Plant Physiology

edited by Malcolm B Wilkins

Regius Professor of Botany, University of Glasgow

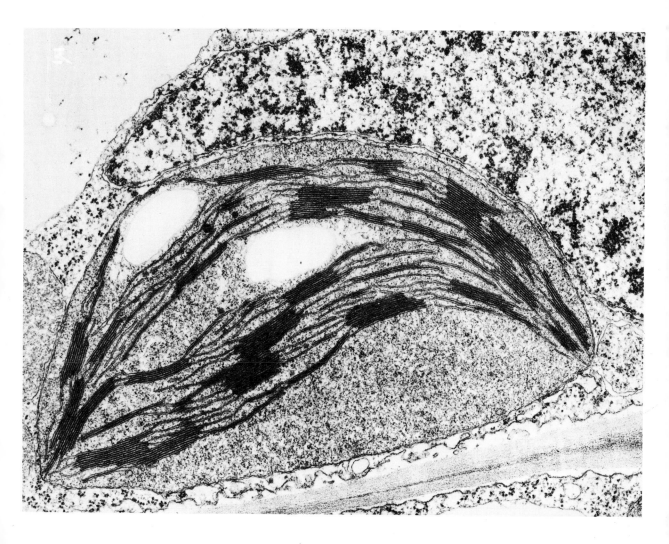

Longman Scientific & Technical
Longman Group UK Limited,
Longman House, Burnt Mill, Harlow,
Essex CM20 2JE, England
and Associated Companies throughout the world

Copublished in the United States with
John Wiley & Sons, Inc., 605 Third Avenue, New York, NY 10158

© Malcolm B. Wilkins 1984
Cover and title page photograph © A D Greenwood 1984

First published in Great Britain by Pitman Publishing Ltd 1984
Second impression 1985
Third impression by Longman Scientific & Technical 1987
Reprinted 1989, 1990

British Library Cataloguing in Publication Data

Advanced plant physiology.
  1. Plant physiology
  I. Wilkins, Malcolm B.
  581.1   QK711.2

ISBN 0-582-01595-2

Produced by Longman Singapore Publishers Pte Ltd
Printed in Singapore

Contents

Preface

Plant physiology is at the very center of both pure and applied botany; its importance cannot therefore be overestimated. It provides an essential link between the biochemical and ecological aspects of the subjects, and a thorough understanding of physiological processes is essential for a proper appreciation of the various other branches of the subject such as biochemistry, morphology, pathology, genetics and ecology. For these reasons plant physiology must find a significant and central place in all courses on plant biology, botany, agricultural botany, horticulture, and agriculture. The purpose of this book is to provide students studying these subjects with an authoritative and up-to-date text.

It is no longer possible for one person to write an advanced textbook on the physiology of plants. The size and diversity of the subject, the sophisticated technology now used in many of its branches, the complexity of the ideas and hypotheses advanced, and the difficulty of evaluating and placing in perspective newly-published results are well beyond the intellectual capacity of the individual scientist. It is to overcome this problem that some twenty-seven contributors, each an internationally recognized authority in his or her field, have been invited to produce this book under my editorship.

Plant physiology contains a number of topics in which progress will have far-reaching consequences. The endogenous mechanisms regulating growth and development are of critical importance in the productivity of crops; the identification and role of naturally-occurring chemical regulators, and their interactions with one another in determining the relative growth and development of different plant parts, in directing nutrient reserves into the harvested parts, and in the regulation of the rate of development and post-harvest characteristics of the product, are all aspects of major importance to the food-producing industry and to the problems of world food supply. Similarly, the environmental factors regulating growth and development are also of critical importance to productivity. Light quality, intensity and photoperiod, the availability of water and inorganic nutrients, and the ability to fix and utilize atmospheric nitrogen are of major importance in determining crop yield. A thorough understanding of the main fundamental aspects of plant physiology is, therefore, an essential element in the training of all applied plant biologists, agriculturalists and horticulturists who will be concerned with problems of practical, economic and commercial importance. In addition, for the fundamental botanist, plant physiology is a subject of substantial intrinsic interest having many unresolved and important problems each presenting a formidable intellectual and technological challenge. It is to provide an advanced fundamental background in plant physiology that this book has been written, and written in such a way that attention is particularly focused on aspects of the subject where there is need for new ideas and experimental work involving the application of new technology.

The editor has been solely responsible for selecting both the topics to be included in this book and the contributors. The contributors have been responsible for the precise content of their chapters and, within broad editorial guidelines, the way in which the topics have been approached. In selecting the topics for inclusion, the editor was guided solely by what he thought ought to be included in an advanced course in plant physiology. Some of the topics dealt with in this book have received scant attention in other works. It is inevitable that some topics have been omitted, but no advanced textbook can be encyclopedic, and other editors would have chosen a somewhat different selection of topics, with equal justification, and omitted some included in this book. Plant physiology must not be studied in isolation, but rather in conjunction with plant biochemistry and plant biophysics, morphology and ecology into which it merges imperceptibly at its boundaries. The topics which are included in this book encompass all those that should, in the editor's view, find a place in an advanced plant physiology course being taken by students of botany, agricultural botany, horticulture and agriculture, and forestry.

The editor is greatly indebted to his friends and fellow-scientists who so readily agreed to contribute to this book, and who submitted their manuscripts so promptly. The care with which the manuscripts were prepared according to the editorial instructions made editing this book a singular pleasure. He is particularly indebted to Drs E. M. Beyer, Jr, P. W. Morgan and S. F. Yang who so readily agreed to write Chapter 5 at short notice after the tragic death of Dr Maurice Lieberman who had originally accepted the editor's invitation to be a contributor. Chapter 5 is dedicated to the memory of Dr Lieberman, a most distinguished scientist and student of ethylene as a plant growth regulator. The editor is also greatly indebted to his colleague, Dr M. F. Hipkins, who also wrote Chapter 11 at very short notice, and to his colleague Professor J. R. Hillman for a number of most helpful discussions.

Glasgow   Malcolm B Wilkins
November 1983

Contributors

Professor D. Baker
Department of Biological Sciences, University of London, Wye College, Nr Ashford, Kent, TN25 5AH, U.K.

Professor R. Bandurski
Department of Botany & Plant Pathology, Michigan State University, East Lansing, Michigan 48824, U.S.A.

Dr A. M. M. Berrie
Department of Botany, University of Glasgow, University Avenue, Glasgow G12 8QQ, U.K.

Dr E. M. Beyer Jr
Agricultural Chemicals Department, Experimental Station, E. I. du Pont de Nemours & Co., Wilmington, Delaware 19898, U.S.A.

Professor M. J. Canny
Department of Botany, Monash University, Clayton, Victoria 3168, Australia.

Dr D. T. Clarkson
ARC Letcombe Laboratory, Wantage, Oxon OX12 9JT, U.K.

Dr D. S. Dennison
Department of Biological Sciences, Dartmouth College, Hanover, New Hampshire 03755, U.S.A.

Dr G. P. Findlay
School of Biological Sciences, Flinders University of South Australia, Bedford Park, South Australia 5042, Australia.

Professor J. R. Hillman
Department of Botany, University of Glasgow, University Avenue, Glasgow G12 8QQ, U.K.

Dr M. F. Hipkins
Department of Botany, University of Glasgow, University Avenue, Glasgow G12 8QQ, U.K.

Dr R. Horgan
Department of Botany & Microbiology, University College of Wales, Aberystwyth, SY23 3DA, U.K.

Dr R. L. Jones
Department of Botany, University of California, Berkeley, California 94720, U.S.A.

Professor J. MacMillan
School of Chemistry, University of Bristol, Cantock's Close, Bristol, BS8 1TS, U.K.

Professor T. A. Mansfield
Department of Biological Sciences, University of Lancaster, Bailrigg, Lancaster LA1 4YQ, U.K.

Professor B. V. Milborrow
School of Biochemistry, University of New South Wales, P.O. Box 1, Kensington, N.S.W. 2033, Australia.

Dr P. W. Morgan
Department of Soil and Crop Sciences, Texas A & M University, College Station, Texas 77843, U.S.A.

Dr H. M. Nonhebel
Department of Botany & Plant Pathology, Michigan State University, East Lansing, Michigan 48824, U.S.A.

Dr D. Vince-Prue
Physiology & Chemistry Division, Glasshouse Crops Research Institute, Worthing Road, Rustington, Littlehampton, West Sussex, BN16 3PU, U.K.

Professor P. Schopfer
Institut für Biologie II, Albert-Ludwigs Universität, Schanzlestrasse 1, D-7800 Freiburg, Germany

Dr R. Sexton
School of Biological Sciences, University of Stirling, Stirling, FK9 4LA, U.K.

Dr P. J. Snaith
Department of Biological Sciences, University of Lancaster, Bailrigg, Lancaster, LA1 4YQ, U.K.

Professor Pill-Soon Song
Department of Chemistry, Texas Tech University, Box 4260, Lubbock, Texas 79409, U.S.A.

Dr J. I. Sprent
Department of Biological Sciences, University of Dundee, Dundee, DD1 4HN, U.K.

Dr B. Thomas
Physiology & Chemistry Division, Glasshouse Crops Research Institute, Worthing Road, Rustington, Littlehampton, West Sussex, BN16 3PU, U.K.

Professor M. B. Wilkins
Department of Botany, University of Glasgow, University Avenue, Glasgow G12 8QQ, U.K.

Professor H. W. Woolhouse
John Innes Institute, Colney Lane, Norwich NR4 7UH, Norfolk, U.K.

Dr Shang Fa Yang
Department of Vegetable Crops, University of California, Davis, California, 95616, U.S.A.

Auxins

1

Robert S Bandurski and Heather M Nonhebel

Introduction

A guiding hypothesis

A peculiarity of the apparently limited number of hormones in plants is that a hormone such as IAA can cause different effects in different plants, or even different effects in the same species of plant at different times. Examples of the varied responses of plants to IAA include: stimulation of cell division, stimulation of shoot growth, inhibition of root growth, control of vascular system differentiation, control of tissue culture differentiation, control of apical dominance, delay of senescence, promotion of flowering and fruit setting and ripening.

To explain these varied effects, and for purposes of this chapter, it is assumed that IAA has one, and only one, receptor. The multiplicity of effects can be explained with current knowledge of second messengers. The IAA–receptor complex could act through a transducer on a calcium gate which in turn could affect intracellular calcium and calmodulin-controlled enzymes, as for example, the plant NAD kinase as studied by Anderson, Charbonneau, Jones, McCann and Cormier (1980). The physiological events observed following application of IAA would then depend on what steps are most limiting in that particular plant in pathways mediated by calmodulin. Thus, one species could elongate, and a different species form roots. It should be emphasized that the hypothesis outlined above is speculative, and is based almost entirely on knowledge of animal systems. However, it is consistent with current knowledge. Valuable discussions of the multiplicity of effects induced by a single hormone may be found elsewhere (Berridge, 1980; Chock, Rhee and Stadtman, 1980; Klee, Crouch and Richman, 1980). The physiological response may be many steps from the IAA binding receptor and thus a Heisenberg uncertainty principle operates in hormone research. Use of *in vitro* systems (Buckhout, Young, Low and Morre, 1981), plant cell cultures and mutant cell lines may provide answers to questions concerning the multiplicity of IAA effects.

Methods of biological and chemical assay for auxin

The bioassays

Only a bioassay can detect the physiological activity of a substance and thus bioassays provided the beginning for all hormone work. The discovery of an assay for auxin was closely linked to the experiments which led to the plant growth hormone concept and to the discovery of IAA. Once it was known that some influence diffused from the tip into the growing region—there to promote growth—it became possible to develop a quantitative assay for the substance. The reader is referred to Went and Thimann (1937), for a detailed description of the early experiments and of several bioassays. Boysen-Jensen demonstrated that the growth stimulus from the tip would diffuse across a wound covered with gelatin and realized that the substance could diffuse out of the tissue and into a collecting gel. It was Went, however, who first used agar blocks and demonstrated that the amount of substance collected from coleoptile tips placed on the agar blocks increased when more tips were used. Thus, Went made the assay quantitative and this assay led to the discovery of IAA.

Avena curvature test Seeds are germinated in darkness using dim red light for necessary manipulation and to suppress mesocotyl growth. Shortly prior to use the plants are decapitated to deplete their hormone reserves. A block of agar, containing the growth substance to be assayed, is placed asymmetrically on the decapitated stump and after one to two hours the curvature resulting from greater growth under the block is measured. Curvature is arithmetically proportional to IAA in the block using concentrations in the region of five micromolar. For a substance to be active in this assay it must permeate the tissue, be reasonably stable to enzymatic destruction, be transported to the growth site *and* be physiologically active. Which auxins are discovered may be restricted by these requirements.

The Avena straight growth assay This assay is less sensitive and will respond to compounds not as readily transported as IAA, but is simpler to perform. Decapitated coleoptiles are incubated in the solution to be assayed and changes in length during a 12-h period are measured. The growth is usually only a few millimeters, so great accuracy is required, but by stringing several sections together average values for changes in length are obtained; mechanical and electrical recording and amplifying devices may also be used (Penny, Penny and Marshall, 1974). In this assay growth is proportional to the log of the IAA concentration, unlike the curvature test, which responds arithmetically to IAA.

Other bioassays A compendium of more than 75 bioassays has been prepared by Mitchell and Livingston (1968). Ruge (1961) has reviewed some of the amplifying devices used in detecting rapid changes and Evans (1974) has provided a review of rapid responses. Some very convenient and rapid bioassays have recently become available (Meudt and Bennett, 1978) and these together with good transduction and recording devices vastly extend the utility of the bioassays.

Chemical and physicochemical assays

Chemical assays can measure a fixed single substance whereas bioassays measure a stimulus, or inhibition, or both simultaneously (*i.e.,* a net effect), and not necessarily a discrete compound.

Colorimetric and fluorometric assays Chemical and physicochemical assays have recently been reviewed (McDougal and Hillman, 1978; Morgan and Durham, 1983). Owing to the chemical instability of IAA, the very small quantities in tissues, and the presence in tissue of relatively large amounts of reactive phenolic compounds, such assays must be used with care. The best of the colorimetric reagents is a combination of the Van Urk and Salkowski reagent as described by Ehmann. This reagent was originally used as a spray reagent but it has been adapted to colorimetric procedures since it yields a stable color (Percival and Bandurski, 1976). To the IAA-containing sample in a volume of 50 μl of 50% ethanol is added a reagent composed of one part Salkowski reagent and one part Ehrlich reagent. After heating at 45°C for 30 min a stable color develops which is diluted to 1 ml with 50% ethanol and may be estimated quantitatively at 615 nm against an appropriate blank. A full scan of the color detects interfering compounds. The method may be scaled down, and sensitivities to a fraction of a microgram are possible. Other indoles interfere, as do some phenolics and phenylpropanes, but only when present in 100- or 1000-fold excess.

Other methods are based upon the intrinsic fluorescence of IAA or upon the 2-methylindolo-2-pyrone formed from IAA cyclized with acetic anhydride in the presence of an acid catalyst. Many precautions must be followed and the interested reader is referred to Mousdale (1978). The method is applicable to 1–5 g fresh weight of tissue and employs ^{14}C-labeled IAA as an internal standard.

GC–MS methods Mass spectrometry was first applied to the assay and certain identification of IAA by Igoshi, Yamaguchi, Takahashi and Hirose, 1971, and other applications are reviewed by McDougall and Hillman (1978) and by Reeve and Crozier (1980). IAA labeled in a stable position with deuterium is a desirable internal standard. To this end Magnus, Bandurski and Schulze (1980), synthesized IAA with deuterium in the 4, 5, 6 and 7 positions of the indole ring. The deuterium in these positions, unlike those in position 2 of the side chain, is not exchanged in alkaline solution so the d_4-IAA may be used as an internal standard even in assays involving alkaline hydrolysis. Further, the incorporation of four deuterium obviates the background owing to the normal abundancy of heavy isotopes. Following addition of a known amount of d_4-IAA and sufficient ^{14}C-IAA to permit ready monitoring, the IAA is purified by DEAE-Sephadex and reverse phase HPLC. The sample is then methylated with diazomethane and chromatographed on a GLC column of intermediate polarity, preferably a wall-coated quartz capillary. The IAA coming from the plant may now be monitored at its molecular ion, 189, and at its major fragment ion, 130. The d_4-IAA is measured at mass 193 and at 134. Since the amount of d_4-IAA added is known, the amount of IAA in the plant may be calculated. If the percentage of d_4-IAA found at the two masses does not agree, it is a sign of a contaminating ion at one of the masses monitored.

The gc-sim-ms method is sensitive in the low nanogram region, but it does require access to a mass spectrometer. It provides the greatest precision and assurance as to the identity of the substance being assayed and it is strongly recommended that simpler methods be validated by gc-sim-ms. Increases in sensitivity are potentially attainable by using negative ion mass spectrometry or tandem mass spectrometry.

A double-standard isotope dilution assay has recently been described by Cohen and Schulze (1981). This assay greatly simplifies determination of specific activity and can be adapted to almost any detector.

Immunoassays Lately a radio-immunoassay (RIA) has been described by Pengelly, Bandurski and Schulze (1981). This method is convenient, rapid and inexpensive. The antibodies are prepared by coupling IAA to serum albumin with formaldehyde. The method was applicable to measurement of free IAA in corn shoots but was in error when used for the assay of IAA in tissue extracts which had been subjected to alkaline hydrolysis. A response versus concentration curve would, however, have detected the error.

An immobilized enzyme assay has been described by Weiler (1982). Literally hundreds of samples can

be assayed per day and almost any degree of sensitivity can be attained. Undoubtedly RIA and immobilized enzyme assays will be convenient when large numbers of assays are to be made of a particular plant.

Chemical structure of the auxins

Discovery of IAA

The extraordinary story of the development of an assay for indole-3-acetic acid (IAA) by Went, and its isolation and characterization by Kögl and Haagen-Smit, is described in detail by Went and Thimann (1937). Figure 1.1 shows a brief pictorial survey of this history. The horticulturist Ciesielski, working with roots, and Charles and Francis Darwin, working with shoots, found that the tip of the root or shoot controlled the rate of growth of the growing region located some distance from the tip (Heslop-Harrison, 1980). Ciesielski and Darwin reasoned correctly that some 'influence' had to move from the tip to the growing region. Boysen-Jensen found that this influence could diffuse through a gelatin junction across a wound and Went discovered that this substance could be collected in an agar block. Placing the block asymmetrically on a decapitated plant stump caused that stump to grow faster on the side under the block. The resultant curvature could be measured and the degree of curvature was a function of how many tips had been permitted to diffuse into the agar. This provided Kögl and Haagen-Smit with a quantitative assay for the 'influence' and they soon isolated and characterized the substance IAA whose structure is shown in Fig. 1.2.

Structure–activity relationships

Knowledge of the structure of IAA led to a search for compounds of related structure having auxin activity (e.g. Thompson, Swanson and Norman, 1946). This had important consequences in that it led to our current knowledge of the structural requirements for auxin activity and to the development of synthetic weed killers such as 2,4-dichlorophenoxyacetic acid (2,4-D). This serves as an example of the practical worth of basic research, since the

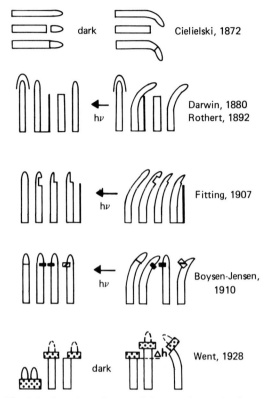

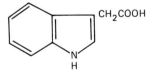

Fig. 1.1 Summary of some of the experiments leading to the discovery of plant growth hormones and the structure of auxin. References, and a description of these experiments may be found in Went and Thimann (1937). Readers should compare the conclusions they draw from these experiments with those drawn by the original research workers. (The drawings are adapted from: A. Ehmann (1973) *'Indole compounds in seeds of Zea mays'*, p. 3, Ph.D. Thesis, Michigan State University, East Lansing.)

Fig. 1.2 The structure of indole-3-acetic acid.

phenoxyacetic acids as selective herbicides is difficult since some of the work was conducted under conditions of the military secrecy of World War II. The names of the pioneers would include Kraus, Zimmerman, Hitchcock, Mitchell, Templeman, Sexton and Quastal (Norman, 1946; Thimann, 1974; and E. J. Kraus, personal communication).

Initially it seemed that the major requisites for biological activity were an aromatic ring, a carboxyl group at least one carbon removed from the ring, a free *ortho* position or a strong electron-withdrawing substituent. Later, Thimann and Leopold (1955) developed a charge-separation theory indicating that there must be a negative charge on the carboxyl and a positive charge on the ring system which must be 5.5 Å (0.55 nm) from the negative charge.

A further important observation was made by Hansch and Muir (1950). These workers studied the kinetics of the IAA-induced growth response and observed that growth rate increased as a function of added IAA in the test media up to a maximum of about 10^{-5} M. Above this concentration growth was inhibited and reached almost zero at 10^{-3} M. To explain the peculiar kinetics these workers suggested a two-point attachment theory. In this theory, when the concentration of IAA was too high then the growth site would be occupied by two molecules of IAA, one attached by the carboxyl and a second through a free *ortho* position forming an inactive $(IAA)_2$ complex with the growth-promoting site. The two-point attachment theory is perhaps not correct but the theory plus the data on structure versus activity have had theoretical importance since they focused attention on the IAA-binding site.

The skilful work on structure–activity led to many new synthetic auxin-like growth regulators. It also led to attempts, as described above, to obtain profiles of electron density in compounds active as auxins. However, there were always important exceptions in structure–activity correlations. Recently Katekar and Geissler (1982), developed a highly empirical, but apparently effective, stochastic method. Using this method the domains necessary for auxin activity can be assigned. The method has confirmed the necessity of a carboxyl domain and of at least two domains called Ar_1 and Ar_2 in the aromatic ring. The method shows potential and with use of computer modeling may be able to draw a three-dimensional picture of the IAA-binding site.

world-wide use of selective herbicides saves sufficient money in one year to pay for all plant physiological research since the time of Aristotle. Assignment of priority to the workers who first recognized the possibilities of use of chlorinated

A precautionary remark must be made and this concerns the use of any bioassay to detect structure–activity relationships or to diagram binding sites. For a substance to show activity in a bioassay it must: (a) permeate the cell cuticle and plasma membrane; (b) be transported from the site of application to the growth promoting site; (c) have reasonable stability; and (d) have the correct shape and properties for biological activity. Thus, current bioassays may restrict discovery to only certain types of auxins. Current procedures in bioassay may be the equivalent of trying to cure a vitamin deficiency in a rabbit by rubbing coenzymes on the rabbit's stomach. It must be emphasized that bioassays can only find certain kinds of hormones and will miss others, and that structure–activity relationships may be complicated by factors other than auxin binding, for example by penetration or metabolism.

Naturally-occurring auxins other than IAA

Phenylacetic acid Wightman and Lighty (1982) have demonstrated that phenylacetic acid is a naturally-occurring compound that fulfills the definition of a non-indolylic auxin. Its activity is usually lower than that of IAA but it is often present in concentrations much higher than IAA. It has been found in a range of crop plants including tomato, tobacco, sunflower, pea, barley and corn. The probable precursor of phenylacetic acid would be phenylalanine and this, of course, is ubiquitous. Phenylalanine would be converted to phenylpyruvic acid by a non-specific aminotransferase and this could then be oxidatively decarboxylated. For both IAA and phenylacetic acid it would be helpful if mutant techniques could be used to prove that plants lacking IAA or phenyl-acetic acid have disturbances in their growth characteristics.

Chlorinated IAAs The chlorinated indole, 4-chloro-3-indoleacetic acid has been reported to occur in immature seeds of *Pisum*, *Lathyrus* and *Vicia* (*e.g.*, Engvild, Egsgaard and Larsen, 1980; Gander and Nitsch, 1967). The chlorinated IAA is more active than IAA in the *Avena* straight growth test. Immature pea seeds also contain chlorinated tryptophan derivatives (Marumo and Hattori, 1970). It seems possible that these compounds are made by chloroperoxidase or peroxidase (Hollenberg, Rand-Meir and Hager, 1974).

Biosynthesis of IAA

Routes of biosynthesis

Multiplicity of IAA sources There are several routes to the synthesis of IAA. For example it is possible to begin with the shikimate pathway (Miflin and Lea, 1977) and derive the aromatic indole ring *de novo*. Such *de novo* synthesis must occur in growing, seed producing, autotrophic green plants. There is indirect evidence that *de novo* synthesis occurs in the leaf and something, possibly tryptophan, is transported to and stored in the seed where it may be converted to other indoles (Mockaitis, Kivilaan and Schulze, 1973). The tryptophan may then be converted to IAA by removal of the amino and carboxyl group. Most commonly it is conversion of tryptophan to IAA which is regarded as IAA synthesis. In the old literature (Cholodny, 1935) and more recently (Cohen and Bandurski, 1982) the hydrolysis of IAA conjugates to yield free IAA has been considered as IAA synthesis and as a major source of IAA in seedling plants. There is uncertainty as to how much of a plant's IAA is derived from hydrolysis of a conjugate and how much from tryptophan or even total *de novo* synthesis. Lastly, it should be emphasized that, for purposes of growth control, transport of IAA from one point to another can be the equivalent of synthesis as can conversion of IAA from a transportable form to an active form (Van Overbeek, 1941). It is hoped this section serves to emphasize the role of multiplicity of IAA sources in determining amounts of IAA in the tissue.

Tryptophan as a precursor The biosynthesis of tryptophan in plants has been reviewed by Miflin and Lea (1977) and in bacteria by Crawford and Stauffer (1980). The tryptophan operon is the best-studied anabolic gene in enteric bacteria. *In vivo* studies with labeled precursors and *in vitro* studies with enzymes—in a limited number of plants, mainly carrot and tobacco tissue cultures—indicate that plants make tryptophan as do bacteria. The tryptophan synthase [L-serine hydrolase (adding indole EC 4.2.1.20)] consists of two components, like the bacterial tryptophan synthase, in contrast to the single-component fungal enzyme. Both an A and B component are required to form tryptophan from indole-3-glycerol phosphate and serine, whereas B catalyzes the serine-indole condensation and A catalyzes the conversion of indole-3-glycerol phosphate to indole. Antibody neutralization experiments indicate that the B protein is more closely related to that of *Escherichia coli* than is the A protein. Of very great importance is the fact that pea embryos, as well as 3- to 4-day-old corn roots already have anthranilate synthetase, phosphoribosyl-transferase, indole-glycerol phosphate synthetase and phosphoribosylanthranilate isomerase. These enzymes were shown to be present in excess of any possible microbial contamination (Hankins, Largen and Mills, 1976). Young plants may have the capability of synthesizing tryptophan, despite the large amount of tryptophan in the seed, to provide an alternative supply in the event of damage by a predator. Mitra, Burton and Varner (1976) have attempted to determine when a seedling plant ceases heterotrophic metabolism and begins a true autotrophic existence. Young plants are mainly heterotrophic. More studies will be required to determine the relative importance of *de novo* synthesis and utilization of seed reserves.

The relationship between auxin autotrophy and tryptophan has been studied. Widholm (*see* Widholm, 1981) has cultured cell lines of carrot and tobacco which mutate, with a frequency of about 1 in 10^6 to strains which will grow in the presence of higher concentrations of 5-methyltryptophan (5MT) and form an anthranilate synthetase enzyme resistant to inhibition by L-tryptophan and by 5-methyltryptophan. Such cells form a tryptophan pool which is 30 times the normal pool. About ten different tryptophan analogs inhibit carrot and tobacco cell tryptophan synthase *in vivo* and *in vitro*. The cells form the corresponding tryptophan analog when grown on 4-fluoroindole, 5-hydroxyindole and 5-methoxyindole or 5-methylindole. No studies have been made of the IAA content of these high tryptophan cultures or whether a substituted IAA is made. A total of 5 of 10 tryptophan-accumulating carrot lines were found to be auxin autotrophic in that they did not require 2,4-D for growth. By contrast, tobacco lines which oversynthesize tryptophan are not auxin autotrophic. Obviously IAA determinations on cell lines with altered tryptophan metabolism are required.

The pathway from tryptophan to IAA There is evidence that tryptophan is converted to IAA by

loss of the carboxyl carbon and the amino nitrogen. Oxidative decarboxylation to indole-3-acetamide followed by hydrolysis to yield ammonia and IAA is a bacterial reaction that proves convenient for radiological synthesis of IAA (Michalczuk and Chisnell, 1982). Long ago, Thimann had fed tryptophan to *Rhizopus* (*see* Went and Thimann, 1937) and showed an increased auxin formation and, in fact, this evidence aided in the acceptance of the structure of IAA. Formation of IAA from tryptophan has been studied *in vivo* and *in vitro* in more than 20 plant species; 10 enzyme preparations; and many studies of the conversion of ^{14}C-labeled tryptophan to IAA (Schneider and Wightman, 1978; Sembdner, Gross, Liebisch and Schneider 1980). There are many difficulties in studies of the synthesis of IAA from tryptophan. First, the amounts of tryptophan are very high relative to that of IAA (Schneider and Wightman, 1978) and, secondly, simply drying radiolabeled tryptophan yields IAA in 30% yield (Epstein, Cohen and Bandurski, 1980). Lastly, microbial conversions are possible (Heerkloss and Libbert, 1976a; Sembdner *et al.*, 1980). Double-labeling experiments, however, establish that plants can convert tryptophan to IAA and that the pathway does not pass through indole (Heerkloss and Libbert, 1976; Sembdner *et al.*, 1980).

The possible pathways from tryptophan to IAA are shown in Fig. 1.3. First, indole-3-pyruvic acid can be formed by transamination (Schneider and Wightman, 1978) and Wightman and his colleagues have studied this reaction using α-ketoglutarate as acceptor and measuring the glutamate formed. The characterization of indole-3-pyruvate is based on characterization of the 2,4-dinitrophenylhydrazone derivative in incubation mixtures. The instability of indole-3-pyruvic acid has precluded its isolation from plant tissue (Schneider and Wightman, 1978). Studies from this laboratory on wine (Ehmann, 1976), using mass spectrometry, showed the presence of indole-3-lactic acid and its ethyl ester as a common component of wine; these compounds were probably formed from tryptophan in the grape juice by yeast. A related transamination from phenylalanine to form phenylacetic acid has been carefully studied by Wightman and Lighty (1982). The α-keto acid derived from tryptophan or phenylalanine can then be converted to the corresponding aldehyde. The alcohol-to-aldehyde conversion is the only reversible step in the synthesis of IAA from

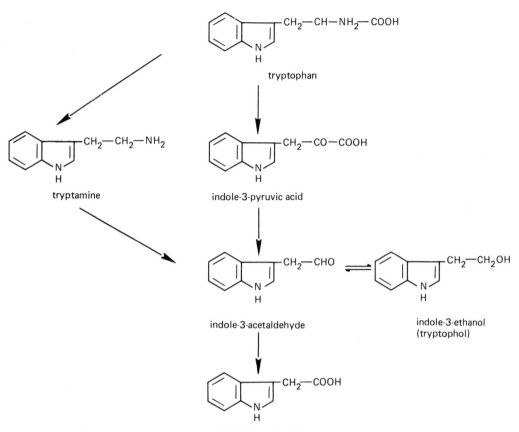

Fig. 1.3 Intermediates in the conversion of tryptophan to indole-3-acetic acid. (Adapted from: G. Sembdner, D. Gross, H. W. Liebisch and G. Schneider (1980) Biosynthesis and metabolism of plant hormones, p. 284 in *Encyclopedia of Plant Physiology*, ed. J. MacMillan, Springer-Verlag, Berlin.)

the amino acid (Purves, personal communication). Purves and his colleagues consider tryptamine, indole-3-ethanol and indole-3-acetaldehyde to have been convincingly demonstrated to occur in plant tissues.

The tryptamine pathway Tryptamine may also be converted to indole-3-acetaldehyde. Tryptamine has a more limited occurrence (Schneider and Wightman, 1978) in plants. Its pool size in *Zea mays* is small (*see* Epstein *et al.*, 1980) but it does occur as *N*-coumaryl and ferulyl tryptamine. Amine oxidases occur in plants but none are specific for tryptamine (Schneider and Wightman, 1978).

The indole-3-ethanol pathway Indole-3-ethanol is probably the product of a side reaction (Schneider and Wightman, 1978). None the less, indole-3-ethanol has been identified as a naturally-occurring component by Rayle and Purves (1967) using combined gas chromatography–mass spectrometry. The indole-3-acetaldehyde reductase is associated with a microsomal fraction and inhibited by high NADPH. The aldehyde reduction is a pyridine-nucleotide-dependent enzyme and, most importantly, the indole-3-ethanol oxidase is subject to feed-back regulation by IAA (Percival, Purves and Vickery, 1973).

Thus, the biosynthesis of IAA from tryptophan is

a well established route. The problems are that the pool of tryptophan is so large relative to IAA that regulation of tryptophan levels hardly seems an appropriate control mechanism. The intermediates between tryptophan and IAA would be more reasonable control points. Work on tissue cultures, and use of mutants and derepressed strains, coupled with careful assays of IAA levels, will be required to understand how the rate of synthesis of IAA from tryptophan controls IAA levels.

Hydrolysis of IAA conjugate to yield IAA Amide-linked and ester conjugates of IAA are widespread in seeds and vegetative tissues (Cohen and Bandurski, 1982). Conjugates usually compose 50 to 90% of the IAA of the tissue, as for example IAA-aspartate in the leaves of 2-week-old pea plant (Law and Hamilton, 1982); IAA esters in the case of *Zea mays* kernels (*see* Cohen and Bandurski, 1982) or IAA-1-aspartate in the seeds of *Glycine max* (Cohen, 1982).

Conjugates can be hydrolyzed to yield free IAA as shown by the following evidence: Cholodny (1935) showed that moistened bits of conjugate-rich endosperm tissue cause seedling growth; conjugates function as 'slow release forms' of IAA for tissue cultures (Hangarter and Good, 1981); labeled conjugates are hydrolyzed following application to *Zea* endosperm (Nowacki and Bandurski, 1980); IAA is formed and destroyed in *Zea* endosperm at a rate commensurate with the rate of conjugate disappearance (Epstein *et al.*, 1980); and lastly, there is an enzyme in *Zea* seedlings which hydrolyzes IAA-*myo*-inositol (Hall and Bandurski, 1981). The question is to what extent does conjugate hydrolysis provide a source of IAA for the growing tissue. It has been proposed that conjugates provide a major source of IAA for seedlings and further that there is a homeostatic system involving IAA and its conjugates (Bandurski, 1979)

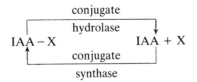

where X is the conjugating moiety. Some of the elements of proof have been provided as discussed above.

Summary It is clear that no single plant tissue has been studied well enough to know what determines the IAA content of the tissue. We must know, and measure, *all* of the routes of IAA synthesis, transport, use, and catabolism in one species, and at one age, to be able to evaluate relative contributions.

Metabolism of IAA

As early as 1935 it was suggested that the metabolism of auxin might be important in controlling its endogenous level (*e.g.*, Van Overbeek, 1935). Since that time, three main approaches have been used in the study of IAA metabolism: (1) Enzymes capable of destroying IAA have been studied and the reaction products characterized; (2) Radiolabeled IAA has been applied to plant tissues and the resultant metabolites analyzed; and (3) Endogenous indolic compounds have been extracted and identified.

IAA metabolites include compounds formed by irreversible oxidation, as well as conjugates in which the intact IAA molecule is attached *via* an ester or amide linkage to a conjugating moiety. The chemical structures of a selection of IAA metabolites are shown in Fig. 1.4.

IAA catabolism

IAA decarboxylation In 1947, Tang and Bonner found an enzyme capable of catalyzing IAA oxidation. 'IAA oxidases' which usually have the characteristics of peroxidases are now known to be widespread in higher plants (Schneider and Wightman, 1974). The major products of IAA-oxidases are 3-hydroxymethyloxindole (Hinman and Lang, 1965) and indole-3-aldehyde (*e.g.*, Ricard and Job, 1974). These compounds are readily oxidized to 3-methyleneoxindole and 3-methyloxindole or indole-3-carboxylic acid, respectively (Sembdner *et al.*, 1980).

Owing to the widespread occurrence of 'IAA oxidases' and the commercial availability of enzymes such as horseradish peroxidase, the *in vitro* oxidation of IAA has been widely studied. Most, although not all, IAA-oxidases require Mn^{2+} as a cofactor. Monohydric phenols such as *p*-coumaric acid are also often needed as cofactors. On the other hand, *o*- and *p*-dihydric phenols (*e.g.*, caffeic acid)

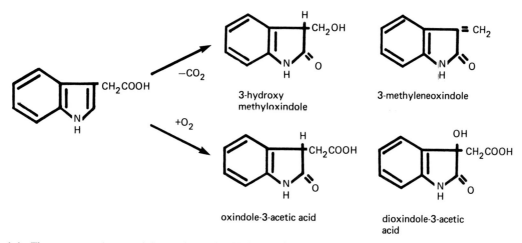

Fig. 1.4 The structure of some of the products of oxidative catabolism of indole-3-acetic acid.

and polyphenols often act as inhibitors (Sembdner et al., 1980).

The mechanism of the enzymatic oxidation of IAA has been studied extensively using spectroscopic methods. Several pathways have been proposed (see Sembdner et al., 1980, for details). In general, the route involves the production of the IAA free radical and requires the presence of H_2O_2. The stoichiometry is one mol of oxygen consumed per mol of IAA destroyed. The products formed vary according to the reaction conditions (Ricard and Job, 1974). High [substrate]/[enzyme] ratios and a pH of 6 favor production of 3-methyleneoxindole whereas indole-3-aldehyde is the major product at pH 4 and with stoichiometric proportions of IAA and peroxidase.

Despite the large amount of research on the *in vitro* destruction of IAA in the presence of IAA oxidases, evidence for the *in vivo* oxidation of IAA to form compounds such as 3-methyleneoxindole, indole-3-aldehyde and related compounds is scant. A number of workers have reported the evolution of substantial quantities of radioactive CO_2 following the application of IAA labeled with ^{14}C in the 1-position of the side chain (e.g., Wilkins, Cane and McCorquodale, 1972). The IAA decarboxylation products indole-3-aldehyde, indole-3-methanol and indole-3-carboxylic acid have been reported in pea stem sections (Magnus, Iskric and Kueder, 1971), 3-methyleneoxindole in oat coleoptiles (Menschick and Hild, 1976) and 3-hydroxylmethyl-

oxindole in etiolated pea seedlings. Definitive mass spectral identification of these compounds is lacking.

Catabolism without decarboxylation Recently the compound oxindole-3-acetic acid, which retains both carbon atoms of the IAA side chain, has been identified by mass spectrometry as an endogenous constituent of *Zea mays* endosperm and shoot tissue (Reinecke and Bandurski, 1981, 1983). This compound has also been demonstrated following the application of IAA-1-^{14}C to the endosperm and would account for at least 26% of the IAA catabolized during the incubation period. The methyl esters of oxindole-3-acetic acid and the related compounds dioxindole-3-acetic acid, 5-hydroxyoxindole-3-acetic acid and 5-hydroxydioxindole-3-acetic acid have also been extracted from rice bran (Kinashi, Suzuki, Takeuchi and Kawarada, 1976). In further recent work using maize root tissue, Nonhebel et al. (1983) found that although over 60% of the IAA was metabolized during a 2-h incubation period, negligible decarboxylation had taken place. Instead a number of products were formed, the most prominent of which co-chromatographed with oxindole-3-acetic acid. Similar results were also obtained for coleoptile tissue (Nonhebel, 1982).

Thus at present, the data indicate two pathways of IAA catabolism, peroxidative decarboxylation, and oxidation to oxindole-3-acetic acid. It is possible that the major pathway of IAA catabolism may vary between species although some reports of decarboxylation may be attributable to the route of IAA application. There is evidence that decarboxylation of IAA may take place at cut surfaces (Andreae and Collet, 1968; Waldrum and Davies, 1981). Also, evolution of $^{14}CO_2$ following application of carboxyl-labeled IAA could be due to further metabolism of oxindole-3-acetic acid. Further studies will be required to determine the relative importance of these two catabolic pathways.

IAA conjugates

Endogenous IAA conjugates The IAA conjugates of *Zea mays* kernels have been extensively studied (Cohen and Bandurski, 1982). Approximately one half are high-molecular-weight esters, mainly a $\beta1\rightarrow4$ glucan. The remainder are esters of IAA and *myo*-inositol, *myo*-inositol-arabinose, and *myo*-inositol-galactose with traces of IAA glucose. IAA-*myo*-inositol has also been found in vegetative tissues of *Zea mays* (Chisnell and Bandurski, 1982) and in seeds of *Oryza sativa* (Hall, 1980). In *Avena* seeds approximately 80% of the IAA is conjugated to a β 1-4, 1-3-glucoprotein (Percival and Bandurski, 1976). IAA-1-aspartate has been identified as a major endogenous constituent of soybean seeds (Cohen, 1982). The structures of some of the IAA conjugates are shown in Fig. 1.5.

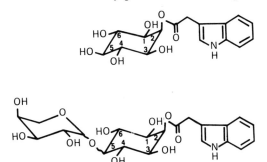

Fig. 1.5 The structure of indole-3-acetyl-2-*O*-*myo*-inositol (upper) and indole-3-acetyl-2-*O*-*myo*-inositol-5-*O*-arabinoside.

Enzymatic synthesis of IAA conjugates

The *in vitro* synthesis of IAA esters using an enzyme from *Zea mays* caryopses has been demonstrated

(*see* Corcuera *et al.*, 1982; Corcuera and Bandurski, 1982). The reaction sequence is as follows:

IAA + UDP-glucose
$$\rightleftharpoons IAA\text{-}1\text{-}O\text{-}\beta\text{-}D \text{ glucose} + UDP$$
IAA-glucose + *myo*-inositol
$$\rightarrow IAA\text{-}myo\text{-inositol} + glucose$$
IAA-*myo*-inositol + UDP-galactose
$$\rightarrow IAA\text{-}myo\text{-inositol-galactoside} + UDP$$
or
IAA-*myo*-inositol + UDP-arabinose
$$\rightarrow IAA\text{-}myo\text{-inositol-arabinoside} + UDP$$

The *in vitro* synthesis of an IAA-amide has not been observed.

Formation of IAA conjugates following application of exogenous IAA Conjugation of IAA and, in fact, any applied growth regulator is an important factor in the use of growth regulators as herbicides (Cohen and Bandurski, 1982). The first conjugate to be identified was indoleacetyl-aspartic acid (Andreae and Good, 1955). Later Klämbt (1961) and Zenk (1961) found that some plants formed IAA-glucose following the application of IAA. Thus, there are at least two conjugation routes for the detoxification of applied growth regulators. IAA aspartate formation occurs after a lag of about 2 h and reaches a maximum rate in 6 to 8 h (Venis, 1972). There is no lag in the formation of IAA-glucose and this reaches its maximum rate in about 12 h. There is specificity in the detoxification route since a weak aromatic acid, such as benzoic acid, is conjugated to form benzoylmalate. Further, pretreatment of a plant with IAA, 2,4-D, and naphthaleneacetic acid, but not benzoic acid, will abolish the lag in IAA aspartate formation.

The ability to identify and measure the auxin conjugates and their oxidative catabolites should make it possible to understand the basis for selective herbicide toxicity.

Physiological role of IAA conjugation

The formation of IAA conjugates is non-destructive and it is known that in some cases it is reversible. During germination of kernels of *Zea mays* some 90% of the conjugates are hydrolyzed to yield free IAA (*see* Cohen and Bandurski, 1982). Some of the IAA-*myo*-inositol is transported into the shoot and there hydrolyzed to yield free IAA (Nowacki and

Bandurski, 1980). Thus, the conjugate can function, first, as a storage form of IAA, stored during seed ripening and used during germination, and secondly, as a transport form during movement of IAA from kernel to shoot. A third function has been reported by Cohen and Bandurski (1978) to be the protection of IAA against peroxidative attack, since it could be demonstrated that IAA conjugates are not attacked by peroxidase preparations. Lastly, evidence has been adduced that when seedling growth is inhibited free IAA disappears and the IAA ester content is increased (Bandurski, Schulze and Cohen, 1977). On this basis it has been postulated that conjugate formation and hydrolysis may constitute a homeostatic mechanism for regulating IAA levels in the plant.

Conclusion

In view of the high rates of oxidation and conjugation, it is apparent that metabolism can be a major determinant of the amount of IAA in the tissue. Two phenomena are being considered, however—the normal metabolic transformations of endogenous IAA, and the pathways used in detoxification. Careful regulation of the amount and route of IAA application should permit distinction between these pathways. Finally, it is likely that the rate and pathway of IAA metabolism may vary significantly between different groups of cells. Substantial differences in pool sizes of IAA and its conjugates in cortical and stelar tissues have been reported (Pengelly, Hall, Schulze and Bandurski, 1982) and differences in IAA metabolism of the cortical and stelar tissues of the maize root have already been shown (Greenwood, Hillman, Shaw and Wilkins, 1973; Nonhebel, 1982).

IAA transport

Polar transport

The polar transport of auxin was noted as early as 1928 by Went. Using segments of *Avena* coleoptiles and agar blocks containing material that had diffused from coleoptile tips, he showed that auxin movement was greatest in the basipetal direction. This observation was later repeated with synthetic

IAA and in shoot segments from several different species (*see* Went and Thimann, 1937). The polar transport of IAA in shoots is in a basipetal direction although following tropic stimulation there is evidence that lateral transport is possible (*see* Gardner, Shaw and Wilkins, 1974 and Chapters 7 and 8). In root segments polar transport is in an acropetal direction (Wilkins and Scott, 1968).

IAA transport in intact plants has also been studied. Labeled IAA applied to the apex of pea seedlings is transported at a rate of $11\,mm\,h^{-1}$ (*see* Morris and Thomas, 1978), *i.e.*, at a similar velocity to polar transport in sections. Transport is inhibited by 2,3,5-triiodobenzoic acid—an inhibitor of polar IAA transport. The specific polar transport of IAA, therefore, takes place in intact plants.

Characteristics of polar transport

Polar transport of IAA in both roots and shoots requires metabolic energy since IAA is not transported under anaerobic conditions (see Wilkins and Whyte, 1968). The polar transport system is specific for active auxins, and inactive weak acids such as benzoic acid are not transported. α-Naphthaleneacetic acid is transported rapidly in a polar direction whereas the growth-inactive β-naphthaleneacetic acid is not transported (*e.g.*, Hertel, Evans, Leopold and Sell, 1969). IAA metabolites do not move in the polar transport system of *Zea mays* coleoptiles (*e.g.*, Nonhebel, 1982).

The time to transport a given amount of IAA increases approximately linearly with distance rather than with the distance squared as would be the case for diffusion (Goldsmith, 1977). When supplied in concentrations of 0.5 to 5×10^{-3} mol m^{-3}, ^{3}H-IAA moves as a discrete pulse down a *Zea mays* coleoptile section with a velocity of about $12\,mm\,h^{-1}$ (Goldsmith, 1982). Similar velocities have been reported in other plants.

The involvement of a saturable site in IAA transport was demonstrated by Goldsmith (1982). When a pulse of ^{3}H-IAA was applied to coleoptile sections equilibrated with 1×10^{-2} mol m^{-3} of IAA, the rate of transport was reduced and the labeled IAA no longer moved as a discrete pulse.

The mechanism of IAA polar transport

The classical theory of IAA polar transport suggests

that IAA moves in the cytoplasm and is secreted at the basal end of shoot cells (the apical end of root cells) by a carrier-mediated, energy-dependent mechanism (see Goldsmith, 1977). This mechanism is attractive since IAA would be required to permeate the plasma membrane only once and thereafter would pass through the plasmodesmata and remain in the symplast. A carrier would, of course, be required to provide the requisite specificity. It is not clear how direction of transport would be established. This mechanism is reviewed briefly by Goldsmith (1977) and a number of references to earlier reviews are provided.

More recently a second theory of auxin transport has been proposed by Rubery and Sheldrake and is extensively reviewed by Goldsmith (1977). This theory also suffers from lack of a proven mechanism for establishing the direction of transport. It is based upon the facts that the pH of the interior of the cells is higher than the pH of the solution bathing the cell walls and that acids enter cells in the undissociated form and not as an anion (Overton, 1902; Simon and Beevers, 1951). The pH of the cell wall solution of *Avena* coleoptiles is 5 (Cleland, 1976) whereas the pH of the cytoplasm and vacuole are 7.1 and 5.5, respectively (Roberts, Ray, Wade-Jardetzky and Jardetzky, 1980). Thus, IAA inside the cell would be ionized to a greater degree than that outside the plasmalemma and the cells would accumulate IAA. Polarity could then be attained if one end of the cell were more permeable to the dissociated acid than the other end.

Cellular location of IAA polar transport

In maize roots it is possible to separate the cortical and stelar tissues. Taking advantage of this property, Wilkins and co-workers (e.g., Shaw and Wilkins, 1974) were able to show that polar transport of IAA took place predominantly in the stele. Studies of ^{14}C-IAA transport from the apical bud in intact pea seedlings using dissection techniques and autoradiography indicated that transport took place in the cambium and differentiated vascular tissue. As labeled IAA could not be tapped by aphids the mature sieve elements did not appear to be involved (see Morris and Thomas, 1978).

IAA transport in the phloem

The presence of IAA in the phloem sap has been demonstrated by mass spectrometry (see Allen and Baker, 1980). Measurements of concentration varied from 4 to 13 ng l^{-1}. IAA supplied to mature exporting leaves of several species is transported at rates similar to that of [^{14}C] glucose (see Morris and Thomas, 1978) and accumulates at metabolic sinks. Following application of labeled IAA, radioactivity can be collected by aphids feeding on the phloem sap. The velocity of transport ranges from 100 to 240 mm h^{-1} and there is no fixed polarity (see Goldsmith, 1977). This pathway of IAA transport appears to be separate from the polar transport system since IAA applied to the apical bud does not appear to enter the phloem sap (see Morris and Thomas, 1978).

Transport of IAA conjugates

Transport from seed to shoot Cholodny (1935) recognized that the seed was a rich source of auxin for germinating shoots. He demonstrated this by excising small blocks of endosperm, moistening the blocks, and applying them asymmetrically to growing shoots. The shoots responded by growing faster on the side to which the block had been applied. A physiological study was made by Skoog in 1937 to characterize better this 'seed auxin precursor'. Skoog found that something diffused upward from seed to shoot which was not itself an auxin but which could be converted to an auxin by prolonged contact with the shoot tissue. He found that substances like tryptamine had the approximate properties of the seed auxin precursor.

Recently there has been renewed interest in the seed auxin precursor. High-specific-activity labeled tryptophan is available commercially and labeled IAA and IAA-*myo*-inositol are readily prepared from tryptophan (Michalczuk and Chisnell, 1982) or chemically synthesized (Nowacki, Cohen and Bandurski, 1978). Thus, it is possible, with at least these three compounds, to determine whether they can serve as the seed auxin precursor and move from the endosperm to the vegetative shoot. Application of labeled IAA, tryptophan and IAA-*myo*-inositol to the endosperm led to the appearance of 0.015, 0.15 and 6.3 pmol shoot^{-1} h^{-1}, respectively, of IAA or IAA ester, in the shoot. Thus, the seed can provide some of the IAA required by the vegetative shoot as IAA-*myo*-inositol.

Summary Understanding how IAA is moved about in the plant is important in understanding the determinants of IAA concentration in the tissue. Further, the agronomic use of growth-regulating chemicals often depends upon their selective upward or downward transport, so knowledge of hormone transport is important. There is still no general agreement as to how the rapid and specific downward polar transport of IAA is accomplished and there are only the beginnings of knowledge concerning upward transport from seed to shoot. Transport- and carrier-deficient mutants, together with improved analytic capabilities for IAA and its conjugates, and the use of photosensitive probes should enable increased progress to be made.

Table 1.1 Physiological responses to exogenous auxin.

Phenomenon	Effect
(a) Effects at the cellular level	
Cell division	Stimulates cambial cell division
Cell elongation	Stimulates shoot growth
	Inhibits root growth
Cell differentiation	Stimulates xylem differentiation
	Stimulates phloem differentiation
	Promotes root initiation from cuttings
	Regulates callus tissue morphogenesis
(b) Effects at the organ and whole plant levels	
Seedling morphology	Reverses red-light inhibition of mesocotyl elongation (*see* Chapter 17)
Geotropism	IAA is transported to lower side of shoot (*see* Chapter 8)
Phototropism	IAA is transported to dark side of shoot (*see* Chapter 7)
Apical dominance	Replaces apical bud (*see* Chapter 6)
Leaf senescence	Delays senescence (*see* Chapter 20)
Leaf abscission	Auxin applied to leaf inhibits abscission
	Auxin applied proximal to the abscission layer promotes abscission (*see* Chapter 20)
Flowering	May promote flowering (*see* Chapter 18)
Fruit setting	Allows development of parthenocarpic fruit
Fruit ripening	Delays ripening

Physiological effects

Table 1.1 lists a number of responses induced by IAA. These responses involve cell division, cell enlargement and cell differentiation. Since the responses are induced by exogenous application of IAA, they may be pharmacological in nature, involving abnormal amounts of IAA. It is uncertain to what extent these responses mirror processes under the control of endogenous IAA. There are, however, a few examples of physiological activity which do correlate with the amount of IAA in the tissue. For example, stamen filaments of *Gaillardia grandiflora* elongate following application of IAA and can also be shown to peak in endogenous IAA content when they are elongating (Koning, 1983). Photo-inhibition of mesocotyl growth induces a decrease in free IAA, and an increase in ester IAA, which is commensurate with the inhibition of growth (Bandurski *et al.*, 1977).

Time course of action

As discussed in the introduction, there may be many target sites for IAA action or a single target site which unleashes a cascade of events. The events may occur in seconds (*see* Evans, 1974) or about 10 min, as in coleoptile elongation, or in days, as in leaf abscission (Addicott and Lynch, 1951).

Mechanism of IAA action

For the purpose of this chapter it is suggested that the sequence of events following addition of IAA to a plant system is more or less as described below. The initial event would be adsorption of IAA to a hormone-specific binding site. The IAA-binding site complex would then initiate a cascade (Chock *et al.*, 1980) of reactions evidenced as (a) membrane phenomena leading to media acidification, and (b) nucleic acid-related phenomena leading to longer-range enzyme changes involved in growth and morphogenesis. One of the concomitants of (a) and (b) would be changes in the plasticity of the wall such that growth could occur. This increase in plasticity would require changes (1) in the protein matrix of the wall, (2) in the cellulosic matrix and (3) in the hemicellulosic matrix. This working hypothesis is illustrated in Fig. 1.6. Each of these steps will be considered in turn.

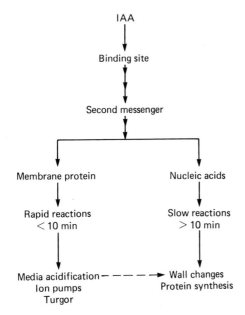

Fig. 1.6 A working hypothesis for the sequence of reactions involved in IAA action. IAA, once adsorbed to its specific binding site, would by means of a second messenger initiate the cascade of reactions called growth.

IAA binding

This has been critically reviewed (*see* Rubery, 1981) with the conclusion that no plant hormone receptor has been isolated with the certainty that the binding protein is involved in the hormone's action. However, by analogy with animal hormones, it seems certain that the hormone reacts with a macromolecular cellular component to exert its effect. The most widely used method to identify IAA-binding sites is to add the labeled hormone to a particular cell fraction and then measure how much of it is displaced by addition of unlabeled hormone. These results serve to measure a dissociation constant. The

site must then be evaluated for its structural specificity, its kinetic parameters and its tissue specificity.

Recently methods have been developed for the synthesis of photosensitive, IAA and cytokinin probes (Melhado, Jones and Leonard, 1982; Sussman and Kende, 1977). Perhaps by use of these probes it will be possible to link IAA irreversibly to its binding site so that it may be isolated with some assurance that it is in fact the binding site.

Membrane and ion movements leading to media acidification

Incubation of auxin-sensitive tissue in solutions of low pH will promote growth (Bonner, 1934). The pH optimum for acid-induced growth is 4.5 to 5 for tissues whose cuticle has been removed (Rayle, 1973). Acid-induced and IAA-induced growth have similar Q_{10} and similarly affect cell wall extensibility (*see* Cleland, 1982). These observations led Hager, Menzel and Krauss (1971) to propose the 'acid-growth' hypothesis. They suggested that IAA activated a plasma membrane ATPase, thereby stimulating active proton efflux from the cell. The resultant lowering of the pH of the solution bathing the cell walls might then activate enzymes capable of hydrolyzing wall polysaccharides to soften the wall and allow cell extension. They further showed that an uncoupler such as carboxyl cyanide m-chlorophenyl hydrazone (CCCP) prevented both cell extension and media acidification. Subsequently Rayle showed that KCN, dinitrophenol, valinomycin, abscisic acid and cycloheximide all inhibited growth and media acidification.

Even isolated walls will respond to solutions of low pH. For example, Rayle and Cleland (*see* Cleland, 1982) repeatedly froze and thawed *Avena* coleoptile sections and then subjected the flaccid walls to a deforming force. At pH 7, little deformation occurred, whereas with an acidic buffer the walls were stretched by that same force. The acid-growth hypothesis is also supported by the finding that neutral buffers will prevent IAA-induced growth; thus, it is certain that media acidification is somehow related to IAA-enhanced growth.

Mechanism of the media acidification

The mechanism of the media acidification is not known. It is known that the reaction has a requirement for Ca^{2+} as shown by Cohen and Nadler (1976). It is necessary carefully to deplete the plant section of Ca^{2+} before this requirement can be demonstrated. There appears to be a decrease in the ATP/ADP ratio (Trewavas, Johnston and Crook, 1967) and an increase in respiration (Anderson, Lovrien and Brenner, 1981). There have also been reports that ATP can restore IAA-induced growth under anaerobic conditions (Hager *et al.*, 1971). Interestingly, the rapidly-growing side of a tropically-stimulated stem is electropositive relative to the slower growing side and shows enhanced media acidification (Jaffe and Nuccitelli, 1977).

Conclusion It seems certain that media acidification is closely linked to IAA-induced growth. The finding that acid can induce short term growth, and that buffers stop the growth, argues that media acidification is more than just a concomitant of growth but rather is part of the growth phenomenon. It is difficult to see how cell expansion, media acidification and ion movements integrate into a system which must ultimately show cell division and cell differentiation, but these turgor-related changes may be only a small early part of the cascade released by IAA treatment.

Nucleic acid and protein changes

The observation that IAA stimulation of growth may be preceded by an increase in RNA synthesis (Silberger and Skoog, 1953) early led to the idea that IAA might act by depressing certain genes, thus causing altered RNA and protein synthesis. It was found that inhibitors of RNA and protein synthesis would inhibit auxin-induced growth and Key and Shanoren observed an IAA-stimulated incorporation of radiolabeled nucleotides into RNA (*see* Key, 1969).

Recently, Zurfluh and Guilfoyle (1982) and Theologis and Ray (1982) demonstrated the appearance of specific mRNA sequences following application of IAA to pea and soybean stem segments. The technique used involved autoradiography of two-dimensional electrophoresis gels on which [35S]methionine-labeled translation products from polyA+-mRNA were separated. Both increases and decreases in certain proteins were observed and some changes were apparent after 20 min, while others appeared only after 1–4 h.

There is some question as to whether these nucleic acid and protein changes are fast enough to be part of the early growth process. However, the effect of cycloheximide on growth (15–20 min) as observed by Evans and Ray (1969) and the earlier studies of Noodén and Thimann (1965) on growth-associated protein synthesis must be explained.

Conclusion There seems no doubt that enhanced protein and nucleic acid synthesis are concomitants of IAA-induced growth. It now seems within reach to determine what enzymes are encoded in the sequences of mRNA which become active after the plant is treated with IAA. The lack of synchrony of cell extension, and changes in RNA and protein, may reflect the complexity of the sequence of events triggered by IAA treatment.

Changes in wall plasticity

Since plant cells are encased in a cell wall it is obligatory that the wall stretches, or grows, as the protoplast enlarges. The cell may increase in size some 50-fold with little or no increase in dry weight. It was at first believed that the primary site of IAA action was upon the wall and as early as 1930 Heyn measured the effect of auxin on the stretching properties of the wall. When the wall softens the cell will expand owing to the turgor of the protoplast. The present view is that the changes in mechanical properties of the wall, although they occur very rapidly, are an essential and rapid process involved in growth but that the wall is not the primary target of IAA.

The reader is referred to Went and Thimann (1937) for a discussion of the early experiments on wall structure and wall stretching. Briefly, the early studies demonstrated that the cellulosic microfibrils were almost at right angles to the long axis of the plant prior to cell expansion but that the microfibrils become more nearly parallel to the plant's axis after expansion. What determines the rigidity of the wall and how the sliding of the points of attachment of the microfibrils occurs are still the subject of active research. In essence, there are four polymeric substances which have been most frequently considered as determinants of wall rigidity. They are the wall protein, the cellulosic β 1-4 glucan itself, a β 1-3 callose-type glucan, and a lichenin-type glucan having mixed β 1-3, 1-4 linkages.

Wall protein Pure, primary cell walls contain about 5% protein (Kivilaan, Beaman and Bandurski, 1959) and some of these proteins are enzymatically active including peroxidases and many hydrolytic enzymes. However, there is also a hydroxyproline-rich glycoprotein, somewhat like the collagen of animals, which has been named extensin by Lamport (1970). There is evidence that this structural protein is covalently linked to the carbohydrate components of the wall by an acid-labile galactose-serine linkage. Such cross-linking would enhance wall rigidity.

Wall carbohydrates Effects of IAA on both the cellulosic and hemicellulosic constituents of the wall have been reported (Zeroni and Hall, 1980). MacLachlan and his colleagues early showed a correlation between IAA stimulation of growth and cellulase activity (Fan and MacLachlan, 1966). Masuda (1968) observed an increase in β 1-3 glucanase activity induced by IAA and found that cell walls could be softened by glucanase treatment. Recently Nevins and his associates (Huber and Nevins, 1981) have solubilized wall proteins and used these proteins as antigens to immunize rabbits. The antibodies so obtained will reduce IAA-induced growth. To date, no antibody to a purified glucanase will reduce growth.

Conclusions The plant cell wall is a highly-ordered organelle with a complex structure of cellulosic microfibrils held in a matrix of hemicellulose and protein. Probably there is no single determinant of the structural rigidity of the wall. When IAA-induced growth occurs, the solution bathing the cell wall becomes more acid and the wall becomes more easily deformed. This change in the plasticity of the wall seems to be due not simply to the lower pH but rather to enzyme-induced changes in the wall. Whether this enhanced enzymatic activity is attributable to more enzyme secreted into the wall or to enhanced activity of enzymes already in the wall is not known.

Acknowledgement. We are indebted to A. Schulze and to Marianne La Haine for aid in manuscript preparation. This work was supported by the US National Science Foundation, PCM 79-04637 and the National Aeronautic and Space Administration, NAGW-97-ORD-25796. This is Journal Article 10727 from the Michigan Agricultural Experiment Station.

Selected further reading

Cohen, J. D. and Bandurski, R. S. (1982). Chemistry and physiology of the bound auxins. *Ann. Rev. Plant Physiol.* **33**, 403–30.

Sembdner, G., Gross, D., Liebisch, H. W. and Schneider, G. (1980). Biosynthesis and metabolism of plant hormones. In *Encyclopedia of Plant Physiology*, **9**, ed. J. MacMillan, Springer-Verlag, Berlin, pp. 281–390.

Thimann, K. V. (1969). The auxins. In *The Physiology of Plant Growth and Development*, Ed. M. B. Wilkins, McGraw-Hill, New York, pp. 3–36.

Thimann, K. V. (1977). *Hormone Action in the Whole Life of Plants.* Univ. of Mass. Press, Amherst, MA.

Went, F. W. and Thimann, K. V. (1937; reprinted 1978). *Phytohormones.* Allanheld, Osmun/Universe Books. Montclair, New Jersey.

Zeroni, M. and Hall, M. A. (1980). Molecular effects of hormone treatment on tissue. In *Encyclopedia of Plant Physiology*, **9**, ed. J. MacMillan, Springer-Verlag, Berlin, pp. 511–68.

References

Addicott, F. T. and Lynch, R. S. (1951). Acceleration and retardation of abscission by indoleacetic acid. *Science* **114**, 688–9.

Allen, J. R. F. and Baker, D. A. (1980). Free tryptophan and IAA levels in the leaves and vascular pathways of *Ricinus communis* L. *Planta* **148**, 69–74.

Anderson, J. M., Charbonneau, H., Jones, H. P., McCann, R. O. and Cormier, M. J. (1980). Characterization of the plant nicotinamide adenine dinucleotide kinase activator protein and its identification as calmodulin. *Biochemistry* **19**, 3113–20.

Anderson, P. C., Lovrien, R. E. and Brenner, M. L. (1981). Energetics of the response of maize coleoptile tissue to indoleacetic acid. *Planta* **151**, 499–505.

Andreae, W. A. and Collet, A. (1968). The effect of phenolic substances on the growth activity of indoleacetic acid applied to pea, root or stem sections. In *Biochemistry and Physiology of Plant Growth Substances*, eds F. Wightman and G. Setterfield, Runge Press, Ottawa, pp. 553–61.

Andreae, W. A. and Good, N. E. (1955). The formation of indoleacetylaspartic acid in pea seedlings. *Plant Physiol.* **30**, 380–2.

Bandurski, R. S. (1979). Homeostatic control of concentrations of indole-3-acetic acid. In *Plant Growth Substances 1979*, ed. F. Skoog, Springer-Verlag, Berlin, pp. 37–49.

Bandurski, R. S., Schulze, A. and Cohen, J. D. (1977). Photoregulation of the ratio of ester to free indole-3-

acetic acid and its derivatives in plants. *Biochem. Biophys. Res. Commun.* **79**, 1219–23.

Berridge, M. J. (1980). Receptors and calcium signalling. *Trends in Pharm. Sci.* **1**, 419–23.

Bonner, J. (1934). The relation of hydrogen ions to the growth rate of the *Avena* coleoptile. *Protoplasma* **21**, 406–23.

Buckhout, T. J., Young, K. A., Low, P. S. and Morré, D. J. (1981). *In vitro* promotion by auxins of divalent ion release from soybean membranes. *Plant Physiol.* **68**, 512–15.

Chisnell, J. R. and Bandurski, R. S. (1982). Isolation and characterization of indol-3-yl-acetyl-*myo*-inositol from vegetative tissue of *Zea mays*. *Plant Physiol.* **69** (S), 55.

Chock, P. B., Rhee, S. G. and Stadtman, E. R. (1980). Interconvertible enzyme cascades in cellular regulation. *Ann. Rev. Biochem.* **49**, 813–43.

Cholodny, N. G. (1935). Uber das Keimungshormon von Gramineea. *Planta* **23**, 289–312.

Cleland, R. E. (1976). Kinetics of hormone-induced H⁺ excretion. *Plant Physiol.* **58**, 210–13.

Cleland, R. E. (1982). The mechanism of auxin-induced proton efflux. In *Plant Growth Substances 1982*, ed. P. F. Waring, Academic Press, London, pp. 23–31.

Cohen, J. D. (1982). Identification and quantitative analysis of indole-3-acetyl-L-aspartate from seeds of *Glycine max* L. *Plant Physiol.* **70**, 749–53.

Cohen, J. D. and Bandurski, R. S. (1978). The bound auxins: protection of indole-3-acetic acid from peroxidase-catalyzed oxidation. *Planta* **139**, 203–8.

Cohen, J. D. and Bandurski, R. S. (1982) Chemistry and physiology of the bound auxins. *Ann. Rev. Plant Physiol.* **33**, 403–30.

Cohen, J. D. and Nadler, K. D. (1976). Calcium requirement for indoleacetic acid-induced acidification by *Avena* coleoptiles. *Plant Physiol.* **57**, 347–50.

Cohen, J. D. and Schulze, A. (1981). Double standard isotope dilution assay. I. Quantitative assay of indole-3-acetic acid. *Anal. Biochem.* **112**, 249–57.

Corcuera, L. J. and Bandurski, R. S. (1982). Biosynthesis of indol-3-yl-acetyl-*myo*-inositol in kernels of *Zea mays* L. *Plant Physiol.* **70**, 1664–6.

Corcuera, L. J., Michalczuk, L. and Bandurski, R. S. (1982). Enzymic synthesis of indol-3-ylacetyl-*myo*-inositol galactoside. *Biochem. J.* **207**, 283–90.

Crawford, I. P. and Stauffer, G. V. (1980). Regulation of tryptophan biosynthesis. *Ann. Rev. Biochem.* **49**, 163–95.

Ehmann, A. (1976). Indoles in wine I. Chloroform-soluble non-volatile indoles. *Plant Physiol.* **57**(S), 30.

Engvild, K. C., Egsgaard, H. and Larsen, E. (1980). Determination of 4-chloroindole-3-acetic acid methyl ester in *Lathyrus*, *Vicia* and *Pisum* by gas chromatography-mass spectrometry. *Physiol. Plant* **42**, 499–503.

Epstein, E., Cohen, J. D. and Bandurski, R. S. (1980). Concentration and metabolic turnover of indoles in germinating kernels of *Zea mays* L. *Plant Physiol.* **65**, 415–21.

Evans, M. E. (1974). Rapid responses to plant hormones. *Ann. Rev. Plant Physiol.* **25**, 195–223.

Evans, M. L. and Ray, P. M. (1969). Timing of the auxin response in coleoptiles and its implications regarding auxin action. *J. Gen. Physiol.* **53**, 1–20.

Fan, D. F. and MacLachlan, G. A. (1966). Control of cellulase activity by indole-acetic acid. *Can. J. Bot.* **44**, 1025–34.

Gardner, G. K., Shaw, S. and Wilkins, M. B. (1974). IAA transport during the phototropic responses of intact *Zea* and *Avena* coleoptiles. *Planta* **121**, 237–51.

Gander, J. C. and Nitsch, C. (1967). Isolement de l'ester methylique d'un acid chloro-3-indolylacetique a partir de graines immature de pois *Pisum sativum* L. *C. R. Acad. Sci. D.* **265**, 1795–8.

Goldsmith, M. H. M. (1977). The polar transport of auxin. *Ann. Rev. Plant Physiol.* **28**, 439–78.

Goldsmith, M. H. M. (1982). A saturable site responsible for polar transport of indole-3-acetic acid in sections of maize coleoptiles. *Planta* **155**, 68–75.

Greenwood, M. S., Hillman, J. R., Shaw, S. and Wilkins, M. B. (1973). Localization and identification of auxin in roots of *Zea mays*. *Planta* **109**, 369–74.

Hager, A., Menzel, H. and Krauss, A. (1971). Versuche und Hypothese zur Primarwirkung des Auxins beim Streckungswachstum. *Planta* **100**, 47–75.

Hall, P. J. (1980). Indole-3-acetyl-*myo*-inositol in kernels of *Oryza sativa*. *Phytochemistry* **19**, 2121–23.

Hall, P. J. and Bandurski, R. S. (1981). Hydrolysis of ³H-IAA-*myo*-inositol by extracts of *Zea mays*. *Plant Physiol.* **67** (S), 2.

Hangarter, R. P. and Good, N. E. (1981). Evidence that IAA conjugates are slow-release sources of free IAA in plant tissues. *Plant Physiol.* **68**, 1424–7.

Hankins, C. N., Largen, M. T. and Mills, S. E. (1976). Some physical characteristics of the enzymes of L-tryptophan biosynthesis in higher plants. *Plant Physiol.* **57**, 101–4.

Hansch, C. and Muir, R. M. (1950). The ortho effect in plant growth regulators. *Plant Physiol.* **25**, 389–93.

Heerkloss, R. and Libbert, E. (1976). On the question of β-indolyl-acetic acid synthesis from indole without a tryptophan intermediate. *Planta* **131**, 299–302.

Hertel, R., Evans, M. L., Leopold, C. A. and Sell, H. M. (1969). The specificity of the auxin transport system. *Planta* **85**, 238–49.

Heslop-Harrison, J. (1980). Darwin and the movement of plants: a retrospect, in *Plant Growth Substances 1979*, ed. F. Skoog, Springer-Verlag, Berlin, pp. 3–14.

Heyn, A. N. J. (1930). On the relationship between growth and extensibility of the cell wall. *Proc. Kon.*

Akad. Wetensch. Amsterdam **33**, 1045–58.

Hinman, R. L. and Lang, S. (1965). Peroxidase-catalyzed oxidation of indole-3-acetic acid. *Biochemistry* **4**, 144–58.

Hollenberg, P. F., Rand-Meir, T. and Hager, L. P. (1974). The reaction of chlorite with horseradish peroxidase and chloroperoxidase. *J. Biol. Chem.* **249**, 5816–25.

Huber, D. J. and Nevins, D. J. (1981). Wall protein antibodies as inhibitors of growth and of autolytic reactions of isolated cell wall. *Physiol Plant.* **53**, 533–9.

Igoshi, M., Yamaguchi, I., Takahashi, N. and Hirose, K. (1971). Plant growth substances in the young fruit of *Citrus unshiu. Agric. Biol. Chem.* **35**, 629–31.

Jaffe, L. F. and Nuccitelli, R. (1977). Electrical controls of development. *Ann. Rev. Biophys. Bioeng.* **6**, 445–76.

Katekar, G. F. and Geissler, A. E. (1982). Auxins II: the effect of chlorinated indolylacetic acids on pea stems. *Phytochemistry* **21**, 257–60.

Key, J. L. (1969). Hormones and nucleic acid metabolism. *Ann. Rev. Plant Physiol.* **20**, 449–74.

Kinashi, H., Suzuki, Y., Takeuchi, S. and Kawarada, A. (1976). Possible metabolic intermediates from IAA to β-acid in rice bran. *Agric. Biol. Chem.* **40**, 2465–70.

Kivilaan, A., Beaman, T. C. and Bandurski, R. S. (1959). A partial chemical characterization of maize coleoptile cell walls prepared with the aid of a continually renewable filter. *Nature* **184**, 81–2.

Klämbt, H. D. (1961). Wachstums Induktion und Wuchsstoffmetabolismus in Weizenkoleoptilzylinder. II. Stoffwechselprodukte der Indol-3-Essigsaure und der Benzoesaure. *Planta* **56**, 618–31.

Klee, C. B., Crouch, T. H. and Richman, P. G. (1980). Calmodulin. *Ann. Rev. Biochem.* **49**, 489–515.

Koning, R. E. (1983). The role of auxin, ethylene and acid growth in filament elongation in *Gaillardia grandiflora* (*Asteraceae*). *Am. J. Bot.* **70**, 602–10.

Lamport, D. T. A. (1970). Cell wall metabolism. *Ann. Rev. Plant Physiol.* **21**, 235–70.

Law, D. M. and Hamilton, R. H. (1982). A rapid isotope dilution method for analysis of indole-3-acetic acid and indoleacetyl aspartic acid from small amounts of plant tissue. *Biochem. Biophys. Res. Commun.* **106**, 1035–41.

Magnus, V., Bandurski, R. S. and Schulze, A. (1980). Synthesis of 4,5,6,7 and 2,4,5,6,7 deuterium-labeled indole-3-acetic acid for use in mass spectrometric assays. *Plant Physiol.* **66**, 775–81.

Magnus, V., Iskric, S. and Kueder, S. (1971). Indole-3-methanol, a metabolite of indole-3-acetic acid in pea seedlings. *Planta* **97**, 116–25.

Marumo, S. and Hattori, H. (1970). Isolation of D-4-chlorotryptophan derivatives as auxin related metabolites from immature seeds of *Pisum sativum. Planta* **90**, 208–11.

Masuda, Y. (1968). Role of cell wall degrading enzymes in cell-wall loosening in oat coleoptiles. *Planta* **83**, 171–84.

McDougall, J. and Hillman, J. R. (1978). Analysis of indole-3-acetic acid using GC-MS techniques. In *Isolation of Plant Growth Substances*, ed. J. R. Hillman, Camb. Univ. Press, Cambridge.

Melhado, L. L., Jones, A. M. and Leonard, N. J. (1982) Photolysis of azido auxins *in vitro* and *in vivo*. *Plant Physiol.* **69** (S), 56.

Menschick R. and Hild, V. (1976). Decarboxylation of IAA in coleoptiles in relation to phototropism, *Abstracts 9th International Conference on Plant Growth Substances*, ed. P. E. Pilet, pp. 252–4.

Meudt, H. and Bennett, H. W. (1978). Rapid bioassay for auxin. *Physiol. Plant.* **44**, 422–8.

Michalczuk, L. and Chisnell, J. R. (1982). Enzymatic synthesis of 5-^{3}H-indole-3-acetic acid and 5-^{3}H-indole-3-acetyl-*myo*-inositol from 5-^{3}H-L-tryptophan. *J. Labelled Comp. Radiopharm.* **19**, 121–8.

Miflin, B. J. and Lea, P. J. (1977). Amino acid metabolism. *Ann. Rev. Plant Physiol.* **28**, 299–329.

Mitchell, J. W. and Livingston, G. A. (1968). Methods of studying plant hormones and growth-regulating substances. *Agric. Handb.* **336**, 131 pp., U.S. Dept. of Agric., Washington, D.C.

Mitra, R., Burton, J. and Varner, J. E. (1976). Deuterium oxide as a tool for the study of amino acid metabolism. *Anal. Biochem.* **70**, 1–17.

Mockaitis, J. M., Kivilaan, A. and Schulze, A. (1973). Studies of the loci of indole alkaloid biosynthesis and alkaloid translocation in *Ipomoeae violacea* plants. *Biochem. Physiol. Pflanz.* **164**, 248–57.

Morgan, P. W. and Durham, J. I. (1983). Strategies for extracting, purifying and assaying auxins from plant tissues. *Bot. Gaz.* **144**, 20–31.

Morris, D. A. and Thomas, A. G. (1978). Microautoradiographic study of auxin transport in the stem of intact pea seedlings (*Pisium sativum* L.). *J. Exp. Bot.* **29**, 147–57.

Mousdale, D. M. A. (1978). Spectrophotofluorimetric methods of determining indole-3-acetic acid. In *Isolation of Plant Growth Substances*, ed. J. R. Hillman, pp. 27–41, Cambridge University Press, Cambridge.

Nonhebel, H. M. (1982). Metabolism of indole-3-acetic acid in seedlings of *Zea mays* L. Ph.D. Thesis, Univ. of Glasgow, Scotland.

Nonhebel, H. M., Crozier, A. and Hillman, J. R. (1983). Analysis of [^{14}C]indole-3-acetic metabolites from the primary roots of *Zea mays* seedlings using reverse-phase high-performance liquid chromatography. *Physiol. Plant* **57**, 129–34.

Noodén, L. D. and Thimann, K. V. (1965) Inhibition of protein synthesis and auxin-induced growth by chloramphenicol. *Plant Physiol.* **40**, 193–207.

Norman, A. G. (1946). Studies on plant growth regulating substances. *Bot. Gaz.* **107**, 475.

Nowacki, J. and Bandurski, R. S. (1980). *Myo*-inositol esters of indole-3-acetic acid as seed auxin precursors of *Zea mays* L. *Plant. Physiol.* **65**, 422–7.

Nowacki, J., Cohen, J. D. and Bandurski, R. S. (1978). Synthesis of ^{14}C-indole-3-acetyl-*myo*-inositol. *J. Labelled Comp. Radiopharm.* **15**, 325–9.

Overton, E. (1902). Beitrage zur allgemeinen Muskel- und Nerven Physiologie. *Pflugers Arch. ges Physiol.* **92**, 115.

Pengelly, W. L., Bandurski, R. S. and Schulze, A. (1981). Validation of a radioimmunoassay for indole-3-acetic acid using gas chromatography-selected ion monitoring-mass spectrometry. *Plant Physiol.* **68**, 96–8.

Pengelly, W. L., Hall, P. J., Schulze, A. and Bandurski, R. S. (1982). Distribution of free and ester indole-3-acetic acid in the cortex and stele of the *Zea mays* mesocotyl. *Plant Physiol.* **69**, 1304–7.

Penny, D., Penny, P. and Marshall, D. C. (1974) High resolution measurement of plant growth. *Can. J. Bot.* **52**, 959–69.

Percival, F. W. and Bandurski, R. S. (1976). Esters of indole-3-acetic acid from *Avena* seeds. *Plant Physiol.* **58**, 60–7.

Percival, F. W., Purves, W. K. and Vickery, L. E. (1973). Indole-3-ethanol oxidase. *Plant Physiol.* **51**, 739–43.

Rayle, D. L. (1973). Auxin-induced hydrogen ion secretion in *Avena* coleoptiles and its implications. *Planta* **114**, 63–73.

Rayle, D. L. and Purves, W. K. (1967). Conversion of indole-3-ethanol to indole-3-acetic acid in cucumber seedling shoots. *Plant Physiol.* **42**, 1091–3.

Reeve, D. R. and Crozier, A. (1980). Quantitative analysis of plant hormones. In *Encyclopedia of Plant Physiol.*, ed. J. MacMillan, Springer-Verlag, Berlin, pp. 41–78.

Reinecke, D. M. and Bandurski, R. S. (1981). Metabolic conversion of ^{14}C-indole-3-acetic acid to ^{14}C-oxindole-3-acetic acid. *Biochem. Biophys. Res. Commun.* **103**, 429–33.

Reinecke, D. M. and Bandurski, R. S. (1983). Oxindole-3-acetic acid, an indole-3-acetic acid catabolite in *Zea mays*. *Plant Physiol.* **71**, 211–13.

Ricard, J. and Job, D. (1974). Reaction mechanisms of indole-3-acetate degradation by peroxidases. A stopped-flow and low-temperature spectroscopic study. *Eur. J. Biochem.* **44**, 359–74.

Roberts, J. K. M., Ray, P. M., Wade-Jardetzky, N. and Jardetzky, O. (1980). Estimation of cytoplasmic and vacuolar pH in higher plant cells by ^{31}P NMR. *Nature* **283**, 870–2.

Rubery, P. H. (1981). Auxin Receptors. *Ann. Rev. Plant Physiol.* **32**, 569–96.

Ruge, U. (1961). Methoden der Wachstummessung. In *Encyclopedia of Plant Physiology*, ed. W. Ruhland, vol. 14, Springer-Verlag, Berlin, pp. 47–67.

Schneider, E. A. and Wightman, F. (1974). Metabolism of auxin in higher plants. *Ann. Rev. Plant Physiol.* **25**, 487–513.

Schneider, E. A. and Wightman, F. (1978). Auxins. In *Phytohormones and Related Compounds—a Comprehensive Treatise*, eds D. S. Letham, P. B. Goodwin and T. J. V. Higgins, Elsevier, North Holland, 1, 29–105.

Sembdner, G., Gross, D., Liebisch, H. W. and Schneider, G. (1980). Biosynthesis and metabolism of plant hormones. In *Encyclopedia of Plant Physiology*, ed. J. MacMillan, Springer-Verlag, Berlin, 9, 281–390.

Shaw, S. and Wilkins, M. B. (1974) Auxin transport in roots. X. Relative movement of radioactivity from IAA in the stele and cortex of *Zea* root segments. *J. Exp. Bot.* **25**, 199–207.

Silberger, J. and Skoog, F. (1953). Changes induced by indoleacetic acid in nucleic acid contents of tobacco pith tissue. *Science* **118**, 443–4.

Simon, E. W. and Beevers, H. (1951). Quantitative relation between pH and the activity of weak acids and bases in biological experiments. *Science* **114**, 124–6.

Skoog, F. (1937). A deseeded *Avena* test method for small amounts of auxin and auxin precursors. *J. Gen. Physiol.* **20**, 311–34.

Sussman, M. R. and Kende, H. (1977). The synthesis and biological properties of 8-azido-N^6-benzyladenine, a potential photoaffinity reagent for cytokinin. *Planta* **137**, 91–6.

Tang, Y. W. and Bonner, J. (1947). The enzymatic inactivation of indoleacetic acid. I. Some characteristics of the enzyme contained in pea seedlings. *Arch. Biochem.* **13**, 11–25.

Theologis, A. and Ray, P. M. (1982). Changes in messenger RNAs under the influence of auxin. In *Plant Growth Substances 1982*, ed. P. F. Wareing, pp. 43–57, Academic Press, London.

Thimann, K. V. (1974). Fifty years of plant hormone research. *Plant Physiol.* **54**, 450–3.

Thimann, K. V. and Leopold, H. C. (1955). In *The Hormones*, eds G. Pincus and K. V. Thimann, Academic Press, New York, 3, 1–56.

Thompson, H. E., Swanson, C. P. and Norman, A. G. (1946). New growth regulating compounds. I. Summary of growth inhibitory activities of some organic compounds as determined by three tests. *Bot. Gaz.* **107**, 476–507.

Trewavas, A. J., Johnston, I. R. and Crook, E. M. (1967). The effects of some auxins on the levels of phosphate esters in *Avena sativa* coleoptile sections. *Biochim. Biophys. Acta* **136**, 307–11.

Van Overbeek, J. (1935). The growth hormone and the dwarf type of growth in corn. *Proc. Nat. Acad. Sci.* **21**, 292–9.

Van Overbeek, J. (1941). A quantitative study of auxin and its precursor in coleoptiles. *Am. J. Bot.* **28**, 1–10.

Venis, M. A. (1972). Auxin-induced conjugation system in peas. *Plant Physiol.* **49**, 24–7.

Waldrum, J. D. and Davies, E. (1981). Subcellular localization of IAA oxidase in peas. *Plant Physiol.* **68**, 1303–7.

Weiler, E. W. (1982). Plant hormone immunoassay. *Physiol. Plant.* **54**, 230–4.

Went, F. W. and Thimann, K. V. (1937). *Phytohormones.* MacMillan, New York.

Widholm, J. M. (1981). Utilization of indole analogs by carrot and tobacco cell tryptophan synthase *in vivo* and *in vitro. Plant Physiol.* **67**, 1101–4.

Wightman, F. and Lighty, D. G. (1982). Identification of phenylacetic acid as a natural auxin in the shoots of higher plants. *Physiol. Plant.* **55**, 17–24.

Wilkins, M. B., Cane, A. R. and McCorquodale, I. (1972). Auxin transport in roots. IX. Movement, export, resorption and loss of radioactivity from IAA by *Zea* root segments. *Planta* **106**, 291–310.

Wilkins, M. B. and Scott, T. K. (1968). Auxin transport in roots. *Nature* **219**, 1388–9.

Wilkins, M. B. and Whyte, P. (1968). Relationship between metabolism and the lateral transport of IAA in corn coleoptiles. *Plant Physiol.* **43**, 1435–42.

Zenk, M. H. (1961). 1-(indole-3-acetyl)-β-D-glucose, a new compound in the metabolism of indole-3-acetic acid in plants. *Nature* **191**, 493–4.

Zeroni, M. and Hall, M. A. (1980). Molecular effects of hormone treatment on tissue. In *Encyclopedia of Plant Physiology*, vol. 9, ed. J. MacMillan, Springer-Verlag, Berlin, pp. 511–68.

Zurfluh, L. L. and Guilfoyle, T. J. (1982). Auxin induced changes in the population of translatable messenger RNA in elongating sections of soybean hypocotyl. *Plant Physiol.* **69**, 332–7.

Gibberellins

2

Russell L Jones and Jake MacMillan

Introduction

The gibberellins (GAs) are a large family of diter-
pene acids. They were originally isolated as metabolites of the fungus *Fusarium moniliforme*, the imper-
fect stage of *Gibberella fujikuroi*, and were shown to
cause a wide range of often spectacular growth
responses when applied to intact plants. The GAs
are now known to be of widespread, and probably
universal, occurrence in higher plants where they
are generally accepted to function as hormones. In
this discussion of the GAs, an attempt is made to
draw together the vast literature on this subject and
to provide a concise and balanced overview. The
dangers in oversimplification, however, are readily
acknowledged.

The discovery of the GAs as plant hormones is a
fascinating story which has been related in many
publications. The early history is described in a
review by Stowe and Yamaki (1957) and in the
Source Book on Gibberellins 1828–1957 by Stodola (1958). The latter booklet makes compelling
reading and it is most regrettable that it had such
limited circulation. Recently Phinney (1983) has
provided a more personal and anecdotal account of
the history up to the time of the discovery of the
GAs in higher plants. Tamura (1969, 1977) has
discussed the earlier work from a Japanese point of
view; unhappily these fascinating accounts are not
generally available in English translation (but *see*
Phinney, 1983).

Occurrence and distribution

At the time of this writing 62 individual GAs are
known. Of these, 25 have been isolated from the
fungus *Gibberella fujikuroi*, 51 from higher plants
and 14 are common to both sources. The chemical
structures are shown in Figs 2.1 and 2.2 where they
are annotated F (for fungus) and P (for higher
plant). The different structural types are discussed
later. Gibberellin A_3 and mixtures of GA_4 and GA_7
are produced commercially from *G. fujikuroi*. Many
other fungi have been examined for GA production
but only recently has a fungus other than *G. fujikuroi* been shown to produce GAs. The pathogen,
Sphaceloma manihoticola, also an ascomycete which
causes the 'super-elongation disease' of cassava
(CIAT Annual Reports, 1972, 1975), produces GA_4
(Rademacher and Graebe, 1979).

The GAs from higher plants have been chemically
identified in 28 species representing 11 families,
predominantly angiosperms. An even wider distribution is indicated by the presence of GA-like
biological activity in plant extracts. Katsumi (1969)
records GA-like substances in 130 species of angiosperms and 9 species of gymnosperms. Murakami
(1972) found GA-like substances in 75 of 85 species
from 82 different families. Gibberellin-like activity
has also been reported in extracts from mosses,
algae and ferns, and the methyl ester of GA_9 has
been found in prothallia of *Lycopodium japonicum*
(Yamane, Takahashi, Takeno and Furuya, 1979).
Thus the GAs are indeed of widespread occurrence
in plants.

Gibberellin-like activity has been obtained from

all parts of higher plants. In general, reproductive organs such as immature seeds contain more (several μg per g fresh weight) than vegetative tissue (*ca.* 1–10 ng per g fresh weight). Most of the GAs which have been isolated in pure form from higher plants have therefore been obtained from seeds. However, analytical methods are now available by which GAs can be identified without isolation in the pure state (Gaskin and MacMillan, 1978; Yokota, Murofushi and Takahashi, 1980; Reeve and Crozier, 1980) and radio-immunoassays have now been developed (Atzorn and Weiler, 1982).

Most plants contain groups of GAs. For example 10 have been identified in seeds of *Phaseolus coccineus*. Answers to the question as to why there are so many GAs in plants are beginning to emerge from the metabolic studies discussed later.

In addition to the GAs listed in Figs 2.1 and 2.2, several GA derivatives occur in higher plants (for leading references *see* relevant chapters in MacMillan, 1980). These derivatives include alkyl esters, such as *n*-propyl esters of GA_1 and GA_3 and the methyl ester of GA_{17}, and GA-conjugates. The latter derivatives appear to be of widespread occurrence since enzymatic hydrolysis of water-soluble fractions of extracts from a wide range of plants gives GAs or GA-like compounds. Naturally-occurring conjugates of defined structure are listed in Fig. 2.3. Conjugates with glucose are of two types, glucosyl esters and glucosyl ethers; an example of each structural type is shown in Fig. 2.3.

Chemical structures

All GAs have the same basic ring-structure (*ent*-giberellane) in which the carbon atoms are numbered as shown in the first structure of Fig. 2.1. The GAs are subdivided into C_{20}-GAs, containing all 20 carbon atoms of their diterpenoid precursors, and the C_{19}-GAs which have lost carbon-20. Individual GAs are referred to as GA_x in the series GA_1-GA_n (MacMillan and Takahashi, 1968). The structures of the known C_{20}-GAs and C_{19}-GAs are shown in Fig. 2.1 and Fig. 2.2 respectively in a manner which classifies them into structural types. These structural types are discussed later in relation to present knowledge of the GA-metabolic pathways.

Metabolism

Metabolic studies have been conducted to establish the pathways by which the GAs are biosynthesized and then deactivated, and to characterize the enzymes which catalyze the individual steps in the pathways. With this information the control of GA metabolism, and hence GA-regulated plant development, can be investigated in relation to environmental and genetic factors. Progress towards these objectives is discussed in the following three sections. The subject has been extensively reviewed (Hedden, MacMillan and Phinney, 1978; Graebe and Ropers, 1978; Sembdner, Gross, Liebisch and Schneider, 1980; and relevant chapters in MacMillan, 1980 and in Crozier, 1983).

Metabolic pathways

The pathway shown in Fig. 2.4 from mevalonic acid to GA_{12}-7-aldehyde has been established in enzyme preparations from seeds of *Cucurbita maxima* (originally misidentified as *C. pepo*) (Graebe, Bowen and MacMillan, 1972), *Marah macrocarpus* (formerly *Echinocystis macrocarpa*) (Graebe, Dennis, Upper and West, 1965; Lew and West, 1971; West, 1973), and *Pisum sativum* (Ropers, Graebe, Gaskin and MacMillan, 1978) and in the fungus *G. fujikuroi* (West, 1973). The branch from *ent*-7α-hydroxy-kaurenoic acid to *ent*-6α,7α-dihydroxykaurenoic acid (Fig. 2.4) has been observed in enzyme preparations from *C. maxima* and cultures of *G. fujikuroi*. The pathway to GA_{12}-7-aldehyde, the first intermediate with the gibberellin ring structure, is probably common to all plants.

Gibberellin A_{12}-7-aldehyde is a branch-point to the various GAs. Several pathways have been established which differ mainly in the position and sequence of hydroxylation. More than one pathway from GA_{12}-7-aldehyde can occur in the same plant. The most detailed information on these pathways has been obtained using enzyme preparations from *C. maxima* endosperm (Fig. 2.5), in enzyme preparations from embryos and in developing seeds of *P. sativum* (Fig. 2.6), and in cultures of the fungus *Gibberella fujikuroi* (Fig. 2.7).

In the *C. maxima* system a complex metabolic grid has been established from GA_{12}-7-aldehyde *via* GA_{12} to GA_{43} and GA_4 (Graebe, 1980; Graebe, Hedden and Rademacher, 1980). Only the main

I. 10−Methyl

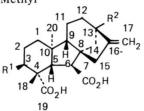

		R^1	R^2
GA_{12}	(F,P)	H	H
GA_{14}	(F)	OH	H
GA_{53}	(P)	H	OH
GA_{18}	(P)	OH	OH

II. 10−Hydroxymethyl

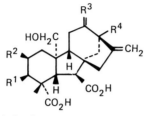

		R^1	R^2	R^3	R^4
GA_{15}	(F,P)	H	H	H_2	H
GA_{37}	(F,P)	OH	H	H_2	H
GA_{44}	(P)	H	H	H_2	OH
GA_{38}	(P)	OH	H	H_2	OH
GA_{27}	(P)	OH	OH	H_2	H
GA_{52}	(P)	OH	OH	$H,a−OH$	OH

isolated as

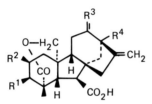

III. 10−Aldehyde

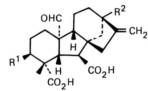

		R^1	R^2
GA_{24}	(F,P)	H	H
GA_{36}	(P)	OH	H
GA_{19}	(P)	H	OH
GA_{23}	(P)	OH	OH

IV. 10−Carboxylic acid

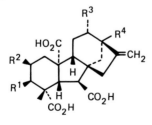

		R^1	R^2	R^3	R^4
GA_{25}	(F,P)	H	H	H	H
GA_{13}	(F,P)	OH	H	H	H
GA_{46}	(P)	H	OH	H	H
GA_{17}	(P)	H	H	H	OH
GA_{43}	(P)	OH	OH	H	H
GA_{39}	(P)	OH	H	OH	H
GA_{28}		OH	H	H	OH

V. 16−Hydroxylated

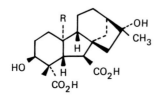

		R
GA_{41}	(F)	CO_2H
GA_{42}	(F)	CH_3

Fig. 2.1 Structures of C_{20}-GAs. F = fungal origin; P = higher plant origin.

I.   Non-hydroxylated

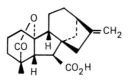

GA$_9$ (P,F)

II.   1β–Hydroxylated

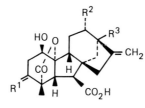

		R^1	R^2	R^3
GA$_{33}$	(P)	O	OH	H
GA$_{54}$	(P,F)	H,β–OH	H	H
GA$_{55}$	(P,F)	H,β–OH	H	OH
GA$_{60}$	(P)	H$_2$	H	OH
GA$_{61}$	(P)	H$_2$	H	H

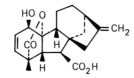

GA$_{62}$ (P)

III.   1a–Hydroxylated

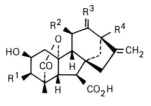

		R^1	R^2
GA$_{16}$	(P,F)	OH	H
GA$_{57}$	(F)	OH	OH

IV.   2β–Hydroxylated

		R^1	R^2	R^3	R^4
GA$_{51}$	(P)	H	H	H$_2$	H
GA$_{29}$	(P)	H	H	H$_2$	OH
GA$_{34}$	(P)	OH	H	H$_2$	H
GA$_8$	(P)	OH	H	H$_2$	OH
GA$_{48}$	(P)	OH	H	H,β–OH	H
GA$_{49}$	(P)	OH	H	H,a–OH	H
GA$_{50}$	(P)	OH	OH	H$_2$	H
GA$_{26}$	(P)	OH	H	O	H

Fig. 2.2 Structures of C$_{19}$-GAs. F = fungal origin; P = higher plant origin.

V.   2a–Hydroxylation

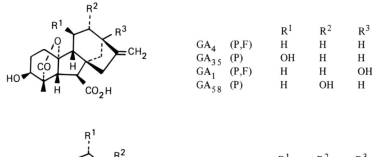

		R^1	R^2
GA_{40}	(F)	H	H
GA_{47}	(F)	OH	H
GA_{56}	(F)	OH	OH

VI.   3β–Hydroxylated

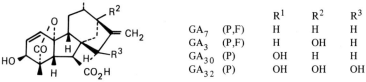

		R^1	R^2	R^3
GA_4	(P,F)	H	H	H
GA_{35}	(P)	OH	H	H
GA_1	(P,F)	H	H	OH
GA_{58}	(P)	H	OH	H

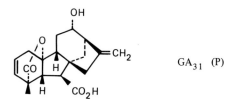

		R^1	R^2	R^3
GA_7	(P,F)	H	H	H
GA_3	(P,F)	H	OH	H
GA_{30}	(P)	OH	H	H
GA_{32}	(P)	OH	OH	OH

VII.   11β–Hydroxylation – all included in Groups IV and VI.

VIII.   12–Hydroxylation, included in Groups II, IV and VI except for

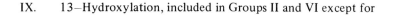

GA_{31} (P)

IX.   13–Hydroxylation, included in Groups II and VI except for

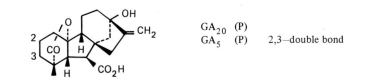

GA_{20} (P)

GA_5 (P)   2,3–double bond

Fig. 2.2 (contd)

X.   15β–Hydroxylation, included in Groups II and VI except for

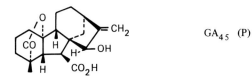

GA$_{45}$ (P)

XI.   16a–Hydroxylated

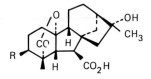

	R
GA$_{10}$ (F)	H
GA$_2$ (F)	OH

XII.   18–Oxygenation

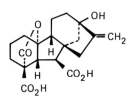

GA$_{21}$ (P)

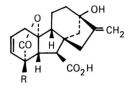

	R
GA$_{22}$ (P)	CH$_2$OH
GA$_{59}$ (P)	CO$_2$H

XIII.   Epoxides

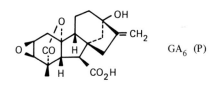

GA$_6$ (P)

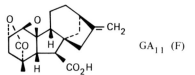

GA$_{11}$ (F)

Fig. 2.2 (*contd*)

GA$_1$ 3—O—β—D—Glucopyranosyl ether
GA$_3$ 3—O—β—D—Glucopyranosyl ether
GA$_8$ 2—O—β—D—Glucopyranosyl ether
GA$_{26}$ 2—O—β—D—Glucopyranosyl ether
GA$_{27}$ 2—O—β—D—Glucopyranosyl ether
GA$_{29}$ 2—O—β—D—Glucopyranosyl ether
GA$_{35}$ 11—O—β—D—Glucopyranosyl ether

GA$_1$ β—D—Glucopyranosyl ester
GA$_4$ β—D—Glucopyranosyl ester
GA$_9$ β—D—Glucopyranosyl ester
GA$_{37}$ β—D—Glucopyranosyl ester
GA$_{38}$ β—D—Glucopyranosyl ester

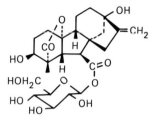

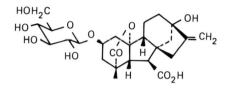

GA$_{29}$ 2—O—β—D— glucosyl ether

GA$_1$ β—D—glucosyl ester

Fig. 2.3 Higher plant conjugates of GAs and glucose.

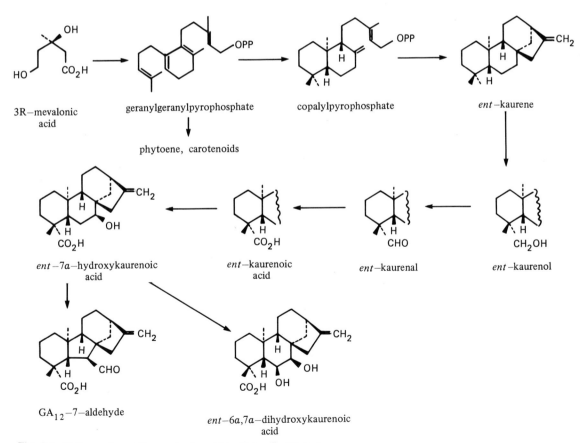

3R—mevalonic acid

geranylgeranylpyrophosphate

phytoene, carotenoids

copalylpyrophosphate

ent—kaurene

ent—kaurenol

ent—kaurenal

ent—kaurenoic acid

ent—7a—hydroxykaurenoic acid

GA$_{12}$—7—aldehyde

ent—6a,7a—dihydroxykaurenoic acid

Fig. 2.4 Pathway from 3R-mevalonic acid to GA$_{12}$-7- aldehyde.

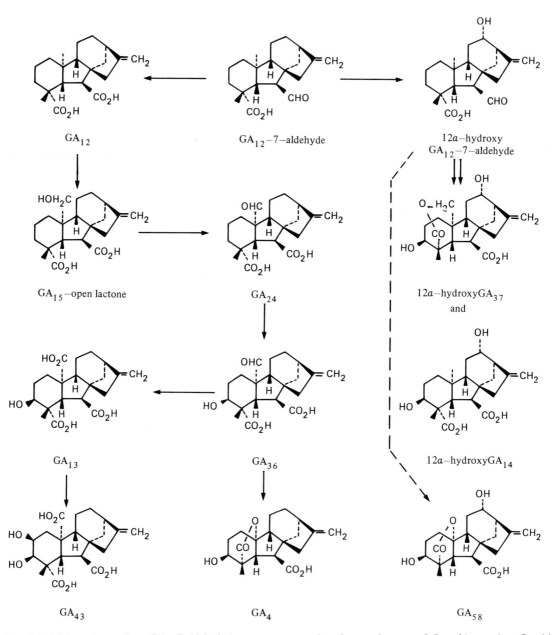

Fig. 2.5 Main pathways from GA_{12}-7-aldehyde in enzyme preparations from endosperm of *Cucurbita maxima*. Step(s) shown by dashed arrow are not established but GA_{58} is native to seeds.

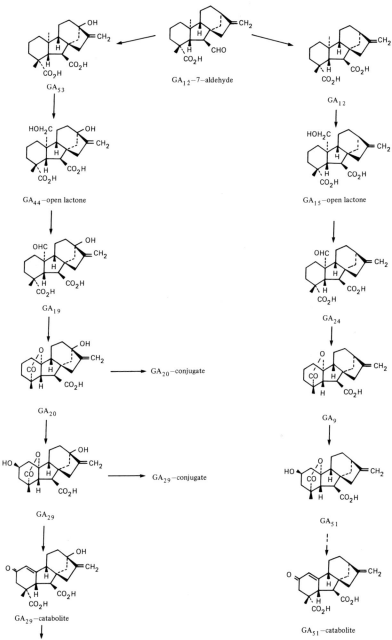

Fig. 2.6 Pathways from GA_{12}-7-aldehyde in seeds of *Pisum sativum*. The step shown by dashed arrow has not been established but the GA_{51}-catabolite is native to seeds.

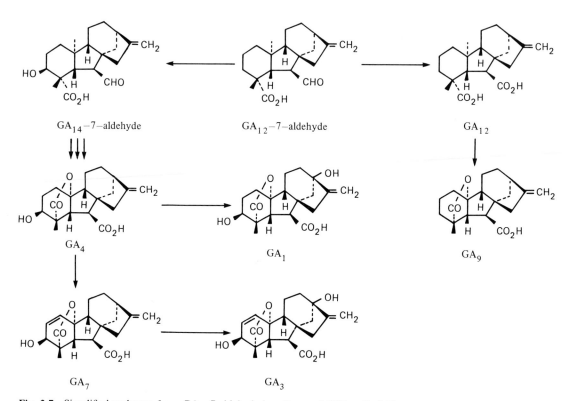

Fig. 2.7 Simplified pathways from GA_{12}-7-aldehyde in cultures of *Gibberella fujikuroi*.

pathway to these two GAs, native to *C. maxima* seeds, is shown in Fig. 2.5. In addition there is a branch from GA_{12}-7-aldehyde to 12α-hydroxy-GA_{12}-7-aldehyde (Fig. 2.5), which is further converted to 12α-hydroxyGA_{14} and 12α-hydroxyGA_{37}. This pathway presumably leads to GA_{58} which is native to *C. maxima* seeds but the steps have yet to be established (Hedden, Graebe, Beale, Gaskin and MacMillan, 1983).

In *P. sativum* seeds two pathways from GA_{12}-7-aldehyde also occur (Fig. 2.6). The steps from GA_9 and GA_{20} were first established *in vivo* in developing seeds of the cultivar Progress No. 9 (Frydman and MacMillan, 1975; Sponsel and MacMillan, 1977, 1978, 1980). Recently the steps from GA_{12}-7-aldehyde to GA_{51} and GA_{29} were established *in vitro* using enzyme preparations from the embryos of the cultivar Grosser Schnabel (Ropers *et al.*, 1978; Kamiya and Graebe, 1983). All C_{19}-GAs and the catabolites of GA_{29} and GA_{51} are native to

seeds of *P. sativum*. The pathway to GA_{29} is the major one.

In the fungus *G. fujikuroi* two pathways have been established using wild-type strains (Evans and Hanson, 1975) and mutants blocked for GA synthesis (Bearder, MacMillan and Phinney, 1975). Only the outlines of these pathways are shown in Fig. 2.7. The pathway to GA_3 is the major one.

In addition to these three systems, fragmentary evidence is available for seeds of *Phaseolus* species. In enzyme preparations from suspensors of *P. coccineus* the conversion *ent*-7α-hydroxykaurenoic acid to GA_1, GA_5 and GA_8 has been realized (Ceccarelli, Lorenzi and Alpi, 1981). In developing seeds of *P. vulgaris*, the metabolism of labeled C_{19}-GAs native to the seed has been investigated (Yamane, Murofushi and Takahashi, 1975; Yamane, Murofushi, Osada and Takahashi, 1977). The results, taken together with the known endogenous GAs, suggest three convergent pathways to

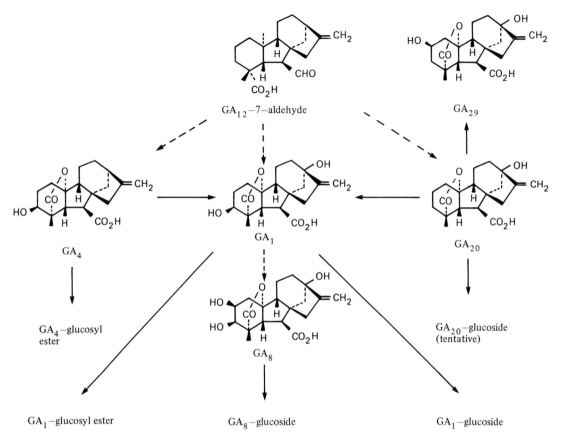

Fig. 2.8 Pathways in seeds of *Phaseolus*.

GA₁ as outlined in Fig. 2.8. The origins of the native GA₅, GA₆, and GA₃₄ have yet to be directly established.

Most of the detailed evidence for the metabolic pathways in higher plants has therefore been obtained from seeds, particularly using enzyme preparations. The reasons are technical. Seeds contain higher levels of GAs than other plant parts. Enzyme preparations provide a convenient *in vitro* system. They can be dialyzed to reduce or remove endogenous GAs and their precursors, thus minimizing dilution of radioactively-labeled substrates. As a result [¹⁴C] can be used as a radiolabel to follow metabolism by radio-counting and as a heavy isotope to determine the percentage incorporation of label by mass spectrometry. Thus metabolites can be identified and their specific radioactivity can be

determined simultaneously by gas liquid chromatography–mass spectrometry, requiring less than a nanogram of metabolite (Bowen, MacMillan and Graebe, 1972; Kamiya and Graebe, 1983). However, there are also disadvantages of *in vitro* systems, particularly the loss of compartmentalization and of correlative interactions in the whole plant.

Metabolic studies *in vivo* present a special problem. Since the pathways are hormonal, and since the objective is to determine the normal pathways and their control, it is important that the experimental methods should disturb the *in vivo* system as little as possible. An instructive example is the study of the metabolism of the C₁₉-GAs in developing seeds of *P. sativum* cultivar Progress No. 9 (Fig. 2.6). The native GAs were first identified (Frydman and

MacMillan, 1973) and then quantified throughout seed maturation (Frydman, Gaskin and MacMillan, 1974). The conversions of GA_{20} to GA_{29} and GA_{29} to GA_{29}-catabolite were established under the following conditions. Firstly, the labeled GAs were injected into the cotyledons of developing seeds at the predetermined levels. Secondly, the labeled GAs were applied and their metabolites were identified at the stage of seed development when the native compounds were near maximal concentration. Thirdly, using stable isotopes, it was shown by time-course experiments that the applied labeled GA and the native GA were metabolized in the same qualitative and quantitative manner (Sponsel and MacMillan, 1978, 1980). In view of the experimental design it is reasonable to conclude that the pathway $GA_{20} \rightarrow GA_{29} \rightarrow GA_{29}$-catabolite, is a natural sequence in these seeds. Recently Sponsel (1983b) has shown that the 2β-hydroxylation of GA_{20} to GA_{29} occurs in both the cotyledons and testas, but that the conversion of GA_{29} to GA_{29}-catabolite occurs exclusively in the testas. During desiccation and germination of dry seeds, GA_{29}-catabolite and unidentified compounds of its further degradation are transported back to embryos.

In vitro and *in vivo* studies of GA-metabolism in vegetative tissues and roots are fragmentary and have yielded little definitive information. Such studies present a major challenge for future work but a beginning has been made (*see* later).

Enzymes of the metabolic pathways

Geranylgeranyl pyrophosphate (GGPP) is the precursor of all diterpenes. The enzymes which catalyze its formation from mevalonic acid (for leading references *see* Porter and Spurgeon, 1981) have not been studied in detail in relation to GA metabolism.

Geranylgeranyl pyrophosphate is at a branch-point leading either to diterpenes (and hence GAs) or to tetraterpenes (phytoene, carotenoids) (Fig. 2.4). The cyclization of GGPP to *ent*-kaurene (Fig. 2.4) has therefore been investigated as a possible site of regulation of GA biosynthesis. The cyclization is catalyzed by *ent*-kaurene synthetase, which is composed of two activities. Activity-A catalyzes the cyclization of GGPP to copalyl pyrophosphate (CPP) and activity-B catalyzes the cyclization of CPP to *ent*-kaurene. The overall cyclization is catalyzed by activity-AB. The enzyme has been

purified from several plant sources (West, 1981) and requires a divalent metal ion, preferably Mg^{2+}. The A and B activities from endosperm of *M. macrocarpus* have been partly resolved by ion exchange chromatography (Duncan and West, 1981) and have similar properties. Duncan and West (1981) found that CPP derived from GGPP is more readily converted to *ent*-kaurene than exogenously supplied CPP and have suggested that the overall activity-AB is catalyzed by an enzyme complex of the free, but co-operating A- and B-enzymes. It is surprising therefore that B-activity, but no A-activity, has been obtained from some plants (Simcox, Dennis and West, 1975; Yafin and Schechter, 1975). The A-activity may be less stable but it may be under regulation.

There are some indications that *ent*-kaurene synthetase may be a site of regulation. For example, Coolbaugh and Moore (1969) have reported that, in cell-free preparations from seed of *P. sativum* cv. Alaska, the synthesis of *ent*-kaurene increased throughout the early stages of development, and Ecklund and Moore (1974) have reported a correlation between the rate of *ent*-kaurene synthesis in extracts of young shoots of the same plant and the rate of growth. Hedden and Phinney (1979) found that cell-free preparations from etiolated shoots of the GA-responding d-5 mutant of *Zea mays* produced only minor amounts of *ent*-kaurene and large amounts of *ent*-isokaurene when incubated with mevalonic acid, GGPP or CPP. The reverse was found for the normal genotype. The results indicate that the genetic lesion in the d-5 mutant affects the B-activity of *ent*-kaurene synthetase. Finally, some synthetic growth retardants, such as AMO-1618 and Phosphon D, strongly inhibit the A-activity in a non-competitive way, and others, such as SKF-3301, inhibit both activities (Dennis, Upper and West, 1965; Frost and West, 1977); the growth-retardant properties of these synthetic compounds may thus be explained by this inhibition of *ent*-kaurene synthetase.

The enzymes which catalyze the steps from mevalonic acid to *ent*-kaurene (Fig. 2.4) are soluble. Those which catalyze the steps from *ent*-kaurene to GA_{12}-7-aldehyde appear to be membrane-bound and mixed function oxidases. They require oxygen and a reduced pyridine nucleotide (preferably NADH) and appear to be dependent on cytochrome P450 with electron-transport systems like those found in liver (for leading references *see* West, 1980). Evidence for an *ent*-kaurene carrier protein in cotyledons of *P. sativum* (Moore, Barlow and Coolbaugh, 1972) has been questioned (Graebe and Ropers, 1978), although there is also evidence for such a protein in the mycelia of *G. fujikuroi* (Hanson, Willis and Parry, 1980). The growth-retardant, ancymidol, and synthetic triazoles such as Pachlobutazole inhibit the oxidation between *ent*-kaurene and *ent*-kaurenoic acid (Fig. 2.4).

The formation of GA_{12}-7-aldehyde from *ent*-7α-hydroxykaurenoic acid is a single oxidation step that is accompanied by the formation of *ent*-6α,7α-dihydroxykaurenoic acid (Fig. 2.4). The mechanism of both reactions in cell-free preparations from endosperm of *C. maxima* has been investigated in detail (Graebe *et al.*, 1972; Graebe and Hedden, 1980; Graebe, 1980). The enzyme activities for ring contraction and 6β-hydroxylation of *ent*-7α-hydroxykaurenoic acid could not be separated. They have the same cofactor requirements, the same temperature and pH optima and the same kinetics. The branch from *ent*-7α-hydroxykaurenoic acid to GA_{12}-7-aldehyde, although an attractive candidate, may not therefore be a point of regulation of GA biosynthesis. The enzyme activity for the further oxidation appears to be located on the endoplasmic reticulum (ER) (Hafemann, quoted by Graebe, 1980; Hafemann, Froneberg and Graebe, 1982).

The further oxidation of GA_{12}-7-aldehyde by enzyme preparations from *C. maxima* endosperm to GA_{12} and 12α-hydroxy GA_{12}-7-aldehyde (Fig. 2.5) is also catalyzed by microsomal enzymes. The 12α-hydroxylase has a low pH optimum (6.2) compared to that (7.0–7.5) for the enzyme which catalyzes the formation of GA_{12} (Hedden *et al.*, 1983). In *P. sativum* enzyme preparations, the oxidations of GA_{12}-7-aldehyde to GA_{12} and GA_{53} and of GA_{12} and GA_{53} (Fig. 2.6) are also catalyzed by microsomal enzymes and GA_{12} appears to be a better substrate for the 13-hydroxylase than GA_{12}-7-aldehyde (Kamiya and Graebe, 1983). The enzyme activity for the oxidation of GA_{12} appears to be located on the ER (Hafemann *et al.*, 1982).

The later hydroxylating enzymes in GA-metabolic pathways are soluble. In enzyme preparations from *C. maxima* endosperm, the steps from GA_{12} and 12α-hydroxyGA_{12}-7-aldehyde (Fig. 2.5) are catalyzed by soluble enzymes which require 2-oxoglutarate and Fe^{2+} (Hedden and Graebe, 1982).

In seeds of *P. sativum* similar soluble enzymes with the same requirements, which are characteristic of dioxygenases, have been shown to catalyze the steps from GA_{12} (Fig. 2.6) (Kamiya and Graebe, 1983; Hoad, MacMillan, Smith, Sponsel and Taylor, 1982). The soluble enzyme from germinating seeds of *P. vulgaris* (Patterson, Rappaport and Breidenbach, 1975) has been purified and shown to be a dioxygenase-type enzyme (Hoad *et al.*, 1982). The unfractionated soluble enzyme preparations from *C. maxima* endosperm and *P. sativum* embryos have not been fractionated, but preliminary evidence suggests that there is more than one 2β-hydroxylating activity in seed of *P. sativum* (Hoad *et al.*, 1982).

Significance of metabolic studies

With increasing information on the pathways, the status of the plethora of GAs is becoming clearer and the number of GAs which may be hormones *per se* is becoming smaller.

Of the C_{20}-GAs (Fig. 2.1) groups I, II and III represent intermediates in the sequential steps from GA_{12}-7-aldehyde to the C_{19}-GAs. Those which show biological activity probably do so by their metabolism to C_{19}-GAs. Those in Group IV are probably side products from the pathway to C_{19}-GAs.

The function of the various structural types of C_{19}-GAs (Fig. 2.2) is less well-defined. Some distinction between the structures of fungal and plant GAs can be made. Gibberellins hydroxylated at the 1β-, 2β- and 15β-positions (groups II, IV and X) and at the 11- and 12-positions occur in plants; the only exceptions are the trace fungal GAs, GA_{54} and GA_{55}. Hydroxylations at the 1α-, 2α- and 16α-positions (groups III, V and XI) are characteristic of fungal GAs; the only exception is GA_{16}. 3β- and 13-hydroxylated GAs (groups VI and IX) are common to both fungus and higher plants. Structure–activity relationships based upon plant bioassays show that non-hydroxylation (group I), 3β-hydroxylation (group VI) and 13-hydroxylation (group IX) are associated with high biological activity. However these conclusions may be misleading without detailed knowledge of the metabolic fate of the applied GAs in the bioassay system (*see* later). The significance of hydroxyl groups at carbons-1, -11, -12, -15, and -18 is not known. Most attention has centered on the significance of 2β-hydroxylation.

2β-Hydroxylated C_{19}-GAs have low or no biological activity (*see* Sponsel, Hoad and Beeley, 1977). They are commonly observed as metabolites of bioactive GAs which have been applied to plants (for leading references *see* Sembdner *et al.*, 1980) and the conversions of GA_{20} to GA_{29} and GA_9 to GA_{51} are natural processes in developing seeds of *P. sativum* (see earlier). 2β-Hydroxylation of C_{19}-GAs may therefore be a natural and irreversible process regulating the deactivation of hormonal GAs. 2β-Hydroxylated GAs are either oxidized further to unsaturated ketones in which the lactone ring is opened (e.g. GA_{29}-catabolite, Fig. 2.6) or are conjugated to give 2-O-β-D-glucosides (Fig. 2.3). Both processes can occur in the same plant, but conjugation is the minor pathway in *P. sativum* seeds (Fig. 2.6) and the major one in seeds of *P. vulgaris* (Fig. 2.8). The role of these further metabolites of 2β-hydroxyGAs is unknown but they cannot be transportable or sequestered forms of bioactive GAs since they are metabolites of the bio-inactive 2β-hydroxyGAs (for leading references *see* Sponsel, 1980, 1983a). The functions of the conjugates of the bioactive GAs, such as the glucosyl esters and ethers of GA_1, GA_3 and GA_4 (Fig. 2.3) are not known.

Although detailed information on the GA-pathways in higher plants is restricted to studies with seeds, the pathways in Figs 2.5, 2.6 and 2.8 provide useful models for other tissues. From a knowledge of the native GAs, speculative pathways can be constructed and rational experiments can be devised to test the hypothesis and explore the effects of environmental and genetic factors. Two examples illustrate this approach.

In shoots of *Spinacia oleracea*, six GAs were identified (Metzger and Zeevaart, 1980a) and a pathway based upon the pathway in *P. sativum* seeds (Fig. 2.6) was considered plausible. In fact, a more likely pathway is shown in Fig. 2.9. The changes in levels of these GAs, except for GA_{53}, were then determined in relation to photoperiod. Long-day conditions caused a five-fold decrease in the level of GA_{19} and a dramatic increase in the levels of GA_{20} and GA_{29}, while the levels of GA_{17} and GA_{44} did not change significantly (Metzger and Zeevaart, 1980b). Taken in conjunction with the subsequent demonstration in cell-free preparations from *P. sativum* embryos that GA_{19} is the precursor of GA_{20}, the results indicate that the step GA_{19} to GA_{20} is under photoperiodic control.

The second example concerns dwarfism. Seven GAs have been identified in maize tassels (Hedden, Phinney, Heupel, Fujii, Cohen, Gaskin, MacMillan and Graebe, 1982), allowing the construction of a possible pathway (Fig. 2.9) again by analogy with the pathway in seeds of *P. sativum*. Using this pathway as a working hypothesis, bioassay data and metabolic studies have provided evidence that the recessive single gene mutant dwarfs-1, -2, and -3 block the steps indicated in the pathway shown in Fig. 2.9 (Phinney and Spray, 1982 and 1983). The results of their studies indicate that there is only one pathway that controls elongation growth in maize and that GA_1 is the only GA in this pathway which is bioactive *per se*. The other GAs in the pathway are only bioactive because they are metabolized to GA_1. The latter conclusion has interesting implications for the interpretation of bioassay results using dwarf mutants of higher plants. Current studies with dwarf mutants of *P. sativum* (Potts, Ingram, Reid and Murfet, 1982) are leading to similar conclusions.

The action of gibberellins

Although the GAs were discovered because of the dramatic response elicited from certain higher plants following treatment with GA_3, the effect of this class of hormones is not confined to the angiosperms. In many conifer species, GAs promote shoot elongation, an increase in the diameter of stems, and flowering. In ferns, algae and fungi GAs have also been shown to influence the growth and development of a number of species. In yeast (*Saccharomyces ellipsoideus*), for example, GA can promote sporulation as well as growth (Kamisaka, Masuda and Yanagishima, 1967). It is the responses of higher plants to GAs that have been most intensively studied, however, and it is experiments with these organisms that have provided insight into the mechanism of GA action. The multitude of responses the GAs can elicit when applied to plants and the ways the GAs are used in the production of food and fiber will not be discussed here. The interested reader is referred to numerous books and reviews which provide comprehensive accounts of the effects of GAs on plant growth and development (Crozier, 1983; Jones, 1973; Krishnamoorthy, 1975; Skoog, 1980) and the application of these

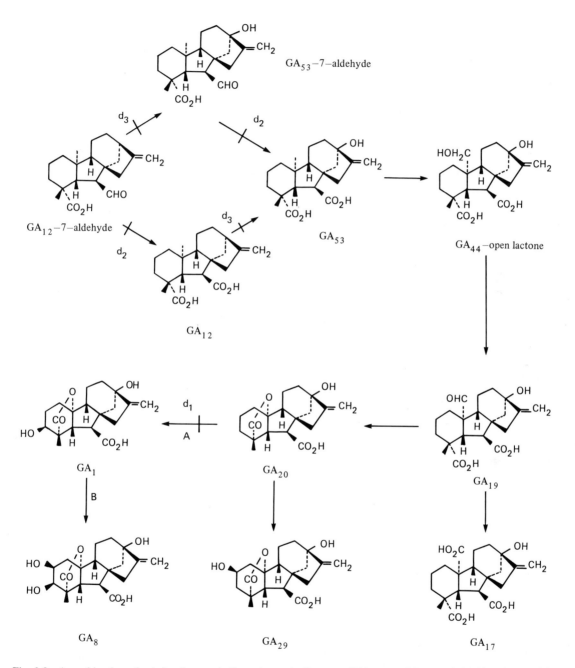

Fig. 2.9 A working hypothesis for the metabolic pathways in *Zea mays* (Phinney and Spray, 1982a, b) and, except for steps A and B, in *Spinacia oleracea* (Metzger and Zeevaart, 1980a, b). The symbols d_1, d_2 and d_3 denote the single gene mutants, dwarf-1, dwarf-2 and dwarf-3, of *Zea mays*. (It is not known whether GA_{12}-7-aldehyde is converted to GA_{53} *via* GA_{53}-7-aldehyde.)

compounds in agriculture (Skoog, 1980; Weaver, 1972). Rather, our discussion will be confined to experiments that have been designed to elucidate the mechanisms whereby the gibberellins modify plant growth and development.

Research into the action of GAs in higher plants has advanced along two distinct lines in the past decade. One avenue has emphasized the role of GA in growth, and the work has been biophysical in nature, while the other has focused on the capacity of GA to promote RNA and protein synthesis in germinating seeds and seedlings. This chapter will deal with the effects of GA on growth and protein synthesis separately; however, it should not be inferred that the molecular basis of GA action is distinctly different for these two processes. Rather, the effects of GA on growth and metabolism are discussed separately only because they represent two distinct experimental approaches to the examination of the mechanism of GA action.

GA and growth

The role of cell division It has been argued that GA can promote growth of plants by affecting either cell expansion or cell division or both. It must be emphasized that cell division alone cannot result in the growth of an organism (*see* Green, 1976), growth being defined as an irreversible increase in volume. The distinction between growth and cell division can be exemplified by the early stages of blastula development in animals. Cleavage of the fertilized egg occurs without a concomitant increase in volume. Thus the blastula is more highly differentiated than the fertilized egg, but growth has not occurred because there has been no increase in volume.

Cell division can contribute to growth only by producing more cells which can undergo expansion. There is evidence that GA affects the process of cell division in higher plants, and it does so in two ways. In the subapical region of both rosette and caulescent plants, GA_3 increases the size of the meristematic region and also increases the proportion of cells which are undergoing division (Loy, 1977). These effects of GA_3 on cell division can be readily accounted for by an effect on the cell cycle. Jacqmard (1968) has proposed that one of the effects of GA_3 is to promote the onset of DNA synthesis in cells which are arrested in the G_1 phase of the cell

cycle. This notion is supported by the data of Liu and Loy (1976) on the effect of GA_3 on the cell cycle of watermelon seedlings (Fig. 2.10). GA_3 reduces the duration of the cell cycle by nearly 30%, and it does so primarily by reducing the length of G_1 by 30% and that of S by 36%.

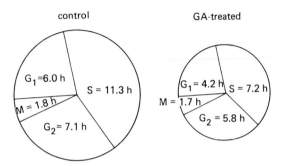

Fig. 2.10 The effect of GA on the duration in hours of the G_1, S, G_2 and M phases of the cell cycle of dwarf watermelon seedlings. Diagram drawn from the data of Liu and Loy (1976).

Unlike cytokinins, which promote cell division in cultured cells, GAs do not have a pronounced effect on cell number in suspension-cultured plants. GA_3 treatment does not promote cell division in cultured cells whose capacity to synthesize their own GAs has been prevented either by a single gene mutation or by chemical inhibitors of GA_3 biosynthesis (Rappaport, 1980). Although these observations appear to contradict the evidence from intact plants that implicates GAs in the control of cell division, it should be borne in mind that plant cells grown in culture in the absence of plant hormones are poorly differentiated, whereas cells in the shoot meristem show a high degree of specialization. It is easy to envisage that the capacity of cells to respond to GA_3 by undergoing division is a function of their state of differentiation.

Cell expansion: its biophysical basis Plant cell expansion can be described in biophysical terms by the following equation, formulated by Lockhart (1965) and modified by Cosgrove (1981), whereby steady state growth rate, V_s, is given by:

$$V_s = \frac{L\varphi}{L + \varphi}(\sigma \Delta \pi - Y)$$

where L is the hydraulic conductivity of the cell; φ, cell wall extensibility; σ, solute reflection coefficient; $\Delta\pi$, difference in osmotic potential across the plasma membrane; and Y, the yield threshold, defined as the minimum turgor pressure required to initiate cell expansion. The components of this equation contribute either to water uptake into the cell (L, σ and $\Delta\pi$) or to the capacity of the cell wall to extend (φ and Y). Plant growth regulators have been shown to influence each of these parameters; however, emphasis has been placed on changes in the osmotic potential (π) of cells and tissues and on the extensibility properties (φ) of the cell wall.

The role of osmotic potential One of the first proposals advanced to explain the biophysical mechanism of GA-stimulated growth stated that GA_3 lowered π (i.e. making ψ_s more negative, *see* Chapter 14) by increasing the solute concentration of the cell sap. This proposal was based on circumstantial evidence that correlated GA-stimulated growth with the disappearance of starch from the elongating tissues. A mechanism for GA_3 action in growth involving the conversion of starch to sugar was inferred by analogy with the known effects of GA_3 in regulating starch metabolism in cereal seeds (Paleg, 1965). Support for this idea was provided by experiments with cucumber hypocotyls which respond to both auxin and GA_3 by increasing in length (Cleland, Thompson, Rayle and Purves, 1968). Following auxin treatment there was an increase in cell wall extensibility, but GA-treatment did not result in a change in cell-wall yielding properties. Since GA_3-induced elongation was not correlated with an increase in cell-wall yielding properties, it was inferred that growth must be brought about by changes in π.

Measurement of the solute concentration of cell sap from control and GA_3-treated lettuce hypocotyl sections shows clearly that GA_3 does not promote growth by increasing the solute concentration of the cell sap (Fig. 2.11). Rather, the data show that during the most rapid phase of cell elongation the solute concentration of the tissue declines (Stuart and Jones, 1977). That growth can be independent of π is also indicated by the results of experiments which show that increased solute concentration of cells resulting from enhanced ion uptake does not always lead to an increase in growth rate. It is only when tissue has been treated with GA that changes

in π as a result of ion uptake lead to an increase in growth rate (Fig. 2.11 and Stuart and Jones, 1977). If the hydraulic conductivity (L) and solute reflection coefficient (σ) of the lettuce hypocotyl tissue are unaffected by GA, then it must be concluded that elongation is governed by changes in wall yielding properties.

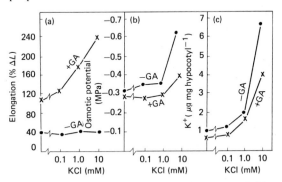

Fig. 2.11 The effect of KCl concentration on growth (a), osmotic potential (b) and K^+ uptake (c) of light-grown lettuce hypocotyl sections incubated in light in GA_3 ($\times$) or H_2O ($\bullet$) for 48 h. (Reproduced from Stuart and Jones, 1977.)

Extensibility changes in GA-treated tissues Although the early experiments of Cleland *et al.* (1968) suggested that GA_3 did not affect the extensibility properties of the cucumber hypocotyl cell wall, evidence from several laboratories using many different plant species indicates that GA_3 has a marked influence on cell-wall extensibility. In *Avena* internode sections, for example, GA_3 causes a large increase in wall extensibility as determined by Instron extensometry (Fig. 2.12 and Adams, Montague, Tepfer, Rayle, Ikuma and Kaufman, 1975). These changes in wall extensibility brought about by GA_3 in *Avena* parallel closely the hormone-induced changes in growth rate (Fig. 2.12). Extensometric measurements of the hypocotyl of intact lettuce also show that following GA_3 treatment there is an increase in cell wall extensibility relative to untreated control seedlings (Kawamura, Kamisaka and Masuda, 1976).

Using the osmotic method developed by Green and Cummins (1974) to determine the extensibility of tissue, Stuart and Jones (1977) examined the effects of light, GA_3 and KCl on the wall yielding properties of lettuce hypocotyl sections. Light

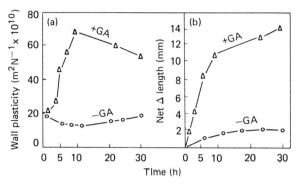

Fig. 2.12 Cell wall plasticity determined by Instron extensometry (a) and elongation (b) of *Avena* internode segments incubated with ($\triangle$) and without ($\bigcirc$) GA$_3$ (30 μM), (Redrawn from Kaufman and Dayanadan, 1983.)

reduced the extensibility of hypocotyl tissue while darkness and GA$_3$ overcame this inhibition (Fig. 2.13). Potassium chloride, on the other hand, had no effect on extensibility even though it stimulated growth of GA-treated or dark-grown tissue (Fig. 2.13). This result is in keeping with the observation that KCl treatment changes the π (ψ_s) of lettuce

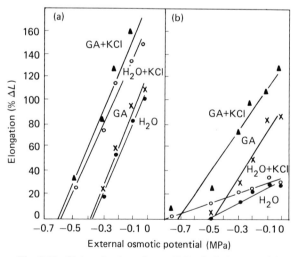

Fig. 2.13 Determination of extensibility in dark-grown (a) and in light-grown (b) lettuce hypocotyl sections treated with KCl in the presence and absence of GA using the osmotic method of Green and Cummins (1974). Tissue extensibility is proportional to the slope of the plot relating elongation to external osmotic potential. (Redrawn from Stuart and Jones, 1977.)

hypocotyl tissue (Fig. 2.11). This increased solute concentration, however, leads to an increase in growth rate of lettuce hypocotyl sections only when the inhibitory effects of light on extensibility are countered by GA$_3$ treatment or darkness. Thus, in lettuce, increased cell wall extensibility brought about by GA$_3$ treatment or darkness is the principal driving force for growth.

Although cell expansion can be described in biophysical terms, the events leading to altered cell wall extensibility are dependent on metabolism. It is well established, for example, that cell expansion is rapidly inhibited by respiratory poisons. Several hypotheses have been advanced to explain how the plant cell wall yields, but the two most widely held ideas are that either newly synthesized wall polymers are inserted into the wall, or cell-wall loosening factors break bonds within or between the cell wall polymers. These two ideas need not be mutually exclusive. Indeed, Albersheim (1976) has proposed that transglycosylation occurs in the cell wall and results in the cleavage of a glycosylic bond which causes weakening of the wall followed by the transfer of the polysaccharide terminus to a new position where the formation of a new glycosylic bond is catalyzed. Transglycosylases could therefore serve the dual purposes of glycosylic bond cleavage and synthesis (Albersheim, 1976).

There have been numerous demonstrations that in GA-treated tissue elongation is paralleled by an increase in the synthesis of cell wall polymers (Kaufman and Dayanandan, 1983; Montague and Ikuma, 1975; Srivastava, Sawhney and Taylor, 1975). In *Avena* internode segments, incorporation of [^{14}C] glucose into a cell wall fraction was promoted by GA$_3$ after one hour of incubation in the hormone, and the time course of incorporation closely matched the growth of the segments (Fig. 2.14 and Montague and Ikuma, 1975). Although these data and the experiments of others (Srivastava *et al.*, 1975) establish a relationship between cell wall synthesis and elongation, it is not known whether these changes in polysaccharide synthesis are causally related to growth. Careful measurement of the kinetics of GA-induced growth in lettuce hypocotyls shows that the hormone causes an increase in growth rate within 10 min of its addition to the incubation medium (Moll and Jones, 1981a). If cell expansion resulting from increased wall extensibility were dependent on the synthesis of new cell

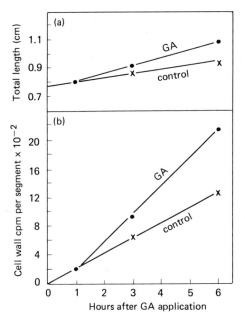

Fig. 2.14 The time course of elongation (a) and [^{14}C] glucose incorporation into cell wall polymers (b) of *Avena* internode sections incubated in the presence (●) and absence (×) of GA$_3$. (Reproduced with permission from Montague and Ikuma, 1975.)

wall polymers, then enhanced cell wall synthesis, as measured by radiolabel incorporation, should precede the earliest detectable growth response.

The polysaccharide composition of the cell wall that is synthesized during elongation of *Avena* internodes (Montague and Ikuma, 1975) or lettuce hypocotyls (Jones, 1983) is not affected by GA$_3$. Neither the sugar nor the uronide composition of the lettuce hypocotyl cell wall changes in response to GA$_3$ over a 24-h growth period (Jones, 1983).

The plant cell wall contains proteins and inorganic ions in addition to complex polysaccharides, and there have been proposals that both protein and ions, especially Ca^{2+}, play a role in wall extensibility (Preston, 1979). Calcium has long been known to cause a decrease in wall extensibility (Preston, 1979), and this ion has been implicated in the control of both auxin-induced (Cleland and Rayle, 1977) and GA-induced (Moll and Jones, 1981b) growth. The results of Moll and Jones on lettuce hypocotyls indicate that GA$_3$ promotes uptake of Ca^{2+} into the cytoplasm. It was proposed that in this

tissue Ca^{2+} activity in the cell wall governs tissue extensibility and that extensibility could be regulated by uptake of Ca^{2+} into the cytoplasm (Moll and Jones, 1981b).

Attention has recently been focused on the role of the phenolic constituents of the plant cell wall in growth. Fry (1979, 1980) has proposed that cell wall extensibility can be influenced by the formation of cross-bridges between adjacent ferulic acid moieties in the wall. The formation of diferuloyl bridges could be catalyzed by peroxidase resulting in the gelation of the cell wall (Fry, 1979). In suspension-cultured spinach cells Fry has shown that GA$_3$ promotes cell expansion and inhibits the release of peroxidase from the cell. According to this hypothesis, Fry envisages a less-rigid wall in GA-treated cells because peroxidase is not able to catalyze the formation of the diferuloyl bridge. Although this hypothesis is attractive, it may not explain how extensibility is controlled in dicotyledonous tissues since the primary cell walls of dicots contain negligible amounts of phenolics (Harris and Hartley, 1981).

It is clear from the above that neither changes in the rate of cell wall synthesis nor changes in its composition can be implicated in the control of cell elongation. The most serious limitation of most experimental approaches to these problems concerns the resolution of the growth response to GA$_3$ and the timing of cell-wall synthesis. Before a causal relationship can be established it must be demonstrated that changes in cell-wall synthesis precede the growth response to GA.

At least two cell-wall-loosening factors have been identified and shown to be involved in hormone-induced cell expansion in plants. The role of protons (H$^+$) is described by the acid growth hypothesis which has been shown to account for the early phase of auxin-induced elongation (Chapter 1). This hypothesis is based on the observation that auxin treatment results in rapid proton pumping from elongating tissue and that H$^+$ alone can promote growth of auxin-sensitive tissues (Cleland and Rayle, 1978).

The acid growth hypothesis has also been advanced to explain GA-induced elongation (Hebard, Amantangelo, Dayanandan and Kaufman, 1976). *Avena* internode segments release H$^+$ into the incubation medium following treatment with GA, but the magnitude of this effect is small, the hormone increasing H^+ release by only 17% above controls (Hebard et al., 1976). Protons were also implicated in GA-induced growth of *Avena* because acidic buffers promote elongation at optimal GA_3 concentrations (Hebard et al., 1976). The observation that acidic buffers promote the elongation of GA-treated tissue as well as controls is inconsistent with the hypothesis that GA promotes growth by regulating H^+ release.

Results of experiments with lettuce hypocotyl sections do not support the view that GA-stimulated proton release regulates elongation. GA_3 does not promote proton release from hypocotyl tissue under a wide range of incubation conditions despite the fact that this tissue can be induced to secrete protons rapidly when treated with the fungal toxin fusicoccin (Stuart and Jones, 1978). The growth response of lettuce hypocotyls to acidic media and GA_3 also indicates that the acid growth hypothesis does not apply to GA-induced growth (Stuart and Jones, 1978).

Hydrolytic enzymes have been shown to be involved in the modification of higher plant cell walls, and their activity is thought to be modified by hormones (Huber and Nevins, 1981a, b). Huber and Nevins (1981a) have used a novel approach to identify putative cell wall-modifying proteins which may be involved in auxin-induced elongation of *Zea mays* coleoptiles. They raised antibodies against proteins purified from corn cell walls and showed that auxin-induced growth could be inhibited by this antiserum but not by non-immune serum. The specific antibodies to cell wall proteins also inhibited cell wall autolysis, indicating that the processes of growth and cell-wall digestion are related. No attempt has yet been made to apply these methods to a study of cell-wall proteins in GA-responsive tissues. A similar experimental approach could provide evidence for the involvement of cell wall-loosening enzymes in GA-induced growth.

GA_3 and the synthesis of RNA and protein

It is well established that plant hormones and growth regulators affect the synthesis of RNA and protein (Bewley and Black, 1978; Jacobsen, 1977; Jacobsen, Higgins and Zwar, 1979). Gibberellins in particular have been shown to be involved in the synthesis of RNA and protein in fruits and seeds, and the effect of GAs on enzyme synthesis in cereal aleurone is regarded as a model system for the study of the molecular basis of hormone action in plants.

Some of the earliest work on the effects of GAs on metabolism derives from studies on dormant seeds. Work with dormant hazel (*Corylus avellana*) seed, for example, established a role for GA_3 in germination. In these seeds GA_3 treatment or stratification overcomes dormancy and stimulates RNA and protein synthesis in both embryo axis and cotyledons (Bradbeer, 1968; Jarvis, Frankland and Cherry, 1968; Pinfield and Stobart, 1969). Although GA-induced germination of many dormant and non-dormant seeds is frequently accompanied by changes in the metabolism of stored reserves, it should not be inferred that these events are causally related. Indeed, in a wide variety of dicotyledonous and monocotyledonous seeds, the effect of GA_3 on growth of the embryo (germination) can be spatially and temporally separated from its effects on the metabolism of stored reserves in the endosperm or cotyledons. The germination of some dormant seeds, on the other hand, results from an effect of GA_3 on the elongation of the embryo axis and is not accompanied by marked changes in the metabolism of either embryo axis or cotyledons (Carpita, Nabors, Ross and Petrie, 1979). Because the role of GA_3 in substrate mobilization in seeds has been intensively studied, our discussion of the effects of GA_3 on RNA and protein synthesis will be limited to its involvement in these processes in germinating seeds and seedlings.

The cereal aleurone layer

Although the involvement of the aleurone layer of cereal seeds in the degradation of the starchy endosperm was recognized nearly a century ago (Haberlandt, 1884), it was not until 1960 that GA_3 was shown to substitute for the embryo in the control of this process. Paleg in Australia and Yomo in Japan showed simultaneously that endosperm breakdown in de-embryonated barley caryopses could be stimulated by the addition of GA_3 to the embryo-less endosperm. It was subsequently shown by bioassay (Radley, 1967) that the barley embryo produces GAs and it has now been demonstrated by gas chromatography–mass spectroscopy (Gaskin, Gilmour, Lenton, MacMillan and Sponsel, 1982) and immunochemical methods (Atzorn and Weiler, 1982) that a large number of GAs occur in the barley grain. While it is presumed that these GAs function to control the production of hydrolytic enzymes by the aleurone layer, the regulation of enzyme production in whole barley caryopses is not simply by GA_3 produced in the embryo. There are many additional controls on the production of hydrolytic enzymes by the aleurone layer including regulation by the end products of starch hydrolysis (Fig. 2.15A; Briggs, 1973; Jones and Armstrong, 1971).

The demonstration that embryo-less barley

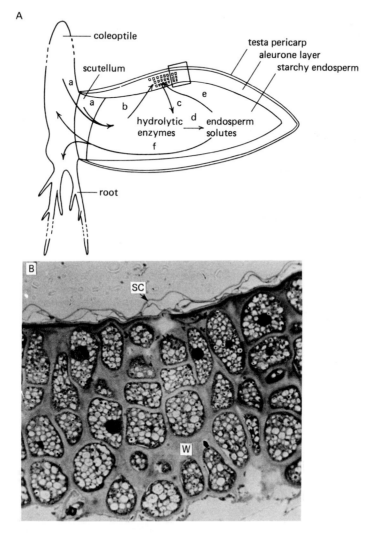

Fig. 2.15 A: The relationship between GA production and hydrolytic enzyme synthesis and release in germinating barley grain. GA produced by the coleoptile and scutellum (a) migrates into the aleurone layer (b) where hydrolytic enzyme production is stimulated. These enzymes are released into the starchy endosperm (c) where they serve to hydrolyze endosperm reserves (d) producing solutes that can inhibit further enzyme production (e) and function to nourish the growing embryo (f). Box indicates area shown in Fig. 2.15B. (Reproduced from Armstrong and Jones, 1971.) B: Light micrograph of aleurone cells from the area indicated by the box in Fig. 2.15A. The prominent, thick cell wall (W) and seed coats (SC) are indicated. (Reproduced from Jones, 1969.)

caryopses respond to exogenously supplied GA_3 was the stimulus for considerable research on endosperm degradation. The suitability of the embryoless endosperm as a research system was further emphasized by the work of Chrispeels and Varner (1967), which demonstrated that the aleurone layer could be removed from the starchy endosperm. The aleurone layer of barley is a uniform tissue consisting of one cell type, the aleurone cell. In barley the aleurone layer is 3–4 cell layers thick (Fig. 2.15B), while in oats, rice, rye, wheat and corn it is only one cell layer thick. Although the aleurone layer responds to GA_3 by producing a wide spectrum of hydrolytic enzymes, it neither grows nor differentiates following hormone treatment.

When isolated aleurone layers are incubated in optimal GA_3 concentrations (*ca.* 10^{-5} to 10^{-4} M) containing 10–20 mM Ca^{2+}, numerous hydrolytic enzymes are produced (Table 2.1). Among the newly synthesized enzymes is α-amylase which makes up 60–70% of the total protein produced by these cells after 24 h incubation in GA_3 (Mozer, 1980a; Higgins, Jacobsen and Zwar, 1982). Most GA-induced enzymes are secreted into the incubation medium although some, notably those involved in fat and lipid metabolism, are not released to the exterior (Table 2.1).

The study of the action of GA in stimulating hydrolase production by cereal aleurone has focused on those molecular events that lead to *de novo* protein synthesis and the structural events that permit the secretion of these proteins. Electron

Table 2.1 GA-induced enzymes in cereal aleurone.

Enzyme	Secreted	Reference*
α-Amylase	yes	(2)
Protease	yes	(5)
Ribonuclease	yes	(2)
β-Glucanase	yes	(12)
Esterase	yes	(4)
Acid phosphatases:	yes	
p-nitrophenyl phosphatase	yes	(1),(9)
ATPase	yes	(10)
Phytase	yes	(9)
Naphthol AS-B1 phosphatase	yes	(1)
GTPase	yes	(10)
Pentosanase	yes	(8)
Endoxylanase	yes	(11)
Xylopyranosidase	yes	(11)
Arabinofuranosidase	yes	(11)
Glucosidase	yes	
Peroxidase	yes	(4)
Catalase	no	(7)
Malate dehydrogenase	no	(7)
Malate synthetase	no	(3),(7)
Isocitrate lyase	no	(3),(7)
Citrate synthetase	no	(7)
Phosphorylcholine glyceride transferase	no	(6)
Phosphorylcholine cytidyl transferase	no	(6)

* References: (1), Ashford and Jacobsen (1974); (2), Chrispeels and Varner (1967); (3), Doig *et al.* (1975); (4), Jacobsen and Knox (1972); (5), Jacobsen and Varner (1967); (6), Johnson and Kende (1971); (7), Jones (1972); (8), McLeod and Millar (1962); (9), Obata and Suzuki (1976); (10), Pollard and Singh (1968); (11), Taiz and Honigman (1976); (12), Taiz and Jones (1970).

microscopy has been used to follow changes in the ultrastructure of the aleurone cell that accompany hydrolase synthesis and secretion (Jacobsen *et al.*, 1979; Jones, 1973; Varner and Ho, 1976). In barley, a marked organization of the endoplasmic reticulum (ER) into stacks in response to GA_3 treatment has been reported by all investigators. There is no evidence for an increase in the total amount of lipid in GA_3-treated barley aleurone (Firn and Kende, 1974), and measurement of total ER membrane also indicates only a small and transient difference between GA_3-treated and control tissue (Jones, 1980b). These data strongly suggest that GA treatment results in a reorganization of pre-existing ER membrane and not the synthesis of additional ER.

The ER is the site of α-amylase synthesis and sequestration in both control and GA_3-treated barley aleurone tissue (Jones, 1980a; Jones and Jacobsen, 1982; Locy and Kende, 1978). In GA_3-treated tissue the ER contains four isoenzymes of α-amylase, while in control tissue it contains only two (Jones and Jacobsen, 1982). Both GA-treated and control aleurone layers secrete α-amylase to the incubation medium. Four isoenzymes of α-amylase are released from GA-treated tissue while only two are released from controls incubated in the absence of the hormone. Although there are clearly differences in the ultrastructure of GA_3-treated and control aleurone layers, it is not known whether organization of the ER into stacks of lamellae is a prerequisite for the synthesis and/or secretion of a specific group of enzymes or isoenzymes.

In the aleurone of wheat the organization of the ER into lamellar stacks is not dependent on the presence of GA_3 (Buckhout, Gripshover and Morré, 1981; Colborne, Morris and Laidman, 1976). In this tissue the ER becomes stacked during the first few hours of imbibition in either H_2O or GA_3. Although α-amylase activity is associated with the ER of wheat aleurone (Gibson and Paleg, 1976), it is not known whether the ER systems of control and GA-treated tissue have different α-amylase isoenzymes associated with them.

The ER is also involved in the transport of α-amylase from barley aleurone cells. Pulse-chase studies with radioactive amino acids show that the labeled α-amylase located in the lumen of the ER is lost from this compartment during a chase with unlabeled amino acid (Chen and Jones, 1974; Jones and Jacobsen, 1982; Locy and Kende, 1978). The

precise path of enzyme transport to the cell exterior has yet to be determined. Despite numerous attempts to isolate and identify an ER- or Golgi-derived transport vesicle there has been no unequivocal demonstration that such organelles participate in enzyme secretion.

There is compelling evidence that GA_3 affects the synthesis of hydrolytic enzymes in cereal aleurone and that it does so by controlling the transcription and possibly translation of new mRNAs. The synthesis of RNA is a prerequisite for the production of GA-induced enzymes. Inhibitors of RNA synthesis, for example cordycepin, inhibit α-amylase synthesis (Ho and Varner, 1974), while the synthesis of poly (A) RNA accompanies GA-treatment of the aleurone layer (Jacobsen and Zwar, 1974). When poly(A)-rich RNA is translated in a wheat germ *in vitro* protein-synthesizing system, there is an increase in the amount of translatable mRNA for several proteins, the most prominent of which is α-amylase (Fig. 2.16A and Higgins, Zwar and Jacobsen, 1976; Higgins *et al.*, 1982). GA_3-treatment does not cause a general increase in the amount of translatable RNA, however. Rather the activity of some RNAs is increased, as is the case for α-amylase mRNA, while the levels of others is decreased (Fig. 2.17). Mozer (1980b) has purified poly(A) RNA from GA-treated barley aleurone using DMSO-formaldehyde sucrose density gradient centrifugation followed by agarose gel electrophoresis and shown that the predominant poly(A) RNA in aleurone tissue treated with GA_3 corresponds to translatable α-amylase mRNA.

Although these experiments suggested that GA_3 treatment influences transcription by stimulating the production of some sequences and inhibiting the production of others, direct evidence for this was lacking. Thus it could be argued that the enhanced translatability of α-amylase mRNA resulted from activation of pre-existing mRNA or from decreased catabolism of the amylase messenger (Higgins *et al.*, 1976). Using cDNA prepared from purified α-amylase mRNA, Bernal-Lugo, Beachey and Varner (1981) have shown that there is an increase in the number of sequences which hybridize with α-amylase cDNA following GA_3 treatment. This conclusion is supported by experiments using recombinant DNA cloning methods (Chandra and Muthukrishnan, 1982; Jacobsen, Chandler, Higgins and Zwar, 1982). These workers have cloned cDNA com-

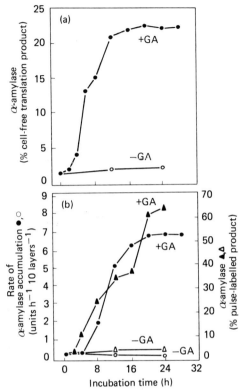

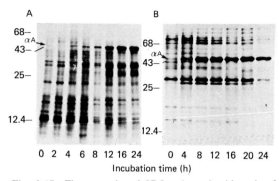

Fig. 2.17 Fluorographs of SDS–polyacrylamide gels of proteins synthesized *in vitro* by poly(A) RNA extracted from GA₃-treated barley aleurone layers (A) and *in vivo* (B). Arrow indicates the position of α-amylase and numerals the positions of molecular weight markers in kilodaltons. (Reproduced with permission from Higgins *et al.*, 1982.)

Fig. 2.16 The proportion of radioactivity incorporated into α-amylase *in vitro* (a) and the proportion incorporated *in vivo* compared with the rate of enzyme accumulation in the medium (b) of barley aleurone layers treated with (●, ▲) or without (○, △) GA₃. (Reproduced with permission from Higgins *et al.*, 1982.)

plementary to α-amylase mRNA isolated from GA₃-treated barley aleurone layers. Using the cloned cDNA as a probe they have shown that GA₃ treatment results in an increase in the number of sequences which produce α-amylase when translated *in vitro*.

Evidence for control of protein synthesis at the translational level comes from a comparison of the *in vivo* protein labeling pattern with the pattern of polypeptides produced when mRNA from aleurone layers is translated *in vitro* (Figs 2.16 and 2.17). α-Amylase represents more than 60% of the total labeled protein produced in aleurone cells incubated in GA₃ for 24 h (Figs 2.16B and 2.17B), but less than 25% of the total labeled protein *in vitro* (Figs 2.16A and 2.17A). Several explanations can account

for this difference. It may represent differences between the translation of mRNAs *in vivo* and *in vitro* or differences in the availability of amino acids or the pool size of other components of the protein-synthesizing system. On the other hand, the difference may represent differences in the efficiency of mRNA translation (Higgins *et al.*, 1982). It might be speculated that those sequences transcribed in the presence of GA₃ are translated more efficiently than those transcribed without hormone. This hypothesis would be consistent with the data shown in Figs 2.16 and 2.17.

The α-amylase of barley produced *in vivo* has an Mr of 44 000 while that produced *in vitro* has an Mr of 46 000 (Higgins *et al.*, 1982). The addition of microsomal membranes from dog pancreas to the *in vitro* system results in the processing of synthesized α-amylase to a polypeptide with Mr 44 000 (Higgins *et al.*, 1982). Similar processing of α-amylase has been described for wheat aleurone (Boston, Miller, Mertz and Burgess, 1982). These workers injected poly(A) RNA from GA-treated wheat aleurone into *Xenopus* oocytes and showed that the α-amylase produced had an Mr of 41 500, whereas α-amylase produced *in vitro* with the wheat germ protein-synthesizing system had an Mr of 43 000. Boston *et al.* (1982) also showed that addition of dog pancreas microsomes to a wheat germ *in vitro* system programmed with poly (A) RNA from GA-treated wheat aleurone produced α-amylase with an Mr of 41 500.

The results from both barley and wheat are consistent with the idea that α-amylase contains a signal peptide (Blobel and Dobberstein, 1975). Since the signal sequence can be cleaved from the α-amylase polypeptide by membranes isolated from canine pancreas or by the protein-synthesizing machinery of *Xenopus* oocytes, it must be presumed that cleavage of the signal sequence occurs on membranes within the aleurone cell. Because *in vivo* labeling of aleurone cell proteins shows that labeled α-amylase is located in the lumen of the ER (Jones and Jacobsen, 1982), it seems probable that co-translational processing of α-amylase occurs on the ER membrane.

The events leading to the synthesis and secretion of GA-induced hydrolases in cereal aleurone can be summarized as follows. The transcription of a specific group of genes is promoted by GA_3 and the newly-transcribed sequences are transported to the cytoplasm and translated on the ER. Following co-translational cleavage of the signal sequence by the ER membrane, the completed α-amylase molecule is transported from the ER to the exterior of the cell by an as yet unidentified route. The evidence that GA_3 influences the transcription process is convincing. However, despite the differences in labeling patterns of proteins produced *in vivo* and *in vitro*, there is no evidence that GA_3 influences translation directly. Similarly, there is no evidence that GA_3 controls post-translational processing of proteins or their transport despite the fact that in barley GA_3 affects the organization of the ER system.

Dicotyledonous fruits and seeds

The effects of GA_3 on germination of dormant and non-dormant dicotyledonous seeds and on the growth of the seedling are well known (Khan, 1977). Investigations of the biochemical basis of GA_3 action in these seeds have concentrated largely on fat-storing seeds, in particular those of almond, castor bean and hazel.

In the endosperm of castor bean (*Ricinus communis*) seedlings GA_3 stimulates the activity of at least 16 enzymes involved in gluconeogenesis, but not the activities of nine other enzymes of intermediary metabolism that are not involved in gluco-

neogenesis (Table 2.2, and Gonzalez and Delsol, 1981). The enzymes whose activities are increased by GA are strategically placed in the pathway of carbon from fatty acids to sucrose and Gonzalez and Delsol (1981) propose that the overall effect of GA_3 is to accelerate the process of sucrose synthesis.

The view that GA_3 selectively induces the activities of specific enzymes in castor bean has been challenged by the recent experiments of Martin and Northcote (1982a, b), who have argued that the effect of GA on protein synthesis is a quantitative one and that no specific induction of mRNA synthesis is observed. They support their argument with experiments which show a general increase in protein synthesis following treatment with GA_3 (Martin and Northcote, 1982a) and by the study of a specific mRNA (Martin and Northcote, 1982b). In the latter study Martin and Northcote purified the mRNA for isocitrate lyase from control and GA_3-treated castor bean endosperm and showed that the relative amount of mRNA for isocitrate lyase is lower in GA_3-treated tissue than in untreated controls (Fig. 2.18A). Because the total mRNA (poly (A) RNA) level is higher in GA_3-treated tissue, however, the absolute amount of isocitrate lyase mRNA is higher in GA_3-treated tissue than in controls (Fig. 2.18B).

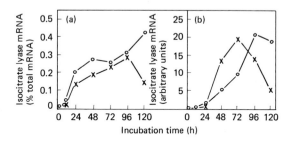

Fig. 2.18 (a) Changes in isocitrate lyase mRNA expressed as a percentage of the total extractable mRNA from endosperm of castor bean seedlings during incubation in the presence ($\times$) and absence ($\bigcirc$) of GA_3. (b) Absolute amounts of isocitrate lyase mRNA activity extracted from the endosperm of seedlings incubated in the presence ($\times$) and absence ($\bigcirc$) of GA_3. (Recalculated and redrawn from Martin and Northcote, 1982b.)

When endogenous GA_3 synthesis was inhibited by AMO 1618, the appearance of isocitrate lyase was further retarded relative to that in GA-treated tissue (Martin and Northcote, 1982b). Furthermore,

Table 2.2   The effect of GA on the activity of enzymes of intermediary metabolism in castor bean seeds.

Enzyme	GA induced	Organelle*	Reference**
Aconitase	yes	M	(1)
Amino transferase	yes	G, M	(1)
Catalase	yes	G	(1),(2)
CDP choline glyceride transferase	yes	ER	(2)
Citrate synthase	yes	M	(1),(2)
Fructose diphosphatase	yes	C	(1)
Fumarase	yes	M	(1)
3-Hydroxy acyl-CoA dehydrogenase	yes	N.D.	(3)
Isocitrate lyase	yes	G	(1),(2),(3),(4)
Malate dehydrogenase	yes	C, G	(1)
Malate synthase	yes	G	(2)
NADH-cytochrome C reductase	yes	ER	(2)
NADH ferricyanide reductase	yes	M	(4)
NADH oxidase	yes	M	(4)
Phosphoenolypyruvate carboxykinase	yes	C	(1)
Succinic dehydrogenase	yes	M	(1)
Acid lipase	no	N.D.	(3)
Alkaline lipase	no	N.D.	(3)
Choline phosphotransferase	no	ER	(4)
Citrate synthase	no	G	(1)
Cytochrome oxidase	no	M	(1)
Enolase	no	C	(1)
Malate dehydrogenase	no	M	(1)
OH pyruvate reductase	no	G	(1)
Pyruvate kinase	no	C	(1)

* Organelles: C, cytoplasm; ER, endoplasmic reticulum; G, glyoxysome; M, mitochondria; N.D., not determined.
** References: (1), Gonzales and Delsol (1981); (2), Gonzales (1978); (3), Marriott and Northcote (1975); (4), Wrigley and Lord (1977).

the total amount of mRNA produced did not change although the peak of mRNA, which appeared in GA_3-treated seedlings at 3 days and in controls at 4 days, was delayed until 8 days following AMO 1618 treatment. These observations led Martin and Northcote to conclude that a pre-programmed pattern of protein synthesis exists in castor bean endosperm and that the qualitative nature of the program cannot be altered by GA. The onset of the program can be hastened by GA, however, and they propose that this can be achieved by an acceleration of RNA synthesis.

The action of GA_3 in the cereal aleurone that involves the induction of the synthesis of some proteins and an inhibition of the synthesis of others seems to conflict with the non-specific effects of GA_3 on RNA and protein synthesis in castor bean endosperm. Additional experiments are needed before it can be concluded that the action of GA_3 in these seeds is indeed different. It is possible that in the castor bean seed high concentrations of pre-formed GAs exist which during imbibition induce enzyme synthesis. The level of pre-formed GAs would not be affected by AMO 1618, for example, since this inhibitor only interferes with the synthesis of GA_3 (Lang, 1970). In the cereal caryopsis, on the other hand, the sites of GA_3 synthesis and storage are the embryo and scutellum and in the experiments described above these tissues are removed from the endosperm before imbibition of the half-seed. The response to GA_3 in the cereal aleurone is obtained because untreated aleurone tissue has low levels of endogenous GA, while the response of castor bean endosperm to GA_3 may be quantitative because the endosperm may always contain high levels of GA. Addition of GA_3 to castor bean therefore serves only to amplify a pre-existing program which was initiated by endogenous GA.

A general mechanism for GA₃ action?

Although the effects of GA_3 on growth, development and metabolism appear to involve the action of GA at several biochemical levels, it is possible that the molecular mechanism whereby this hormone acts is the same. There is now convincing evidence that auxins are involved in the control of the synthesis of specific mRNAs in elongating tissues. In soybean hypocotyls treated with 2:4 dichlorophenoxyacetic acid, new translatable mRNAs are found within 15 min of hormone treatment (Zurfluh and Guilfoyle, 1982), while in pea internodes several new mRNAs appear within 20 min of indoleacetic acid treatment (Theologis and Ray, 1981). While it is not known whether the proteins synthesized in response to auxin treatment are involved in the initiation of the growth response, these experiments establish that the metabolism of mRNA is one site at which auxin acts (see Chapter 1).

Since GAs and auxin promote growth in a similar fashion, GA_3 may also promote changes in RNA synthesis during elongation. The action of GA_3 in all target tissues may therefore be *via* its effect on the transcriptional process. This could be a quantitative effect, where GA_3 could accelerate the synthesis of all classes of RNA, as has been reported for castor bean, or its effect could be qualitative by the induction of specific enzymes, as in the cereal aleurone.

Suggested further reading

Bewley, J. D. and Black, M. (1978). *Physiology and Biochemistry of Seeds. I. Development, Germination and Growth*, Springer-Verlag, Berlin.

Crozier, A., ed. (1983) *The Biochemistry and Physiology of Gibberellins*, Praeger Scientific, New York.

Hedden, P., MacMillan, J. and Phinney, B. O. (1978). The metabolism of the gibberellins. *Ann. Rev. Plant Physiol.* **29**, 149–92.

Jones, R. L. (1973). Gibberellins: their physiological role. *Ann. Rev. Plant Physiol.* **24**, 571–98.

Khan, A. A., ed. (1977). *The Physiology and Biochemistry*

of Seed Dormancy and Germination, Elsevier/North-Holland Biomedical Press, New York.

Krishnamoorthy, H. N., ed. (1975). *Gibberellins and Plant Growth*, John Wiley, New York.

Letham, D. S., Goodwin, P. B. and Higgins, T. J. V., eds (1978). The biochemistry of phytohormones and related compounds. *Phytohormones and Related Compounds—A Comprehensive Treatise*, vol. 1, Elsevier/North-Holland Biomedical Press, New York.

MacMillan, J., ed. (1980). *Hormonal regulation of development. I. Molecular aspects of plant hormones. Encyclopedia of Plant Physiology New Series*, vol. 9, Springer-Verlag, Berlin, Heidelberg, New York.

Skoog, F., ed. (1980). *Plant Growth Substances 1979*, Springer-Verlag, Berlin, Heidelberg, New York.

Weaver, R. H., ed. (1972). *Plant Growth Substances in Agriculture*, Freeman and Co., San Francisco.

References

Adams, P. A., Montague, M. J., Tepfer, M., Rayle, D. L., Ikuma, H. and Kaufman, P. B. (1975). Effects of gibberellic acid on the plasticity and elasticity of *Avena* stem segments. *Plant Physiol.* **56**, 757–60.

Albersheim, P. (1976). The primary cell wall. In *Plant Biochemistry*, 3rd edn, eds J. Bonner and J. E. Varner, Academic Press, New York, pp. 225–74.

Ashford, A. E. and Jacobsen, J. V. (1974). Cytochemical localization of phosphatase in barley aleurone cells: The pathway of gibberellic-acid-induced enzyme release. *Planta* **120**, 81–105.

Atzorn, R. and Weiler, E. W. (1982). The immunoassay of gibberellins. *XIth International Conference on Plant Growth Substances* Abstract 122, p. 19.

Bearder, J. R., MacMillan, J. and Phinney, B. O. (1975). Fungal products. Part XIV. Metabolic pathways from *ent*-kaurenoic acid to the fungal gibberellins in mutant B1-41a of *Gibberella fujikuroi*. *J. Chem. Soc. Perkin Trans.* **1**, 721–6.

Bernal-Lugo, I., Beachy, R. N. and Varner, J. E. (1981). The response of barley aleurone layers to gibberellic acid includes the transcription of new sequences. *Biochem. Biophys. Res. Commun.* **102**, 617–23.

Bewley, J. D. and Black, M. (1978). *Physiology and Biochemistry of Seeds. I. Development, Germination and Growth*, Springer-Verlag, Berlin.

Blobel, G. and Dobberstein, B. (1975). Transfer of proteins across membranes. Presence of proteolytically processed and unprocessed nascent immunoglobulin light chains on membrane bound ribosomes of murine myeloma. *J. Cell Biol.* **67**, 835–51.

Boston, R. S., Miller, T. J., Mertz, J. E. and Burgess, R. R. (1982). In vitro synthesis and processing of wheat α-amylase. *Plant Physiol.* **69**, 150–4.

Bowen, D. H., MacMillan, J. and Graebe, J. E. (1972). Determination of specific radioactivity of[^{14}C]-compounds by mass spectrometry. *Phytochemistry* **11**, 2253–7.

Bradbeer, J. W. (1968). Studies in seed dormancy. IV. The role of endogenous inhibitor and gibberellin in the dormancy and germination of *Corylus avellana* L. seeds. *Planta* **78**, 266–76.

Briggs, D. E. (1973). Hormones and carbohydrate metabolism in germinating cereal grains. In *Biosynthesis and Its Control in Plants*, Ann. Proc. Phytochem. Soc. Number 9, ed. B. V. Milborrow, Academic Press, New York, pp. 219–77.

Buckhout, T. J., Gripshover, B. M. and Morré, D. J. (1981). Endoplasmic reticulum formation during germination of wheat seeds. *Plant Physiol.* **68**, 1319–22.

Carpita, N. C., Nabors, M. W., Ross, C. W. and Petrie, N. L. (1979). The growth physics and water relations of red-light induced germination in lettuce seeds. III. Changes in the osmotic and pressure potential in the embryonic axes of red- and far-red-treated seeds. *Planta* **144**, 217–24.

Ceccarelli, N., Lorenzi, R. and Alpi, A. (1981). Gibberellin biosynthesis in *Phaseolus coccineus* suspensor. *Z. Pflanzenphysiol.* **102**, 37–44.

Chandra, G. R. and Muthukrishnan, S. (1982). Gibberellic acid dependent formation of α-amylase mRNA in barley aleurone cells. *XIth International Conference on Plant Growth Substances*, Abstract 115, p. 16.

Chen, R.-F. and Jones, R. L. (1974). Studies on the release of barley aleurone cell proteins: kinetics of labelling. *Planta* **119**, 193–206.

Chrispeels, M. J. and Varner, J. E. (1967). Gibberellic acid-enhanced synthesis and release of α-amylase and ribonuclease by isolated barley aleurone layers. *Plant Physiol.* **42**, 398–406.

CIAT Annual Report (1972). Centro Internacional de Agricultura Tropical, Cali, Columbia.

CIAT Annual Report (1975). Centro Internacional de Agricultura Tropical, Cali, Columbia.

Cleland, R. E. and Rayle, D. L. (1977). Reevaluation of the effect of calcium ions on auxin-induced elongation. *Plant Physiol.* **60**, 709–12.

Cleland, R. E. and Rayle, D. L. (1978). Auxin, H$^+$-excretion and cell elongation. *Bot. Mag. Tokyo Special Issue* **1**, 125–39.

Cleland, R. E., Thompson, M. L., Rayle, D. L. and Purves, W. K. (1968). Difference in effects of gibberellins and auxins on wall extensibility of cucumber hypocotyls. *Nature* 219, 510–11.

Colborne, A. J., Morris, G. and Laidman, D. L. (1976). The formation of endoplasmic reticulum in aleurone cells of germinating wheat: an ultrastructural study. *J. Exp. Bot.* **27**, 759–68.

Coolbaugh, R. C. and Moore, T. C. (1969). Apparent changes in rate of kaurene biosynthesis during the development of pea seeds. *Plant Physiol.* **44**, 1364–7.

Cosgrove, D. J. (1981). Analysis of the dynamic and steady-state responses of growth rate and turgor pressure to changes in cell parameters. *Plant Physiol.* **68**, 1439–46.

Crozier, A., ed. (1983). *The Biochemistry and Physiology of Gibberellins*. Praeger Scientific, New York.

Dennis, D. T., Upper, C. D. and West, C. A. (1965). An enzymatic site of inhibition of gibberellin biosynthesis by AMO 1618 and other plant growth retardants. *Plant Physiol.* **40**, 948–52.

Doig, R. I., Colborne, A. J., Morris, G. and Laidman, D. L. (1975). The induction of glyoxysomal enzyme activities in aleurone cells of germinating wheat. *J. Exp. Bot.* **26**, 387–98.

Duncan, J. D. and West, C. A. (1981). Properties of kaurene synthetase from *Marah macrocarpus* endosperm. Evidence for the participation of separate but interacting enzymes. *Plant Physiol.* **68**, 1128–34.

Ecklund, P. R. and Moore, T. C. (1974). Correlations of growth rate and de-etiolation with rate of *ent*-kaurene biosynthesis in pea (*Pisum sativum* L.). *Plant Physiol.* **53**, 5–10.

Evans, R. and Hanson, J. R. (1975). Studies in terpenoid biosynthesis. Part XIII. The biosynthetic relationship of the gibberellins in *Gibberella fujikuroi*. *J. Chem. Soc. Perkin Trans.* **1**, 663–6.

Firn, R. D. and Kende, H. (1974). Some effects of applied gibberellic acid on the synthesis and degradation of lipids in isolated barley aleurone layers. *Plant Physiol.* **54**, 911–15.

Frost, R. G. and West, C. A. (1977). Properties of kaurene synthetase from *Marah macrocarpus*. *Plant Physiol.* **59**, 22–9.

Fry, S. C. (1979). Phenolic components of the primary cell wall and their possible role in the hormonal regulation of growth. *Planta* **146**, 343–51.

Fry, S. C. (1980). Gibberellin-controlled pectinic acid and protein secretion in growing cells. *Phytochemistry* **19**, 735–40.

Frydman, V. M., Gaskin, P. and MacMillan, J. (1974). Qualitative and quantitative analyses of gibberellins throughout seed maturation in *Pisum sativum* cv. Progress No. 9. *Planta* **118**, 123–32.

Frydman, V. M. and MacMillan, J. (1973). Identification of gibberellins A$_{20}$ and A$_{29}$ in seed of *Pisum sativum* cv. Progress No. 9 by combined gas chromatography-mass spectrometry. Planta **115**, 11–15.

Frydman, V. M. and MacMillan, J. (1975). The metabolism of gibberellins A$_9$, A$_{20}$ and A$_{29}$ in immature seeds of *Pisum sativum* cv. Progress No. 9. *Planta* **125**, 181–95.

Gaskin, P., Gilmour, S. J., Lenton, J. R., MacMillan, J. and Sponsel, V. M. (1982). Endogenous gibberellins and related compounds in developing grain and germinating seedlings of barley. *XIth International Conference on Plant Growth Substances* Abstract 158, p. 50.

Gaskin, P. and MacMillan, J. (1978). GC and GC-MS techniques for gibberellins. In *Isolation of Plant Growth Substances*, ed. J. Hillman, Cambridge Univ. Press, London, New York, Melbourne, pp. 79–85.

Gibson, R. A. and Paleg, L. G. (1976). Purification of GA_3-induced lysosomes from wheat aleurone cells. *J. Cell Sci.* **22**, 413–26.

Gonzalez, E. (1978). Effect of gibberellin A_3 on the endoplasmic reticulum and on the formation of glyoxysomes in the endosperm of germinating castor bean. *Plant Physiol.* **62**, 449–53.

Gonzalez, E. and Delsol, M. A. (1981). Induction of glyconeogenic enzymes by gibberellin A_3 in endosperm of castor bean seedlings. *Plant Physiol.* **67**, 550–4.

Graebe, J. E. (1980). GA-biosynthesis. The development and application of cell-free systems for biosynthetic studies. In *Plant Growth Substances 1979*, ed. F. Skoog. Springer-Verlag, Berlin, Heidelberg, New York, pp. 180–7.

Graebe, J. E., Bowen, D. H. and MacMillan, J. (1972). The conversion of mevalonic acid into gibberellin A_{12}-aldehyde in a cell-free system from *Cucurbita pepo*. *Planta* **102**, 261-71.

Graebe, J. E., Dennis, D. T., Upper, C. D. and West, C. A. (1965). Biosynthesis of gibberellins. I. The biosynthesis of (−)-kaurene, (−)-kaur-19-ol and trans-geranylgeraniol in endosperm nucellus of *Echinocystis macrocarpa* Greene. *J. Biol. Chem.* **240**, 1847–54.

Graebe, J. E., Hedden, P. and Rademacher, W. (1980). Gibberellin biosynthesis. In *Gibberellins—Chemistry, Physiology and Use*, ed. J. R. Lenton, British Plant Growth Regulators Group, Wantage, pp. 31–47.

Graebe, J. E. and Ropers, H.-J. (1978). Gibberellins. In *Phytohormones and Related Compounds—A Comprehensive Treatise, vol. 1*, eds D. S. Letham, P. B. Goodwin and T. J. V. Higgins, Elsevier/North Holland Biomedical Press, Amsterdam, pp. 107–204.

Green, P. B. (1976). Growth and cell pattern formation on an axis: critique of concepts, terminology, and modes of study. *Bot. Gaz.* **137**, 187–202.

Green, P. B. and Cummins, W. R. (1974). Growth rate and turgor pressure. Auxin effect studies with an automated apparatus for single coleoptiles. *Plant Physiol.* **54**, 863–9.

Haberlandt, G. F. J. (1884). *Physiologische Pflanzenanatomie*, Erste Anflage, Leipzig.

Hafemann, C., Froneberg, M. and Graebe, J. E. (1982). Experiments towards localization of selected reactions in gibberellin biosynthesis. *XIth International Conference on Plant Growth Substances* Abstract 153, p. 49.

Hanson, J. R., Willis, C. L. and Parry, K. P. (1980). The inhibition of gibberellin biosynthesis by *ent*-kauran-16β,17-epoxide. *Phytochemistry* **19**, 2323–5.

Harris, P. J. and Hartley, R. D. (1981). Phenolic constitutents of the cell walls of flowering plants. *Proc. XIII Int. Bot. Congr.* (Report).

Hebard, F. V., Amantangelo, S. J., Dayanandan, P. and Kaufman, P. B. (1976). Studies on acidification of media by *Avena* stem segments in the presence and absence of gibberellic acid. *Plant Physiol.* **58**, 660–74.

Hedden, P. and Graebe, J. E. (1982). The co-factor requirements for the soluble oxidases in the metabolism of the C_{20}-gibberellins. *J. Plant Growth Regulation* **1**, 105–16.

Hedden, P., Graebe, J. E., Beale, M. H., Gaskin, P. and MacMillan, J. (1983). The biosynthesis of 12α-hydroxylated gibberellins in a cell-free preparation from *Cucurbita neoxenia* endosperm. *Phytochemistry*, submitted.

Hedden, P., MacMillan, J. and Phinney, B. O. (1978). The metabolism of the gibberellins. *Ann. Rev. Plant Physiol.* **29**, 149–92.

Hedden, P. and Phinney, B. O. (1979). Comparison of *ent*-kaurene and *ent*-isokaurene synthesis in cell-free systems from etiolated shoots of normal and dwarf-5 maize seedlings. *Phytochemistry* **18**, 1475–9.

Hedden, P., Phinney, B. O., Heupel, R., Fujii, D., Cohen, H., Gaskin, P., MacMillan, J. and Graebe, J. E. (1982). Hormones of young tassels of *Zea mays*. *Phytochemistry* **21**, 391–3.

Higgins, T. J. V., Jacobsen, J. V. and Zwar, J. A. (1982). Changes in protein synthesis and mRNA levels in barley aleurone layers treated with gibberellic acid. *Plant Mol. Biol.* **1**, 191–215.

Higgins, T. J. V., Zwar, J. A., and Jacobsen, J. V. (1976). Gibberellic acid enhances the level of translatable mRNA for α-amylase in barley aleurone layers. *Nature* **260**, 166–9.

Ho, D. T. and Varner, J. E. (1974). Hormonal control of messenger ribonucleic acid metabolism in barley aleurone layers. *Proc. Nat. Acad. Sci. USA* **71**, 4783–6.

Hoad, G. V., MacMillan, J., Smith, V. A., Sponsel, V. M. and Taylor, D. A. (1982). Gibberellin 2-β-hydroxylases and biological activity of 2-β-alkyl gibberellins. In *Plant Growth Substances 1982*, ed. P. F. Wareing, Academic Press, London, New York, 91–100.

Huber, D. J. and Nevins, D. J. (1981a). Wall-protein antibodies as inhibitors of growth and of autolytic reactions of isolated cell wall. *Physiol. Plant.* **53**, 533–9.

Huber, D. J. and Nevins, D. J. (1981b). Partial purification of endo- and exo-β-D-glucanase enzymes from *Zea mays* seedlings and their involvement in cell-wall autohydrolysis. *Planta* **151**, 206–14.

Jacobsen, J. V. (1977). Regulation of ribonucleic acid metabolism by plant hormones. *Ann. Rev. Plant Physiol.* **28**, 537–64.

Jacobsen, J. V., Chandler, P. M., Higgins, T. J. V. and Zwar, J. A. (1982). Control of protein synthesis in barley aleurone layers by gibberellin. In *Plant Growth*

Substances 1982, ed. P. F. Wareing, Academic Press, London, New York, pp. 111–20.

Jacobsen, J. V., Higgins, T. J. V., and Zwar, J. A. (1979). Hormonal control of endosperm function during germination. In *The Plant Seed*, eds I. Rubenstein, B. G. Gegenbach, R. L. Phillips and C. E. Green, Academic Press, New York, pp. 241–62.

Jacobsen, J. V. and Knox, R. B. (1972). Cytochemical localization of gibberellic acid-induced enzymes in the barley aleurone layer. In *Plant Growth Substances 1970*, ed. D. J. Carr, Springer-Verlag, Berlin, pp. 344–51.

Jacobsen, J. V. and Varner, J. E. (1967). Gibberellic acid-induced synthesis of protease by isolated aleurone layers of barley. *Plant Physiol.* **42**, 1596–600.

Jacobsen, J. V. and Zwar, J. A. (1974). Gibberellic acid causes increased synthesis of RNA which contains poly(A) in barley aleurone tissue. *Proc. Nat. Acad. Sci. USA* **71**, 3290–3.

Jacqmard, A. (1968). Early effects of gibberellic acid on mitotic activity and DNA synthesis in the apical bud of *Rudbeckia bicolor*. *Physiol. Vég.* **6**, 409–16.

Jarvis, B. C., Frankland, B. and Cherry, J. H. (1968). Increased DNA template and RNA polymerase associated with the breaking of seed dormancy. *Plant Physiol.* **43**, 1734–6.

Johnson, K. D. and Kende, H. (1971). Hormonal control of lecithin synthesis in barley aleurone cells: regulation of the CDP-choline pathway by gibberellin. *Proc. Nat. Acad. Sci. USA* **68**, 3290–3.

Jones, R. L. (1972). Fractionation of the enzymes of the barley aleurone layer: evidence for a soluble mode of enzyme release. *Planta* **103**, 95–109.

Jones, R. L. (1973). Gibberellins: their physiological role. *Ann. Rev. Plant Physiol.* **24**, 571–98.

Jones, R. L. (1980a). Quantitative and qualitative changes in the endoplasmic reticulum of barley aleurone layers. *Planta* **150**, 70–81.

Jones, R. L. (1980b). The isolation of endoplasmic reticulum from barley aleurone layers. *Planta* **150**, 58–69.

Jones, R. L. (1982). Gibberellin control of cell elongation. In *Plant Growth Substances 1982*, ed. P. F. Wareing, Academic Press, London, New York, pp. 121–30.

Jones, R. L. (1983). The role of gibberellins in plant cell elongation. *CRC Crit. Rev. Plant Sci.* **1**, 23–47.

Jones, R. L. and Armstrong, J. E. (1971). Evidence for osmotic regulation of hydrolytic enzyme production in germinating barley seeds. *Plant Physiol.* **48**, 137–42.

Jones, R. L. and Jacobsen, J. V. (1982). The role of the endoplasmic reticulum in the synthesis and transport of α-amylase in barley aleurone layers. *Planta* **156**, 421–32.

Kamisaka, S., Masuda, Y. and Yanagishima, N. (1967). Yeast sporulation and RNA as affected by gibberellic acid. *Plant Cell Physiol.* **8**, 121–7.

Kamiya, Y. and Graebe, J. E. (1983). The biosynthesis of all major pea gibberellins in a cell-free system from *Pisum sativum*. *Phytochemistry* **22**, 681–89.

Katsumi, M. (1969). Bioassay and distribution of gibberellins. In *Gibberellins: Chemistry and Physiology*, ed. S. Tamura, Univ. Tokyo Press, pp. 221–63.

Kaufman, P. B. and Dayanandan, P. (1983). Gibberellin-induced growth in *Avena* coleoptiles. In *The Biochemistry and Physiology of Gibberellins*, Vol. 2, ed. A. Crozier, Praeger Scientific, New York, in press.

Kawamura, H., Kamisaka, S. and Masuda, Y. (1976). Regulation of lettuce hypocotyl elongation by gibberellic acid. Correlation between cell elongation, stress relaxation properties of the cell wall and wall polysaccharide content. *Plant Cell Physiol.* **17**, 23–34.

Khan, A. A., ed. (1977). *The Physiology and Biochemistry of Seed Dormancy and Germination*, Elsevier/North-Holland Biomedical Press, New York.

Krishnamoorthy, H. N., ed. (1975). *Gibberellins and Plant Growth*. John Wiley, New York.

Lang, A. (1970). Gibberellins: structure and metabolism. *Ann. Rev. Plant Physiol.* **21**, 537–70.

Lew, F. T. and West, C. A. (1971). (-)-Kaur-16-en-7β-19-oic acid, an intermediate in gibberellin biosynthesis. *Phytochemistry* **10**, 2065–76.

Liu, P. B. W. and Loy, J. B. (1976). Action of gibberellic acid on cell proliferation in the subapical shoot meristem of watermelon seedlings. *Am. J. Bot.* **63**, 700–4.

Lockhart, J. A. (1965). An analysis of irreversible plant cell elongation. *J. Theor. Biol.* **8**, 264–75.

Locy, R. and Kende, H. (1978). The mode of secretion of α-amylase in barley aleurone layers. *Planta* **143**, 89–99.

Loy, J. B. (1977). Hormonal regulation of cell division in the primary elongating meristems of shoots. In *Mechanisms and Control of Cell Division*, eds T. L. Rost and E. M. Gifford, Dowden, Hutchinson and Ross, Stroudsburg, Pennsylvania, pp. 92–110.

MacLeod, D. M. and Millar, A. S. (1962). Effects of gibberellic acid on barley endosperm. *J. Inst. Brew.* **68**, 322–32.

MacMillan, J., ed. (1980). *Hormonal Regulation of Development. I. Molecular Aspects of Plant Hormones. Encyclopedia of Plant Physiology New Series, vol. 9*, Springer-Verlag, Berlin, Heidelberg, New York.

MacMillan, J. and Takahashi, N. (1968). Proposed procedure for the allocation of trivial names to the gibberellins. *Nature* **217**, 170–1.

Marriott, K. M. and Northcote, D. H. (1975). The induction of enzyme activity in the endosperm of germinating castor-bean seeds. *Biochem. J.* **152**, 65–70.

Martin, C. and Northcote, D. H. (1982a). The action of exogenous gibberellic acid on protein and mRNA in germinating castor bean seeds. *Planta* **154**, 168–73.

Martin, C. and Northcote, D. H. (1982b). The action of exogenous gibberellic acid on isocitrate lyase-mRNA in germinating castor bean seeds. *Planta* **154**, 174–83.

Metzger, J. D. and Zeevaart, J. A. D. (1980a). Identification of six endogenous gibberellins in spinach shoots. *Plant Physiol.* **65**, 623–6.

Metzger, J. D. and Zeevaart, J. A. D. (1980b). Effect of photoperiod on the levels of endogenous gibberellins in spinach as measured by combined gas chromatography-selected ion current monitoring. *Plant Physiol.* **66**, 844–6.

Moll, C. and Jones, R. L. (1981a). Short-term kinetics of elongation growth of gibberellin-responsive lettuce hypocotyl sections. *Planta* **52**, 442–9

Moll, C. and Jones, R. L. (1981b). Calcium and gibberellin-induced elongation of lettuce hypocotyl sections. *Planta* **152**, 450–6.

Montague, M. J. and Ikuma, H. (1975). Regulation of cell wall synthesis in *Avena* stem segments by GA. *Plant Physiol.* **55**, 1043–7.

Moore, T. C., Barlow, S. A. and Coolbaugh, R. C. (1972). Participation of non-catalytic 'carrier' protein in the metabolism of kaurene in cell-free extracts of pea seeds. *Phytochemistry* **11**, 3225–33.

Mozer, T. J. (1980). Control of protein synthesis in barley aleurone layers by the plant hormones gibberellic acid and abscisic acid. *Cell* **20**, 479–85.

Mozer, T. J. (1980b). Partial purification and characterization of the mRNA for α-amylase from barley aleurone layers. *Plant Physiol.* **65**, 834–7.

Murakami, Y. (1972). A survey of the gibberellins in shoots of angiosperms by rice seedling test. *Bot. Mag. Tokyo* **83**, 312–14.

Obata, T. and Suzuki, H. (1976). Gibberellic acid-induced secretion of hydrolases in barley aleurone layers. *Plant Cell Physiol.* **17**, 63–71.

Paleg, L. G. (1965). Physiological effects of gibberellins. *Ann. Rev. Plant Physiol.* **16**, 291–322.

Patterson, R., Rappaport, L. and Breidenbach, R. W. (1975). Characterization of an enzyme from *Phaseolus vulgaris* seeds which hydroxylates GA_1 to GA_8. *Phytochemistry* **14**, 363–8.

Phinney, B. O. (1983). The history of the gibberellins. In *The Biochemistry and Physiology of Gibberellins*, vol. 1, ed. A. Crozier, Praeger Press, New York, pp. 19–52.

Phinney, B. O. and Spray, C. (1982). Chemical genetics and the gibberellin pathway in *Zea mays* L. In *Plant Growth Substances*, ed. P. F. Wareing, Academic Press, London, New York, pp. 101–10.

Phinney, B. O. and Spray, C. (1983). Gibberellin biosynthesis in *Zea Mays*: the 3-hydroxylation step GA_{20} to GA_1. In *Fifth IUPAC Internat. Congress on Pesticide Chemistry, Kyoto, 1982*. Academic Press, pp. 81–6.

Pinfield, N. J. and Stobart, A. K. (1969). Gibberellin-stimulated nucleic acid metabolism in the cotyledons and embryonic axes of *Coryllus avellana* (L.) seeds. *New Phytol.* **68**, 993–9.

Pollard, C. J. and Singh, B. N. (1968). Early effects of gibberellic acid on barley aleurone layers. *Biochem. Biophys. Res. Commun.* **33**, 321–6.

Porter, J. W. and Spurgeon, S. L. (1981). *Biosynthesis of Isoprenoid Compounds*, vol. 1. Wiley-Interscience, New York, Chichester, Brisbane, Toronto.

Potts, W. C., Ingram, T. J., Reid, J. B. and Murfet, J. C. (1982). Internode length in pea genotypes and the involvement of gibberellins. *Plant Physiol.* (Suppl.) **69**, 25.

Preston, R. D. (1979). Polysaccharide composition and cell wall function. *Ann. Rev. Plant Physiol.* **30**, 55–78.

Rademacher, W. and Graebe, J. E. (1979). Gibberellin A_4 produced by *Sphaceloma manihoticola*, the carrier of the hyperelongation disease of cassava (*Manihot esculenta*). *Biochem. Biophys. Res. Commun.* **91**, 35–40.

Radley, M. (1967). Site of production of gibberellin-like substances in germinating barley embryos. *Planta* **75**, 164–71.

Rappaport, L. (1980). Plant growth hormones: internal control points. *Bot. Gaz.* **141**, 125–30.

Reeve, D. R. and Crozier, A. (1980). Qualitative analysis of plant hormones. In *Hormonal Regulation of Development. I. Molecular Aspects of Plant Hormones. Encyclopedia of Plant Physiology, New Series* vol. **9**, ed. J. MacMillan, Springer-Verlag, Berlin, Heidelberg, New York, pp. 203–80.

Ropers, H.-J., Graebe, J. E., Gaskin, P. and MacMillan, J. (1978). Gibberellin biosynthesis in a cell-free system from immature seeds of *Pisum sativum*. *Biochem. Biophys. Res. Commun.* **80**, 690–7.

Sembdner, G., Gross, D., Liebisch, H.-W. and Schneider, G. (1980). Biosynthesis and metabolism of plant hormones. In *Hormonal Regulation of Development. I. Molecular Aspects of Plant Hormones, Encyclopedia of Plant Physiology, New Series, vol. 9*, ed. J. MacMillan, Springer-Verlag, Berlin, Heidelberg, New York, pp. 281–444.

Simcox, P. D., Dennis, D. T. and West, C. A. (1975). Kaurene synthetase from plastids of developing plant tissues. *Biochem. Biophys. Res. Commun.* **66**, 166–72.

Skoog, F., ed. (1980). *Plant Growth Substances 1979*. Springer-Verlag, Berlin, Heidelberg, New York.

Sponsel, V. M. (1980). Gibberellin metabolism in legume seeds. In *Gibberellins—Chemistry, Physiology and Use, Monograph 5*, ed. J. R. Lenton, British Plant Regulators Group, Wantage, pp. 49–62.

Sponsel, V. M. (1983a). *In vitro* metabolism in higher plants. In *The Biochemistry and Physiology of Gibberellins*, vol. 1, ed. A. Crozier, Praeger Press, New York, pp. 151–250.

Sponsel, V. M. (1983b). The localization and biological activity of gibberellins in maturing and germinating seeds of *Pisum sativum* cv. Progress No. 9. *Planta*, in press.

Sponsel, V. M., Hoad, G. V. and Beeley, L. J. (1977). The biological activities of some new gibberellins (GAs) in six plant bioassays. *Planta* **135**, 143–7.

Sponsel, V. M. and MacMillan, J. (1977). Further studies on the metabolism of gibberellins (GAs) GA$_9$, A$_{20}$, A$_{29}$ in immature seeds of *Pisum sativum* cv. Progress No. 9. *Planta* **35**, 129–36.

Sponsel, V. M. and MacMillan, J. (1978). Metabolism of gibberellin A$_{29}$ in seeds of *Pisum sativum* cv. Progress No. 9: use of [^{2}H] and [^{3}H] GAs and the identification of a new GA catabolite. *Planta* **144**, 69–78.

Sponsel, V. M. and MacMillan, J. (1980). Metabolism of [^{13}C]gibberellin A$_{29}$ to [^{13}C]gibberellin A$_{29}$-catabolite in maturing seeds of *Pisum sativum* cv. Progress No. 9. *Planta* **150**, 46–52.

Srivastava, L. M., Sawhney, V. K. and Taylor, J. E. P. (1975). Gibberellic acid-induced cell elongation in lettuce hypocotyls. *Proc. Nat. Acad. Sci. USA* **72**, 1107–11.

Stodola, F. H. (1958). *Source Book on Gibberellin 1828–1957*. Agricultural Research Service, U.S. Department of Agriculture.

Stowe, B. B. and Yamaki, T. (1957). The history and physiological action of the gibberellins. *Ann. Rev. Plant Physiol.* **8**, 181–216.

Stuart, D. A. and Jones, R. L. (1977). The roles of extensibility and turgor in gibberellin- and dark-stimulated growth. *Plant Physiol.* **59**, 61–8.

Stuart, D. A. and Jones, R. L. (1978). The role of acidification in gibberellic acid- and fusicoccin-induced elongation growth of lettuce hypocotyl sections. *Planta* **142**, 135–45.

Taiz, L. and Honigman, W. A. (1976). Production of cell wall hydrolyzing enzymes by barley aleurone layers in response to gibberellic acid. *Plant Physiol.* **58**, 380–6.

Taiz, L. and Jones, R. L. (1970). Gibberellic acid, β-1,3-glucanase and the cell walls of barley aleurone layers. *Planta* **92**, 73–84.

Tamura, S. (1969). The history of studies on gibberellins. In *Gibberellins: Chemistry and Physiology*, ed. S. Tamura, Univ. Tokyo Press, pp. 13–26.

Tamura, S. (1977). History of plant hormone gibberellins: how it was found. In *Plant Hormones*, ed. S. Tamura, Dainippontosho Co. Ltd., pp. 18–50.

Theologis, A. and Ray, P. M. (1981). Auxin-regulated polyadenylated mRNA sequences in pea epicotyl. *Plant Physiol.* **67**, Suppl. (Abstr.), p. 3.

Varner, J. E. and Ho, D. T. (1976). The role of hormones in the integration of seedling growth. In *The Molecular Biology of Hormone Action*, ed. J. Papaconstantinou, Academic Press, New York, pp. 173–94.

Weaver, R. H., ed. (1972). *Plant Growth Substances in Agriculture*, Freeman and Co., San Francisco.

West, C. A. (1973). Biosynthesis of gibberellins. In *Biosynthesis and Its Control in Plants*, ed. B. V. Milborrow, Academic Press, New York, pp. 472–82.

West, C. A. (1980). Hydroxylases, monooxygenases and cytochrome P-450. In *The Biochemistry of Plants*, vol. **2**, *Metabolism and Respiration*, ed. D. D. Davies, Academic Press, New York, pp. 317–64.

West, C. A. (1981). Biosynthesis of diterpenes. In *Biosynthesis of Isoprenoid Compounds*, vol. **1**, eds J. W. Porter and S. L. Spurgeon, Wiley-Interscience, New York, Chichester, Brisbane, Toronto, pp. 375–411.

Wrigley, A. and Lord, J. M. (1977). The effects of gibberellic acid on organelle biogenesis in the endosperm of germinating castor bean seeds. *J. Exp. Bot.* **28**, 345–53.

Yafin, Y. and Schechter, I. (1975). Comparison between biosynthesis of *ent*-kaurene in germinating tomato seeds and cell-suspension cultures of tomato and tobacco. *Plant Physiol.* **56**, 671–5.

Yamane, H., Murofushi, N., Osada, H. and Takahashi, N. (1977). Metabolism of gibberellins in early immature bean seeds. *Phytochemistry* **16**, 831–5.

Yamane, H., Murofushi, N. and Takahashi, N. (1975). Metabolism of gibberellins in maturing and germinating bean seeds. *Phytochemistry* **14**, 1195–200.

Yamane, H., Takahashi, N., Taneko, K. and Furuya, M. (1979). Identification of gibberellin A$_9$ methyl ester as a natural substance regulating formation of reproductive organs in *Lycopodium japonicum*. *Planta* **147**, 251–6.

Yokota, T., Murofushi, N. and Takahashi, N. (1980). Extraction, purification and identification. In *Hormonal Regulation of Development. I. Molecular Aspects of Plant Hormones. Encyclopedia of Plant Physiology, New Series*, vol. **9**, ed. J. MacMillan, Springer-Verlag, Berlin, Heidelberg, New York, pp. 113–210.

Zurfluh, L. L. and Guilfoyle, T. J. (1982). Auxin-induced changes in the population of translatable messenger RNA in elongating sections of soybean hypocotyl. *Plant Physiol.* **69**, 332–7.

Cytokinins

3

R Horgan

Introduction

Growth in a complex multicellular organism is usually the result of the interrelated processes of cell division and cell expansion. Normal growth and development clearly requires strict spatial and temporal control, and co-ordination of these processes. An attractive hypothesis to account for control and co-ordination in growth and development is that cell division and expansion may be regulated by the distribution within tissues of specific chemical controllers of these processes.

The existence of specific substances which can control cell division in plants was postulated many years before such substances were finally discovered. The concept of chemical controllers of plant cell division can be traced back to at least Wiesner (1892). Haberlandt (1913) is generally credited with having obtained the first experimental evidence for the existence of such substances. He showed that phloem diffusates could cause cell division in potato parenchyma. In 1921 he observed that cell division induced by wounding, in several plant species, could be prevented by rinsing the wound surface and that application of crushed tissue to the wound re-initiated cell division. Haberlandt coined the word 'wundhormone' to describe the putative substance responsible for the cell division inducing activity of the damaged tissue (Haberlandt, 1921).

The work that led to the isolation and identification of the first plant cell division promoting substance derived directly from the studies of Skoog and his co-workers on plant tissue cultures. Investigations into the nutritional requirements for growth of tissue cultures derived from the pith tissue of tobacco stems (*Nicotiana tabacum* cv. Wisconsin No. 38) indicated the existence of specific cell division inducing factors in vascular tissue (Skoog and Tsui, 1948; Jablonski and Skoog, 1954). In the presence of auxin, pith explants showed extensive cell enlargement but no cell division. However, when placed in contact with vascular tissue division of the pith cells was observed. The tobacco pith tissue culture system was thus developed into a bioassay for cell division promoting activity. Many materials were investigated as potential sources of cell division promoting substances. Particularly rich sources, as evidenced by activity in the tobacco pith assay, were coconut milk, malt extract, yeast extract, extracts of vascular tissue and autoclaved DNA (Miller, Skoog, Saltza and Strong, 1955). In 1956, 63 years after Wiesner's original idea, Miller isolated a pure, highly-active, cell-division-inducing factor from autoclaved herring sperm DNA (Miller, Skoog, Okomura, Saltza and Strong, 1956). This compound was identified as 6-(furfurylamino)purine and was called kinetin. Kinetin was not a component of DNA, but was derived from breakdown and rearrangement during autoclaving. Synthetic kinetin was found to be a very potent promoter of cell division in the tobacco pith assay, with detectable activity at 10^{-6} g l^{-1}.

The original generic name of 'kinins' for plant cell division promoters such as kinetin, conflicted with the same term used in animal biochemistry and was replaced with the name cytokinin. This is the generic name now accepted for compounds which promote cytokinesis in plant cells. In practice a cytokinin is defined as a compound which, in the presence of optimal auxin, induces cell division in the tobacco pith or similar assay (*e.g.*, carrot phloem or soybean callus) grown on an optimally defined medium.

Although kinetin did not occur naturally, application of bioassay methods to numerous plant extracts indicated that substances with cell division promoting activity were widely distributed in plants. In 1963, nine years after the discovery of kinetin, Letham isolated 6-(4-hydroxy-3-methylbut-trans-2-enylamino)purine from immature kernels of *Zea mays* and named this compound zeatin (Letham, 1963; Letham, Shannon and MacDonald, 1964). Further studies by Letham on *Zea mays* kernels resulted in the isolation of zeatin riboside, zeatin ribotide (the 5′-monophosphate) and other cytokinins (Letham, 1973).

In 1966, 6-(3-methylbut-2-enylamino)purine was isolated and identified from two serine tRNAs of yeast (Biemann, Tsunakawa, Sonnenbichler, Feldman, Dutting and Zachau, 1966). This compound and its riboside had previously been synthesized and shown to be active cytokinins.

Nature and distribution of cytokinins

Chemical nomenclature

Virtually all the known naturally-occurring cytokinins are substituted purines. In naming cytokinins the accepted numbering of the purine nucleus is used to define the positions of substitution. The normal way of writing the full chemical name of a cytokinin is to express it as a substituted 6-amino-purine or as a N^{-6}-substituted adenine. Readers should be aware that a considerable diversity of ways of naming cytokinins have been used in the past.

Throughout this chapter the abbreviations used for cytokinins will be those based on the established system for nucleic acid derivatives (IUPAC–IUB Commission on Biochemical Nomenclature, 1970). The abbreviations used for some of the common cytokinins are shown in Fig. 3.1.

Free cytokinins in plants

The nine-year gap between the discovery of kinetin and the isolation of zeatin as a free, naturally-occurring cytokinin from a higher plant reflects the extreme technical difficulties faced by the early workers in the field. Although immature *Zea mays* kernels are a particularly rich source of cytokinins, Letham had to extract 60 kg of material to obtain about 1 mg of zeatin. The difficulties of purifying the minute amount of biologically active material from this large quantity of plant material with the very rudimentary purification techniques available at that time were enormous. These, together with the skill required to determine the structure on such a small amount of material using physical techniques of limited sensitivity, made the isolation and identification of zeatin a very notable achievement in the history of plant growth substance research.

There are many reports in the literature describing the existence in extracts of plants of biological activity with chromatographic properties similar to one or more of the known cytokinins. However, the number of cytokinins unambiguously identified in higher plants is relatively few. This is mainly due to technical difficulties of purification and structure determination. These problems will be discussed later. Cytokinins which have been unambiguously identified in plants are listed in Table 3.1.

N,N'-Diphenylurea has been isolated from coconut milk as a cell division inducing compound (Shantz and Steward, 1955). However, its identity as a naturally-occurring cytokinin is questionable as it has not been isolated from a plant source since. Letham (1963) has shown that the cell division promoting activity of coconut milk can be accounted for by the presence of zeatin riboside.

Dihydroconiferyl alcohol has been isolated from the bleeding sap of sycamore, *Acer pseudoplatanus* and from maple syrup (Lee, Purse, Pryce, Horgan and Wareing, 1981). This compound promotes cell division in the soybean and tobacco callus tests and thus warrants consideration as a cytokinin. However, it does not exhibit many of the other biological properties of cytokinins.

Cytokinins in transfer RNA

The identification of i^6Ade as a constituent of two serine tRNA species from yeast by Zachau and co-workers (Zachau, Dutting and Feldman, 1966; Biemann *et al.*, 1966) initiated a major research effort into the occurrence of cytokinins in tRNA. It was found in both tRNA species that the cytokinin

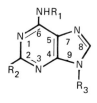

Fig. 3.1 Chemical structures, trivial names and IUPAC–IUB abbreviations for some naturally-occurring and synthetic cytokinins.

Substituents			Trivial name	Abbr.
R_1	R_2	R_3		
CH₃ / CH₂ CH₃ (structure)	H	H	N^6 (Δ^2 – isopentenyl) adenine	i^6Ade
	H	ribofuranosyl	N^6 (Δ^2 – isopentenyl) adenosine	i^6A
	CH_3S	ribofuranosyl	2 methylthio N^6(Δ^2 isopentenyl) adenosine	ms^2i^6A
CH₂OH / CH₂ CH₃ (structure)	H	H	zeatin	io^6Ade
	H	ribofuranosyl	zeatin riboside	io^6A
CH₂OH / CH₂ CH₃ (structure)	H	H	dihydrozeatin	H_2io^6Ade
CH₂– benzyl (structure)	H	H	N^6 (benzyl) adenine	bzl^6Ade
CH₂– furan (structure)	H	H	kinetin	
H	H	ribofuranosyl (structure)	β-D-ribofuranosyl adenosine	

Table 3.1 Selected examples of naturally-occurring cytokinins that have been identified in plant tissues by physicochemical methods.

Zeatin. t-io⁶Ade

Zea mays kernels (Letham, 1967); coconut milk (van Staden and Drewes, 1975); tomato xylem sap (van Staden and Menary, 1976); sycamore sap (Purse, Horgan, Horgan and Wareing, 1976); *Vinca rosea* crown gall tissue (Miller, 1975); *Prunus cerasus* fruit (Young, 1977).

Zeatin riboside. t-io⁶A

Zea mays kernels (Letham 1967, 1973); coconut milk (Letham, 1967); sycamore sap (Horgan, Hewett, Purse, Horgan and Wareing, 1973); *Vinca rosea* crown gall tissue (Miller, 1975); needles of *Picea sitchensis* (Lorenzi, Horgan and Wareing, 1975); tomato xylem sap (van Staden and Menary, 1976); *Prunus cerasus* fruit (Young, 1977).

Dihydrozeatin. H₂-io⁶Ade

Lupinus luteus seed (Koshimizu, Matsubara, Kasaki and Mitsui, 1967); *Phaseolus vulgaris* seed (Krasnuk, Witham and Tegley, 1971); sycamore sap (Purse *et al.*, 1976); culture medium of *Vinca rosea* crown gall tissue (Palni and Horgan, 1982).

Dihydrozeatin riboside. H₂-io⁶A

Phaseolus vulgaris leaves (Wang and Horgan, 1978); culture medium of *Vinca rosea* crown gall tissue (Palni and Horgan, 1982).

Zeatin-9-β-D-glucoside

Zea mays kernels (Summons, Entsch, Letham, Gollnow and MacLeod, 1980); *Vinca rosea* crown gall tissue (Scott, Horgan and McGaw, 1980).

Zeatin-*O*(4)-β-D-glucoside and ribosyl zeatin-*O*(4)-β-glucoside

Vinca rosea crown gall tissue (Morris, 1977); pods of *Lupinus luteus* (Summons, Entsch, Parker and Letham, 1979); *Zea mays* kernels (Summons, Entsch, Letham, Gollnow and MacLeod, 1980).

Dihydrozeatin-*O*(4)-β-D-glucoside

Phaseolus vulgaris leaves (Wang, Thompson and Horgan, 1977); pods of *Lupinus luteus* (Summons, Entsch, Parker and Letham, 1979); *Zea mays* kernels (Summons *et al.*, 1980).

Ribosyl dihydrozeatin-*O*(4)-β-D-glucoside

Pods of *Lupinus luteus* (Summons, Entsch, Parker and Letham, 1979); *Zea mays* kernels (Summons, Entsch, Letham, Gollnow and MacLeod, 1980).

*N*⁶-(*o*-Hydroxybenzyl) adenosine

Populus robusta leaves (Horgan, Hewett, Horgan, Purse and Wareing, 1975).

---

occurred only once in the polynucleotide chain and was in a position adjacent to the 3′ end of the anticodon. Later work established that the cytokinin was associated only with tRNA species corresponding to codons with the initial letter U.

N - [- (β - D - Ribofuranosyl)purine - 6 - carbonoyl] threonine (Ad - CO - thr) has been found in an analogous position in certain tRNA species which respond to codons beginning with A (Nishamura, 1972). This compound is inactive as a cytokinin although more lipophillic ureidopurines do exhibit cytokinin activity.

A summary of the cytokinins that have been found in tRNA is shown in Table 3.2. The methods of tRNA hydrolysis used resulted in these compounds being isolated as ribosides.

Table 3.2 Selected examples of cytokinins which occur in plant tRNA.

---

*N*⁶-isopentenyl)adenosine. i⁶A

Wheat germ (Burrows, Armstrong, Kaminek, Skoog, Bock, Hecht, Danmann, Leonard and Occolowitz, 1970); tobacco callus tissue (Burrows, Skoog and Leonard, 1971); leaves and isolated chloroplasts of *Spinacia oleracea* (Vreman, Thomas, Corse, Swaminathan and Murai, 1978)

cis-Zeatin riboside. c-io⁶A

Wheat germ (Burrows *et al.*, 1970); pea roots (Babcock and Morris, 1970); tobacco callus tissue (Burrows *et al.*, 1971)

2-Methylthio-*N*⁶-(Δ²-isopentenyl)adenosine. ms²i⁶A

Wheat germ (Burrows *et al.*, 1970)

2-Methylthiozeatin riboside. ms²io⁶A

Wheat germ (Burrows *et al.*, 1970); tobacco callus tissue (Burrows *et al.*, 1971); pea shoots (Vreman, Schmitz, Skoog, Playtis, Frihart and Leonard, 1974); leaves and isolated chloroplasts of *Spinacia oleracea* (Vreman *et al.*, 1978)

Other sources of cytokinins

Free cytokinins have been identified in the culture filtrates of several species of bacteria. In particular the plant pathogens, *Corynebacterium fasciens* which causes a 'fasciation' disease in certain dicotyledonous plants, and *Agrobacterium tumefaciens* which causes the formation of crown galls, have been shown to produce cytokinins (Armstrong, Scarborough, Skoog, Cole and Leonard, 1976; Kaiss-Chapman and Morris, 1977; Murai, 1981).

Free cytokinins have also been found in culture filtrates of a number of fungi. Zeatin and zeatin riboside have been identified in cultures of *Rhizopogen roseolus* (Miller, 1967).

There has been considerable speculation as to whether or not cytokinins produced by pathogenic fungi, *e.g.*, the powdery mildew (*Erisiphe graminus*) and the rusts (*Uromyces phaseoli* and *U. fabae*), are involved in the physiology of infection.

The above account of the occurrence of cytokinins is not by any means exhaustive. It is intended to acquaint the reader with the chemical identity of the naturally-occurring cytokinins and to provide some insight into their occurrence. Comprehensive listings of the sources of naturally-occurring cytokinins can be found in Letham (1978) and Bearder (1980).

Synthetic cytokinins

The identification of kinetin as a 6-substituted amino purine led to the synthesis of a very large number of analogues. This work was aimed at producing more potent cell division promoting compounds and at understanding cytokinin structure/activity relationships.

Although the fine details of cytokinin structure/activity relationships are complex, they may be summarized by saying that replacement of the N^6 substituent of kinetin with a variety of different groups leads to compounds with a wide range of cell division promoting activity. In general the most active compounds have a side chain of four to six carbon atoms. The naturally-occurring cytokinins zeatin and i[6]Ade are still the most active cytokinins in cell division tests. One of the most successful synthetic cytokinins is 6-(benzylamino)purine (bz[6]Ade). This compound together with kinetin is the most widely used cytokinin for agricultural and horticultural experiments.

A large number of compounds have been synthe-sized with substituents at different positions on the purine nucleus and with alterations in the nucleus itself. Substituents at different positions on the purine nucleus generally result in lower biological activity. Alterations in the nucleus usually lead to compounds with much reduced activity and often result in complete loss of cytokinin activity. In some cases changes in the purine nucleus have resulted in compounds which reversibly inhibit cytokinin-induced cell division in the tobacco pith assay. These compounds, *e.g.*, 4-cyclopentylamino-2-methyl-thiopyrrolo[2,3-*d*]-pyrimidine (Fig. 3.2), have been called anticytokinins. The structure/activity relationship of anticytokinins have been reviewed by Skoog and Ghani (1981).

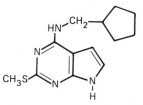

Fig. 3.2 4-Cyclopentylamino-2-methylthiopyrrolo[2,3-*d*]-pyrimidine. A potent anticytokinin.

For a detailed account of cytokinin structure/activity relationships the reader should consult the comprehensive reviews of Letham (1978) and Matsubara (1980).

Identification and measurement of cytokinins

The very small amounts of cytokinins found in most plant tissues create considerable difficulties for the isolation and identification of these compounds. A major problem is the isolation of microgram or even nanogram amounts of cytokinins, from what may be kilograms of plant material, in a sufficiently high degree of purity for structural studies by physico-chemical methods. To a certain extent this problem has been overcome by the use of modern chromatographic techniques such as gas chromatography (GC) and high performance liquid chromatography (HPLC) for sample purification. In addition the enormous increases in sensitivity of instrumental methods of structure determination have meant that much less material is now needed to determine chemical structures. Over a period of 20 years the amount of sample required for mass spectrometric analysis has decreased from milligrams to nanograms. This represents a sensitivity increase of 10^6. In order to understand the importance of advances in this area to plant growth substance research, readers should compare the enormous effort involved in the isolation and identification of zeatin in 1963 with the isolation and identification of a cytokinin using present-day techniques (*e.g.*, Scott *et al.*, 1980).

A detailed exposition of the methods whereby the structure of an unknown cytokinin can be determined are beyond the scope of this chapter. However a brief outline is as follows. Once biological activity has been observed in a plant extract using one of the standard bioassays, the bioassay may be used as a specific detector to monitor the purification of the cytokinin by a series of chromatographic stages. If these have been judiciously chosen, this will result ultimately in the isolation of the biologically active compound in a high state of purity. The two most frequently used techniques for cytokinin structure determination may then be applied. Firstly, ultraviolet spectroscopy may be used to determine the position of substituents on the purine ring. Secondly, mass spectrometry or possibly combined gas chromatography/mass spectrometry may be used to determine the molecular weight and molecular formula of the compound. Detailed analysis of the mass spectrometric fragmentation pattern together with the molecular weight and UV data can frequently lead to a sufficiently good idea of the structure to warrant attempts at chemical synthesis. Confirmation of the proposed structure by unambiguous chemical synthesis usually represents the complete identification of a new cytokinin.

The application of these techniques to a relatively simple problem of cytokinin structure determination is illustrated by the isolation and identification of an unusual cytokinin from *Populus robusta* leaves (Horgan *et al.*, 1975). Successful solutions to more complex problems can be found in Letham (1973) and Morris (1977).

Biological assays

Although by definition cytokinins are promoters of cell division in plant tissue cultures, they exhibit a wide range of other physiological effects on a variety of plants and plant tissues. The physiology of several of these will be discussed later. However, a number of physiological responses to cytokinins have been used as bioassays for the presence of these compounds in plant extracts. The most generally used bioassays for cytokinins are shown in Table 3.3. These bioassays fall into four distinct groups: (1) promotion of cell division; (2) retardation of senescence; (3) promotion of cell expansion; (4) induction of pigment synthesis. In general all the bioassays in Table 3.3 exhibit the typical log/linear dose/response curves shown by other plant growth substance assays. However, they differ markedly in

Table 3.3 Commonly used bioassays for cytokinins.

Bioassay	Time (days)	Min. det. conc. kinetin (μg l^{-1})
Tobacco pith callus (Skoog, Hamzi, Szweykowska, Leonard, Carraway Fujii, Helgeson and Leoppky, 1967)	21	<1
Soybean callus (Miller, 1963)	21	1
Carrot phloem (Letham, 1967)	21	0.5
Barley leaf senescence (Kende, 1965)	2	3
Oat leaf senescence (Varga and Bruinsma, 1973)	4	3
Radish leaf disk expansion (Kuraishi, 1959)	1	2
Radish cotyledon expansion (Letham, 1971)	3	10
Cucumber greening (Fletcher and McCullagh, 1971a)	1	1
Amaranthus betacyanin (Biddington and Thomas, 1973)	2	5

their concentration ranges and the time taken for maximum response. The use of non-cell-division-based bioassays for cytokinins has resulted mainly from the long growth period required for the cell division tests. It should, however, always be remembered that the definition of a cytokinin is a physiological one, based on promotion of plant cell division. Although several of the non-cell-division assays have proved valuable for monitoring the isolation of new cytokinins, care should always be exercised in classifying an unknown compound as a cytokinin on the basis of activity in these assays.

Quantitative measurement of cytokinins

A major underlying assumption of much of plant growth regulator research is that the endogenous levels of plant substances control, in certain circumstances, the growth and development of plants. It is therefore not surprising that a large amount of research effort has been aimed at attempting to correlate these with plant growth and development. The pros and cons of this assumption will be discussed later. This section will examine critically the methodological approaches to the problem of accurately measuring the amounts of cytokinin in plant tissues.

The vast majority of literature reports on the levels of cytokinins in plants are based on the quantitative use of bioassays. Although suitable bioassays will always be the most important means of detecting new plant growth substances their use for quantitative measurement of endogenous plant growth substances is in most cases unacceptable. A typical situation which may be found in a great many publications on cytokinin levels in plants is illustrated in Fig. 3.3. In this case paper chromatography

of a relatively crude plant extract is followed by bioassay of the R_f zones of the chromatogram. Cytokinin levels are calculated by using the bioassay response to the pure compound (shown on the right of the figure) usually a synthetic cytokinin (bz⁶Ade or kinetin) to construct a standard curve. This is used to determine the amount of cytokinin in each of the R_f zones of the chromatogram. Often these are summed to give a value of the total cytokinin concentration in the extract. In spite of the satisfactory dose–response curves of many of the cytokinin bioassays to pure compounds the method outlined above can only lead to very approximate results. There are several major problems associated with use of bioassays for the quantitative determination of cytokinins. Firstly, the large losses that always occur during the purification stages are not accounted for. Secondly, the bioassay response is the resultant of the growth-promoting and growth-inhibiting substances in the various chromatographic zones. Thirdly, the considerable biological variation often present in the plant material used for the bioassay necessitates the use of a considerable degree of replication for statistically satisfactory

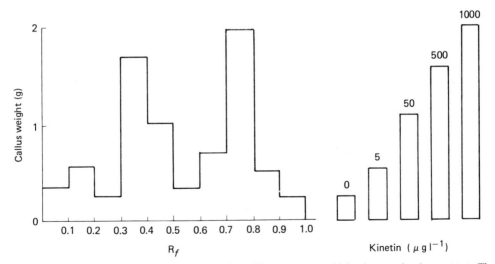

Fig. 3.3 Use of a cell division bioassay to detect and roughly measure cytokinins in a crude plant extract. The crude extract has been run on a paper chromatogram and the chromatogram cut into zones. Each zone has been incorporated into the medium of a callus bioassay. The callus growth corresponding to each zone is shown on the vertical axis. The response of the assay system to known amounts of cytokinin is shown on the right of the figure. Comparison of the growth induced by known amounts of cytokinin with that of the zones of the paper chromatogram enables a very rough estimate of the amount of cytokinin in these zones to be obtained.

results. Unfortunately the degree of replication found in most published bioassay results for cytokinins is very low.

The realization of the low degree of reliability that can be placed on quantitative determinations of cytokinins by bioassay methods has led in the last few years to the development of accurate and precise physical methods of assay. The two most important methods in current use are outlined below.

Radio-immunoassay

Cytokinins can be chemically linked to a carrier protein such as bovine serum albumin by a variety of standard methods. This material can then be used to raise specific antibodies in rabbits. These antibodies may be used as a means of measuring cytokinin levels in crude plant extracts by virtue of their highly specific binding of these compounds. Using the competitive binding between cytokinins in the extract and the same compound added as a radioactive ligand, the cytokinin concentration in the extract may be measured by reference to a suitable standard curve. Although this method can only be used after the cytokinins have been positively identified and suitable antibodies prepared, it has considerable advantages over other methods in terms of low cost and rapidity. Potentially the method can be both precise and accurate although in the case of the cytokinins this has not yet been established unambiguously. The use of this technique for the measurement of cytokinin concentrations in crude extracts can be found in Weiler (1980) and Weiler and Spanier (1981). Because of the possibility of interference by unknown compounds in crude extracts this technique has also been used in conjunction with HPLC (MacDonald, Akiyoshi and Morris, 1981).

Mass spectrometry

The combination of gas chromatography and mass spectrometry (GCMS) can be used as a highly specific method for quantitative estimation of many organic compounds. By focusing the mass spectrometer on one or a few ions known to be derived exclusively from the compound of interest, this compound can be specifically and sensitively detected even in a chemically complex mixture. Provided that a suitable internal standard is available to account for losses during the preliminary purification this technique may be made the basis of an accurate and precise method for measuring cytokinin concentrations in plant extracts. In practice the method is used as follows. A known quantity of the internal standard is added to the crude plant extract. The internal standard should be as similar chemically to the endogenous compound as possible and must be distinguishable from it in the final analysis. Deuterium-labeled (Summons, MacLeod, Parker and Letham, 1977) and ^{15}N labeled cytokinins (Scott and Horgan, 1980) have been used for this purpose since they can be distinguished from the endogenous compounds by their molecular weights. After a suitable purification regime the material is converted to volatile derivatives and examined by GCMS. The instrument is adjusted to monitor simultaneously a specific ion from the endogenous cytokinin and the corresponding ion from the internal standard. The latter ion will be at a higher mass than the ion from the endogenous compound because of its higher mass isotopes. Provided that there is no interference between the two ions the amount of cytokinin present in the original extract can be calculated by

$$(\text{amount of added internal standard}) \times \frac{\text{intensity of ion of endogenous cytokinin}}{\text{intensity of ion of internal standard}}$$

Usually a standard curve is produced by plotting the observed intensity ratio for known ratios of unlabeled/labeled cytokinins. This enables the technique to be used even if there is some interference between the ions. Detailed accounts of the basis of this technique may be found in Millard (1978). Although this method is expensive in terms of the apparatus used, and time consuming when compared to radioimmunoassay, it is at present the standard by which other methods of cytokinin quantification must be assessed.

Biochemistry of cytokinins

Studies of the biosynthesis and metabolism of cytokinins are important for several reasons. Firstly the level or pool size of a physiologically active cytokinin is the result of a dynamic balance between biosynthesis and metabolism. In this context 'physiologically active' cytokinin is intended to mean a cytokinin that is exerting some effect on plant growth and development by virtue of its concentration in the cell or some cellular compartment. Although cytokinin research is not yet at a stage to be able to recognize physiologically active pools, knowledge of the biochemical mechanisms controlling cytokinin turnover can provide considerable insight into the way in which processes which modify growth and development may modulate pool sizes of cytokinins.

Secondly, when cytokinins are applied externally to plant tissues they are extensively metabolized. Thus the observed biological activity of a particular cytokinin is in some way a function of this metabolism. Plant tissues may metabolize cytokinins to inactive compounds or conversely to compounds which may have higher biological activity than the applied compound. Thus any assessment of the biological properties of externally applied cytokinins must take into account their rates and extent of metabolism.

If cytokinins do play direct roles in the coordination of development there must be strict control of their production and distribution within the plant. The following section summarizes what is at present known about the biosynthesis and metabolism of cytokinins and discusses the possible functional significance of some of these processes.

Cytokinin biosynthesis

Two mechanisms have been proposed to account for the production of free cytokinins. The evidence for and against these routes is presented below.

Biosynthesis via tRNA Since certain tRNA species contain cytokinins, turnover of tRNA is a potential route to free cytokinins. Although it is not possible to discount totally tRNA turnover as a source of free cytokinins there are several pieces of evidence that suggest it is not a major source of free cytokinins under normal conditions. Firstly, plant tissue cultures which require an external supply of cytokinins for growth contain cytokinins in their tRNA (Burrows et al., 1971). Secondly, certain cytokinins found as free compounds in plants are not found in the corresponding tRNA (Burrows, 1978). Thirdly, the major zeatin-like cytokinins in plant tRNA have the cis-configuration of the side chain whereas the

most predominantly-found free cytokinins have the trans-side chain configuration.

It has been suggested that the rate of incorporation of ^{14}C from adenine into zeatin riboside in Vinca rosea crown gall tissue is too high to be accounted for by tRNA turnover (Stuchbury, Palni, Horgan and Wareing, 1979). However, this argument assumes a uniform rate of turnover of all tRNA species. It has been shown that in certain animal tumors this assumption is not valid (Borek, Baglia, Gehrke, Kuo, Belman, Troll and Waalkes, 1977).

De novo biosynthesis There are several reports in the literature of the incorporation of radioactive label from adenine into cytokinins in plant tissues (Peterson and Miller, 1976; Chen and Petschow, 1978a; Stuchbury et al., 1979; Wang, Beutelmann and Cove, 1981). These reports, however, need to be read critically, bearing in mind the difficulties of establishing the radioactive purity of the putative cytokinins. Nevertheless, there is now an increasing body of evidence that plant tissues contain enzymes capable of the direct synthesis of free cytokinins. Initial evidence for this pathway came from studies on the slime mold Dictyostelium discoideum (Taya, Tanaka and Nishimura, 1978). A cell-free extract of this organism converted 5'-AMP and Δ^2-isopentenylpyrophosphate (Δ^2-iPP) into i^{6}Ade. Chen and Melitz (1979) have isolated a similar cell-free system from a cytokinin autotrophic strain of tobacco tissue and have partially characterized the enzyme involved. This enzyme has been called Δ^2-isopentenylpyrophosphate:AMP - Δ^2 - isopentenyltransferase. It catalyzes the reaction shown in Fig. 3.4. Adenine and adenosine do not act as substrates for this enzyme. This suggests that free cytokinin biosynthesis in plants occurs at the nucleotide level and confirms some of the previous findings in in vivo systems (Stuchbury et al., 1979).

Biosynthesis of cytokinins in the tRNA Treatment of tRNA species containing cytokinins with permanganate destroys the Δ^2-isopentenyl side chain leaving an adenine residue in place of the cytokinin. Crude enzyme preparations from yeast, rat liver and tobacco pith have been shown to catalyze the reattachment of a Δ^2-isopentenyl side chain (derived from radioactively-labeled mevalonic acid) to these adenine residues in the tRNA. Thus, it appears that

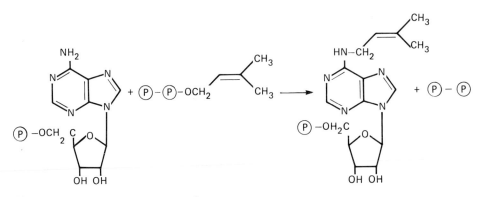

Fig. 3.4 The reaction between 5'-AMP and Δ^2-isopentenylpyrophosphate to produce i^{6}A-5'-monophosphate. This reaction, which is catalyzed by the enzyme Δ^2-isopentenylpyrophosphate:AMP-Δ^2-isopentenyltransferase, is considered to be the key step in free cytokinin biosynthesis.

tRNA cytokinins are produced after the formation of the polynucleotide chain (Fittler, Kline and Hall, 1968; Chen and Hall, 1969).

Thus, the mechanism of cytokinin biosynthesis in plants appears to involve two routes, one leading to tRNA cytokinins and the other to free cytokinins. In addition there is some evidence, more circumstantial than direct, that tRNA does not normally function as a source of free cytokinins.

Cytokinin metabolism

When both synthetic and naturally-occurring cytokinins are applied to plant tissues they are rapidly metabolized to a variety of different compounds. Cytokinin metabolism has been studied extensively during the past eight years and a generalized scheme of the metabolic conversions which cytokinins undergo can now be presented.

It is convenient to distinguish between two broad types of metabolic reactions involving cytokinins: firstly, metabolism in which the essential features of the molecule are retained (the N^6-side chain) with the retention of some degree of biological activity or potential biological activity; secondly, metabolism which results in irreversible loss of biological activity. The first type of metabolism is typified by such reactions as side-chain hydroxylation and saturation and by sugar and amino acid conjugation. The second type of reaction is typified by side-chain cleavage to give biologically inactive products such as adenine.

Side-chain modification Both i^{6}Ade and its nucleoside i^{6}A can be stereospecifically hydroxylated to trans-zeatin and trans-zeatin riboside respectively. This reaction has been shown to occur in several plant tissues (Miura and Hall, 1973; Einset and Skoog, 1973; Laloue, Terrine and Guern, 1977). It is probably of general occurrence in plants which produce zeatin-like cytokinins and would represent the final step in the biosynthesis of zeatin-5'-monophosphate from Δ^2-iPP and 5'-AMP via i^{6}A-5'-monophosphate. In contrast to the above, i^{6}A residues in tRNA are stereospecifically hydroxylated to cis-zeatin riboside.

Hydrogenation of the unsaturated side chain of zeatin has been shown to occur in the axes of *Phaseolus vulgaris* seeds (Sondheimer and Tzou, 1971) and in mature leaves of the same plant (Palmer, Scott and Horgan, 1981). This conversion of zeatin-type cytokinins to dihydrozeatin-type cytokinins may have significance in protecting the molecules from side-chain cleavage.

Side-chain cleavage Adenine and adenosine have been found as metabolites of externally applied cytokinins in many systems. Indeed it is the exception rather than the rule to find that externally applied zeatin and i^{6}Ade are not extensively broken down. An enzyme which could convert i^{6}A to adenosine by cleavage of the Δ^2-isopentenyl group was first detected in tobacco pith tissue by Paces, Werstiuk and Hall (1971). A similar enzyme was partially purified and characterized from *Zea mays* kernels by Whitty and Hall (1974) and Brownlee, Hall and Whitty (1975). This enzyme exhibited high reactivity towards the naturally-occurring cytokinins i[6]Ade, i[6]A, zeatin and zeatin riboside. However, cytokinins lacking a Δ^2-isopentenyl side chain were not substrates. The enzyme required molecular oxygen and hence was termed 'cytokinin oxidase'.

Since side-chain cleavage results in loss of cytokinin activity, cytokinin oxidase could be of importance in the regulation of endogenous cytokinin levels. Paradoxically, however, tissues which contain high endogenous cytokinin levels, e.g., Zea mays kernels and crown gall tissue of Vinca rosea, also contain high levels of cytokinin oxidase activity. This activity is immediately obvious when applying radioactively-labeled cytokinins to these tissues. External application of ^{14}C-labeled zeatin or zeatin-O(4)-glucoside to Vinca rosea crown gall tissue results in a large part of the applied cytokinin being converted to adenine and its derivatives and very little entering the endogenous cytokinin pools (Horgan, Palni, Scott and McGaw, 1981). Thus the relationship between 'cytokinin oxidase' and endogenous cytokinin turnover is complex. More research is needed before the role of this enzyme can be established.

Conjugation of cytokinins Conjugation of cytokinins with sugars (principally glucose and ribose) is a very commonly observed metabolic reaction of externally applied cytokinins. As can be seen from Table 3.1 a number of cytokinin conjugates have been isolated as endogenous compounds. The frequently-reported formation of 9-ribosides and ribosyl-5'-monophosphates of cytokinins is in keeping with the expected metabolism of purines in plant tissues. Several of the common purine metabolizing enzymes have been shown to catalyze the formation of ribosides and ribosyl-5'-monophosphates from cytokinin bases (Chen and Petschow, 1978b; Chen and Eckert, 1977). Laloue, Terrine and Gawer (1974) have shown that cytokinins can be converted to nucleoside di- and tri-phosphates in tobacco and sycamore cell suspension cultures.

The more unusual metabolites of both naturally-occurring and synthetic cytokinins are the variously substituted glucosides. The 7-β-D-glucopyranoside of zeatin was first isolated after applying ^{3}H-zeatin to de-rooted seedlings of Raphanus sativus (Parker, Letham, Cowley and MacLeod, 1972). The corres-

ponding glucoside of i[6]Ade was isolated from tobacco cells grown on ^{14}C-i[6]Ade (Laloue, Gawer and Terrine, 1975).

The 9-β-D-glucopyranoside of zeatin has been found as a metabolite of ^{3}H-zeatin in roots, cultured embryos and intact seedlings of Zea mays (Parker and Letham, 1974; Parker, Wilson, Letham, Cowley and MacLeod, 1973). The synthetic cytokinin benzyladenine is also metabolized to glucosides; the 7,9- and 3-β-D-glucopyranosides of this compound have been isolated from seedlings of Raphanus sativus (Letham, Wilson, Parker, Jenkins, MacLeod and Summons, 1975). An enzyme has been partially purified from cotyledons of Raphanus sativus which synthesizes the 7-β-D-glucopyranoside of zeatin from zeatin and UDP glucose (Entsch, Parker, Letham and Summons, 1979).

The side chain OH-group of zeatin is also a position at which conjugation with glucose occurs. Application of labeled zeatin to cytokinin-requiring soybean callus (Horgan, 1975) and de-rooted seedlings of Lupinus augustifolius (Parker, Letham, Gollnow, Summons, Duke and MacLeod, 1978), resulted in the formation of zeatin-O(4)-glucoside. Zeatin was metabolized in a similar fashion in leaves of Populus alba (Duke, Letham, Parker, MacLeod and Summons, 1977) and Phaseolus vulgaris (Palmer et al., 1981). In these tissues side-chain saturation also occurred leading to ribosyl dihydrozeatin-O(4)-glucoside in the case of Populus and dihydrozeatin-O(4)-glucoside in the case of Phaseolus. A scheme illustrating the known metabolic conversions of zeatin is shown in Fig. 3.5.

Zeatin applied to seedlings of Lupinus augustifolius is also converted into an amino acid conjugate. This metabolite was identified as β-(zeatin-9-yl)-alanine (Parker et al., 1978). Its synthesis has been demonstrated in a cell-free system (Murakoshi, Ikegami, Ookawa, Haginawa and Letham, 1977).

There has been considerable speculation as to the physiological significance of cytokinin metabolism. An early idea was that cytokinin bases such as zeatin needed to be metabolically activated before functioning as promoters of cell division. Although it was originally thought that 7-glucosylation may represent such an activation, later work has not substantiated this idea (Horgan, 1975; Laloue, 1977). It has been suggested that conjugates of cytokinins are storage and/or transported forms of the active compounds. Cytokinin glucosides are often found to

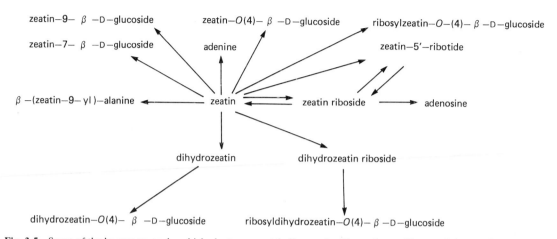

Fig. 3.5 Some of the known routes by which plants can metabolize zeatin. The pathways illustrated do not all occur in any single plant tissue but represent the results of metabolic experiments with several species of plant.

accumulate in tissues and persist after the applied free base has ceased to be detectable. However, this property does not permit them to be called storage compounds. Much more evidence is required as to the compartmentation and turnover of these compounds before functional roles can be assigned to them.

The formation of glycosides and amino acid conjugates is a very common metabolic response of plant tissues to externally applied compounds. In this context many of the metabolic processes involving cytokinins reported in this section may have little to do with the processes involved in the metabolic regulation of endogenous cytokinin levels. Determining the extent to which the metabolism of externally-applied labeled cytokinins (and other plant growth substances) mimics the metabolic fate of the endogenous compounds is a very important problem to be faced by plant physiologists and biochemists. This problem has to be resolved before accurate assessments of turnover rates and control of pool sizes can be made.

Physiology of cytokinins

In addition to their effects on cell division, cytokinins exhibit a wide range of physiological effects when applied externally to whole plants, plant tissues and plant organs. The ability of externally applied cytokinins to influence such processes as

differentiation and senescence had led to considerable speculation amongst plant physiologists that these processes may also be controlled by the endogenous levels of cytokinins. A very large amount of research into cytokinins, and indeed other plant growth substances, has sought evidence of roles for cytokinins as endogenous controllers of plant growth and development. As yet little clear evidence that specific physiological processes in higher plants are under the control of endogenous cytokinins has been obtained. It is not intended to dismiss out of hand the many attempts that have been made to correlate endogenous cytokinin levels with the physiological status of plants. However the reader is urged to take a critical view, at both the technical and the conceptual level, of much of the literature in this area. The technical difficulties of isolation and measurement of cytokinins in plant tissues have already been discussed. Lack of knowledge of the details of the mechanism of action, and of possible inter- and/or intra-cellular compartmentation of sites of production and sites of action, makes interpretation of such experiments extremely difficult. Thus, although cytokinins exert spectacular effects on the growth and development of plants and it would therefore seem obvious to link these compounds with the control of these processes, the reader should note the distinction between the clear-cut effects of externally applied cytokinins and the, as yet, very incomplete evidence for their endogenous roles.

In this section the effects of externally applied cytokinins on several physiological processes will be described, and any evidence for control of these processes by endogenous cytokinins will be critically assessed.

Morphogenesis

One of the most striking responses to cytokinins is the redifferentiation of certain plant tissue cultures to form organs. In combination with auxin, cytokinins exhibit a striking quantitative relationship in the regulation of morphogenesis.

The classic experiments of Skoog and Miller (1957) showed that organ formation in tobacco pith tissue cultures could be controlled by the concentration and ratio of auxin and cytokinin in the medium. At an IAA concentration of $2 \, mg \, l^{-1}$, increasing the kinetin concentration from 0.02 to $10 \, mg \, l^{-1}$, led to spectacular changes in the growth and differentiation of the tissue. A low ratio of cytokinin to auxin led to extensive root formation. As the cytokinin/auxin ratio increased the tissue went through a stage of rapid undifferentiated growth in which large thin-walled cells were produced. At a kinetin concentration of approximately 0.2 to $1.0 \, mg \, l^{-1}$, a large number of buds were produced on the tissue. These buds could be grown into complete tobacco plants under appropriate conditions. At higher kinetin concentrations hard undifferentiated callus was obtained consisting of small thick-walled cells, and finally growth was inhibited at concentrations of kinetin above $10 \, mg \, l^{-1}$.

The ratio of auxin to cytokinin has been shown to affect root and shoot formation in many plant callus cultures. The concentrations and ratios of auxin to cytokinin required to elicit a particular morphogenetic response vary widely with the plant species used, the plant part from which the tissue culture is derived and the constitution of the growth medium. Although this method of organogenesis does not function in all plants it has become a valuable technique for the commercial production of certain plants via tissue cultures. The striking nature of the above phenomenon and its extreme simplicity, in that it involves two known growth regulators, has resulted in the obvious suggestion that the endogenous cytokinin/auxin ratio can control growth and morphogenesis in whole plants. There is no firm evidence yet that this is so. However, two experimental systems in which evidence does exist for correlations between endogenous cytokinin levels and morphogenesis deserve further mention.

The most direct evidence for control of morphogenesis by endogenous cytokinin levels comes from studies with the moss *Physcomitrella patens.* As with several other moss species *Physcomitrella patens* when treated with auxin and cytokinin produces large numbers of buds. Using chemical mutagenesis Cove and his co-workers (Cove, Ashton, Featherstone and Wang, 1979) have isolated mutants of this moss phenotypically resembling the wild type organism which has been treated with a high concentration of cytokinin (*i.e.*, they produce large numbers of buds). These mutants (OVE mutants) fall into three complementation groups, suggesting that at least three genes may be involved in the mutation process. It has now been clearly shown that the enhanced bud formation of the OVE mutants does indeed arise from increased endogenous cytokinin levels. HPLC and MS analysis of the culture medium of the wild type and OVE mutants showed a 25- to 100-fold increase in the level of i^6Ade in the mutants (Wang, Horgan and Cove, 1981). Thus in this instance there is a simple and direct correlation between endogenous cytokinin levels and morphogenesis.

Some evidence for the control of growth and differentiation in plant tissue cultures by endogenous cytokinins comes from studies of crown gall tissues. The soil bacterium *Agrobacterium tumefaciens* causes the formation of crown galls on a variety of dicotyledonous plants. These galls are frequently referred to as plant tumors. A characteristic of crown gall tissue is that it grows in tissue culture without externally supplied cytokinin and auxin (Braun, 1956). It has been clearly demonstrated that crown gall tissues from a variety of plants produce large quantities of cytokinins (Weiler and Spanier, 1981).

The molecular basis of crown gall disease involves the incorporation into the plant cell genome of a small fragment (T-DNA) of a large bacterial plasmid (Ti-plasmid) (Chilton, Drummond, Merlo, Sciaky, Montoya, Gordon and Nester, 1977). The introduction of transposons into the bacterial T-DNA has been shown in certain cases to influence markedly the morphology of the crown galls produced on infection. Some insertions produce galls with a preponderance of shoots and others galls with a preponderance of roots (Ooms, Hooykaas, Moolenar and Schilperoort, 1981; Garfinkel, Simpson, Ream, White, Gordon and Nester, 1981). Since genetic information relating to the production of auxin and cytokinins is clearly carried on the T-DNA it is very likely that certain insertions will interfere with the expression of this information. This may affect the endogenous auxin/cytokinin ratio and thus lead to the production of tumors with roots or shoots. Thus, in this case the redifferentiation of the tissue may be controlled by the absolute levels and relative concentrations of endogenous cytokinins and auxins, and would thus fit the model of Skoog and Miller (1957). Future research will certainly clarify this situation.

The role of cytokinins in crown gall disease is additionally important because of the existence of the T-DNA as a natural genetic vector. Intense interest thus centers on it as a possible means of introducing foreign genes into plants. Since the regeneration of whole plants from tissue cultures depends on the formation of shoots, knowledge of the control of morphogenesis in crown gall tissues by levels of endogenous cytokinins is important to the future of genetic engineering in plants.

Cell enlargement

Cytokinins can cause an increase in size of leaf and cotyledonary tissue by a process involving only cell enlargement. Bean leaf disks floated on a 5×10^{-4} M solution of kinetin showed an increase in diameter of approximately three times that of control disks floated on water (Miller, 1956). Similar effects have been observed with radish leaf disks (Kuraishi, 1959). Excised radish cotyledons show marked cell expansion in the presence of cytokinins (Letham, 1971). This response, which has been used as a bioassay for cytokinins, can be mimicked by NaCl and KCl. The response to KCl and NaCl and cytokinins is additive (Norris, 1976). It is of interest to note that 2-(3-methyl-but-2-enylamino)purine-6-one has recently been isolated from coconut milk as a compound which markedly stimulates cell expansion in the radish cotyledon bioassay. It does not, however, function as a cytokinin in cell division tests (Letham, 1982).

In combination with gibberellins cytokinins are able to modify markedly the shapes of leaves on intact plants. It has been suggested that normal leaf development could be controlled by the gibberellin/cytokinin ratio.

Delay of senescence

A remarkable effect of externally applied cytokinins is their ability to delay the rate of chlorophyll disappearance and protein degradation which usually accompanies the senescence process in leaves (Chapter 20). This effect was first described by Richmond and Lang (1957) in detached leaves of *Xanthium* using kinetin. They were able to maintain cytokinin-treated leaves in a fully green condition for up to 20 days, by which time untreated leaves had lost all their chlorophyll. More recent studies have shown that several processes involved in senescence are influenced by cytokinins. Polysome aggregates are stabilized (Berridge and Ralph, 1969) and the reduction in several membrane-associated activities in the chloroplast which inevitably accompanies senescence is significantly delayed (Thomas, 1975; Woolhouse and Batt, 1976). Cytokinins also suppress senescence-linked changes in respiration rate and mitochondrial coupling (Tetley and Thimann, 1974).

The extent to which senescence in the whole plant may be controlled by endogenous cytokinin levels has been the subject of considerable debate (*see* Stoddart and Thomas, 1982). The influence of the root system on leaf senescence (Chibnall, 1939), the fact that initiation of roots on cuttings or the petioles of excised leaves can delay senescence, together with a certain amount of bioassay-based evidence that root tips (Weiss and Vaadia, 1965; Short and Torrey, 1972) and root exudates contain cytokinins (Skene, 1975), has led to the rather dogmatic view that root-derived cytokinins are responsible for the control of leaf senescence. However, there is little evidence for such a simplistic view. The formation of roots on disbudded cuttings and rooted leaves of *Phaseolus vulgaris* does indeed lead to a massive accumulation of dihydrozeatin-*O*(4)-glucoside in the leaves (Engelbrecht, 1972; Wang, Thompson and Horgan, 1977). However good is this sort of circumstantial evidence that the roots are the source of cytokinins, it should be borne in mind that there is as yet not direct evidence for cytokinin biosynthesis in roots or for the involvement of endogenous cytokinins in the control of senescence.

Hormone-directed transport

In relation to the 'anti-senescence' effects of cytokinins described above, an equally spectacular effect of cytokinins is their ability to direct the movement of numerous substances to areas of the plant treated with these growth substances. Tobacco leaves treated with a single localized spot of kinetin and supplied with ^{14}C-labeled amino acids showed accumulation of radioactivity (determined by autoradiography) totally coincident with the localized area of cytokinin application (Mothes, 1960). As with senescence, the mechanism responsible for the 'directed' accumulation is still highly speculative. It is probably more correct to refer to this phenomenon as 'metabolism-directed' transport since the most important factor in directing material to the site of cytokinin application is probably the higher rate of metabolism in this area as a result of delay of senescence.

Localized application of cytokinins to yellowing leaves produces green areas in which the rate of senescence is markedly reduced. In this respect cytokinins mimic the green islands which are observed in yellowing leaves infected with various pathogens (Wood, 1967; Mothes, 1970). The leaf-mining caterpillars *Stigmella argyropeza* and *Stigmella argentipedella* which produce green islands probably produce cytokinins in their labial glands (Engelbrecht, Orban and Heese, 1969).

Cytokinins and light

Cytokinins can substitute for, or interact with, light in the control of a number of processes. These include seed germination, pigment synthesis and chloroplast development.

Light-sensitive seeds of lettuce, *Lactuca sativa*, exhibit little germination in the dark but show a high level of germination on exposure to a brief period of light. This process has been shown to be under phytochrome control (Chapters 16 and 17); thus, germination is promoted by red light (660 nm) and this effect is reversed by far-red light (730 nm). Cytokinins can substitute for the red light requirement for germination (Miller, 1956) and in addition can act synergistically with light (Miller, 1958).

The synthesis of betacyanin in *Amaranthus* seedlings is likewise under phytochrome control. Cytokinins can markedly stimulate betacyanin production in *Amaranthus* seedlings in the dark (Köhler and Conrad, 1966). This response, which appears to show a high degree of specificity for cytokinins, has been developed into a rapid and convenient bioassay system for these compounds (Biddington and Thomas, 1973). Elliot (1979a,b,c,d) has investigated in detail the ionic regulation and the effects of temperature, water stress and ageing in this assay.

Cytokinins exhibit a rather complex relationship with light in influencing chloroplast development. In particular they appear to stimulate the synthesis of certain chloroplast constituents during the light-dependent etioplast-to-chloroplast transformation in greening seedlings, cotyledons or tissue cultures. In addition they can restore chloroplast structures and restart chlorophyll synthesis in detached yellow leaves of certain plants (Dyer and Osborne, 1971). The most frequently studied effect of cytokinins on chloroplast development has been their effect on chlorophyll accumulation. Stimulation by cytokinins of chlorophyll accumulation during etioplast to chloroplast transformation has been described for a number of plant tissues. In etiolated cucumber cotyledons a 14-h pretreatment in the dark with cytokinin increased the chlorophyll content by 450% after a subsequent 3-h period of illumination (Fletcher and McCullagh, 1971a). This response has found widespread use as a very rapid bioassay for cytokinins. In this system cytokinins eliminate the usual lag period of chlorophyll formation after exposure to light. Although it has been suggested (Fletcher and McCullagh, 1971b) that the cytokinin is acting directly on chlorophyll biosynthesis and chloroplast differentiation, there is little firm evidence for this. As will be discussed later, lack of knowledge of the primary site of action of cytokinins at the molecular level precludes rationalization of their roles in a process as complex as chloroplast development. The major effect of cytokinins on chloroplast structure appears to be an increase in the internal membrane system (Khokhlova, 1977; Farineau and Rousseaux, 1975). The cytokinin effects outlined above appear to be independent of cell division inducing activity (Seyer, Marty, Lescure and Péaud-Lenoel, 1975; Naito, Tsuji and Hatakeyama, 1978). Parthier (1979) has reviewed the effects of cytokinins and other growth substances on chloroplast development.

Cytokinins and stomata

Cytokinins show very little effect on stomatal aperture in the isolated epidermal systems usually used for stomatal studies, i.e., *Commelina*, *Vicia* and *Tridax*. However, in the graminaceous species treatment of whole leaves with cytokinin has been reported to increase transpiration (Meidner, 1967; Cooper, Digby and Cooper, 1972; Biddington and Thomas, 1978). In *Anthephora pubescens* an increase of stomatal aperture of 50% has been observed with a range of naturally-occurring and synthetic cytokinins (Incoll and Whitelam, 1977; Jewer and Incoll, 1980). Similar effects have been observed in detached epidermis of *Kalanchoë daigremontiana* (Jewer and Incoll, 1981).

Mechanism of action of cytokinins

Cytokinin-binding factors

A widespread view regarding the primary mode of action of plant growth regulators is that the initial event regarding expression of biological activity involves binding of the active substance to a specific receptor molecule. Thus considerable research effort has been expended in the search for such receptor molecules.

The first report of a binding factor for cytokinins was by Berridge, Ralph and Letham (1970) who described the non-saturable multisite reversible binding of kinetin to ribosomes isolated from leaves of Chinese cabbage, *Brassica periniensis*. The apparently non-specific, low-affinity binding reported by these workers did not fit the requirements for a receptor site. However, Fox and Erion (1975), in a detailed study of cytokinin binding by wheat germ ribosomes, discovered a high specificity and high affinity cytokinin binding protein. This protein has a molecular weight of 183 000 and consists of four subunits. It has a dissociation constant, K_d, of 6.5×10^{-7} for bzl^{6}Ade and 1.7×10^{-6} for i^{6}Ade. Surprisingly this protein exhibits a very low affinity for zeatin and dihydrozeatin (Fox and Erion, 1977).

There have been several other reports of the isolation of relatively high-affinity cytokinin-binding proteins. A protein very similar to the one described by Fox and Erion has been isolated from wheat embryos by Polya and Davis (1978) and Moore

(1979). A binding factor showing high affinity for bz^{6}Ade ($K_d 1.4 \times 10^{-7}$M) has been detected in extracts of tobacco callus and protonemata of the moss *Funaria hygrometrica* (Sussman and Kende, 1978; Gardner, Sussman and Kende, 1975). Recently a very high-affinity binding factor has been found associated with a mitochondrial fraction from mung bean seedlings, *Phaseolus aureus*. This particulate material exhibits a K_d of 1×10^{-8} for bzl^{6}Ade (Keim, Erion and Fox, 1981). Unfortunately it has not been possible to assign any biological function to the cytokinin binding described above. Thus these binding factors cannot yet be considered to be cytokinin receptors in the strict sense of the term, and their role in relation to cytokinin action remains to be elucidated.

Cytokinins in tRNA

The identification of cytokinins in certain species of tRNA and their specific location adjacent to the 3' end of the anticodon initially suggested that the primary mode of action of cytokinins might be via incorporation into tRNA. There is, however, evidence against this hypothesis. Cytokinin-requiring tissue cultures grown on bzl^{6}Ade have the usual complement of naturally-occurring cytokinins in their tRNA (Burrows *et al.*, 1971). Such cultures do incorporate externally supplied cytokinin into their tRNA but this incorporation is low (about 1 in 10^4 to 10^5 tRNA molecules) and non-specific (Armstrong, Murai, Taller and Skoog, 1976).

Control of transcription and translation

An intuitively satisfying hypothesis for the mode of action of plant growth substances is that involving specific regulation of protein synthesis. Many attempts have been made to discover if cytokinins act by specific transcriptional and/or translational control of certain key proteins.

As might be expected of compounds which exert such dramatic effects on growth and development there are many reports of effects of cytokinins on nucleic acid and protein metabolism. However, there is little evidence for specific effects. A report that kinetin may regulate gene activity by binding to pea bud chromatin (Matthysee and Abrams, 1970) has not been investigated further. With regard to translational control it has been claimed that cytokinins may stimulate protein synthesis by stimulating the recruitment of previously untranslated mRNA into polysomes (Tepfer and Fosket, 1978). Bevan and Northcote (1981) have shown that this stimulation of polysome formation by cytokinins is non-specific and thus does not represent the stimulation of the synthesis of any particular polypeptide.

As has been pointed out by Stoddart and Thomas (1982), there is probably no need to invoke a specific role for cytokinins with regard to protein synthesis in the regulation of senescence. Cytokinins may simply be involved in the general process of protein synthesis in the leaf although their supply to the leaf could in certain circumstances be limiting and thus influence the timing and rate of senescence.

Cytokinins and ion transport

Marre and his co-workers (Marre, Colombo, Lado and Rasi-Caldogno, 1974; Marre, Lado, Ferroni and Ballarin-Denti, 1974) have shown a correlation between proton extrusion and cytokinin-stimulated cell enlargement. However a detailed study on the multiple effects of cytokinins on the growth and development of excised watermelon cotyledons indicates that these effects are probably not the result of a single primary response of proton extrusion (Longo, Longo, Lampugnani, Rossi and Servettaz, 1981).

Cytokinins: future perspectives

In 1907 Arrhenius, speaking of immunity, said: 'The physiological side of the problem will not find a satisfactory solution until the more simple chemical aspect is elucidated'. The same is certainly true at present with regard to research into cytokinins. The pioneering work of Skoog, Miller, Letham and their co-workers laid a firm foundation regarding the chemical nature of cytokinins. The amazing potency of cytokinins, and the diverse and, in many cases spectacular, nature of their biological effects, naturally led many plant physiologists to propose central roles for these compounds as plant hormones and thus as controllers of many phases of plant growth and development. It is unfortunate that at the period when there was the greatest euphoria regarding the possible functions of cytokinins the method-

ology to test the theories was far too crude to provide meaningful results. Modern analytical techniques now provide reliable means of identifying and measuring cytokinins in most situations. Presumably it will not be long before techniques are refined sufficiently to enable accurate estimates of inter- and intracellular pool sizes to be made. As can be seen from this chapter, considerable progress is being made in the area of the biochemistry of the cytokinins. Although a number of important questions still need to be answered (e.g., relationships between metabolism of endogenous and externally applied compounds) an understanding of the control of turnover and pool sizes of cytokinins will certainly emerge in the near future.

In the case of the mechanism of action of cytokinins the future prospects are less certain. It will be obvious from this chapter that in spite of considerable research effort knowledge of the mechanism of cytokinin action is virtually non-existent. There are several reasons for this. Possibly the most important stumbling block to understanding cytokinin action is lack of basic knowledge of the mechanisms of many of the processes influenced by cytokinins. Thus, until we know a lot more about the biochemical control of plant cell division it will be almost impossible to ascertain how cytokinins can cause cell division in certain plant tissue cultures. The same is true of senescence and directed transport. However, there are several reasons to look optimistically on the future of research in cytokinin mode of action. The spectacular recent advances in molecular genetics have provided a powerful tool with which the control of transcription and translation may be investigated. Recombinant DNA technology now provides the means to investigate the control of individual genes, in favorable circumstances. If suitable systems can be found, the use of these techniques could well provide some much needed insight in the primary mode of action of cytokinins. In addition, the use of mutants may lead to an understanding of the relationship between the biochemistry of cytokinins and their physiological role.

Although it is no longer correct to think of cytokinins as plant hormones, certainly not if the term 'hormone' is used in the animal sense, they are unique and highly physiologically active substances. Understanding their roles in plant growth and development represents a great challenge to future research workers.

Further reading

Bearder, J. R. (1980). Plant hormones and other growth substances—their background structures and occurrence. In *Hormonal Regulation of Development I, Molecular Aspects of Plant Hormones, Encyclopaedia of Plant Physiology*, New Series, ed. J. MacMillan, Springer, Berlin, vol. 9, pp. 9–112.

Letham, D. S. (1978). Cytokinins. In *Phytohormones and Related Compounds: A Comprehensive Treatise*, eds D. S. Letham, P. B. Goodwin and T. J. V. Higgins, Elsevier/North-Holland Biomedical Press, Amsterdam, vol. 1, pp. 205–63.

Sembdner, G., Gross, D., Leibisch, H-W. and Schneider, G. (1980). Biosynthesis and metabolism of plant hormones. In *Hormonal Regulation of Development I, Molecular Aspects of Plant Hormones, Encyclopaedia of Plant Physiology*, New Series, ed. J. MacMillan, Springer, Berlin, vol. 9, pp. 281–444.

The above reviews provide a comprehensive coverage of the nature, distribution and biochemistry of the cytokinins.

Brenner, M. L. (1981). Modern methods for plant growth substance analysis. *Ann. Rev. Plant Physiol.* **32**, 511–38.

Horgan, R. (1982). Modern methods for plant hormone analysis. In *Progress in Phytochemistry*, ed. L. Reinhold, J. B. Harborne and T. Swain, Pergamon Press, Oxford, pp. 137–70.

The above two articles provide background information of the techniques currently used for cytokinin analysis.

Everett, N. P., Wang, T. L. and Street, H. E. (1978) Hormone regulation of cell growth and development *in vitro*. In *Frontiers of Plant Cell Tissue Culture 1978*, International Association for Plant Tissue Culture, pp. 307–16

Parthier, B. (1979) Phytohormones and chloroplast development. *Biochem. Physiol. Pflanzen* **174**, 173–214.

Stoddart, J. L. and Thomas, H. (1982) Leaf senescence. In *Nucleic Acids and Proteins in Plants I, Encyclopaedia of Plant Physiology*, New Series, vol. 14A, eds D. Boulter and B. Parthier, Springer, Berlin, pp. 592–636.

The above reviews critically discuss the roles of cytokinins in cell division, chloroplast development and senescence.

Dennis, F. G. Jr. (1977). Growth hormones: pool size, diffusion or metabolism. *Hort. Sci.* **12**, 217–20.

Trewavas, A. (1981). How do plant growth substances work. *Plant, Cell and Environment* **4**, 203–28.

The above articles question the ideas of simple correlations between plant growth substance levels and their physiological effects. The latter article is iconoclastic and should be read very critically.

References

Armstrong, D. J., Murai, N., Taller, B. J. and Skoog, F. (1976). Incorporation of cytokinin N^6-benzyladenine into tobacco callus transfer ribonucleic acid and ribosomal ribonucleic acid preparations. *Plant Physiol.* **57**, 15–22.

Armstrong, D. J., Scarborough, E., Skoog, F., Cole, D. L. and Leonard, N. J. (1976) Cytokinins in *Corynebacterium faciens* cultures. Isolation and identification of 6-(-4-hydroxy-3-methyl*cis*-2-butenylamino)-2-methylthiopurine. *Plant Physiol.* **58**, 749–52.

Babcock, D. F. and Morris, R. O. (1970). Quantitative measurements of isoprenoid nucleosides in tRNA. *Biochemistry* **9**, 3701–5.

Bearder, J. R. (1980). Plant hormones and other growth substances—their background, structures and occurrence. In *Hormonal Regulation of Development I, Molecular Aspects of Plant Hormones, Encyclopaedia of Plant Physiology*, New Series, ed. J. MacMillan, Springer, Berlin, vol. 9, pp. 9–112.

Berridge, M. V. and Ralph, R. K. (1969). Some effects of kinetin on floated Chinese cabbage leaf discs. *Biochim. Biophys. Acta* **182**, 266–9.

Berridge, M. V., Ralph, R. K. and Letham, D. S. (1970). The binding of cytokinins to plant ribosomes. *Biochem. J.* **119**, 75–84.

Bevan, M. and Northcote, D. H. (1981). Subculture-induced protein synthesis in tissue cultures of *Glycine max* and *Phaseolus vulgaris*. *Planta* **152**, 24–31.

Biddington, N. L. and Thomas, T. H. (1973). A modified *Amaranthus* betacyanin bioassay for the rapid determination of cytokinins in plant extracts. *Planta* **111**, 183–6.

Biddington, N. L. and Thomas, T. H. (1978). Influence of different cytokinins on the transpiration and senescence of oat leaves. *Physiol. Plant.* **42**, 369–74.

Biemann, K., Tsunakawa, S., Sonnebichler, J., Feldman, H., Dutting, D. and Zachau, H. G. (1966). The structure of an odd nucleoside from serine-specific transfer RNA, *Angew. Chem.* **5**, 590–1.

Borek, E., Baglia, B. S., Gehrke, C. W., Juo, G. W., Belman, S., Troll, W. and Waalkes, T. P. (1977). High turnover rate of transfer RNA in tumor tissue. *Cancer Res.* **37**, 3362–6.

Braun, A. C. (1956). The activation of two growth substance systems accompanying the conversion of normal cells to tumor cells in crown gall. *Cancer Res.* **16**, 53–6.

Brownlee, B. G., Hall, R. H. and Whitty, D. (1975). 3-Methyl-2-butenal: An enzymatic degradation product of the cytokinin, N^6-(Δ^2-isopentenyl)adenine. *Can. J. Biochem.* **53**, 37–41.

Burrows, W. J. (1978). Evidence in support of biosynthesis *de novo* of free cytokinins. *Planta* **138**, 53–7.

Burrows, W. J., Armstrong, D. J., Kamínek, M., Skoog, F., Bock, R. M., Hecht, S. M., Dammann, L. G., Leonard, N. J. and Occolowitz, J. (1970). Isolation and identification of four cytokinins from wheat germ transfer RNA. *Biochemistry* **9**, 1867–72.

Burrows, W. J., Skoog, F. and Leonard, N. J. (1971). Isolation and identification of cytokinins located in the transfer ribonucleic acid of tobacco callus grown in the presence of 6-benzylaminopurine. *Biochemistry* **10**, 2189–94.

Chen, C. M. and Eckert, R. L. (1977). Phosphorylation of cytokinin by adenosine kinase from wheat germ. *Plant Physiol.* **59**, 443–7.

Chen, C. M. and Hall, R. H. (1969). Biosynthesis of N^6-(Δ^2-isopentenyl)adenosine in transfer ribonucleic acid of cultured tobacco pith tissue. *Phytochemistry* **8**, 1687–95.

Chen, C. M. and Melitz, D. K. (1979). Cytokinin biosynthesis in a cell-free system from cytokinin-autotrophic tobacco tissue cultures. *FEBS Lett.* **107**, 15–20.

Chen, C. M. and Petschow, B. (1978a). Cytokinin biosynthesis in cultured rootless tobacco plants. *Plant Physiol.* **62**, 861–5.

Chen, C. M. and Petschow, B. (1978b). Metabolism of cytokinins. Ribosylation of cytokinin bases by adenosine phosphorylase from wheat germ. *Plant Physiol.*, **62**, 871–4.

Chibnall, A. C. (1939). *Protein Metabolism in the Plant*, Yale University, New Haven, pp. 1–306.

Chilton, M-D., Drummond, H. J., Merlo, D. J., Sciaky, D., Montoya, A. L., Gordon, M. P. and Nester, E. W. (1977). Stable incorporation of plasmid DNA into higher plant cells: the molecular basis of crown gall tumorigenesis. *Cell* **11**, 263–71.

Cooper, M. J., Digby, J. and Cooper, J. P. (1972). Effects of plant hormones on the stomata of barley: a study of the interaction between abscisic acid and kinetin. *Planta* **105**, 43–9.

Cove, D. J., Ashton, N. W., Featherstone, D. R. and Wang, T. L. (1979). The use of mutant strains in the study of hormone action and metabolism in the moss *Physcomitrella patens*. In *The Plant Genome, Proceedings of the IV John Innes Symposium*, eds D. R. Davis and D. A. Hopwood, John Innes Charity, Norwich, pp. 231–41.

Duke, C. C., Letham, D. S., Parker, C. W., MacLeod, J. K. and Summons, R. L. (1979). The structure and synthesis of cytokinin metabolites IV. The complex of *O*-glucosylzeatin derivatives formed in *Populus* species. *Phytochemistry* **18**, 819–24.

Dyer, T. A. and Osborne, D. J. (1971). Leaf nucleic acids. II Metabolism during senescence and the effect of kinetin. *J. Exp. Bot.* **22**, 552–60.

Einset, J. W. and Skoog, F. (1973). Biosynthesis of cytokinins in cytokinin autotrophic tobacco callus. *Proc. Nat. Acad. Sci. USA* **70**, 658–60.

Elliot, D. C. (1979a) Ionic regulation for cytokinin-dependent beta cyanin synthesis in *Amaranthus* seedlings. *Plant Physiol.* **63**, 264–8.

Elliot, D. C. (1979b). Temperature-dependent expression of betacyanin synthesis in *Amaranthus* seedlings. *Plant Physiol.* **63**, 277–9.

Elliot, D. C. (1979c). Analysis of variability in the *Amaranthus* bioassay for cytokinins. *Plant Physiol.* **63**, 269–73.

Elliot, D. C. (1979d). Analysis of variability in the *Amaranthus* betacyanin assay for cytokinins. *Plant Physiol.* **63**, 274–6.

Engelbrecht, L. (1972). Cytokinins in leaf cuttings of *Phaseolus vulgaris* L. during their development. *Biochem. Physiol. Pflanzen* **163**, 335–43.

Engelbrecht, L., Orban, U. and Heese, W. (1969). Leaf miner caterpillars and cytokinins in the 'green islands' of autumn leaves. *Nature* **223**, 319–21.

Entsch, B., Parker, C. W., Letham, D. S. and Summons, R. E. (1979). Preparation and characterisation, using high-performance liquid chromatography, of an enzyme forming glucosides of cytokinins. *Biochim. Biophys. Acta* **570**, 124–39.

Farineau, N. and Rousseaux, J. (1975). Influence de la 6-benzylaminopurine sur la différenciation plastidiale dans les cotylédons de concombre. *Physiol. Plant.* **33**, 194–202.

Fittler, F., Kline, L. K. and Hall, R. H. (1968). N^6-(Δ^2-isopentenyl)adenosine: biosynthesis *in vitro* by an enzyme extract from yeast and rat liver. *Biochem. Biophys. Res. Commun.* **31**, 571–6.

Fletcher, R. A. and McCullagh, D. (1971a). Cytokinin induced chlorophyll formation in cucumber cotyledons. *Planta* **101**, 88–90.

Fletcher, R. A. and McCullagh, D. (1971b). Benzyladenine as a regulator of chlorophyll synthesis in cucumber cotyledons. *Can. J. Bot.* **49**, 2197–201.

Fox, J. E. and Erion, J. L. (1975). A cytokinin binding protein from higher plant ribosomes. *Biochem. Biophys. Res. Commun.* **64**, 694–700.

Fox, J. E. and Erion, J. L. (1977). Cytokinin binding proteins in higher plants. In *Plant Growth Regulation*, ed. P. E. Pilet, Springer, Berlin, pp. 139–46.

Gardner, G., Sussman, M. R. and Kende, H. (1975). Cytokinin binding to particulate fractions from moss protonemata. *Plant Physiol. Suppl.* **58**, 28.

Garfinkel, D. J., Simpson, R. B., Ream, L. W., White, F. F., Gordon, M. P. and Nester, E. W. (1981). Genetic analysis of crown gall: fine structure map of the T-DNA by site directed mutagenesis. *Cell* **27**, 143–53.

Haberlandt, G. (1913). Zur Physiologie der Zellteilungen. *Sitzungsber. K. Preuss. Acad. Wiss.* 318–45.

Haberlandt, G. (1921). Wundhormone als Erreger von Zellteilungen. *Beitr. Allg. Bot.* **2**, 1–53.

Horgan, R. (1975). A new cytokinin metabolite. *Biochem. Biophys. Res. Commun.* **65**, 358–63.

Horgan, R., Hewett, E. W., Horgan, J. M., Purse, J. G. and Wareing, P. F. (1975). A new cytokinin from *Populus robusta*. *Phytochemistry* **14**, 1005–8.

Horgan, R., Hewett, E. W., Purse, J. G., Horgan, J. M. and Wareing, P. F. (1973). Identification of a cytokinin in sycamore sap by gas chromatography-mass spectrometry. *Plant Sci. Lett* **1**, 321–4.

Horgan, R., Palni, L. M. S., Scott, I. M. and McGaw, B. (1981). Cytokinin biosynthesis and metabolism in *Vinca rosea* crown gall tissue. In *Metabolism and Molecular Activities of Cytokinins*, eds J. Guern and C. Péaud-Lenoel, Springer, Berlin, pp. 56–65.

Incoll, L. D. and Whitelam, G. C. (1977). The effect of kinetin on stomata of the grass *Anthephora pubescens* Nees. *Planta* **137**, 243–5.

IUPAC-IUB Commission on Biochemical Nomenclature (1970). Abbreviations and symbols for nucleic acids, polynucleotides and their constituents. *J. Biol. Chem.* **245**, 5171–6.

Jablonski, J. and Skoog, F. (1954). Cell enlargement and cell division in excised tobacco pith tissue. *Physiol. Plant.* **7**, 16–24.

Jewer, P. C. and Incoll, L. D. (1980). Promotion of stomatal opening in the grass *Anthephora pubescens* Nees by a range of natural and synthetic cytokinins. *Planta* **150**, 218–21.

Jewer, P. C. and Incoll, L. D. (1981) Promotion of stomatal opening in detached epidermis of *Kalanchoë daigremontiana* Hamet et Perr. by natural and synthetic cytokinins. *Planta* **153**, 317–18.

Kaiss-Chapman, R. W. and Morris, R. O. (1977). Trans-zeatin in culture filtrates of *Agrobacterium tumefaciens*. *Biochem. Biophys. Res. Commun.* **76**, 453–9.

Keim, P., Erion, J. and Fox, J. E. (1981). Cytokinin: the current status of cytokinin binding moieties. In *Metabolism and Molecular Activities of Cytokinins*, eds J. Guern and C. Péaud-Lenoel. Springer, Berlin, pp. 179–90.

Kende, H. (1965). Kinetin-like factors in the root exudate of sunflowers. *Proc. Nat. Acad. Sci. USA* **53**, 1302–7.

Khokhlova, V. A. (1977). The effect of cytokinin on plastid formation in excised pumpkin cotyledons in the light and in the dark. *Fiziol. Rastenii (Moscow)* **24**, 1189–92.

Köhler, K-H. and Conrad, K. (1966). Ein quantitativer Phytokinin-Test. *Biol. Rdsch.* **4**, 36–7.

Koshimizu, K., Matsubara, S., Kasaki, T. and Mitsui, T. (1967). Isolation of a new cytokinin from immature yellow lupin seeds. *Agric. Biol. Chem.* **31**, 795–801.

Krasnuk, M., Witham, F. H. and Tegley, J. R. (1971). Cytokinins extracted from Pinto bean fruit. *Plant Physiol.* **48**, 320–4.

Kuraishi, S. (1959). Effect of kinetin analogs on leaf growth. *Scientific Papers of the College of General Education, University of Tokyo*, **9**, 67–104.

Laloue, M. (1977). Cytokinins: 7-glucosylation is not a prerequisite of the expression of their biological activity. *Planta* **134**, 273–5.

Laloue, M., Gawer, M. and Terrine, C. (1975). Modalites de l'utilisation des cytokinines exogènes par les cellules de tabac cultivés en milieu liquide agité. *Physiol. veg.* **13**, 781–96.

Laloue, M., Terrine, C. and Gawer, M. (1974). Cytokinins: formation of the nucleoside-5'-triphosphate in tobacco and *Acer* cells. *FEBS Lett.* **46**, 45–50.

Laloue, M., Terrine, C. and Guern, J. (1977). Cytokinins: metabolism and biological activity of N^6-(Δ^2-isopentenyl)adenine in tobacco cells and callus. *Plant Physiol.* **59**, 478–83.

Lee, T. S., Purse, J. G., Pryce, R. J., Horgan, R. and Wareing, P. F. (1981). Dihydroconiferyl alcohol—a cell division factor from *Acer* species. *Planta* **152**, 571–7.

Letham, D. S. (1963). Zeatin, a factor inducing cell division from *Zea mays*. *Life Sci.* **8**, 569–73.

Letham, D. S. (1967). Regulators of cell division in plant tissues. A comparison of the activities of zeatin and other cytokinins in five bioassays. *Planta* **74**, 228–42.

Letham, D. S. (1971). Regulators of cell division in plant tissues XII. A cytokinin bioassay using excised radish cotyledons. *Physiol. Plant.* **25**, 391–6.

Letham, D. S. (1973). Cytokinins from *Zea mays*. *Phytochemistry* **12**, 2445–55.

Letham, D. S. (1974). The cytokinins of coconut milk. *Physiol. Plant.* **32**, 66–70.

Letham, D. S. (1978). Cytokinins. In *Phytohormones and Related Compounds: A Comprehensive Treatise*, eds D. S. Letham, P. B. Goodwin and T. J. V. Higgins, Elsevier/North-Holland Biomedical Press, Amsterdam, vol. 1, pp. 205–63.

Letham, D. S. (1982). A 6-oxypurine with growth promoting activity. *Plant Sci. Lett.* **26**, 241–9.

Letham, D. S., Shannon, J. C. and MacDonald, I. R. C. (1964). The structure of zeatin, a (kinetin-like) factor inducing cell division. *Proc. Chem. Soc. London* 230–1.

Letham, D. S., Wilson, M. M., Parker, C. W., Jenkins, I. D., MacLeod, J. K. and Summons, R. E. (1975). Regulators of cell division in plant tissues. XXIII. The identity of an unusual metabolite of 6-benzylaminopurine. *Biochim. Biophys. Acta* **399**, 61–70.

Longo, C. P., Longo, G. P., Lampugnani, M., Rossi, G.

and Servettaz, S. (1981). Light and fusicoccin as tools for discriminating among responses of cotyledons to cytokinins. In *Metabolism and Molecular Activities of Cytokinins*, eds J. Guern and C. Péaud-Lenoel, Springer, Berlin, pp. 261–6.

Lorenzi, R., Horgan, R. and Wareing, P. F. (1975). Cytokinins in *Picea sitchensis* Carriere—identification and relation to growth. *Biochem. Physiol. Pflanzen* **168**, 333–40.

MacDonald, E. M. S., Akiyoshi, D. E. and Morris, R. O. (1981). Combined high-performance liquid chromatography-radioimmunoassay for cytokinins. *J. Chromatog.* **214**, 101–9.

Marre, E., Colombo, R., Lado, P. and Rasi-Caldogno, F. (1974). Correlation between proton extrusion and stimulation of cell enlargement. III. Effects of fusicoccin and cytokinins on leaf fragments and isolated cotyledons. *Plant Sci. Lett.* **2**, 139–50.

Marre, E., Lado, P., Ferroni, A. and Ballarin-Denti, A. (1974). Transmembrane potential increase induced by auxin, benzyladenine and fusicoccin. Correlation with proton extrusion and cell enlargement. *Plant Sci. Lett.* **2**, 257–65.

Matsubara, S. (1980). Structure–activity relationships of cytokinins. *Phytochemistry* **19**, 2239–53.

Matthysee, A. G. and Abrams, M. (1970). A factor mediating interaction of kinins with the genetic material. *Biochim. Biophys. Acta* **199**, 511–18.

Meidner, H. (1967). The effect of kinetin on stomatal opening and the rate of intake of CO_2 in mature primary leaves of barley. *J. Exp. Bot.* **18**, 556–61.

Millard, B. J. (1978). *Quantitative Mass Spectrometry*. Heyden.

Miller, C. O. (1956). Similarity of some kinetin and red light effects. *Plant Physiol.* **31**, 318–19.

Miller, C. O. (1958). The relationship of the kinetin and red light promotions of lettuce seed germination. *Plant Physiol.* **33**, 115–17.

Miller, C. O. (1963). Kinetin and kinetin-like compounds. In *Modern Methods of Plant Analysis*, vol. 6, eds H. F. Linskens and M. V. Tracey, Springer, Berlin, pp. 194–202.

Miller, C. O. (1967). Zeatin and zeatin riboside from a mycorrhizal fungus. *Science* **157**, 1055–7.

Miller, C. O. (1975). Cell division factors from *Vinca rosea* L. crown gall tumor tissue. *Proc. Nat. Acad. Sci. USA* **72**, 1883–6.

Miller, C. O., Skoog, F., Okomura, F. S., Saltza, M. H. von and Strong, F. M. (1956). Isolation, structure and synthesis of kinetin, a substance promoting cell division. *J. Am. Chem. Soc.* **78**, 1345–50.

Miller, C. O., Skoog, F., Saltza, M. H. von and Strong, F. M. (1955). Kinetin a cell division factor from deoxyribonucleic acid. *J. Am. Chem. Soc.* **77**, 1329.

Miura, G. A. and Hall, R. H. (1973). Trans-ribosylzeatin. Its biosynthesis in *Zea mays* endosperm and the mycorrhizal fungus *Rhizopogon roseolus*. *Plant Physiol.* **51**, 563–9.

Moore, H. (1979). Kinetin binding protein from wheat germ. *Plant Physiol.*, Suppl., **59**, 17.

Morris, R. O. (1977). Mass spectroscopic identification of cytokinins. Glucosyl zeatin and glucosyl ribosylzeatin from *Vinca rosea* crown gall. *Plant Physiol.* **59**, 1029–33.

Mothes, K. (1960). Uber das Altern Blatter und die Moglichkert iher Wiederverjungung. *Naturwissenschaften* **47**, 337–51.

Mothes, K. (1970). Uber grune Inslen. *Leopoldina* **15**, 171–2.

Murai, N. (1981). Cytokinin biosynthesis and its relationship to the presence of plasmids in strains of *Corynebacterium fascians*. In *Metabolism and Molecular Activities of Cytokinins*, eds J. Guern and C. Péaud-Lenoel, Springer, Berlin, pp. 17–26.

Murakoshi, I., Ikegami, F., Ookawa, N., Haginiwa, J. and Letham, D. S. (1977). Enzymatic synthesis of lupinic acid, a novel metabolite of zeatin in higher plants. *Chem. Pharm. Bull.* **25**, 520–2.

Naito, K., Tsuji, H. and Hatakeyama, I. (1978). Effect of benzyladenine on RNA, DNA, protein and chlorophyll contents in intact bean leaves: differential responses to benzyladenine according to the leaf age. *Physiol. Plant.* **43**, 367–71.

Nishamura, A. (1972). Minor components in transfer RNA: their characterisation, location and function. *Prog. Nucleic Acid Res. Mol. Biol.* **12**, 49–85.

Norris, R. F. (1976). Expansion of radish cotyledons: an interaction between NaCl and cytokinins. *Plant Sci. Lett.* **6**, 63–8.

Ooms, G., Hooykaas, P. J. J., Moolenaar, G. and Schilperoort, R. A. (1981). Crown gall tumors of abnormal morphology, induced by *Agrobacterium tumefaciens* carrying mutated octopine Ti plasmids; analysis of T-DNA functions. *Gene* **14**, 33–50.

Paces, V., Westiuk, E. and Hall, R. H. (1971). Conversion of N^6-(Δ^2-isopentenyl)adenosine to adenosine by enzyme activity in tobacco tissue. *Plant Physiol.* **48**, 775–8.

Palmer, M. V., Scott, I. M. and Horgan, R. (1981). Cytokinin metabolism in *Phaseolus vulgaris* L. II. Comparative metabolism of exogenous cytokinins by detached leaves. *Plant Sci. Lett.* **22**, 187–95.

Palni, L. M. S. and Horgan, R. (1982). Cytokinins from the culture medium of *Vinca rosea* crown gall tumour tissue. *Plant Sci. Lett.* **24**, 327–34.

Parker, C. W. and Letham, D. S. (1974). Regulators of cell division in plant tissues XVIII. Metabolism of zeatin in *Zea mays* seedlings. *Planta* **115**, 337–44.

Parker, C. W., Letham, D. S., Cowley, D. E. and MacLeod, J. K. (1972). Raphanatin, an unusual purine derivative and a metabolite of zeatin. *Biochem. Biophys. Res. Commun.* **49**, 460–6.

Parker, C. W., Letham, D. S., Gollnow, B. I., Summons, R. E., Duke, C. C. and MacLeod, J. K. (1978). Regulators of cell division in plant tissues. XXV. Metabolism of zeatin by lupin seedlings. *Planta* **142**, 239–51.

Parker, C. W., Wilson, M. M., Letham, D. S., Cowley, D. E. and MacLeod, J. K. (1973). The glycosylation of cytokinins. *Biochem. Biophys. Res. Commun.* **55**, 1370–6.

Parthier, B. (1979). Phytohormones and chloroplast development. *Biochem. Physiol. Pflanzen* **174**, 173–214.

Peterson, J. B. and Miller, C. O. (1976). Cytokinins in *Vinca rosea* L. crown gall tumor tissue as influenced by compounds containing reduced nitrogen. *Plant Physiol.* **57**, 393–9.

Polya, G. M. and Davis, A. W. (1978). Properties of a high affinity cytokinin binding protein from wheat germ. *Planta* **139**, 139–47.

Purse, J. G., Horgan, R., Horgan, J. M. and Wareing, P. F. (1976). Cytokinins of sycamore spring sap. *Planta* **132**, 1–8.

Richmond, A. and Lang, A. (1957). Effect of kinetin on protein content and survival of detached *Xanthium* leaves. *Science* **125**, 650–1.

Scott, I. M. and Horgan, R. (1980). Quantification of cytokinins by selected ion monitoring using ^{15}N-labelled internal standards. *Biomed. Mass Spectrometry* **7**, 446–9.

Scott, I. M., Horgan, R. and McGaw, B. A. (1980). Zeatin-9-glucoside a major endogenous cytokinin of *Vinca rosea* L. crown gall tissue. *Planta* **149**, 472–5.

Seyer, P., Marty, D., Lescure, A. M. and Péaud-Lenöel, C. (1975). Effect of cytokinin on chloroplast cyclic differentiation in cultured tobacco cells. *Cell Differentiation* **4**, 187–97.

Shantz, E. M. and Steward, F. C. (1955). The identification of compound A from coconut milk as 1,3-diphenylurea. *J. Am. Chem. Soc.* **77**, 6351–3.

Short, K. C. and Torrey, J. G. (1972). Cytokinins in seedling roots of pea. *Plant Physiol.* **49**, 155–60.

Skene, K. G. M. (1975). Cytokinin production by roots as a factor in the control of plant growth. In *The Development and Function of Roots*, eds J. G. Torrey and D. T. Clarkson, Academic Press, London, pp. 366–9.

Skoog, F. and Ghani, A. K. B. A. (1981). Relative activities of cytokinins and antagonists in releasing lateral buds of *Pisum* from apical dominance compared with their relative activities in the regulation of growth of tobacco callus. In *Metabolism and Molecular Activities of Cytokinins*, eds J. Guern and C. Péaud-Lenoel, Springer, Berlin, pp. 140–50.

Skoog, F., Hamzi, H. Q., Szweykowska, A. M., Leonard, N. J., Carraway, K. L., Fujii, T., Helgeson, J. P. and Leoppky, R. N. (1967). Cytokinins: structure/activity relationships. *Phytochemistry* **6**, 1169–92.

Skoog, F. and Miller, C. O. (1957). Chemical regulation of growth and organ formation in plant tissues cultured *in vitro. Symp. Soc. Exp. Biol. Med.* **11**, 118–31.

Skoog, F. and Tsui, C. (1948). Chemical control of growth and bud formation in tobacco stem segments and callus cultured *in vitro. Am. J. Bot.* **35**, 782–7.

Sondheimer, E. and Tzou, D. S. (1971). The metabolism of hormones during seed germination and dormancy. II. The metabolism of 8-^{14}C-zeatin in bean axes. *Plant Physiol.* **47**, 516–19.

Stoddart, J. L. and Thomas, H. (1982). Leaf senescence. In *Nucleic Acids and Proteins in Plants I, Encyclopaedia of Plant Physiology*, New Series, vol. 14A, eds D. Boulter and B. Parthier, Springer, Berlin, pp. 592–636.

Stuchbury, T., Palni, L. M. S., Horgan, R. and Wareing, P. F. (1979). The biosynthesis of cytokinins in crown gall tissue of *Vinca rosea. Planta* **147**, 97–102.

Summons, R. E., Entsch, B., Letham, D. S., Gollnow, B. I. and MacLeod, J. K. (1980). Regulators of cell division in plant tissues. XXVIII. Metabolites of zeatin in sweet corn kernels: purifications and identifications using high-performance liquid chromatography and chemical-ionisation mass-spectrometry. *Planta* **147**, 422–34.

Summons, R., Entsch, B., Parker, C. W. and Letham, D. S. (1979). Mass spectrometric analysis of cytokinins in plant tissues. III. Quantitation of the cytokinin glycoside complex of lupin pods by stable isotope dilution. *FEBS Lett.* **107**, 21–5.

Summons, R. E., MacLeod, J. K., Parker, C. W. and Letham, D. S. (1977). The occurrence of raphanatin as an endogenous cytokinin in radish seed. Identification and quantitation by gas chromatographic-mass spectrometric analysis using deuterium labelled standards. *FEBS Lett.* **82**, 211–14.

Sussman, M. R. and Kende, H. (1978). *In vitro* cytokinin binding to a particulate fraction of tobacco cells. *Planta* **140**, 251–9.

Taya, Y., Tanaka, Y. and Nishimura, S. (1978) 5'-AMP is a direct precursor of cytokinins in *Dictyostelium discoideum. Nature* **271**, 545–7.

Tepfer, D. A. and Fosket, D. E. (1978). Hormone-mediated translational control of protein synthesis in cultured cells of *Glycine max. Develop. Biol.* **62**, 486–97.

Tetley, R. M. and Thimann, K. V. (1974). The metabolism of oat leaves during senescence. I. Respiration, carbohydrate metabolism and the action of cytokinins. *Plant Physiol.* **54**, 859–62.

Thomas, H. (1975). Regulation of alanine aminotransferase in leaves of *Lolium temulentum* during senescence. *Z. Pflanzenphysiol.* **74**, 208–18.

van Staden, J. and Drewes, S. E. (1975). Identification of zeatin and zeatin riboside in coconut milk. *Physiol. Plant.* **34**, 106–9.

van Staden, J. and Menary, R. C. (1976). Identification of

cytokinins in the xylem sap of tomato. *Z. Pflanzenphysiol.* **78**, 262–5.

Varga, A. and Bruinsma, J. (1973). Effect of different cytokinins on the senescence of detached oat leaves. *Planta* **111**, 91–3.

Vreman, H. J., Schmitz, R. Y., Skoog, F., Playtis, A. J., Frihart, C. R. and Leonard, N. J. (1974). Synthesis of methylthio-*cis*- and *trans*-ribosylzeatin and their isolation from *Pisum* tRNA. *Phytochemistry* **13**, 31–7.

Vreman, H. J., Thomas, R., Corse, J., Swaminathan, S. and Murai, N. (1978). Cytokinins in tRNA obtained from *Spinacia oleracia* L. leaves and isolated chloroplasts. *Plant Physiol.* **61**, 296–306.

Wang, T. L., Beutelmann, P. and Cove, D. J. (1981). Cytokinin biosynthesis in mutants of the moss *Physcomitrella patens*. *Plant Physiol.* **68**, 739–44.

Wang, T. L. and Horgan, R. (1978). Dihydrozeatin riboside, a minor cytokinin from the leaves of *Phaseolus vulgaris*. *Planta* **140**, 151–3.

Wang, T. L., Horgan, R. and Cove, D. J. (1981). Cytokinins from the moss *Physcomitrella patens*. *Plant Physiol.* **68**, 735–8.

Wang, T. L., Thompson, A. G. and Horgan, R. (1977). A cytokinin glucoside from the leaves of *Phaseolus vulgaris* L. *Planta* **135**, 285–8.

Weiler, E. W. (1980). Radioimmunoassay for trans-zeatin and related cytokinins. *Planta* **149**, 155–62.

Weiler, E. W. and Spanier, K. (1981). Phytohormones in the formation of crown gall tumors. *Planta* **153**, 326–37.

Weiss, C. and Vaadia, Y. (1965). Kinetin activity in root apices of sun flower plants. *Life Sci.* **4**, 1323–6.

Whitty, C. D. and Hall, R. H. (1974). A cytokinin oxidase in *Zea mays*. *Can. J. Biochem.* **52**, 789–99.

Wiesner, J. (1892). *Die Elementarstruktur und das Wachstum der lebenden Substanz*. A. Holder, Vienna.

Wood, R. K. S. (1967). *Physiological Plant Pathology*, Blackwell, Oxford and Edinburgh, pp. 393–7.

Woolhouse, H. W. and Batt, T. (1976). The nature and regulation of senescence in plastids. In *Perspectives in Experimental Biology*, ed. N. Sunderland, Pergamon Press, Oxford, pp. 163–75.

Young, H. (1977). Identification of cytokinins from natural sources by GCMS. *Anal. Biochem.* **79**, 226–33.

Zachau, H. G., Dutting, D. and Feldman, H. (1966). Nucleotide sequences of two serine-specific transfer ribonucleic acids. *Angew. Chem. Int. Ed.* **5**, 422.

Inhibitors

4

B V Milborrow

Introduction

The plant hormones that have been isolated and characterized were first detected by using the responses of sensitive pieces of plant to the presence of active compounds in extracts. By far the easiest response to detect and measure is an effect on growth and it is not surprising, therefore, that all the plant hormones known at present are regulators of growth processes. Plant hormones differ from those in animals in that they control a broader spectrum of processes than does an individual animal hormone, and movement from their sites of production may not occur. A considerable amount of work is necessary to establish that a compound acts as an inhibitory regulator of growth and, sadly, after the first brief description, work on most of the putative inhibitors has been abandoned.

It is becoming increasingly clear that most plants contain from a few to many secondary products, polyketide, terpenoid or alkaloidal compounds, which often have potent pharmacological effects. Their function is obscure but many are now considered to comprise a static chemical defense against insect or fungal attack. Others are classified as phytoalexins: these are compounds which are synthesized rapidly in response to attack by a pathogenic organism or damage to a tissue. These compounds of the 'active defense' can be considered to be analogous to the mammalian antibody system and it is hardly surprising to find that many cause drastic changes in a plant's physiology. It is often difficult to discriminate between such compounds and those that have a genuine regulatory function in the plant.

Secondary products in plants

Two main strategies of experimentation have been adopted to investigate whether or not a given compound regulates a process: (1) apply the compound and observe its effects; and (2) adjust the environmental conditions and relate the amount of hormone present to the response of the plant. Both of these are unsatisfactory. The applied hormone may be inappropriate for the tissue at the time chosen, it may not penetrate into the correct cellular compartment, it may be applied in supraoptimal amounts, or the response may not be proportional to concentration. Experimentation with growth inhibitors is particularly hazardous because almost any compound will cause inhibition of growth, toxicity and death if applied at a sufficiently high concentration.

Phenolic acids and related compounds

Extracts of higher plants invariably contain a considerable amount of a number of growth-inhibitory phenolic acids and the proportions in any plant could be the result of the particular affinities of that plant's hydroxylating enzymes for the various substrates (Stafford, 1974).

Phenolic compounds and IAA metabolism

The aspect of the action of phenolic acids that has excited most interest among physiologists has been their interaction with IAA oxidase. Goldacre, Galston and Weintraub (1953) found that some substituted phenols stimulated the activity of IAA oxidase in cell-free extracts of plants. The idea of growth regulation by modifying the activity of IAA oxidase is an attractive one, but in the absence of a precise quantitative background it appears that there is a vast excess of potential modifying phenols to IAA oxidase and an even greater disproportion to the amount of free IAA in cells of growing plants. Again, we seem to have fallen into the trap of assuming that the substrate fed to, and attacked by, an enzyme preparation is (*a*) the natural substrate and (*b*) the only substrate and (*c*) that the enzyme and substrate are present together in undamaged cells. While IAA may be oxidized by a tissue homogenate, the reaction may be brought about by an unspecific oxidase released from ruptured tissues. Generally, monophenolic acids such as *p*-coumaric acid increase the activity of IAA oxidase whereas diphenolic acids, such as caffeic acid, inhibit its

activity. This is illustrated in Fig. 4.1 where the increase in growth caused by 10^{-4} M caffeic acid in the presence of $10\,\mu$M IAA is attributed to increased IAA concentrations which, in turn, are attributed to decreased IAA oxidase activity.

Unfortunately, most of the data on the interaction of phenolics and IAA do not stand up to a rigorous, critical examination. The enzyme has not been isolated in pure form and so the interaction between IAA as substrate and a phenolic material as cofactor, activator or deactivator has not been established. Rubery (1972) has found that malate stimulates IAA oxidase activity, so the reaction is not specific for phenolic acids. The IAA has been assayed by its net effect over a period of hours (usually 12 to 24) on the elongation of a coleoptile section and so the phenolics could have affected growth by inhibiting IAA destruction, inhibiting degenerative changes in the cells, assisting penetration of IAA or other indirect effects. This is not to say that the phenols do not have clear, reproducible effects; Zenk and Müller (1963) have measured the decarboxylation of added [1-^{14}C]IAA by oat coleoptiles (Table 4.1) and found that the rate of

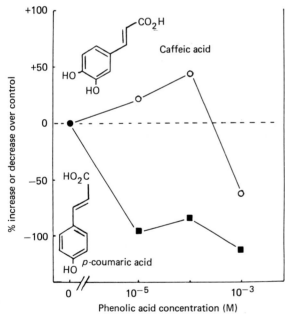

Fig. 4.1 Caffeic acid, a diphenolic acid, promotes the growth of oat mesocotyl segments incubated in the presence of $10\,\mu$M IAA. In contrast, *p*-coumaric, a monophenol, inhibits the growth. The dotted line represents the control value in the presence of $10\,\mu$M IAA. Above the line represents % growth stimulation, below the line % growth inhibition by the phenolic acids in relation to IAA alone. Caffeic acid, ●; *p*-coumaric acid ■. (Adapted from Nitsch and Nitsch, 1962.)

Table 4.1 Effect of various phenolic acids on the formation of [^{14}C]CO$_2$ from [^{14}C]IAA by 3-mm sections of oat (*Avena sativa*) coleoptiles: IAA 0.1 mM, phenolic acids 0.1 mM

Monophenolic acids	[^{14}C]CO$_2$ produced as a % of control
o-Coumarate	104
m-Coumarate	117
p-Coumarate	181
p-Hydroxybenzoate	113
Ferulate	126
Diphenolic acids	
Caffeate	79
Chlorogenate	73

Data from Zenk and Müller (1963).

[^{14}C]CO$_2$ production was increased by monophenols and reduced by diphenols, as expected from their effects on growth. The function of the phenols remains uncertain; they may cause tanning of proteins and so render the damaged cells less liable to attack by saprophytes, or the free phenolics and polyphenols formed may be toxic to pathogens.

Overall, the rôle of 'IAA oxidase' *in vivo* must be judged, at best, as not proven.

At present the involvement of the phenolic acids in the regulation of growth must also be judged as 'not proven'.

Abscisic acid

The discovery of abscisic acid

The presence of growth-inhibitory materials in extracts of plants has been known for many years, but the detailed investigation of the causative agents did not begin until Bennet-Clark and his colleagues used paper chromatography to try and rediscover the mythical auxins A and B. Bennet-Clark, Tambiah and Kefford (1952) chromatographed the acid fraction of alcoholic extracts of a number of plants and subjected segments of the chromatograms to bioassays using oat coleoptiles. The areas near the origin showed the presence of a growth accelerator, α, while a zone at about R_f 0.6 carried potent growth inhibitor activity which was named inhibitor β (Bennet-Clark and Kefford, 1953).

The investigations of the β-inhibitor were pursued in several laboratories during the 1950s, Radley and Hemberg correlated the amount of the inhibitor with the capacity of potato tubers to sprout, Barlow found that the amounts in plum shoots varied seasonally and Van Steveninck (1959) found that the abortion of fertilized yellow lupin (*Lupinus luteus*) pods was correlated with the amount of inhibitory material present. The inhibitory activity appeared to be responsible for the self-thinning of the immature lupin pods, any one of which could develop to fertile maturity if the remainder were removed. This work was pursued by Rothwell and Wain (1964) who extracted an active fraction from which ABA was later isolated (Cornforth, Milborrow, Ryback, Rothwell and Wain, 1966). Wareing found that an inhibitor was present in dormant buds of deciduous trees such as sycamore (*Acer pseudoplatanus*) and birch (*Betula pendula*) and that the amounts were correlated with the depth of bud dormancy. (*see* Milborrow, 1966a.)

The work of Robinson and Wareing (1964) was followed up by Cornforth, Milborrow, Ryback and Wareing (1965) who isolated the crystalline inhibi-

tor, known as 'dormin', from sycamore leaves. Meanwhile Addicott and his colleagues in California had found an inhibitor in the gibberellin fraction from cotton (*Gossypium hirsutum*) bolls and had isolated nine milligrams of crystalline material (Okhuma, Lyon, Addicott and Smith, 1963). This was sufficient for them to obtain a nuclear magnetic resonance spectrum, and to propose a structure for the compound which they believed to be an abscission accelerator and which they named 'Abscisin II' (Okhuma, Addicott, Smith and Thiessen, 1965). The Milstead group compared the infra-red spectrum of their dormin with that of Abscisin II and found them to be identical; they confirmed the proposed structure by synthesizing it 11 days later (Cornforth, Milborrow, Ryback and Wareing, 1965). The natural, optically active Abscisin II, now renamed abscisic acid (ABA) (1), melts at 161°C whereas the synthetic material is composed of mirror-image pairs that form a more stable crystal structure which melts at 190°C.

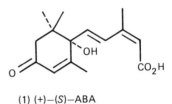

(1) (+)−(*S*)−ABA

(+)-Abscisic acid is one of the most intensely optically active compounds known and this feature in particular, together with chromatographic and biological properties, enabled Milborrow (1968) to prepare extracts from a number of plants whose inhibitor-β fractions had been reported previously and to account for most of the inhibitory activity in terms of their (+)-ABA content.

Bioassay and physical methods of measurement

The method used to detect ABA before it was characterized chemically was, perforce, bioassay. Abscission of petiolar stumps of cotton (Addicott, Carns, Lyon, Smith and McMeans, 1964), inhibition of wheat embryo germination (Cornforth *et al.*, 1965), amylase synthesis in barley aleurone (Aspinall, Paleg and Addicott, 1967), closure of stomata in epidermal strips (Jones and Mansfield, 1970), have all been used. While bioassays are still used occasionally, the sensitivity and specificity of several physical methods are far superior.

The first of these made use of the compound's intense and characteristic optical rotatory dispersion spectrum (ORD) (Milborrow, 1968) (Fig. 4.2). The disadvantages of the method are that the ABA has to be purified to a considerable degree and such purification always causes considerable losses.

assay for ABA has been developed that can detect amounts of ABA as small as 10 fg (Weiler, 1980). The antibody to the (+) ABA appears to be specific while the antibody to the racemate also appears to react with ABA glucose ester. Nevertheless, radio-immunoassay may be the method of choice in the future for extremely small amounts of ABA.

The main failing of all these methods of analysis is

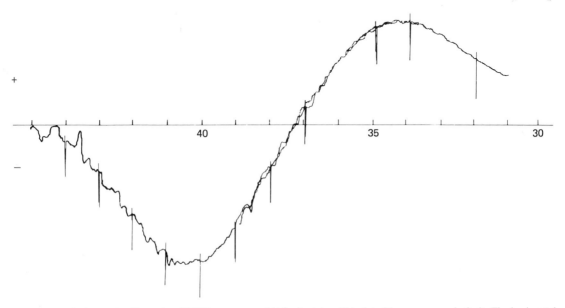

Fig. 4.2 Optical rotatory dispersion (ORD) spectrum of (+)-abscisic acid isolated from an avocado fruit. The horizontal line represents zero rotations, distance above the line is proportional to (+) rotation, distance below the line is proportional to (−) rotation. The horizontal scale is marked in wave numbers where 40 (250 nm) is in the ultraviolet, 30 (333 nm) is the blue end of the visible spectrum. The (+) extremum at 34.6 wave numbers (289 nm) is the most convenient for quantitative measurements. This spectrum was produced by 6.4 μg (+)-ABA/ml in acidic methanol.

Gas chromatography of a volatile derivative of ABA, such as the methyl ester, has been used successfully but some purification is usually required. Flame ionization detectors (FID) give a similar response for a similar weight of any carbon compound but electron capture detectors (EC) are some 6000 times more sensitive to ABA than to compounds lacking the conjugated dienoic acid grouping, so Me ABA stands out from a background of most plant constituents (Seeley and Powell, 1970). The method is extremely sensitive and detectable signals are obtained with as little as 100 fg (1×10^{-13} g) (Fig. 4.3).

Recently a highly-sensitive radio-immunological

that they require some purification of the sample which causes losses and so the amount of ABA detected is less than the true value.

Several attempts have been made to overcome this. Lenton, Perry and Saunders (1971) added a known amount of the 2-*trans* isomer of ABA (2) to

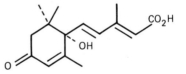

(2) 2–*trans*–ABA

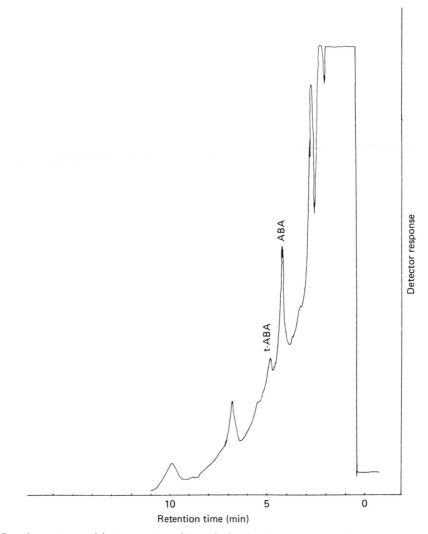

Fig. 4.3 Gas chromatogram (electron capture detector) showing the occurrence of trace amounts of 2-*trans*-ABA although the extraction was carried out in darkness. The injection is equivalent to 0.1 g tomato leaf. Ordinate = detector response; abscissa = retention time in minutes from injection, 190°C.

an extract so that when the Me ABA peak was detected by glc the size of the Me *trans* ABA peak, which eluted shortly after Me ABA, could be used to calculate how much had been lost and, therefore, how much ABA had been lost. The method is a dangerous one because 2-*trans*-ABA does occur naturally in plants and ABA is reversibly isomerized by light to a 1:1 mixture with the 2-*trans* isomer;

thus any exposure of the solution to light will distort the result. Recently, the ethyl ester of ABA has been added to the extract as an internal standard. It is separated from Me ABA by glc and would be expected to suffer the same losses as the methylated, endogenous ABA.

The ORD method for (+)-ABA was elaborated to account for losses during extraction: the racemate dilution method (Milborrow, 1968). A known amount of synthetic, racemic, optically-inactive ABA was added to a tissue extract, the total amount of ABA in a highly purified sample was measured by ultraviolet absorption and then the proportion of natural (+) present was determined by ORD. These two measurements enabled the original ABA content of the tissue to be calculated. The method is little used now because it demands relatively large amounts of plant material, rigorous purification and access to a sensitive spectropolarimeter.

Once ^{14}C and stably ^{3}H labeled ABA became available it was possible to monitor losses during isolation by measuring the amount of label recovered. It must be stressed, however, that the detection of Me ABA by glc is considerably more sensitive than a labeling method, unless ABA of extremely high specific activity is used.

The most accurate and, at the same time, reasonably sensitive method is gas chromatography–mass spectrometry with selected ion monitoring (GC/MS SIM)(Rivier, Milon and Pilet, 1977). The glc separation of Me ABA enables small samples of fairly impure solutions to be used, and identifies the Me ABA by retention time. The mass spectrum gives unambiguous confirmatory identification and the use of stably deuteriated ABA enables the true endogenous concentration to be calculated. A known quantity of stably deuteriated ABA is added to the extract and then the mixture of unlabeled and deuterium-labeled ABA is methylated. 3-Trideuteriomethyl ABA gives a parent-ion peak three mass units greater than that of the unlabeled, endogenous ABA, and the ratio between the two parent ions is used to calculate the amount of ABA originally present. The method can be used to detect as little as 1×10^{-10} g (Netting, Milborrow and Duffield, 1981).

Distribution in plants

A variety of methods has been used to show that ABA is present in all vascular plants in which it has been sought, these include dicotyledons, monocotyledons, gymnosperms (Yew, *Taxus baccata*; Silverfir, *Abies alba*), ferns (Bracken, *Pteridium aquilinium*), and horsetails (*Equisetum arvensis*).

Except for two fungi (*see* later), ABA has not been found in fungi, bacteria or algae; it is absent from liverworts but the situation in mosses is still uncertain. It will be necessary to grow mosses on defined media in culture before it is certain that any ABA they appear to contain has not been derived from a higher plant source; contamination of mosses by aphid exudate, known to contain ABA, or the leachate of taller vegetation, will have to be excluded.

ABA has been found in all higher plant tissues examined; these include roots, xylem of tree trunks, xylem sap, phloem sap, pollen, petals, fruits and seeds (Milborrow, 1974). Concentrations vary widely from $3-5\,\mu g\,kg^{-1}$ in aquatic plants to $10\,mg\,kg^{-1}$ in the avocado fruit mesocarp. The amounts in the leaves of temperate crop plants are usually between 50 and $500\,\mu g\,kg^{-1}$. ABA is transported quite rapidly in the phloem and from cell to cell, so the presence of ABA in a tissue is no guarantee that it is formed there. The incorporation of the labeled precursor mevalonic acid into ABA by an isolated tissue is a clear indication that biosynthesis has occurred, provided that the ABA has been freed of all radiochemical impurities. The most convenient method to purify ABA is to isolate the free acid by thin-layer chromatography, methylate, rechromatograph, reduce to 1′,4′-diols with sodium borohydride, chromatograph and determine the label in each (Milborrow and Noddle, 1970). The diols can be reoxidized to ABA. If, at the end of this procedure the label still co-chromatographs with ABA, it can be safely assumed to have been incorporated into ABA.

With these methods it has been shown that roots of sunflower and avocado, stems, leaves, fruit mesocarp and seeds of avocado, and embryos, endosperm and leaves of wheat can incorporate labeled mevalonate into ABA (Milborrow and Robinson, 1973).

Effects of ABA on growth

Inhibition Went's dictum—'There is no growth without auxin' has sometimes been tacitly extended to embrace the concept that growth is proportional to the supply of the various growth-promoting substances. Clearly, if this were so then there would be no function for the growth inhibitors. Mature organs do not grow, even when supplied with a variety of growth substances, so there is an overriding effect imposed by the tissues. Even rapidly growing juvenile tissues contain some ABA; does it have a slight inhibitory activity on their growth? It is very difficult to answer this question without some method of removing or blocking the action of ABA and, so far, no competitive inhibitor or agent that can prevent its synthesis has been discovered.

The general observation that applications of growth-promoting phytohormones, such as IAA (Chapter 1) or GA (Chapter 2), to growing tissues do cause more rapid growth could be interpreted as a swamping of the effect of endogenous ABA, but it could as easily be attributed to less than optimal concentrations of available IAA and GA. Rothwell and Wain (1964) introduced a novel approach to such problems when they investigated the response of coleoptiles treated with IAA and ABA, using the methods of enzyme kinetics. They obtained evidence for the presence of an active, endogenous IAA fraction but their data were not sufficiently extensive to allow for an effect of endogenous ABA to be identified. This approach is an interesting one and it should be re-examined. It is possible that the assumptions on which mathematical analyses are based may not be fulfilled by growth regulators if penetration rates mask the true values.

Apart from the clearly-demonstrable inhibitory action of ABA when applied to growing organs and the presence of ABA even in rapidly growing organs, there is little evidence that it does act as a slight brake on normal growth (Milborrow, 1974; Walton, 1980).

Correlative inhibition The role of ABA as a correlative inhibitor (Chapter 6), i.e., one that regulates the growth of lateral buds in response to changes in the apical system, appears uncertain at present (Dorffling, 1976). Although it is present in inhibited lateral buds there is no marked change in the amount present when lateral buds are released from apical dominance (White and Mansfield, 1977). Exposure of whole tomato plants to far-red irradiation (700–800 nm) after decapitation caused the lateral buds to remain inhibited while those of untreated plants grew out (Tucker, 1977). The ABA content of far-red-treated plants increased, but the amounts in their buds were not determined.

Growth promotion The few experiments in which ABA has been found to promote a growth process include the stimulation of parthenocarpic seed development (Jackson and Blundell, 1966), rooting of

cuttings (Chin, Meyer and Beevers, 1969), soybean callus growth in the presence of kinetin (Blumenfeld and Gazit, 1970), elongation of cucumber hypocotyls (Aspinall *et al.*, 1967) and excised root tips of peas (Gaither, Lutz and Forrance, 1975). While some of these effects were marked, McWha and Jackson (1976) suggest that the stimulatory effects of ABA are not as uncommon as has been suggested and that at very low concentrations (*e.g.*, 1×10^{-9} M) (Fig. 4.4) stimulation of growth processes was not unusual. They found that the frond number of aseptically-grown greater duckweed (*Lemna polyrrhiza*) was increased by concentrations of ABA between 1×10^{-10} to 1×10^{-8} M after seven days.

It is extremely difficult to discover which is the primary action of ABA when such a complex, interrelated and cyclic process as cell growth is affected with an excess of one component. When ABA has been applied to non-growing organs it has been found to inhibit the synthesis of proteins and nucleic acids, to affect membranes and the amounts

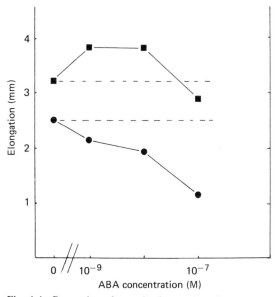

Fig. 4.4 Promotion of growth of oat coleoptile segments by low concentrations of (±)-ABA. Apical segments (8 mm) with the tip intact ■, or with 3 mm of the tip removed ●, were incubated for 24 h. The ordinate shows the elongation in mm. The dotted line represents the control value without ABA. (Data from McWha and Jackson, 1976.)

of the other phytohormones present, and it may induce degenerative changes such as abscission and senescence: an embarrassment of metabolic riches.

Geotropism The involvement of ABA in root geotropism is covered in Chapter 8 and will not be discussed here.

Abscission

As its name bears historical witness, abscisic acid was originally isolated as an abscission-accelerating substance (Ohkuma *et al.*, 1965), but subsequent work has shown that it has only a slight effect when applied to intact plants. The original isolation was guided, in the early stages, by a cotton seedling explant assay in which ABA stimulates the abscission of the petiolar stumps; explants of *Coleus* and *Phaseolus vulgaris* (Osborne, Jackson and Milborrow, 1972) have given similar responses. However, these tissues respond more rapidly and sensitively to ethylene (*see* Chapter 5).

Dormancy

The original suggestion that ABA was involved in the induction and maintenance of dormancy (*see* Chapter 19) was based on the observations that dormant potato tubers and resting buds of deciduous trees contained more ABA than those germinating or growing. Furthermore, gibberellins and cytokinins antagonized the action of ABA in a number of bioassays and broke dormancy. Later work has established that there is usually a correlation of ABA contents with dormancy but this is not exact. El Antably, Wareing and Hillman (1967) reported that applications of ($\pm$)-ABA to leaves and terminal buds of *Betula pendula* induced the formation of dormant buds at the apices, but attempts to reproduce these experiments have failed, despite having been attempted on a number of deciduous species (Hocking and Hillman, 1975).

Exposure of woody plants to short days was believed to initiate dormancy by Lenton, Perry and Saunders (1972) but Alvim, Thomas and Saunders (1978) and Powell (1976) failed to detect significant differences between the ABA contents of plants exposed to these treatments and those kept in long days.

It is difficult to envisage how changes in the endogenous ABA contents could regulate dormancy in the face of the large amounts of ABA that are synthesized in leaves whenever they become wilted, especially since a considerable proportion of this extra ABA is transported to the rest of the plant (Zeevaart, 1977). Attempts to establish a pre-eminant rôle for ABA in the induction and maintenance of bud dormancy in woody plants have thus been inconclusive, and the same can be said for its role in other types of dormant organs such as potato tubers (Rappaport and Wolf, 1969). It may be that ABA has an important function in dormancy phenomena but as a necessary component of a system rather than as a master switch. It is also well to bear in mind that the dormancy of different plant species may involve a changing pattern of gibberellins and cytokinins, and changes in tissue sensitivity (Trewavas, 1981).

The case for a rôle for ABA in seed dormancy is better established than for buds, but the correlation is not exact (Wareing and Saunders, 1971) and the effects of vernalization treatments appear to be brought about by increasing cytokinin or gibberellin production rather than by decreasing the amount of ABA present (Khan, 1971). Numerous studies have shown that ABA inhibits the germination of seeds, if used at sufficiently high concentration, but the effect is transitory.

Karssen (1982) suggested that ABA has an indirect effect on the germination of lettuce (*Lactuca sativa* cv. Grand Rapids) because applications of synthetic ABA abolished their red-light-induced ability to germinate. However, it is not clear how this differs from the usual effect of ABA on seeds. Braun and Khan (1975) measured the endogenous ABA levels in lettuce seeds and reported that during the 24 h after soaking the initial level of $120-140 \, \text{ng} \, \text{g}^{-1}$ of seed fell faster under conditions that promote germination (25°C and light) than under those that delay germination (35°C and light; or 25°C and darkness). They noted that germination was not always correlated with a decrease in ABA content.

Durand, Thévenot and Côme (1975) and Le Page-Degivry (1973) have been able to induce dormancy in yew (*Taxus baccata*) embryos in culture by ABA treatment, and found that a later application of gibberellin, or a cold vernalization treatment, was necessary before the embryos could germinate. Applications of ABA to viviparous mutants of maize (*Zea mays*) prevented the spontaneous germination and caused the seeds to develop normally (Smith, McDaniel and Lively, 1978).

Fronds of greater duckweed (*Lemna polyrrhiza*) bud continuously under long-day conditions but form dense, vegetative propagules known as turions under short days. Frond growth of *Lemna minor* is highly susceptible to inhibition by ABA (50% inhibition occurs at about 50 μg l^{-1}) (Van Overbeeck, Loeffler and Mason, 1967) but ABA induces turion formation even under long-day conditions (Stewart, 1969; Perry and Byrne, 1969). A ratio of 1:1 fronds to turions was obtained at concentrations of 50 μg l^{-1} for *Lemna* and 10 μg l^{-1} for *Spirodela*.

Before dismissing the ability of ABA to induce dormancy in woody species, it is as well to examine the experimental protocol. The attempts to induce resting buds have been carried out under constant, long-day conditions and it is possible that foliar applications of ABA cannot swamp the degradative capacity of the leaves and reach the apical buds in sufficient amount to overcome the effects of the gibberellins and cytokinins present. It could be that a less extreme long-day regime would allow this to happen. For example, *Myriophyllum verticillatum* was found by Weber and Nooden (1976) to form resting buds (turions) under marginal long days (8 and 12 h) but not under the normal, 16 h, long-day treatment at 15°C. At 20°C no turions were formed even under the 8-h days.

Fruit growth

Fruit ripening and development Ripening fruit tissues include the richest source of (+)-ABA, yet application of ABA to fruit has little or no effect. The exception to this general finding is the ripening grape berry (Coombe and Hale, 1973) where ABA, alone of the compounds tested, has the capacity to hasten the ripening and coloring of the fruit.

Sondheimer, Tzou and Galson (1968) concluded from analyses of ash (*Fraxinus americana*) seeds that the ABA in the fruit coat does not affect the germination or dormancy of the seed and Dorffling's data (1970) indicate that the amounts of ABA in tomato fruits are insufficient to inhibit germination. On the other hand the viviparous maize mutant develops normal seed when treated with ABA.

ABA is present in fairly constant amounts throughout the development of the seed, although several workers have detected a small rise followed by a fall in the ABA content at mid development (Davis and Addicott, 1972) with a large final rise as the seeds mature. This is illustrated for wheat (Fig. 4.5) but in this plant the mid-term rise is slight

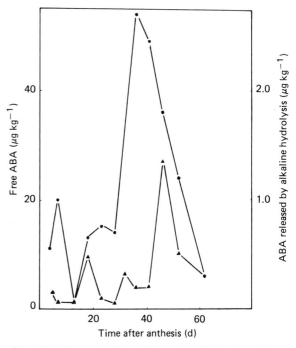

Fig. 4.5 Changes in free ABA and in ABA released by alkaline hydrolysis during the growth and maturation of wheat grains. From day 42 the dry weight remained constant while the seeds dried out, losing 34% of their total weight. Free ABA (●) left-hand ordinate, ABA released by alkaline hydrolysis (▲) right-hand ordinate. N.B. The weights of ABA released by hydrolysis are plotted on a scale 20 times that of ABA. (Adapted from King, 1976, recalculated on a fresh weight basis.)

(King, 1976) and there is no sign of it in the data for barley published by Naumann and Dorffling (1982). The amount of inhibitory material (later shown to be ABA) in aborted fruitlets of yellow lupin (*Lupinus luteus*) is very high, and Van Steveninck (1959) produced convincing evidence that the inhibitor was responsible for the self-thinning of the excess, fertilized fruits. Removal of some of the yellow lupin fruit soon after fertilization in a variety of patterns established that all were capable of development to maturity but normally only the two basal whorls survived. Whether ABA production and export by the more mature fruitlets can induce a massive, suicidal synthesis of ABA by the newly fertilized fruitlets, or whether the ABA is transported into them in sufficient quantities to cause abortion, is unknown. Whatever the mechanism of self-thinning, the 'June drop' is an important one commercially and should be amenable to experimentation now that [^{14}C]- and [^{3}H]ABA are available and the amounts of all the hormones present can be determined by sensitive physico-chemical methods.

Pathenocarpy Jackson and Blundell (1966) found that applications of ABA to emasculated fruit of wild rose (*Rosa sherardii*) were able to initiate some parthenocarpic development. Arditti, Flick and Jeffrey (1971), in contrast to this, found that orchid (*Cymbidium*) flowers underwent several post-pollination changes when treated with ABA (between 0.001 and 1 μg per flower) but their stigmata did not close nor did their columns swell.

Flowering Abscisic acid applications have a very slight promoting effect on flower growth in short-day plants. It failed to affect flower initiation in *Plumbago indica* but promoted flower development, even on internodal explants (Nitsch, 1968). High concentrations of ABA inhibit or delay flowering in a number of species but this effect is probably a reflexion of an inhibitory effect on growth.

Stress and threshold of leaf water potentials

Wright (1969) made the important observation that the amount of a growth-inhibiting material in wheat shoots increased dramatically when they lost water and wilted and Wright and Hiron (1969) later identified the inhibitory agent as ABA. They were able to measure the increase in ABA in considerable detail and show that radiant heating of leaves (bean, *Phaseolus vulgaris*, tomato, *Lycopersicon esculentum*, and wheat, *Triticum aestivum*), or waterlogging of roots, also caused the amount of ABA to rise. Since these first observations, chilling and salinity have also been found to cause an increase in ABA content but they are not as effective as wilting. It might be supposed that all these treatments act by causing incipient loss of turgor and so operate indirectly via the wilting response. However, this

does not appear to be correct because stress conditions such as chilling, heating, waterlogging and salinity can be adjusted so that leaf water potentials remain below the critical point for wilting but the ABA content nevertheless increases. In support of this view, Mizrahi and Richmond (1972) found that the transfer of tobacco (*Nicotiana rustica*) plants growing in half-strength Hoagland solutions to distilled water caused the amount of ABA in the leaves to rise progressively for up to seven days (Fig. 4.6) although their water status was unimpaired. Although the content of ABA had doubled by day two and quadrupled by day four, only thereafter did the growth rate and the leaf chlorophyll content decline. Mizrahi, Blumenfeld, Bittner and Richmond (1971) found in a similar experiment that transfer of plants to a 6 g l^{-1} NaCl solution caused the ABA content to increase (Table 4.2) but only when the leaves were in dry air. They concluded that it is the NaCl in the leaf that induces the change rather than an osmotic effect on the roots.

Influence of turgor

Zabadal (1974) sampled leaves of two species of *Ambrosia* during a 23-h period, measuring their water potential, the fresh and dry weights and ABA content after transferring rooted plants to a desiccating environment. During the first eight hours the leaf water potentials remained at −0.9 MPa (*A. artemisifolia*) or decreased from −0.4 MPa (*A. trifida*) and the ABA content remained at the initial level (Fig. 4.7). When the water potential of the leaves fell below −1.0 MPa the amount of ABA increased and, thereafter, the ABA content was inversely proportional to the water potential. Beardsell and Cohen (1975) and Wright (1977) also found a sudden increase in the ABA content when leaf water potentials fall below a critical value. There are slight differences in the position of the threshold value between plants and species; they appear to be related to the inherent drought tolerance of the leaf and its pretreatment. Furthermore, the more gradual response found in sorghum, in comparison with maize, is in accord with the former's greater capacity to resist drought (Fig. 4.8). The sudden increase in the amount of ABA at a critical water potential coincides with wilting, and Pierce and Raschke (1978) have proposed that the critical factor appears to be turgor pressure. In cockle burr (*Xanthium*

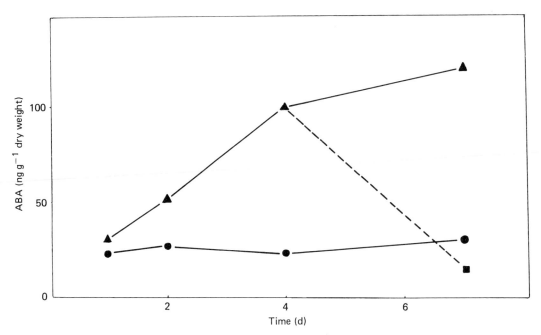

Fig. 4.6 Changes in the amounts of ABA in tobacco plants transferred from half-strength Hoagland solution to distilled water (▲). Some plants were returned to the Hoagland solution from day 4 (■). Plants maintained in Hoagland solution are shown (●). (Data from Mizrahi and Richmond, 1972.)

Table 4.2 Effect of relative humidity and salinity on the water saturation deficit and transpiration of the leaves of tobacco plants over 24 h. The relative units of ABA were calculated from bioassay results. (Adapted from Mizrahi *et al.*, 1971.)

	High relative humidity			Low relative humidity		
	Water saturation deficit (%)	Transpiration (ml d^{-1} plant^{-1})	ABA (relative units)	Water saturation deficit (%)	Transpiration (ml d^{-1} plant^{-1})	ABA (relative units)
Plant in sodium chloride solution 6 g l^{-1}	4.6	93	6.6	12.1	116	3.6
No sodium chloride	1.4	95	5.4	7.3	183	4.0

strumarium) the onset of ABA accumulation occurred exactly at zero turgor, but pre-adaptation of cotton to semi-drought conditions altered the water potential at which ABA was accumulated.

Effects of ABA on stomata

The closure of stomata by ABA as its level in the leaf rises during wilting makes an attractive regulatory mechanism, but the detailed, quantitative aspects of this phenomenon in the whole plant have not received adequate attention. Overall, foliar levels of ABA frequently rise from about $20 \, \mu g \, kg^{-1}$ initial fresh weight to $500 \, \mu g \, kg^{-1}$ (*i.e.*, from about 1×10^{-7} to $2 \times 10^{-6} M$). The stomata are in the epidermal layers which synthesize little or no ABA when wilted. Furthermore, the amount of ABA in the epidermis is negligible and the stomata in strip-

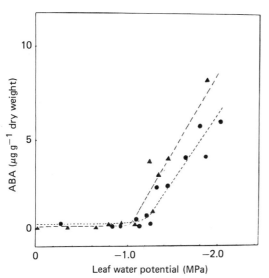

Fig. 4.7 Relationship between the ABA content and the leaf water potential in intact plants of *Ambrosia* during a period of progressive dehydration. *A. trifida*, ▲; *A. artemisifolia*, ●. (From Zabadal, 1974.)

ped epidermis respond rapidly and proportionally to changes in concentration (Tucker and Mansfield, 1971). Stomata usually comprise less than 5% of the epidermis, so there is a great excess of ABA in the wilted leaf over the amount necessary to affect the guard cells.

Resistance to injury

An application of ABA to growing, mesophytic plants enables them to withstand a subsequent, sudden desiccation (Wilson, 1976), and ABA also appears to protect plants from toxicity. Oat (*Avena sativa*) mesocotyl segments which were killed by $50 \, \text{mg} \, \text{l}^{-1}$ IAA appeared healthy and grew at the same rate as controls, even in solutions containing $100 \, \text{mg} \, \text{l}^{-1}$ IAA when ABA ($2 \, \text{mg} \, \text{l}^{-1}$) was present (Milborrow, 1966b).

Treatment of whole plants of *Acer negundo* (Irvine and Lanphear, 1968), apple (Holubowicz and Boe, 1969), alfalfa (*Medicago sativa*) (Rikin, Waldman, Richmond and Dobrat, 1975), and cucumber (*Cucumis sativus*) (Rikin, Blumenfeld

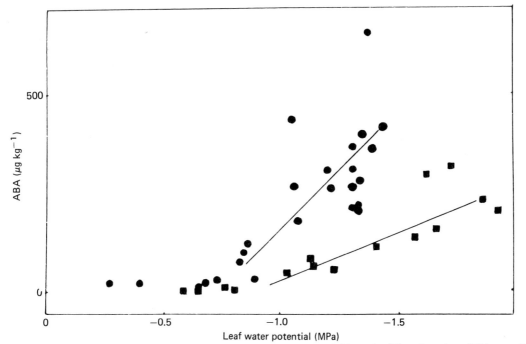

Fig. 4.8 The relationship between the ABA content and leaf water potential in maize (●) and sorghum (■) leaves. The ordinate values were calculated from the data for maize. (Data from Beardsell and Cohen, 1975.)

and Richmond, 1976) has been found to make them more resistant to chilling injury.

Fast and slow reactions

Natural (+)-abscisic acid is optically active, having one center of asymmetry at C-1′. ABA synthesized chemically is racemic and composed of equal amounts of the (+)- and (−)-enantiomers (Fig. 4.9).

These two mirror-image forms have been separated and, when the separate enantiomers were

form inhibits a fast reaction and a slow reaction, while the (−) form inhibits only the slow reaction. The two sites of action must differ in their steric requirements so the fast reaction is not a necessary forerunner of the slow reaction (Milborrow, 1980).

Transport, intracellular compartmentation and distribution

Heilmann, Hartung and Gimmler (1979) carried out a series of measurements of ABA permeation rates between different cellular compartments and inter-

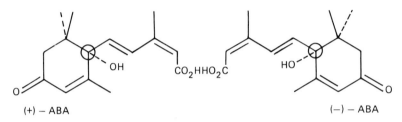

Fig. 4.9   The two mirror-image forms of abscisic acid. The single center of asymmetry (C-1′) is ringed. If the unnatural (−) enantiomer is turned over, as if the page were turned back, the downwards projecting hydroxyl group, shown with a dotted line, would project upwards.

tested in a wheat embryo bioassay, the (−) was found to be as active as the (+) (Milborrow, 1968). This is an extremely unusual result because one of a pair of opposite antipodes is usually inactive or may even be inhibitory. In some experimental systems the two mirror-image forms of ABA appeared to penetrate into the tissues and to be metabolized at different rates (Sondheimer, Galson, Chang and Walton, 1971) but even this did not prejudice the strong activity of the (−).

Some phytohormones, particularly IAA, appear to act in two ways: a fast response which occurs within ten minutes and before protein synthesis begins, followed by a slow reaction beginning after about 30 min that involves protein synthesis. Normally both occur together but the two reactions have recently been separated in the response of coleoptiles to IAA (Vanderhoef and Dute, 1981). The question becomes: 'Is the fast reaction a necessary precursor of the slow reaction or are they quite separate?'

Natural (+)-ABA causes stomatal closure within minutes of application while the unnatural (−) has little effect. Both (+) and (−) forms inhibit seed germination and protein synthesis. Thus the (+)

preted the distribution of ABA in terms of the permeability of membranes to the unionized form and the pH within the organelles. They found that ABA is synthesized in the cytosol of spinach leaf cells and is degraded there but not in the chloroplasts. ABA is concentrated within the chloroplasts because the pH of the stroma is higher than that of the rest of the cell, particularly when illuminated, and so they accumulate ABA in the ionized form (Fig. 4.10). This distribution pattern applies to the free ABA only and a considerable proportion of the unconjugated compound appears to be present in a bound form (i.e., not conjugated but attached to a macromolecule in some way) within chloroplasts isolated from normal, turgid leaf tissue (Loveys, 1977). The chloroplasts isolated from wilted tissue contain far less bound ABA, e.g.,

ABA in turgid leaf chloroplasts   95%
ABA in wilted leaf chloroplasts   15%

Distribution within plants

Knowledge of how ABA is transported within the plant is essential to an understanding of how and

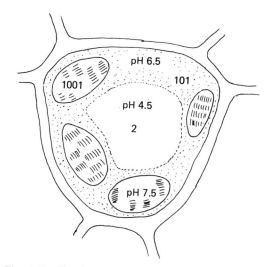

Fig. 4.10 The distribution of ABA within a mesophyll cell. ABA is a weak acid (pKa 4.5) and the numbers refer to the relative amounts of ABA that would be present at equilibrium in the chloroplasts, the vacuole and the cytosol if these compartments had equal volumes and the pH values shown.

where it produces its effects, and yet the topic has received little attention until recently. It was shown by Lenton, Bowen and Saunders (1968) that ABA was detectable in xylem sap, and present at much higher concentrations in phloem sap and in aphid honeydew derived from it. Hocking, Hillman and Wilkins (1972) found that $(\pm)$-[^{14}C]ABA applied to a leaf of a pea seedling was widely distributed within the plant within 24 h and 18% was present in the root nodules. The movement of ^{14}C was both upwards and downwards but did not pass downwards through a steam-girdled zone. A very similar pattern of distribution has been observed by Shindy, Asmundson, Smith and Kumamato (1973) in cotton (*Gossypium hirsutum*) seedlings and, although this species lacks nodules, half of the [^{14}C]ABA had been transported out of a treated leaf during eight days and was recovered as unmetabolized ABA from the roots. Thus ABA moves rapidly in the phloem as well as in the parenchymatous cells of short sections of stems and petioles. In this latter system IAA shows strong polar movement while ABA shows little or no polarity in its movement. In bean petiolar segments, basipetal transport of [^{14}C]ABA was some 2–3 times faster than in the

acropetal direction (Milborrow, 1968) but Dorffling and Bottger (1968) found that polarity in *Coleus* petioles was less than this in young segments and absent in segments from old petioles. Similarly, in bean (*Phaseolus coccineus*) roots Astle and Rubery (1980) found there was no polarity of ABA movement.

Rubery and co-workers (Astle and Rubery, 1980) have been able to identify two mechanisms for the entry of ABA into cells of bean root segments. First, there is a passive entry of the uncharged molecule (written as ABAH), presumably by partitioning into the lipid phase of the membranes and then diffusing into the cytosol. The second method of entry of ABA into cells of roots is brought about by a saturable carrier which is specific for ABA, does not accept benzoic acid or IAA, and is restricted to the elongating zone. The most probable mechanism for the ABA carrier is an ABA$^-$/H$^+$ symport (Rubery and Astle, 1982; Goldsmith, 1977).

Solute movement

Malek and Baker (1978) have proposed that the loading of sucrose into phloem sieve tubes is brought about by a H$^+$/K$^+$ exchange between the phloem sap and the apoplast such that the pH gradient between the sieve elements and the apoplast provides the energy for the proton co-transport of sucrose taken into the phloem. It also accounts for the high K$^+$ concentration and high pH of sieve elements. Malek and Baker perfused 2% sucrose solutions through hollow petioles of castor oil (*Ricinus communis*) leaves and measured the uptake of [^{14}C]sucrose under the influence of ABA, IAA, KCl and the fungal toxin fusicoccin: a compound that stimulates proton efflux.

Perfusion with a 20 mM KCl solution was followed by a fall in the pH of the extracellular solution of about one pH unit and this was increased and accelerated by fusicoccin and inhibited by ABA. The ABA inhibited the absorption of K$^+$ into the petioles and also inhibited the accumulation of sucrose. Malek and Baker incorporated their results into the model mechanism shown in Fig. 4.11.

On the other hand, ABA has been found to stimulate the accumulation of ions by roots (Collins and Kerrigan, 1974; Karmoker and Van Steveninck, 1978; Van Steveninck 1972) and also to inhibit this process (Shaner, Mertz and Arntzen, 1975). Pitman

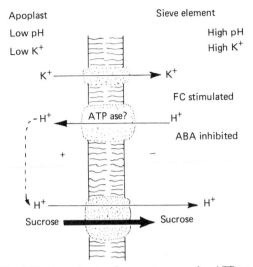

Fig. 4.11 A mechanism for a proton-pumping ATPase at the plasma membrane of a sieve element coupled with a K^+ influx. Sugars are co-transported down the resulting proton gradient. Fusicoccin (FC) increases the proton gradient and increases the amount of sucrose taken in, while ABA has the opposite effect. (From Malek and Baker, 1978.)

and colleagues (Pitman, Luttge, Lauchli and Ball, 1974) found that the promotion or inhibition depended on the conditions under which the plants had been grown and the temperature during the experiment. When they wilted and then rewatered barley plants the turgor and transpiration rate rapidly returned to normal while ion uptake was promoted and export of ions from the roots was reduced for several hours longer. They attribute the effect to the ABA produced during the wilting period.

Glinka and Reinhold (1971) found that ABA increased the permeability of carrot disks to tritiated water, both into and out of the cells, and the effect was reversible. The authors conclude that the effect of ABA in lowering the permeability of the membranes could account for the stimulation of exudation observed when excised tomato roots were treated with ABA (Tal and Imber, 1971).

Yet another complication is that Karmoker and Van Steveninck (1978) have found that the effect of ABA on ion uptake was different on excised root system of french bean (*Phaseolus vulgaris*) compared with that in the intact plant. ABA stimulated

the uptake of Cl^- and Na^+ in the former, while the influx of these ions into the roots of intact seedlings was stimulated only transiently.

Effects on enzyme activity and synthesis

ABA has not been found to alter the catalytic activity of an enzyme to any appreciable extent but it has been found to affect enzyme production. The most studied example is the formation of α-amylase by barley aleurone cells (Chispeels and Varner, 1966; Varner and Johri, 1968) which is powerfully stimulated by GA (*see* Chapter 2) and inhibited by ABA, but the synthesis of other enzymes is also inhibited (Varner and Ho, 1976). Ho and Varner (1974) proposed that the inhibition of α-amylase synthesis is caused, at least in part, by an effect on translation because ABA still inhibited the formation of α-amylase at 12 h when cordycepin (an inhibitor of RNA synthesis) no longer had an effect. Cordycepin abolished the inhibitory effects of ABA at 12 h when both were added together. A possible explanation for this was that ABA might interfere with the production of a regulatory protein.

Ho (1979) and Jacobsen, Higgins and Zwar (1980) detected new peptides when aleurone layers were treated with ABA; they turned over rapidly, and the amounts formed were reduced by inhibitors of transcription and translation. The function of these peptides is unknown but the experimental system is obviously of great promise because it responds to a stimulatory and an inhibitory hormone *in vitro*.

Biosynthesis

The carotenoid and the direct synthesis pathways The carbon skeleton of abscisic acid resembles the terminal parts of some carotenoids, even to the positions of oxygen atoms (Scheme 4.1). Simpson and Wain (1961) observed that the amounts of growth-inhibitory material increased when dark-grown plants were illuminated and that the action spectrum resembled the absorption spectrum of carotene. This suggested that a carotenoid could give rise to ABA by photolytic cleavage and so Taylor and Smith (1967) exposed the carotenoid fraction from nettle (*Urtica dioica*) leaves to sunlight on damp filter paper and found that a potent growth inhibitor had been formed. They showed that violaxanthin (3)(Scheme 4.1) gave rise to the

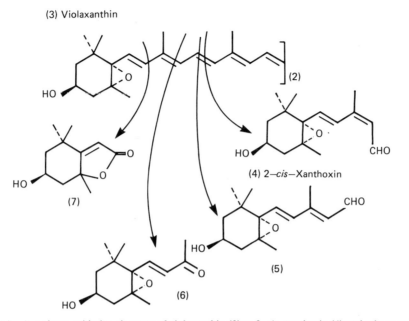

Scheme 4.1 The *in vitro* photo-oxidative cleavage of violaxanthin (3) to 2-*cis*-xanthoxin (4) and other products (5, 6 and 7). These compounds are also formed in similar proportions by the soybean lipoxygenase. All four of the photo-oxidation products have been isolated from green leaves.

inhibitor which they named 2-*cis*-xanthoxin (4) (Taylor and Burden, 1970).

2-*Cis*-xanthoxin is rapidly metabolized to ABA and has been isolated from shoots and leaves of a number of plants (Firn, Burden and Taylor, 1972). However, although cleavage of some violaxanthin to ABA by light may occur it is unlikely to be the main mechanism of ABA biosynthesis because wheat leaves wilted in darkness and avocado fruit supplied with [^{14}C]mevalonate in darkness were able to synthesize ABA.

In another experiment the late Dr Robinson fed [^{14}C]phytoene (a precursor of carotenoids) to avocado fruit with [^{3}H]mevalonic acid (Scheme 4.2). The ABA contained ^{3}H but no ^{14}C, the carotene contained ^{14}C and ^{3}H. If ABA had been formed via a carotenoid it would have been expected to contain ^{14}C. The experiment can be criticized because intact tissue was present and it is possible that the [^{14}C]phytoene failed to reach the cellular compartment in which the [^{3}H]ABA was formed.

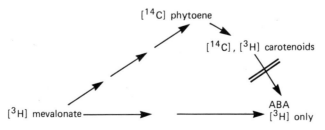

Scheme 4.2 An experiment designed to discover whether ABA could be biosynthesized without the involvement of a carotenoid. [^{3}H]Mevalonate was incorporated into ABA but although [^{14}C]phytoene was converted into cyclic carotenoids, no ^{14}C was found in ABA.

Three other compounds formed photolytically from violaxanthin *in vitro* (5, 6 and 7 in Scheme 4.1) are also present in leaves suggesting that photolytic cleavage of violaxanthin does occur *in vivo*. If light caused the breakdown of violaxanthin then the ratio of the inactive 2-*trans*-isomer should increase as the 2-*cis*-xanthoxin is converted into ABA. No such accumulation of the 2-*trans*-isomer has been observed.

Firn and Friend (1972) have introduced another complication; they found that a lipoxygenase enzyme from soybean leaves cleaves violaxanthin to give *cis*-xanthoxin and a similar range of products to those formed by photolysis, so this system could give rise to ABA by yet another route.

Biosynthetic mechanisms in the terpenoid pathway
Mevalonic acid (8 in Scheme 4.3) is the precursor of all the terpenoid material in plants and animals and its metabolic conversions have been thoroughly explored (*see* Chapter 2). One of the first experiments to try and discriminate between the carotenoid pathway and ABA formation by direct synthesis involving intermediates with no more than 15 carbon atoms was carried out using 4R and 4S

(8) Mevalonic acid

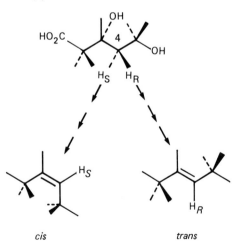

Scheme 4.3 The sources of the hydrogen atoms retained from C-4 of mevalonic acid on *cis* and *trans* double bonds. The six-carbon mevalonic acid (top) provides the five-carbon skeleton of the isoprene unit (shown in heavy lines). The 4-*pro-S*-hydrogen atom is retained on the *cis* double bond, the 4-*pro-R* on the *trans*.

tritiated mevalonate. Popjak and Cornforth (1966) had shown that when a *trans* (E or Entgegen in modern nomenclature) double bond is formed between what has been C-3 and C-4 of mevalonate, the 4-*pro-R* hydrogen atom is retained at C-4 (Scheme 4.3). In contrast to this the 4-*pro-S* hydrogen is retained when a *cis* (Z or Zusammen) double bond is formed, as in rubber hydrocarbon. If mevalonic acid containing ^{14}C and 4R tritium atoms was incorporated into ABA then two tritium atoms would be expected to be retained in ABA if the double bond at C-2 was formed *trans*(E) and isomerized at a later stage with the retention of its hydrogen atom. No tritium would be retained from 4S tritiated, [^{14}C]mevalonate. If the C-2 double bond was formed *cis*(Z) then one 4R tritium atom would have been retained per three ^{14}C atoms.

The analogous double bond in carotenoids is formed *trans* and so the retention of a 4S tritium in ABA would have proven that a carotenoid could not have been an intermediate in its formation. The results (Table 4.3) showed that the 4-*pro-R* hydrogen atom was retained (Robinson and Ryback, 1969) and so the labeling pattern could not be used to discriminate between the two pathways.

Formation of ABA by a fungus Assante, Merlini and Nasini (1977) reported that the pathogenic fungus *Cercospora rosicola* biosynthesized ABA in culture. Since then Neill, Horgan, Walton and Lee (1982) have identified the penultimate precursor as 1'-desoxy ABA (9) so the insertion of the 1'-hydroxyl

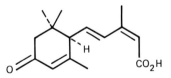

(9) 1'—desoxy ABA

group appears to be the final biosynthetic step. It now remains for 1'-desoxy ABA to be established as an intermediate in plants for the carotenoid pathway to be disproved. However, xanthoxin does occur in leaves and so a small proportion of the ABA may be derived from violaxanthin when leaves are exposed to strong sunlight.

The function of the ABA in the parasitism of the rose leaves by *C. rosicola* is unknown at present. It would seem likely that the ageing effect of ABA and

Table 4.3 ^{3}H/^{14}C ratios in ABA biosynthesized by avocado fruit halves from $3R$-[2-^{14}C,4-R-4-^{3}H${}_1$]-and $3R$-[2-^{14}C,4S-4-^{3}H${}_1$]mevalonates. The removal of one $4R$ tritium atom of mevalonate by base-catalyzed exchange from ABA is consistent with its being attached to C-5′ of ABA. All ratios are normalized to the substrate. The predicted ratios are calculated on the expectation that the $4R$ hydrogen atom is retained or C-2 and C-5. (Data from Robinson and Ryback, 1969.)

	ABA from $4R$-4-^{3}H${}_1$ mevalonate		ABA from $4S$-4-^{3}H${}_1$ mevalonate	
	Expected ratio if ^{3}H from $4R$ retained in ABA	Observed ratio	Expected ratio if ^{3}H from $4R$ retained in ABA	Observed ratio
ABA as methyl ester	2.0:3.0	1.93:3	0.0:3	0.05:3
ABA after base-catalyzed exchange	1.0:3	0.92:3	0.0:3	0.02:3

possibly its inhibition of protein synthesis may favor the growth of the fungus; it is also possible that ABA could prevent the manifestation of the phytoalexin response. More recently another plant pathogenic fungus (*Botrytis cinerea*) has been found to make ABA (Marumo, Katayama, Komori and Yozaki, 1982).

Inactivation of ABA

Nomenclature Free ABA turns over, in other words there is continuous biosynthesis balanced by the same rate of deactivation. It is usually convenient to measure the rate of turnover as the rate of destruction and for ABA, as for the other classes of plant hormones, this occurs by two processes: oxidation and conjugation. 'Conjugation' is used here as proposed by Sembdner, Weiland, Schneider, Schreiber and Focke (1972), namely that a covalent link, such as glucoside or ester bond, should be formed between the hormone and another (usually small) molecule such as a monosaccharide or an amino acid. The definition of 'bound ABA' encompasses those forms of the hormone in which it is held by a macromolecule so that it cannot be extracted by washing with the aqueous medium and yet the attachment is by non-covalent bonding.

Oxidative inactivation of ABA The first attempt to isolate a polar, acidic metabolite of ABA gave colorless crystals which showed properties similar to, but clearly different from those of ABA. It was referred to as 'Metabolite C'; nowadays it is 6′S-hydroxymethyl ABA (10). The metabolite later isomerized to a known compound, phaseic acid, that

had been isolated from french beans (*Phaseolus vulgaris*) by MacMillan and Pryce (1968).

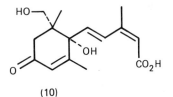

(10)

The structure that had been proposed for phaseic acid (11) was incompatible with isomerization from a hydroxylated form of ABA and an alternative structure for phaseic acid (12) was proposed (Milborrow, 1969). The difference between the two structures

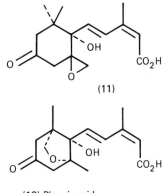

(11)

(12) Phaseic acid

hinges on the origin of the methyl signals in the nuclear magnetic resonance spectrum (NMR). In the spectrum of the original structure the two sharp methyl signals were attributed to the 6' geminal methyl groups while in the new structure (12) one signal would come from the remaining 6' methyl and the other from the 2' methyl group. It is possible to exchange the hydrogen atoms of the 2'-methyl group of ABA and replace them with deuterium atoms which do not produce signals in the proton NMR. A 2'-Me-deuteriated sample of ABA was fed to tomato plants and metabolized to phaseic acid (PA); two sharp methyl signals would be expected in the NMR spectrum if the original structure was correct, only one signal would be present if the new structure was correct. The methyl signal was clearly absent in the spectrum of phaseic acid; this is seen in Fig. 4.12. Thus the revised structure (12) is correct. Phaseic acid is formed by the attack of the 6'-hydroxymethyl group on C-2' to make an ether bridge and to saturate the ring double bond.

After phaseic acid has been formed it is reduced to the two epimeric dihydrophaseic acids with the 4'-hydroxyl group projecting upwards or downwards respectively (13,14) (Tinelli, Sondheimer,

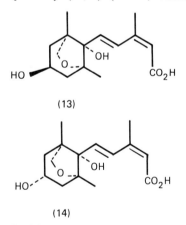

(13)

(14)

Walton, Gaskin and MacMillan, 1973; Zeevaart and Milborrow, 1976). Enzymes normally form just one of the two possible diastereoisomers at each position so the reduction of the 4'-ketone group of phaseic acid to the two epimers with hydroxyl groups 'up' and 'down' is extremely unusual. Two possible mechanisms can be considered for this: the ring of phaseic acid could fit into the enzyme's active site in two ways (Fig. 4.13), or two enzymes could be involved. Preliminary results suggest that the latter possibility is the correct one. The reverse reaction was used to investigate the mechanism, *i.e.*, the two

dihydrophaseic acids were oxidized to phaseic acid by driving the reaction backwards with excess NAD (alcohol dehydrogenases are quite readily re-

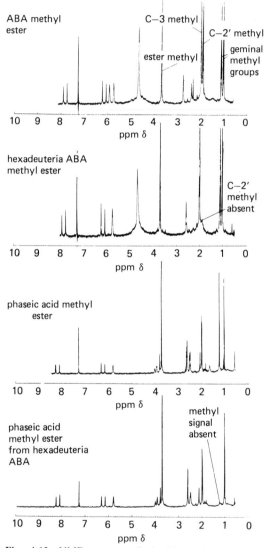

Fig. 4.12 NMR spectra of abscisic and phaseic acid methyl esters. The ring methyl (C-2') signal of ABA at 1.9δ is missing in the deuteriated sample and when this was fed to tomato plants the phaseic acid formed from it lacked a methyl signal at 1.2δ. This confirmed that the latter methyl signal of PA arose from what had been the C-2' methyl of ABA, thereby confirming the new structure. (Milborrow, 1971.)

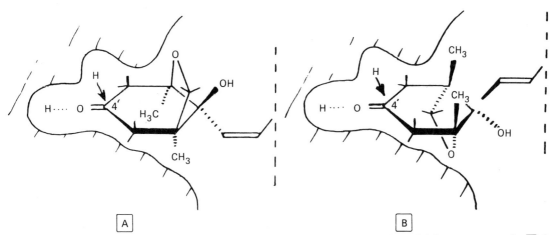

Fig. 4.13 A possible mechanism for the formation of both DPA and *epi*-DPA from PA by one enzyme. In Ⓐ the hydrogen atom is being added to C-4′ to form DPA, in Ⓑ it is being added to the opposite face of the ring to form *epi*-DPA.

versible). ^{14}C-labeled 4′-*epi*-dihydrophaseic acid (4′-*epi*-DPA) was incubated with a cell-free preparation in the presence of different amounts of unlabeled DPA, and ^{14}C-DPA with varying amounts of unlabeled 4′-*epi*-DPA. If one enzyme was involved, competition between the DPA and *epi*-DPA should have been observed. No competition was detected in either experiment so it is assumed that two different enzymes are responsible. Dihydrophaseic acid is converted into a 4′-β-D-glucoside (Milborrow and Vaughan, 1982) and then the glucoside is converted into an even more polar conjugate whose structure has not yet been determined. The claim that carboxyl-labeled ABA fed to apple (*Pyrus malus*) gave rise to $[^{14}C]CO_2$ (Rudnicki and Czapski, 1974) has been shown to be attributable to bacterial contamination (Milborrow and Vaughan, 1979). Sterile apple seeds

supplied with $(\pm)$-$[2$-$^{14}C]$ABA failed to give rise to $[^{14}C]CO_2$ during 40 days incubation and contained the usual pattern of labeled metabolites. $[^{14}C]CO_2$ was present in flasks containing $(\pm)$-$[2$-$^{14}C]$ABA and non-sterile seeds.

ABA metabolism can be considered to involve two functions: inactivation of the active hormone by destruction or masking of an essential feature, and a change to give it highly polar, hydrophilic properties such as by combination with a glucosyl or similar residue.

The polar conjugates may be accumulated in the vacuoles and the proportion of ABA released by alkaline hydrolysis (ABAGS and ABAGE) to free ABA is higher in the cell debris and protoplasm fraction of citrus fruit than in the expressed vacuolar sap (Table 4.4). Now that vacuoles can be isolated this distribution could be measured directly.

Table 4.4 Free and conjugated ABA in vacuolar sap. Vesicles of ripe fruit were removed and ruptured, the vacuolar sap was collected immediately and analyzed by gas chromatography for free and conjugated ABA. The higher ratio of conjugated to free ABA in the sap suggests that a larger proportion of the conjugated ABA is present in the vacuolar sap than in the protoplasm.

Fruit	Free ABA in cell debris (mg kg^{-1})	Conjugated ABA in cell debris (mg kg^{-1})	Ratio (conjugated: free)	Free ABA in vacuolar sap (mg kg^{-1})	Conjugated ABA in sap (mg kg^{-1})	Ratio (conjugated: free)
Mandarin	0.75	1.79	2.4	0.14	0.48	3.4
Grapefruit	3.95	6.91	1.75	0.77	2.68	3.5

At present the physiological effects of ABA metabolites are uncertain. Phaseic and dihydrophaseic acids have virtually no effect on growth but the difficulty of obtaining sufficient material for experimentation is considerable and they have been examined in very few tests. Phaseic acid has been claimed to inhibit photosynthesis while ABA did not (Kriedemann, Loveys and Downton, 1975); Sharkey and Raschke (1980), however, found that crystaline PA was inactive and the effects were probably caused by contaminants in the solvents used. They did find that PA was as effective as ABA in causing the stomata of *Commelina communis* to close but PA had no effect on the stomata of broad beans (*Vicia faba*). No stomata responded to DPA.

Conjugation of ABA The first conjugate of ABA isolated was the glucose ester (15) (ABAGE) (Koshimizu, Inui and Fukui, 1968). Its inhibitory activity was followed by bioassay and the pure compound had almost exactly half the effect, on a weight basis, as ABA; in other words the glucose ester appears to be hydrolyzed to ABA. ABAGE was also characterized (Milborrow, 1970) as a derivative of [^{14}C]ABA in tomato shoots.

Recently a second conjugate of ABA has been

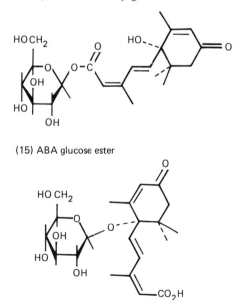

(15) ABA glucose ester

characterized (16); this is the 1′-*O*-glucoside (ABAGS) in which the β-D-glucosyl unit is attached to the tertiary 1′-hydroxyl group of ABA (Loveys and Milborrow, 1981). The 1′-*O*-glucoside and the glucose ester are susceptible to hydrolysis by weak alkali so any attempts to measure the amounts of free ABA and conjugates have to be carried out with great care. Additionally, neutral and, particularly, alkaline methanol attack the glucose ester to release glucose and form Me ABA (Milborrow and Mallaby, 1975) and the 1′-*O*-glucoside undergoes slow spontaneous rearrangement to form ABAGE. Consequently slightly acidic acetone should be used to extract plant material and the conjugates should be separated as soon as is feasible. Most early determinations of the amounts of free ABA and conjugated ABA are of dubious validity.

A number of experiments carried out with wilting plants have revealed a rise in the amount of free ABA of up to 40-fold, while the conjugated ABA fraction remained constant. This suggests that the conjugates do not act as a source of free ABA. An experiment to detect hydrolysis directly made use of *Beta vulgaris* plants that had been fed [^{14}C]ABA and whose conjugated ABA fraction had become labeled (Milborrow, 1978). When subsamples of these plants were wilted the amount of free ABA rose considerably but the amount of free [^{14}C]ABA present fell slightly. The ^{14}C in the conjugates did not fall (it rose slightly) showing that free ABA was being glycosylated but the conjugates were not being hydrolyzed (Table 4.5). The result appears clear-cut for tomato shoots but whether or not the result applies to other plant tissues remains to be established.

(+)-ABA is readily converted into phaseic acid, the (−) enantiomer is oxidized much less readily, and the 2-*trans*-isomer of (+)-ABA is also oxidized much less efficiently. On the other hand, formation of both glucose conjugates proceeds faster with the (−) than with the (+) and 2-*trans*-ABA, and a number of related compounds such as xanthoxin acid (17), are conjugated with ease. It

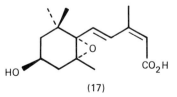

(17)

(16) ABA glucoside

Table 4.5 Silverbeet shoots (*Beta vulgaris*) fed with $(\pm)[2\text{-}^{14}C]ABA$ (2.83×10^3 Bq, 3.77×10^4 Bq mol^{-1}) were kept for 3 days until most of the labeled ABA had been metabolized. Batches were either maintained turgid or wilted (Milborrow, 1978).

Treatment	Original fresh weight (g)	Free acids (Bq × 60)			Acids released by alkaline hydrolysis (Bq × 60)		
		ABA	PA	DPA	ABA	PA	DPA
5-h Turgid control	8.1	2560	499	236	10 484	1424	1235
8-h Turgid control	9.4	2370	357	217	11 151	1262	1857
2-h Wilt	9.0	1862	323	169	11 764	1677	1288
5-h Wilt	10.0	1977	344	230	9 834	1434	1291
8-h Wilt	10.5	2171	453	260	9 878	1159	957

appears that oxidation is a specific inactivation mechanism while conjugation is a general, unspecific one.

Phaseic acid and dihydrophaseic acid are also converted into glucose esters and glucosides and a number of other metabolites can be detected after [^{14}C]ABA is supplied to plants; they have been detected by autoradiography of chromatograms or by monitoring HPLC elution profiles, but they have not yet been identified. They may not all be naturally occurring because racemic [^{14}C]ABA was used. If the metabolites contain a second optically active center then those formed from $(+)$-ABA will be separable chromatographically from those of $(-)$-ABA.

The use of $(\pm)$-[^{14}C]ABA, therefore, potentially doubles the number of metabolites that may be found. The reason for this is as follows. The mirror-image forms of an optically-active compound cannot be separated by normal chromatographic procedures. However, if a second optically-active center is introduced then two kinds of molecules are formed. This can be illustrated with $(\pm)$-ABA and D-$(+)$-glucose, ABA glucoside will comprise two forms:

$(+)$-ABA . . . $(+)$ glucose
$(-)$-ABA . . . $(+)$ glucose.

These are not equivalent (*i.e.*, they are not mirror-images) and so have different solubility properties which may allow them to be separated readily.

Another possible source of unnatural metabolites could arise from labeled impurities or breakdown products of the [^{14}C]ABA supplied. Methods for isolating ^{14}C-containing compounds are highly

sensitive so that metabolites of ^{14}C-labeled impurities in the synthetic material may be detected. On the other hand the pattern varies from species to species and so the identification of a range of new metabolites can be expected. Hirai, Fukui and Koshimizu (1978) have found the hydroxymethyl glutaryl adduct (18) of the unstable intermediate hydroxymethyl ABA in the seeds of false acacia (*Robinia pseudacacia*).

Tietz, Dorffling, Wohrle, Erxleben and Leimann (1979) found that pea plants converted [^{14}C]ABA into a slightly more polar compound which they identified as 4'-deoxy ABA. It has now been shown that ABA is reduced to the 1',4'-*trans* diol of ABA (19) (Milborrow, 1983) and it is this compound that is present in the peas. It dehydrates on heating during glc–mass spectrometry and so gives a mass spectrum of 4'-deoxy ABA. The amounts of endogenous diol present and its metabolic significance are unknown.

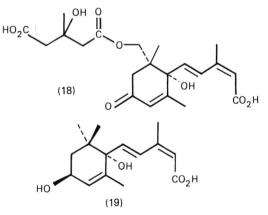

Isomerization Light isomerizes ABA to a 1:1 mixture with its 2-*trans* isomer (2)(*t*-ABA) and this relatively small change in structure eliminates activity. Some reports have suggested that *t*-ABA is inhibitory to growth but the experiments were carried out in normal light over several days (Roberts, Heckman, Hege and Bellin, 1968) and so the active isomer would have been formed. When the *trans* isomer was tested in darkness it had less than 2% of the potency of ABA.

Many determinations of ABA by gas chromatography have shown that a small proportion of 2-*trans* isomer is present in leaves although every care has been taken to minimize isomerization during the work-up (Fig. 4.14). In one experiment the isomerization during isolation was monitored by adding (±)-[^{14}C]ABA to the extraction medium,

Abscisic acid

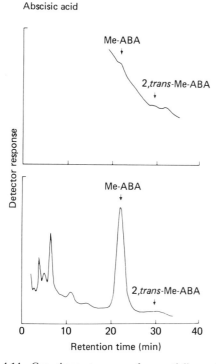

Fig. 4.14 Gas chromatograms of a partially purified, methylated extract of birch buds using, upper, a flame ionization detector, and, lower, an electron capture detector. The samples used were equivalent to 0.25 g of tissue for the flame ionization detector and 0.025 g for the electron capture detector. (Data from Harrison and Saunders, 1975.)

then measuring the total amount of natural (+) *t*-ABA by its optical activity (ORD) and then the amount of optically inactive (±)-[^{14}C]ABA that had isomerized during the course of the experiment (1%). An additional 4% of the total, endogenous (+)-[^{14}C]ABA was present as *trans* isomer. The (+)-*t*-ABA, therefore, must have been present in the leaves before they were extracted (Milborrow, 1970).

The biological inactivity of *t*-ABA, the failure of plants to convert it into ABA, and its absence from seeds of avocado (Milborrow, 1970) and from maize (*Zea mays*) roots grown in darkness (Rivier and Pilet, 1981), suggest that it is not a precursor of ABA and that it is not formed enzymatically. It seems probable that *t*-ABA is a photolytic degradation product. The equilibrium value of 1:1 for ABA:*t*-ABA is not approached in leaves because the *trans* isomer is conjugated with glucose preferentially (some 20 times faster than ABA) and is thus rapidly removed from the pool of free acid. ABA 'turns over' with a half-time of about three hours in wilted leaves of french beans (Harrison and Walton, 1975) and so there is continual production of ABA and preferential removal of *t*-ABA.

Structural requirements for activity The chemical synthesis and testing of hundreds of analogs of ABA have shown that almost any change to the molecule reduces activity considerably and most alterations abolish it (Milborrow, 1974). Essential features include the carboxyl group, the tertiary hydroxyl group and the 2-*cis* and ring double bond, and it is perhaps noteworthy that the three main degradation products, ABAGE, ABAGS and phaseic acid, are formed with the loss of one of these essential groups. The increased polarity of the conjugates may favor their accumulation in vacuoles where they would be less likely to undergo hydrolysis than in the metabolic milieu of the cytoplasm.

Turnover Harrison and Walton (1975) measured the amount and specific activity of [^{14}C]ABA and the [^{14}C]PA and [^{14}C]DPA formed from it by french bean leaves while the leaves remained turgid or were kept wilted. They found that the endogenous ABA content of 40 μg kg^{-1} rose to 500 μg kg^{-1} on wilting and the [^{14}C]ABA that had been fed to the leaves was continuously degraded to phaseic and then to dihydrophaseic acids while the plants remained wilted (Fig. 4.15). The rate of formation of phaseic acid (destruction of ABA) was calculated to be 150 μg kg^{-1} h^{-1}. The recovery of [^{3}H]ABA added to the extracts showed that half was lost during isolation, so, if the recovery of PA was similar, then the rate of ABA turnover would be double. These data indicate that there is continuous and rapid synthesis and destruction of ABA while the leaves are wilted.

Other compounds

Lunularic acid

While abscisic acid is present in flowering plants, gymnosperms, ferns and horsetails (Milborrow,

1974), it has not been detected in liverworts. Its place seems to be taken by a growth-inhibitory compound known as lunularic acid: a dihydrostilbene carboxylic acid (20). This compound not only inhibits growth of the thallus but it occurs, and is effective, at 'hormonal concentrations'. It is conveniently bioassayed by using the gemmae of *Lunularia cru-*

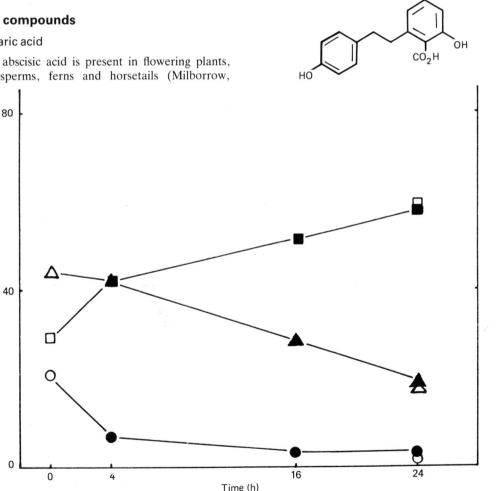

(20) Lunularic acid

Fig. 4.15 The turnover of (+)-[2-^{14}C]ABA in wilted bean leaves. The labeled ABA was supplied for 6h and the samples were either wilted or left turgid. The ABA was degraded rapidly and the ^{14}C in DPA rose as the amounts in PA fell. ABA in turgid leaves, ○; wilted leaves, ●. PA in turgid leaves, △; wilted leaves, ▲. DPA in turgid leaves, □; wilted leaves, ■. The ordinate indicates the percentage of ^{14}C in these compounds, and the abscissa, time in hours from wilting. (Data from Harrison and Walton, 1975.)

ciata, the liverwort from which it was first isolated (Valio, Burden and Schwabe, 1969). These are small (*ca.* 1 mm diameter) vegetative propagules formed in moon-shaped cups on the upper surface of the liverwort's thallus. Schwabe and Valio (1970a, b) obtained evidence which suggests that the gemmae are prevented from growing by the high concentrations of lunularic acid they contain while on the mother plant. When separated from the mature thallus and placed in a damp environment the lunularic acid leaches out and, provided that the light regime is suitable, growth begins.

Lunularic acid has been detected in several other genera of liverwort (Pryce, 1971, 1972) but, although its methyl ester can be readily detected by glc, this technique alone is insufficient to demonstrate its presence. Another, quite different, compound has almost identical chromatographic properties and is readily confused with lunularic acid methyl ester (Schwabe, personal communication).

The significant feature of lunularic acid in *Lunularia* is that the amounts present change with alterations in daylength: in short-day conditions (16 h dark, 8 h light) the liverwort grows rapidly and lunularic acid content is low. On transfer to long days the mature plants stop growing and show an increase in their lunularic acid content. The quality of light is also important, red light (650 nm) stimulates growth while far-red (735 nm) causes inhibition of growth. The effects of each waveband are overcome, as in higher plants (Table 4.6), by a subsequent exposure to the other. This is the characteristic pattern of processes involving phytochrome (*see* Chapters 16 and 17).

Lunularia thalli maintained under long days and rich in the inhibitor are able to withstand extreme drought for many months and recover and grow again when provided with water and short daylength conditions, while plants growing rapidly under short-day light regimes are killed by desiccation.

Unfortunately, it has not been possible to induce drought tolerance in *Lunularia* by applying high concentrations of lunularic acid to growing plants because the high concentrations apparently necessary to stop growth under short days are highly toxic. Perhaps a borderline day : night regime could be found under which high, but non-toxic, concentrations of lunularic acid can induce drought tolerance. It appears that lunularic acid in *Lunularia* fulfils the role played by ABA in vascular

Table 4.6 Effect of daily red and far-red radiation treatment (3 min) on growth in area of *Lunularia cruciata* gemmalings, after 10 days. Figures in table show the mean area of 16 plantlets (mm²). (Data from Schwabe and Valio, 1970b.)

Treatment	Control	Treatment given at beginning of dark period	Treatment given in middle of dark period
SD (8-h photo-period)	2.46	—	—
Continuous light	1.24	—	—
Red		2.52	1.85
Far-red		1.78	2.32
Red/far-red		1.72	2.44
Far-red/red		2.65	2.06

L.S.D. 5% 0.24.

plants. Growth of *Lunularia* is inhibited by low concentrations of ABA in the medium.

The production of one of the enzymes required by the proposed pathway of formation of lunularic acid, phenyl ammonia lyase (PAL), is known to be regulated by phytochrome and two others, cinnamic acid 4-hydroxylase and *P*-coumarate-CoA ligase have also been shown to be affected similarly by light regimes, so these enzymes also, if involved in the synthesis of lunularic acid, could be two further points at which light could affect biosynthesis. [^{14}C]Lunularic acid is decarboxylated by *Lunularia* thallus to lunularin (21), which is also present

(21) Lunularin

naturally (Pryce, 1972). The lunularic acid, therefore, can be inactivated by the plant or lost to the environment by leaching. Pryce found no evidence for the decarboxylation reactions being affected by light.

Pryce and Linton (1974) extracted an enzyme from a related liverwort (*Conocephalum conicum*) that was capable of decarboxylating lunularic acid and they measured some of its properties. They did not establish whether or not it was specific for lunularic acid or whether it decarboxylated a variety of acids. Schwabe and Valio (1970b) found that lunularic acid is deactivated at similar rates in light and darkness and have evidence which suggests that the rate of biosynthesis is affected by photoperiod. The internal content, therefore, appears to be regulated by the adjustment of the rate of biosynthesis rather than by alteration in the rate of degradation.

Batatasins

The yam plant (*Dioscorea batatus*) forms vegetative, reproductive structures (bulbils) which arise by the swelling of aerial, lateral buds. The bulbils germinate but not until they have undergone a period of after-ripening or stratification (Hasegawa and Hashimoto, 1975). Hashimoto and Tamura (1969) found that the depth of dormancy was correlated with the abundance of a substituted phenanthrene named Batatasin I (22) (Letcher, 1973). The

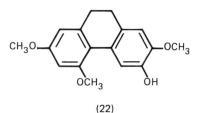

(22)

amounts of batatasins increased as the bulbils matured and decreased as the fully dormant bulbils underwent stratification (Hasegawa and Hashimoto, 1973, 1975; Hashimoto and Hasegawa, 1974), while control, unstratified bulbils retained an approximately constant amount of batatasins (Fig. 4.16). Dormancy was deepened by applications of batatasin but abscisic acid not only remained constant during stratification but did not prevent germination when applied to stratified bulbils. In contrast, GA_3 caused the batatasin content to rise and suppressed sprouting, but had no clear effect on ABA content (Hasegawa and Hashimoto, 1974). The batatasins are concentrated in the skin of the bulbils and are absent from the core, whereas ABA was distributed evenly throughout (Hasegawa and Hashimoto, 1973).

Batatasin I is a hydroxy, trimethoxyphenanthrene while Batatasin III (Hashimoto, Hasegawa, Yamaguchi, Saito and Ishimoto, 1974) is a bibenzyl analog

(23) with potentially the same substitution pattern. It is possible that it is a precursor of batatasin I but this is unlikely on chemical grounds. More probably, both are derived from a common, stilbene intermediate. These observations, when taken together, provide strong evidence that dormancy and germination processes of yam bulbils are regulated by batatasins but their occurrence in other plants has not been reported.

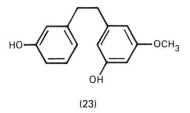

(23)

Jasmonic acid and methyl jasmonate

The methyl ester of jasmonic acid (JA) (24) was isolated from the leaves of wormwood (*Artemisia absinthium*) by Ueda and Kato (1980) who detected its senescence-promoting activity in a bioassay using oat leaf segments. Methyl jasmonate accelerated the loss of chlorophyll from the leaves and its action antagonized that of kinetin. Methyl jasmonate and a number of related compounds had been isolated previously from oil of jasmine and the free acid (25) has also been shown to occur widely in plants

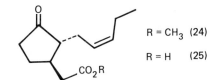

R = CH₃ (24)

R = H   (25)

(Dathe, Ronsch, Preiss, Schade, Sembdner and Schreiber, 1981) although the concentrations do not appear to be related to senescence (Table 4.7). Methyl jasmonate was more active than ABA in causing breakdown of chlorophyll and the half effective concentration was about $0.25 \, \text{mg} \, l^{-1}$, within the range of what is usually considered a hormonal concentration.

The natural material is optically active, (−), while the synthetic compound, which is composed of an equal mixture of (−) with the (+), has between 0.5 and 0.2 of the biological activity of the natural compound. This may mean that the unnatural, (+)

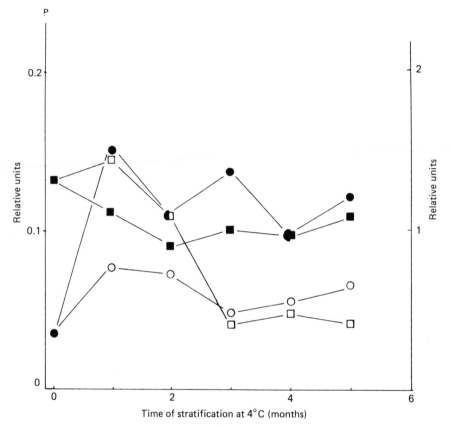

Fig. 4.16 The influence of stratification on the amounts of batatasin I and ABA in yam bulbils. Those stored at 4°C (□, batatasin I; ○, ABA) showed a marked decrease in batatasin content while those at 23°C (■, batatasin I; ●, ABA) showed little progressive change. The ordinates are in relative units. (Data from Hasegawa and Hashimoto, 1975.)

Table 4.7 Levels of jasmonic acid (JA) and ABA in the developing pericarp of broad bean (*Vicia faba*). (Data from Dathe *et al.*, 1981.)

Weight of pericarp (g)	JA (μg kg^{-1} fresh weight)	ABA (μg kg^{-1} fresh weight)
0.5	Not detectable	4
1.8	3100	8
2.4	900	1
2.8	500	2
3.3	400	2

enantiomer interferes slightly with the action of the natural (−).

Methyl jasmonate (Me JA) is slightly volatile and

it could be the volatile senescence factor postulated by Thimann and Satler (1979). Obviously much work is required before these compounds can be established as phytohormones.

Observations which suggest that jasmonic acid does not have a hormonal role are provided by measurements of the amounts in leaves. The first isolation gave 4.4 mg from 9.7 kg of plant material (*i.e.*, approximately 0.5 mg kg^{-1}). This was for a large-scale extraction and these seldom give better than a 20% recovery so the endogenous level is rather high for a potential hormone. The amounts of JA in developing bean pericarp decreased with time (Table 4.7): a result at variance with its projected rôle as a senescence hormone.

Jasmonic acid inhibited the elongation of the

second leaf sheath in rice to approximately the same degree as (±)-ABA: 50% inhibition at about $2\,mg\,ml^{-1}$. On the other hand, (−)-JA at $10\,mg\,l^{-1}$ had little effect on the growth of lettuce seedlings, while ABA was considerably more potent.

Camellia pollen was prevented from germinating by immersion in a solution of JA ($100\,mg\,l^{-1}$) and the grains regained most of their ability to germinate 120 min after being transferred to JA-free medium; thus JA is not an irreversible toxin (Yamane, Takagi, Abe, Yokota and Takahashi, 1981).

Jasmonic acid is closely related structurally to cucurbic acid (26) and it and its 5-O-glucoside and

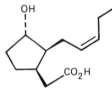

(26)

5-O-glucoside methyl ester of the latter acid have also been isolated from pumpkin (*Cucubita pepo*) seeds (Koshimizu, Fukui, Usuda and Mitsui, 1974), The last compound was the more inhibitory to elongation growth in a rice second leaf-sheath bioassay.

Seselin

A compound which shows some structural similarity to the *Eucalyptus* inhibitors (*see* below) has been isolated from roots of citrus trees (orange, lime and grapefruit) (Tomer, Goren and Monseliser, 1969) and named Seselin (27). Goren and Tomer (1971) suggest that it can act as a cofactor for IAA oxidase.

(27)

They demonstrated this satisfactorily *in vitro* using IAA oxidase from roots of orange seedling but there is no evidence for its acting in this way *in vivo*.

Avocado inhibitor

Avocado fruit contain an inhibitor of wheat coleoptile elongation and soybean callus growth (28) which was isolated by Blumenfeld and Gazit (1969). So far it has not been found in any other species but chromatograms of extracts of some other fruits show biological activity at the same R_f. No regulatory rôle for the inhibitor has been defined beyond the observation that it increases in the developing fruit mesocarp and embryo while the growth rate decreases; the ABA level remains constant. It is not toxic in that tissues resumed their growth after transfer to fresh, control solutions (Bittner, Gazit and Blumenfeld, 1971).

Eucalyptus inhibitors

Extracts of mature leaves and bark of *Eucalyptus grandis* contain a series of compounds (G inhibitors, 29, 30, 31) that inhibit growth and, in particular, prevent the rooting of cuttings. The suppression of root formation has been demonstrated on cuttings of *E. grandis* and *E. glupta* as well as mung beans (*Phaseolus mungo*).

Juvenile tissues of *E. grandis* form roots readily while the mature parts, which contain high concentrations of the inhibitors, do not form roots. Paton and Willing (1974) have noted that young leaves of *E. grandis* wilt under drought conditions while the mature leaves appear to be unaffected, they suggest that the G inhibitors are involved in the resistance of mature leaves to the effects of drought. They failed to detect ABA in the leaves, but another species *E. haemastoma* (Netting, Milborrow and Duffield, 1981) contains abundant amounts which rise even in mature leaves when they suffer a water deficit. The G inhibitors may affect rooting but there is no necessity to postulate their playing the rôle of ABA under conditions of water deficit.

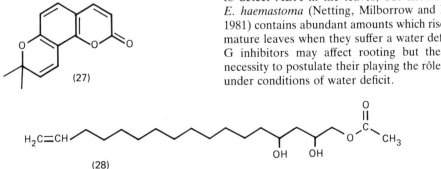

(28)

The inhibitors are closely related; two are epimers at one position, and the third differs only in the lack of a methylene group at the same point (Sterns, 1971; Nicholls, Crow and Paton, 1972).

The function of these G inhibitors (29, 30, 31) remains obscure; they are not translocated and show only very small changes in amount in response to

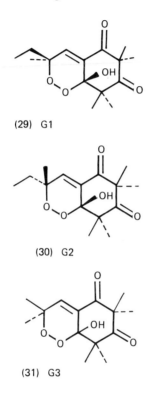

(29) G1

(30) G2

(31) G3

environmental factors. Paton and colleagues (Paton, Dhawan and Willing, 1980) reported that a mixture of the three compounds had a slight effect on the transpiration rate when fed as an aqueous solution to cut shoots of mung beans.

Postscript

When the role of inhibitors in the control of normal plant growth is examined it is necessary to consider how metabolism is regulated. The analysis soon brings into question the strategy adopted in the experimental investigation of control mechanisms

and the validity of the conclusions that can be drawn from the usual kinds of experimental results.

The bioassay that has to be used to detect a given biological activity has to be reasonably specific and sensitive. It is perhaps dangerous to use highly responsive assay systems to investigate the physiological role of hormones once they are available in pure form because the very responsiveness of the tissue may depend on an artificially-induced absence of the hormone in question. It is one of the fundamental characteristics of living systems that they resist change: chemical and physical systems respond in accordance with the pressures exerted on them, whereas living systems are self-regulating. The law of mass action does not apply when a negative feed-back system is in operation, and yet the thinking behind many experiments is that the more compound that is applied to a living tissue the greater the effect that may be expected. Even if a particular cellular process does respond in proportion to the amount of a hormone present, then some other factor has perceived a stimulus, and responded by raising the amount of the hormone present. Simple experiments designed to analyze complex, interlocking systems may be useless or even misleading. This is not a policy of despair, advocating abandonment of experimentation, it is a plea for thorough experimental design. The next decades may see the computer modeling of integrated metabolic systems so that a few crucial experiments can be carried out to determine the important parameters. Such research will be able to give answers to questions of how hormones work, questions that have evaded solution for more than half a century.

Further reading

Letham, D. S. (1978). Naturally-occurring plant growth regulators other than the principal hormones of higher plants. In *Phytohormones and Related Compounds: a comprehensive treatise*, vol. I, *The Biochemistry of Phytohormones and Related Compounds*, eds D. S. Letham, P. B. Goodwin and T. J. V. Higgins, Elsevier/North-Holland, Amsterdam, pp. 349–465.

Milborrow, B. V. (1974). The chemistry and physiology of abscisic acid. *Ann. Rev. Plant. Physiol.* **25**, 259–307.

Walton, D. C. (1980). Biochemistry and physiology of abscisic acid. *Ann. Rev. Plant Physiol.* **31**, 453–89.

References

Åberg, B. and Johannsson, I. (1969). Studies on plant growth regulators, XXIV: Some phenolic compounds. *K. Lantbrukshogsk. Ann.* **35**, 3–27.

Addicott, F. T., Carns, H. R., Lyon, J. L., Smith, O. E. and McMeans, J. L. (1964). On the physiology of abscission. In *Regulateurs naturels de la croissance végétale,* Paris, C.N.R.S.

Alvim, R., Thomas, S. and Saunders, P. F. (1978). Seasonal variation in the hormone content of willow. II. Effect of photoperiod on growth and abscisic acid content of trees under field conditions. *Plant Physiol.* **62**, 779–80.

Arditti, J., Flick, B. and Jeffrey, D. (1971). Post-pollination phenomena in orchid flowers. II. Induction of symptoms by abscisic acid and its interactions with auxin, gibberellic acid and kinetin. *New Phytol.* **70**, 333–4.

Aspinall, D., Paleg, L. G. and Addicott, F. T. (1967). Abscisin II and some hormone-regulated plant responses. *Aust. J. Biol. Sci.* **20**, 869–82.

Assante, G., Merlini, L. and Nasini, G. (1977). (+)-Abscisic acid, a metabolite of the fungus *Cercospora rosicola. Experientia* **33**, 1556–7.

Astle, M. C. and Rubery, P. H. (1980). A study of the abscisic acid uptake by apical and proximal root segments of *Phaseolus coccineus* L. *Planta* **150**, 312–20.

Beardsell, M. F. and Cohen, D. (1975). Relationship between leaf water status, abscisic acid levels, and stomatal resistance in maize and sorghum. *Plant Physiol.* **56**, 307–12.

Bennet-Clark, T. A. and Kefford, N. P. (1953). Chromatography of the growth substances in plant extracts. *Nature* **171**, 645–8.

Bennet-Clark, T. A., Tambiah, M. S. and Kefford, N. P. (1952). Estimation of plant growth sustances by partition chromatography. *Nature* **169**, 452–3.

Bittner, S., Gazit, S. and Blumenfeld, A. (1971). Isolation and identification of a plant growth inhibitor from avocado. *Phytochemistry* **10**, 1417–21.

Blumenfeld, A. and Gazit, S. (1969). An endogenous inhibitor of auxins and kinetin. *Israel J. Bot.* **18**, 217–19.

Blumenfeld, A. and Gazit, S. (1970). Interaction of kinetin and abscisic acid in the growth of Soybean callus. *Plant physiol.* **45**, 535–6.

Braun, W. and Khan, A. A. (1975). Endogenous abscisic acid levels in germinating and non-germinating lettuce seeds. *Plant Physiol.* **56**, 731–3.

Chin, T. Y., Meyer, M. M. and Beevers, L. (1969). Abscisic acid-stimulated rooting of stem cuttings. *Planta* **88**, 192–6.

Chispeels, M. J. and Varner, J. E. (1966). Inhibition of gibberelic acid-induced formation of α-amylase by abscisin II. *Nature* **212**, 1066–7.

Collins, J. C. and Kerrigan, A. P. (1974). The effect of kinetin and abscisic acid on water and ion transport in isolated maize roots. *New Phytol.* **73**, 309–14.

Coombe, B. G. and Hale, C. R. (1973). Hormone content of ripening grape berries and the effects of growth substances. *Plant Physiol.* **51**, 6629–34.

Cornforth, J. W., Milborrow, B. V., Ryback, G., Rothwell, K. and Wain, R. L. (1966). Identification of the yellow lupin inhibitor as (+)-abscisin II ((+)-dormin). *Nature* **211**, 742–3.

Cornforth, J. W., Milborrow, B. V., Ryback, G. and Wareing, P. F. (1965). Chemistry and physiology of 'Dormins' in sycamore. Identity of sycamore 'Dormin' with abscisin II. *Nature* **205**, 1269–70.

Coster, H. G. L., Stendle, E. and Zimmermann, U. (1977). Turgor pressure sensing in plant cell membranes. *Plant Physiol.* **58**, 636–43.

Dathe, W., Rönsch, H., Preiss, A., Schade, W., Sembdner, G. and Schreiber, K. (1981). Endogenous plant hormones of broad bean, *Vicia faba* L. (−)-jasmonic acid, a plant growth inhibitor in pericarp. *Planta* **153**, 530–5.

Davis, L. A. and Addicott, F. T. (1972). Abscisic acid-correlations with abscission and with development in cotton fruits. *Plant Physiol.* **49**, 644–8.

Dörffling, K. (1970). Quantitative changes in the abscisic acid content during fruit development in *Solanum lycopersicum* (tomato). *Planta* **93**, 233–42.

Dörffling, K. (1976). Correlative bud inhibition and abscisic acid in *Acer pseudoplatanus* and *Syringa vulgaris. Physiol. Plant.* **38**, 319–22.

Dörffling, K. and Böttger, M. (1968). Transport von Abscisinsäure in Explanten. Blattstiel und Unternodialsegmenten von *Coleus rheneltianus. Planta* **80**, 299–308.

Durand, M., Thévenot, C. and Côme, D. (1975). Rôle des cotylédons dans l'acide abscissique. *Physiol. Vég.* **13**, 603–10.

El Antably, H. M., Wareing, P. F. and Hillman, J. R. (1967). Some physiological responses to d,l-abscisin II (dormin). *Planta* **73**, 73–90.

Firn, R. D., Burden, R. S. and Taylor, H. F. (1972). The detection and estimation of the growth inhibitor xanthoxin in plants. *Planta* **102**, 115–26.

Firn, R. D. and Friend, J. (1972). Enzymatic production of the plant growth inhibitor, xanthoxin. *Planta* **103**, 263–6.

Gaither, D. H., Lutz, D. H. and Forrance, L. E. (1975). Abscisic acid stimulates elongation of excised per root tips. *Plant Physiol.* **55**, 948–9.

Gazit, S. and Blumenfeld, A. (1972). Inhibitor and auxin activity in the avocado fruit. *Physiol. Plant.* **27**, 77–8.

Glinka, Z. and Reinhold, L. (1971). Abscisic acid raises the permeability of plant cells to water. *Plant Physiol.* **48**, 103–5.

Goldacre, P. L., Galston, A. W. and Weintraub, R. L.

(1953). The effect of substituted phenols on the activity of the indoleacetic acid oxidase of peas. *Arch. Biochem.* **43**, 358–73.

Goldschmidt, E. E., Goren, R. and Monselise, S. P. (1967). The IAA-oxidase system of citrus roots. *Planta* **72**, 213–22.

Goldsmith, M. H. M. (1977). The polar transport of auxin. *Ann. Rev. Plant Physiol.* **28**, 439–78.

Goren, R. and Tomer, E. (1971). Effects of sesilin and coumarin on growth, indoleacetic acid oxidase, and peroxidase, with special reference to cucumber (*Cucumis sativus* L.) radicles. *Plant Physiol.* **47**, 312–16.

Harrison, M. A. and Saunders, P. F. (1975). The abscisic acid content of dormant birch buds. *Planta* **123**, 291–8.

Harrison, M. H. and Walton, D. C. (1975). Abscisic acid metabolism in water-stressed bean leaves. *Plant Physiol.* **56**, 250–4.

Hasegawa, K. and Hashimoto, T. (1973). Quantitative changes in batatasins and abscisic acid in relation to the development of dormancy in yam bulbils. *Plant Cell Physiol.* **14**, 369–77.

Hasegawa, K. and Hashimoto, T. (1974). Gibberellin-induced dormancy and batatasin content in yam bulbils. *Plant Cell Physiol.* **15**, 1–6.

Hasegawa, K. and Hashimoto, T. (1975). Variation in abscisic acid and batatasin content of yam bulbils—effects of stratification and light exposure. *J. exp. Bot.* **26**, 757–64.

Hashimoto, L. and Hasegawa, K. (1974). Batatasin I: its structure and possible involvement in yam bulbil dormancy. In *Plant Growth Substances 1973*, Hirokawa Publishing Co., Tokyo, pp. 150–6.

Hashimoto, T., Hasegawa, K., Yamaguchi, H., Saito, M. and Ishimoto, S. (1974). Structure and synthesis of batatasins, dormancy-inducing substances of yam bulbils. *Phytochemistry* **13**, 2849–52.

Hashimoto, T. and Tamura, S. (1969). Effects of abscisic acid on the sprouting of aerial tubers of *Begonia eransiana* and *Dioscorea batatas*. *Bot. Mag. Tokyo* **82**, 69–75.

Heilmann, B., Hartung, W. and Gimmler, H. (1979). The distribution of abscisic acid between chloroplasts and cytoplasm of leaf cells and the permeability of the chloroplast envelope for abscisic acid. *Zeit. Pflanz.* **97**, 67–78.

Hirai, N., Fukui, H. and Koshimizu, K. (1978). Isolation of a novel abscisic acid metabolite from seeds of *Robinia pseudacacia*. *Phytochemistry* **17**, 1625–8.

Ho, D. T. H. (1979) On the mode of action of abscisic acid in barley aleurone cells. *Plant Physiol.* **63**, 80 (Suppl.).

Ho, D. T. and Varner, J. E. (1974). Hormonal control of messenger ribonucleic acid metabolism in barley aleurone layers. *Proc. Nat. Acad. Sci. USA* **71**, 4783–6.

Hocking, T. J., Hillman, J. R. (1975). Studies on the role of abscisic acid in the initiation of bud dormancy in *Alnus glutinosus* and *Betula pubescens*. *Planta* **125**, 235–42.

Hocking, T. J., Hillman, J. R. and Wilkins, M. B. (1972). Movement of abscisic acid in *Phaseolus vulgaris* plants. *Nature (N.B.)* **235**, 124–25.

Holubowicz, T. and Boe, A. A. (1969). Development of cold hardiness in apple seedlings treated with gibberellic acid and abscisic acid. *J. Am. Soc. Hort. Sci.* **94**, 661–4.

Irvine, R. M. and Lanphear, F. O. (1968). Regulation of cold hardiness in *Acer negundo*. *Plant Physiol.* **43**, 9–13.

Jackson, G. A. D. and Blundell, J. B. (1966). Effect of dormin on fruit-set in *Rosa*. *Nature* **212**, 1470–1.

Jacobsen, J. V., Higgins, T. J. V. and Zwar, J. A. (1980). Hormonal control of endosperm function during germination. In *The Plant Seed—Development, Preservation and Germination*, ed. J. Rubenstein, Academic Press, New York.

Jones, R. L. and Mansfield, T. A. (1970). Suppression of stomatal opening in leaves treated with abscisic acid. *J. exp. Bot.* **21**, 714, 719.

Karmoker, J. L. and Van Steveninck, R. F. M. (1978). Stimulation of volume flow and ion flux by abscisic acid in excised root systems of *Phaseolus vulgaris* L. cv. Redland Pioneer. *Physiol. Plant.* **45**, 453–9.

Karssen, C. M. (1982). Indirect effect of abscisic acid on the induction of secondary dormancy in lettuce seeds. *Physiol. Plant.* **54**, 258–66.

Kefeli, V. I. and Kadyrov, C. S. (1971). Natural growth inhibitors, their chemical and physiological properties. *Ann. Rev. Plant Physiol.* **22**, 185–96.

Khan, A. A. (1971). Cytokinins: permissive role in seed germination. *Science* **171**, 853–9.

King, R. W. (1976). Abscisic acid in developing wheat grains and its relationship to grain growth and maturation. *Planta* **132**, 43–51.

Koshimizu, K., Fukui, H., Usuda, S. and Mitsui, T. (1974). Plant growth inhibitors in seeds of pumpkin. In *Plant Growth Regulators 1973*, Hirokawa Publishing Co., Tokyo, pp. 86–92.

Koshimizu, K., Inui, M. and Fukui, H. (1968). Isolation of (+)-abscisyl-β-D-glucopyranoside from immature fruit of *Lupinus luteus*. *Agric. Biol. Chem.* **32**, 789–91.

Kriedemann, P. E., Loveys, B. R. and Downton, W. J. S. (1975). Internal control of stomatal physiology and photosynthesis II Photosynthetic responses to phaseic acid. *Aust. J. Plant. Physiol.* **2**, 553–67.

Lenton, J. R., Bowen, M. R. and Saunders, P. F. (1971). Detection of abscisic acid in the xylem sap of Willow (*Salix viminalis* L.) by gas-liquid chromatography. *Nature* **220**, 86–7.

Lenton, J. R., Perry, V. M. and Saunders, P. F. (1971). The identification and quantitative analysis of abscisic acid in plant extracts by gas liquid chromatography. *Planta* **96**, 271–80.

Lenton, J. R., Perry, V. M. and Saunders, P. F. (1972). Endogenous abscisic acid in relation to photoperiodically induced bud dormancy. *Planta* **106**, 13–22.

Le Page-Degivry, M. T. (1973). Influence de l'acide abscissique sur le développement des embryons de *Taxus baccata* L. cultivés *in vitro*. *Z. Pfl. Physiol.* **70**, 406–13.

Letcher, R. M. (1973). Structure and synthesis of the growth inhibitor Batatasin I from *Dioscorea batas*. *Phytochemistry* **12**, 2789–90.

Loveys, B. R. (1977). The intracellular location of abscisic acid in stressed and non-stressed leaf tissue. *Physiol. Plant.* **40**, 6–10.

Loveys, B. R. and Milborrow, B. V. (1981). Isolation and characterisation of l′-*O*-abscisic acid-*β*-D-glycopyranoside from vegetative tomato tissue. *Aust. J. Plant Physiol.* **8**, 571–89.

MacMillan, J. and Pryce, R. J. (1968) Phaseic acid, a putative relative of abscisic acid, from seeds of *Phaseolus multiflorus*. *Chem. Commun.* 124–6.

McWha, J. A. and Jackson, D. L. (1976). Some growth promotive effects of ABA. *J. exp. Bot.* **27**, 1004–8.

Malek, T. and Baker, D. A. (1978). Effects of fusicoccin on proton co-transport of sugars in the phloem loading of *Ricinus communis* L. *Plant Sci. Lett.* **11**, 233–9.

Marumo, S., Katayama, M., Komori, E. and Yozaki, H. (1982). Microbiological production of abscisic acid by *Botrytis cinerea*. *Agric. Biol. Chem.* **46**, 1967–8.

Milborrow, B. V. (1966a). Identification of (+)-Abscisin II [(+)-Dormin] in plants and measurements of its concentrations. *Planta* **76**, 93–113.

Milborrow, B. V. (1966b). The effects of synthetic dl-Dormin (Abscisin II) on the growth of the oat mesocotyl. *Planta* **70**, 155–71.

Milborrow, B. V. (1968). Identification and measurement of (+)-Abscisic acid in plants. In *Biochemistry and Physiology of Plant Growth Substances*, ed. F. Wightman and G. Setterfield, Runge Press, Ottawa, pp. 1531–45.

Milborrow, B. V. (1969). Identification of 'Metabolite C' from abscisic acid and a new structure for phaseic acid. *Chemical Commun.*, 966–7.

Milborrow, B. V. (1970a). The metabolism of abscisic acid. *J. exp. Bot.* **21**, 17–29.

Milborrow, B. V. (1970b). Abscisic acid. In *Aspects of Terpenoid Chemistry and Biochemistry*, ed. T. W. Goodwin, Academic Press, London, pp. 137–51.

Milborrow, B. V. (1974). The chemistry and physiology of abscisic acid. *Ann. Rev. Plant Physiol.* **25**, 259–307.

Milborow, B. V. (1978). The stability of conjugated abscisic acid during wilting. *J. exp. Bot.* **29**, 1059–66.

Milborrow, B. V. (1979). Anti transpirants and the regulation of abscisic acid content. *Aust. J. Plant Physiol.* **6**, 249–54.

Milborrow, B. V. (1980). A distinction between the fast and slow responses to abscisic acid. *Aust. J. Plant Physiol.* **7**, 749–54.

Milborrow, B. V. (1983). The reduction of (±)-[2-
^{14}C]abscisic acid to the 1′,4′-*trans* diol by pea seedlings and the formation of 4′-desoxy ABA as an artefact. *J. exp. Bot.* **34**, 303–8.

Milborrow, B. V. and Mallaby, R. (1975). Occurrence of methyl (+)-abscisate as an artefact of extraction. *J. exp. Bot.* **26**, 741–8.

Milborrow, B. V. and Noddle, R. C. (1970). Conversion of 5-(1,2-epoxy-2,6,6-trimethylcyclohexyl)-3-methylpenta-*cis*-2-*trans*-4-dienoic acid into abscisic acid in plants. *Biochem. J.* **119**, 727–34.

Milborrow, B. V. and Robinson, D. R. (1973). Factors affecting the biosynthesis of abscisic acid. *J. exp. Bot.* **24**, 537–48.

Milborrow, B. V. and Vaughan, G. (1979). The long term metabolism of (±)-[2^{14}C]abscisic acid by apple seeds. *J. exp. Bot.* **30**, 983–95.

Milborrow, B. V. and Vaughan, G. T. (1982). Characterisation of dihydrophaseic acid 4′-*O*-*β*-D-glucopyranoside as a major metabolite of abscisic acid. *Aust. J. Plant. Physiol.* **9**, 361–72.

Mizrahi, Y., Blumenfeld, A., Bittner, S. and Richmond, A. E. (1971). Abscisic acid and cytokinin contents of leaves in relation to salinity and relative humidity. *Plant Physiol.* **48**, 752–5.

Mizrahi, Y. and Richmond, A. E. (1972). Abscisic acid in relation to mineral deprivation. *Plant Physiol.* **50**, 667–70.

Naumann, R. and Dörffling, K. (1982). Variation of free and conjugated abscisic acid and phaseic acid and dihydrophaseic acid levels in ripening barley grains. *Plant Sci. Lett.* **27**, 111–17.

Neill, S. J., Horgan, R., Walton, D. C. and Lee, T. S. (1982). The biosynthesis of abscisic acid in *Cerospora rosicola*. *Phytochemistry* **21**, 61–5.

Netting, A. G., Milborrow, B. V. and Duffield, A. M. (1981). Determination of abscisic acid in *Eucalyptus haemastoma* leaves using gas chromatography, mass spectrometry and deuterated internal standards. *Phytochemistry* **21**, 385–9.

Nicolls, W., Crow, W. D. and Paton, D. M. (1972). Chemistry and physiology of rooting inhibitors in adult tissues of *Eucalyptus grandis*. In *Plant Growth Substances 1970*, ed. D. J. Carr, Springer-Verlag, Berlin, pp. 324–9.

Nitsch, C. (1968). Induction de la floraison chez *Plumbago indica* L. *Ann. Sci. Naturelles Botanique Paris* **9**, 64–92.

Nitsch, J. P. and Nitsch, C. (1962). Composés phenolique et croissance végétale. *Ann. Physiol. Vég.* **4**, 211–25.

Okhuma, K., Lyon, J. L., Addicott, F. T. and Smith, O. E. (1963). Abscisin II, an abscission accelerating substance from young cotton fruit. *Science* **142**, 1592–3.

Okhuma, K., Addicott, F. T., Smith, O. E. and Thiessen, W. E. (1965). The structure of abscisin II. *Tetrahedron Lett.* **29**, 2529–35.

Osborne, D. J., Jackson, M. B. and Milborrow, B. V.

(1972). Physiological properties of abscission accelerator from senescent leaves. *Nature New Biology* **240**, 98–101.

Paton, D. M., Dhawan, A. K. and Willing, R. R. (1980). Effect of *Eucalyptus* growth regulators on the water loss from plant leaves. *Plant Physiol.* **66**, 254–6.

Paton, D. M. and Willing, R. R. (1974). Inhibitor transport and ontogenetic age in *Eucalyptus grandis*. In *Plant Growth Substances 1973*, Kirokawa Publishing Co., Tokyo, pp. 126–32.

Perry, T. O. and Byrne, O. R. (1969). Turion induction in *Spirodela Polyrrhiza* by abscisic acid. *Plant Physiol.* **44**, 784–5.

Pierce, M. and Raschke, K. (1978). The relationship between abscisic acid and leaf turgor. *Plant Physiol.* **61**, (Suppl.) 25.

Pitman, M. G., Luttge, U., Lauchli, A. and Ball, E. (1974). Effect of previous water stress on ion uptake and transport in barley seedlings. *Aust. J. Plant Physiol.* **1**, 377–85.

Popják, G. and Cornforth, J. W. (1966). Substrate stereochemistry in squalene biosynthesis. *Biochem. J.* **101**, 553–68.

Powell, L. E. (1976). Effect of photoperiod on endogenous abscisic acid in *Malus* and *Betula*. *Hort. Sci.* **11**, 498–9.

Pridham, J. B. (1965). Low molecular weight phenols in higher plants. *Ann. Rev. Plant Physiol.* **16**, 13–36.

Pryce, R. J. (1971). The biosynthesis of lunularic acid—a dihydrostilbene endogenous growth inhibitor of liverworts. *Phytochemistry* **10**, 2679–85.

Pryce, R. J. (1972). Metabolism of lunularic acid to a new plant stilbene by *lunularia cruciata*. *Phytochemistry* **11**, 1355–64.

Pryce, R. J. (1976). Lunularic acid, a common, endogenous growth inhibitor of liverworts. *Planta* **97**, 354–7.

Pryce, R. J. and Linton, L. (1974). Lunularic acid decarboxylase from the liverwort *Conocephalum conicum*. *Phytochemistry* **13**, 2497–501.

Rappaport, L. and Wolf, N. (1969). Problem of dormancy in potato tubers and related structures. *Symp. Soc. Exp. Biol.* **23**, 219–40.

Rikin, A., Blumenfeld, A. and Richmond, A. E. (1976). Chilling resistance as affected by stressing environments and abscisic acid. *Bot. Gaz.* **137**, 307–12.

Rikin, A. and Richmond, A. E. (1976). Amelioration of chilling injuries in cucumber seedlings by abscisic acid. *Physiol. Plant.* **38**, 95–7.

Rikin, A., Waldman, M., Richmond, A. E. and Dobrat, A. (1975). Hormonal regulation of morphogenesis and cold resistance. I. Modifications by abscisic acid and by gibberelleic acid in alfalfa (*Medicago sativa* L.) seedlings. *J. exp. Bot.* **26**, 175–83.

Rivier, L., Milon, H. and Pilet, P. E. (1977). Gas-chromatography–mass spectrometric determinations of abscisic acid levels in the cap and the apex of maize roots. *Planta* **134**, 23–7.

Rivier, L. and Pilet, P. E. (1981). Abscisic acid levels in the root tips of seven varieties of *Zea mays*. *Phytochemistry* **20**, 17–19.

Roberts, D. L., Heckman, R. A., Hege, B. P. and Bellin, S. A. (1968). Synthesis of (*RS*)-abscisic acid. *J. Org. Chem.* **33**, 3566–9.

Robinson, D. R. and Ryback, G. (1969). Incorporation of tritium from [(4*R*)-4-^{3}H] mevalonate into abscisic acid. *Biochem. J.* **113**, 895–7.

Robinson, P. M. and Wareing, P. F. (1964). Chemical, native and biological properties of the inhibitor varying with photoperiod in sycamore (*Acer pseudoplatanus*). *Physiol. Plant.* **17**, 314–23.

Rothwell, K. and Wain, R. L. (1964). Studies on a growth inhibitor in yellow lupin (*Lupinus luteus* L.). In *Regulateurs naturels de la croissance végétale*, Paris: C.N.R.S.

Rubery, P. H. (1972). Studies on indoleacetic acid oxidation by liquid medium from crown gall tissue culture cells: the role of malic acid and related compounds. *Biochim. Biophys. Acta* **261**, 21–34.

Rubery, P. H. and Astle, M. C. (1982). The mechanism of transmembrane abscisic acid transport and some of its implications. In *Plant Growth Substances 1982*, ed. P. F. Wareing, Academic Press, London, pp. 353–62.

Rudnicki, R. and Czapski, J. (1974). The uptake and degradation of 1'-^{14}C-abscisic acid by apple seeds during stratification. *Ann. Bot. New Series* **38**, 189–92.

Schwabe, W. W. and Valio, I. F. M. (1970a). Growth and dormancy in *Lunularia cruciata* (L.) Dum. V. The control of growth by a natural endogenous inhibitor. *J. exp. Bot.* **21**, 112–21.

Schwabe, W. W. and Valio, I. F. M. (1970b). Growth and dormancy in *Lunularia cruciata* (L.) Dum. VI. Growth regulation by daylength, by red, far red, and blue light, and by applied growth regulators and chelating agents. *J. exp. Bot.* **21**, 122–37.

Seeley, S. D. and Powell, L. E. (1970). Electron capture gas chromatography for sensitive assay of abscisic acid. *Anal. Biochem.* **35**, 530–3.

Sembdner, G., Weiland, J., Schmeider, G., Schreiber, K. and Focke, I. (1972). Recent advances in the metabolism of gibberellins. In *Plant Growth Substances 1970*, ed. D. J. Carr, Springer-Verlag, Berlin, pp. 143–50.

Shaner, D. L., Mertz, S. M. and Arntzen, C. J. (1975). Inhibition of ion accumulation in maize roots by abscisic acid. *Planta* **122**, 79–90.

Sharkey, T. D. and Raschke, K. (1980). Effects of phaseic acid and dihydrophaseic acid on stomata and the photosynthetic apparatus. *Plant Physiol.* **65**, 291–7.

Shindy, W. W., Asmundson, C. M., Smith, O. E. and Kumamoto, J. (1973). Absorption and distribution of high specific radioactivity 2-^{14}C-abscisic acid in cotton seedlings. *Plant Physiol.* **52**, 443–7.

Simpson, G. M. and Wain, R. L. (1961). A relationship between gibberellic acid and light in the control of intermode extension in dwarf pea (*Pisum sativum*). *J. exp. Bot.* **12**, 207–16.

Smith, J. D., McDaniel, S. and Lively, S. (1978). Regulation of embryo growth by abscisic acid *in vitro*. *Maize Genet. Newslett.* **52**, 107–8.

Sondheimer, E., Galson, E. C., Chang, Y. P. and Walton, D. C. (1971). Asymmetry, its importance to the action and metabolism of abscisic acid. *Science* **174**, 829–31.

Sondheimer, E., Tzou, D. S. and Galson, E. C. (1968). Abscisic acid levels and seed dormancy. *Plant Physiol.* **43**, 1443–7.

Smith, O. E., Lyon, J. L. and Addicott, F. T. (1963). Abscisin II, an abscission accelerating substance from young cotton fruit. *Science* **142**, 1592–3.

Stafford, H. A. (1974). Activation of 4-hydroxycinnamate hydroxylase in extracts from Sorghum. *Plant Physiol.* **54**, 686–9.

Sterns, M. (1971). Crystal and molecular structure of a root inhibitor from *Eucalyptus grandis*, 4-ethyl-L-hydroxy - 4,8,8,10,10 - pentamethyl - 7,9 - dioxo - 2,3 - dioxabicyclo[4.4.0]-decene-5. *J. Cryst. Molec. Struct.* **1**, 373–81.

Stewart, G. R. (1969). Abscisic acid and morphogenesis in *Lemna polyrrhiza*. *Nature* **221**, 61–2.

Tal, M. and Imber, D. (1971). Abnormal stomatal behaviour and hormonal imbalance in *flacca*, a wilty mutant of tomato. III. Hormonal effects on the water status in the plant. *Plant Physiol.* **47**, 849–50.

Taylor, H. F. and Burden, R. S. (1970). Identification of plant growth inhibitors produced by photolysis of violaxanthin. *Phytochemistry* **9**, 2217–23.

Taylor, H. F. and Smith, T. A. (1967). Production of plant growth inhibitors from xanthophylls: a possible source of dormin. *Nature* **215**, 1513–14.

Thimann, K. V. and Satler, S. O. (1979). Relation between leaf senescence and stomatal closure: senescence in light. *Proc. Nat. Acad. Sci. USA* **76**, 2295–8.

Tietz, D., Dörffling, K., Wohrle, D., Erxleben, I. and Liemann, F. (1979). Identification by combined gas chromatography–mass spectrometry of phaseic acid and dihydrophaseic acid and characterisation of further abscisic acid metabolites in pea seedlings. *Planta* **147**, 168–73.

Tinelli, E. T., Sondheimer, E., Walton, D. C., Gaskin, P. and MacMillan, J. (1973). Metabolites of 2-^{14}C-abscisic acid. *Tetrahedron Letts.* 139–40.

Tomer, E., Goren, R. and Monselise, S. P. (1969). Isolation and identification of sesilin in *Citrus* roots. *Phytochemistry* **8**, 1315–16.

Trewavas, A. J. (1981). How do plant growth substances work? *Plant Cell Env.* **4**, 203–28.

Tucker, D. J. (1977). The effects of far-red light on lateral bud outgrowth in decapitated tomato plants and the associated changes in the levels of auxin and abscisic acid. *Plant Sci. Lett.* **8**, 339–44.

Tucker, D. J. and Mansfield, T. A. (1971). A simple bioassay for detecting 'antitranspirant' activity of naturally occurring compounds such as abscisic acid. *Planta* **98**, 157–63.

Ueda, J. and Kato, J. (1980). Isolation and identification of a senescence-promoting substance from wormwood (*Artemisia absinthum* L.). *Plant Physiol.* **66**, 246–9.

Valio, I. F. M., Burden, R. S. and Schwabe, W. W. (1969). New natural growth inhibitor in the liverwort, *Lunularia cruciata* (L.) Dum. *Nature* **223**, 1176–8.

Van Overbeeck, J., Loeffler, J. E. and Mason, M. I. R. (1967). Dormin (Abscisin II), inhibitor of plant DNA synthesis? *Science* **156**, 1497–8.

Van Steveninck, R. F. M. (1959). Abscission accelerators in lupins (*Lupinus luteus* L.). *Nature* **183**, 1246–8.

Van Steveninck, R. F. M. (1972). Abscisic acid stimulation of ion transport and alteration in K$^+$ Na$^+$ selectivity. *Z. Pflanzenphysiol.* **67**, 282–6.

Van Steveninck, R. F. M. (1976). Effect of hormones and related substances on ion transport. In *Transport in Plants II, Part B*, eds U. Lüttge and M. G. Pitman, Springer-Verlag, Berlin, pp. 307–42.

Vanderhoef, L. N. and Dute, R. R. (1981). Auxin-regulated wall loosening and sustained growth in elongation. *Plant Physiol.* **67**, 146–9.

Varner, J. E. and Ho, D. T. H. (1976). The role of hormones in the integration of seedling growth. In *The Molecular Biology of Hormone Action*, Academic Press, New York, pp. 173–94.

Varner, J. E. and Johri, M. M. (1968). Hormonal control of enzyme synthesis. In *Biochemistry and Physiology of Plant Growth Substances*, eds F. Wightman and G. Setterfield, Runge Press, Ottawa, pp. 793–814.

Walton, D. C. (1980). Biochemistry and physiology of abscisic acid. *Ann. Rev. Plant Physiol.* **31**, 453–89.

Wareing, P. F. and Saunders, P. F. (1971). Hormones and dormancy. *Ann. Rev. Plant Physiol.* **22**, 261–88.

Weber, J. A. and Nooden, L. D. (1976). Environmental and hormonal control of turion formation in *Myriophyllum verticillatum*. *Plant Cell Physiol.* **17**, 721–32.

Weiler, E. (1980). Plant hormone immunoassay. *Physiol. Plant.* **54**, 230–4.

White, J. C. and Mansfield, T. A. (1977). Correlative inhibition of lateral bud growth in *Pisum sativum* L. and *Phaseolus vulgaris* L. Studies of the role of ascisic acid. *Ann. Bot.* **41**, 1163–70.

Wilson, J. M. (1976). The mechanisms of chill- and drought-hardening of *Phaseolus vulgaris* leaves. *New Phytol.* **76**, 257–70.

Wright, S. T. C. (1969). An increase in the 'inhibitor-β' content of detached wheat leaves following a period of wilting. *Planta* **86**, 10–20.

Wright, S. T. C. (1977). The relationship between leaf water potential (ψ leaf) and the levels of abscisic acid and ethylene in excised wheat leaves. *Planta* **134**, 183–9.

Wright, S. T. C. and Hiron, R. W. P. (1969). (+)-Abscisic acid, the growth inhibitor induced in detached leaves by a period of wilting. *Nature* **224,** 719–20.

Yamane, H., Takagi, H., Abe, H., Yokota, T. and Takahashi, N. (1981). Identification of jasmonic acid in three species of higher plants and its biological activities. *Plant Cell Physiol.* **22,** 689–97.

Zabadal, T. J. (1974). A water potential threshold for the increase of abscisic acid in leaves. *Plant Physiol.* **53,** 125–7.

Zeevaart, J. A. D. (1977). Sites of abscisic acid synthesis and metabolism in *Ricinus communis* L. *Plant Physiol.* **59,** 788–91.

Zeevaart, J. A. D. and Milborrow, B. V. (1976). Metabolism of abscisic acid and the occurrence of *epi*-dihydrophaseic acid in *Phaseolus vulgaris. Phytochemistry* **15,** 493–500.

Zenk, M. H. and Müller, G. Z. (1963). *In vivo* destruction of exogenously applied indoleacetic acid as influenced by phenolic acids. *Nature* **200,** 761–3.

Ethylene

5

Elmo M Beyer, Jr, Page W Morgan and Shang Fa Yang

This chapter is dedicated to the memory of Dr Morris Lieberman and to his many contributions to ethylene biology. Dr Lieberman had agreed to write this chapter but died before he could begin the work.

Introduction

Ethylene (C_2H_4) is produced by all higher plants and in trace amounts interacts with the other plant hormones, especially auxin, to coordinate and regulate a wide variety of growth and developmental processes. While at times the effects of ethylene are subtle, often they are spectacular as in the case of fruit ripening, abscission, breaking of dormancy, flowering, and modification of sex expression. Ethylene as a plant hormone is unique in its structural simplicity and gaseous nature. These qualities make it the easiest plant hormone to detect and measure. Ethylene evolved from plant tissues can be measured quickly with considerable sensitivity and accuracy by gas chromatography.

Discovery

The field of ethylene biology can be traced back to the turn of the century and historically only the auxins predate ethylene. The Russian botanist Neljubow (1901) is believed to have been the first to recognize the growth-regulatory properties of ethylene. In 1901 he reported that ethylene was the agent in illuminating gas which most effectively caused the 'triple response' in dark-grown pea seedlings (Fig. 5.1). Subsequently, much of the earlier literature describing the growth-regulatory effects of illuminating gas, various fumes, and smoke took on new meaning. These effects have now been attributed largely to the presence of ethylene.

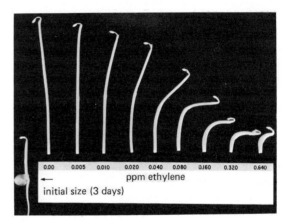

Fig. 5.1 The effect of ethylene (μl l^{-1} or ppm) on the growth and gravitropism of etiolated pea epicotyls after 48 h exposure. (Photograph courtesy of J. D. Goeschl.)

By 1930 ethylene was recognized to have a wide variety of interesting effects on plants. It was not until 1934, however, that Gane in England first obtained positive proof that ethylene was a natural plant product. Shortly thereafter, scientists at the Boyce Thompson Institute (Crocker, Hitchcock and Zimmerman, 1935) proposed that ethylene is a fruit-ripening hormone and also acts as a regulator in vegetative plant organs. This hypothesis was supported by Hansen (1943) and Kidd and West (1945). However, Biale, Young and Olmstead (1954), using the available but insensitive chemical assay procedures, could find no evidence that fruits produced sufficient ethylene prior to ripening to induce the response. Thus, the role of ethylene in fruit ripening remained unsettled until the adaptation of gas chromatography to ethylene analysis (Burg and Stolwijk, 1959; Burg and Thimann, 1959). This major development rapidly advanced the field of ethylene research and by the mid-1960s ethylene had been clearly identified as an endogenous regulator of fruit ripening. With the resurgence of interest that followed, ethylene rapidly gained in hormonal stature.

Ethylene production in plants

Two discoveries, separated by more than a decade, have contributed enormously to our current understanding of ethylene biosynthesis in higher plants. The first was the realization that ethylene is derived from carbons 3 and 4 of methionine (Lieberman and Mapson, 1964), and the second was the discovery by Adams and Yang (1979) that this conversion proceeds through the stable intermediate, 1-amino-cyclopropane-1-carboxylic acid (ACC). Recent advances stemming from this latter discovery have led to a much better appreciation of the regulatory mechanisms involved in plant responses to ethylene.

Scope of ethylene production

Ethylene is produced by essentially all parts of higher plants (Burg, 1962; Abeles, 1973). The rate of this production varies with both the type of tissue and its stage of development. In germinating seeds, high rates of ethylene production usually occur when the radical starts to penetrate the seed coat and during the period when the seedling forces its way through soil. Generally, meristematic tissue and nodal regions are the most vigorous sites of ethylene synthesis. In flowers, maximum rates of ethylene production usually occur at the initial stages of fading. Possibly the highest rate of ethylene production recorded in higher plants is $3400\,nl\,g^{-1}h^{-1}$ from fading blossoms of *Vanda* orchids.

Of all plant tissues, the ethylene production of fruits is the best documented (Biale and Young, 1981). Fruits are categorized into climacteric and non-climacteric types depending on the pattern of respiration after harvest. The climacteric fruits are characterized by a low rate of ethylene production during the pre-climacteric or unripe stage. This is followed by the climacteric, a sudden increase in

ethylene production and respiration, accompanied by other biochemical changes of ripening. Apple, banana, tomato, avocado, mango and cherimoya belong to this group. Ethylene production by these fruits during the climacteric peak ranges from 1 for mango to $320\,nl\,g^{-1}h^{-1}$ for cherimoya fruit (Table 5.1). In non-climacteric fruits, such as orange, lemon and cherry, there is no surge in respiration

Table 5.1 Ethylene production rates by various plant tissues.

Tissue	Production rate ($nl\,g^{-1}\,h^{-1}$)	Conversion constant[a] ($\mu l\,l^{-1}/ nl\,g^{-1}\,h^{-1}$)
Climacteric fruits[b]		
Avocado	0.02–130	0.7
Banana	0.02–20	1.9
Cherimoya	0.10–320	1.3–0.4
Mango	0.01–1	3.8
Cantaloupe	0.02–80	2.5
Tomato	0.02–20	2.0
Non-climacteric fruits		
Lemon	0.02–0.1	1.8
Orange	0.02–0.1	4.0
Leaf		
Tobacco	0.50–2	0.1
Stem		
Bean	2	0.4
Flower		
Carnation[b]	0.05–30	
Fungi[c]		
Penicillium digitatum	6000	
Fusarium oxysporum f. sp. *tulipae*	3800	

(a) The ratio of internal ethylene concentration ($\mu l\,l^{-1}$) to the ethylene production rate ($nl\,g^{-1}\,h^{-1}$) is expressed as a conversion constant. Thus, when the production rate and the conversion constant are known, it is possible to estimate the internal ethylene concentration.

(b) The range of production rates represents those at preclimacteric stage and at climacteric peak stage.

(c) Maximal ethylene production rate in pure culture.

and ethylene production is usually stable and very low, about $0.05\,nl\,g^{-1}\,h^{-1}$.

Higher plants are not the only biological source of ethylene. Biale and, independently, Miller, Winston and Fisher in 1940 discovered that *Penicillium digitatum*, a common green mold on citrus fruits, produced ethylene at a high rate. Since then many workers have reported ethylene production by a variety of microorganisms including *Fusarium oxysporum*, *Aspergillus clavatus*, *Aspergillus flavus*, *Mucor hiemalis*, *Escherichia coli*, and many other soil microbes. Ethylene thus produced may play an important ecological role in seed germination, seedling growth, and some host–pathogen interactions.

Pathway of ethylene biosynthesis

The task of unraveling the biochemistry of ethylene formation has not been easy. Many compounds were once proposed as precursors of ethylene including linolenic acid, propanal, ethanol, acrylic acid, β-alanine, β-hydroxypropionic acid, fumaric acid and methionine. Methionine, the only confirmed precursor, was first proposed by Lieberman and Mapson (1964) based on work with a model chemical system. Subsequently Lieberman, Kunishi, Mapson and Wardale (1966) demonstrated *in vivo* conversion of ^{14}C-labeled methionine to [^{14}C] ethylene in apple tissue. It is now established that methionine indeed serves as a precursor of ethylene in all higher plant tissues and the biochemical mechanism has been discovered. The CH_3S group of methionine was found to be retained by apple tissue. Biologically this is important since the level of methionine in plant tissues is normally very low compared to their rates of ethylene production. Adams and Yang (1977) demonstrated that the sulfur of methionine is recycled to maintain a steady rate of ethylene production. The CH_3S group of methionine is released from S-adenosylmethionine (SAM) as methylthioadenosine (MTA), then rapidly hydrolyzed to methylthioribose (MTR), and the CH_3S group of MTR is recycled back into methionine (Fig. 5.2). More recently Yung, Yang and Schlenk (1982) and Wang, Adams and Lieberman (1982) showed that the ribose moiety of MTR is directly incorporated into the 2-aminobutyrate moiety of methionine along with the CH_3S unit. Thus, the overall result of this cycle is that the CH_3S group is conserved for the continued synthesis of

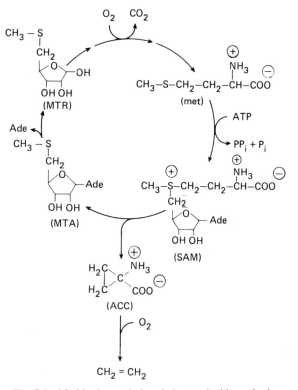

Fig. 5.2 Methionine cycle in relation to the biosynthesis of ethylene in apple tissue. (From Yung *et al.*, 1982.)

methionine while the 3,4-carbon moiety from which the ethylene molecule is derived is ultimately replenished from the ribose moiety of ATP.

Ethylene production ceases when plant tissue is placed under anaerobic conditions, but a surge of ethylene production occurs upon re-exposure of the tissue to air. This suggests that an intermediate accumulates during anaerobic incubation which is subsequently converted to ethylene upon exposure to oxygen. Using apple tissue, Adams and Yang (1979) compared the metabolism of ^{14}C-labeled methionine in air and nitrogen. In air, methionine was efficiently converted to ethylene; however, in nitrogen, methionine was metabolized not to ethylene but to MTR and an unknown compound which was subsequently identified as 1-aminocyclopropane-1-carboxylic acid (ACC). In air ACC was rapidly converted to ethylene. This indicates that ACC is the immediate precursor of ethylene and that the conversion is oxygen-dependent. Thus, the pathway of ethylene biosynthesis is as follows: methionine → SAM → ACC → ethylene. Soon after the pathway for ethylene biosynthesis became known, ACC synthase, which catalyzes the conversion of SAM to ACC, was demonstrated in tomato extracts. Since the enzyme requires pyridoxal phosphate for maximum activity and is inhibited by aminoethoxyvinylglycine (AVG) and aminooxyacetic acid (AOA), well-known inhibitors of pyridoxal phosphase enzymes, it is thought that ACC synthase is a pyridoxal enzyme (Boller, Herner and Kende, 1979; Yu, Adams and Yang, 1979).

When ACC was applied to various plant organs, including root, stem, leaf, inflorescence, and fruit, a marked increase in ethylene production was observed. This indicates that the enzyme converting ACC to ethylene is present in most plant tissues. This enzyme, however, has not yet been identified, but it is known to be very labile and is assumed to be membrane-bound. In retrospect, it should be noted that ACC was isolated more than two decades ago from apples and pears by Burroughs (1957). Although he was unable to assign a specific role to ACC, he noted that it was detectable only in ripe fruit.

In addition to its conversion to ethylene, ACC is also metabolized to *N*-malonyl-ACC, and the enzyme responsible for this conjugation appears to be present in most plant tissues (Amrhein, Schneebeck, Skorupka, Tophof and Stöckigt, 1981; Hoffman, Yang and McKeon, 1982). Since this conversion is irreversible and the conjugate does not release ethylene readily, *N*-malonyl-ACC is thought to be an inactive end-product rather than a storage form of ACC. Thus, the formation of *N*-malonyl-ACC represents another important mechanism for regulation of ethylene biosynthesis.

Induction of ethylene production by auxin

The initial discovery that auxin regulates ethylene production was made by Zimmerman and Wilcoxon (1935), who observed that application of the newly discovered 'heteroauxin' (IAA) to plant shoots promoted epinasty of other untreated plants enclosed in the same container. This observation, plus the ability of auxins and ethylene to cause a number of similar plant responses, suggested that some responses previously attributed to auxin might be

due to ethylene produced in response to the auxin treatment. This idea went largely unnoticed until Morgan and Hall (1964) found the same parallel relationship between ethylene and auxin responses and independently rediscovered the capacity of auxin to promote ethylene synthesis. The common occurrence of auxin induction of ethylene synthesis was soon confirmed in other laboratories, and it was shown to be involved in many plant responses to auxin (Table 5.2). Induction of ethylene production by auxin has come to be recognized as one of the most pervasive plant hormone interactions.

Table 5.2 Plant responses to ethylene[a][b].

Growth inhibition**[c]	Flower initiation**
Growth promotion	Flower sex shifts**
Geotropism modification**	Fruit growth stimulation**
Tissue proliferation**	Fruit degreening
Root and root hair initiation**	Fruit ripening**
Aeronchyma development	Respiratory changes
Leaf epinasty**	Storage product hydrolysis
Leaf movement inhibition**	Latex and gum exudation promotion**
Formative growth	Protein synthesis promotion
Hook formation	RNA synthesis promotion
Chlorophyll destruction	Abscission and dehiscence promotion**
Anthocyanin synthesis promotion	Seed and bud dormancy release
Anthocyanin synthesis inhibition**	Apical dominance release

(a) Adapted from Morgan (1976) which contains detailed references.

(b) Under suitable conditions most of the responses can be induced by auxins as well as ethylene. However, the last three, abscission, seed dormancy and apical dominance are usually affected differently.

(c) ** Indicates early examples where auxin-stimulated ethylene production was implicated in plant responses to auxin, as reviewed by Morgan (1976).

Auxins are by far the most effective stimulators of ethylene production. In vegetative tissues the rate of ethylene production is thought to be regulated by the internal level of free auxin. Accordingly, higher rates of ethylene production are consistently associated with those tissues, such as the growing apex and younger parts of organs, which contain higher amounts of auxin (Abeles, 1973). However, senescing tissue may be an exception to the general rule that high ethylene production corresponds to high auxin content (Roberts and Osborne, 1981).

The mechanism of auxin-induced ethylene biosynthesis has been studied in mungbean and pea seedlings where IAA stimulates ethylene production several hundred-fold. A variety of studies indicate that auxin stimulates ethylene production by enhancing the conversion of SAM to ACC. Induction of enzyme synthesis appears to be the mechanism involved since treatment of the tissue with cycloheximide, a protein synthesis inhibitor, blocked ACC synthase activity and IAA-induced ethylene production (Yoshii and Imaseki, 1981).

Regulation of ethylene production in ripening fruits

It is now recognized that ethylene plays an essential role in the ripening of climacteric fruits. To study the regulation of ethylene production during ripening, Hoffman and Yang (1980) examined the changes in the internal ACC content, as well as the effect of exogenous ACC on ethylene production. Although the ACC content in preclimacteric avocado fruits was very low ($< 0.1 \, \text{nmol g}^{-1}$), a large increase occurred at the onset of rapid ethylene evolution and ripening. The relationship between the change in ACC content and ethylene production rate is illustrated in Fig. 5.3. ACC increased to $45 \, \text{nmol g}^{-1}$ in the later stage of the climacteric rise, then declined to $5 \, \text{nmol g}^{-1}$, and later increased. This second increase in ACC content was not accompanied by a rise in ethylene production; the ability to convert ACC to ethylene was impaired. These data indicate that the failure of preclimacteric fruit tissue to produce ethylene is due to its inability to form ACC. However, since application of ACC resulted in only a slight increase in ethylene production, preclimacteric fruit tissues must also lack the ability to convert ACC to ethylene.

When ethylene is applied to fruits at the preclimacteric stage, the ripening process is triggered and is accompanied by a massive synthesis of ethylene. This capability of tissues to synthesize large quantities of ethylene in response to the application of low concentrations of ethylene is referred to as 'autocatalytic' ethylene production. It is common

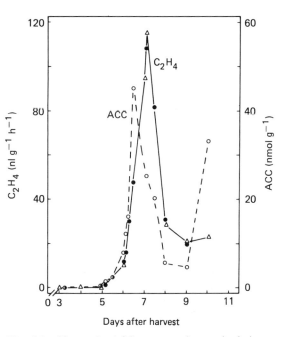

Fig. 5.3 Change in ACC content of avocado fruit at various stages of ripeness. (From Hoffman and Yang, 1980.)

not only of fruits, but also of other senescing organs. In citrus leaf tissue autocatalysis has been observed when the tissue is exposed to ethylene for 24h or longer. This ethylene effect has been found to be due to the enhancement of ACC formation and its conversion to ethylene (Riov and Yang, 1982).

Regulation of stress-induced ethylene production

Many other factors and conditions, in addition to auxin and ethylene, have been found to promote ethylene synthesis. These include cytokinins, ABA, and numerous stresses (Abeles, 1973). The tendency of stress to promote the production of ethylene by plant tissue has proven to be so universal that the term 'stress ethylene' was coined. Various kinds of stress, such as presence of certain chemicals, chilling temperature, drought, flooding, radiation, insect damage, disease or mechanical wounding, promote ethylene production (Abeles, 1973). Stress ethylene is of metabolic origin and is produced by stressed but living cells. Severe stress results in cell death and cessation of ethylene production. The physiological functions of stress ethylene are recognized in a few cases, such as accelerated senescence and abscission (*see* Chapter 20). Thus, stress ethylene appears to act as a 'second-messenger'; it communicates the effect of stress to the plant in a way that allows a response. Stress ethylene may also be involved in wound healing, production of phytoalexins, and increased disease resistance. However, these responses do not appear to be general phenomena.

The biosynthesis of stress ethylene in response to chilling, drought, flooding, and mechanical wounding has been studied in a number of plant tissues. In all cases the methionine-to-ACC pathway is functional. In each case both ethylene production and ACC content are low before stress treatment but increase rapidly following stress. As in auxin-induced ethylene production, the conversion of SAM to ACC is a key reaction controlling the production of stress ethylene. This view is supported by the observation that application of AVG, a potent inhibitor of ACC synthase, eliminates the increase in ACC formation and the production of stress ethylene. Furthermore, pretreatment with cycloheximide, an inhibitor of protein synthesis, eliminates stress-induced ACC accumulation and ethylene production. Apparently, stress induces the synthesis of ACC synthase, which in turn causes accumulation of ACC and the onset of stress-ethylene production.

It has long been recognized that flooding creates anaerobic conditions in the root zone and brings about elevated synthesis of ethylene in the shoots of many plants. Bradford and Yang (1980) studied the regulation of ethylene biosynthesis in waterlogged tomato plants. They found that ACC is synthesized in anaerobic roots in response to flooding and is then transported through the xylem to the shoot where it is converted aerobically to ethylene. Thus, while ethylene itself exerts its physiological effects at or near its site of production, ACC is readily transported and can liberate ethylene some distance from its site of synthesis.

The current understanding of the pathway and regulation of ethylene biosynthesis is summarized in Fig. 5.4. An important conclusion is that ACC synthase, which converts SAM to ACC, is the main site of control. The synthesis of ACC synthase accompanies certain developmental events such as fruit ripening and flower senescence and is induced

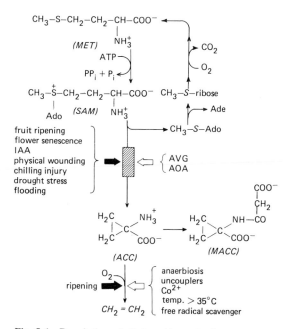

Fig. 5.4   Regulation of ethylene biosynthesis. →, catalytic reaction; ▨, this reaction is normally suppressed and is the rate-limiting step in the pathway; ➤, induction of synthesis of the enzyme; ▷, inhibition of the reaction. Ado, Ade, AVG and AOA stand for adenosine, adenine, aminoethoxyvinylglycine, and aminooxyacetic acid, respectively. (From Yang, 1981.)

by external factors, including the application of auxins or ethylene, physical wounding, chemical injury, root anaerobiosis, drought, and chilling injury. Learning how such a diverse set of stimuli can induce the synthesis of the same key enzyme, ACC synthase, is a challenging problem for future work.

Plant responses to ethylene

Range of responses

The range of plant responses to ethylene is illustrated in Table 5.2 (p. 114). Some of the effects of ethylene are positive such as growth promotion and flower initiation while others appear to be negative such as growth inhibition and abscission. Ethylene influences all phases of plant development, from seed germination to senescence, in all parts of the plant, from roots to shoot tips; hence, ethylene must be viewed as a substance with broad regulatory activity. However, any one process is presumably influenced by the balance of the existing complement of plant hormones. Many of the responses of plants to ethylene can also be produced by auxins and, as explained in the previous section, this is related to the ability of auxin to stimulate ethylene synthesis.

Criteria for hormonal involvement

A correlation between the capacity to make ethylene, even at an accelerated rate, and the occurrence of a symptom that can be induced by exogenous ethylene does not prove that a cause and effect relationship exists. This applies to natural regulatory roles as well as stress situations. Five criteria must be met before ethylene can be assumed to be involved in any response: (1) the kinetics of ethylene synthesis should be consistent with a causal relationship; (2) added ethylene should produce the same effect more rapidly than the inducing agent, if one is involved; (3) saturating levels of exogenous ethylene should reduce, mask, or remove the response to the inducing agent; (4) reduction of the internal concentration of ethylene by hypobaric treatment should reduce or delay the response; and (5) application of ethylene antagonists such as CO_2 and Ag^+ (*see* next section) or ethylene biosynthesis inhibitors such as AVG (*see* previous section), should delay or reduce the response.

Examples of ethylene roles in plant development

Fruit ripening Fruit ripening was the first plant response which was clearly shown to be regulated by ethylene. It was an ideal experimental system because some ripening fruits produce very high levels of ethylene and ethylene will hasten the ripening of a wide variety of fruits. Further, the cavity of many fruits is sufficiently large that an air sample can be removed with a syringe, making it possible to measure internal ethylene levels directly rather than measuring production rates from which internal levels must be estimated. This is important for the

obvious reason that it is the internal ethylene, not the excess which has escaped from the tissue, that must produce the physiological response.

As discussed in the Introduction, the generally-accepted hypothesis that ethylene is a fruit ripening hormone was cast into doubt by the failure of manometric methods to demonstrate that sufficient ethylene was produced prior to ripening actually to induce the response. Gas chromatography increased the sensitivity of the manometric method approximately a million-fold. This allowed Burg and Burg (1962) to demonstrate convincingly that the rise in ethylene production precedes the climacteric rise in respiration and ripening. Internal ethylene levels in several fruits were found always to reach concentrations adequate to stimulate ripening before the climacteric began. Pratt and Goeschl (1968) obtained additional convincing data by monitoring ethylene levels in the seed cavities of honeydew melons during their development. There was a 50- to 100-fold increase prior to the rise in respiration. The rise in ethylene began at least eight days prior to the preclimacteric minimum at which time the internal levels in fruits approached $3 \mu l \, l^{-1}$, a level well in excess of the minimum necessary to hasten and promote the climacteric and ripening. Also, the Burgs (1962) found that ripening was delayed by the ethylene-antagonist CO_2 or hypobaric removal of ethylene from fruit. Further confirmation of ethylene's hormonal role has come from the observed inhibition of both ethylene synthesis and fruit ripening by AVG and the similar patterns of ACC and ethylene production during ripening (Fig. 5.3).

Some questions offer challenges for future studies. The natural agent which delays or prevents ripening of fruit while they are attached to the tree still has not been identified. Also unknown is the agent or condition which initiates ethylene production prior to the climacteric events of ripening.

Early seedling growth The early investigations of effects of ethylene on plants involved the behavior of etiolated pea seedlings. They show a characteristic 'triple' response to ethylene (Fig. 5.1). Elongation growth is inhibited, radial growth is promoted, and orientation of shoots to gravity is disturbed.

Pea seedlings demonstrate several regulatory roles of ethylene. The epicotyl forms a natural hook with the apical bud folded over back toward the shoot; this hook is the site of elevated ethylene production. Light, specifically red light, reduces the rate of production (Goeschl, Pratt and Bonner, 1967). A transient red-light exposure produces a transient decrease in ethylene production and an increase in plumule growth, expressed as an opening of the hook. Exogenous ethylene will reclose the hook even in light. Other diagnostic tests, partial vacuum, Ag^+ or high CO_2 levels, promote hook opening in the dark. Thus, the hook is a growth response regulated by ethylene.

When pea seedling shoots emerge through soil and encounter a barrier, such as a crust, elongation growth slows, radial growth increases, and the shoot grows horizontally. This is the characteristic 'triple' response. Goeschl, Rappaport and Pratt (1966) showed that physical resistance to growth increases ethylene production enough to cause this altered growth habit. This type of plant behavior has survival value since an increase in the shoot diameter increases its lifting capacity thereby facilitating emergence of the seedling. Horizontal growth might bring the shoot tip to a crack in the crust where light or the absence of physical resistance would turn off elevated ethylene production and allow a resumption of vertical growth.

The 'triple' response in pea also demonstrates an effect of ethylene on cell shape (Fig. 5.1). The gas inhibits cell elongation and growth but promotes cell radial growth and stem swelling so that fresh weight is unchanged. Thus, ethylene acts as a regulator of cell shape and seedling behavior rather than strictly as a growth inhibitor.

Abscission Abscission is the separation of organs from the plant; autumn leaf fall is the most familiar example (Chapter 20). The process becomes important in agriculture where abscission or non-abscission of flowers, fruits and leaves influences yields and the efficiency of harvesting operations. Both ethylene's mechanism of action and hormonal role in abscission have been studied extensively.

Since auxin counteracts the promotive action of ethylene in abscission, the initial act of ethylene might be to lower the auxin supply from the leaf to the abscission zone. Ethylene was proposed to act by (1) inhibition of auxin synthesis, (2) promotion of auxin destruction, (3) promotion of auxin conjugation, (4) inhibition of auxin transport, or (5) promotion of auxin binding. Interestingly enough, all of

these processes except the last have been shown by various assays to be influenced by ethylene. IAA levels in leaves of ethylene-treated plants are reduced; thus, ethylene does reduce the auxin supply. The exact *in vivo* mechanism, however, has not been identified.

The ethylene-mediated events in the abscission zone occur as a dissolution of the middle lamella, hydrolysis of the cell walls, and a localized cell enlargement which facilitates separation by providing shearing action. Ethylene has been shown to promote the *de novo* synthesis of cellulase as well as the secretion of this enzyme into the cell wall (Abeles, 1973). The promotion of both cell enlargement and cellulase synthesis by ethylene in the abscission zone is unique to that isolated zone of cells (Osborne and Sargent, 1976). For example, cells surrounding this layer enlarge in response to auxin but not ethylene. Thus, the separation layer cells are target cells for ethylene.

Hall (1952) was the early advocate of the idea that plant leaves make ethylene which results in their own abscission. Almost two decades later ethylene production was shown to increase in tissue adjacent to the abscission zone in bean leaf explants prior to abscission (Jackson and Osborne, 1970). The rates of production were sufficient to develop internal levels that promote abscission. Ethylene production rates and internal levels increased with age in the leafy cotyledons of cotton plants, especially as they

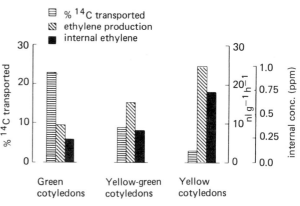

Fig. 5.5 Basipetal auxin transport capacity, ethylene production rate, and internal ethylene concentration of three physiological ages of cotton cotyledons. Natural abscission occurs within about 10 d of visible senescence. (From Beyer and Morgan, 1971.)

became visibly senescent (Beyer and Morgan, 1971) (Fig. 5.5). The rise in ethylene production was inversely correlated with a decline in auxin transport. Similar levels of ethylene applied to non-senescent cotyledons inhibited auxin transport and promoted abscission. Ethylene levels were about 6-fold higher in cotton petioles than in leaf blades. The findings with beans and cotton have been confirmed in other species. The other diagnostic tests for ethylene involvement have also been positive: CO_2, Ag^+, hypobaric treatment, and AVG all delay or prevent abscission. Thus, leaves make ethylene which initiates their abscission. Similar evidence indicates that ethylene is involved in flower bud, flower, and young fruit abscission as well as fruit dehiscense.

Ethylene in normal growth and development

Accumulated evidence indicates that ethylene regulates a variety of processes ranging from release of seed dormancy and early seedling behavior to the events associated with ageing such as leaf abscission and fruit ripening. These changes can be viewed as isolated events during which ethylene production is turned on, the ethylene has its effect, and the synthetic system is subsequently turned off. However, some lines of evidence indicate that ethylene, which is a common part of the hormone complement at all times, may modify development even when its production rate is at a very low, base level. Zobel (1974) found a tomato mutant whose shoots and roots grew horizontally; it also had thin stems without large secondary xylem vessels, hyponastic leaf orientation, and primary roots without lateral branching. The new growth in this mutant could be returned to normal morphology by surprisingly low levels of ethylene (5 nl l^{-1}). Tracheid differentiation in lettuce explants has been shown to require 100 nl l^{-1} or less of ethylene (Zobel and Roberts, 1978). These observations suggest that ethylene participates broadly in all phases of plant behavior, but more work is needed.

Mode of action

Uncovering the mechanism(s) by which ethylene induces the diverse and complex changes discussed above continues to be a major challenge. One of the main obstacles has been the lack of an isolated subcellular system that responds to ethylene in a manner that clearly reflects its *in vivo* action. The lack of such a system has made it necessary to use intact plants or tissue segments where the ethylene response measured usually reveals little about the underlying biochemical mechanism. Nevertheless, several important concepts concerning the mechanism of ethylene action have emerged and suggest that the general features of this mechanism are very similar in all plants. The way in which the initial changes triggered by ethylene are perceived and expressed appears to account for the multiplicity of ethylene responses.

Ethylene antagonists

CO_2 At concentrations of 5 to 10%, CO_2 can rather specifically prevent or delay many ethylene responses (Abeles, 1973). Accordingly, CO_2 has been used as a diagnostic test for ethylene action. CO_2 is not, however, a universal ethylene antagonist. While effective under conditions of low ethylene concentrations, the inhibitory property is lost as the ethylene concentration approaches or exceeds 1 μl l^{-1}. CO_2 probably does not function as a natural ethylene antagonist except when excessive amounts of CO_2 accumulate in the intercellular tissue spaces as in certain fruits. Such accumulation is thought to account for the higher than normal thresholds for ethylene-induced responses in some fruits (Burg and Burg, 1967). CO_2 is used practically in controlled atmosphere storage of fruits where high CO_2 levels help to delay the ripening action of ethylene, making prolonged storage possible (Dilley, 1979). The mechanism by which CO_2 acts is not known, but it has been suggested to be a competitive inhibitor of ethylene action (Burg and Burg, 1965).

Ag^+ In 1976, Beyer reported that a foliar spray of $AgNO_3$ prevented a wide variety of ethylene-induced plant responses including growth inhibition, abscission, senescence, and changes in sex expression. These observations have since been confirmed by others, and today Ag^+ is widely used diagnostically and also commercially (Fig. 5.6). Ag^+ is generally a more potent inhibitor of ethylene action than is CO_2. When applied at the appropriate concentration, the antiethylene effect of Ag^+ is specific, persistent, and non-phytotoxic. No other metal is

$[Ag(S_2O_3)_2]^{3-}$ extends vase-life

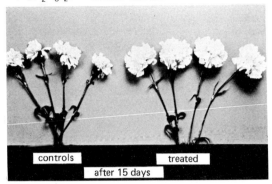

controls   treated

after 15 days

Fig. 5.6 Treatment of cut carnations with Ag^+ as $[Ag(S_2O_3)_2]^{3-}$ to extend vase-life. (Photograph courtesy of H. Veen.)

known to have these properties. A scavenging effect of Ag^+ has been ruled out as a possible mechanism of action. As with CO_2, the exact mechanism by which Ag^+ blocks or reduces ethylene action is unknown.

Dose–response relationships

The dose–response relationships for over 40 ethylene-mediated phenomena indicate that concentrations of 0.01, 0.1 and $10 \mu l \, l^{-1}$ represent threshold, half-maximal, and saturating doses, respectively (Abeles, 1973). In assessing such data, it is important to consider the exposure time and treatment conditions since, if the response is measured too quickly or if the conditions for ethylene action are unfavorable, high threshold values may result. For example Nichols (1968) found that, while $0.2 \mu l \, l^{-1}$ ethylene applied for 6 h had no noticeable effect on carnation flowers, the same concentration induced significant senescence after 24 h. Considering differences in the physiological state of the tissue, internal ethylene concentrations, and treatment conditions, the dose–response relationships observed for most ethylene responses are remarkably similar. This suggests that a common basic mechanism underlies most, if not all, of the physiological processes mediated by the gas.

Structure-activity studies

In a detailed analysis of the structural requirements for ethylene-like action in the pea stem bioassay,

Burg and Burg (1965, 1967) concluded that (a) only unsaturated compounds possess ethylene-like activity, (b) double bonds confer much greater activity than triple bonds, (c) activity requires a terminal carbon atom adjacent to an unsaturated bond, (d) activity is inversely related to molecular size, and (e) substituents, like halogens, which lower the electron density in the unsaturated position decrease activity. Since these requirements for ethylene-like activity closely resemble those for the binding of olefins to metals, the Burgs proposed that the *in vivo* ethylene receptor site contains a metal. Today this idea is generally accepted and seems highly plausible based on well-established principles of organometallic chemistry. Copper (Cu^+) and several other metals have been suggested as likely candidates but direct evidence is still lacking.

Ethylene action model

From their pioneering work, the Burgs (1965, 1967) proposed that ethylene acts by reversible binding to a metal whose receptivity depends on molecular oxygen. The O_2 requirement was deduced from observations that ethylene action was inhibited at reduced oxygen concentrations which had no apparent effect on respiration. A reversible mechanism was considered essential since most plant responses to ethylene are readily reversible, provided ethylene does not induce its own production ('autocatalytic ethylene'). Although based on circumstantial evidence, this model has markedly influenced subsequent studies of ethylene action.

Deuterated ethylene

Abeles, Ruth, Forrence and Leather (1972) and Beyer (1972) first used deuterated ethylene to investigate changes occurring at the ethylene receptor site. It was reasoned that if ethylene forms a reversible metal complex, this might result in the double bond rotation of ethylene and thus provide a convenient means of monitoring receptor activation. To test this idea, *cis*-dideuteroethylene ($C_2H_2D_2$) was applied to pea seedlings and after several days the gas phase was analyzed for the presence of the *trans* isomer. The possibility that carbon-to-hydrogen bond splitting or hydrogen exchange occurs was also explored by applying tetradeuteroethylene (C_2D_4) and assaying the gas phase for C_2HD_3, $C_2H_2D_2$, or

C_2H_3D; no evidence could be found for either of these changes. Also, C_2D_4 and C_2H_4 were found to have identical biological activities indicating that the deuterium atoms did not significantly alter receptor-site activation. This lack of an isotopic effect, plus other considerations, led to the conclusion that bond splitting or metabolism probably does not occur during ethylene action. Evidence to the contrary has been obtained (*see* following section).

Ethylene metabolism

Although several early studies had reported that a small but significant amount of ^{14}C- or ^3H-labeled ethylene was incorporated into both fruit and vegetative tissues, these studies were not convincing since adequate precautions had not been taken to insure radiochemical purity of the ethylene or to prevent microbial contamination (Abeles, 1973). These criticisms, plus the failure to detect any metabolism in some of the early studies, led to the general conclusion that ethylene was metabolically inert in plants.

General features of ethylene metabolism Contrary to the existing opinion, Beyer (1975) was able to demonstrate convincingly, using ultrapure ^{14}C-labeled ethylene under aseptic exposure conditions, that ethylene is metabolized by plants. In dark-grown pea seedlings ethylene was converted to two gaseous metabolites that were identified as CO_2 and ethylene oxide (Beyer, 1980). In addition, ethylene was found to be converted to a number of soluble metabolites including ethylene glycol and its glucose conjugate (Blomstrom and Beyer, 1980). A model for these conversions is presented in Fig. 5.7. Heat killing, homogenization, anaerobiosis, or metabolic inhibitors severely inhibit or completely block metabolism (Beyer and Blomstrom, 1980). Collectively, these results indicate that ethylene metabolism in peas is a natural metabolic function of healthy tissue.

Subsequent studies with other ethylene-sensitive tissues including tomato fruit, carnation and morning glory flowers, cotton abscission zone segments, and *Vicia faba* cotyledons revealed important features of ethylene metabolism (Beyer, 1980; Dodds and Hall, 1982). First, the maximum rate of ethylene metabolism varies greatly in different tissues. *Vicia faba* cotyledons have the highest re-

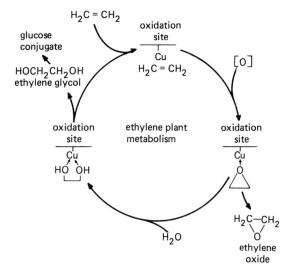

Fig. 5.7 A model for the metabolism of ethylene to ethylene oxide, ethylene glycol and its glucose conjugate.

ported rate. Second, the metabolic system present in all tissues examined appears to be similar in that the primary metabolites are simple oxidation products of ethylene such as CO_2, ethylene oxide, and ethylene glycol. Third, the rate of ethylene metabolism, like natural ethylene production, often changes dramatically during development. Fourth, the rate of conversion of ethylene to CO_2 and ethylene oxide is both tissue and time dependent. This final point is illustrated in Table 5.3. As seen in this table, the major gaseous product formed in peas during the first 6 h of $^{14}C_2H_4$ incubation is $^{14}CO_2$, whereas after 24 h of incubation, it is [^{14}C]ethylene oxide. In *Vicia faba* [^{14}C]ethylene oxide is the only gaseous product detected, whereas in morning glory flowers and cotton abscission zones, only $^{14}CO_2$ appears. The significance of these differences is not known.

Jerie and Hall (1978) first recognized the importance of ethylene oxide as a major product of ethylene metabolism. They used *Vicia faba* cotyledons which for unknown reasons convert ethylene to ethylene oxide at much higher rates than all other tissues examined. Dodds, Musa, Jerie and Hall (1979) isolated a cell-free, particulate fraction from these cotyledons capable of oxidizing ethylene to ethylene oxide, albeit at low rates compared to intact tissues. This system has a requirement for molecular oxygen, a high affinity for ethylene (K_m

Table 5.3   Ethylene conversion to CO_2 and ethylene oxide by various plant tissues.

Tissue	Exposure period[a] (h)	CO_2[b] (%)	Ethylene oxide[b] (%)	Unidentified[b] (%)
Pea epicotyl tips	6	90	10	0
	24	40	60	0
Vicia faba cotyledons	6	0	100	0
	24	0	100	0
Morning glory flowers	6	98	0	2
	24	99	0	1
Cotton abscission zones	6	100	0	0
	24	100	0	0

(a) From Beyer (1980). Tissue exposed in sealed flasks to $5 \mu l \ l^{-1}$ $^{14}C_2H_4$ ($110 \, mCi \, mmol^{-1}$).
(b) Percent of total ^{14}C volatile products recovered at end of 6- or 24-h exposure periods.

of $4.2 \times 10^{-10} M$ or $0.1 \, \mu l \ l^{-1}$ in gas phase), and is inhibited by several enzyme inhibitors. Their suggestion that an oxygenase is involved is supported by the recent finding that ethylene metabolism by *Vicia faba* cotyledons in the presence of $^{18}O_2$ produces labeled ethylene oxide (Beyer, unpublished data).

Physiological role of ethylene metabolism Metabolism might be expected to function as a mechanism for ethylene removal or inactivation as observed with other plant hormones. Based on the ability of ethylene to diffuse freely from most tissues and the relatively small amounts of ethylene metabolized by most tissues, Beyer (1979, 1980) has reasoned that the ethylene-metabolic system most probably serves some other function; namely, that this system is an integral part of the ethylene-action mechanism. The principal support for this hypothesis comes from demonstrated correlations between changes in ethylene metabolism and tissue responsiveness to ethylene. For example, in peas the inhibition of both ethylene action and metabolism by Ag^+ diminished in strikingly similar patterns as the ethylene concentration increased from 0.2 to $20 \, \mu l \ l^{-1}$ (Fig. 5.8). Additional work is needed to confirm this relationship.

Ethylene binding

Sisler (1979) and Bengochea, Dodds, Evans, Jerie, Niepel, Shaari and Hall (1980) independently discovered ethylene binding in plants. The discovery by Bengochea *et al.* arose unexpectedly from attempts to explain why certain plants accumulate ^{14}C-labeled

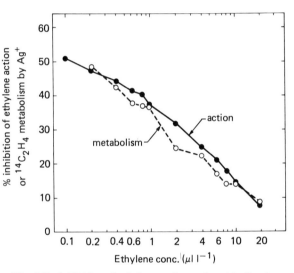

Fig. 5.8   Inhibition of ethylene action and metabolism by Ag^+. (From Beyer, 1979.)

ethylene (Jerie, Shaari and Hall, 1979). This accumulation, called 'compartmentation', resulted in the retention of ethylene in amounts far in excess of that which could be explained simply in terms of solubility. Furthermore, if the tissue were steam-killed, this capacity disappeared. Curiously, once bound, hours were required for the fixed ethylene to be released at ambient temperature; but when heated to 60°C, all of it was immediately lost.

General characteristics of ethylene binding The discovery of ethylene binding probably could not have been made if it were not for the very slow rate at which ethylene dissociates from its binding site. This characteristic allows the 'background', non-bound radioactive gas to be removed from the tissue by ventilation, thus leaving the bound fraction. Binding in isolated cell-free systems parallels binding characteristics demonstrated in vivo—a fact which has greatly facilitated detailed studies. In isolated systems from *Phaseolus vulgaris* and *Phaseolus aurens*, the binding site appears to be associated with the endoplasmic reticulum system or Golgi bodies. Binding is heat-labile and only partially destroyed by proteolytic enzymes suggesting that the site may be buried within a membrane. Several protein-reactive reagents inhibit binding, whereas a variety of cations and the other plant hormones do not.

In those cases studied (Sisler, 1979, 1980; Hall, 1983) the ethylene-binding site has been found to have a high affinity and specificity for ethylene and to exhibit saturation kinetics. Scatchard analyses indicate that a single class of binding sites exists with an apparent K_D of about 0.1 μl l^{-1}. This is the concentration needed to fill one half the sites and is remarkably similar to the half-maximal concentration of ethylene required to induce many ethylene responses. Also, a reasonably good correlation has been observed between the ability of ethylene analogs to compete with ethylene for the binding sites and to mimic plant responses to ethylene (Table 5.4).

Physiological significance of ethylene binding The high affinity and specificity of the binding site for ethylene suggest that it may either regulate the internal status of ethylene or constitute a true hormone-receptor site (Sisler and Goren, 1981; Hall, 1983). The first suggestion cannot be dismissed in the light of the finding that *Phaseolus* cotyledons bind enough ethylene to account for at least 200 h of normal emanation (Jerie *et al.*, 1979). Whether or not ethylene is ever bound or released at a rate sufficient to affect development needs to be resolved, but the occurrence of binding does call into question the validity of simply equating ethylene biosynthesis with the amount of ethylene escaping from the tissue. Thus far, investigators have been unable to establish a clear relationship between ethylene binding and the onset of characteristic changes in the plant. Therefore, the functional nature of the ethylene receptor, as is the case for other putative plant hormone receptors, will remain obscure until this obstacle can be overcome.

Relationship between ethylene binding and metabolism Could the ethylene-binding phenomenon represent a rather stable metal–ethylene complex where oxidation or metabolism occurs when other rate-limiting components become available? Could this site serve multiple functions such as regulating the ethylene status of the plant, or functioning as the true physiological receptor in ethylene action? While answers to such questions must await future investigations, there are some data that suggest that ethylene binding and metabolism may be linked. For example, the kinetic parameters for ethylene

Table 5.4 Relative ethylene-like activity and binding inhibition of structural analogs of ethylene.

Compound	Structure	Relative activity[a]	Relative binding inhibition[b]
Ethylene	$CH_2{=}CH_2$	1	1
Propylene	$CH_3CH{=}CH_2$	100	128
Vinyl chloride	$CH_2{=}CHCl$	1 400	466
Carbon monoxide	CO	2 700	1 068
Acetylene	$CH{\equiv}CH$	2 800	1 013
Vinyl fluoride	$CH_2{=}CHF$	4 300	1 139
Propyne	$CH_3C{\equiv}CH$	8 000	2 651
Vinyl methyl ether	$CH_2{=}CH{-}O{-}CH_3$	100 000	136 196
1-Butene	$CH_3CH_2CH{=}CH_2$	270 000	601 227
Carbon dioxide	CO_2	300 000	141 104

(a) From Burg and Burg (1967). Relative concentration required for half maximal inhibition of pea stem growth.
(b) From Bengochea *et al.* (1980). Extracts from *Phaseolus vulgaris* cotyledons were treated with $^{14}C_2H_4$ at various concentrations in the presence or absence of the analogs. The K_i or inhibitor constant was calculated from Lineweaver–Burk plots of the data and used as a basis for relative binding inhibition.

metabolism in *Vicia faba* and for binding in *Phaseolus vulgaris* are very similar; and a key metabolite in *Vicia* and *Pisum* is ethylene oxide.

Commercial importance

The problems associated with use of ethylene gas to regulate plant responses of commercial and practical value were largely overcome with the discovery of the ethylene-releasing properties of 2-chloroethyl-phosphonic acid ($ClCH_2CH_2PO_3H_2$) (Maynard and Swan, 1963) and its subsequent commercialization under the tradename Ethrel® (de Wilde, 1971). Today this 'liquid-ethylene', referred to generically as ethephon, is the most important and versatile ethylene-release agent sold and is registered for more than 20 crops. Ethrel® is used for improving the quality or promoting ripening of fruits such as tomatoes, apples, coffee berries and grapes; facilitating harvesting of cherries, walnuts and cotton by accelerating abscission or fruit dehiscence; increasing rubber production by prolonging latex flow in rubber trees; increasing sugar production in sugarcane; synchronizing flowering in pineapple; and accelerating senescence of tobacco leaves. It is often desirable to delay plant responses mediated by ethylene; therefore, ventilation, absorption, or destruction of ethylene during the storage or transit of fruits, vegetables and flowers have all been attempted. Today, depending on the commodity, a wide variety of specific preservation techniques are used in commerce to prevent ethylene effects (Dilley, 1978). Examples include the use of controlled atmospheric or hypobaric storage and the treatment of cut carnations with Ag^+ formulations to extend shelf-life (Veen and Van de Geijn, 1978). Future work may lead to a commercially-useful ethylene-synthesis inhibitor, thus increasing the flexibility to delay ethylene effects.

Further reading

Yang, S. F. (1981). Biosynthesis of ethylene and its regulation. In *Recent Advances in the Biochemistry of Fruits and Vegetables*, eds J. Friend and M. J. C. Rhodes, pp. 89–106, Academic Press, London.

Beyer, E. M., Jr. (1981). Ethylene action and metabolism. In *Recent Advances in the Biochemistry of Fruits and*

Vegetables, eds J. Friend and M. J. C. Rhodes, pp. 107–21, Academic Press, London.

Lieberman, M. (1979). Biosynthesis and action of ethylene. *Ann. Rev. Plant Physiol.* **30**, 533–91.

References

Abeles, F. B. (1973). *Ethylene in Plant Biology*, p. 302, Academic Press, New York.

Abeles, F. B., Ruth, J. M., Forrence, L. E. and Leather, G. R. (1972). Mechanisms of hormone action. *Plant Physiol.* **49**, 669–71.

Adams, D. O. and Yang, S. F. (1977). Methionine metabolism in apple tissue: implication of *S*-adenosylmethionine as an intermediate in the conversion of methionine to ethylene. *Plant Physiol.* **60**, 892–6.

Adams, D. O. and Yang, S. F. (1979). Ethylene biosynthesis: identification of 1-aminocyclopropane-1-carboxylic acid as an intermediate in the conversion of methionine to ethylene. *Proc. Natl Acad. Sci. USA* **76**, 170–4.

Amrhein, N., Schneebeck, D., Skorupka, H., Tophof, S. and Stöckigt, J. (1981). Identification of a major metabolite of the ethylene precursor 1-aminocyclopropane-1-carboxylic acid in higher plants. *Naturwissenschaften* **68**, 619–20.

Bengochea, T., Acaster, M. A., Dodds, J. H., Evans, D. E., Jerie, P. H. and Hall, M. A. (1980). Studies on ethylene binding by cell free preparations from cotyledons of *Phaseolus vulgaris* L. *Planta* **148**, 407–11.

Bengochea, T., Dodds, J. H., Evans, D. E., Jerie, P. H., Niepel, B., Shaari, A. R. and Hall, M. A. (1980). Studies on ethylene binding by cell free preparations from cotyledons of *Phaseolus vulgaris* L. *Planta* **148**, 397–406.

Beyer, E. M., Jr. (1972). Mechanism of ethylene action. *Plant Physiol.* **49**, 672–5.

Beyer, E. M., Jr. (1975). ^{14}C-ethylene incorporation and metabolism in pea seedlings. *Nature* **255**, 144–7.

Beyer, E. M., Jr. (1976). A potent inhibitor of ethylene action in plants. *Plant Physiol.* **58**, 268–71.

Beyer, E. M., Jr. (1979). Effect of silver ion, carbon dioxide, and oxygen on ethylene action and metabolism. *Plant Physiol.* **63**, 169–73.

Beyer, E. M., Jr. (1980). Recent advances in ethylene metabolism. In *Aspects and Prospects of Plant Growth Regulators*, DPGRG/BPGRG, Monogr. **6**, 27–38.

Beyer, E. M., Jr. and Blomstrom, D. C. (1980). Ethylene metabolism and its possible physiological role in plants. In *Proc. Tenth Int. Conf. Plant Growth Subs. (1979)*, ed. F. Skoog, pp. 208–18, Springer-Verlag, Berlin.

Beyer, E. M., Jr. and Morgan, P. W. (1971). Abscission: the role of ethylene modification of auxin transport. *Plant Physiol.* **48**, 208–12.

Biale, J. B. (1940). Effect of emanations from several species of fungi on respiration and color development of citrus fruits. *Science* **91**, 458–9.

Biale, J. B. and Young, R. E. (1981). Respiration and ripening of fruits—retrospect and prospect. In *Recent Advances in the Biochemistry of Fruits and Vegetables*, eds J. Friend and M. J. C. Rhodes, pp. 1–39, Academic Press, London.

Biale, J. B., Young, R. E. and Olmstead, A. J. (1954). Fruit respiration and ethylene production. *Plant Physiol.* **29**, 168–74.

Blomstrom, D. C. and Beyer, E. M., Jr. (1980). Plants metabolise ethylene to ethylene glycol. *Nature* **283**, 66–8.

Boller, T., Herner, R. C. and Kende, H. (1979). Assay for and enzymatic formation of an ethylene precursor, 1-aminocyclopropane-1-carboxylic acid. *Planta* **145**, 293–303.

Bradford, K. J. and Yang, S. F. (1980). Xylem transport of 1-aminocyclopropane-1-carboxylic acid, an ethylene precursor, in waterlogged tomato plants. *Plant Physiol.* **65**, 322–6.

Burg, S. P. (1962). The physiology of ethylene formation. *Ann. Rev. Plant. Physiol.* **13**, 265–302.

Burg, S. P. and Burg, E. A. (1962). Role of ethylene in fruit ripening. *Plant Physiol.* **37**, 179–89.

Burg, S. P. and Burg, E. A. (1965). Ethylene action and the ripening of fruits. *Science* **148**, 1190–6.

Burg, S. P. and Burg, E. A. (1967). Molecular requirements for the biological activity of ethylene. *Plant Physiol.* **42**, 144–52.

Burg, S. P. and Stolwijk, J. A. A. (1959). A highly sensitive katharometer and its application to the measurement of ethylene and other gases of biological importance. *J. Biochem. Microbiol. Technol. Eng.* **1**, 245–59.

Burg, S. P. and Thimann, K. V. (1959). The physiology of ethylene formation in apples. *Proc. Natn. Acad. Sci. USA* **45**, 335–44.

Burroughs, L. F. (1957). 1-Aminocyclopropane-1-carboxylic acid: a new amino acid in perry pears and cider apples. *Nature* **179**, 360–1.

Crocker, W., Hitchcock, A. E. and Zimmerman, P. W. (1935). Similarities in the effects of ethylene and the plant auxins. *Contrib. Boyce Thompson Inst.* **7**, 231–48.

de Wilde, R. C. (1971). Practical applications of (2-chloroethyl) phosphonic acid in agricultural production. *HortSci.* **6**, 364–70.

Dilley, D. R. (1978). Approaches to maintenance of postharvest integrity. *J. Food Biochem.* **2**, 235–42.

Dodds, J. H. and Hall, M. A. (1982). Metabolism of ethylene by plants. *Int. Rev. Cytol.* **76**, 299–325.

Dodds, J. H., Musa, S. K., Jerie, P. H. and Hall, M. A. (1979). Metabolism of ethylene to ethylene oxide by cell-free preparations from *Vicia faba* L. *Plant Sci. Lett.* **17**, 109–14.

Gane, R. (1934). Production of ethylene by some ripening fruits. *Nature* **134**, 1008.

Goeschl, J. D., Pratt, H. K. and Bonner, B. A. (1967). An effect of light on the production of ethylene and the growth of the plumular portion of etiolated pea seedlings. *Plant Physiol.* **42**, 1077–88.

Goeschl, J. D., Rappaport, L. and Pratt, H. K. (1966). Ethylene as a factor regulating the growth of pea epicotyls subjected to physical stress. *Plant Physiol.* **41**, 877–84.

Hall, M. A. (1983). Ethylene receptors. In *Receptors in Plants and Cellular Slime Moulds*, eds C. M. Chadwick and D. R. Garrod, Marcel Dekker, New York (in press).

Hall, W. C. (1952). Evidence on the auxin–ethylene balance hypothesis of foliar abscission. *Bot. Gaz.* **113**, 310–22.

Hansen, E. (1943). Relation of ethylene production to respiration and ripening of premature pears. *Proc. Am. Soc. Hort. Sci.* **43**, 69–72.

Hoffman, N. E. and Yang, S. F. (1980). Changes of 1-aminocyclopropane-1-carboxylic acid content in ripening fruits in relation to their ethylene production rates. *J. Am. Soc. Hort. Sci.* **105**, 492–5.

Hoffman, N. E., Yang, S. F. and McKeon, T. (1982). Identification of 1-(malonylamino) cyclopropane-1-carboxylic acid as a major conjugate of 1-aminocyclopropane-1-carboxylic acid, an ethylene precursor in higher plants. *Biochem. Biophys. Res. Commun.* **104**, 765–70.

Jackson, M. B. and Osborne, D. J. (1970). Ethylene, the natural regulator of leaf abscission. *Nature* **225**, 1019–22.

Jerie, P. H. and Hall, M. A. (1978). The identification of ethylene oxide as a major metabolite of ethylene in *Vicia faba* L. *Proc. R. Soc. Lond.* B, **200**, 87–94.

Jerie, P. H., Shaari, A. R. and Hall, M. A. (1979). The compartmentation of ethylene in developing cotyledons of *Phaseolus vulgaris* L. *Planta* **144**, 503–7.

Kidd, F. and West, C. (1945). Respiratory activity and duration of life of apples gathered at different stages of development and subsequently maintained at a constant temperature. *Plant Physiol.* **20**, 467–504.

Lieberman, M. and Mapson, L. W. (1964). Genesis and biogenesis of ethylene. *Nature* **204**, 343–5.

Lieberman, M., Kunishi, A., Mapson, L. W. and Wardale, D. A. (1966). Stimulation of ethylene production in apple tissue slices by methionine. *Plant Physiol.* **41**, 376–82.

Maynard, J. A. and Swan, J. M. (1963). Organophosphorous compounds. *Aust. J. Chem.* **16**, 596–608.

Miller, E. V., Winston, J. R. and Fisher, D. F. (1940). Production of epinasty by emanations from normal and decaying citrus fruits and from *Penicillium digitatum*. *J. Agric. Res.* **60**, 269–77.

Morgan, P. W. (1976). Effects on ethylene physiology. In *Herbicides: Physiology, Biochemistry, Ecology*, 2nd edn. ed. L. J. Audus, Academic Press, New York and London, vol. **1**, 255–80.

Morgan, P. W. and Hall, W. C. (1964). Accelerated release of ethylene by cotton following application of indolyl-3-acetic acid. *Nature* **201**, 99.

Neljubow, D. (1901). Über die horizontale Nutation der Stengel von *Pisum sativum* und einiger Anderer. *Pflanzen. Beih. Bot. Zentralbl.* **10**, 128–38.

Nichols, R. (1968). Response of carnations (*Dianthus caryophyllus*) to ethylene. *J. Hort. Sci.* **43**, 335–49.

Osborne, D. J. and Sargent, J. A. (1976). The positional differentiation of abscission zones during the development of leaves of *Sambucus nigra* and the response of the cells to auxin and ethylene. *Planta* **132**, 197–204.

Pratt, H. K. and Goeschl, J. D. (1968). The role of ethylene in fruit ripening. In *Biochemistry and Physiology of Plant Growth Substances*, eds F. Wightman and G. Setterfield, The Runge Press, Ottawa, Canada, pp. 1295–302.

Riov, J. and Yang, S. F. (1982). Effects of exogenous ethylene on ethylene production in citrus leaf tissue. *Plant Physiol.* **70**, 136–41.

Roberts, J. A. and Osborne, D. J. (1981). Auxin and the control of ethylene production during the development and senescence of leaves and fruits. *J. Exp. Bot.* **32**, 875–87.

Sisler, E. C. (1979). Measurement of ethylene binding in plant tissue. *Plant Physiol.* **64**, 538–42.

Sisler, E. C. (1980). Partial purification of an ethylene-binding component from plant tissue. *Plant Physiol.* **66**, 404–6.

Sisler, E. C. and Goren, R. (1981). Ethylene binding—the basis for hormone action in plants? *What's New in Plant Physiol.* **12**, 37–40.

Veen, H. and Van de Geijn, S. C. (1978). Mobility and ionic form of silver as related to longevity of cut carnations. *Planta* **140**, 93–6.

Wang, S. Y., Adams, D. O. and Lieberman, M. (1982). Recycling of 5′-methylthioadenosine-ribose carbon atoms into methionine in tomato tissue in relation to ethylene production. *Plant Physiol.* **70**, 117–21.

Yoshii, H. and Imaseki, H. (1981). Biosynthesis of auxin induced ethylene: effects of indole-3-acetic acid, benzyladenine and abscisic acid on endogenous levels of 1-aminocyclopropane-1-carboxylic acid (ACC) and ACC synthase. *Plant Cell Physiol.* **22**, 369–79.

Yu, Y. B., Adams, D. O. and Yang, S. F. (1979). 1-Aminocyclopropanecarboxylate synthase, a key enzyme in ethylene biosynthesis. *Arch. Biochem. Biophys.* **198**, 280–6.

Yung, K. H., Yang, S. F. and Schlenk, F. (1982). Methionine synthesis from 5-methylthioribose in apple tissue. *Biochem. Biophys. Res. Commun.* **104**, 771–7.

Zimmerman, P. W. and Wilcoxon, F. (1935). Several chemical growth substances which cause initiation of roots and other responses in plants. *Contrib. Boyce Thompson Inst.* **7**, 209–29.

Zobel, R. W. (1974). Control of morphogenesis in the ethylene-requiring tomato mutant diageotropica. *Can. J. Bot.* **52**, 735–41.

Zobel, R. W. and Roberts, L. W. (1978). Effects of low concentrations of ethylene on cell division and cytodifferentiation in lettuce pith explants. *Can. J. Bot.* **56**, 987–90.

Apical dominance

6

J R Hillman

Introduction

Terrestrial multicellular plants typically have a sessile mode of life and have evolved strategies to exploit and adapt to their changing environment. They are continuously developing organisms and their shape is a product of the differential activity of regions in the apices where the cells remain in an embryonic condition—the meristems. The non-determinate growth habit of most perennial plants and vegetative monocarpic plants reflects an interplay between gene expression in the cells derived from the meristem and responses to the environment.

Plants are not apparently a weakly controlled complex of individual meristems. Each apex influences the development and positioning of lateral structures derived from the same or different apices. Thus, the constituent parts of a plant are inter-related, not only by virtue of the fact that the cytoplasm of adjacent cells is often continuous *via* intercellular connexions (plasmodesmata), but by the possession of physiological processes to correlate or integrate centers of cell growth and differentiation. The study of growth correlations has been dismembered somewhat into several areas of investigation and this has tended to mask the essential unity of the various correlative phenomena in the whole plant. Modern plant physiology now emphasizes the spatial and temporal organization of growth and development, analyzing the role of positional signalling of various types, and the nature of positional information.

The phenomenon of apical dominance is broad-ranging, encompassing at least four themes:

(i) Complete or partial inhibition of development of lateral (usually axillary, but occasionally adventitious) buds by an actively-growing apical region on the same or a different shoot axis. The apex may also control the initiation of buds. Domination of the apical region over the lateral buds is termed 'correlative inhibition'. It is seen especially in etiolated shoots but also in leafy and modified shoots (tubers, rhizomes, stolons, bulbs, corms etc).
(ii) The suppressive influence of one dominant shoot upon one or more subordinate shoots.
(iii) Influence of the apex on the development and positioning of leaves, axillary shoots, stolons, tubers, rhizomes and roots. An inflorescence apex can modify the positioning and development of the floral and derived structures.
(iv) Influence of the apex on the transport of nutrients and cellular differentiation in the stem or root axis.

The correlative inhibition of lateral buds by the apex has been investigated experimentally for over a century and an awareness of the dominating role of the apex has undoubtedly, if unwittingly, influenced gardening, horticultural and pruning practices. There is good evidence that the phenomenon is widespread throughout all those members of the plant kingdom that bear apices and lateral buds, although the controlling mechanisms may not be the same.

Compared with the seedlings of herbaceous dicotyledonous species, relatively little attention has been paid to apical dominance in woody perennial species, monocotyledonous plants, lower plants, roots and inflorescences. This state of affairs is not an oversight by plant physiologists, merely a reflexion of the complexity of the phenomenon. Apical dominance should be considered in association with other developmental phenomena in the whole plant, particularly bud dormancy; senescence of leaves, fruits and apices; phase change and flowering; environmental effects; tropisms; epinasty and hyponasty.

This chapter is concerned mainly with the physiology of correlative inhibition in the shoots of herbaceous dicotyledonous plants, especially leguminous seedlings. The text references cited are either review articles or representative research papers. It is not the intention to assume that interpretations of experiments performed on one species or even cultivar can be applied *a priori* to another in the search for a common mechanism. Nevertheless, common themes are highlighted together with the main theories and possible areas of further study.

Correlative inhibition

Following the formation of the embryo, the apical regions influence the development of lateral structures and this continues throughout the life of the plant. The apex is a self-determining region controlling differentiation. It affects those processes that culminate in the appearance of lateral bud meristems which usually arise in the axils of the leaf primordia, and also influences, along with other internal and environmental factors, subsequent bud development.

The studies of Wardlaw (1952) on ferns have shown that microsurgical techniques can modify the positioning of primordia on the apical dome. There appear to be transportable chemical 'evocators' which determine the site of primordial initiation. Bud initiation in aseptic tissue cultures of various species including *Nicotiana, Daucus* and *Phaseolus* can be controlled by a balance in levels of indole-3-acetic acid, cytokinins and nutrients. This indicates that a complex of various components could be the correlative determinant for bud initiation rather than the single bud-inducing factor envisaged by the distinguished nineteenth-century plant physiologists.

Normally, the apical meristem together with the recently-formed stem and leaf tissues above the uppermost unfolded leaf constitute the 'apical' (terminal) bud. Removal of this bud or the shoot above the lateral bud of interest is termed 'decapitation'. A meristem is also found in the axils of the leaf primordia and gives rise to a lateral (axillary) bud analogous and apparently similar to the apical bud.

After initiation, lateral-bud development in the intact plant may be curtailed, depending on the variety and position on the axis, and conditions of cultivation. In some species, inhibited or quiescent buds may be fairly rudimentary whereas in others the buds may be well-formed with readily discernible leaves, an internode and even floral primordia. In most species, the inhibited lateral buds are morphologically and anatomically heterogeneous. They contain cells of several types in differing phases of differentiation and development. Reactivation of the bud occurs when correlative inhibition is removed.

The term 'dormancy' is not usually applied nowadays to buds held under correlative inhibition. Detailed studies show that correlatively inhibited buds do show barely perceptible growth and in contrast to buds which show winter dormancy, more rapid growth commences on removal of the shoot apex.

Correlative inhibition is patently an excellent experimental model for the study of spatial organization of developmental activities in the plant (Phillips, 1975). The possession of quiescent lateral buds may be regarded as an important survival adaptation for the supply of replacement apices. However, the significance of studies on apical control extend beyond fundamental aspects of morphogenesis, correlative signals and ecological adaptations, to encompass agricultural and horticultural practice. Considerable expense is incurred in the control of lateral growth in certain crop and ornamental plants. In species such as *Lycopersicon esculentum*, a strong leader habit (lateral shoot growth suppressed or eliminated) is encouraged. On the other hand, lateral growth is often desirable in cereals, where the laterals are termed 'tillers', and a 'bushy' growth habit is frequently required in many dicotyledonous fruit-bearing crops as well as plantation crops such as *Hevea brasiliensis*. Thus the position and outgrowth of lateral buds in effect determine the shape of the plant and ultimately its productivity. Factors which lead to axillary outgrowth on roots and shoots of weeds will also assist in the uptake of herbicides.

Historical perspectives and development of theories on apical dominance

Goebel (1880) first introduced the term 'growth correlations' to describe mutual, growth-modifying influences between stems, buds and leaves, influences which determine the properties of these parts within the intact plant. Investigators around the beginning of this century usually interpreted correlative phenomena in terms of nutrient availability and regarded competition for nutrients between the shoot apex and lateral structures as the determinant (see Jost, 1907). Apical meristems, when active, would consume all available nutrients or growth factors, thereby depriving the later-formed lateral meristems and associated primordial organs. The lateral buds would then remain quiescent and the apical meristematic areas would act as 'metabolic sinks' to which nutrients would be transported. This theory is referred to as the 'nutritive' theory (Phillips, 1969).

Evidence in support of the theory came from Goebel (1900), Loeb (1915, 1917a, 1918, 1923) and Dostál (1926) who noted that the removal of organs led to 'compensation' growth in the remaining parts of the plants. Confirmation of compensatory growth effects in apical dominance (Jacobs and Bullwinkel, 1953) and the undeniable effects of the nutritive state of the plant on bud growth point towards nutrition as a factor in the mechanism of apical dominance.

Von Sachs, the formidably intellectual and industrious Professor of Botany in the University of Würzburg, reasoned that root formation in plants results from the downward travel of root-forming substances (von Sachs, 1882). At that time, the transmission of stimuli involved in plant movements occupied the attention of several eminent plant physiologists and Pfeffer (see Vines, 1886) suggested that plasmodesmata may act as channels for the conduction of stimulating, possibly chemical, influences. Jost (1913), without experimental evidence, considered that these internal stimuli could indeed be chemical. This view was reinforced by the observation of Haberlandt (1913) that wound-induced cork formation is dependent on the formation of a thermostable 'wound hormone'. The word 'hormone' in the context of plant physiology was first used by Fitting (1909, 1910) in his researches on orchids.

The early experiments on apical dominance in *Phaseolus* (McCallum 1905a, b; Child and Bellamy, 1920; Harvey, 1920) and *Faba* (also known as *Vicia*) (Mogk, 1913), in conjunction with the suggestions of Errera (1904), Dostál (1909) and Loeb (1917b), led to the general view that inhibitory substances might be responsible for correlative inhibition. Snow (1925) discovered that the influence inhibitory to lateral-bud growth could pass through a zone of stem internode from which all tissues outside the xylem had been removed. He further showed the passage of inhibition through pith alone and xylem alone, and also confirmed the result of Harvey (1920) that living tissue is required for the transmission of inhibition. Somewhat surprisingly, he claimed to have shown the movement of inhibition across a moist protoplasmic discontinuity (water-filled gap).

The conclusion that a diffusible inhibitor was involved in correlative inhibition was seemingly justified by the discovery of the effects of auxin (indole-3-acetic-acid, IAA), then known as 'growth substance' or 'heteroauxin'. Thimann and Skoog (1933, 1934), in their classic and pioneer studies, found that auxin (detected by the *Avena* curvature bioassay) from excised shoot apices of *Faba* could be collected in agar. Similar amounts of auxin isolated from cultures of the fungus *Rhizopus*, applied to the cut-stem surfaces of decapitated plants, could partially replace the shoot tip with respect to the inhibition of bud growth. This result agreed with the observations of Laibach (1933) and Müller (1935) who used extracts of orchid pollinia and urine. Thimann and Skoog also noted that auxin was produced by rapidly growing buds and developing leaves, but not by inhibited buds or older organs. Bud outgrowth was associated with the capacity of buds to synthesize their own supply of auxin. In addition, pure heteroauxin (IAA) could inhibit bud outgrowth on decapitated plants with the amount of bud growth dependent on auxin concentration (Skoog and Thimann, 1934). From these distinguished scientists came the suggestion that auxin synthesized in the shoot apex reaches the lateral buds and inhibits the localized production of IAA.

Observing that bud inhibition by IAA was not accompanied by compensation growth elsewhere in the plant, and considering the differential effects of IAA on roots, buds and stems, Thimann (1937) proposed that the responses of these parts of the plant show dose–response curves with different optima, such that auxin concentrations that were optimal for stem growth would be supraoptimal for bud growth. This reasoning was supported by Skoog (1939) and gave rise to the 'direct' theory of auxin action in correlative inhibition. Were the theory valid then the following statements could be confirmed experimentally:

(a) auxin from the tissues that are the source of inhibition enters the bud;
(b) the transport of auxin exactly parallels that of the correlative inhibitor;
(c) direct applications of auxin to buds inhibit bud growth;
(d) the kinetics of bud growth inversely reflect the amount of IAA entering the bud;
(e) specific inhibitors of auxin biosynthesis, action and distribution affect the pattern of bud growth.

To date, none of these points has been demonstrated satisfactorily and (a), (c), and (d) have been refuted.

In a long series of papers noted for their perceptive arguments but unsound experimental design, Snow developed what has become known as the 'correlative inhibitor' or 'indirect' theory (see Snow, 1940). The gist of his theory is that, as a result of auxin action, perhaps by process related to growth, an inhibitor is formed which moves into lateral buds. Although the inhibitor is present in both stem and lateral buds, its effect is manifest only in buds because the stem is protected by basipetally moving auxin. There is little evidence to support this notion, for Snow's experiments were largely designed to disprove others. He was a virulent antagonist of the 'direct' theory, attempting to prove, without the benefit of proper controls and radioactively-labeled compounds, that the inhibitor could travel where auxin apparently could not.

Went (1936) suggested a modification of the 'nutritive' theory, the 'nutrient-diversion' theory, in which endogenous or exogenous auxin creates in some way a flow of growth factors towards the point of auxin production (presumably the active apex) or application, such that the lateral buds become starved in a manner akin to that envisaged in the 'nutritive' theory. Went (1938, 1939) obtained evidence that root-produced factors were required for growth and that factors promotory to bud growth

accumulate in IAA-treated internodes. Thus the 'nutrient-diversion' theory refers to the diversion of root-produced bud-growth factors from the lateral buds in response to a stimulus from apically synthesized auxin, a theory which provided a foundation for further hypotheses.

One such hypothesis blending elements of the 'direct' and 'nutrient-diversion' theories, is the 'vascular-connection' theory where auxin and/or the correlative inhibitor prevents the entry of factors into the lateral buds by an effect on the vascular connexions between bud and stem (Overbeek, 1938), perhaps by an effect on vascular differentiation (Sorokin and Thimann, 1964). Recent analyses of bud growth, however, indicate that a correlation does not always exist between vascular development and the loss of the quiescent state.

With the identification of gibberellins, cytokinins, abscisic acid and ethylene as endogenous plant growth substances and the frequent demonstrations of their effects on bud growth, the possibility that a balance of hormonal factors controling the inhibition and stimulation of bud development, possibly acting in sequence, cannot be discounted. Yet evidence in favor of this complicated 'hormonal-balance' theory is fragmentary, despite considerable interest in the rôle of auxin and the cytokinins.

A derivation of the 'nutrient-diversion' theory arises from the demonstration of the long-distance diversion of nutrients and growth factors by apices, exogenous auxins and other growth substances (Phillips, 1975; Patrick, 1982). Hormone-directed metabolite transport could help explain the correlative mechanisms, and there is evidence in support of the following points which are essential to the acceptance of the process:

(a) active meristems or associated tissues are sites of auxin synthesis or release, possibly indicated by elevated auxin levels or metabolism;
(b) auxin induces long-distance transport of metabolites;
(c) nutrients and growth factors accumulate in active apices.

Phillips (1975) highlighted the enigmatic nature of the mechanism of hormone-directed transport and queried its significance in correlative inhibition. There are no definitive experiments yet published which distinguish unequivocally between an effect of auxin and the other growth substances on local metabolite demand (formation of a metabolic sink) and a direct effect on one or more components of phloem transport. Such experiments are confounded by the vagaries of using exogenous compounds, inhibitors of questionable specificity, and inadequately sensitive techniques of gauging metabolic-sink activity. Interpretation, in any case, must also be speculative while phloem physiology and endogenous hormonal relationships remain problematical areas of study.

Patterns of bud growth

Using intact seedlings of *Phaseolus vulgaris* cv. Canadian Wonder (Fig. 6.1) as an example it is apparent that an exquisite degree of correlative control exists in the sequential development of leaves, internodes and lateral buds (Fig. 6.2). Their growth patterns exhibit the sigmoid curve typical of biological systems, with the leaf or internode entering its grand or rapid phase of growth just as the preceding leaf or internode completes its rapid growth phase.

Bud growth in *Phaseolus* can be measured non-destructively by determining internodal or total bud length rather than destructively by assaying fresh or dry weight. The larger of the axillary buds in the axils of the opposite primary leaves shows little growth whereas the buds in the axils of the later-formed trifoliate leaves show slow but continuous growth. In common with most species, the rate and duration of *Phaseolus* bud growth varies markedly between seed batches, the time of year, position on

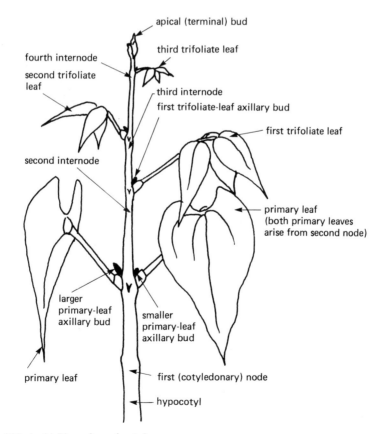

Fig. 6.1 Shoot of 25-d-old *Phaseolus vulgaris* L.

the stem, cultivation conditions and plant age. Transformation from the monopodial to sympodial growth habit in many plants occurs at the time of flower initiation or the attainment of maturity.

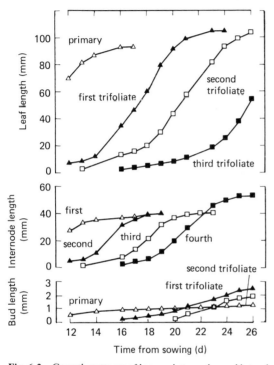

Fig. 6.2   Growth patterns of leaves, internodes and lateral buds in intact plants of *Phaseolus vulgaris*. Growth of leaves was determined by measuring the length of the laminae (central laminae in the case of trifoliate leaves). Bud growth was determined by measuring the first internode length of the bud. (Adapted from Yeang, 1980.)

Removal of dominance

Excision of the apical portion of the stem is a simple method to eliminate correlative inhibition. If, for example, seedlings are decapitated at the third internode then vigorous lateral bud growth ensues (Fig. 6.3). Usually the greatest growth is shown by the highest bud but the presence of lower growing buds retards its outgrowth. There are at least fourteen ways in which the growth of inhibited buds can be promoted (Table 6.1) but it is not known if they share the same mechanism of bud activation.

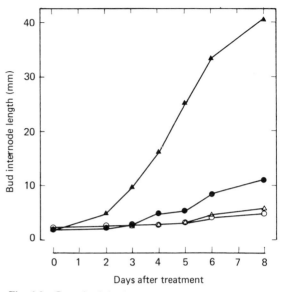

Fig. 6.3   Growth of the first-trifoliate-leaf lateral bud and the larger lateral bud in the axils of the primary leaves of intact and decapitated *Phaseolus vulgaris*. Bud growth was determined by measuring the first internode length of the bud. ($\triangle$) first-trifoliate-leaf lateral bud, intact plant; ($\blacktriangle$) first-trifoliate-leaf lateral bud, decapitated plant; ($\bigcirc$) primary-leaf lateral bud, intact plant; ($\bullet$) primary-leaf lateral bud, decapitated plant. (Adapted from Yeang, 1980.)

Short-term changes in bud growth and metabolism

The time lag between removal of the apex and subsequent detection of growth changes is crucial to an understanding of the mechanism of correlative inhibition. Inhibited buds of *Phaseolus* grow, albeit slowly, at $2 \mu m \ h^{-1}$ in the case of primary-leaf axillary buds. Thus, any changes occurring on removal of correlative inhibition must be carefully compared with the normal events in the inhibited bud. It is possible to measure changes in many parameters which indicate growth, e.g. gross extension, cell expansion, cell division, cytological changes, biochemical changes in the levels or activities of selected metabolites and enzymes etc. Great technological problems accompany each of these approaches, especially as the limited size and structural complexity of the bud, coupled with expansion of bud leaves, may obscure any small but significant differences.

Table 6.1 Summary of treatments which promote growth of lateral buds held under correlative inhibition.

(1) Excision of apical portion of stem (decapitation).
(2) Removal of young developing leaves.
(3) Physical restriction of apical growth.
(4) Isolation of the apical position of the stem by disease, bark-ringing, or steam-girdling.
(5) Infection by pathogens causing witches'-broom (Hexenbesen).
(6) Elevated carbon dioxide levels.
(7) Quantity and spectral quality of light.
(8) Humidity.
(9) Water and nutrient supply to root system.
(10) Gravimorphic treatments.
(11) Induction of the reproductive condition.
(12) Application of auxin-transport inhibitors, abscisic acid, ethephon, ethylene or May & Baker 25–105 to tissues above bud.
(13) Application of chemical pruning agents to shoot.
(14) Application of indole-3-acetic acid or cytokinins to bud.

Considerable variation has been noted in the length of time between removal of the shoot apex and the onset of measurable growth. In *Pisum*, a lag of 6 to 10h has been recorded by Wardlaw and Mortimer (1970). In the same species, Nagao and Rubinstein (1976) found that active growth of the young leaves in the axillary bud preceded extension growth of the bud internode by about four hours following removal of the shoot apex; they calculated the time lags for the two events to be approximately 8 and 12h, respectively. Couot-Gastelier (1978) using *Faba vulgaris* (*Vicia faba*) reported that following shoot decapitation, leaf growth in the bud could be detected after four hours whereas active growth in the bud internode had a lag period of eight hours. Couot-Gastelier (1979) also noted that one of the first events following decapitation of *Faba* was mitotic activation in the cambium of the stem subapex after one hour.

Working with *Phaseolus*, Hall and Hillman (1975) observed enhanced outgrowth of the entire larger primary-leaf axillary bud within 30min of decapitation. These measurements included both the bud internode and partially developed leaves. Measurements of internodal growth of the smaller first-trifoliate-leaf bud in the same variety (Yeang and Hillman, 1981a) showed that accelerated growth

following decapitation could be detected after a lag period of three to five hours. Anatomical analyses by Yeang and Hillman revealed that cell extension could account for internodal extension for one day after decapitation; presumably this applies to leaf expansion too. As bud development proceeds, other factors such as cell division are involved. Thus bud growth may be regarded in at least two phases—an initial release from dominance involving cell expansion, followed by the rapid establishment of the lateral shoot. In *Faba vulgaris*, in the period 12 to 48h following decapitation, there was an increase in epidermal cell numbers in the bud internode (Couot-Gastelier, 1978) which again indicates the part played by cell division and differentiation at a later stage in bud outgrowth.

Short-term changes in bud metabolism after the release from apical control offer exciting potential for research into the control of correlative inhibition. Perhaps a specific group of cells is activated which in turn switches on metabolic activity elsewhere in the bud. It may be argued that longer-term studies monitoring bud growth after the switching-on of active growth may tell us little about correlative inhibition itself but may be of relevance to correlative control of subordinate shoots. Likewise, cytological and biochemical changes that can be detected after the release from inhibition may be the result rather than the cause of bud activation.

Correlative inhibition is associated with a complete or partial retardation of mitotic activity in the apical meristem of the lateral bud. Increased cell divisions can be detected after removal of the shoot apex (Rubinstein and Nagao, 1976). In *Glycine max*, an increase in cell division was noted from about 25h after shoot decapitation or complete removal of the lateral bud from the plant (Ali and Fletcher, 1970). In *Pisum*, enhanced mitotic division was noted within 12h of decapitation, shortly after a discernible increase in lateral-bud length (Nagao and Rubinstein, 1976).

Some studies, principally with *Pisum* and *Tradescantia*, have been made of nuclear development in the apical meristems of lateral buds during correlative inhibition and its release. In *Pisum* the majority of nuclei in the inhibited bud are found in the presynthetic phase of the cell cycle (G_1) with a 2C DNA (*i.e.*, diploid DNA content) level. These cells have a low rate of DNA synthesis and cell division. Release from dominance is associated with mitosis of a few G_2 nuclei that are already present, and entry into the S phase of nuclei previously in G_1 (Nougarède and Rondet, 1975). Microdensitometric studies reveal that the lysine-rich fraction of the total histone present in these nuclei increases after release, while the argenine fraction decreases with increasing mitotic activity (Tobin, Yun and Naylor, 1974; Nougarède and Rondet, 1977).

A rapid effect of release from correlative inhibition has been noted in *Cicer,* where cell division in the bud meristem was detected one hour after treatment (Guern and Usciati, 1972). Such a rapid effect led Guern and Usciati to postulate that DNA synthesis in inhibited nuclei is blocked at the G_2 phase.

Dwividi and Naylor (1968) using *Tradescantia* noted accelerated cell division in leaf primordia and the activation of an ovoid group of cells at the center of the bud apex beneath a single layer of cells. This group of cells possessed limited genetic activity in the inhibited state but following decapitation their DNA content doubled, they underwent cell division and exhibited a decreased DNA:histone ratio.

One structural change which may be linked to the release from dominance is an increase in volume and in pyroninophily of the nucleolus which at the G_1 phase becomes irregularly coated with granular material (Nougarède, 1977a).

The 2C nuclei of inhibited buds contain dispersed chromatin which becomes condensed during the release from dominance. In *Tradescantia*, this condensation is observed before the first mitosis but only after histones have been synthesized (Dwividi and Naylor, 1968; Booker and Dwividi, 1973).

A relationship may exist in lateral buds between the predominance of the argenine-rich fraction and the low condensation of the chromatin of the 2C inhibited nuclei (Booker and Dwividi, 1973; Nougarède, 1977b). In reactivated nuclei of *Pisum* increased chromatin condensation seems to be related to a decrease in nuclear volume and an increase in the lysine-rich fraction. This latter fraction is invariably found in active nuclei of reactivated lateral buds as well as the shoot apex (Nougarède and Rondet, 1976). Inhibited nuclei could therefore lack a specific lysine-rich histone fraction necessary for chromatin development and mitotic activity.

Not surprisingly, active RNA synthesis is required for lateral-bud growth following decapitation. Substantial inhibition of *Nicotiana* buds following shoot

decapitation was achieved with the inhibitor 6-methylpurine and its effects could be partially reversed by application of the appropriate metabolite uridine (Sharpe and Schaeffer, 1970). Similar experiments monitoring short-term effects on protein and nucleic acid metabolism immediately after eliminating correlative inhibition would be interesting.

One of the most rapid changes taking place after decapitation is the apparent accumulation of K^+ ions in the stem tissue around the nodes of *Bidens* (Kramer, Desbiez, Garrec, Thellier, Fourcy and Bossy, 1980). X-ray micro-analysis revealed K^+ accumulation within five minutes of decapitation, possibly indicating that ion-transport processes could play a fundamental rôle in bud activation.

In summary, there is still no clear-cut instance of where cytological and biochemical changes point to the signal and primary mechanism by which the apex governs the growth of the lateral bud. Further aspects of short-term effects are considered in the subsection dealing with growth substances.

Bud growth and vascular connexions

Some studies have implicated the vascularization of lateral buds as a point of control for apical dominance. Anatomical studies in *Linum* and *Faba* indicated a close correlation between the extent of bud growth following the loss of apical control and the amount of vascular tissue connecting the stem and bud (Gregory and Veale, 1957; Panigrahi and Audus, 1966). Sorokin and Thimann (1964) noted that in *Pisum* bud growth following release from correlative inhibition coincided with the establishment of xylem continuity between bud and stem.

More recent studies in various species indicate that bud growth can be detected many hours before increased vascular interconnexions become manifest (Rubinstein and Nagao, 1976). What is more, in species such as *Glycine* and *Phaseolus*, inhibited vascular buds appear to possess functional xylem and phloem connexions with the stem (Peterson and Fletcher, 1973; Hall and Hillman, 1975; Yeang, 1980). Vascular elements connecting the stem with the primary-leaf axillary buds (White, 1973) and first-trifoliate-leaf bud (Mullins, 1970) have been detected in *Phaseolus*. In addition, ^{14}C photoassimilates from $^{14}CO_2$ supplied to the primary leaf of decapitated *Phaseolus* plants (with or without IAA applied to the cut stem) were recovered from the subtended bud within one hour after decapitation, too short a period for substantial differentiation of additional vascular tissue (Hall and Hillman, 1975).

Notwithstanding the debate about the rôle of vascular connexions in apical dominance, their development is crucial to continued growth of the lateral shoot, and active buds in turn induce vascular differentiation (Wetmore and Sorokin, 1955). Even so, it is unlikely that the primary control of bud development resides in the development or functioning of the xylem and phloem interconnexions.

Source of bud inhibition in correlative inhibition

Surgical experiments have shown that the shoot apical region appears to be the main source of lateral-bud inhibition although other regions may contribute directly or indirectly as a source of growth factors (Went, 1939), specific inhibitors or even inhibitor-precursors (Libbert, 1955).

Internodal development (Phillips, 1971; Zieslin and Halevy, 1976), cotyledons (Dostàl, 1909; Champagnat, 1951) and mature leaves (Goebel, 1880; Weiskopf, 1927; Zieslin and Halevy, 1976) have all been noted to influence bud development quantitatively rather than qualitatively. However, the inhibitory role of mature leaves has been disputed (White, Medlow, Hillman and Wilkins, 1975).

The presence of developed roots does not appear to be a prerequisite for bud inhibition or lateral-shoot growth in the initial stages. Correlative inhibition is usually maintained in de-rooted cuttings. It is conceivable though that root initials may play a correlative role. Moreover, in the ancestral potato, *Solanum andigena*, the transformation of stolons (lateral shoots) into normal orthogravitropic shoots following shoot decapitation requires the presence of roots or cytokinins (Woolley and Wareing, 1972a,b). The significance of roots in correlative phenomena in the shoot have been noted by Goebel (1908) and Libbert (1955) and indirect effects of root treatments on shoot-growth relationships are extensively catalogued in the scientific literature.

Libbert (1955) has suggested that roots and mature leaves might be the source of inhibitor-precursors for correlative inhibition, in which case their removal from the plant need not have a demonstrable effect if the plant has already achieved a saturable level of such substances.

In dicotyledonous plants the major inhibitory influence on lateral buds is known to originate in the young expanding leaves of the shoot apex (Weiskopf, 1927; Snow, 1929; White *et al.*, 1975). Some of these leaves are capable of inhibiting lateral-bud growth in the absence of other leaves (Fig. 6.4) and careful removal of the apical meristem and leaf primordia reveals that these zones are not the source of bud inhibition. Physical restriction of the growth of the young leaves is as effective as decapitation (Hillman and Yeang, 1979).

Little is known of the role of the apical region in the Gramineae. Leopold (1949) noted that in *Hordeum* and *Teosinte* destruction of the apex increased tillering. Laidlaw and Berrie (1974) observed that correlative inhibition during the vegetative stage of *Lolium* was apparently due to expanding leaves in the vegetative apex, but in the flowering plant the source of bud inhibition was in the inflorescence or elongating stem. Tiller-bud growth tends to be suppressed during the reproductive phase when rapid stem elongation commences (Langer, 1974).

The inhibitory leaves of herbaceous and certain woody angiosperms are thought to have a basipetal polarity of action (Thimann, 1939). Growing lateral buds inhibit other lateral buds also with a basipetal polarity, presumably by virtue of their young expanding leaves. This polarity effect discounts a purely nutritive theory of apical dominance, although some of the inhibition could be the result of compensatory growth effects. The generally better growth of the uppermost buds on decapitated plants would be expected on the basis of a primarily downward-acting inhibition. In the woody plants *Betula* and *Syringa* the inhibitory action of leaves extends to buds nearer the apical meristem (Champagnat, 1955), in which case the correlative inhibition can act upwards and downwards in the same axis.

Little is known about the correlative relationships of shoots which may be regarded as competing with each other for nutritive and growth factors. In *Phaseolus vulgaris* the buds in the axils of the opposite primary leaves are unequal in size. About two weeks after decapitation, the growth of the shorter bud may be completely suppressed by the longer shoot which develops from the initially bigger bud. Eventually the shorter shoot may die, an effect first noted by Mogk (1913). Further investigations by Snow (1931, 1937) with *Pisum* revealed that one of two equal shoots could be rendered more susceptible to inhibition by darkening or excision of its young leaves. He also noted that this susceptibility could be overcome by exogenous IAA.

Environmental factors

Higher plants have highly efficient photosynthetic carbon metabolism, conservative mineral nutrition (especially with respect to nitrogen), and special systems for the uptake and transport of water.

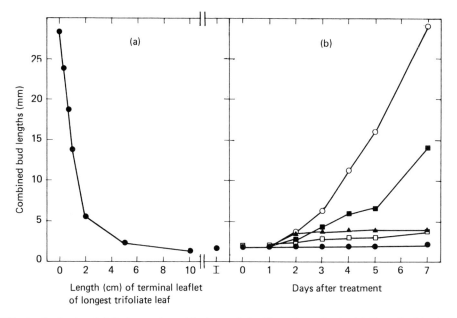

Fig. 6.4 Effects of selective defoliation on lateral-bud growth in *Phaseolus vulgaris*. (a) Growth of both primary-leaf axillary buds seven days after commencing defoliation treatments to 19-d-old intact plants. All trifoliate leaves (with associated buds) greater than a particular length were continually excised over the seven-day period. A main shoot bearing leaves with a maximum terminal-leaflet length of five centimeters inhibited lateral-bud growth to the same extent as an intact main shoot. More extensive defoliation led to progressively greater bud outgrowth, indicating that the apex itself is not the source of inhibition. I = intact plants. (b) Growth of both primary-leaf axillary buds on 21-d-old decapitated plants with all (●) or none (○) of the trifoliate leaves, or with only the first (■), second (□), or third (▲) trifoliate leaf remaining. The rapidly growing second- and third-trifoliate leaves strongly inhibited bud growth. (Adapted from White *et al.*, 1975.)

Although the inhibition of shoots resembles the inhibition of lateral buds, the death of inhibited shoots of *Pisum* (Sachs, 1966) and *Phaseolus* occurs within 20 to 30 days of the onset of inhibition, whereas inhibited buds of many species survive for much longer periods, and may die with the rest of the plant without ever being released from correlative inhibition. Generally then, there are basic physiological differences between inhibited buds and subordinate shoots.

Their growth and reproductive activities are affected by predictable and non-predictable changes in the environment. The successful existence of a plant depends on its sensitivity and responses to environmental changes, including adaptive changes in correlative relationships.

Inorganic nutrients, water, light, temperature, gaseous atmosphere and gravity have been shown to influence, singly and in combination, apical-dominance phenomena. Nevertheless, relatively few experiments have been designed to test whether these environmental factors play a direct or modulating role in the underlying mechanism.

Inorganic nutrients

The availability of nitrogen has long been recognized as an important environmental factor affecting bud growth. Raising the nitrogen status of a low-nitrogen growth medium usually stimulates the growth of lateral buds and shoots and this effect is frequently manifest within 48 h (Phillips, 1975; Rubinstein and Nagao, 1976).

In *Tradescantia*, the cytological changes in buds released by either decapitation or elevated nitrogen availability were closely similar (Yun and Naylor, 1973). Because increases in bud total-nitrogen accompany increases in bud elongation and dry weight, McIntyre (1977) postulated that nitrogen is a limiting factor in bud growth. Yet it is feasible that elevated endogenous nitrogen levels could equally be the result rather than the cause of bud growth. Nitrogen, phosphorus and potassium levels (calculated on a dry-weight basis) in the apical five millimeters of inhibited *Phaseolus* buds were as high as in the equivalent regions of released buds (Phillips, 1968).

Although there are variations between species in their responses to phosphorus, potassium and calcium, particularly in field experiments, increased availability of these elements is normally associated with enhanced bud growth (Phillips, 1969; Rubinstein and Nagao, 1976). Only with the advent of sophisticated microanalyses has it been possible to show that K^+ accumulation around the nodes is one of the earliest changes occurring after decapitation (Kramer *et al.*, 1980). Perhaps similar studies of ion distribution may yield important clues as to the part played by inorganic ions in growth responses.

Water

In their natural environments, most terrestrial plants would normally be expected to tolerate varying degrees of water-stress. Active plant meristems have the competitive ability to obtain water at the expense of mature parts of the plants, including fruits. McIntyre (1977) has pointed out that competition for water may play a critical role in correlative inhibition, since bud growth in *Agropyron*,

Helianthus, *Phaseolus*, and *Pisum* is promoted by water availability and high air humidity.

Light and temperature

The quality, quantity and temporal distribution of light markedly influences correlative inhibition. With few exceptions, the degree of dominance is greater at lower light than at higher light intensities (Rubinstein and Nagao, 1976). Lower light intensities can apparently re-establish the non-dividing central zone in *Tradescantia* buds (Yun and Naylor, 1973). Nevertheless, Gregory and Veale (1957) pointed out that in *Linum* correlative inhibition resulting from nitrogen supply was independent of light intensity.

The spectral quality of irradiation affects lateral-bud outgrowth in intact plants of several species including *Xanthium* (Tucker and Mansfield, 1972) and *Lycopersicon* (Tucker, 1975). When the photoperiod is extended with a brief period of illumination rich in the far-red wavelengths, correlative inhibition is significantly increased in *Nicotiana* (Kasperbauer, 1971), *Xanthium* (Tucker and Mansfield, 1972) and *Lycopersicon* (Tucker, 1975), but there is little direct effect on bud growth in *Phaseolus* (White and Mansfield, 1978).

In an extensive investigation of branching in *Pisum*, Nakamura (1965) demonstrated that light intensity, photoperiod and temperature, in addition to nutrients, markedly affected the pattern of lateral-bud growth. Photoperiodic and temperature effects have also been noted in other species, *e.g.*, *Phalaris* (Scurfield, 1963) and *Solanum* (Woolley and Wareing, 1972a,b). A fairly general observation in photoperiodically sensitive species is that apical dominance is favored under long daylengths whereas short-days tend to promote lateral-shoot growth (Phillips, 1969). Fluctuating temperatures promote sprouting of *Eucalyptus* lignotubers after decapitation (Blake, 1972). Without critical research it is not possible to state whether true thermoperiodic control (Went, 1953) of branching exists. By way of contrast, neither variations in photoperiod nor varying day and night temperatures modify correlative inhibition in *Phaseolus* (White and Mansfield, 1978).

The effects of light have been interpreted as being mediated *via* carbohydrate metabolism and distribution in the shoot, a view which is reinforced by the fact that bud growth in *Pisum* can be promoted by high irradiance and elevated carbon dixode levels (Andersen, 1976). Indeed, McIntyre (1977) tentatively ascribed the effects of low light intensity to a limitation of carbohydrate for bud growth, noting that decapitation of *Agropyron* rhizome segments aged for seven days in water will not lead to lateral-bud growth except in the presence of sucrose. Similar results can be obtained with other species such as *Pisum* (Wickson and Thimann, 1958).

There are, however, at least four observations which raise doubts about the validity of this idea. Firstly, although the dry weights of *Phaseolus* buds from plants grown under high light intensities are greater than those under low light intensities, the amounts of carbohydrates analyzed as a proportion of the bud dry weight were the same under both light treatments (McIntyre, 1973). Secondly, during the early growth of *Pisum* buds following release from correlative inhibition there was no significant increase in bud total carbohydrate (Wardlaw and Mortimer, 1970). Thirdly, treatment of tubers, non-decapitated plants and stem segments with solutions containing relatively low concentrations of sucrose or mineral salts does not modify bud outgrowth (Rubinstein and Nagao, 1976). Lastly, even under apparently optimal environmental conditions, bud growth on intact plants of many species is much less than that occurring after decapitation.

The nutritive theory and the environmental control of bud growth

To what extent can nutritional and environmental control be applied to an unifying theme of correlative regulation? While accepting that current knowledge is far too incomplete to provide unqualified answers, three caveats seem to preclude making a firm commitment to direct nutritional control:

(a) Lateral buds are predisposed to growth which involves complex ultrastructural, biochemical and physiological changes. Buds at different morphological positions may respond differently to changes in the same environmental factor (Gregory and Veale, 1957). This could partially relate to the phase of reproductive development, the stage of anatomical and morphological development, or other correlative effects.

(b) Water, inorganic nutrients and light are essential to autotrophic plant growth. Augmentation of these factors is likely to aid lateral-bud growth. The mechanisms underlying their utilization are largely unresolved.

(c) Environmental factors modify the effectiveness of exogenous growth-regulating chemicals (White and Mansfield, 1978), and affect the synthesis and metabolism of endogenous growth substances.

Nutrients may well play a fundamental role in correlative inhibition possibly *via* hormone-directed metabolite transport, in which case the nutrients would not act as a correlative signal. It seems likely that the basis for apical dominance resides in the *control* of water and nutrient distribution.

Composition of the atmosphere

Increasing the carbon dioxide levels of the air diminishes apical control in *Pisum* (Andersen, 1976). The effects of ethylene are discussed in a separate subsection of this chapter.

Gravity and shoot orientation

The training or enforced curvature of shoots from their normal negatively orthogravitropic habit gives rise to an example of the phenomenon termed 'gravimorphism'. This is defined as those gravity-induced morphogenetic or developmental effects in addition to those of gravitropism (Wareing and Phillips, 1981). In the case of lateral buds, correlative inhibition is reduced and usually eliminated in the highest upwardly-directed bud (Wareing and Nasr, 1958; Phillips, 1969). If the whole stem is secured horizontally, lateral buds closest to the roots are the most strongly stimulated.

Champagnat (1954) investigated the effects of gravity on bud growth in woody species. Horizontal branches of *Sambucus* were reversed so that the lowermost buds become uppermost. When the reversal took place in summer, then during the following spring, after winter dormancy, the uppermost (ex-lowermost) buds produced the largest branches. If, on the other hand, the reversal was delayed into the winter period, the lowermost (ex-uppermost) buds produced the largest branches. When *Syringa* and *Prunus* shoots are bent horizontally during the summer there is a loss of apical dominance in the following spring, whereas if the treatment is delayed until winter the dominance is retained. Champagnat proposed that there are 'propriétés fixées' induced by gravity which determine the pattern of bud growth.

Gravity, therefore, influences bud growth and buds can be conditioned to respond to reorientation. Gravimorphic effects have been interpreted in terms of gravity-induced alterations to auxin (Audus, 1959) or cytokinin distribution (Hillman, 1968; Phillips, 1975), although in *Coleus* (Abeles and Gahagan, 1968) and *Pinus, Pyrus* and *Prunus* (Leopold, Brown and Emerson, 1972) gravimorphic treatments led to enhanced endogenous ethylene production which could affect outgrowth.

Plant growth substances

Correlative control in higher plants can be regarded as a form of positional signalling. Since there is no differentiated nervous system in plants, relatively little research has been devoted to biophysical signalling although publications discussing electrical conduction systems have appeared for more than 120 years. Most attention has been directed towards the chemical-signalling roles of the plant growth substances, particularly IAA. Whether or not these compounds can be regarded as *hormones* is not a point of semantics but in reality a central issue in formulating experiments. Regardless of their modes of action at the molecular level, their role as integrating factors in growth and differentiation cannot be accepted unreservedly.

A greater part of research on growth substances has been carried out on the effects and metabolism of exogenous compounds (*i.e.,* synthesized artificially). This approach is pertinent to apical dominance studies, for exogenous auxins, cytokinins, gibberellins, abscisic acid and ethylene, singly and in combination, affect bud growth. The relevance of this 'effectology' experimentation is difficult to assess without comprehensive knowledge of the physiology and biochemistry of the equivalent endogenous compounds. Yet, as in all areas of developmental physiology, research on endogenous growth substances in apical dominance is still at an elementary stage. The following two points outline the chief aims of research programmes in this field of scholarship:

(1) *Identification and quantification* Confirmation of chemical identity should be obtained with unambiguous physicochemical methods such as combined gas chromatography–mass spectrometry (GC–MS). The quantification techniques should have known parameters of sensitivity (amount needed to elicit a response), selectivity (extent of differential responses to different compounds), accuracy (correctness with respect to standards), and precision (variability). Losses during purification should be monitored by internal standards.

(2) *Related studies* Measurements of 'levels' of a compound in a tissue or organ are in themselves of little significance without consideration of (a) its metabolism (low levels could indicate rapid turnover); (b) its location and binding in specific cell organelles, cells, tissues or organs; (c) characterization of enzyme systems and precursors as well as degradative and conservative metabolites; (d) environmental effects and possible diurnal changes in production; and (e) the temporal relationship between alterations in some aspect of the formation and metabolism of the endogenous compound and a change in bud growth.

Technical limitations have so far precluded a complete study of the role of a single growth substance in correlative inhibition and much of the earlier research is now outmoded. Furthermore, the proof of a hormone-like role *in vivo* would require, in addition to the details outlined above, demonstration of a site of synthesis and intercellular transport to the site of action where demonstrable effects take place (Huxley, 1935).

In the absence of strong contradictory evidence, the growth substances seem to be strong contenders as correlative factors in bud growth. Evaluation of their involvement is outlined for each substance, and possible multi-hormone systems are discussed.

Indole-3-acetic acid

In the section 'Historical perspectives' IAA was considered as having either a direct or indirect rôle in correlative inhibition. The most widely cited evidence supporting the 'direct' theory is the observation that exogenous IAA will replace the shoot apex with respect to bud inhibition. This is misleading in the light of recent investigations.

White *et al.* (1975) estimated by mass spectrometry that IAA levels in *Phaseolus* shoot tips were in the order of 0.1 to 0.7 ng per shoot tip, but over one thousand times this level of exogenous IAA is required to maintain bud inhibition by application to the cut stem stump. The IAA-induced inhibition of bud growth is associated with cell division and expansion in the treated internode, processes which would be expected to give rise to a metabolic sink and a concomitant diversion of nutrients and growth factors from the buds. In addition, there is a readily detectable time-lag between replacing the apex with a source of exogenous IAA and the reimposition of dominance, an observation confirmed by monitoring the importation of ^{14}C-labeled photosynthetic assimilates (Hall and Hillman, 1975).

To exert a direct effect on lateral-bud growth, IAA must be in or close to the lateral bud. The basipetal polarity of correlative inhibition is in accord with the known transport characteristics of radioactivity from radiolabeled exogenous IAA in shoots. GC–MS analyses demonstrate conclusively that IAA is recovered in the agar receiver blocks used to collect radiolabeled compounds transported basipetally through *Zea* coleoptile segments supplied with radioactive IAA to their morphological apical end (Nonhebel, 1982). This could indicate that exogenous IAA may be transported in an unaltered form and might reach lateral buds. Despite the attempts of Snow and others it is not known whether the transport of IAA is identical to that of the correlative inhibitor.

Five inhibitors of exogenous-IAA transport in plants, triiodobenzoic acid (TIBA) (Panigrahi and Audus, 1966), morphactin (White and Hillman,

1972), naptalam (Morgan, 1964), DPX 1840 (Beyer, 1972), and phenacylidene phthalide hydrate (Brown, Johansen and Sasse, 1972) can effectively release buds basipetal to the point of treatment. These results appear to be firm evidence in favor of a central role for IAA in the phenomenon. There are, nonetheless, contra-indications. The specificity of action and relevance to correlative inhibition of both TIBA and morphactin can be called into question given their effects in promoting the abscission of young leaves, the known source of bud inhibition (White and Hillman, 1972).

Further doubt was cast on the significance of IAA transport by Morris (1977) who was unable to detect radioactivity derived from IAA passing from the inhibitory to the inhibited branches of two-branched pea seedlings. Similarly, Hall and Hillman (1975) could not detect IAA movement into inhibited buds of *Phaseolus*.

The sheer problem of harvesting large quantities of lateral buds has effectively limited research on endogenous growth substances at the site of correlative inhibition. Early studies indicated that auxin levels tend to rise rather than fall upon their release from inhibition (Overbeek, 1938). More recently Hillman, Math and Medlow (1977) analyzed the IAA levels of 56 000 *Phaseolus* lateral buds using GC–MS and found that 24 h after decapitation there was a consistent rise in endogenous IAA content. The IAA levels varied between 2.9 and 29.4 pg per bud with mean levels of 6.4 and 15.6 pg per bud in intact and decapitated plants respectively (Table 6.2). Unfortunately, there are no GC–MS studies of changes in IAA levels during the few hours following decapitation.

Table 6.2 Endogenous IAA levels of the paired lateral buds in axils of *Phaseolus* primary leaves. Decapitation was performed 24 h prior to harvesting. IAA identification and quantification was by GC–MS with internal standards. (Hillman *et al.*, 1977.)

Experiment	Age of plant (d)	No. of plants per treatment	μg IAA (g dry wt)$^{-1}$ Intact	Decapitated	pg IAA (bud)$^{-1}$ Intact	Decapitated
1	18	2500	0.08	0.13	2.9	3.9
2	18	5000	0.06	0.14	10.8	29.4
3	20	1500	0.02	0.03	3.3	6.0
4	20	1500	0.02	0.03	5.0	13.0
5	25	1500	0.04	0.13	8.0	26.0
6	25	2000	0.06	0.08	8.2	15.0

Further evidence against the 'direct' theory comes from the observation that application of IAA to the bud stimulates bud release (Yeang and Hillman, 1982) but this growth may differ from normal bud release (see Ethylene).

The 'indirect' theory of auxin inhibition accommodates some of the objections raised against the 'direct' theory. Apically-derived IAA could modify other factors which in turn modulate bud growth. Examples of indirect action include hormone-directed transport and inhibition of vascular connexions but, as described earlier, in neither instance is there a solid experimental base to substantiate these concepts.

Cytokinins

The fact that applications of synthetic and naturally-occurring cytokinins will release buds from correlative inhibition in a wide range of species had led to the view that these compounds may be limiting for bud growth. Stimulatory effects have been reported in several species including Cicer, Coleus, Faba, Glycine, Helianthus, Nicotiana, Pisum, Scabiosa (Rubinstein and Nagao, 1976), Phaseolus (Hillman, 1968) and Lycopersicon (Aung and Byrne, 1978). The response to cytokinin has a time-lag of about 6 h as determined by cytological, biochemical and bud-growth changes (Usciati, Codaccioni and Guern, 1972; Nagao and Rubinstein, 1976).

It is commonplace for the stimulatory effects of cytokinins to be short-lived, lasting for about two days, but the duration of development can be prolonged by the application of IAA following the initial response to cytokinin, possibly indicating sequential roles of growth substances in bud outgrowth.

Cytokinin action may require supplementation from a stem factor because isolated lateral buds respond poorly to benzyladenine unless a stem segment is excised along with the bud (Peterson and Fletcher, 1975).

Hadacidin, an inhibitor of purine synthesis, has been used to investigate the role of endogenous cytokinins in correlative inhibition. Bud growth following decapitation can be inhibited by hadacidin, an effect which can be reversed by cytokinin (Lee, Kessler and Thimann, 1974). The inhibitor may not be specific, however, as adenine could not reverse the hadacidin-induced inhibition.

There are no detailed reports of changes in endogenous cytokinins of lateral buds following decapitation. Woolley and Wareing (1972a,b) proposed that the buds are deficient in cytokinins because apically-synthesized auxin inhibits acropetal transport of these compounds. This theory presupposes that whereas outgrowth of inhibited lateral buds requires cytokinins, these compounds are not necessary for growth of the main axis.

Root-derived cytokinins seem to be of limited significance since roots do not appear to be essential for the initiation of bud outgrowth in decapitated plants. As regards cytokinins elsewhere in the plant, Palmer (1980) noted dihydrozeatin-O-glucoside accumulation in leaves of disbudded Phaseolus plants but this accumulation was prevented when the buds were allowed to develop. To date, there is no evidence that IAA or other hormones influence the form of cytokinins entering lateral structures.

The site of cytokinin action is not known. Skoog and Abdul Ghani (1980) found that a range of cytokinins as well as compounds which show anti-cytokinin activity in bioassay are equally effective in promoting bud growth.

Thus the cytokinins offer a promising line of enquiry into the hormonal control of bud growth but a considerable amount of basic knowledge about endogenous cytokinin biosynthesis, metabolism and distribution is needed urgently.

Gibberellins

Gibberellins have dramatic stimulatory effects on the growth of released buds whether applied to the bud or stem (Phillips, 1969), but they may not play a major role in the release from correlative inhibition. When applied to quiescent buds of Pisum (Sachs and Thimann, 1964), Lycopersicon (Catalano and Hill, 1969) and Glycine (Ali and Fletcher, 1971), the effects were minimal and became evident only after treatment with cytokinin.

Gibberellins applied in combination with auxins to the cut stem of decapitated plants may either stimulate (Phillips, 1969, Hillman, 1970) or inhibit (Scott, Case and Jacobs, 1967) bud growth. The discrepancy has been explained by differences in the age of the internode to which the hormones were applied. Gibberellin applied near the bud in decapitated Phaseolus multiflorus promoted its growth whereas application to a younger internode further away reduced growth, perhaps indicating compensation-growth effects arising from gibberellin-induced extension of the younger internode (Phillips, 1975).

As with the cytokinins, there are no reports on the rigorous analysis of endogenous gibberellins in quiescent and newly-released lateral buds. Discussions on the involvement of these compounds in the release of correlative inhibition are therefore highly speculative.

Abscisic acid

Largely as a result of bioassays of inhibitory activity in dormant buds, a number of attractive hypotheses have been proposed implicating ABA as an inhibitor of bud growth (*e.g.* Tucker and Mansfield, 1973).

Direct applications of ABA to lateral buds suppressed growth of the buds in *Pisum* (Bellandi and Dörffling, 1974), *Lycopersicon* (Tucker, 1977) and *Phaseolus* (White and Mansfield, 1977). In the latter species, ABA applied on its own to the decapitated stem slightly promoted axillary-bud growth, but increased inhibition marginally when applied in combination with IAA and kinetin (Hillman, 1970). Very low doses of ABA, however, promote the growth of lateral buds (Hartung and Steigerwald, 1977).

Studies on the ABA relations of *Phaseolus vulgaris* (Onckelen, Horemans and De Greef, 1981) revealed that decapitation of seedlings reversed the build-up of endogenous ABA in cotyledons. The elevated ABA levels were normally associated with cotyledonary senescence and the minimal ABA level coincided with the onset of visible cotyledonary growth. Once correlative inhibition was re-established by an axillary shoot, ABA accumulation in the cotyledons was detected and they became senescent. Similar indications of an inhibitory role of ABA were found by Tamas and associates (1979, 1981) and Everat-Bourboloux and Charnay (1981).

There are, however, findings that call into question a fundamental role of ABA in bud-growth inhibition. In *Acer* and *Syringa* the absolute quantities of ABA present in active lateral buds following decapitation and defoliation of the shoot remained unchanged, even though there was a decrease in 'concentration' due to the increase in bud fresh weight (Dörffling, 1976). Taylor and Rossall (1982) were unable to detect a difference in ABA content

between the *lateral suppressor* mutant of *Lycopersicon* and the branching wild type.

In a detailed and thorough analysis of endogenous ABA in *Phaseolus vulgaris*, Knox (1982) has shown that the decline in ABA levels of the bud following shoot decapitation occur *after* visible growth commences. While it is possible to argue that the techniques employed were not adequate to detect subtle cellular changes in a complicated structure such as a bud, the weight of evidence is against ABA having a primary role in bud inhibition. It cannot be deemed to share the properties of the elusive correlative inhibitor described by Snow. The promotion of lateral-bud growth by treating the apex with ABA (Aung and Byrne, 1978) can be explained as an inhibition of apical growth, tantamount to decapitation, causing compensation growth in the buds. Similar explanations could be applied equally to the effects of chemical pruning agents (Cathey and Steffens, 1968), May and Baker 25–105 (Yeang and Hillman, 1981b) and the auxin-transport inhibitors.

Ethylene

Unlike the other growth substances, relatively little work has been carried out on the involvement of ethylene in bud growth. Nevertheless, recent studies indicate that this compound may well have a complex role in the phenomenon.

Hall, Truchelut, Leinweber and Herrero (1957) reported the loss of apical dominance in field-grown *Gossypium*; they attributed this to atmospheric pollution by ethylene. Inhibited buds of *Petunia* (Burg, 1973) and *Solanum* (Catchpole and Hillman, 1976) supplied with a pulse of ethylene were activated when the ethylene was subsequently withdrawn. Applications of the ethylene-generating agent, ethephon, to a wide range of ornamental and rhizomatous species (Morgan, Meyer and Merkle, 1969; De Wilde, 1971) and to *Hevea* (Leong, Leong and Yoon, 1976) leads to outgrowth of lateral buds.

Effective removal of correlative inhibition is obtained by physical restriction of apical growth. Branching in *Hevea* can be induced by covering the apical portion of the shoot or by folding and securing the uppermost whorl of leaves over the shoot apex (Leong *et al.*, 1976). Likewise confinement of the developing apical shoot of *Phaseolus* in sealed or ventilated tubes inhibits its growth and promotes the vigorous growth of buds beneath the enclosure

(Hillman and Yeang, 1979). These developments in *Phaseolus* are thought to be brought about by ethylene because firstly, the concentration of ethylene in gases removed by vacuum extraction (internal ethylene) and the rate of ethylene emanation are significantly increased in the treated zone, and secondly, similar responses are observed in plants treated apically with either ethylene gas or ethephon (Yeang and Hillman, 1981b).

Endogenous ethylene could influence bud growth at or near the bud. Burg and Burg (1968a) proposed that since ethylene evolution in etiolated *Pisum* was greater at the nodes than the internodes, then inhibition of the lateral buds was due to endogenous ethylene. However, in an article the same year they noted that although ethylene emanation from the node decreased upon stem decapitation, this decrease was not observed when the scale leaves at the node were removed (Burg and Burg, 1968b). Both articles attributed the inhibition of bud development in isolated nodal segments by IAA to the production of auxin-induced ethylene, despite measuring ethylene emanation 21 h after excision of the segments. Bud growth is now recognized to take place within 6 to 12 h.

Yeang and Hillman (1982) examined the role of ethylene in the nodal region of *P. vulgaris* in relation to bud inhibition. Analyses of endogenous ethylene were carried out within 30 min of excising the tissues, at a time before the burst in wound-induced ethylene synthesis. Both ethylene emanation and internal ethylene levels in nodal tissues declined when the plants were decapitated, reflecting the oft-observed decrease in the whole shoot. If bud development were inhibited by ethylene in or near the bud then inhibitors of ethylene biosynthesis and action would be expected to overcome this inhibition. Yet aminoethoxyvinyl glycine (AVG) or silver ions (Ag$^+$) applied to the bud inhibited rather than promoted bud outgrowth in decapitated plants. Ethephon had a similar effect. These three substances were ineffective in the buds of intact plants, so it can be argued that ethylene is essential in maintaining the growth of released buds but the absence of bud development in the intact plant is not due to a lack of free ethylene.

These authors further demonstrated (Yeang and Hillman, 1982) that inhibition of lateral-bud growth in decapitated plants by IAA applied to the cut stem was not ascribable to auxin-induced ethylene at the nodal regions. Application of a single dose of IAA directly to the axillary buds of intact *Phaseolus* plants gave rise to a transient increase in bud growth. The enhancement of growth was completely annulled by a low concentration of AVG supplied to the bud at the same time. This dosage of AVG did not affect bud outgrowth due to shoot decapitation or the slow bud growth of the intact plant—only the additional growth due to IAA was prevented. An explanation for this could be that the response to IAA is ethylene-dependent and IAA-induced bud growth is dissimilar to the growth arising from shoot decapitation.

Another justification for considering ethylene as a bud-growth factor comes from the fact that certain environmental treatments which stimulate bud growth (carbon dioxide, gravimorphic treatments, light quality) are also known to affect ethylene production and/or action.

Thus ethylene could have a basic rôle in the control of bud growth when, for example, apical growth is affected. Its seemingly poor transport (Chapter 5) may prevent it having a truly hormonal rôle as a correlative signal unless it is transported in a modified form.

Conclusions

The correlative inhibition of lateral-bud growth is the most extensively-investigated aspect of apical dominance in higher plants. Some form of positional signalling occurs between the main source of bud inhibition, the young developing leaves, and the quiescent bud.

At least fourteen types of treatment (Table 6.1) will activate growth in the quiescent bud, but there is no site or basic process which is clearly defined as the locus of inhibition. The lag period between removal of apical influence and various detectable changes in the lateral bud can be less than six hours.

Six main theories have been advanced to explain the phenomenon, namely the 'nutritive' theory, the 'direct' theory of auxin action, the 'correlative inhibitor' theory where auxin has an indirect action, the 'nutrient-diversion' theory, the 'vascular-connection' theory and the 'hormonal-balance' theory. None is satisfactory.

All of the plant growth substances affect bud growth, directly or indirectly, and there is evidence to suggest that IAA, cytokinins and, under certain circumstances, ethylene could have basic roles in correlative inhibition.

The correlative signal has not been characterized. It could operate by controlling the distribution of ions and water to strategic sites in the bud.

Potentially rewarding lines of future research include monitoring short-term changes in fine structure, nucleic acids, proteins, growth substances and ions in the various tissues following removal of bud inhibition. The precise path and nature of the correlative signal and its production by the developing leaves merit careful scrutiny.

Further reading

Audus, L. J. (1959). Correlations. *J. Linn. Soc.* (Bot.) **56**, 177–87.

Brown, C. L., McAlpine, R. G. and Kormanik, P. P. (1967). Apical dominance in woody plants: a reappraisal. *Am. J. Bot.* **54**, 153–62.

McIntyre, G. I. (1977). The role of nutrition in apical dominance. In The Society for Experimental Biology Symposium 31. *Integration of Activity in the Higher Plant*, ed. D. H. Jennings, Cambridge University Press, Cambridge, pp. 251–73.

Phillips, I. D. J. (1969). Apical dominance. In *The Physiology of Plant Growth and Development*, ed. M. B. Wilkins, McGraw-Hill, London, pp. 165–202.

Phillips, I. D. J. (1975). Apical dominance. *Ann. Rev. Plant Physiol.* **26**, 341–67.

Rubinstein, B. and Nagao, M. A. (1976). Lateral bud outgrowth and its control by the apex. *Bot. Rev.* **42**, 83–109.

References

Abeles, F. B. and Gahagan, H. E. (1968). Accelerated abscission of *Coleus* petioles by placing plants in a horizontal position. *Life Science* **7**, 653–5.

Ali, A. and Fletcher, R. A. (1970). Hormonal regulation of apical dominance in soybeans. *Can. J. Bot.* **48**, 1989–94.

Ali, A. and Fletcher, R. A. (1971). Hormonal interaction in controlling apical dominance in soybeans. *Can. J. Bot.* **49**, 1717–31.

Andersen, A. S. (1976). Regulation of apical dominance by ethephon, irradiance and CO_2. *Physiol. Plant.* **37**, 303–8.

Audus, L. J. (1959). Correlations. *J. Linn. Soc.* (Bot.) **56**, 177–87.

Aung, L. H. and Byrne, J. M. (1978). Hormones and young leaves control development of cotyledonary buds in tomato seedlings. *Plant Physiol.* **62**, 276–9.

Bellandi, D. M. and Dörffling, K. (1974). Effect of abscisic acid and other plant hormones on growth of apical and lateral buds of seedlings. *Physiol. Plant.* **32**, 369–72.

Beyer, E. M. Jr (1972). Auxin transport: a new synthetic inhibitor. *Plant Physiol.* **50**, 322–7.

Blake, T. J. (1972). Studies on the lignotubers of *Eucalyptus obliqua* L'Herit. 3. The effects of seasonal and nutritional factors on dormant bud development. *New Phytol.* **71**, 327–34.

Booker, C. E. and Dwividi, R. S. (1973). Ultrastructure of meristematic cells of dormant and released buds of *Tradescantia paludosa*. *Exp. Cell Res.* **82**, 255–61.

Brown, B. T., Johansen, O. and Sasse, W. H. F. (1972). New inhibitors of auxin transport. *Experimentia* **28**, 1290–1.

Burg, S. P. (1973). Ethylene in plant growth. *Proc. Nat. Acad. Sci. USA* **70**, 591–7.

Burg, S. P. and Burg, E. A. (1968a). Auxin stimulated ethylene formation: its relationship to auxin inhibited growth, root geotropism and other plant processes. In *Biochemisty and Physiology of Plant Growth Substances*, ed. F. Wightman and G. Setterfield, The Runge Press Ltd., Ottawa, pp. 1275–94.

Burg, S. P. and Burg, E. A. (1968b). Ethylene formation in pea seedlings; its relation to the inhibition of bud growth caused by indole-3-acetic acid. *Plant Physiol.* **43**, 1069–74.

Catalano, M. and Hill, T. A. (1969). Interaction between gibberellic acid and kinetin in overcoming apical dominance, natural and induced by IAA, in tomato (*Lycopersicum esculentum* Mill, Cultivar Potentate). *Nature* **222**, 985–6.

Catchpole, A. H. and Hillman, J. R. (1976). The involvement of ethylene in the coiled-sprout disorder of potato. *Ann. Appl. Biol.* **83**, 413–23.

Cathey, H. M. and Steffens, G. L. (1968). Relation of the structure of fatty acid derivatives to their action as chemical pruning agents. In *Plant Growth Regulators*. S.C.I. Monograph No. 31, Society of Chemical Industry, London, pp. 224–35.

Champagnat, P. (1951). Action du cotylédon de *Bidens pilosus* L. var. radiatus sur son bourgeon axillaire cas d'inhibition et cas de stimulation. Dosages auxiniques. *C.R. hebd. Séanc, Acad. Sci. (Paris)* **145**, 1371–3.

Champagnat, P. (1954). Les corrélations sur le rameau d'un an des végétaux Ligneux. *Phyton* **4**, 1–102.

Champagnat, P. (1955). Les corrélations entre feuilles et bourgeons de la pousse herbacée du Lilas. *Rev. Gén. Bot.* **62**, 325–72.

Child, C. M. and Bellamy, A. W. (1920). Physiological isolation by low temperature in *Bryophyllum*. *Bot. Gaz.* **70**, 249–67.

Couot-Gastelier, J. (1978). Etude de quelques modalités de la croissance des bourgeons axillaires de la fève *Vicia faba* L. libérés de la contrainte apicale. *Z. Pflanzenphysiol.* **89**, 189–206.

Couot-Gastelier, J. (1979). Activité mitotique et croissance des bourgeons axillaires provoquées par la décapitation de jeunes plants de fève (*Vicia faba*). *Can. J. Bot.* **57**, 2478–88.

De Wilde, R. C. (1971). Practical application of (2-chloroethyl) phosphonic acid in agricultural production. *Hort. Science* **6**, 364–70.

Dörffling, K. (1976). Correlative bud inhibition and abscisic acid in *Acer pseudoplatanus* and *Syringa vulgaris*. *Physiol. Plant.* **38**, 319–22.

Dostál, R. (1909). Die Korrelations beziehung zwischen dem Blatt und seiner Axillarknospe. *Ber. dtsch. Bot. Ges.* **27**, 547–54.

Dostál, R. (1926). Über die wachstumsregulierende Wirkung des Laubblattes. *Act. Soc. Sci. Nat. Moravicae* **3**, 83–209.

Dwividi, R. S. and Naylor, J. M. (1968). Influence of apical dominance on the nuclear proteins in cells of the lateral bud meristem in *Tradescantia paludosa*. *Can. J. Bot.* **46**, 289–98.

Errera, L. (1904). Conflicts de préséance et excitations inhibitoires chez les végétaux. *Bull. Soc. Roy. Bot. Belgique* **42**, 27–43.

Everat-Bourbouloux, A. and Charnay, D. (1981). Endogenous abscisic acid levels in stems and axillary buds of intact or decapitated broad-bean plants (*Vicia faba* L.). *Physiol. Plant.* **54**, 440–5.

Fitting, H. (1909). Die Beeinflussing der Orchideenblüten durch die Bestaubund und durch andere Umstände. *Z. Bot.* **1**, 1–86.

Fitting, H. (1910). Weitere Entwicklungsphysiologische Untersuchungen an Orchideenblüten. *Z. Bot.* **2**, 225–66.

Goebel, K. (1880). Beiträge zur Morphologie und Physiologie des Blattes. *Botan. Zeitung* **38**, 800 (cited by Snow, 1929).

Goebel, K. (1900). *Organography of Plants especially of the Archegoniatae and Spermaphyta. Part I. General Organography.* Clarendon Press, Oxford.

Goebel, K. (1908). Einleitung in die experimentelle Morphologie der Pflanzen. Leipzig (cited by Snow, 1925).

Gregory, F. G. and Veale, J. A. (1957). A reassessment of the problem of apical dominance. *Symp. Soc. Exp. Biol.* **11**, 1–20.

Guern, J. and Usciati, M. (1972). The present status of the problem of apical dominance. In *Hormonal Regulation of Plant Growth and Development*, ed. H. Kaldewey and Y. Vardar, Verlag Chemie, Weinheim, Federal Republic of Germany, pp. 383–400.

Haberlandt, G. (1913). Zur Physiologie der Zellteilung. *Sitzungsber. K. Preuss. Adad. Wiss.* 318–45.

Hall, S. M. and Hillman, J. R. (1975). Correlative inhibition of lateral bud growth in *Phaseolus vulgaris* L. Timing of bud growth following decapitation. *Planta* **123**, 137–43.

Hall, W. C., Truchelut, G. B., Leinweber, C. L. and Herrero, F. A. (1957). Ethylene production by the cotton plant and its effects under experimental and field conditions. *Physiol. Plant.* **10**, 306–17.

Hartung, W. and Steigerwald, F. (1977). Abscisic acid and apical dominance in *Phaseolus coccineus* L. *Planta* **134**, 295–9.

Harvey, E. N. (1920). An experiment on regulation in plants. *American Naturalist* **54**, 362–7.

Hillman, J. R. (1970). The hormonal regulation of bud outgrowth in *Phaseolus vulgaris* L. *Planta* **90**, 222–9.

Hillman, J. R., Math, V. B. and Medlow, G. C. (1977). Apical dominance and the levels of indole acetic acid in *Phaseolus* lateral buds. *Planta* **134**, 191–3.

Hillman, J. R. and Yeang, H. Y. Y. (1979). Correlative inhibition of lateral bud growth in *Phaseolus vulgaris* L. Ethylene and the physical restriction of apical growth. *J. Exp. Bot.* **30**, 1079–83.

Hillman, S. K. (1968). *Translocation in Plants with Special Reference to the Role of Growth Hormones*, Ph.D. Thesis, University of Wales.

Huxley, J. S. (1935). Chemical regulation and the hormone concept. *Biol. Rev.* **10**, 427–41.

Jacobs, W. P. and Bullwinkel, B. (1953). Compensatory growth in *Coleus* shoots. *Am. J. Bot.* **40**, 385–91.

Jost (1907). *Lectures on Plant Physiology,* Translated by R. J. Harvey Gibson, Clarendon Press, Oxford.

Jost (1913). *Lectures on Plant Physiology,* Translated by R. J. Harvey Gibson, Supplement, Clarendon Press, Oxford.

Kasperbauer, M. J. (1971). Spectral distribution of light in a tobacco canopy and effects of end-of-day light quality on growth and development. *Plant Physiol.* **47**, 775–8.

Knox, J. P. (1982). *Hormonal Aspects of Apical Dominance*, Ph.D. Thesis, University of Wales.

Kramer, D., Desbiez, M-O., Garrec, J. P., Thellier, M., Fourcy, A. and Bossy, J. P. (1980). The possible role of potassium in the activation of axillary buds of *Bidens pilosus* L. after decapitation of the apex. An examination by X-ray microanalysis. *J. Exp. Bot.* **31**, 771–6.

Laibach, F. (1933). Wuchsstoffversuche mit lebenden Orchideenpollinen. *Ber. dtsch. Bot. Ges.* **51**, 336–40.

Laidlaw, A. S. and Berrie, A. M. M. (1974). The influence of expanding leaves and the reproductive stem apex on apical dominance in *Lolium multiflorum*. *Annls Appl. Biol.* **78**, 75–82.

Langer, R. H. M. (1974). Control of tiller bud growth in the Gramineae. *Proc. 12th int. Grassld Cong.* Moscow 1974, 178–90.

Lee, P.K-W., Kessler, B. and Thimann, K. V. (1974). The

effect of hadacidin on bud development and its implications for apical dominance. *Physiol. Plant.* **31**, 11–14.

Leong, W., Leong, H. T. and Yoon, P. K. (1976). *Some Branch Induction Methods for Young Buddings*, Rubber Research Institute of Malaysia, Kuala Lumpur, Malaysia.

Leopold, A. C. (1949). The control of tillering in grasses by auxin. *Am. J. Bot.* **36**, 437–40.

Leopold, A. C., Brown, K. M. and Emerson, F. H. (1972). Ethylene in the wood of stressed trees. *Hort. Science* **7**, 715.

Libbert, E. (1955). Nachweis und chemische Trennung des Korrelationshemmstoffes und seiner Hemmstoffvorstufe. *Planta* **45**, 405–25.

Loeb, J. (1915). Rules and mechanism of inhibition and correlation in the regeneration of *Bryophyllum calycinum*. *Bot. Gaz.* **60**, 249–76.

Loeb, J. (1917a). The chemical basis of regeneration and geotropism. *Science* **46**, 115–18.

Loeb, J. (1917b). The chemical basis of axial polarity in regeneration. *Science* **46**, 547–51.

Loeb, J. (1918). Chemical basis of correlation. I. Production of equal masses of shoots by equal masses of sister leaves in *Bryophyllum calycinum*. *Bot. Gaz.* **65**, 150–74.

Loeb, J. (1923). Theory of regeneration based on mass action. *J. Gen. Physiol.* **5**, 831–52.

McCallum, W. B. (1905a). Regeneration in plants. 1. *Bot. Gaz.* **40**, 97–120.

McCallum, W. B. (1905b). Regeneration in plants. 2. *Bot. Gaz.* **40**, 241–63.

McIntyre, G. I. (1973). Environmental control of apical dominance in *Phaseolus vulgaris*. *Can. J. Bot.* **51**, 293–9.

McIntyre, G. I. (1977). The role of nutrition in apical dominance. In The Society for Experimental Biology Symposium 31. *Integration of Activity in the Higher Plant*, ed. D. H. Jennings, Cambridge University Press, 251–73.

Mogk, W. (1913). Untersuchungen uber Korrelationen von Knospen und Sprossen. *Arch. Entwick. Org.* **38**, 584–681.

Morgan, D. G. (1964). Influence of α-naphthylpthalamic acid on the movement of indolyl-3-acetic acid in plants. *Nature* **201**, 476–7.

Morgan, P. W., Meyer, R. E. and Merkle, M. G. (1969). Chemical stimulation of ethylene evolution and bud growth. *Weed Sci.* **17**, 353–5.

Morris, D. A. (1977). Transport of exogenous auxin in two-branched pea seedlings (*Pisum sativum* L.). *Planta* **136**, 91–6.

Müller, A. M. (1935). Über den Einfluss von Wuchstoff auf das Austreiben der Seitenknospen und auf die Wurzelbildung. *Jahr, wiss. Bot.* **81**, 497–540.

Mullins, M. G. (1970). Transport of ^{14}C-assimilates in seedlings of *Phaseolus vulgaris* L. in relation to vascular anatomy. *Ann. Bot.* **34**, 889–96.

Nagao, M. A. and Rubinstein, B. (1976). Early events associated with lateral bud growth of *Pisum sativum* L. *Bot. Gaz.* **137**, 39–44.

Nakamura, E. (1965). Studies on the branching in *Pisum sativum* L. *Special Rep. Lab. Hort.*, Shiga Agric. Coll., Japan.

Nonhebel, H. M. (1982). *Metabolism of Indole-3-acetic Acid in Seedlings of Zea mays* L., Ph.D. Thesis, University of Glasgow.

Nougarède, A. (1977a). On the infrastructural localization of basic proteins in the nucleus and nucleolus of cells in pea axillary meristems submitted to or released from apical dominance. *Protoplasma* **93**, 341–56.

Nougarède, A. (1977b). Infrastructure des axillaires cotylédonaires du *Pisum sativum* L. (var. nain hatif d'Annonay) durant le blocage en phase G_1 (état inhibé) et après la reprise d'activité. *C.R. hebd. Séanc. Acad. Sci. (Paris)* **284**, 25–8.

Nougarède, A. and Rondet, P. (1975). Evolution des index mitotiques et des teneurs en DNA nucléaire dans le méristème axillaire de la feuille de rang 6, lors de la levée de dominance apicale, provoquée par ablation de l'axe principal, chez le *Pisum sativum* L. *C.R. hebd. Séanc. Acad. Sci. (Paris)* **280**, 973–6.

Nougarède, A. and Rondet, P. (1976). Durée des cycles cellulaires du méristème terminal et des méristèmes axillaires du *Pisum sativum* L. *C.R. hebd. Séance. Acad. Sci. (Paris)* **282**, 715–18.

Nougarède, A. and Rondet, P. (1977). Les histones nucléaires et leurs variations dans les méristèmes axillaires de rang 6 du pois (*Pisum sativum* L. var. nain hâtif d'Annonay) soumis ou non à la dominance apicale. *C.R. hebd. Séance. Acad. Sci. (Paris)* **284**, 623–6.

Onckelen, H. A. van, Horemans, S. and De Greef, J. A. (1981). Functional aspects of abscisic acid metabolism in cotyledons of *Phaseolus vulgaris* L. seedlings. *Plant Cell Physiol.* **22**, 507–15.

Overbeek, J. van (1938). Auxin distribution in seedlings and its bearing on the problem of bud inhibition. *Bot. Gaz.* **100**, 133–66.

Palmer, M. V. (1980). *Cytokinin Relations in the Whole Plant*, Ph.D. Thesis, University of Wales.

Panigrahi, B. M. and Audus, L. J. (1966). Apical dominance in *Vicia faba*. *Ann. Bot.* **30**, 457–73.

Patrick, J. W. (1982). Hormonal control of assimilate transport. In *Plant Growth Substances 1982*, ed. P. F. Wareing, Academic Press, London, pp. 669–78.

Peterson, C. A. and Fletcher, R. A. (1973). Apical dominance is not due to a lack of functional xylem and phloem in inhibited buds. *J. Exp. Bot.* **24**, 97–103.

Peterson, C. A. and Fletcher, R. A. (1975). Lateral bud growth on excised stem segments: effect of the stem. *Can. J. Bot.* **53**, 243–8.

Phillips, I. D. J. (1968). Nitrogen, phosphorus and potassium distribution in relation to apical dominance in dwarf bean (*Phaseolus vulgaris* c.v. Canadian Wonder). *J. Exp. Bot.* **19**, 617–27.

Phillips, I. D. J. (1969). Apical dominance. In *The Physiology of Plant Growth and Development,* ed. M. B. Wilkins, McGraw-Hill London, pp. 165–202.

Phillips, I. D. J. (1971). Factors influencing the distribution of growth between stem and axillary buds in decapitated bean plants. *J. Exp. Bot.* **22**, 465–71.

Phillips, I. D. J. (1975). Apical dominance. *Ann. Rev. Plant Physiol.* **26**, 341–67.

Rubinstein, B. and Nagao, M. A. (1976). Lateral bud outgrowth and its control by the apex. *Bot. Rev.* **42**, 83–113.

Sachs, J. von (1882). Stoff und Form der Pflanzenorgane. *Arb. Bot. Inst. Würzburg* (**II**)3, 452–88.

Sachs, T. (1966). Senescence of inhibited shoots of peas and apical dominance. *Ann. Bot.* **30**, 447–56.

Sachs, T. and Thimann, K. V. (1964). Release of lateral buds from apical dominance. *Nature* **201**, 939–40.

Scott, T. K., Case, D. B. and Jacobs, W. P. (1967). Auxin-gibberellin interaction in apical dominance. *Plant Physiol.* **42**, 1329–33.

Scurfield, G. (1963). The effects of temperature on the early vegetative growth of *Phalaris canariensis* L. and *P. tuberosa* L. *Aust. J. Agric. Res.* **14**, 165–79.

Sharpe, F. T. Jr. and Schaeffer, G. W. (1970). Methylpurine inhibition and benzyladenine stimulation of axillary bud growth. *Am. J. Bot.* **57**, 629–32.

Skoog, F. (1939). Experiments on bud inhibition with indole-3-acetic acid. *Am. J. Bot.* **26**, 702–7.

Skoog, F. and Abdul Ghani, A. K. B. (1980). Relative activities of cytokinins and antagonists in releasing lateral buds of *Pisum* from apical dominance compared with their relative activities in the regulation of growth of tobacco callus. In *Metabolism and Molecular Activities of Cytokinins,* ed. J. Guern and C. Péaud–Lenoël, Springer-Verlag, pp. 140–50.

Skoog, F. and Thimann, K. V. (1934). Further experiments on the inhibition of the development of lateral buds by growth hormone. *Proc. Nat. Acad. Sci. USA* **20**, 480–5.

Snow, R. (1925). The correlative inhibition of the growth of axillary buds. *Ann. Bot.* **39**, 841–59.

Snow, R. (1929). The young leaf as the inhibiting organ. *New Phytol.* **28**, 345–58.

Snow, R. (1931). Experiments on growth and inhibition. 2. New phenomena of inhibition. *Proc. Roy. Soc. Lond. B.* **108**, 305–16.

Snow, R. (1937). On the nature of correlative inhibition. *New Phytol.* **36**, 283–300.

Snow, R. (1940). A hormone for correlative inhibition. *New Phytol.* **39**, 177–84.

Sorokin, H. P. and Thimann, K. V. (1964). The histological basis for inhibition of axillary buds in *Pisum sativum* and effects of auxins and kinetin on xylem development. *Protoplasma* **59**, 326–50.

Tamas, I. A., Ozbun, J. L., Wallace, D. H., Powell, L. E. and Engels, C. J. (1979). Effect of fruits on dormancy and abscisic acid concentration in axillary buds of *Phaseolus vulgaris* L. *Plant Physiol.* **64**, 615–19.

Tamas, I. A., Engels, C. J., Kaplan, S. L., Ozbun, J. L. and Wallace, D. H. (1981). Role of indole acetic acid and abscisic acid in the correlative control by fruits of axillary bud development and leaf senescence. *Plant Physiol.* **68**, 476–81.

Taylor, I. B. and Rossall, S. (1982). The genetic relationship between the tomato mutants, flacca and lateral suppressor, with reference to abscisic acid accumulation. *Planta* **154**, 1–5.

Thimann, K. V. (1937). On the nature of inhibitions caused by auxin. *Am. J. Bot.* **24**, 407–12.

Thimann, K. V. (1939). Auxins and the inhibition of plant growth. *Biol. Rev.* **14**, 314–37.

Thimann, K. V., Sachs, T. and Mathur, K. N. (1971). The mechanism of apical dominance in *Coleus. Physiol. Plant.* **24**, 68–72.

Thimann, K. V. and Skoog, F. (1933). Studies on the growth hormone of plants. 3. The inhibiting action of the growth substance on bud development. *Proc. Nat. Acad. Sci. USA* **19**, 714–16.

Thimann, K. V. and Skoog, F. (1934). On the inhibition of bud development and other functions of growth substance in *Vicia faba. Proc. Roy. Soc. Lond.* B**114**, 317–39.

Tobin, R. S., Yun, K-B, and Naylor, J. M. (1974). Nuclear proteins of quiescent and mitotically active cells in shoot meristems of *Tradescantia paludosa. Can. J. Bot.* **52**, 2049–53.

Tucker, D. J. (1975). Far-red light as a suppressor of side shoot growth in the tomato. *Plant Sci. Lett.* **5**, 127–30.

Tucker, D. J. (1977). Hormonal regulation of lateral bud outgrowth in the tomato. *Plant Sci. Lett.* **8**, 105–11.

Tucker, D. J. and Mansfield, T. A. (1972). Effects of light quality on apical dominance in *Xanthium strumarium* and the associated changes in endogenous levels of abscisic acid and cytokinins. *Planta* **102**, 140–51.

Tucker, D. J. and Mansfield, T. A. (1973). Apical dominance in *Xanthium strumarium*. A discussion in relation to current hypotheses of correlative inhibition. *J. Exp. Bot.* **24**, 731–40.

Usciati, M., Codaccioni, M. and Guern, J. (1972). Early cytological and biochemical events induced by a 6-benzylaminopurine application on inhibited axillary buds of *Cicer arietinum* plants. *J. Exp. Bot.* **23**, 1009–20.

Vines, S. H. (1886). *Lectures on the Physiology of Plants,* Cambridge University Press.

Wardlaw, C. W. (1952). *Morphogenesis in Plants,* Methuen, London.

Wardlaw, I. F. and Mortimer, D. C. (1970). Carbohydrate movement in pea plants in relation to axillary bud growth and vascular development. *Can. J. Bot.* **48**, 229–37.

Wareing, P. F. and Nasr, T. A. A. (1958). Gravimorphism in trees. Effects of gravity on growth, apical dominance and flowering in fruit-trees. *Nature* **182**, 379–81.

Wareing, P. F. and Phillips, I. D. J. (1981). *Growth and Differentiation in Plants*, 3rd edn, Pergamon, Oxford.

Weiskopf, B. (1927). *Publ. biol. École haut., Études vét.* (Brno) **6**, No. 4 (cited by Snow, 1929).

Went, F. W. (1936). Allgemeine Betrachtungen über das Auxinproblem. *Biol. Zbl.* **56**, 449–63.

Went, F. W. (1938). Specific factors other than auxin affecting growth and root formation. *Plant Physiol.* **13**, 55–80.

Went, F. W. (1939). Some experiments on bud growth. *Am. J. Bot.* **26**, 109–17.

Went, F. W. (1953). The effect of temperature on plant growth. *Ann. Rev. Plant Physiol.* **4**, 347–62.

Wetmore, R. H. and Sorokin, S. (1955). On the differentiation of xylem. *J. Arnold Arbor.* **36**, 305–17.

White, J. C. (1973). *Apical Dominance in Leguminous Plants*, Ph.D. Thesis, University of Glasgow.

White, J. C. and Hillman, J. R. (1972). On the use of morphactin and triiodobenzoic acid in apical dominance studies. *Planta* **107**, 257–60.

White, J. C. and Mansfield, T. A. (1977). Correlative inhibition of lateral bud growth in *Pisum sativum* L. and *Phaseolus vulgaris* L.: studies of the role of abscisic acid. *Ann. Bot.* **41**, 1163–70.

White, J. C. and Mansfield, T. A. (1978). Correlative inhibition of lateral bud growth in *Phaseolus vulgaris* L.—Influence of the environment. *Ann. Bot.* **42**, 191–6.

White, J. C., Medlow, G. C., Hillman, J. R. and Wilkins, M. B. (1975). Correlative inhibition of lateral bud growth in *Phaseolus vulgaris* L. Isolation of indole acetic acid from the inhibitory region. *J. Exp. Bot.* **26**, 419–24.

Wickson, M. E. and Thimann, K. V. (1958). The antagonism of auxin and kinetin in apical dominance. *Physiol. Plant.* **11**, 62–74.

Woolley, D. J. and Wareing, P. F. (1972a). The role of roots, cytokinins and apical dominance in the control of lateral shoot form in *Solanum andigena*. *Planta* **105**, 33–42.

Woolley, D. J. and Wareing, P. F. (1972b). The interaction between growth promoters in apical dominance. II. Environmental effects on endogenous cytokinin and gibberellin levels in *Solanum andigena*. *New Phytol.* **71**, 1015–25.

Yeang, H. Y. (1980). *Ethylene and the Control of Axillary Bud Growth in* Phaseolus vulgaris *L.*, Ph.D. Thesis, University of Glasgow.

Yeang, H. Y. and Hillman, J. R. (1981a). Internodal extension in the first trifoliate leaf axillary bud of *Phaseolus vulgaris* L. following shoot decapitation. *Ann. Bot.* **48**, 25–32.

Yeang, H. Y. and Hillman, J. R. (1981b). Control of lateral bud growth in *Phaseolus vulgaris* L. by ethylene in the apical shoot. *J. Exp. Bot.* **32**, 395–404.

Yeang, H. Y. and Hillman, J. R. (1982). Lateral bud growth in *Phaseolus vulgaris* L. and the levels of ethylene in the bud and adjacent tissue. *J. Exp. Bot.* **33**, 111–17.

Yun, K. and Naylor, J. M. (1973). Regulation of cell reproduction in bud meristems of *Tradescantia paludosa*. *Can. J. Bot.* **51**, 1137–45.

Zieslin, N. and Halevy, A. H. (1976). Components of axillary bud inhibition in rose plants. I. The effect of different plant parts (correlative inhibition). *Bot. Gaz.* **137**, 291–6.

Phototropism

7

David S Dennison

Introduction

History

The history of phototropism investigations has several intertwining themes, as was nicely detailed by Curry (1969) in an earlier book. Darwin (1881) used continuous unilateral illumination in his phototropic studies, whereas Blaauw (1909) was interested in the effect of short flashes of light. This dichotomy of method persists to the present, with corresponding differences in result and difficulty of interpretation. Also, Darwin was the first to conclude that the stimulus can be perceived at the apex of a seedling and subsequently cause bending in the lower part of the plant. Alternatively, in simpler systems such as the sporangiophore of *Phycomyces*, the effect seems to be mainly local, with little evidence for transmission of a signal from a site of perception to a site of expression. The existence of stimulus transmission has of course been the major impetus for the development of our present understanding of auxins. It has been long and widely held that phototropism is based on the differential growth between the illuminated and shaded sides of the organ. Blaauw strengthened the concept of growth as the basis of phototropism by demonstrating that both *Avena* coleoptiles and *Phycomyces* sporangiophores show a light-growth response (the former a decrease and the latter an increase in growth rate) after an increase in the intensity of uniformly distributed light (Blaauw, 1919). The Cholodny–Went theory finally tied these two ideas together by proposing that light causes the lateral redistribution and longitudinal translocation of auxin and that the resulting auxin concentration gradient brings about differential growth.

Definition

What, precisely, is the phototropic response? Defined narrowly, it is the developing curvature of a growing organ in response to some asymmetric external light source. It must be narrowed further to specify that the bending of the organ is oriented in relation to the *direction* of the illumination, usually approximately towards it or away from it. Excluded therefore is polarotropism, in which the bending is oriented in relation to the electric vector of the polarized light stimulus. In laboratory situations, of course, the organ may respond to light intensity distributions having no simple relation to an external light source; these also would be examples of phototropism. As to the mechanism of bending, the most common is that of differential elongation on two sides of the organ; for example, the shaded side of a coleoptile may elongate more rapidly than the illuminated side, causing curvature towards the light direction. Alternatively, the growth may be maximal at the point of maximum illumination; this functions in a relatively opaque organ, such as the moss protonema, to cause the cell wall to 'grow out' towards the light.

Experimental conditions

The literature of phototropism contains a diversity of experimental organisms and experimental conditions. While the former is as inescapable as the diversity of life itself, the latter is more subject to control. In the hope that it will provide useful guidance to the reader, a brief discussion of some of this experimental diversity is offered.

Response The response consists of the change in orientation of the phototropic organ with time. Ideally, the data should consist of the recording with time of the shoot axis angle in three-dimensional space. Interestingly, Darwin (1881) did take the trouble to record the movement of shoots and coleoptiles as seen from above. Rather few studies since then have followed this example, perhaps because of the complex repertoire of nutational movements Darwin discovered. Most phototropic data are derived from angles projected onto a plane, under the reasonable but usually unproven assumption that little significant information is thereby lost.

How long should this bending be followed? In oat or corn coleoptiles starting from a vertical position, interference by the geotropic response becomes apparent after 100 min or so. One simple but arbitrary approach is therefore to record the angle at 100 min. The difficulty is that the method will not distinguish the speed of bending from the maximum accumulated angle of curvature. Since phototropic organs frequently differ in both of these measures, it is important to use consistent methods in making comparisons. Another attractive but experimentally more difficult approach is to use a clinostat to eliminate geotropism altogether (Pickard, Dutson, Harrison and Donegan, 1969). Using this method, both the rate of bending and the total curvature in response to a discrete stimulus may be measured with considerable confidence.

Stimulus There is even less agreement about what constitutes a standard stimulus. As mentioned above, Blaauw was one of the first to abandon the more natural conditions used by Darwin and investigate the effect of short light flashes. Blaauw was interested in the validity of the Bunsen–Roscoe reciprocity law, and in fact he found that to obtain a threshold response from *Avena coleoptiles*, a constant energy fluence (= fluence rate × time) was required, and that this constancy held over a range of exposure times from 43 h to 1 ms (Blaauw, 1909). This simplified view no longer prevails, however, and long exposure times are now known to trigger

processes that lead to qualitatively different results. For example, the classical third positive curvature (area 'C' in Fig. 7.2, p. 152) does not obey the reciprocity law and is probably due to the effect of blue light in altering the sensitivity of the system. This 'tonic' effect evidently is time limited instead of energy limited.

Pretreatment There are many variations in the handling of experimental material prior to the beginning of the experiment, although in the majority of cases dark-grown seedlings, handled in 'dim' green light, are used. One of the most persistent problems is that the effect of a red light pretreatment is overlooked. Not only is it risky to compare results from dark-grown seedlings with those from seedlings exposed to a red pretreatment, it is also questionable to assume that 'white' or 'blue' phototropic stimulus sources have too little contaminating red light to affect the results. A system so complex in dose–response relations, in spectral sensitivity, and in time sensitivity should not be trusted to behave in a simple manner!

Description of phenomenon

Fluence–response curve

Typical fluence (dose)–response curves (Fig. 7.1) were obtained by Zimmerman and Briggs (1963a,b) using initially vertical *Avena* coleoptiles pretreated with 2 h of red light. They exposed the apical 3 mm to three fluence rates (intensities) of monochromatic blue light at 436 nm, and at each fluence rate the total fluence was varied by varying the exposure time from 0.7 s to 37 min. The curvatures were measured at 100 min after the end of the exposure.

The curve at the highest fluence rate (curve 3 in Fig. 7.1) shows the characteristic features noted by earlier workers. The initial rising part of the curve and the first peak are termed the first-positive curvature, the subsequent minimum is termed the first-negative curvature, and the second rising part is termed the second-positive curvature. Although not shown in Fig. 7.1, it is possible to have another minimum after the second-positive curvature, and another peak, called the third-positive curvature.

The first-positive region is constant for the three

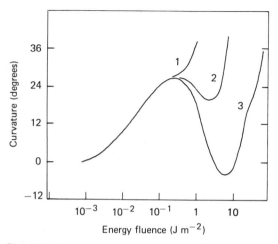

Fig. 7.1 Fluence–response curve for phototropism in *Avena* coleoptiles. Curves 1, 2, and 3 are at energy fluence rates of 3.84×10^{-4}, 3.84×10^{-3} and 3.84×10^{-2} Wm^{-2}, respectively, at a wavelength of 436 nm. (Redrawn from Zimmerman and Briggs, 1963a.)

fluence rates of Fig. 7.1, in accordance with the reciprocity law, but this is not true for the other regions. As the fluence rate is reduced, the region of the second-positive curvature shifts towards lower fluences (curves 2 and 1), and in the process the first-negative region becomes obliterated. For each 10-fold reduction in fluence rate, the fluence required to produce a given second-positive curvature, say 36°, is likewise reduced 10-fold. In other words, this curvature is characteristic of a certain exposure time rather than a particular fluence, and hence reciprocity does not hold.

Older ideas

Zimmerman and Briggs (1963b) proposed an empirical kinetic model in which first-positive and first-negative curvatures are each explained by simple two-step reactions with a pair of photosensitive forward rate constants. By appropriate adjustments of the four arbitrary constants the authors explained their experimental results in the region of the first-positive and first-negative curvature. In the region of the second-positive curvature, they proposed a very complex scheme, comprising a photoequilibrium step followed by a thermochemical step, so that at suitably long times the amount of product

accumulated becomes a linear function of time, but not intensity.

The multiplicity of light-driven steps in this model and the aberrant behavior of the second-positive curvature suggest multiple photoreceptor pigments. If this were so, it would then be expected that the shape of the fluence–response curve might be wavelength-dependent, reflecting different action spectra for each pigment. Thimann and Curry (1961) compared the dose–response curve at 436 nm (blue) with that at 365 nm (near UV), wavelengths corresponding to separate peaks in the action spectrum. The fluence–response curves for the first-positive and first-negative curvatures were found to be essentially identical and similar to those of Fig. 7.1; this result is evidence against a multiplicity of photoreceptors for first-positive and first-negative curvatures. The different behavior of the second-positive curvature might again argue for a difference in photoreceptor pigment systems involved in this response. However, the action spectrum for the second-positive curvature in *Avena* is not different from that for the first-positive curvature (Everett and Thimann, 1968), and it must be concluded that all regions of the fluence–response curve derive from excitation of a single photoreceptor. Although the complexities of the kinetic model of Zimmerman and Briggs do not correspond to complexities in the primary photoreceptor system of the phototropic response, this of course in no way invalidates the model, since its features could well be the result of many other kinds of processes, removed from the primary photo-event.

Response dependence on time and fluence rate

Fluence–response surface One of the difficulties in understanding the fluence–response relations of coleoptiles is that the measured curvature is a function of two variables: the fluence rate and the duration of the exposure. The fluence–response relationship must therefore be visualized as a three-dimensional surface, rather than a two-dimensional curve.

Blaauw and Blaauw-Jansen (1970a) have measured curvature of *Avena* coleoptiles as a function of fluence rate from 10^{-5} to 12.6 W m^{-2} and as a function of exposure time from 0.01 s to 40 min. The coleoptile was pretreated with red light, and the curvature was measured 100 min after the end of the exposure. The data have the usual appearance when plotted either as a function of fluence rate or as a function of time, holding the other factor constant in each case. However, when the curvature is displayed as a function of two variables at once, a three-dimensional surface results (Fig. 7.2). Although it has a rather unconventional appearance, this surface may be the most meaningful way to comprehend the fluence–response relation as a whole.

Fig. 7.2 Phototropic curvature of *Avena* coleoptiles (*ordinate*) as a function of blue-light fluence rate and exposure time. From *back* to *front* are increasing energy fluence rates and from *left* to *right* are increasing exposure times. A red pretreatment was used. (From Blaauw and Blaauw-Jansen, 1970a.)

To relate the topography of Fig. 7.2, to the more conventional appearance of curve 3 in Fig. 7.1, the reader should imagine the surface sectioned by vertical planes running parallel to the time axis. These sections yield response as a function of time and differ markedly when made at different fluence rates. At the highest fluence rates (12.6 W m^{-2}) the curve is quite complex, showing three maxima known classically as first-positive, second-positive, and third-positive curvature, and designated by Blaauw and Blaauw-Jansen as areas A, B, and C, respectively. As sections are made at progressively lower fluence rate values, areas

A and B shift to longer exposure times, as does the negative region between A and B (the classical first-negative response). This shift is in accordance with the reciprocity law; for example, the peak curvature of area A corresponds to a constant energy fluence (given by the authors as 0.12 J m^{-2}) equal to the product of time and fluence rate. Area C does not shift along the time axis; rather, its peak and slope increase with decreasing fluence rates, showing that reciprocity does not hold.

This pattern of changing curve shapes has the intriguing result that at progressively lower fluence rates area B becomes merged with, and finally obliterated by, the rising flank of area C. The negative region, too, becomes swallowed up and finally vanishes at a fluence rate of about 10^{-3} W m^{-2}. This strongly implies that low intensity fluence–response curves such as those of Fig. 7.1 (Zimmerman and Briggs, 1963a) do not correspond to a classical first-positive, second-positive response, but, on the contrary, correspond to a classical first-positive, third-positive response, the second-positive response (Blaauw and Blaauw-Jansen's area B) having vanished at this intensity.

Fluence–response curves at constant exposure times can be visualized as vertical sections running from back to front of the surface in Fig. 7.2. At shorter exposure times the sections do in fact correspond to the classical first-positive, second-positive curvatures, areas A and B respectively. For example, imagine the surface of Fig. 7.2 sectioned by a vertical plane at an exposure time of 15 s. Although this would look like curve 3 in Fig. 7.1, the second peak would be due to the B area and not the C area.

The fluence–response surface relationship is a convenient way to synthesize a great variety of experimental results into a single pattern. It also allows the disentangling of the second-positive and third-positive responses, and it indicates what conditions of energy fluence rate and exposure time are best suited for their individual study.

Tonic effect of blue light The complexity of the fluence–response relationship implies that each energy increment of the phototropic stimulus has two effects. It triggers a bending response, and it also changes the state of the organ so that its sensitivity to a subsequent energy increment is now altered. The latter is a tonic, or non-directional effect of blue light.

In *Avena*, the first stimulus can be an effective desensitizer even when given vertically from above. Blaauw and Blaauw-Jansen (1970b) found that the C-type (third-positive) curvature obtained with a 10-min exposure to blue light (fluence rate 1.48×10^{-2} W m^{-2}) could also be elicited if as much as the first 9 min of the exposure was given vertically instead of horizontally. Also, the vertical exposure could be shortened to 1 min, provided that a dark period of at least 2 min intervened before the subsequent horizontal exposure. This dark period could be as long as 40 min without diminishing the positive (C-type) curvature in response to the second exposure.

Since the horizontal 1-min exposure taken by itself is in the first-negative region, the increased curvature can be interpreted as a reactivation of the first-positive response caused by a lowered sensitivity. Thus the C-type curvature is not actually an independent response at all, but the result of desensitization of the first-positive response system. This desensitization requires from 5 to 10 min to become fully developed after which it gradually dies out. This process can be visualized as the movement of the A-region, or first-positive curvature, towards higher fluence rates, reaching a maximum shift 5 to 10 min after the first (desensitizing) exposure, and then moving gradually back to its original position.

Effect of red light

Curry (1957, cited in Curry, 1969) discovered that the first-positive region in *Avena* was shifted to higher fluence rates by red light. These results were confirmed by Zimmerman and Briggs (1963a), who showed that the regions of first-positive and first-negative curvature were shifted to higher fluence rates (10-fold shift) by a red-light exposure immediately prior to blue-light exposure. They found that the second-positive region was shifted towards lower fluence rates (3-fold shift) by this same red-light pretreatment. Thus the red pretreatment lowers the sensitivity of the first-positive and first-negative response, and raises the sensitivity of the second-positive response (the C-type curvature of Blaauw and Blaauw-Jansen, 1970a). In terms of the fluence-response surface (Fig. 7.2), red pretreatments shift the A-area curvature toward higher fluences and the C-area curvature toward lower fluences which amounts to an increase in C-area

curvature (Blaauw and Blaauw-Jansen, 1970a). A different situation exists for area B, which disappears completely if no red pretreatment is used, leaving only a broad indifferent region between the two minima (Blaauw and Blaauw-Jansen, 1970a).

In addition to an increase in the second-positive (C-type) curvature (Fig. 7.2), red light enhances the rate of elongation near the coleoptile tip (Curry, Thimann and Ray, 1956). This change in elongation distribution may be related to the effect of red light on the localization of curvature. When blue stimuli in the fluence range of the C-type curvature were used, red pretreatment shifted the locus of maximum curvature towards the coleoptile tip (Blaauw and Blaauw-Jansen, 1970b).

This distribution of red-light sensitivity along the coleoptile is fairly well correlated with the distribution of phytochrome: both were most heavily concentrated in the apical 3 mm (Briggs and Chon, 1966). Fluence-response comparison of the two red-light effects had quite unexpected results, however. The sensitivity of the physiological effect on phototropism is a factor of 10^3 or 10^4 higher than that of the effect on spectrophotometrically detectable phytochrome. This enormous difference in sensitivity indicates that if both effects are indeed due to phytochrome, the pigment must be divided into two pools: a large spectrophotometrically detectable one, and a small spectrophotometrically undetectable one that is, however, physiologically active. Further data collected by Briggs and Chon (1966), concerning the decay kinetics of P_{fr} and P_r and the shape of the fluence-response curves, indicate that if the same phytochrome pigment is responsible for both effects, the pools must be functionally distinct, and the photochemical rate constants for P_r–P_{fr} interconversion must be different within each pool.

Tip and base responses

Much has been made of the separation of *Avena* phototropism into the so-called tip and base responses (Thimann and Curry, 1960; Curry, 1969). According to their view, only the tip is sensitive to low fluence stimuli, and only the tip is responsible for the first-positive curvature, which is slowly propagated down the length of the coleoptile. The sensitivity to larger stimuli is not confined to the tip, but rather is distributed throughout the entire coleoptile and is hence called the base response. This base response is thought to be triggered only by large stimuli (in the fluence region of the second-positive curvature), and it is not propagated, but develops at the same time all along the coleoptile. Support for this distinction is largely morphological, in that a coleoptile with a pure tip response is bent relatively sharply near the apex, whereas one with a pure base response is bent more or less uniformly along the coleoptile (Thimann and Curry, 1960; Briggs, 1964). More recently, Blaauw and Blaauw-Jansen (1970b) have maintained that the distinction between tip and base curvatures is spurious, and arises from uncontrolled red-light effects. They found that without a red pretreatment the curvature was uniformly distributed, and that red-light shifts the curvature towards the coleoptile tip. Further, they report that stimuli in both the first-positive and the second-positive regions produce the same tip curvature.

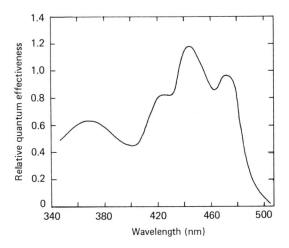

Fig. 7.3 Action spectrum for phototropism in *Avena* coleoptiles, showing the quantum effectiveness for a curvature of 10°, relative to the response at 436 nm. (After Thimann and Curry, 1960.)

Perception of stimulus

Action spectra

The analysis of a photobiological process requires the identification of the primary photoreceptor pigment, and the first step in such identification is the determination of an action spectrum. The action spectra for the blue-light type of response (in which red is phototropically ineffective), typical of coleoptiles, *Phycomyces* sporangiophores and many other systems, will be described first.

An *Avena* action spectrum, with quantum effectiveness (the inverse of the quantum requirement for a given curvature) related to a curvature of 10°, is shown in Fig. 7.3 (Thimann and Curry, 1960; Curry and Thimann, 1961). There are two prominent peaks and a shoulder in the blue region (475, 450, and 420 nm), and there is a peak in the near-ultraviolet region (370 nm). The *Avena* action spectrum in the ultraviolet region shows a prominent peak at about 290 nm (Fig. 7.4, Curry *et al.*, 1956). Since the curve in Fig. 7.4 was obtained using coleoptiles with caps over the tips (apical 2–4 mm shielded), it is not possible to draw any conclusions from the relative peak heights in Figs. 7.3 and 7.4.

Phycomyces action spectra are determined most

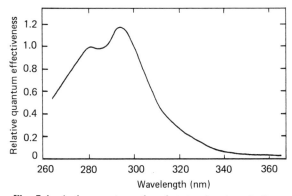

Fig. 7.4 Action spectrum for phototropism in apically-shielded *Avena* coleoptiles, showing the quantum effectiveness for a curvature of 8.6°, relative to the response at 289 nm. (After Curry *et al.*, 1956.)

advantageously with a balance method, since the adaptive mechanism makes it impossible to obtain the usual fluence–response curve. One method (Curry and Gruen, 1959) balances a standard stimulus against the monochromatic stimulus, each source being located at one end of a 2-m plank. Cultures are placed along the plank, and the point of balance is determined. Action spectra for the visible (Fig. 7.5), in which the standard stimulus is a source with peak wavelength at 460 nm, and for the ultraviolet

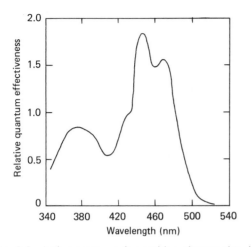

Fig. 7.5 Action spectrum for positive phototropism in *Phycomyces* sporangiophores. The determinations were made by balancing the phototropic effectiveness of a standard source at 436 nm against that of a test source. On the *ordinate* are the reciprocals of the quanta needed for balance at each wavelength, relative to the quanta needed at the standard wavelength. (After Curry and Gruen, 1959.)

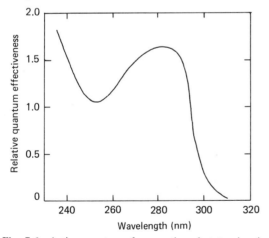

Fig. 7.6 Action spectrum for negative phototropism in *Phycomyces* sporangiophores. The standard source was at 254 nm. Otherwise as in Fig. 7.5. (After Curry and Gruen, 1959.)

(Fig. 7.6), in which the standard stimulus is a 254-nm germicidal lamp, are at first glance similar to those for *Avena* (Figs 7.3 and 7.4, respectively). However, in contrast to the behavior of coleoptiles,

Phycomyces sporangiophores show negative curvatures at ultraviolet wavelengths. This reversal of *Phycomyces* phototropism is due to the strong ultraviolet absorption by the sporangiophore. There is an indifferent region, where neither positive nor negative curvatures occur, between 310 nm and 334 nm. If a single standard is used in both the ultraviolet and visible ranges, the resulting curves can be quantitatively compared. Using this method, Delbrück and Shropshire (1960) found that the UV peak at 280 nm is about four-fold higher than the main peak at 455 nm. Here also, the reason lies in the strong absorption by the cell at 280 nm, leading to an exaggerated intensity gradient across the cell. The growth–response action spectrum is closely similar, except that the magnitude of its 280-nm peak does not exceed that of the 455-nm peak. The light–growth response is positive at all wavelengths.

A strikingly different action spectrum was found for the chloronemal filaments of protonemata of the moss, *Physcomitrium turbinatum* (Nebel, 1968). These structures grow at the apex, and the initiation of a phototropic bend occurs by a 'growing out' mechanism, in which the axial growing point shifts towards the light source. Subsequent growth follows this new direction. The action spectrum (Fig. 7.7) was obtained in an aqueous medium using the balance method and a standard source of wavelength 665 nm. Blue light is almost totally without

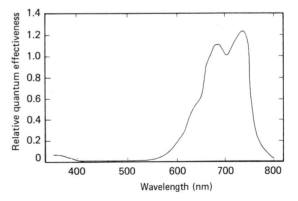

Fig. 7.7 Action spectrum for phototropism in protonemata of *Physcomitrium turbinatum*. Determinations were made by balancing the phototropic effectiveness of a standard source at 665 nm against that of a test source at each wavelength. Otherwise as in Fig. 7.5. (After Nebel, 1968.)

phototropic effect, although there is a slight suggestion of activity at 365 nm. Phototropism in moss protonemata thus uses a different photoreceptor pigment than in *Phycomyces* and most higher plants. When a correction is made to the action spectrum for the screening action of chlorophyll, the result resembles the absorption spectrum of phytochrome in the P_{fr} form (*see* Chapters 16, 17).

Identity of the photoreceptor pigment

Phototropic action spectrum *Avena* coleoptiles and *Phycomyces* sporangiophores have very nearly identical action spectra, and this strongly indicates that both organisms have the same phototropic photoreceptor pigment. Many other action spectra are similar and are probably due to the same blue-absorbing pigment. This similarity suggests that there may be a unity in phototropic mechanism for these systems, and that such unity is built upon a unity of photoreceptor pigment. Is this pigment a flavin or a carotenoid?

There is little basis for choice in the visible region, where the absorption spectra of both flavins and carotenoids are generally a good fit with the action spectra. Carotenoids have a characteristic triple-peak structure in the region around 450 nm, and although riboflavin absorption spectra shown in earlier reports (Thimann and Curry, 1960) are devoid of fine structure in this region, spectra obtained in ethanol at 77 K (Sun, Moore and Song, 1972) do indeed have the requisite shape. More pronounced discrepancies show up in the near ultraviolet, however, where the riboflavin absorption peak in ethanol at 77 K is shifted toward shorter wavelengths by about 30 nm, as compared to the action spectrum. The carotenoid absorption spectrum typically lacks peaks altogether in the 300–400 nm range. Both pigments generally have another peak near 280 nm, although the magnitudes differ. Thus neither flavins nor carotenoids can be ruled out conclusively on the basis of these spectral comparisons.

Growth action spectrum Action spectra have been obtained for the *Phycomyces* light–growth response, the transient change of growth rate as a result of a change in the steady-state light intensity. Using a null point method, Delbrück and Shropshire (1960) obtained a growth–response action spectrum essentially identical to the phototropic spectrum,

except that the size of the 280 nm peak was different, due to the screening effect of UV-absorbing substances. This work has been extended by Delbrück, Katzir and Presti (1976) in the region from 575 to 630 nm, using a tunable laser. The authors discovered a small shoulder at about 600 nm, whose magnitude is about 10^{-9} relative to the main peak at 455 nm. This shoulder is attributed to the 'forbidden' triplet excited state of riboflavin, known from spectroscopic data (Sun *et al.*, 1972) to be centered at 600 nm. Another possibility is that a red-sensitive pigment is involved in the light-growth system.

Albino mutants Meissner and Delbrück (1968) analyzed the carotenoids of an albino *Phycomyces* strain with normal phototropism and were unable to detect any β-carotene. Using methods 10 times more sensitive, Presti, Hsu and Delbrück (1977) have detected no β-carotene in albino *Phycomyces* mutants doubly blocked for β-carotene synthesis but with normal phototropism. Thus, the photoreceptor in *Phycomyces* is probably not β-carotene but a flavin.

Red-absorbing photoreceptors As mentioned above, moss protonemata are phototropically responsive to red light, suggesting the involvement of a red-absorbing pigment such as phytochrome, rather than the more common blue-absorbing photoreceptor pigment.

It has recently been discovered that *Phycomyces* shows phototropism to red light, with a very weak but unmistakable maximum in the action spectrum at 650 nm. Although this red peak is similar to that found in the growth action spectrum, comparison of the phototropic action spectra of wild type and 'night blind' mutants suggests that the red phototropic peak is due to a second photoreceptor pigment and not to the excitation of a flavin triplet state in the primary pigment (Galland, 1982). An interesting red–blue interaction has also been described in *Phycomyces* by Löser and Schäfer (1980). When red light (605 nm) is combined with blue light (450 nm), the phototropic response is less than that for blue light alone. Thus it appears that red light has a dual effect; it can either trigger phototropism when acting alone, or it can inhibit the phototropic response to blue light. Although this could be the result of a single photochromic pigment like phytochrome, the mutant evidence favors two independent pigments.

It is not known whether such red-sensitivity may exist in coleoptile phototropism. Since the size of the red peak in the *Phycomyces* action spectrum is about 10^8-fold smaller than the main blue peak, a comparable red peak may well have gone undetected in previous work with coleoptiles.

Location and orientation of photoreceptor

The locus of maximum phototropic sensitivity of coleoptiles is thought to be the apex. This sensitivity is not confined to the extreme tip, however, but extends downwards for a considerable distance. For example, Meyer (1969a) obtained phototropic responses by stimulating either the apical 350 μm of the coleoptile or the next 350 μm-segment with 5-s blue light stimuli. Although curvatures were twice as large when the apical segment was stimulated, the fluence–response curve was not shifted along the fluence axis, but was merely changed in amplitude. This suggests that the sensitivity of the more basal section cannot be attributed to scattered light reaching the apical section indirectly, but must instead be due to the existence of a pigment in the more basal section.

If continuous illumination is used instead of short flashes, significant phototropic curvatures can be obtained even if as much as 5 mm of the apex are covered by an opaque cap. Although Curry observed only 'small' curvatures in capped coleoptiles after 100 min (Curry, 1969), much greater curvatures can be obtained after several hours, especially if geotropism is suppressed by the use of a clinostat (Franssen, Firn and Digby, 1982).

The use of microbeam stimuli has shown that the phototropic sensitivity of *Phycomyces* sporangiophores extends from 0.1 to about 2 mm below the sporangium (Cohen and Delbrück, 1959; Delbrück and Varjú, 1961). Page and Curry (1966) found that the phototropic sensitivity of young sporangiophores of the fungus *Pilobolus kleinii* was confined to the terminal 50 μm of the cell. Illumination of the more basal region, which has a high concentration of a carotenoid pigment, produced little or no phototropic curvature. Thus in these cells the photoreceptor is much more localized than in *Phycomyces*. An extremely localized receptor is also found in the alga *Vaucheria geminata*, which grows only at the extreme tip and responds to a phototropic stimulus by a lateral shift of the growth center. Kataoka (1975)

gradually increased the fraction of the hemispherical tip illuminated by the half field and found that bending reached only 50% of normal when all but the terminal 20% was illuminated. This shows that most of the photosensitivity is concentrated in the extreme tip of this cell, perhaps the apical 10 μm.

The existence of ordered, dichroic photoreceptors in *Phycomyces* sporangiophores has been well established. Under conditions of high optical density, Jesaitis (1974) found that transversely polarized light was more effective than longitudinally polarized light in eliciting the growth response. He further concluded that the orientation of the blue and ultraviolet dipole moments of the receptor molecule are 5.5° and 33.5°, respectively, from the transverse (hoop) axis. Dichroic, oriented photoreceptors have also been postulated in *Phycomyces* sporangiophores by Hertel (1980) and by Wulff (1974), but their measurements of the effect of the plane of polarization on the time of onset of phototropism seem to support a radial, rather than a transverse orientation. In moss protonemata, the finding that the phototropic effectiveness of red or far-red light is maximal when the plane of polarization of the 45° incident beam is vertical, leads to the conclusion that the receptors are near the cell surface and the preferred dipole orientation is parallel to it (Nebel, 1969).

Signal processing after perception

Intensity distribution

The phototropic response is directed toward or away from a source of light, and this response must be based on the distribution of light intensity set up within that organ by purely optical mechanisms. Such a distribution will not be a simple gradient of intensity across the organ if it is created principally through optical refraction by the cell. On the other hand, transmission losses due to internal scattering or absorption will lead to a simple gradient.

Focusing The lens property of the sporangiophore of *Phycomyces* has been shown to be the principal means of establishing the intensity distribution. This was proved by the reversal of the direction of phototropism when the cell is immersed in a medium of higher refractive index (*e.g.*, paraffin oil), but Shropshire (1962) demonstrated it still more elegantly by showing that the illumination of a sporangiophore with a diverging ray bundle also reverses phototropism. The receptors are apparently able to detect this intensity distribution by virtue of their circulation around the periphery of the cell in synchrony with the cell wall's spiral growth. The introduction of an artificial rotation in the opposite direction greatly reduces the strength of the phototropic response (Dennison and Foster, 1977).

Phototropism in *Avena* coleoptiles is also reversed by immersion in paraffin oil, but it is not clear whether the mechanism is that of a focusing effect. Meyer (1969b) found that soaking in oil prior to unilateral illumination in air also produces reversal. She found that reversal occurs only when this oil treatment is given at least 1 min before the 5-s light stimulus. Penetration of the oil into the intercellular spaces and into the upper $300\,\mu m$ of the central cavity was observed microscopically. Although Meyer concluded that the reversal was due to physical changes in the cell wall structure, it also seems possible that optical effects may play a role at the cellular level in coleoptiles. Direct evidence for a focusing mechanism in *Avena* was provided by Shropshire (1975), who reported that coleoptiles showed negative phototropism when exposed in air to a diverging beam of light, using a fluence in the region of first-positive curvature. A converging beam at the same fluence produced positive phototropism.

Screening In the absence of focusing effects, the light gradient across the cell will be determined by light absorption and scattering due to the cell contents.

Although in *Phycomyces* the focusing effect is dominant, there are experimental conditions under which the screening becomes large and phototropism reverses sign. The reversal in ultraviolet light is due to the high concentration of gallic acid in the cell which is strongly absorbant at these wavelengths (Delbrück and Shropshire, 1960). In addition, a mutant strain with a superabundant amount of β-carotene has an optical density about ten times normal and shows negative phototropism in blue light (Jesaitis, 1974).

Screening is important in moss protonemata, in which the cell apex contains high concentrations of chlorophyll. Apparently the lens effect is of no significance in this case, and phototropism is caused by a 'growing out' phenomenon on the illuminated side of the filament apex. In support of this are the observations that light stimulates the growth rate and that there is no reversal of phototropism in paraffin oil (Nebel, 1968).

Photoelectric potentials

Following the electronic excitation of the photoreceptor molecule, but before changes occur in the distribution of auxin, there must intervene a sequence of signal-processing events. A clue to these events may be found in studies of light-induced biopotential changes, although there is no easy way to distinguish signals that relate directly to the causal sequence from those that are merely byproducts.

Coleoptiles develop transverse electric potentials when illuminated unilaterally, but the phenomenon is probably not related in any way to signal processing. Johnsson (1965) found that with a unilateral stimulus of up to 10 min the potential across a corn coleoptile does not appear until 30 min after the stimulus, suggesting that the potential may be a result of the auxin redistribution rather than its cause. A light-induced signal of short latency in *Avena* coleoptiles has been briefly reported by Briggs (1976). The potential between the tip and the base of the coleoptile changed within 1 s of the beginning of a blue-light stimulus and could well be related to early intracellular signal processing.

Interesting light-induced biopotential changes have been observed by Hartmann and Schmid (1980) in etiolated bean hooks. Biopotential changes occurred within 1–5 s after the beginning of the light pulse and reached a maximum within 2 min. Although not phototropism, hook opening is nevertheless a light-induced growth phenomenon, and it has a typical blue-light action spectrum. Thus it can be expected that future work will confirm the existence of comparable biopotentials in coleoptiles.

Metabolic effects

Although there have been several reports of light-induced changes in ATP and cyclic AMP (cAMP) levels in *Phycomyces* sporangiophores, these findings have been difficult to confirm (Shropshire and

Bergman, 1968; Cohen, 1974; Leutwiler and Brandt, 1983). Cyclic AMP has also been implicated in *Vaucheria* phototropism by Kataoka (1977), who found that it greatly reduced phototropic bending without affecting the linear growth rate. Kataoka proposed that the cAMP functions in the cell's 'steering mechanism', which can be swamped by an excess of the substance.

The use of low temperature is an attractive approach to the study of signal processing because in this way it may be possible to separate the primary photo-event from the subsequent processes.

In *Avena* coleoptiles, a 45-min unilateral blue-light stimulus delivered at 2°C was found to remain in 'cold storage' for at least 450 min of additional time in the cold (Pickard, Dutson, Harrison and Donegan, 1969). Data of these authors also suggest that some signal processing occurs even at 2°C: at 25°C a 10-min exposure produced a 19° curvature, whereas at 2°C a 30-min exposure at the same fluence rate was required for the same curvature. Thus at least one step subsequent to the primary photochemical one continues in the cold, but at a lower rate. Other steps must be blocked almost completely at 2°C.

Lateral auxin redistribution

Lateral transport

Briggs (1963) used the standard *Avena* curvature test to determine the auxin content of agar blocks placed beneath the basal end of illuminated corn coleoptile tips. In the range of either the first-positive or the second-positive curvatures, there was no significant difference between the auxin yields of the illuminated tips and those of the dark controls, thus indicating that under the usual conditions for phototropism no significant photo-destruction of the total diffusible auxin occurs. When coleoptile tips were used that had been sectioned longitudinally from the base up to, but not including, the extreme apex, and into which thin glass barriers had been inserted between the sectioned portions, it was found that the shaded half of the tip delivered significantly more auxin into the receiver block. If the tips were completely divided

and the halves totally separated by a thin glass barrier, no auxin differences were noted, thus indicating that some substance must move laterally in the intact part of the apex.

Lateral asymmetry of indoleacetic acid (IAA) has been demonstrated by Pickard and Thimann (1964), who applied ^{14}C-IAA to the extreme apex of excised corn coleoptiles in which the basal 1 mm was partitioned by razor blades. When a unilateral white light fluence corresponding to the first-positive curvature was administered, the radioactivity recovered in the basal agar blocks was in the ratio of 25:75, for the illuminated and shaded sides, respectively. Under similar conditions the radioactivity of the illuminated and dark halves of the coleoptile tissue itself had the ratio of 35:65. Further, there was no evidence for any effect of illumination on the total radioactivity recovered, ruling out a photo-inactivation mechanism. Thus, in this system at least, the IAA applied to the apex becomes asymmetrically redistributed, establishing a 'concentration' gradient in the tissue itself, and finally leading, via basipetal transport, to a concentration difference in the agar blocks. Further, since in these experiments the unilateral light stimulus was administered before the ^{14}C-IAA was applied, the light clearly does not act on the applied IAA itself, but rather sets up in the tissue a lateral polarity, which subsequently leads to an asymmetric distribution of IAA.

Several workers have noted rather variable results when the IAA application was not rigorously limited to the extreme coleoptile tip, suggesting that the site of origin of the asymmetric redistribution is highly localized at the apex (Pickard and Thimann, 1964; Gardner, Shaw and Wilkins, 1974). Another approach to this problem is that of a clearly off-center IAA application, which gives an unequivocal test of lateral transport. Using this method, Gardner *et al.* (1974) found good evidence for lateral transport: intact corn coleoptiles exposed to blue light producing 11.6° (first-positive) curvature in controls had 31.1% of the radioactivity on the side opposite the point of application when the point of application was illuminated and 15.8% on the opposite side when the point of application was in darkness (Fig. 7.8). These results indicate a 'net lateral movement' of 15.3% from the illuminated side to the dark side. Unfortunately, this technique did not always give consistent quantitative correlations between radioactivity differentials and curvatures, notably when comparing intact and excised tips and when introducing a red pretreatment. This may be due to a problem with the technique, or to the existence of hormones other than IAA capable of causing differential growth.

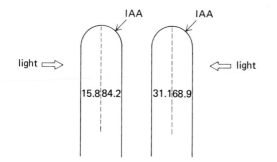

Fig. 7.8 Lateral transport of radioactive IAA applied asymmetrically to intact *Zea* coleoptile tips. Numbers show percentages of radioactivity recovered from the two sides of the uppermost 25 mm of the coleoptile. (Data from Gardner *et al.*, 1974.)

In excised third internodes of pea seedlings (*Pisum sativum*), Kang and Burg (1974) were able to show that ^{3}H-IAA applied to the apical end of the section in lanolin paste became laterally asymmetrically distributed in the ratio of 35:65 (light:dark) under conditions where a phototropic curvature of 24° developed. Remarkably, a red pretreatment that nearly doubled the curvature did not significantly change the lateral IAA gradient, again showing that curvature does not correlate quantitatively with IAA gradients.

In light-grown sunflower seedlings (*Helianthus annuus* L.), Bruinsma, Karssen, Benschop and Van Dort (1975) found no detectable gradient of IAA between the concave and convex flanks of phototropically bending plants when the flanks were extracted and analyzed for IAA by a spectrophotometric method. The authors found that the cut basal end of the hypocotyl secreted considerable auxin into a water or buffer solution, as measured by *Avena*-test activity. However, when spectroscopic IAA determinations were made on these diffusates, only a few per cent of this auxin activity could be accounted for. Thus, IAA is probably not responsible for differential growth in *Helianthus*. In the same organism, Phillips (1972) found a ratio of 12:88 (light:dark) in the diffusible gibberellins extracted from unilaterally illuminated excised hypocotyls. Hormone activity was measured on chromatographed extracts using a lettuce hypocotyl elongation test. Thus it seems likely that *Helianthus* phototropism involves a hormone distinct from IAA, possibly a gibberellin.

Polar transport

Light inhibition of longitudinal polar IAA transport was shown by Shen-Miller, Cooper and Gordon (1969), who applied ^{14}C-IAA solutions to *Avena* coleoptile tips and measured the effect of light on the subsequent recovery of radioactive IAA from basal sections. They found that the inhibition of longitudinal IAA transport, relative to dark controls, had approximately the same relationship to fluence and wavelength as did the phototropic response. Thornton and Thimann (1967) studied the transport of ^{14}C-IAA applied to the apical ends of 4-mm sections of *Avena* coleoptiles. The transport rate, measured as increased radioactivity in basal receiver blocks, showed a 12 to 17% transient decrease following a 15-min blue-light stimulus. Photo-inhibition of polar auxin transport in corn coleoptiles was also found by Hager and Schmidt (1968), who introduced ^{14}C-IAA to the apex in a block of agar and measured the accumulation of radioactivity in a receiver block at the base of the section. Light inhibition of auxin transport was observed, even though the light dose was completed before the ^{14}C-IAA block was added. Inhibition was observed with monochromatic illumination at 450 nm, but not at 660 nm.

Further reading

Dennison, D. S. (1979). Phototropism. In *Encyclopedia of Plant Physiology*, New Series, vol. 7, eds W. Haupt and M. E. Feinleib, Springer-Verlag, Berlin, Heidelberg, New York, pp. 506–66.

Firn, R. D. and Digby, J. (1980). The establishment of tropic curvatures in plants. *Ann. Rev. Plant Physiol.* **31**, 131–48.

Gressel, J. and Horwitz, B. (1982). Gravitropism and phototropism. In *Molecular Biology of Plant Development*, eds. H. Smith and D. Grierson, University of California Press, pp. 405–33.

Hertel, R. (1980). Phototropism of lower plants. In *Photoreception and Sensory Transduction in Aneural Organisms*, eds F. Lenci and G. Colombetti, Plenum Press, New York, pp. 89–105.

Russo, V. E. A. and Galland, P. (1980). Sensory physiology of *Phycomyces blakesleeanus*. In *Molecular Structure and Sensory Physiology*, ed. P. Hemmerich, Springer-Verlag, Berlin, Heidelberg, New York, pp. 71–110.

Schmidt, W. (1980). Physiological bluelight reception. In *Molecular Structure and Sensory Physiology*, ed. P. Hemmerich, Springer-Verlag, Berlin, Heidelberg, New York, pp. 1–44.

References

Blaauw, A. H. (1909). Die Perzeption des Lichtes. *Rec. trav. bot. neerl.* **5**, 209–372.

Blaauw, A. H. (1919). Licht und Wachstum. III. *Meded. Landbouwhoogeschool Wageningen* **15**, 89–204.

Blaauw, O. H. and Blaauw-Jansen, G. (1970a). The phototropic responses of *Avena* coleoptiles. *Acta Bot. Neerl.* **19**, 755–63.

Blaauw, O H. and Blaauw-Jansen, G. (1970b). Third positive (c-type) phototropism in the *Avena* coleoptile. *Acta Bot. Neerl.* **19**, 764–76.

Briggs, W. R. (1963). Mediation of phototropic responses of corn coleoptiles by lateral transport of auxin. *Plant Physiol.* **38**, 237–47.

Briggs, W. R. (1964). Phototropism in higher plants. In *Photophysiology*, vol. 1, ed. A. C. Giese, Academic Press, New York, pp. 223–71.

Briggs, W. R. (1976). The nature of the blue light photoreceptor in higher plants and fungi. In *Light and Development*, ed. H. Smith, Butterworth, London, pp. 7–18.

Briggs, W. R. and Chon, H. P. (1966). The physiological versus the spectrophotometric status of phytochrome in corn coleoptiles. *Plant Physiol.* **41**, 1159–66.

Bruinsma, J., Karssen, C. M., Benschop, M. and Van Dort, J. B. (1975). Hormonal regulation of phototropism in the light-grown sunflower seedling, *Helianthus annuus* L.: immobility of endogenous indoleacetic acid and inhibition of hypocotyl growth by illuminated cotyledons. *J. Exp. Bot.* **26**, 411–18.

Cohen, R. and Delbrück, M. (1959). Photoreactions in *Phycomyces*. Growth and tropic responses to the stimulation of narrow test areas. *J. Gen. Physiol.* **42**, 677–95.

Cohen, R. J. (1974). Cyclic AMP levels in *Phycomyces* during a response to light. *Nature* **251**, 144–6.

Curry, G. M. (1957). *Studies on the Spectral Sensitivity of Phototropism*. Ph.D. thesis, Harvard University, USA.

Curry, G. M. (1969). Phototropism. In *The Physiology of Plant Growth and Development*, ed. M. B. Wilkins, McGraw-Hill, New York, pp. 241–73.

Curry, G. M. and Gruen, H. E. (1959). Action spectra for the positive and negative phototropism of *Phycomyces* sporangiophores. *Proc. Nat. Acad. Sci. USA* **45**, 797–804.

Curry, G. M. and Thimann, K. V. (1961). Phototropism: the nature of the photoreceptor in higher and lower plants. In *Progress in Photobiology*, eds. B. C. Christensen and B. Buchmann, Elsevier, New York, pp. 127–34.

Curry, G. M., Thimann, K. V. and Ray, P. M. (1956). The base curvature of *Avena* to the ultraviolet. *Physiol. Plant.* **9**, 429–40.

Darwin, C. and Darwin, F. (1881). *The Power of Movement in Plants*, D. Appleton and Co., New York.

Delbrück, M., Katzir, A. and Presti, D. (1976). Responses of *Phycomyces* indicating optical excitation of the lowest triplet state of riboflavin. *Proc. Nat. Acad. Sci. USA* **73**, 1969–73.

Delbrück, M. and Shropshire, W., Jr (1960). Action and transmission spectra of *Phycomyces*. *Plant Physiol.* **35**, 194–204.

Delbrück, M. and Varjú, D. (1961). Photoreactions in *Phycomyces*. Responses to the stimulation of narrow test areas with ultraviolet light. *J. Gen. Physiol.* **44**, 1177–88.

Dennison, D. S. and Foster, K. W. (1977). Intracellular rotation and the phototropic response of *Phycomyces*. *Biophys. J.* **18**, 103–23.

Everett, M. and Thimann, K. V. (1968). Second positive phototropism in the *Avena* coleoptile. *Plant Physiol.* **43**, 1786–92.

Franssen, J. M., Firn, R. D. and Digby, J. (1982). The role of the apex in the phototropic curvature of *Avena* coleoptiles. I. Positive curvature under conditions of continuous illumination. *Planta* **155**, 281–6.

Galland, P. (1983). Action spectra of photogeotropic equilibrium in *Phycomyces* wild type and three behavioral mutants. *Photochem. Photobiol.* **37**, 221–8.

Gardner, G., Shaw, S. and Wilkins, M. B. (1974). IAA transport during the phototropic responses of intact *Zea* and *Avena* coleoptiles. *Planta* **121**, 237–51.

Hager, A. and Schmidt, R. (1968). Auxintransport und Phototropismus. I. Die lichbedingte Bildung eines Hemmstoffes für den Transport von Wuchsstoffen in Koleoptilen. *Planta* **83**, 247–71.

Hartmann, E. and Schmid, K. (1980). Effects of UV and blue light on the biopotential changes in etiolated hypocotyl hooks of dwarf beans. In *The Blue Light Syndrome*, ed. H. Senger, Springer-Verlag, Berlin, Heidelberg, New York, 221–37.

Hertel, R. (1980). Phototropism of lower plants. In *Photoreception and Sensory Transduction in Aneural Organisms*, eds. F. Lenci and G. Colombetti, Plenum Press, New York, pp. 89–105.

Jesaitis, A. J. (1974). Linear dichroism and orientation of the *Phycomyces* photopigment. *J. Gen. Physiol.* **63**, 1–21.

Johnsson, A. (1965). Photoinduced lateral potentials in *Zea mays. Physiol. Plant.* **18**, 574–6.

Kang, B. G. and Burg, S. P. (1974). Red light enhancement of the phototropic response of etiolated pea stems. *Plant Physiol.* **53**, 445–8.

Kataoka, H. (1975). Phototropism in *Vaucheria geminata* I. The action spectrum. *Plant Cell Physiol.* **16**, 427–37.

Kataoka, H. (1977). Phototropic sensitivity in *Vaucheria geminata* regulated by 3',5'-cyclic AMP. *Plant Cell Physiol.* **18**, 431–40.

Leutwiler, L. S. and Brandt, M. (1983). Absence of significant light-induced changes in cAMP levels in sporangiophores of *Phycomyces blakesleeanus, J. Bacteriol.* **153**, 555–7.

Löser, G. and Schäfer, E. (1980). Phototropism in *Phycomyces*: a photochromic sensor pigment? In *The Blue Light Syndrome*, ed. H. Senger, Springer-Verlag, Berlin, Heidelberg, New York, pp. 244–50.

Meissner, G. and Delbrück, M. (1968). Carotenes and retinal in *Phycomyces* mutants. *Plant Physiol.* **43**, 1279–83.

Meyer, A. M. (1969a). Versuche zur 1. positiven und zur negativen phototropischen Krümmung der *Avena*-koleoptile: I. Lichtperception und Absorptionsgradient. *Z. Pflanzenphysiol.* **60**, 418–33.

Meyer, A. M. (1969b). Versuche zur 1. positiven und zur negativen phototropischen Krümmung der *Avena*-koleoptile: II. Die Inversion durch Paraffinöl. *Z. Pflanzenphysiol.* **61**, 129–34.

Nebel, B. J. (1968). Action spectra for photogrowth and phototropism in protonemata of the moss *Physcomitrium turbinatum. Planta* **81**, 287–302.

Nebel, B. J. (1969). Responses of moss protonemata to red and far-red polarized light: evidence for disc-shaped phytochrome photoreceptors. *Planta* **87**, 170–9.

Page, R. M. and Curry, G. M. (1966). Studies on phototropism of young sporangiophores of *Pilobolus kleinii. Photochem. Photobiol.* **5**, 31–40.

Phillips, I. D. J. (1972). Diffusible gibberellins and phototropism in *Helianthus annuus. Planta* **106**, 363–7.

Pickard, B. G., Dutson, K., Harrison, V. and Donegan, E. (1969). Second positive phototropic response patterns of the oat coleoptile. *Planta* **88**, 1–33.

Pickard, B. G. and Thimann, K. V. (1964). Transport and distribution of auxin during tropistic response. II. The lateral migration of auxin in phototropism of coleoptiles. *Plant Physiol.* **39**, 341–50.

Presti, D., Hsu, W.-J. and Delbrück, M. (1977). Phototropism in *Phycomyces* mutants lacking β-carotene. *Photochem. Photobiol.* **26**, 403–5.

Shen-Miller, J., Cooper, P. and Gordon, S. A. (1969). Phototropism and photoinhibition of basipolar transport of auxin in oat coleoptiles. *Plant Physiol.* **44**, 491–6.

Shropshire, W., Jr. (1962). The lens effect and phototropism of *Phycomyces. J. Gen. Physiol.* **45**, 949–58.

Shropshire, W., Jr. (1975). Phototropism. In *Progress in Photobiology*, ed. G. O. Schenk, Springer-Verlag, Berlin, Heidelberg, New York, pp. 1–5.

Shropshire, W., Jr. and Bergman, K. (1968). Light induced concentration changes of ATP from *Phycomyces* sporangiophores: a reexamination. *Plant Physiol.* **43**, 1317–18.

Sun, M., Moore, T. A. and Song, P.-S. (1972). Molecular luminescence studies of flavins. I. The excited states of flavins. *J. Am. Chem. Soc.* **94**, 1730–40.

Thimann, K. V. and Curry, G. M. (1960). Phototropism and phototaxis. In *Comparative Biochemistry: a Comparative Treatise*, vol. 1, *Sources of Free Energy*, eds. M. Florkin and H. S. Mason, Academic Press, New York, pp. 243–309.

Thimann, K. V. and Curry, G. M. (1961). Phototropism. In *Light and Life*, eds W. D. McElroy and B. Glass, The Johns Hopkins Press, Baltimore, pp. 646–70.

Thornton, R. M. and Thimann, K. V. (1967). Transient effects of light on auxin transport in the *Avena* coleoptile. *Plant Physiol.* **42**, 247–57.

Wulff, U. (1974). *Untersuchungen über lokalisierte Adaptation und Orientierrung der Photorezeptoren bei* Phycomyces. Staatsexamensarbeit, Univ. Freiburg, W. Germany.

Zimmerman, B. K. and Briggs, W. R. (1963a). Phototropic dosage-response curves for oat coleoptiles. *Plant Physiol.* **38**, 248–53.

Zimmerman, B. K. and Briggs, W. R. (1963b). A kinetic model for phototropic responses of oat coleoptiles. *Plant Physiol.* **38**, 253–61.

Gravitropism

8

Malcolm B Wilkins

Introduction

To have any chance of survival, a germinating seed must be able to direct the growth of its shoot upwards into the light, so that photosynthesis can begin before the stored food reserves are exhausted, and its root downwards to reach a reliable supply of water and inorganic ions, as well as to provide the seedling with a secure anchorage and mechanical support, as quickly as possible. If, as in agricultural and horticultural practice, and in some natural situations, the seed is randomly oriented and germinating below the soil surface, gravity is the only environmental factor available with which such an orientation of the primary organs can be achieved.

In rapidly growing organs such as primary roots and shoots, and in some rhizomes, which are stems that grow horizontally beneath the soil surface, the precise control of the direction of growth is, in reality, a gravity-sensing guidance system. In all guidance systems there must be at least three components: (1) a sensory mechanism to detect whether or not the organ is on course; (2) a reaction mechanism which, on receipt of the appropriate signal from the sensory mechanism that the organ is off-course, can initiate growth changes in the organ to ensure that it is brought back on course; and (3) a communication mechanism to conduct the signals from the sensory mechanism to the reaction mechanism. Clearly, the course upon which an organ grows will have been set during the process of evolution to be the most advantageous for the successful survival of the seedling.

In addition to the guidance systems in rapidly growing organs, there exists in many non-growing, mature shoots, and in non-growing mature parts of growing shoots, a capacity for reorientation so that the shoot can regain its preferred orientation with respect to gravity should this have been disturbed. Such a mechanism is found in the leaf sheath bases (nodes) of the shoots of grasses and cereals, and is of importance in that its provides the fully grown shoots with the capacity to regain a vertical orientation, and hence to be harvested, if the shoot has been knocked over in bad weather.

Because of space limitation, consideration will be given in this chapter only to the gravity-sensing guidance systems found in primary roots and shoots of young seedlings, and to the reorientation mechanism found in the nodes of grasses and cereals. Thus, attention will be confined to organs whose normal orientation or direction of growth is vertical; such organs are termed *orthogravitropic* (*orthogeotropic*). Other organs, such as lateral shoots and roots, grow at an angle to the gravity force vector and are called *plagiogravitropic* (*plagiogeotropic*) while those which grow horizontally, such as the rhizomes of *Agropyron repens* and *Aegopodium podograria*, are termed *diagravitropic* (*diageotropic*).

The final point to be raised by way of introduction is that while gravitropism has received continuous study over at least the last century, and a large body of information has been accumulated, much remains uncertain and the view the reader takes of the mechanisms involved depends upon the critical assessment of experimental procedure and results, and the balance of probabilities, since at the present time no published evidence establishes beyond doubt the precise nature either of the gravity-sensing mechanism or of the reaction (growth control) mechanism. The subject is therefore one in which there is opportunity for new ideas and critical experimental work.

Historical background

Dodart (1703) and Austruc (1709) appear to have been the first to draw attention to the vertical orientation of many plant organs and to recognize that this was related to gravity. It was a century later, however, before Knight (1806) established experimentally that gravity was the causal factor. Knight attached seedlings to the edge of a wheel rotating about a vertical axis and noted that the centrifugal acceleration, which would cause mass acceleration in the same way as gravity, caused the shoots to turn inwards towards the center of the wheel, and the roots to bend away from the axis. Any notion that roots curve downward because of the weight of their tips is ruled out by the fact that roots will bend down into a more dense medium, such as mercury, if their bases are held securely (von Sachs, 1887). Frank (1868) was the first to point out that gravitropic curvatures were related to the growth of an organ and it was he who introduced the term *geotropism* which has now been replaced by the term *gravitropism*. Later, von Sachs (1887) elaborated these early studies, showing that the development of gravitropic curvature was confined to the growing zone of plant organs. Ciesielski (1872) and Darwin (1880) experimented on the location of the gravity-sensing mechanism in roots and found that it lay in the extreme apex, or cap.

The two most important landmarks in the history of gravitropism research were the enunciation of the 'starch-statolith' hypothesis of gravity perception independently in 1900 by Haberlandt and Němec, and the hypothesis involving the lateral transport and hence asymmetric distribution of auxin as the underlying cause of the development of curvature, which was advanced independently in 1926 by Cholodny and Went. Both hypotheses have dominated thinking in the subject, indeed, they still do, and have provided enormous stimulation for generations of scientists. Designing experiments to test the validity of these hypotheses has been a major

activity since their introduction, and it is the results of such experiments that will be discussed in this chapter.

The gravity-sensing mechanism

The nature of the gravity-sensing mechanism in plants has been the subject of much speculation since Noll (1892) first suggested that plants might perceive gravity in the same way as do animals, that is, by means of statocyst-like sense organs. He believed the organs in question were ultramicroscopic structures located in the cytoplasm. However, the careful observations of Haberlandt (1900) and Němec (1900) led them to suggest that the statocysts (statocytes) in plants are, in fact, entire cells and that there are a number of statocytes in each organ. Haberlandt and Němec further proposed that within each *statocyte* (cell) there were a number of *statoliths* which sedimented to the lowermost side and that these statoliths were in fact starch grains. The statocytes are, therefore, cells characterized by the presence of sedimentable starch grains. It was the presence of statocytes in all organs that have the capacity to respond to gravity that led Haberlandt and Němec to propose the starch-statolith hypothesis of gravity perception. This hypothesis found support in early experimental studies, since removal of the extreme apex (cap) of a root, to which the statocytes are usually confined (Němec, 1900), led to the loss of gravitropic responsiveness (Ciesielski, 1872; Darwin, 1880).

The way in which gravity affects a plant organ requires careful consideration since gravity does not act as a unilateral stimulus on a plant organ in the same way as light does in phototropism (Chapter 7). In the latter, the *organ* is unilaterally stimulated; the side nearest to the source of radiation receives a larger stimulus than the shaded side. Thus, it is perhaps not surprising that the lighted side behaves differently from the shaded side and a curvature response develops. With gravity, every cell of an organ, whatever its orientation, receives exactly the same stimulus. In a horizontal root, for example, cells on the upper side are receiving exactly the same stimulus as those on the lower side (Fig. 8.1), and the organ itself is therefore not subjected to unilateral stimulation. The only way that a stimulus such as gravity can be detected is for it to induce the movement of particles within the component cells of the organ so that an asymmetry is established. It is from this asymmetry that it is possible for the organ to establish which side is uppermost and which lowermost (Fig. 8.1). Plant organs always develop their curvatures along the plane of symmetry. The

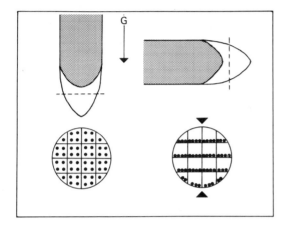

Fig. 8.1 The asymmetry established in a root cap by the sedimentation of amyloplasts following placement of the root in the horizontal position. The lower (circular) diagrams show the distribution of amyloplasts (black dots) in the cells as seen in a transverse section through the root cap taken along the dotted lines and maintained in the orientation with respect to gravity shown by the roots above. The plane of symmetry in the horizontal root is shown by the tail-less arrows. Note the lateral polarity established in the horizontal root in any vertical row of cells. Gravity acts in the direction of the arrow G.

sedimentation of particles (amyloplasts) in the statocytes leads, in effect, to the unilateral stimulation of the cells because of the presence in the lower half of the cell of bodies which are absent from the upper half. Unilateral stimulation does therefore occur in gravitropism, but it takes place at the cellular level and is due to the movement of cellular components by the gravity vector force.

A plant organ detects a change in its orientation with respect to gravity very rapidly, often in only a few seconds. The rapidity of detection is determined by measuring the *presentation time*, which is defined as the minimum time of stimulation required to elicit a just-detectable curvature response. This is usually achieved by placing a root or shoot in the horizontal position for various times of between 1 s and 5 min

and then immediately placing it in a klinostat which rotates it horizontally along its long axis so as to abolish any further effect of gravity. The minimum time in the horizontal position which leads to curvature is then determined. The use of a klinostat is fraught with difficulty and the speed of rotation is a critical factor. Much more precise measurement of the presentation time could be achieved in the near zero-g environment of a space laboratory using a centrifuge to impart the stimulus. The other time which is often referred to in the literature is the *reaction time*, which is the time from the onset of the gravitropic stimulus to the appearance of curvature. This can vary from 10 min to several hours. It must be understood that the *presentation time* has nothing to do with the *reaction time* since the latter measures the time for the *perceived stimulus* to induce growth rate changes, whereas the former measures the time for which the stimulus must be given to be perceived by the organ.

Because plant organs can detect a change in their orientation with respect to gravity within a few seconds, the gravity-sensing mechanism operates rapidly, and if the latter involves the sedimentation of cellular particles, then this sedimentation must occur very rapidly indeed. This finding places a severe limitation on the size and nature of the subcellular particles that could operate as statoliths. Audus (1962, 1969) has given detailed consideration to this question and has concluded that of the various subcellular particles (nucleus, starch grains, mitochondria, microsomes, oil droplets, etc.) the starch grain is the only one with the mass and density to move across one half the width of the statocyte (the average lateral distance moved by the statolith in a horizontally-placed cell) within the observed presentation time. Having arrived at this conclusion on theoretical considerations, Griffiths and Audus (1964) examined roots of *Vicia faba* with the electron microscope and found that the starch grains (amyloplasts) were indeed the only subcellular bodies which underwent sedimentation to lie in the lowermost part of the cell. There is now evidence that endoplasmic reticulum and some other components may become redistributed following gravitropic stimulation (which, in experimental practice, means placing the root or shoot in the horizontal position), but these movements are thought to be a consequence of the starch grain sedimentation (*see* later).

What is now clear is that each statolith is not simply a 'naked' starch grain but rather a group of starch grains enclosed within a membrane (Fig. 8.2), such a structure being called an *amyloplast*. There may be from 1 to 8 starch grains in each amyloplast and perhaps from 4 to 12 amyloplasts in each statocyte.

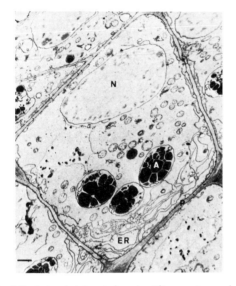

Fig. 8.2 Lateral statocyte in a *Lepidium* root cap, showing the asymmetric distribution of the endoplasmic reticulum (ER) along the outer periclinal wall (P). An amyloplast (A) and the nucleus (N) are also labeled. The scale bar indicates 1 μm. For the location of this statocyte in the root cap see marked cells (white dots) in Fig. 8.4. (Reproduced with permission from Volkmann and Sievers, 1979.)

Thus the starch-statolith hypothesis of gravity perception involves the sedimentation of amyloplasts in specialized cells called statocytes. The statocytes are very localized in roots, being usually confined to the root cap, and more particularly to the central core of the cap, the columella (Němec 1900). In shoots they are not so localized in the sense that they are found all along the shoot, most frequently in the cells that surround the vascular bundles (the bundle sheath), in specific groups located close to vascular bundles, or in a specific layer of cells within the stem such as the inner cortical layer or endodermis (Haberlandt, 1900). For a full discussion of this matter see Drummond's

(1914) translation of Haberlandt's book entitled *Physiological Plant Anatomy*.

In the 84 years which have elapsed since the starch-statolith hypothesis was proposed many attempts have been made to test its validity. This has proved to be unusually difficult, and it is still not possible to find unequivocal evidence upon which to base a definitive answer. Over the years a large body of evidence has been accumulated which is entirely consistent with the validity of the hypothesis. Had the results of virtually all the experiments turned out differently there is no doubt the hypothesis would have had to have been abandoned. However, because the evidence is all essentially of a correlative nature, it does not prove beyond doubt that the hypothesis is correct. It is, nevertheless, substantial, and it is perhaps for the reader to decide whether it is sufficient to be convincing.

The evidence supporting the starch-statolith hypothesis is as follows:

(1) There is a close correlation between the presence of sedimentable amyloplasts in organs and their ability to detect their orientation in a gravitation field and hence respond gravitropically. This evidence has been reviewed by Audus (1962) and by Wilkins (1966). If there were well-documented cases of organs with a gravitropic response but with no statoliths, then the hypothesis would be called seriously into question. There are plants which do not synthesize storage starch (*e.g.*, onion), but these do have sedimentable amyloplasts. In some lower groups of plants the statoliths are not starch grains, but other substances. In the gravitropically-sensitive rhizoids of the alga *Chara fragilis* it has now been established that the statoliths are granules of barium sulfate (Schröter, Läuchli and Sievers, 1975). In the fungi, too, the statolith bodies may not be composed of starch.

(2) There is a very close correlation between the times taken for the amyloplasts to sediment and for the stimulus to be perceived in plant organs. In other words, the time required for the starch grains to fall to the lowermost side of cells in an organ transferred from the vertical to the horizontal position is directly related to the presentation time. Hawker (1933) found that the rate of fall of the statoliths in the stems of the sweet pea *Lathyrus odoratus* varied with

temperature, due presumably to changes in the viscosity of the cytoplasm. The presentation time in *Lathyrus* stems changed in an exactly similar manner at temperatures between 10° and 40°C (Fig. 8.3). Between 10° to 30°C the increased rate of fall of statoliths is accompanied by a shortening of the presentation time, while

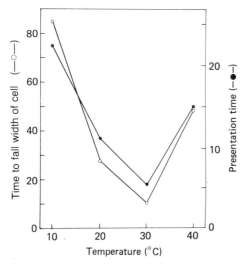

Fig. 8.3 The relationship between the presentation time (min.) and the time (min.) required for statoliths to fall the width of a statocyte in a gravitropically stimulated stem of *Lathyrus odoratus* seedlings at different temperatures. (From Audus, 1962 after Hawker, 1933.)

at 40°C the decreased rate of movement of the statoliths is accompanied by an increase in the presentation time. These findings are obviously what would be expected on the basis of the starch-statolith hypothesis; had the two parameters not been correlated the hypothesis would have been untenable. However, the result does not prove that the hypothesis is correct because it does not establish a causal relationship between sedimentation rate and presentation time.

(3) There is a close correlation between the presence of amyloplasts in the statocytes and of geotropic responsiveness. Starch grains can be removed from the cells of roots and coleoptiles by incubating the organs in a solution containing a relatively high concentration of 6-furfuryl-aminopurine (kinetin) and gibberellic acid

(GA_3) at 30°C for about 35 h (Iversen, 1969). After such treatment, caps of *Lepidium sativum* (cress) roots are completely free of amyloplasts, the roots continue to grow at a rate only slightly less than untreated roots, but they are totally unable to detect their orientation in a gravitation field; they grow straight in any direction. When the treatment is terminated, and the roots are placed in water in the light, the amyloplasts re-form after 20–24 h and there is a simultaneous reappearance of the gravitropic responsiveness. Similar results have been obtained with roots of both red and white clover (*Trifolium pratense* L. and *T. repens* L.) (Iversen, 1974). In these species the amyloplasts reformed 10 h after the treatment ended and again there was a simultaneous reappearance of the root's ability to detect its orientation in the gravitational field. Essentially similar results have been obtained with coleoptiles of wheat (*Triticum durum*) (Iversen, 1974).

Since Iversen carried out his first studies it has been shown that the cap is the source of at least one growth-regulating compound upon which the gravitropic response of the root depends (Gibbons and Wilkins, 1970; Shaw and Wilkins, 1973). As a result of this finding, Iversen's important experimental results might be explained without attributing the loss of gravitropic sensitivity to the loss of amyloplasts. The loss of the gravitropic response after the hormone-incubation treatment might be attributable to the cessation of production, or inactivation, of the inhibiting substance upon which the gravitropic response is dependent, or to interference with its transport. These possibilities could be tested experimentally; if one half of the cap were removed from the destarched root, the continued production and basipetal transport of a growth regulator would be immediately obvious from the development of curvature towards the remaining half-cap.

(4) When the cap is detached from the root the statocytes are removed and there is, as might be expected, a concomitant loss of gravitropic responsiveness (Darwin 1880; Juniper, Groves, Landau-Schachar and Audus, 1966; Gibbons and Wilkins, 1970). However, since the cap is also the source of the growth-regulating compounds necessary for a gravitropic response (*see* above),

loss of gravitropic responsiveness cannot un-equivocally be ascribed solely to the removal of the statocytes.

After removal of the root cap and the conse-quent loss of gravitropic responsiveness, meri-stematic activity in the root apex leads to the regeneration of a new cap and the reappearance of gravitropic responsiveness. Detailed kinetic studies of the regeneration of the cap and the reappearance of gravitropic responsiveness have shown that the two processes do not occur simultaneously; the gravitropic responsiveness reappears some hours before cap regeneration can be clearly distinguished. At first sight this poses a serious obstacle to acceptance of the starch-statolith hypothesis since it implies that a response to gravity can occur in the absence of amyloplastic starch grains. This objection has, however, been obviated by careful micro-scopical studies (Grundwag and Barlow, 1973; Barlow, 1974; Barlow and Grundwag, 1974). Immediately after removal of the cap, starch grains begin to develop from proplastids in the cells of the quiescent center, immature xylem and cortex of the root apex. These starch grains are very prominent after 24 h, by which time the roots are gravitropically responsive despite there being no detectable regeneration of a cap. It is possible, therefore, that starch grains in the *apex* of the decapped root could form the basis of a graviperception mechanism and account for the reappearance of gravitropic sensitivity be-fore the development of a new cap. If this were the case, however, the starch grains might be expected to sediment to the lowermost part of the cell, as do those normally present in the root cap. Recently Hillman and Wilkins (1982) have shown that in decapped roots of *Zea mays* gravitropic responsiveness returns quite sud-denly between 12 and 24 h following detach-ment of the cap, the time varying considerably between individual roots. No sedimentation of the starch grains formed in the root apex was observed in roots which had not regained their gravitropic responsiveness, whereas sedimenta-tion was observed in at least some cells of those which had regained the capacity to respond gravitropically. There being no substantial dif-ference in the size of the amyloplasts in the root apex cells 12 and 24 h after removal of the cap, it

was inferred that sedimentation after the longer time perhaps reflected some change in the physi-cal characteristics of the cytoplasm. Certainly no movement of the amyloplasts could be induced in roots 10 h after removal of the cap even after 4 h centrifugation at 25 g. Thus, in decapped roots, there is a close coincidence between the return of gravitropic responsiveness and the ability of the newly formed amyloplasts in the root apex to sediment to the lowermost side of the stato-cytes.

(5) In genetic mutants of *Zea mays* there also appears to be a close correlation between gravi-tropic responsiveness and either the size or the presence or absence of amyloplasts. Hertel, de la Fuente and Leopold (1969) used coleoptiles of a mutant in which the amyloplasts were very much smaller than those of the wild type. These amyloplasts sedimented to a lesser extent than did those in the wild type, and there was a corresponding decrease in both the lateral trans-port of indoleacetic acid (IAA) (*see* later) and upward gravitropic curvature. Although of con-siderable interest, this investigation did not answer the really critical question of whether the sedimentation time and the presentation time were closely correlated in both the mutant and wild types.

Another interesting nuclear mutant of *Zea mays* (hcf-3) is non-photosynthetic. The stato-cytes in the leaf sheath bases do not contain starch after the endosperm reserves are ex-hausted and the gravitropic responsiveness is lost. If the leaf sheath cells are provided with sucrose, however, the amyloplasts re-form and the shoot becomes gravitropically responsive once again (Miles, 1981).

Taking into consideration all the evidence pre-sently available, there appears to be very little doubt that the sedimentation of amyloplasts is an integral part of the gravity-perception mechanism in the roots and shoots of higher plants. Very little is known, however, about the other components of the mechanism. In an animal sensory organ, the stato-lith sediments to lie on a sensory surface and as a result electrical signals are emitted which ultimately lead to a righting movement by the animal's limbs. In plants, the statocytes may have on their lateral walls sensory surfaces which can detect the presence or absence of the statoliths, but there is no evidence that this is the case. It is not difficult to envisage mechanisms that do not involve specialized sensory areas along the plasmalemma of the lateral walls of the statocytes but which could nevertheless give rise to a lateral polarity between neighboring statocyte cells. It is possible, but perhaps unlikely, that the amyloplasts may have a specific metabolic activity that becomes concentrated only in the lowermost part of the statocytes. Alternatively, the amyloplast membrane may carry an electrical charge which may cause a polarity between the upper and lower surfaces of neighboring statocytes following sedimentation. Evidence that this is so has recently been published by Sack, Priestley and Leopold (1983). A number of electrically charged particles lying in the lowermost part of the cell might affect the permeability and transport properties of the adjacent plasmalemma. On the other hand, the amyloplasts may be metabolically inert and devoid of significant electrical charge, but their sedimentation could initiate displacement of other, metabolically-active, cell constituents away from the vicinity of the plasmalemma in the lowermost part of the statocyte. Many metabolically active subcellular components might be affected, the mitochondria, dictyosomes, ribosomes and endoplasmic reticulum being the most prominent. In principle, such an effect could create a higher metabolic activity in the upper half than in the lower half of each statocyte of a vertical series and result in a metabolic gradient between two adjacent cells (*see* vertical columns of statocytes in the horizontal root, Fig. 8.1). This gradient might induce the polar movement of specific substances from the upper to the lower cell via a specific carrier mechanism. These suggestions are, in practice, extremely difficult to study experimentally. The last suggestion, however, does not necessarily mean that there would be an obvious difference in the number of mitochondria, dictyosomes or other subcellular inclusions in the top and bottom halves of a particular statocyte. What would be critical is that few would be located in the cytoplasm immediately adjacent to the lowermost plasmalemma because the amyloplasts would have displaced them from this part of the cell.

Griffiths and Audus (1964) found evidence that endoplasmic reticulum is displaced to the uppermost part of statocytes when roots of *Vicia faba* are placed horizontally. In *Zea mays* roots, Juniper and French (1973) found the endoplasmic reticulum to be normally oriented parallel to the nuclear membrane and the cell wall in the statocytes and to be symmetrically distributed. After horizontal orientation, the endoplasmic reticulum was pushed away from the lowermost part of the cell by the amyloplasts and accumulated predominantly in the apical region of the statocyte where the amyloplasts were originally located in the vertical root. On the other hand, the distribution of endoplasmic reticulum in statocytes in root caps of *Pisum sativum* (Grundwag, 1971) and of *Lepidium sativum* (Sievers and Volkmann, 1972; Volkmann and Sievers, 1975) is somewhat different from that found in *Zea mays* in that it is normally located in the apical part of the cell underneath the amyloplasts. On reorientation into the horizontal position the amyloplasts move towards the lateral, but now lowermost, part of the cell, leaving the endoplasmic reticulum in its original position.

Sievers and Volkmann (1972) and Volkmann and Sievers (1979) have developed an explanation of graviperception in *Lepidium* roots which involves the precise shape of the statocytes and the asymmetric location of the endoplasmic reticulum within the apical part of the cells (*see* Fig. 8.2). In the vertical root the *pressure* on the endoplasmic reticulum will be equal in the two cells illustrated in Fig. 8.4, whereas in the horizontal position the amyloplasts will exert a pressure on the endoplasmic reticulum only in the lowermost cell. It is the *pressure* which the amyloplasts exert that Sievers and Volkmann think is important because of the very small amount of spatial movement which could be achieved by an amyloplast in the very short presentation times in *Lepidium* roots (12 s). The pressure which can be exerted by the amyloplasts in a *Lepidium* root statocyte cannot, however, much exceed $2-4 \times 10^{-1}$ N m^{-2} (Audus, 1962). Precisely what the amyloplast pressure does to the endoplasmic reticulum has not been established. Volkmann and Sievers (1979) point out that dramatic sedimentation of amyloplasts is not necessary to achieve gravity perception with his hypothesis, because relatively little movement of the amyloplasts need occur. However, what is clear is that they must be free to sediment, otherwise they could not exert a pressure. In experimental study of gravitropism, organs are usually transferred from the vertical to the horizontal position and substantial sedimentation of amyloplasts takes place. In the normal operation of the guidance system in a root growing in the soil, small deviations of, say, 10° from the predetermined course can be detected. With such a small angle of deviation from the vertical the *movement* of the amyloplasts would be small, and Volkmann and Sievers (1979) particularly emphasize that the response to a small angle of deviation can most satisfactorily be accounted for by their hypothesis of differential pressure changes on the endoplasmic reticulum in the statocytes on either side of the root cap (Fig. 8.4).

extent of the sedimentation necessary for gravity perception is not clear; slight movement to exert pressure on nearby membranes may be adequate in roots of some species, but in others the membranous structures are themselves mobile and change their orientation. In shoot statocytes, there is at present little evidence for amyloplast–membrane interaction. Further investigation is required on the nature of the interaction of the statoliths with other cellular components, but really critical studies of this kind are very difficult to make since they have to

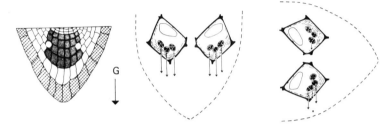

Fig. 8.4   *Left:* The location of the lateral statocytes (white circles) in the columella (heavily shaded) of the *Lepidium* root cap, the ultrastructure of which is illustrated in Fig. 8.2. *Center:* the equal pressure exerted by the amyloplast statoliths on the asymmetrically distributed ER complexes in these statocytes in vertical roots. *Right:* the asymmetric pressure, exerted only on the ER in the lowermost statocyte, following placement of the root in the horizontal position. The arrow labeled G indicates gravity and the smaller arrows emanating from the amyloplasts indicate the direction and magnitude of the pressures they exert on the ER. (Reproduced with permission from Volkmann and Sievers, 1979.)

The hypothesis of Volkmann and Sievers is thus most attractive but there are some difficulties with its general application. As Audus (1975) points out, it cannot adequately account for changes in the geotropic response patterns of pre-inverted roots. Further, the particular pattern of endoplasmic reticulum arrangement in *Lepidium* statocyte cells is, to some extent, found in other species such as *Lens culinaris*, *Daucus carota* and *Allium cepia* (Volkmann, 1974), but, as has already been mentioned, is not found in the statocytes of roots of *Zea mays* (Juniper, 1976), *Vicia faba* (Griffiths and Audus, 1964), or in the statocytes of stems, such as those in grass nodes (Osborne and Wright, 1977; Wright and Osborne, 1977).

Excellent reviews of the gravity perception mechanism can be found in the papers by Audus (1975), Juniper (1976) and Volkmann and Sievers (1979). In summary, there seems to be no doubt that amyloplasts are required for gravity perception, and they must be able to sediment in the statocyte. The

take place within a highly pressurized cell of no more than about 20 μm diameter.

Reaction mechanisms in growing organs

Ever since von Sachs (1887) established, in *Vicia faba* roots, that gravitropic curvature is confined to the sub-apical elongation zone, it has been clear that the development of curvature depended upon the fine regulation of the rate of cell extension in the upper and lower halves of the organ. Following the discovery that the apex of a coleoptile was the source of a growth-promoting substance, auxin (*see* Boysen-Jensen, 1936; Went and Thimann, 1937), Cholodny (1926) and Went (1926) simultaneously proposed the classical hypothesis that now bears their names to account for the growth-rate changes upon which curvature depends. The hypothesis states that in vertical shoots auxin moves basipetally from the apex to the growing zone in a symmetrical manner. Following placement in the horizontal position, however, the auxin becomes asymmetrically redistributed in favor of the lower side by means of a lateral polar transport of the molecule from the upper to the lower half of the organ. This mechanism would give rise to a redistribution of growth; the lower half should increase by approximately the amount that the upper half decreases, and there should be no overall change in the growth rate of the organ. The validity of this hypothesis will be examined separately for coleoptiles, dicotyledonous shoots and roots.

Coleoptiles

The hypothesis has received a substantial measure of experimental support in coleoptiles. The earliest growth-rate measurements showed that in coleoptiles of *Hordeum vulgare* (barley) (Weber, 1931) and *Avena sativa* (oat) (Navez and Robinson, 1933) upward gravitropic curvature was achieved by the growth rate of the lower half increasing and that of the upper half decreasing by approximately the same amount as compared with the growth rate of the vertical control. The validity of the Cholodny–Went hypothesis must depend, however, on establishing firstly, the chemical identity of the growth-promoting hormone in the apex and, secondly, whether or not it undergoes downward lateral transport and becomes asymmetrically distributed in horizontal coleoptiles.

Unequivocal identification of the auxin present in coleoptile tips of *Zea mays* has been achieved using high-resolution mass spectroscopy (Greenwood, Shaw, Hillman, Ritchie and Wilkins, 1972). The diffusate from 15 000 coleoptile apices was purified chromatographically.and the eluate of the R_f zone at which indole-3-acetic acid (IAA) would be expected to occur was injected directly into a mass spectrometer. The fractionation pattern of the molecule was identical to that of authentic IAA (Fig. 8.5) and the high-resolution molecular mass of the sample was 175.0633 while the calculated, theoretical molecular mass of authentic IAA was 175.0628. The auxin in *Zea* coleoptile apices is therefore IAA, and since IAA regulates the rate of extension of the cells in the growing zone of *Zea* coleoptiles at low concentrations, its redistribution laterally in a horizontal organ could obviously give rise to differential growth and hence curvature.

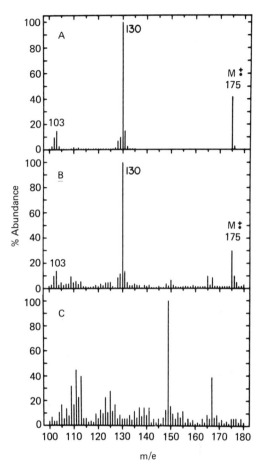

Fig. 8.5 Mass spectra of: A, authentic IAA at a sample concentration of 50 μg in 0.05 ml of methanol and a probe temperature of 280°C; B, the violet fluorescent spot from the IAA R_f obtained after the chromatographic purification of the diffusate from 15 000 *Zea* coleoptile tips, probe temperature 270°C; and C, the same zone of chromatogram as used in B except that the extracts were made of blank agar blocks; probe temperature 220°C. (From Greenwood *et al.*, 1972.)

Although Dolk (1929) had shown with the *Avena* curvature test that the net growth-promoting activity which diffuses out of the base of a horizontal, detached tip of an *Avena* coleoptile was asymmetrically distributed (Fig. 8.6A), there was no evidence that this was wholly or even partly due to an asymmetry in the distribution of a growth-promoting molecule (auxin) being established in favor of

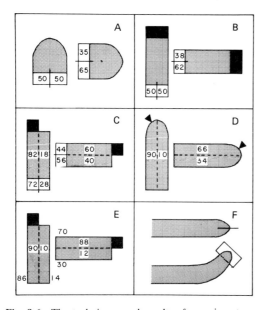

Fig. 8.6 The techniques and results of experiments concerned with attempts to establish whether or not a lateral transport of auxin (IAA) occurs in gravitropically-stimulated coleoptiles. Coleoptile tips or segments are shaded, donor agar blocks containing IAA are shown in black and receiver agar blocks are shown in white. The numbers indicate % distribution of IAA between the blocks or between the two halves of the tissue. The lines separating the receiver blocks in A, B and C are impermeable barriers, which are also used in F to insert into the apex of the coleoptile either horizontally (top) or vertically (bottom). For further explanation see text. (The experiments are those of A, Dolk (1929); B, Gillespie and Thimann (1963); C, Goldsmith and Wilkins (1964); D, Shaw, Gardner and Wilkins (1973); E, Wilkins and Whyte (1968); F, Brauner and Appel (1960).)

the lower half. The result could equally be indicative of the upward redistribution of a growth-inhibiting molecule modifying the effect of a uniformly distributed growth-promoting compound. Bear in mind that the *Avena* curvature test does *not* estimate the amount of auxin diffusing from a coleoptile tip but rather the *net* growth-regulating effect of *all* the substances emerging from the basal end of the tissue.

With the advent of radioactive IAA it was possible to establish whether or not an asymmetry was established in the distribution of this growth-pro-

moting compound in horizontal coleoptiles. Using coleoptile sections of *Avena* and *Zea*, and the technique of Dolk (1929) illustrated in Fig. 8.6B, Gillespie and Thimann (1961, 1963) established that IAA became asymmetrically redistributed in favor of the lower half of the organ, but this procedure cannot elucidate *how* the asymmetry is established for the reasons pointed out by Wilkins (1977). The first unequivocal demonstration of downward lateral transport of IAA within horizontal shoots was made by Goldsmith and Wilkins (1964). Donor blocks were supplied *asymmetrically* to the apical ends of *Zea* coleoptile segments as shown in Fig. 8.6C and the relative distribution of total radioactivity was determined between the two halves of segments that were oriented either vertically or horizontally. The important point of this experiment is that radioactive IAA found in the non-donated half of the coleoptile *must* have come from the half in contact with the donor block because this is the *only* source of the radioactive hormone. If the proportion of the total radioactivity found in the non-donated half of the segment differs when the segment is in different orientations with respect to gravity, then a change in the lateral movement of IAA is established with certainty. That this is so is obvious from Fig. 8.6C; virtually twice as much IAA moves to the lower half of a horizontal segment supplied with an upper donor as moves across in a vertical segment. The downward lateral transport of IAA in horizontal coleoptile segments is thus established, and it has subsequently been shown (Fig. 8.6E) to be entirely dependent upon metabolic energy (Wilkins and Whyte, 1968).

Shaw, Gardner and Wilkins (1973) assessed lateral transport of IAA in intact undamaged coleoptiles of *Zea mays* following geotropic stimulation using a micro-application technique to supply tritiated radioactive IAA (5-^{3}H-IAA) of very high specific activity. A pulse of IAA was applied asymmetrically to the apex of a coleoptile and its movement was followed as a function of time and of orientation in the gravitational field. The very small amount of tissue damage at the point of penetration of the pipette appeared to have no effect on the coleoptile. A strongly polarized, downward, lateral transport of IAA was detected (Fig. 8.6D).

The Cholodny–Went hypothesis can thus be regarded as valid to the extent that apically-synthesized IAA (auxin) undergoes downward lateral transport in horizontal *Zea mays* and *Avena sativa* coleoptiles, and the asymmetric distribution of this growth-promoting compound in the organ will lead to differential growth and hence upward curvature. Despite its validity, however, the hypothesis may represent only a part of the overall response mechanism (Wilkins, 1978, 1979). There are a number of reasons for believing this to be so. Firstly, Cane and Wilkins (1969) demonstrated that, in addition to the initiation of lateral transport of IAA in horizontal coleoptiles of *Zea mays*, there were also induced changes in the basipetal transport of IAA in the coleoptile such that the transport along the lower half was greater than that along the upper half. This differential longitudinal transport of IAA was quite independent of the lateral transport of the compound, and may well affect the distribution of IAA between the upper and lower halves of the growing zone of the organ. Secondly, it has not been established whether natural growth-regulating compounds other than IAA are also involved in the induction of differential growth. For example, Railton and Phillips (1973) found gibberellin-like activity to diffuse into agar blocks at the basal end of detached apices of *Zea mays* coleoptiles, and that when the Dolk (1929) technique was used (Fig. 8.6A) more activity was found in the lower than in the upper block. No evidence for a lateral transport of gibberellins has been detected (Wilkins and Nash, 1974; Webster and Wilkins, 1974). All these studies involved rather long transport times of about 20 and 24 h, however, and it must be remembered that the gravitropic response is complete in 6–8 h. The involvement of inhibitors of cell extension and their possible upward lateral transport in horizontal coleoptiles has not received enough attention to exclude their participation in the response. Certainly the lateral transport of growth-regulating molecules must be involved in part in the gravitropic response of coleoptiles since it is difficult to see how else the results of Brauner and Appel (1960) can be explained. They found that when a barrier to lateral transport was inserted into the apex of a horizontal coleoptile, gravitropic curvature was abolished or greatly diminished, whereas a barrier which did not prevent downward lateral transport did not inhibit curvature (Fig. 8.6F).

Further objection to the acceptance of the Cholodny–Went hypothesis has recently been raised by Digby and Firn (1979) and Hall, Digby and Firn

(1980). They observed that when a *Zea* coleoptile is transferred into the horizontal position, the growth rate of the lower side approximately doubles whereas that of the upper side decreases sharply to zero (Fig. 8.7). This finding contrasts with smaller changes in growth rate observed by Weber (1931)

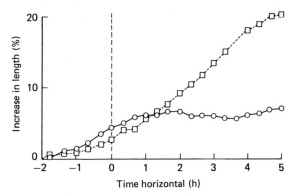

Fig. 8.7 Growth of the two sides of an etiolated shoot of a *Zea mays* seedling before and after gravitropic stimulation at time 0. The lengths of the upper side (○) and lower side (□) of the coleoptile and mesocotyl were measured. (Reproduced with permission from Digby and Firn, 1979.)

and Navez and Robinson (1933), and is not what would be expected on the basis of the Cholodny–Went hypothesis unless there were a really massive movement of auxin from the upper to the lower half of the organ. Hall *et al.* (1980) have therefore questioned whether the magnitude of the asymmetry in IAA distribution which has been demonstrated in *Zea* coleoptiles is adequate to account for the observed changes in the growth rate, bearing in mind the well known optimum-type relationship between the increase in length of coleoptile segments and the logarithm of the *external* concentration of a solution of IAA. There is no doubt that the changes in the growth rate of the upper and lower sides of a gravitropically-responding coleoptile are rather greater than might be expected from the apparent change in 'concentration' of IAA in the upper and lower halves of the responding organ, at least on the basis of the simplistic external concentration–growth rate relationship. The difficulty in assessing the strength of this criticism is that the *concentration* of a compound in a cell or organ cannot be measured, only the *amount* is determined and this cannot be converted to a concentration unless its distribution throughout the volume of the cell or organ is known precisely. The relationship between the external concentration of IAA and growth rate is of no value in assessing the relationship between the *amount* of IAA in an organ and the growth rate that it would induce. External application of IAA to coleoptiles of *Zea* leads to substantial metabolism of the molecule to at least nine metabolites within about 1 h (Nonhebel, Crozier, Hillman and Wilkins, 1983) and there is evidence that the extent of the metabolism increases with applied concentration, thus emphasizing the need for caution when basing a view on the external concentration–growth rate relationship. There can be no way of assessing the criticism of Hall *et al.* (1980) until more is known about the relationship between the *amount* of IAA in the growing zone of a coleoptile and the growth rate of the organ.

At the present time, while it is known that IAA *does* undergo lateral transport towards the lower side of an intact horizontal *Zea* or *Avena* coleoptile and thus becomes asymmetrically distributed (a conclusion recently confirmed by Mertens and Weiler (1983) using radio-immunoassay procedures), it has not been established beyond doubt that this process alone could initiate the growth rate changes which are observed.

Dicotyledonous shoots

The growth-regulating mechanism underlying the upward geotropic curvature in dicotyledonous shoots is not understood, and there is little evidence that the Cholodny–Went hypothesis is involved to any appreciable extent. Although a large number of studies have been made on shoots of *Helianthus annuus*, the results which have been obtained appear to differ according to whether or not segments of the hypocotyl or epicotyl have been used, and whether or not the tissue is etiolated or green.

Growth-rate studies of the upper and lower sides of gravitropically-responding hypocotyls of *Helianthus annuus*, whether green or etiolated, show that the rate of the lower side increases markedly while that of the upper side decreases to zero. In etiolated *Cucumis sativus* hypocotyls upward gravitropic curvature appears to be due entirely to a cessation of growth of the upper side with little change in that of the lower (Firn, Digby and Riley, 1978; Digby and Firn, 1979; Firn and Digby, 1980). The cessation of growth of the upper side and a substantial increase in growth of the lower side could only be achieved on the basis of the Cholodny–Went hypothesis if there was a substantial movement of IAA downwards in the horizontal organ, and it is difficult to reconcile the situation in *Cucumis* with the hypothesis at all. Although downward lateral transport of IAA occurs in horizontal etiolated hypocotyl segments of *Helianthus annuus* and *Phaseolus multifloris* (Bridges and Wilkins, unpublished), it is much less than that found in *Zea* coleoptiles and probably inadequate to account for upward gravitropic curvature. This finding is compatible with earlier reports of a ^{14}C gradient being established between the upper and lower halves of segments supplied symmetrically with ^{14}C-IAA (Gillespie and Thimann, 1963). Further Abrol and Audus (1973a,b) found that ^{14}C-2,4-D underwent lateral displacement towards the lower side of segments of light-grown *Helianthus* hypocotyls. Thus, lateral movement of radioactivity from ^{14}C-IAA, ^3H-IAA and ^{14}C-2,4-D in *hypocotyl* segments of *Helianthus* has been observed, and contrasts with the lack of a detectable lateral transport of IAA in segments of internodes above the insertion of the cotyledons (Phillips and Hartung, 1976). While these findings indicate that auxin redistribution may be involved in the gravitropic response of the hypocotyl, the magnitude of the asymmetry observed is small and the question raised by Hall *et al.* (1980) of whether or not it is adequate to account for the growth rate changes is again pertinent. The lateral movement observed in a hypocotyl segment may not, of course, reflect that which occurs in an intact shoot and, as has been pointed out in the section dealing with coleoptiles, the relationship between the *amount* of IAA in a hypocotyl and its growth rate is quite unknown.

Phillips (1972) has considered whether or not the gibberellins are involved in the gravitropic response of dicotyledonous shoots. With green apical-bud preparations of *Helianthus* seedlings a substantial asymmetry was established between the gibberellin activity diffusing into upper and lower agar receiver blocks at the basal end of the horizontal preparation, the asymmetry being as large as 10:1 in favor of the lower block. It is difficult to assess the significance of this finding in relation to the gravitropic response since the gradients were measured after 20 h and the gravitropic response begins after about 1 h in the horizontal position.

Several investigations have focused attention on the importance of the epidermis in regulating the rate of extension of the two halves of a gravitropically-stimulated shoot, and the control of epidermal cell extension by endogenous hormones. Yamamoto, Shinozaki and Masuda (1970) and Tanimoto and Masuda (1971) found auxin-induced extension of pea stem sections to be controlled by regulation of the growth rate of the epidermal cell walls. Auxin was shown to modify the extensibility of the epidermal cell walls.

Iwami and Masuda (1974) studied the gravitropic response in *Cucumis* hypocotyls especially in relation to epidermal cell extension and its control by auxin. Application of auxin to the apical end of intact seedlings promoted elongation of the hypocotyl especially in the uppermost 5-mm portion (Zone I) but also in the next two 5-mm portions (Zones II and III) beneath. In intact, horizontal seedlings gravitropic curvature began within 30 min, commencing in Zone I and moving basipetally. Removal of the apex and the cotyledons did not affect the gravitropic response appreciably, but removal of Zone I of the hypocotyl substantially reduced the response, which could be restored by the application of IAA. The extensibility of the walls of the lower epidermal cells in horizontal hypocotyls was increased significantly after 15 to 30 min of gravistimulation, that is, before the onset of gravitropic curvature. No change in the extensibility of the upper epidermal cells was observed. Iwami and Masuda (1974) conclude that Zone I or the region immediately above Zone I is the source of IAA necessary for the gravitropic response, that it undergoes downward lateral transport on gravitropic stimulation, and that the increased amount of IAA in the lower epidermal cells increases proton secretion (Chapter 1) which in turn gives rise to an increase in cell-wall extensibility, and hence growth.

Although most interest has been centered on the asymmetric distribution of growth-regulating molecules in gravitropically stimulated shoots, there have been several studies of the establishment of asymmetry in other cellular components. In particular, the distribution of the cations K^+ and Ca^{2+} have been investigated in gravitropically stimulated hypocotyls of *Helianthus annuus*. Bode (1959, 1960) found a higher proportion of K^+ in the ash of the lower half of the hypocotyl than in that of the upper

half, and Ca^{2+} showed the reverse trend. These findings have been confirmed by Goswami and Audus (1976) who found that a maximum difference in Ca^{2+} content of about 20% in favor of the upper half of the hypocotyl was attained after about 1 h in the horizontal position, by which time gravitropic curvature was just beginning. Further study of $^{45}Ca^{2+}$ asymmetry showed that it did not develop at 4°C or when the hypocotyls were treated with N-1-naphthylphthalamic acid, which also abolishes the gravitropic response.

The role of these ion asymmetries and the mechanism by which they arise are unclear. They may be a consequence of the asymmetry in IAA which arises in gravitropically-stimulated hypocotyls. Goswami and Audus (1976) established artificial IAA gradients in vertical *Helianthus* hypocotyls and found more $^{45}Ca^{2+}$ in the non-donated side, and more $^{42}K^+$ and ^{32}P in the IAA-rich, donated side of the hypocotyl, as would have been anticipated on the basis of a downward lateral transport of IAA in horizontal hypocotyls. The phenomenon of ion redistribution may thus fall into the category of the 'geoelectric' effect which is also a consequence of the asymmetric redistribution of IAA in horizontal *Zea* coleoptiles and *Helianthus* hypocotyls (Grahm, 1964; Grahm and Hertz, 1962, 1964; Wilkins and Woodcock, 1965: Woodcock and Wilkins, 1969a,c, 1971). Whether or not the redistribution of these ions has a role in regulating differential cell extension on the two sides of a gravitropically responding hypocotyl remains to be established.

Very recently, radio-immunoassay of the amounts of IAA, GA or ABA in the upper and lower halves of horizontal *Helianthus* hypocotyls has not detected an asymmetry within a period of approximately 2 h, by which time a substantial gravitropic response has developed (Mertens and Weiler, 1983).

In summary, therefore, there is justifiable doubt that the Cholodny–Went hypothesis provides an adequate explanation of the gravitropic response in dicotyledonous shoots. These doubts are strengthened by the growth rate measurements of *Cucumis* hypocotyls in which upward curvature is apparently achieved solely by the cessation of growth on the upper side of the organ while the growth rate of the lower side scarcely changes (Digby and Firn, 1979). There is need for further study of the nature of the gravitropic response in a wider range of dicotyledonous shoots.

Roots

The growth-regulating mechanism underlying the gravitropic response of roots is less well understood than that in coleoptiles. Indeed, scarcely anything is known about the mechanisms controlling the normal growth of roots in their natural environment. The effects of hormones and synthetic growth regulators have, for the most part, been studied by their application to the external surface of excised root segments or cultured roots. It is difficult to assess what such studies reveal about the natural mechanisms regulating growth and development, because roots have the capacity to accumulate and metabolize hormones, especially in the tissues of the cortex through which most of the externally applied compounds must pass (Greenwood, Hillman, Shaw and Wilkins, 1973; Nonhebel, *et al.*, 1983). In their natural environment roots derive neither their major organic nutrients nor their hormones from the exterior (they will, after all, grow perfectly well in humid air), but rather are supplied with these compounds or their precursors by other parts of the plant through the sophisticated longitudinal-transport systems in the stelar core. The very existence of growth hormones in roots has been a matter of speculation for a long time, but it is now certain that IAA, cytokinins, gibberellins and abscisic acid are present, though their physiological functions are unclear. The transport of IAA is highly polarized towards the root tip and takes place only in the stele (Scott and Wilkins, 1968; Wilkins and Scott, 1968; Bowen, Wilkins, Cane and McCorquodale, 1972; Shaw and Wilkins, 1974). Substances which inhibit growth are also present and at least one of these originates in the root cap. Of particular interest here is the evidence which indicates that an inhibitor from the root cap is involved in bringing about the downward gravitropic response of the primary root.

Removal of the cap abolishes the gravitropic response of *Zea mays* roots (Juniper *et al.*, 1966). Since the cap is spatially separated from the extension zone in which curvature develops, there must be a communication mechanism between the former and latter regions of the organ and this probably involves a chemical regulator of growth. On removing one half of the cap of the primary root of *Zea mays* seedlings, the root develops a large curvature towards the side upon which the remaining half-cap was located (Fig. 8.8A), regardless of the orientation of the root with respect to gravity (Gibbons and Wilkins, 1970). The cap is thus the source of a net growth-inhibiting influence and for ease of

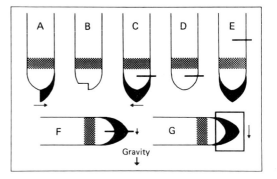

Fig. 8.8   The various treatments to which *Zea mays* roots were subjected by Shaw and Wilkins (1973), and to which reference is made in the text. The root cap is shown in black and the growing zone of the root is shaded. Barriers inserted into the sides of vertical roots are shown in black (C, D and E) and the orientation of the apical barriers in the horizontal roots is horizontal in F and vertical in G. The direction of any curvature which developed as the result of the treatments is shown by the black arrows at the apex of the roots. The small arrow in F indicates a very much smaller curvature than that which was developed in G. (From Wilkins, 1966.)

description this will be referred to as a growth inhibitor. More than one compound may be involved, but the simplest hypothesis to account for the established observations requires no more than one growth-inhibiting compound to be present.

Detailed studies of the transport of the root-cap inhibitor in primary roots of both *Zea* and *Pisum* have been made by Shaw and Wilkins (1973). Although removal of one half of the cap always resulted in curvature of the root towards the remaining half-cap (Fig. 8.8A), excision of one half of the extreme apex of a root from which the cap had been entirely removed did not give rise to curvature (Fig. 8.8B) indicating that the root apex (as distinct from the cap) is not the source of a growth regulator, either of a promoting or inhibiting nature. The insertion of small pieces of mica or metal foil into the side of the root cap, parallel to the long axis of the root, did not induce curvature. It is unlikely, therefore, that surgical trauma is responsible for the curvatures induced by the removal of one half of the cap. Indeed, this possibility is eliminated by the fact that replacing the half-cap immediately after its removal prevents the development of curvature (Pilet, 1973).

Insertion of mechanical barriers to longitudinal transport of growth regulators on one side of a root has shown that the only growth regulators involved in the gravitropic response of the primary root are moving basipetally from the cap to the growing zone. Barriers inserted on the apical side of the extension zone induce large curvatures in roots with intact caps (Fig. 8.8C) and none in those from which the caps have been completely removed (Fig. 8.8D). On the other hand, similar barriers inserted into the side of the root behind the growing zone (Fig. 8.8E) are totally without effect on the behavior of roots.

If an inhibitor arising in the cap is responsible for the induction of gravitropic curvature in horizontally-placed roots, then an asymmetry has to be established in its distribution between the upper and lower halves. There are obviously many ways in which an asymmetry in inhibitor distribution could arise in a horizontal root. One possibility is that downward lateral transport of inhibitor occurs, and Shaw and Wilkins (1973) have shown that this is the case by comparing the curvatures developed by roots which have had either one half of their caps removed, or their caps left intact but a barrier to longitudinal transport of inhibitor inserted into either the upper or the lower half of the root between the cap and the extension zone. These treatments are illustrated in Fig. 8.9. Root A developed a larger *downward* curvature than root B. The only explanation for this difference is that more inhibitor reaches the lower half of the growing zone in A than in B and the only source of this extra inhibitor is the upper half-cap. If this deduction is correct, then the opposite result would be expected in the cases of roots C and D, that is, C must bend *upwards* less than D because lateral transport of the inhibitor will occur in C and not in root D. In root D where there is no lower half-cap into which the inhibitor can be laterally transported, more will move basipetally into the extending zone on the upper side of the root than in root C, and give rise to a greater curvature. This is found to be the case. The validity of this argument is further strengthened by

the observation that insertion of barriers to the lateral movement of growth regulators in the cap decreases the gravitropic response. The experimental treatments are shown in Fig. 8.8F and G.

In summary, therefore, there appears to be strong evidence that at least one inhibitor of growth arises

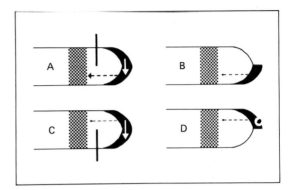

Fig. 8.9 A–D. The differences in curvature developed by roots of *Zea mays* into which a barrier has been inserted or from which one half of the cap has been removed. Root caps are shown in black and broken arrows indicate the movement of inhibitor in the root. The lateral movement of inhibitor takes place in the cap, as shown by the vertical white arrows. The growing zone of the root is shaded. Root A bends down to a greater extent than root B, while root C bends up to a smaller extent than root D. (After Shaw and Wilkins, 1973.)

in the root cap and undergoes downward lateral transport in horizontal roots to give rise, at least in part, to the positive gravitropic response. Whether or not other mechanisms contribute to the establishment of differential growth has yet to be resolved.

The chemical identity of the inhibiting compounds in the root cap obviously needs to be established. According to the Cholodny–Went hypothesis the compound involved in the gravitropic response of roots was auxin (IAA) but recent evidence makes this unlikely. The presence of IAA in roots has only recently been unequivocally established with mass spectrometry (Bridges, Hillman and Wilkins, 1973; Elliott and Greenwood, 1974). In *Zea* roots Bridges *et al.* (1973) found that IAA was virtually confined to the stele, very little occurring in the cortex and in the root apex. Auxin transport is strongly polarized from the base towards the apex in roots (Scott and Wilkins, 1968) and the polar transport is confined to the tissues of the stele in which IAA naturally occurs (Bowen et al., 1972; Shaw and Wilkins, 1974). Thus the polarity of IAA transport is in the wrong direction for this compound to be involved in the geotropic response of roots, and Shaw and Wilkins (unpublished), using a microapplication technique, were unable to find evidence of its downward lateral transport in gravitropically stimulated Zea mays roots. Analytical studies have not led to a unanimity of view on the presence of IAA in root caps. Kundu and Audus (1974) found an inhibitor of growth to be present in root caps of Zea. It occurred at an R_f closely similar to that of abscisic acid (ABA), and it was certainly not IAA. Wilkins and Wain (1974) also examined Zea root caps using physico-chemical methods and a wheat root bioassay. ABA and two other, as yet unidentified, inhibitors were present in the caps of roots which had been exposed to light, but not in those of roots kept in darkness. They found no evidence of IAA. On the other hand Rivier and Pilet (1974) have detected IAA in root caps of Zea by mass spectroscopy. Further study is required to resolve these conflicting reports.

There is, however, unanimity about the presence of ABA in root caps of Zea (Kundu and Audus, 1974; Wilkins and Wain, 1974; Rivier et al., 1977) and the question arises of whether or not ABA is the critical growth inhibitor involved in bringing about gravitropic curvature. There appears to be a body of evidence to suggest that this is so, and it may be summarized as follows:

(1) ABA at concentrations between 10^{-8} and 10^{-4} M inhibits cell elongation in Zea roots and, when supplied asymmetrically at 10^{-8} M, causes substantial curvature. Asymmetric application of IAA did not induce curvature (Pilet, 1975).

(2) The primary roots of some varieties of Zea mays do not have a positive gravitropic response in the dark, but develop the response only after exposure to light (Scott and Wilkins, 1969). Wilkins and Wain (1974) showed that caps of dark-grown Zea mays roots lack ABA, whereas the compound is present following exposure to light.

(3) Light inhibits the elongation of intact Zea mays roots, and this is due solely to an effect of light on the root cap, not on the root apex (Wilkins and Wain, 1974). Similarly the onset of gravitropic curvature following illumination is due solely

to the effect of irradiation on the cap (Wilkins and Wain, 1975a).

(4) Transfer of a cap from dark-grown roots of the Anjou variety of Zea mays, which is gravitropically responsive in the dark, to dark-grown decapped roots of the Kelvedon 33 variety, which is light-requiring, leads to a gravitropic response in the dark (Pilet, 1976).

(5) If ABA at 10^{-9} M is supplied in solution externally to horizontal dark-grown Zea roots in darkness, they develop a gravitropic response, whereas untreated roots do not. The same treatment given to decapped roots does not lead to curvature (Wilkins and Wain, 1975b).

This evidence supports the view that ABA is the growth inhibitor which is present in the root cap and upon which the gravitropic response depends. The response appears to involve the lateral transport of ABA in the cap and the subsequent basipetal movement of the asymmetrically distributed ABA into the zone of cell extension of the root where differential growth and hence curvature occurs.

Despite the evidence supporting this view and its inherent attractiveness over the older notion involving supra-optimal 'concentrations' of IAA (Went and Thimann, 1937) for which there is no evidence whatever, two further pieces of information are required before it can be accepted. Firstly, it must be shown that ABA does undergo downward lateral transport in horizontally placed roots, and this has not been achieved. Hartung (1976, 1981) was unable to demonstrate either an asymmetric distribution or lateral transport of radioactively-labeled ABA supplied by microapplication to root caps of Phaseolus coccineus, Vicia faba or Zea mays. Secondly, it must be established that an asymmetric distribution of naturally occurring ABA develops in favor of the lower half of the horizontal organ. On this question there are conflicting reports. Suzuki, Kondo and Fujii (1979) could not detect such an asymmetry in ABA distribution in horizontal Zea mays roots either in the light or dark using a bioassay procedure. However, there was also in the cap of Zea an unidentified inhibitor of root growth, and this was strongly asymmetrically distributed in favor of the lower half in the light, but not in the dark. Suzuki et al. (1979) conclude that ABA is not the inhibitor involved in the gravitropic response of Zea roots, but suggest that the second unidentified inhibitor may be the compound upon which the development of curvature depends. Mertens and Weiler (1983) have also failed to detect an asymmetry in ABA (or IAA) distribution in horizontal *Zea* roots. In contrast, Pilet and Rivier (1981) have shown that a slight asymmetry in ABA distribution occurs in horizontal roots which have developed a downward curvature, but not in those which have failed to respond, in the time of the experiment. The differences are only just significant and the question naturally arises of whether or not they are sufficient to account for curvature.

Feldman (1981) has recently examined the inhibitors in *Zea* root caps in the light and darkness. He found ABA and a second acidic inhibitor to be present in the light, but no ABA to be present in the dark, the variety of *Zea mays* used, Merit, having a gravitropic response only in the light. In the neutral fraction only one inhibitor is present in the dark, but an additional one appears following exposure to light. If the caps are detached and placed in a culture medium, illumination does not cause the appearance of ABA but the additional neutral inhibitor still appears. The absence of ABA in caps detached in darkness and cultured in light or dark is of considerable interest in that light-cultured, detached caps can induce bending when used to replace the caps of dark-grown roots. This finding indicates that it is the neutral inhibitor and not ABA upon which the gravitropic response depends. Such a conclusion agrees with the suggestions of Suzuki *et al.* (1979) that it was the asymmetric distribution of an unidentified inhibitor rather than ABA that was correlated with the gravitropic responsiveness in their experiments.

If it is an unidentified inhibitor, rather than ABA, which undergoes lateral redistribution in a gravitropically stimulated root, the problem arises of how ABA is able to induce gravitropic responsiveness in horizontal *Zea* roots in the dark (Wilkins and Wain, 1975b). A possible explanation of this finding is that ABA may normally be necessary for the appearance of the unidentified inhibitor by acting either as a precursor or as a controlling factor in its synthesis or release. However, since in detached caps cultured in the light the unidentified inhibitor appears but not ABA (Feldman, 1981), light may also have a more direct influence on the appearance of the unidentified compound, as well as exerting control through ABA. It is of considerable importance that the

chemical structure of the unidentified inhibitors in the cap be established, if only to determine whether or not any of those reported by Feldman (1981), but particularly the neutral fraction one, is identical with that found by Suzuki *et al.* (1979). Published evidence suggests that this may not be the case since the inhibitor reported by Suzuki *et al.* (1979) occurred in the acid fraction of their extract.

Grass nodes

The other important gravitropic response of considerable economic significance is that found in the mature culms of cereals and grass stems. The leaf sheath base (node) of wheat is a structure with the potential for growth which does not occur if the stem remains in the vertical position (Arslan and Bennet-Clark, 1960). The stem within the leaf sheath is quite supple and relies on the surrounding leaf sheath for support. When placed in the horizontal position the lowermost side only of the leaf sheath-base begins to grow (Bridges and Wilkins, 1973a) and this gives rise to upward curvature of the stem and the restoration of the apex to the vertical position (Fig. 8.10). The response involves the *initiation* of cell growth only in that part of the node in which the amyloplasts are lying against the wall of the statocytes nearest to the epidermis. There must, therefore, be differences in the sensitivities of the lateral walls of the statocytes to the presence of statoliths.

The gravitropic response of wheat nodes does not depend upon the redistribution of hormonal or other molecules from one half of the organ to the other, because growth is inititated in node tissue that has been bisected longitudinally merely by placing a half node in the horizontal position with the epidermis directed downwards; growth is not induced in a horizontal half node if the epidermis is directed upward. Furthermore, small pieces of excised node tissue behave in the same way. Growth occurs as long as the tissue is held in the horizontal position (epidermis down) and stops abruptly when returned to either the vertical position (Fig. 8.11) or left horizontal but turned so that the epidermis is uppermost (Bridges and Wilkins, 1973a).

Of the known growth regulators, only IAA initiated growth in a vertical node (Bridges and Wilkins, 1973a), but its presence in an active form in vertical tissue must be minimal in view of the absence of

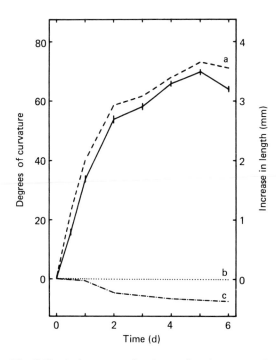

Fig. 8.10 A time course for the gravitropic response in the leaf sheath base of *Triticum* (wheat) (———), and for changes in the lengths of the upper (–·–·, c) and lower (–––, a) surfaces of the horizontal organs as compared with those observed in vertical control segments (......, b). (From Bridges and Wilkins, 1973a.)

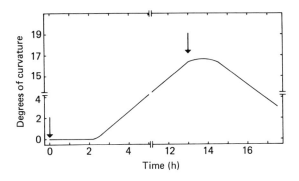

Fig. 8.11 Time course for the initiation and cessation of gravitropic curvature in the leaf sheath base of *Triticum*. Preparations were displaced to the horizontal position at the first arrow and returned to the vertical position at the second. (From Bridges and Wilkins, 1973a.)

growth in this position. A synthesis, release or activation of the molecule might occur upon gravitropic stimulation, especially since the first detectable curvature takes place after a latent period of about 2.3 h following placement in the horizontal position (Fig. 8.11). It is unlikely that a pool of growth regulator is built up in this time because of the fact that growth ceases very shortly after the tissue is returned to the vertical position (Fig. 8.11). There may, however, be a very rapid turnover of IAA in the stimulated tissue. Bridges and Wilkins (1973a) could find no evidence, using bioassay or mass-spectrometry techniques, for the presence of

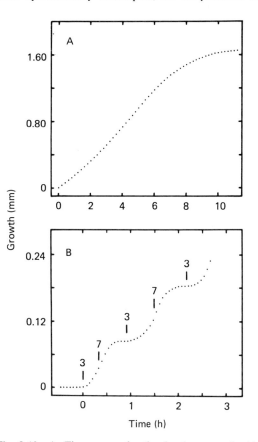

Fig. 8.12 A. Time course for the development of acid-induced growth in wheat leaf-sheath bases on transfer to a buffer at pH 3 at time zero. B. The effect of alternating exposure to pH 3 and pH 7 in inducing and inhibiting the growth of the leaf-sheath base. (From Bridges and Wilkins, 1973b.)

IAA in the node tissue either before or after gravitropic stimulation.

Later investigations on the gravitropic responses of cereal (wheat, barley and rye; Dayanandan, Hebard and Kaufman, 1976) and grass (*Echinochloa colonum*; Osborne and Wright, 1977) nodes have confirmed, for the most part, the results of Arslan and Bennet-Clark (1960) and Bridges and Wilkins (1973a,b, 1974). Osborne and Wright (1977) estimated the amount of IAA in the separated upper and lower halves of a horizontally-placed *Echinochloa* node and obtained values of $5-10$ ng g^{-1} and up to 30 ng g^{-1} fresh mass respectively. The amount of IAA in the lower half increased by at least 40% within 30 min after the tissue was placed in the horizontal position indicating that growth initiation may be associated with the increased amounts of IAA. In this respect the results for *Echinochloa* appear to differ from those for *Triticum* where IAA was not detected (Bridges and Wilkins, 1973a).

Growth can be initiated in the leaf sheath bases of wheat by lowering the pH of the tissue to 3 (Fig. 8.12A) and terminated by raising the pH to 5 or 7 (Bridges and Wilkins, 1973b). This control of growth can be operated repetitively (Fig. 8.12B). Whether or not a proton secretion mechanism is involved in the induction of the growth that underlies the gravitropic response is not known at present, but the observation that growth can be initiated by decreasing the pH of the tissue makes such a mechanism possible. The gravity-induced secretion of protons by cells oriented with their outermost walls directed downward could undoubtedly explain satisfactorily the gravity-induced growth underlying the gravitropic response of the node. If all the cells in the node possessed statoliths, it is possible that each cell could detect its orientation and initiate its own proton secretion. However, only certain cells in the node, those located just inside the vascular bundles, have obvious sedimentable amyloplasts (Wright and Osborne, 1977); it seems likely, therefore, that those cells must send signals to the other cells in the tissue to initiate growth. Clearly, since IAA is known to initiate proton secretion in cells and hence promote growth (Chapter 1), IAA could be either released or synthesized in or near the statocytes, and move to the remainder of the tissue. This view would be in accord with the observation that IAA increases in the tissue following gravitropic stimulation (Wright and Osborne, 1977).

Another possibility, however, is that gravitropic stimulation may induce the secretion of protons from the statocytes without the involvement of IAA, and the protons may themselves travel throughout the apoplast reducing the pH of the walls and initiating growth. It is also possible that the signal may be propagated throughout the tissue

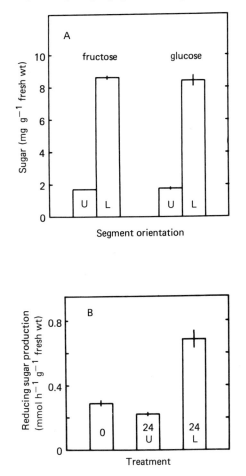

Fig. 8.13   A. Reducing sugar content of the upper (U) and lower (L) portions of the leaf-sheath base of wheat after 24 h in the horizontal position. B. The hydrolysis of sucrose to reducing sugars by homogenates of half nodes (leaf-sheath bases) 24 h after being placed in the horizontal position with their outer epidermis directed either upwards (24U) or downwards (24L). Activity in vertical half nodes is shown by 0. Vertical lines are $2 \times$ S.E. of mean. (From Bridges and Wilkins, 1974.)

electrically (Chapter 9). The latter suggestions imply that gravity can modify the metabolic behavior of statocytes and their neighboring cells without the necessity of growth hormone involvement, and there is evidence that this may be so, since gravitropic stimulation has a large effect on at least one enzyme system in wheat nodes. The reducing sugar content of the lower half of a horizontal, gravitropically-stimulated wheat node increased to 3–4 times that of the upper half (Fig. 8.13A; Arslan and Bennet-Clark, 1960; Bridges and Wilkins, 1974). This increase resulted from an hydrolysis of sucrose, and occurred even in old nodes at the base of the stem which had lost the ability to grow and thus respond gravitropically—it cannot therefore be a result of growth. Crude extracts of the lower halves of gravitropically-stimulated nodes showed a very marked increase in extractable invertase activity, while those of the upper half remained low and similar to that found in the vertical tissue (Fig. 8.13B; Bridges and Wilkins, 1974). This finding raises the interesting possibility that there may be a direct linkage between a mechanism controlling the activity, synthesis or release of an enzyme and the gravity-sensing mechanism in a plant organ, a possibility which, surprisingly, does not appear to have attracted much attention. Obviously a similar effect might occur on enzymes associated with the mechanism of proton extrusion through the plasmalemma, and provide a growth-controlling system which is independent of the presence of plant growth regulators.

References

Abrol, B. K. and Audus, L. J. (1973a). The lateral transport of 2,4-dichlorophenoxyacetic acid in horizontal hypocotyl segments of *Helianthus annuus. J. Exp. Bot.* **24**, 1209–23.

Abrol, B. K. and Audus, L. J. (1973b). The effects of *N*-1-naphthythalamic acid and (2-chloroethyl)-phosphonic acid on the gravity-induced lateral transport of 2,4-dichlorophenoxyacetic acid. *J. Exp. Bot.* **24**, 1224–30.

Arslan, N. and Bennet-Clark, T. A. (1960). Geotropic behaviour of grass nodes. *J. Exp. Bot.* **11**, 1–12.

Audus, L. J. (1962). The mechanism of the perception of gravity by plants. *Symp. Soc. Exp. Biol.* **16**, 197–226.

Audus, L. J. (1969). Geotropism. In *Physiology of Plant Growth and Development*, ed. M. B. Wilkins, McGraw-Hill, London, pp. 204–42.

Audus, L. J. (1975). Geotropism in roots. In *The Development and Function of Roots*. eds J. Torrey and D. T. Clarkson, Academic Press, London, pp. 327–63.

Austruc, J. (1709). Conjecture sur le redressement des plantes inclinées a l'horizon. *Mem. Acad. r. Sci. Paris*, 463–70.

Barlow, P. W. (1974). Recovery of geotropism after removal of the root cap. *J. Exp. Bot.* **25**, 1137–46.

Barlow, P. W. and Grunwag, M. (1974). The development of amyloplasts in cells of the quiescent centre of *Zea* roots in response to removal of the root cap. *Z. Pflanzenphysiol.* **73**, 56–64.

Bode, H. R. (1959). Uber den Einfluss des Heteroauxins auf die Kationenzusammensetzung der Blattasche der Tomate. *Planta* **53**, 212–18.

Bode, H. R. (1960). Uber den Einfluss der Geoinduktion auf die Kationenverteilung im Hypokotyl von *Helianthus annuus. Planta* **54**, 15–33.

Bowen, M. R., Wilkins, M. B., Cane, A. R. and McCorquodale, I. (1972). Auxin transport in roots. VIII. The distribution of radioactivity in the tissues of *Zea* root segments. *Planta* **105**, 273–92.

Boysen-Jensen, P. (1936). *Growth Hormones in Plants*. Transl. by G. S. Avery and P. R. Buckholder, McGraw-Hill, New York and London, pp. 4–5.

Brauner, L. and Appel, E. (1960). Zum Problem der Wuchsstoffquerverschiebung bei der geotropischen Induktion. *Planta* **55**, 226–34.

Bridges, I. G., Hillman, J. R. and Wilkins, M. B. (1973). Identification and localisation of auxin in primary roots of *Zea mays* by mass spectroscopy. *Planta* **115**, 189–92.

Bridges, I. G. and Wilkins, M. B. (1973a). Growth initiation in the geotropic response of the wheat node. *Planta* **112**, 191–200.

Bridges, I. G. and Wilkins, M. B. (1973b). Acid-induced growth and the geotropic response of the wheat node. *Planta* **114**, 331–9.

Bridges, I. G. and Wilkins, M. B. (1974). The role of reducing sugars in the geotropic response of the wheat node. *Planta* **117**, 243–50.

Cane, A. R. and Wilkins, M. B. (1969). Independence of lateral and differential longitudinal movement of indoleacetic acid in geotropically stimulated coleoptiles of *Zea mays. Plant Physiol.* **44**, 1481–7.

Cholodny, N. (1926). Beitrage zur Analyse der geotropischen reaktion. *Jahrb. wiss. Bot.* **65**, 447–59.

Ciesielski, T. (1872). Untersuchung uber die Abwartskrummung der Wurzel. *Beitr. Biol.* **1**, 1–30.

Darwin, F. (1880). *The Power of Movement in Plants*, John Murray, London, pp. 523–47.

Dayanandan, P., Hebard, F. V. and Kaufman, P. B. (1976). Cell elongation in the grass pulvinus in response to geotropic stimulation and auxin application. *Planta* **131**, 245–52.

Digby, J. and Firn, R. D. (1979). An analysis of the changes in growth rate occurring during the initial stages of geocurvature in shoots. *Plant, Cell and Environment*, **2**, 145–8.

Dodart, D. (1703). Sur l'affectation de la perpendiculaire, remarkable dans toutes les tiges, dans plusiers racines, et autant qu'il est possible dans toutes les branches des plantes. *Mem. Acad. r. Sci. Paris*, **1700**, 47–63.

Dolk, H. (1929). Uber die Wukung der Schwerkraft auf Koleoptilen von *Avena sativa. Proc. K. Akad. Wetensch. Amsterdam* **32**, 40–7.

Elliott, M. C. and Greenwood, M. S. (1974). Indol-3yl-acetic acid in roots of *Zea mays. Phytochemistry* **13**, 239–41.

Feldman, L. J. (1981). Light-induced inhibitors from intact and cultured caps of *Zea* roots. *Planta* **153**, 471–5.

Firn, R. D., Digby, J. and Riley, H. (1978). Shoot geotropic curvature—the location, magnitude and kinetics of the gravity induced differential growth in horizontal sunflower hypocotyls. *Ann. Bot.* **42**, 465–8.

Firn, R. D. and Digby J. (1980). The establishment of tropic curvatures in plants. *Ann. Rev. Plant Physiol.* **31**, 131–48.

Frank, A. B. (1868). *Beitrage zur Pflanzenphysiologie. I. Ueber die durch Schwerkeft verursachte Bewegung von Pflanzentheilen*, W. Engelmann, Leipzig, 167 pp.

Gibbons, G. S. B. and Wilkins, M. B. (1970). Growth inhibitor production by root caps in relation to geotropic responses. *Nature* **226**, 558–9.

Gillespie, B. and Thimann, K. V. (1961). The lateral transport of indoleacetic acid-^{14}C in geotropism. *Experientia* **17**, 126–9.

Gillespie, B. and Thimann, K. V. (1963). Transport and distribution of auxin during tropistic response. I. The lateral migration of auxin in geotropism. *Plant Physiol.* **38**, 214–25.

Goldsmith, M. H. M. and Wilkins, M. B. (1964). Movement of auxin in coleoptiles of *Zea mays* L. during geotropic stimulation. *Plant Physiol.* **39**, 151–62.

Goswami, K. K. A. and Audus, L. J. (1976). Distribution of calcium, potassium and phosphorous in *Helianthus* hypocotyls and *Zea mays* coleoptiles in relation to tropic stimuli and curvatures. *Ann. Bot.* **40**, 49–64.

Grahm, L. (1964). Measurement of geoelectric and auxin-induced potentials in coleoptiles with a refined vibrating electrode technique. *Physiologia Pl.* **17**, 231–61.

Grahm, L. and Hertz, C. H. (1962). Measurements of the geoelectric effect in coleoptiles by a new technique. *Physiologia Pl.* **15**, 96–114.

Grahm, L. and Hertz, C. H. (1964). Measurement of the geo-electric effect in coleoptiles. *Physiologia Pl.* **17**, 186–201.

Greenwood, M. S., Shaw, S., Hillman, J. R., Richie, A. and Wilkins, M. B. (1972). Identification of auxin from *Zea* coleoptile tips by mass spectrometry. *Planta* **108**, 179–83.

Greenwood, M. S., Hillman, J. R., Shaw, S. and Wilkins, M. B. (1973). Localization and identification of auxin in roots of *Zea mays. Planta* **109**, 369–74.

Griffiths, H. J. and Audus, L. J. (1964). Organelle distribution in the statocyte cells of the root-tip of *Vicia faba* in relation to geotropic simulation. *New Phytol.* **63**, 319–33.

Grundwag, M. (1971). Personal communication to, and quoted in, Juniper and French (1973).

Grundwag, M. and Barlow, P. W. (1973). Changes in the nucleolar ultrastructure in cells of the quiescent centre after removal of the root cap. *Cytobiologie* **8**, 130–9.

Haberlandt, G. (1900). Ueber die Perzeption des geotropischen Reizes. *Ber. dtsch. Bot. Ges.* **18**, 261–72.

Haberlandt, G. (1914). *Physiological Plant Anatomy* (1928). Transl. M. Drummond, MacMillan, London, pp. 595–612.

Hall, A. B., Digby, J. and Firn, R. D. (1980). Auxins and shoot tropisms—tenuous connection. *J. Biol. Education* **14**, 195–9.

Hartung, W. (1976). Der basipetale [2-^{14}C] Abscisinsauretransport in Wurzeln intakter Bohnenkeimlinge und seine Bedeutung fur den Wurzelgeotropismus. *Planta* **128**, 59–62.

Hartung, W. (1981). The effect of gravity on the distribution of plant growth substance in plant tissues. *The Physiologist* **24**, S-29-32.

Hawker, L. E. (1933). The effect of temperature on the geotropism of seedlings of *Lathyrus odoratus. Ann. Bot.* **47**, 503–15.

Hertel, R., de la Fuente, R. K. and Leopold, A. C. (1969). Geotropism and the lateral transport of auxin in the corn mutant amylomaize. *Planta* **88**, 204–14.

Hillman, S. K. and Wilkins, M. B. (1982). Gravity perception in decapped roots of *Zea mays. Planta* **155**, 267–71.

Iversen, T-H. (1969). Elimination of geotropic responsiveness in roots of cress (*Lepidium sativum*) by removal of statolith starch. *Physiol. Pl.* **22**, 1251–62.

Iversen, T-H. (1974). Experimental removal of statolith starch. In *The roles of statoliths, auxin transport and auxin metabolism in root geotropism, K. norske. vidensk. Selsk. Mus. Miscellanea* **15**. Trondheim.

Iwami, S. and Masuda, Y. (1974). Geotropic response of cucumber hypocotyls. *Plant Cell Physiol.* **15**, 121–9.

Juniper, B. E. (1976). Geotropism. *Ann. Rev. Plant Physiol.* **27**, 385–406.

Juniper, B. E. and French, A. (1973). The distribution and redistribution of endoplasmic reticulum (ER) in geoperceptive cells. *Planta* **109**, 211–24.

Juniper, B. E., Groves, S., Landau-Schacher, B. and Audus, L. J. (1966). Root cap and the perception of gravity. *Nature* **209**, 93–4.

Knight, T. A. (1806). On the direction of the radicle and germen during the vegetation of seeds. *Phil. Trans. R. Soc.* 99–108.

Kundu, K. K. and Audus, L. J. (1974). Root growth inhibitors from root cap and root meristem of *Zea mays*. *J. Exp. Bot.* **25**, 479–89.

Mertens, R. and Weiler, E. W. (1983). Kinetic studies of the redistribution of endogenous growth regulators in gravi-reacting plant organs. *Planta* **158**, 339–48.

Miles, D. (1981). The relationship of geotropic response and statolith in a nuclear mutant maize. In *Abstracts. XIII International Botanical Congress, Sydney, Australia*, ed. D. J. Carr, p. 250.

Navez, A. E. and Robinson, T. W. (1933). Geotropic curvature of *Avena* coleoptiles. *J. Gen. Physiol.* **16**, 133–45.

Němec, B. (1900). Ueber die Art der Wahrnehmung des Schwekraftreizes bei den Pflanzen. *Ber. Dtsch. Bot. Ges.* **18**, 241–5.

Noll, F. (1892). *Ueber heterogene Induktion*, Leipzig, 42 pp.

Nonhebel, H., Crozier, A. and Hillman, J. R. (1983). Analysis of ^{14}C-indole-3-acetic acid metabolites from primary roots of *Zea mays* seedlings using reverse-phase, high-performance, liquid chromatography. *Phys. Plant.* (in press).

Nonhebel, H., Crozier, A., Hillman, J. R. and Wilkins, M. B. (1983). Metabolism of ^{14}C-indole-3-acetic acid by coleoptiles of *Zea mays* L. *Planta* (in press).

Osborne, D. J. and Wright, M. (1977). Gravity induced cell elongation. *Proc. R. Soc. Lond. B* **199**, 551–64.

Phillips, I. D. J. (1972). Endogenous gibberellin transport and biosynthesis in relation to geotropic induction of excised sunflower shoot tips. *Planta* **105**, 234–45.

Phillips, I. D. J. and Hartung, W. (1976). Longitudinal and lateral transport of [3,4-^{3}H]gibberellin A$_1$ and 3-indole(acetic acid-2-^{14}C) in upright and geotropically responding green internode segments from *Helianthus annuus*. *New Phytol.* **76**, 1–9.

Pilet, P-E. (1973). Growth inhibitor from the root cap of *Zea mays*. *Planta* **111**, 275–8.

Pilet, P-E. (1975). Abscisic acid as a root growth inhibitor: physiological analyses. *Planta* **122**, 299–302.

Pilet, P-E. (1976). The light effect on the growth inhibitors produced by the root cap. *Planta* **130**, 245–9.

Pilet, P-E. (1977). Root georeaction: gravity effect on hormone balance. In *Life Sciences Research in Space*, eds W. R. Burke and T. D. Guyenne, Noordwijk. The Netherlands: E.S.A. Scientific and Technical Publications Branch, pp. 223–7.

Pilet, P-E. and Rivier, L. (1981). Abscisic acid distribution in horizontal maize root segments. *Planta* **153**, 453–8.

Railton, I. D. and Phillips, I. D. J. (1973). Gibberellins and geotropism in *Zea mays* coleoptiles. *Planta* **109**, 121–6.

Rivier, L. and Pilet, P-E. (1974). Indolyl-3-acetic acid in cap and apex of maize roots: identification and quantification by mass fragmentography. *Planta* **120**, 107–12.

Rivier, L., Melon, H. and Pilet, P-E. (1977). Gas chromatography-mass spectrometric determinations of abscisic acid levels in the cap and apex of maize roots. *Planta* **134**, 23–7.

von Sachs, J. (1887). *Lectures on the Physiology of Plants*, transl. H. M. Ward, The Clarendon Press, Oxford, pp. 689–92.

Sack, F. D., Priestley, D. A. and Leopold, A. C. (1983). Surface charge on isolated maize coleoptile amyloplasts. *Planta* **157**, 511–17.

Schröter, K., Läuchli, A. and Sievers, A. (1975). Mikroanalytische Identifikation von Bariumsulfat-Kristallen in den Statolithen der Rhizoide von *Chara fragilis* Desv. *Planta* **83**, 323–34.

Scott, T. K. and Wilkins, M. B. (1968). Auxin transport in roots. II. Polar flux of IAA in *Zea* roots. *Planta* **83**, 323–34.

Scott, T. K. and Wilkins, M. B. (1969). Auxin transport in roots. IV. Effects of light on IAA movement and geotropic responsiveness in *Zea* roots. *Planta* **87**, 249–58.

Shaw, S., Gardner, G. and Wilkins, M. B. (1973). The lateral transport of IAA in intact coleoptiles of *Avena sativa* L. during geotropic stimulation. *Planta* **115**, 97–111.

Shaw, S. and Wilkins, M. B. (1973). The source and lateral transport of growth inhibitors in geotropically stimulated roots of *Zea mays* and *Pisum sativum*. *Planta* **109**, 11–26.

Shaw, S. and Wilkins, M. B. (1974). Auxin transport in roots. X. Relative movement of radioactivity from IAA in the stele and cortex of *Zea* root segments. *J. Exp. Bot.* **25**, 199–207.

Sievers, A. and Volkmann, D. (1972). Verursacht differentieller Druck der Amyloplasten auf ein Komplexes Endomembransystem die Geoperzeption in Wurzeln? *Planta* **102**, 160–72.

Suzuki, T., Kondo, N. and Fujii, T. (1979). Distribution of growth regulators in relation to the light-induced geotropic responsiveness in *Zea* roots. *Planta* **145**, 323–9.

Tanimoto, E. and Masuda, Y. (1971). Role of the epidermis in auxin-induced elongation of light grown pea stem segments. *Plant Cell Physiol.* **12**, 663–73.

Volkmann, D. (1974). Amyloplasten und Endomembranen: das Geoperzeptionssystem der Primar-wurzeln. *Protoplasma* **79**, 159–83.

Volkmann, D. and Sievers, A. (1975). Wirkung der Inversion auf die Anordnung des Endoplasmatischen Reticulum und die Polaritat von Statocyten in Wurzeln von *Lepidium sativum*. *Planta* **127**, 11–19.

Volkmann, D. and Sievers, A. (1979). Graviperception in multicellular organs. In *Encyclopedia of Plant Physiology*, New Series, eds W. Haupt and M. E. Feinleib, **7**, Springer-Verlag, Berlin, Heidelberg, pp. 573–600.

Weber, U. (1931). Wachstum und Krummung einzelner Zonen geotropisch gereizter Gerstenkeimlinge. *Jahrb. wiss. Bot.* **75**, 312–76.

Webster, J. and Wilkins, M. B. (1974). Lateral movement of radioactivity from [^{14}C]gibberellic (GA$_3$) in roots and coleoptiles of *Zea mays* L. seedlings during geotropic stimulation. *Planta* **121**, 303–8.

Went, F. W. (1926). On growth accelerating substances in the coleoptile of *Avena sativa. Proc. K. Akad. Wet. Amsterdam* **30**, 10–19.

Went, F. W. and Thimann, K. V. (1937). *Phytohormones,* Macmillan, New York.

Wilkins, H. and Wain, R. L. (1974). The root cap and control of root elongation in *Zea mays* L. seedlings exposed to white light. *Planta* **121**, 1–8.

Wilkins, H. and Wain, R. L. (1975a). The role of the root cap in the response of the primary roots of *Zea mays* L. seedlings to white light and gravity. *Planta* **123**, 217–22.

Wilkins, H. and Wain, R. L. (1975b). Abscisic acid and the response of the roots of *Zea mays* seedlings to gravity. *Planta* **126**, 19–23.

Wilkins, M. B. (1966). Geotropism. *Ann. Rev. Plant Physiol.* **17**, 379–408.

Wilkins, M. B. (1975). The role of the root cap in root geotropism. *Curr. Adv. Plant Sci.* **13**, 317–28.

Wilkins, M. B. (1976a). Identification of auxin in growing plant organs and its role in geotropism. In *Perspectives in Experimental Biology*, vol. 2, ed. N. Sutherland, Oxford, New York, Pergamon, pp. 77–8.

Wilkins, M. B. (1976b). Gravity-sensing guidance systems in plants. *Science Progress, Oxford* **63**, 187–217.

Wilkins, M. B. (1977). Gravity and light-sensing guidance systems in primary roots and shoots. In *Symp. Soc. Exp. Bot.* vol. 31, *Integration of Activity in the Higher Plants*, ed. D. H. Jennings, London, New York, Cambridge University Press, pp. 275–335.

Wilkins, M. B. (1978). Gravity-sensing guidance systems in roots and shoots. *Bot. Mag. Tokyo, Special Issue No. 1*, 255–77.

Wilkins, M. B. (1979). Growth control mechanisms in gravitropism. In *Encyclopedia of Plant Physiol.* vol. **7**, eds W. Haupt and M. E. Feinleib, Springer-Verlag, Berlin, Heidelberg, pp. 601–26.

Wilkins, M. B. and Nash, L. J. (1974). Movement of radioactivity from [^{3}H]GA$_3$ in geotropically stimulated coleoptiles of *Zea mays. Planta* **115**, 245–51.

Wilkins, M. B. and Scott, T. K. (1968). Auxin transport in roots. III. Dependence of the polar flux of IAA in *Zea* roots upon metabolism. *Planta* **83**, 335–46.

Wilkins, M. B. and Whyte, P. (1968). Relationship between metabolism and the lateral transport of IAA in corn coleoptiles. *Plant Physiol.* **43**, 1435–42.

Wilkins, M. B. and Woodcock, A. E. R. (1965). Origin of the geo-electric effect in plants. *Nature* **208**, 990–2.

Woodcock, A. E. R. and Wilkins, M. B. (1969a). The geoelectric effect in plant shoots, I. The characteristics of the effect. *J. Exp. Bot.* **20**, 156–69.

Woodcock, A. E. R. and Wilkins, M. B. (1969b). The geoelectric effect in plant shoots. II. Sensitivity of concentration chain electrodes to reorientation. *J. Exp. Bot.* **20**, 687–97.

Woodcock, A. E. R. and Wilkins, M. B. (1970). The geoelectric effect in plant shoots. III. Dependence upon auxin concentration gradients and aerobic metabolism. *J. Exp. Bot.* **21**, 985–6.

Woodcock, A. E. R. and Wilkins, M. B. (1971). The geoelectric effect in plant shoots. IV. Inter-relationship between growth, auxin concentration and electrical potentials in *Zea* coleoptiles. *J. Exp. Bot.* **22**, 512–25.

Wright, M. and Osborne, D. J. (1977). Control of cell growth and hormone levels by gravity. In *Life Science Research in Space,* ed. W. R. Burke and T. D. Gruyenne, European Space Agency (ESA SP130), Paris, pp. 229–33.

Yamamoto, R., Shinozaki, K. and Masuda, Y. (1970). Stress relaxation properties of plant cell walls with special reference to auxin action. *Plant Cell Physiol.* **11**, 947–56.

Nastic movements

9

G P Findlay

Introduction

Nastic movements are a class of plant movements produced in response to a stimulus of some kind, but whose direction is determined by the anatomy of the moving parts, rather than by the nature and direction of the stimulus. These movements involve either elastic changes in cell walls of moving tissue, or plastic changes. Movements involving plastic changes in cell walls constitute growth, and will only be considered briefly in this chapter. Nastic movements involving elastic changes in cell walls may be classified according to the nature of the stimulus and the type of movement. Thus nyctinastic movements, from the Greek meaning 'night-closure', are the up and down movements of leaves in response to the daily rhythm of light and darkness, with the leaves arranged in a vertical orientation in darkness, and usually a horizontal position in light, and seismonastic movements are the sudden movements in leaves, in plants such as *Mimosa*, produced by mechanical stimulation. Nastic movements are defined as those which are usually reversible and repeatable; an obvious example is the diurnal movement of leaves which may undergo many identical cycles of opening and closing. Thus 'one-shot' processes such as spore dispersal, splitting of seed cases and the like, are not included as nastic movement, although these movements are often caused by elastic changes within the cellular structure.

There are two types of nastic movement, oscillatory movements, with periods ranging from 24 h down to a few minutes, and single-event movements. The single-event movements consist of a fast movement followed by a slow recovery movement with the fast movement ranging in time from a few minutes down to a few milliseconds. Some plants, such as *Mimosa* and *Stylidium*, exhibit both oscillatory and single-event movements. Examples of nastic movements are listed in Fig. 9.1 showing their relationship to other types of plant movement. Passive movements result from forces not generated directly by biological processes; active movements arise from cellular activity. Specific details of oscillatory movements in *Samanea* and of single-event movements in *Mimosa*, *Drosera*, *Dionaea*, and *Stylidium* are shown in Fig. 9.2.

Nastic movements appear as an angular displacement of part of the plant in relation to the whole plant, usually caused by changes in dimensions of special motor cells with respect to neighboring cells. The resultant bending of the motor tissue produces the angular movement of whatever is attached to it—a leaf, for instance. This type of movement is analogous to animal movement caused by contractions in skeletal muscle. In both movements, a small linear movement of a motor cell or muscle produces a magnified angular displacement of the organ; these systems act as levers with mechanical advantage less than unity (Hill and Findlay, 1981).

As plant cells generally change volume by osmosis, there seems no reason, in principle, why plant motor cells should function in any fundamentally different way. Certainly the evidence for osmotically driven movement in the simplest of motor systems—the stomatal pore, consisting of two directly observable motor cells—is very strong indeed. In an osmotically driven movement, net transport of a

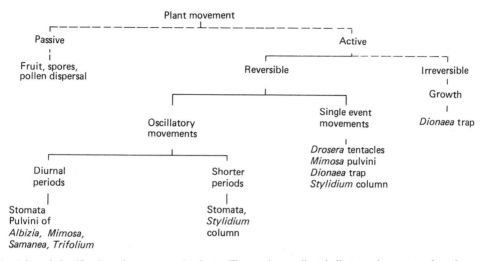

Fig. 9.1 A broad classification of movements in plants. The continuous lines indicate various types of nastic movements discussed in this chapter. (Adapted from Hill and Findlay, 1981.)

solute across the membrane of a motor cell changes the internal osmotic pressure and water potential, thus causing a flow of water. The net volume flow across the cell membrane is accommodated by a stretching or shrinking of the cell wall, and a consequent change in internal hydrostatic pressure. The idea that nastic movements might be driven by osmosis is not new; see Darwin (1880) and Hill and Findlay (1981).

In this chapter, the experimental evidence pertaining to a number of nastic movements is described. These movements have been chosen to give a broad coverage of the physiological information available, knowledge of which is necessary if the phenomenon of nastic movement is to be understood as a problem of cellular physiology; indeed, as a problem of membrane control of electrolyte and water flow between plant motor cells and their external environment.

Oscillatory movements

Trifolium repens

The leaves of clover consist of three leaflets, a terminal leaflet and two lateral leaflets, attached by pulvini to a common petiole. The leaflets undergo

cyclic movement of diurnal period, with the terminal leaflet moving through an angle of about 180° between the closed state of the leaf at night, and the open state during the day. As well, the leaflets exhibit circadian rhythms (see Chapter 10), with the oscillations continuing when the plant is placed either in continuous light or continuous darkness; the endogenous period of the oscillations, τ, is 25–27 h. The oscillation of the leaflet can also be entrained to light/dark cycles with periods ranging from 0.7 to 1.4 τ.

Scott and Gulline (1972) have examined the movement of the terminal leaflet of white clover, *Trifolium repens*, as a function of various light regimes. For a 12L:12D cycle, the motion of the leaflet is periodic but non-sinusoidal (Fig. 9.3a). Scott and Gulline argue that the most likely mechanism for this and other biological oscillators is an over-corrected negative feedback loop, and they have put forward a mathematical model of an oscillating system whose natural oscillation and response to forced oscillations resemble those of the leaf. The oscillatory movement of the *Trifolium* leaflets, as distinct from that in systems such as *Albizia*, *Samanea* and *Mimosa*, is not phytochrome mediated, and is produced by a bending motion of the pulvinus, brought about by periodic swelling and shrinking of the extensor motor cells together with similar changes, but of less magnitude, and 180° out

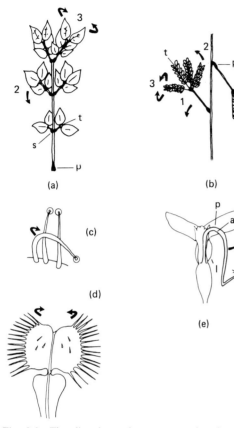

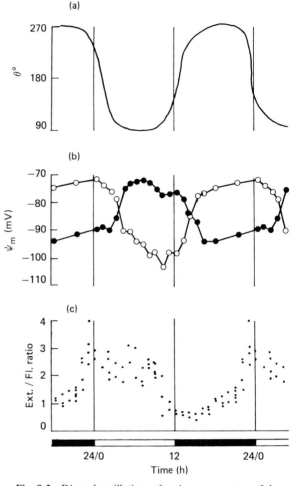

Fig. 9.2 The directions of movement of various plant organs. (a) The movement of parts of the leaf of *Samanea*. (b) The movement of parts of the leaf of *Mimosa*. In these two, p, s and t show primary, secondary and tertiary pulvini respectively, and 1, 2 and 3 the direction of movements produced by bending of these pulvini. (c) The bending of tentacles in *Drosera*. (d) The closing of the trap lobes of *Dionaea*. (e) The movement of the *Stylidium* column; l is the labellum, a and p the anterior and posterior sides of the motor tissue. (Adapted from Hill and Findlay, 1981.)

Fig. 9.3 Diurnal oscillations of various parameters of the pulvinus of *Trifolium repens*. (a) The angular displacement, θ, of the terminal leaflet; the angle is large when the leaf is closed. (b) The membrane potential difference (p.d.), ψ_m in extensor motor cells ($\bullet$) and flexor motor cells ($\bigcirc$). (c) Ratio of K^+ influx into extensor tissue (Ext.) to influx into flexor (Fl.) tissue. The extensor motor cells are defined as those cells within the pulvinus which increase in turgor pressure when the leaflet opens. (Part (a) adapted from Scott and Gulline, 1972, part (b) and (c) adapted from Scott, Gulline and Robinson, 1977.)

of phase, in flexor motor cells. Scott, Gulline and Robinson (1977) have shown that K^+, together with accompanying anions, A^-, whose identity is not known with any certainty, is the major component of the osmoticum, the periodic changes of which, in the cells of separate flexor and extensor motor tissues, produce by osmosis the reversible changes in volume of the motor cells. When the leaflet is open,

the potassium content of the extensor tissue, 138 mmol kg^{-1} fresh weight, is higher than that in the flexor tissue, 114 mmol kg^{-1} fresh weight. When the leaflet is closed the extensor tissue contains 77 mmol kg^{-1} fresh weight and the flexor tissue contains 137 mmol kg^{-1} fresh weight. Thus there is a net movement of potassium back and forth between flexor and extensor regions as the leaflet opens and closes. However, the total K$^+$ content of the pulvinus is significantly higher in the open state, and so there must be some exchange of K$^+$ between the adjacent tissue and the pulvinar cells. Scott *et al.* (1977) measured, with ^{42}K as radioactive tracer, the average influx of potassium to extensor and flexor cells over a two-hour period at various times during the opening and closing cycle. The ratio of potassium influxes to the extensor and flexor regions shows obvious cyclic behavior (Fig. 9.3c). The peaks and troughs of this graph precede the mid-times of the light and dark phases, corresponding to mid-times of leaf open and leaf closed, as shown in Fig. 9.3a, by about three hours. Sodium and calcium influx show no cyclic behavior.

Correlated with the oscillations in potassium influx are oscillations in the membrane potential difference (p.d.) of both extensor and flexor cells (Fig. 9.3b). These oscillations in p.d. are circadian (endogenous) and not responses just to the light/dark (LD) regime. The membrane p.d. is most negative in both extensor and flexor cells when these cells are in their swollen state, but becomes less negative, with the p.d. changing most rapidly, about 9 h before the cells shrink. The less-negative p.d. is maintained while the cells are losing K$^+$ and water, and then swings back to its more negative state again about 9 h before the next phase of swelling begins. It is obvious that while all the rhythms have a period of 24 h, there are distinct separations in phase between ψ_m, Ext./Fl., θ, and the LD cycle (Fig. 9.3). The leaflet oscillation which reflects directly the oscillation in the motor cell volume, leads the LD cycle by about 3 h, while the flux oscillation leads the oscillation in ψ_m in extensor cells by 4–6 h.

There is, however, no direct measure of K$^+$ efflux during the cycle, a quantity that needs to be known if the net flux of K$^+$ is to be determined, because it is the net flux of K$^+$ that determines the concentration of osmoticum within the motor cells, and thus determines the osmotic water flow to or from the cells. The passive fluxes of K$^+$ will be determined by the electrochemical gradient for K$^+$ across the motor cell membrane. Anything which tends to depolarize the membrane (i.e., make ψ_m less negative) will increase the passive efflux of cations, and

decrease the passive influx, and vice versa. By assuming a Goldman model of a membrane Scott *et al.* (1977) calculated the changes in passive efflux of K$^+$ as ψ_m oscillated between -90 and -70 mV. This efflux oscillates in phase with ψ_m, but out of phase with the influx. Thus there will be an oscillation of the net K$^+$ flux. The motor cells will swell when the net flux is inward, and shrink when it is outward (Fig. 9.4).

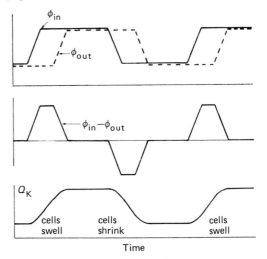

Fig. 9.4 The phase relationships between the measured influx, φ_{in}, and the calculated efflux, φ_{out}, and the net flux ($\varphi_{in} - \varphi_{out}$) of K$^+$ in cells on one side of the pulvinus of *Trifolium repens*. The total K$^+$ content Q_K which will be proportional to the integrated net flux is shown at the bottom. (From Scott *et al.*, 1977.)

The above analysis does not provide any information about the transport of anions, or give any explanation of the causes of the cyclic changes in ψ_m. For the motor cells to swell, work must be done to change the cell volume. Thus there must be active transport of solutes into the motor cells, with energy supplied from cellular metabolism, during this time. The nature of this active transport has not been established, but it is possible that proton extrusion is the primary active transport, as will be discussed later.

Albizia, Samanea, Mimosa

The nyctinastic movements of the leaves of these plants are complex. Each plant possesses double compound leaves with primary pulvini connecting petiole to stem, secondary pulvini connecting pinnae to petiole, and tertiary pulvini connecting pinnules (or leaflets) to pinna. The opening of the leaves in light is brought about by bending movement in tertiary and secondary pulvini. In all three species during opening, the pinnules and the pinnae, which in the dark lie together in pairs, and adjacent to rachilla and rachis respectively, move apart. However, in the dark, the pinnae in *Samanea* are bent down towards the base of the leaf, whereas the pinnae in the other two species, and the pinnules in all three species are bent in the opposite direction as shown in Fig. 9.2a.

In the normal LD cycle, leaflet pairs begin to open before or at the end of the dark period, and close before the end of the light period. *Albizia*, *Samanea* and *Mimosa* as well as undergoing diurnal cycles of opening and closing of the leaves, also show circadian rhythms (Satter, 1979). In *Mimosa*, the pulvini are also involved in more rapid movement in response to various forms of stimulation—mechanical, thermal and chemical—as well as in responses to physical damage to the leaflets. These movements will be discussed later.

The pulvini of these species are essentially cylindrical in shape, with a central vascular core, surrounded by layers of parenchyma cells, forming the cortex. The motor cells are located in the outer region of the cortex. As in the pulvini of clover, the movement of the secondary and tertiary pulvini in *Albizia*, *Samanea* and *Mimosa* is brought about by an alternating swelling and shrinking of motor cells in extensor and flexor regions of the pulvini. These movements almost certainly result from changes in turgor pressure caused by osmosis. The major osmoticum is $K^+ A^-$, where A^- may be Cl^-, but usually not exclusively (Satter and Galston, 1981). Data obtained from the secondary pulvinus of *Samanea* by Campbell, Satter and Garber (1981) show clearly that the movements of the secondary pulvinus of *Samanea* are associated with the movement of K^+ and Cl^- between extensor and flexor regions of the pulvinus (Fig. 9.5). The major route for K^+ and Cl^- between extensor and flexor regions is through the apoplast because during the dark-induced closure of the leaves there is a large increase in ratio of K^+ content in the wall to that in the protoplasts of extensor cells, and a corresponding decrease in this ratio for the flexor cells (Fig. 9.5).

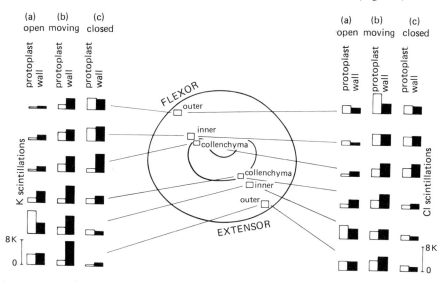

Fig. 9.5   Distribution of K^+ and Cl^-, measured by electron-probe X-ray analysis, in frozen sections of the secondary pulvinus of *Samanea* when (a) the leaflets were open in white light and extensor cells expanded, (b) the leaflets were closing after transfer to darkness, and (c) the leaflets were closed in the dark. The scale is the number of scintillations measured by the X-ray detector in 150s. (Adapted from Satter and Galston, 1981, and reproduced, with permission, from the *Annual Review of Plant Physiology*, Volume 32 (c) 1981 by Annual Review Inc.)

A major problem in any investigation of moving systems such as pulvini where the movement is controlled from within the system by a circadian oscillator (Chapter 10) is the difficulty in distinguishing the responses of the endogenous system, or the 'clock', from the actual physiological machinery of the movement. For instance, changes in light regimes affect the state of phytochrome and a blue light-absorbing pigment in the pulvini. Opening and closing movements of the pulvini in *Albizia*, *Samanea* and *Mimosa* are affected by these two photosystems, see Satter (1979), Satter and Galston (1981) and Chapters 16 and 17 of this book.

Single-event movements

Drosera, Dionaea, Aldrovanda

These three genera of insectivorous plants belong to the family *Droseraceae*.

In *Drosera*, the leaves have numerous glandular, multicellular, hair-like structures, or tentacles, attached to their inner surfaces. At the head of each tentacle is a drop of sticky liquid. An insect which alights on the leaf and becomes stuck to one of the glandular hairs, stimulates the hair by its attempt to escape and after a few seconds the hair begins to bend (Fig. 9.2c). The bending movement starts several seconds after stimulation and may be completed about 3 min later. The response of the hair depends on its location on the leaf. The outer hairs, when touched by an insect, usually bend rapidly inwards, the bending occurring at the base thus carrying the insect to the center of the leaf. The movement is then followed by slow bending movement of other tentacles, and eventually the insect is enclosed in a mesh of tentacles. The insect is then digested and absorbed by the tentacles over a period of several days. On the other hand, if no prey is caught, the hairs straighten out again after several hours, and may subsequently undergo further bending in response to stimuli.

Williams and Pickard (1972a,b) and Williams and Spanswick (1976) have shown that the response of a tentacle to various types of stimuli—chemical, electrical and mechanical—is the production, in cells at the head and neck of the hair, of a receptor potential that slowly rises until it reaches a critical level at which a series of action potentials is produced. Receptor potentials are graded responses, dependent on the magnitude of the stimulus, whereas action potentials have a shape largely independent of the stimulus; for a detailed description of action potentials in plant cells, see Findlay and Hope (1976). The action potentials in the head cells of the tentacle are propagated along the hair to the cells within the base and cause bending of the tentacle. All hair cells appear to be excitable, i.e., capable of producing action potentials. Williams and Pickard (1972a) have shown that a single stimulus, such as an electrical current pulse, usually did not cause the tentacle to bend. A second stimulus, producing a second action potential, caused bending, but not when the interval between the first and second action potentials was too long. This type of response ensures that bending occurs only after the insect is securely stuck to the tentacle.

Williams and Spanswick (1976) have made intracellular recordings of membrane p.d. in the stalk cells, and showed that with the tentacle bathed in a solution containing, among other things, 0.1 mM K^+, the resting membrane p.d. was -110 to -150 mV; action potentials in both inner and outer stalk cells were similar, with peaks at -20 to -80 mV. The shapes of the action potentials were rather variable, with some showing more than one peak, and some showing a plateau. The duration of the action potential was 10–30 s. As yet, little is known about the ionic basis of the action potential in *Drosera*.

In *Dionaea*, the well-known Venus' fly trap, a leaf modification provides a two-lobed structure, joined at the mid-rib, with spines arranged along the periphery of the lobes (Fig. 9.2d). When stimulated by appropriate mechanical displacement of one or more of six sensitive hairs, three on each of the inside surfaces of the lobes, the lobes quickly close together in about 300 ms, with the peripheral spines intermeshing. Insects which bend the sensitive hairs may thus be trapped by the lobes. If this happens the lobes then move more closely together, squeezing the insect, and various digestive enzymes and acid are secreted from glands in the inside surfaces of the lobes. The products of digestion are absorbed into the plant, and after about two weeks the trap re-opens. If the insect is not caught, the trap does not undergo the tighter closing, and re-opens in about 24 h (Jacobson, 1965).

Mechanical stimulation of the sensitive hairs is followed by the appearance of action potentials propagated within cells of the lobes (Burdon-Sanderson, 1882; Stuhlman and Darden, 1950; Siboaka, 1966). These action potentials always precede the closure of the trap (Di Palma, Mohl and Best, 1961). Jacobson (1965) and Benolken and Jacobson (1970) have shown that bending of a hair produces a receptor potential in sensitive cells having special anatomy and located at an indented region near the base of the hair. The receptor potentials are of two types, transient negative-going and transient positive-going. The amplitudes of both depend on the stimulus amplitude and shape. The negative transient is produced by fast displacements of the sensitive hair, the positive transient by slower displacements. The negative response rarely initiates an action potential. The positive-going receptor potential, when it reaches a threshold value, initiates an action potential. Action potentials, once initiated in a sensory cell, are then propagated into the cells of the trap lobe with amplitude of 100 mV, duration of about 1 s, and a propagation rate over the lobe of about 100 mm s^{-1} (Sibaoka, 1966, 1969) (Fig. 9.6a).

Iijima and Sibaoka (1981, 1982) have measured action potentials in the cells of the lobes of the *Aldrovanda* trap. The *Aldrovanda* trap is entirely aquatic, and its structure, while resembling that of *Dionaea*, is simpler. Each lobe is three cells thick at the center and two cells thick at the margins. There are about 20 sensitive hairs on the inside surface of each lobe. All cells of the lobe are excitable, either electrically or by movement of a sensitive hair. By recording from a lobe cell near the base of a sensitive hair, Iijima and Sibaoka were able to show the existence of a receptor potential. The propagation velocity of action potentials in the lobe was about 74 mm s^{-1}.

Mimosa pudica

Mimosa pudica is known as the 'sensitive plant' because it can respond to various stimuli applied at any point on the plant. The overall response is a single-event movement of part or all of a leaf or leaves caused by the movement of pulvini, and initiated by signals propagated from the point of stimulation. The bulk of the published literature on movement in *Mimosa* has been about the various

modes of production and propagation of these signals and has been reviewed by Sibaoka (1966, 1969). Briefly, the responses of the plant are of two main

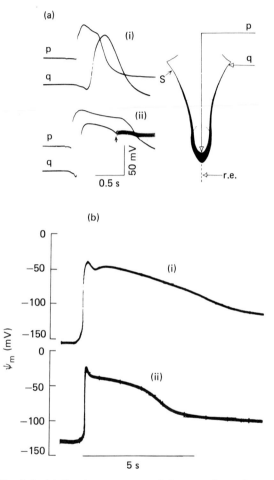

Fig. 9.6 (a) Simultaneous extracellular recordings of action potentials from two points on the surface of a *Dionaea* trap, at the midrib, p, and at the edge, q. Two electric stimuli, 20 s apart, were applied at S. The responses to the first stimulus are shown in (i), and to the second in (ii). In (ii) the arrow shows where movement of the trap, in response to the second stimulus, caused the trap surface to lose contact with the recording electrode. The reference electrode is shown as r.e. (Adapted from Sibaoka, 1966.) (b) Intracellular recordings of the action potential in (i) a flexor cell, and (ii) an extensor cell, in the primary pulvinus of *Mimosa pudica*. (Adapted from Abe and Oda, 1976.)

kinds, one where the stimulus causes no physical damage to the plant tissue, and the other, the 'wounding' response produced by cutting and burning and possibly the actions of chewing mammals, where the stimuli do produce physical damage. The 'non-wounding' responses can be subdivided into two kinds, responses to mechanical stimuli, and responses to other stimuli, such as electrical, thermal and chemical stimuli. Mechanical stimuli affect the pulvini directly by deformations of the pulvinar tissues, whereas other stimuli produce, in groups of cells in the rachis or petiole, action potentials which are then propagated to the pulvinus. These propagated signals arising from 'non-wounding' stimuli rarely go beyond the nearest pulvini. The signals resulting from 'wounding stimuli' are propagated more slowly than those from non-wounding stimuli, can pass through pulvini, and may cause fast movement of many of the leaves on a plant.

The nature of this 'wounding' response is not well understood. Since the work of Ricca (1916) it has been generally thought that wounding produces a substance that travels in the transpiration stream, and can initiate pulvinar movement. However, there are difficulties with this explanation. When a leaflet is 'wounded' the resulting signal travels down to the secondary and primary pulvini at a rate of about $4\,mm\,s^{-1}$, before it travels up to the other part of the leaf, or the rest of the plant. As the transpiration stream in the xylem generally moves up the leaf, and there appears to be no evidence that part or all of it moves in the opposite direction, the 'wound' substance could not travel by bulk flow of solution in the xylem. Diffusion of the substance in the tissue would be too slow. The other possible route for movement of the 'wound substance' could be the phloem, but here the transfer rates of about 16–$30\,mm\,min^{-1}$ (MacRobbie, 1971) seem also to be too slow. This interesting phenomenon obviously needs closer investigation.

The fast movement in the primary pulvinus is a rapid bending in about 1 to 2 s causing the leaf to fall, followed by a slower recovery to the original position in 10 to 15 min. Preceding and initiating the fast movement are action potentials within the pulvinar cells. These excitable cells are found in both extensor and flexor regions of the pulvinus (Fig. 9.6b) but there is evidence that it is the extensor region which is responsible for the actual movement (Samejima and Sibaoka, 1980). Because excitable cells occur throughout the extensor region, it is assumed that action potentials do in fact occur in the same cells that change volume.

In *Mimosa*, the evidence that action potentials probably do occur in the motor cells, and that there is a net efflux of K^+ and Cl^- during the action potential and associated movement, provides strong support for an osmotic model of movement. Toriyama (1955) has shown by histochemical means that K^+ is lost from the motor cells of the primary pulvinus during the fast movement, and Allen (1969) using ^{42}K has demonstrated a rise in K^+ efflux following stimulation of the pulvinus. The most detailed studies of electrical activity and solute movement in the primary pulvinus have been made by Samejima and Sibaoka (1980, 1982). They have made simultaneous measurements of pulvinar movement, intracellular p.d. in pulvinar (and presumably motor) cells, extracellular Cl^- concentration, and electrical conductance of the pulvinar tissue, during both the fast movement and the subsequent slow recovery. The action potential in the motor cell, initiated by the arrival of an action potential propagated from the point of stimulus on the petiole, is characterized by a rising phase lasting about 100 ms, in which the value of the resting membrane potential changes from as low as $-200\,mV$ to $-50\,mV$, followed by a plateau of 4 to 5 s duration. Accompanying the action potential with a delay of about 70 ms is an increase in the electrical conductance of the tissue, reflecting a measured rise in the extracellular concentration of Cl^-. The movement of the pulvinus commences about 200 ms after the start of the action potential, and is complete within about 2 s. Samejima and Sibaoka showed that the efflux of Cl^- from the extensor also was much greater than that from the flexor cells, and concluded that although the flexor cells were excitable, they played a lesser role in the movement.

The initial fast movement results from the release of potential energy stored within the elastic walls of the motor cells. The recovery, on the other hand, taking 10 to 15 min, and in which work clearly must be done by the motor tissue to raise the leaf, requires metabolic energy. Samejima and Sibaoka (1980) showed that the rate of recovery is light-dependent, and thus dependent on photosynthesis, but can continue at a lower rate in the dark, using respiratory energy.

Stylidium

The flowers of the *Stylidiaceae* have stamens and style fused into one structure, the gynostoemium or column. Where it emerges from the throat of the flower, the column is bent in an inverted U-shape, and as a consequence rests, near its end, against the labellum, a modified fifth petal on the anterior side of the flower (Fig. 9.2e). In a spectacular pollination mechanism the column, when mechanically displaced by nectar-gathering insects, flips through an angle of up to 4 radians in 10–20 ms, the movement being stopped either by the posterior petals or, in some species, by throat appendages. After some minutes in this position the column slowly resets to its original position against the labellum in about 400–600 s. After a refractory period the movement can be repeated. During the fast movement, the anthers or receptive stigma on the head of the column come in contact with the insect and exchange pollen.

The movement of the column is produced by a motor tissue, located at the bend of the column, which changes shape when stimulated. This tissue is between 1 and 2 mm in length, weighs up to 0.5 mg, depending on species, and possesses a specialized and very characteristic anatomy, seen in all but a few of the 140 or so species of *Stylidium* and described by Findlay and Findlay (1975). In most species the bend of the column is flatter and wider than the rest of the column. On either side of a central layer containing stigmatoid and vascular tissue is a layer of thick-walled cells containing a large number of amyloplasts. In cross-section, the cells of the anterior layer, located on the concave side of the bend when the column rests against the labellum, are rectangular in shape, whereas cells in the posterior layer are rounder. Surrounding these layers of tissue are layers of parenchyma cells. During the fast movement, the column moves with approximately constant angular velocity of up to 500 radians s^{-1} and the movement extends through 2 to 4 radians, depending on the species. In contrast, the slow recovery movement is quite non-linear (Fig. 9.7). There do not appear to be specific mechano-receptors on the column (Findlay and Findlay, 1975), fast movement being apparently initiated by movement of the motor tissue as a whole. An angular displacement of the column of less than 0.05 radians is usually sufficient to initiate the fast movement (Findlay, 1982).

The movement of the column is caused by torque produced by the motor tissue. This torque develops very rapidly when the column is moved beyond its threshold angular displacement. Findlay (1982), with the aid of an apparatus which controls angular displacement and measures resultant torque development, has shown that this torque is developed in about 1–2 ms. Furthermore, the stiffness, k, of the motor tissue given by $k = d\tau/d\theta$, where τ is the torque and θ the angular displacement, decreases dramatically from 1.51×10^2 N m (rad)$^{-1}$ to 0.644 N m (rad)$^{-1}$ in about the same time. Measurements of the maximum torque exerted by the column when stopped in mid-flight during its fast and slow movements, as a function of angular displacement, show that in both cases the torque is a linear function of θ, indicating Hookean spring-like behavior, but with non-zero values of the torque at each end of the range of movement of the column in the flower. This indicates that the column moves between two constrained states, prevented from moving further by the labellum on one side of the flower, and the petals on the other side.

Findlay and Pallaghy (1978) with an electron probe X-ray analysis have shown that the motor cells in the set column of *Stylidium graminifolium* have a very high concentration of KCl (up to 600 mM) in the anterior layer of cells, but that the KCl is more evenly distributed throughout the motor tissue during the resetting movement. Some recent measurements (C. H. Findlay, personal communication) have shown that before firing 70% of the total K$^+$ content of the motor tissue is in the anterior layer, and 30% in the posterior layer. Within 1–2 min after the fast movement has occurred, the K$^+$ content in the anterior layer falls, with a concomitant rise in content of the posterior layer, continuing for about 800 s in *Stylidium graminifolium* (the column has in the meantime reset in about 600 s), after which the anterior content again rises (and posterior falls) to the original level within about 3.6 ks.

It is often possible to establish the presence of cellular processes requiring metabolic energy by examining the temperature dependence of these processes. It has been shown by Findlay (1978) that in *Stylidium crassifolium* the angular velocity of the slow movement, but not the fast movement, is temperature-dependent, and by Findlay and Findlay (1981) that the resetting movement is inhibited by anaerobic conditions but the fast movement is not. These results suggest that it is the slow resetting motion, and not the fast firing motion, that requires metabolic energy for the transport of solutes and water into the appropriate motor cells.

One of the interesting aspects of the movement of the column, first observed by Gad (1880), is that when the labellum is cut from the flower, or the calyx and corolla removed, or when the column is prevented from resetting completely, it does not reset but oscillates in position without any fast movement, with periods of 10 to 30 min. Furthermore, when the column is held at a constant angular displacement from the reset position, the motor tissue develops an oscillatory torque.

Correlated with these oscillations in angular displacement or torque are oscillations in electrical properties of the motor tissue. Findlay (unpublished data) has shown that when the column is oscillating the p.d. across the membranes of the motor cells oscillates between about -180 and -50 mV for an extracellular K^+ concentration of 10 mM. The physiology underlying the oscillatory behavior is not understood.

While there seems no difficulty in principle in ascribing the slow recovery movement—or the oscillatory movement—to osmosis in the motor cells, some characteristics of the *Stylidium* motor system require particular consideration. If the movement of the column is produced by differential changes in length of anterior and posterior layers of cells, it would be expected that when the column was in the set position, the posterior layer of cells (which are on the convex side of the bend; Fig. 9.2e) would be longer by virtue of possessing a high

(a)

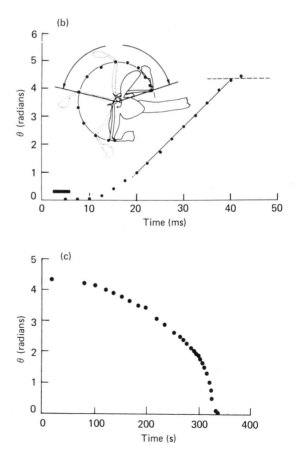

Fig. 9.7 Movement of the column of *Stylidium crassifolium*. (a) A multiple-flash photograph of the fast movement; the time interval between successive flashes was 2 ms. (b) The angular displacement, θ, of the column during fast movement as a function of time after stimulation; the black bar shows the duration of the mechanical stimulus. A tracing of the photographs in (a) is shown in the inset; the dots show the trajectory of the second bend of the column. The column is drawn in solid lines in the initial and final positions, and in dotted lines in the intermediate positions. (c) The angular displacement of the column during its slow recovery movement, as a function of time after stimulation. (Adapted from Findlay and Findlay, 1975.)

turgor pressure. In fact, most of the KCl osmoticum is contained in the anterior cells, and thus in the set column most of the stiffness of the set column would reside in these cells. Findlay (1982) has proposed that the main function of the anterior layer is to provide a holding torque, maintaining the column in its set position against an equal and opposite torque in the posterior layer. On stimulation the anterior cells in about 2–4 ms apparently lose their stiffness, and cease to exert torque, with subsequent movement of the column caused by a torque within cells in the posterior layer. KCl is then lost from the anterior cells, and the process of accumulation of KCl into the posterior cells to produce the resetting movement commences. Finally KCl is re-accumulated into the anterior layer, at which time the column is again able to respond to stimulation.

Models of movement

The preceding survey has considered the details of a range of repeatable plant movements, basically of two types, oscillatory and single-event. The oscillatory movements, endogenously produced, have periods from about 24 h for circadian rhythms in such plants as *Trifolium* and *Samanea*, which in the ordinary course of events are entrained by the day–night cycle, down to about 20 min in the column of *Stylidium*. The single-event movements, on the other hand, have a rapid movement, usually initiated by external stimuli, occurring in times from 1 s in *Mimosa* down to a few milliseconds in *Stylidium*, followed by a very much slower recovery movement, ranging in time from about 1 ks in *Mimosa* to over 100 ks in *Dionaea*.

There is little doubt that the nastic movements of some plants, and particularly the slower diurnal movements, are caused by changes in size of particular motor cells within the motor or moving tissue — changes produced by osmosis in these cells. The interesting question is whether or not nastic movements quite generally are driven by osmosis, and whether it might be possible to construct some generally applicable osmotic model for the movements. Such a model will be put to its severest test in those systems where the movement is very rapid, and where extreme values of appropriate parameters such as the hydraulic conductivity and ionic

conductances of the motor cell membrane may be required.

Assume that only elastic changes in the cell walls of the motor tissue are involved. For a single plant cell with a semi-permeable membrane, $J_v = L_p (\Delta P - \Delta \pi)$ where J_v is the volume flux, L_p the hydraulic conductivity of the membrane, AP and $\Delta \pi$ are hydrostatic and osmotic pressure differences across the membrane. A change in $\Delta \pi$ will cause a water flow across the cell membrane, a change in cell volume, and consequently a change in the internal hydrostatic pressure, P_i, given by $dP_i = \varepsilon (dV/V)$ where ε is the volume elastic modulus of the cell, and is, of course, determined by the elastic properties of the cell wall. For small changes in $\Delta \pi$ it can be shown (Dainty, 1976) that the resultant J_v is an exponential function with the half-time for volume change given by

$$T_{\frac{1}{2}} = \frac{0.693V}{A L_p (\varepsilon + \pi_i)} \qquad (1)$$

where A and V are the surface area and volume of the cell and π_i is the internal osmotic pressure. Thus the important factors affecting $T_{\frac{1}{2}}$ are L_p and ε. For a cell of dimensions $50\,\mu m \times 20\,\mu m \times 20\,\mu m$, $V/A \simeq 4\,\mu m$. Taking the high value of L_p of $3 \times 10^{-12}\,m\,s^{-1}\,Pa^{-1}$ measured for internodal cells of *Nitella flexilis* (Kamiya and Tazawa, 1956) and $\varepsilon + \pi_i = 5 \times 10^6\,Pa$, then $T_{\frac{1}{2}} \simeq 0.5\,s$.

For plant tissue the kinetics of volume changes with changing $\Delta \pi$ are much more complex. However, Equation (1) can probably provide us with some guide as to whether the rates of nastic movements, particularly the fast movements, are consistent with an osmotically driven system. There are clearly no difficulties in accommodating the rates of circadian movement where $T_{\frac{1}{2}}$ is large, of the order of 2–4 ks, or the rate of the slower single-event movements, as in *Drosera* where $T_{\frac{1}{2}}$ is a few seconds. If an increase in $L_p(\varepsilon + \pi_i)$ of 10–20 times is allowed, then rates of movement as fast as that of the *Dionaea* trap can be accommodated. This increase is probably not unreasonable because in a tissue the cells surrounding a particular motor cell will serve to stiffen up the system and increase ε, although Dainty (1976) has expressed doubt about the correctness of values of L_p as high as $3 \times 10^{-12}\,m\,s^{-1}\,Pa^{-1}$. However, the fast movement of the *Stylidium* column appears to be at too high a rate to be driven by osmosis.

For increases in motor cell volume to be caused by osmosis, solute molecules (the osmoticum) must be moved between the cell interior and the extracellular space, across the cell membrane, to do work to stretch the cell walls. Cell shrinkage can be achieved by movement of solute and water outwards down the appropriate potential gradients. One of the most striking features of the physiology of nastic movement is that an electrolyte K^+A^- (where A^- can be Cl^- and other anions) plays the role of almost universal osmoticium. As a consequence the electrophysiology of the motor cell membranes will be intimately involved with the movement of the osmoticum, and if the factors responsible for this movement are to be understood, the active and passive ion transports, the electrochemical gradients, and the ionic conductances in the system must be known in detail.

In broad outline the electrical properties of the plasmalemma of a plant cell may be represented by the equivalent electrical circuit shown in Fig. 9.8.

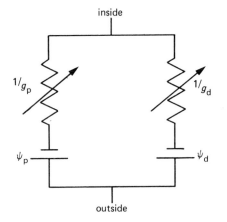

inside

outside

Fig. 9.8 An equivalent electric circuit for a membrane. ψ_p and ψ_d are the equilibrium potentials, and g_p and g_d the conductances, of pump and diffusive pathways, respectively.

The membrane has both pump and diffusive pathways, each with conductance and equilibrium potential. It is generally believed that the pump is an active efflux of H^+, that it is the primary transport directly linked to the biochemical reactions, involving ATP, and that it is electrogenic, transporting net charge across the membrane (Spanswick, 1981). The electrogenic transport in producing a p.d. across the membrane sets up an electrochemical gradient for other ions, such as K^+ and Cl^-, moving in the diffusive pathway. Strong evidence for electrogenic transport is the existence of a hyperpolarized membrane, where the observed membrane p.d. is more negative than any possible ionic diffusion potential given by the Nernst equation, which for K^+, for example, is $\psi_K = 58 \log([K^+]_o/[K^+]_i)\,mV$, where $[K^+]_{o,i}$ are the concentrations of K^+ on the outside and inside of the membrane, respectively (see Chapter 15). The identification of the transported ion as H^+ is more difficult, but in some systems such as *Neurospora* and *Chara* it seems reasonably conclusive (Spanswick, 1981). The evidence for electrogenic proton extrusion in motor cells, however, is less secure. In *Trifolium* motor cells, ψ_m oscillates between about -70 and $-100\,mV$. For swollen extensor motor cells the intracellular potassium concentration. $[K^+]_i$, is approximately $140\,mM$, (Scott *et al.*, 1977). Only if the extracellular concentration is less than $2.64\,mM$ will ψ_K be more negative than $\psi_m(-100\,mV)$, and the membrane be hyperpolarized. However, there is no measurement of extracellular concentration and thus, while values $\leq 2.64\,mM$ are feasible, there is no definite evidence that the membrane is hyperpolarized during any part of the cycle. Similar remarks apply to *Samanea* (Racusen and Galston, 1977). In the motor cells of the primary pulvinus of *Mimosa* the membrane is almost certainly hyperpolarized, because $\psi_m \simeq -160\,mV$, and even though $[K^+]_i$ is unknown, $[K^+]_i$ would have impossibly high values if ψ_K were more negative than $-160\,mV$. The evidence of Findlay (unpublished) that in *Stylidium* motor cells $\psi_m = -180\,mV$ for part of the oscillatory cycle, with the concentration of K^+ in the outside solution $10\,mM$, again suggests that the motor cell membrane is hyperpolarized, at least for part of the cycle.

The representation of the motor cell membrane as shown in Fig. 9.8 has the primary H^+ efflux providing the energy for expansion of the cell with the osmoticum moving through the diffusive pathways. However, there are essentially three electrophysiological processes which might occur in the motor cell, and which could produce changes in flux of K^+ and A^-. There may be changes in pump activity, or changes in the diffusive pathways, or a combination of the two. Any of these if suitably arranged will produce changes in membrane potential and net flux. To establish which of these processes may be occurring we need to know how ψ_p, g_p, ψ_d and g_d vary with time. Unfortunately it appears that none of these variables has yet been measured for oscillatory systems, and only g_d has been measured for the single-event movement in the primary pulvinus of *Mimosa*. By assuming cyclic changes in pump activity and constancy of the diffusive conductances, Scott *et al.* (1977) have produced a feasible description of the behavior of extensor motor cells. Satter and Galston (1981) have constructed a more general model for circadian movement in pulvini. This model attempts to account for phytochrome involvement in light reception, and the control exerted by the endogenous oscillator or 'clock'.

The rapid nastic movements in *Drosera*, *Dionaea*, *Aldrovanda* and *Mimosa* are preceded by action potentials in the motor cells. If these action potentials are similar to the well-characterized action potentials of charophyte plants (Findlay and Hope, 1976) then they will produce a net efflux of solute, probably KCl, from the cells; this is certainly so in *Mimosa*. The loss of solute produced by an increase in diffusive conductances could cause shrinkage of the motor cells, although it is not known whether the magnitude of the efflux is sufficient to cause the observed movement. If we assume that the motor cells in the primary pulvinus of *Mimosa* have dimensions of $60\,\mu m \times 30\,\mu m \times 30\,\mu m$, and suppose that during the action potential the internal solute content declines by 20% from an initial $[K^+]_i$ of 200 mM, then in 3 s (the duration of an action potential) the net flux of K^+ is $\simeq 40\,\mu mol\,m^{-2}\,s^{-1}$, which is about twice the efflux of KCl during an action potential in *Chara corallina* (Hope and Findlay, 1964). In *Dionaea* and *Aldrovanda*, the duration of the action potentials in the motor cells, 0.5 s, is somewhat less than in *Mimosa*, and although there appears to be no information about the ionic components of these action potentials, the net efflux of solute required for movement should not be unreasonably high. Williams and Bennett (1982), who have recently shown that irreversible changes in size of the motor cells of the trap lobes of *Dionaea* are associated with the fast and slow movements of the trap, have suggested that the action potentials, rather than causing loss of osmoticum from the motor cells, trigger a rapid and necessarily active efflux of H^+ that lowers the external pH sufficiently to plasticize the motor-cell walls. However, they do not indicate in any detail how such a change in wall

plasticity would actually cause the movements of the trap.

The very rapid movements of the column of *Stylidium* remain to be considered. For the loss of almost all the stiffness of the motor tissue of *Stylidium* within 1–2 ms of stimulation to be caused by loss of solute and water by osmosis, very high solute effluxes and unbelievably high values of L_p are required (Hill and Findlay, 1981). Findlay (1982) has suggested that the fast movement may be a purely mechanical one—a kind of 'click' mechanism. There may also be another possibility, where the membranes and walls of the anterior motor cells of the column momentarily break down as a result of stress caused by the displacement of the column, causing the loss of some cellular contents and the reduction of the internal hydrostatic pressure to near zero. Following this breakdown, the membrane would re-fuse and the integrity of the cells would be restored. This type of behavior occurs in cells of the marine alga, *Valonia macrophysa*, following osmotic shock (Guggino and Gutknecht, 1982); but there is no direct evidence for it in *Stylidium* motor cells.

Further reading

Bentrup, F. W. (1979). Reception and transduction of electrical and mechanical stimuli. In *Encyclopedia of Plant Physiology*, New Series, vol. 7, *Physiology of Movements*, eds W. Haupt and M. E. Feinleib. Springer-Verlag, Berlin, Heidelberg, New York.

Hill, B. S. and Findlay, G. P. (1981). The power of movement in plants; the role of osmotic machines. *Q. Rev. Biophys.* **14**, 173–222.

Pickard, B. G. (1973). Action potentials in higher plants. *Bot. Rev.* **39**, 172–201.

Roblin, G. (1979). *Mimosa pudica*: a model for the study of excitability in plants. *Biol. Rev.* **54**, 135–53.

Satter, R. L. (1979). Leaf movement and tendril curling. In *Encyclopedia of Plant Physiology*, New Series, vol. 7, *Physiology of Movements*, eds W. Haupt and M. E. Feinleib, Springer-Verlag, Berlin, Heidelberg, New York.

Simons, P. J. (1981). The role of electricity in plant movements. *New Phytol.* **87**, 11–37.

Williams, S. E. (1976). Comparative sensory physiology of the *Droseraceae*—the evolution of a plant sensory system. *Proc. Am. Philosophical Soc.* **120**, 187–204.

Zimmermann, U. (1978). Physics of turgor- and osmoregulation. *Ann. Rev. Plant Physiol.* **29**, 121–48.

References

Abe, T. and Oda, K. (1976). Resting and action potentials of excitable cells in the main pulvinus of *Mimosa pudica*. *Plant Cell Physiol.* **17**, 1343–6.

Allen, R. D. (1969). Mechanism of the seismonastic reaction in *Mimosa pudica*. *Plant Physiol.* **44**, 1101–7.

Benolken, R. M. and Jacobson, S. L. (1970). Response properties of a sensory hair excised from Venus' flytrap. *J. Gen. Physiol.* **56**, 64–82.

Burdon-Sanderson, J. (1882). On the electromotive properties of the leaf of *Dionaea* in the excited and unexcited states. *Phil. Trans. R. Soc. Lond.* B **173**, 1–53.

Campbell, N. A., Satter, R. L. and Garber, R. C. (1981). Apoplastic transport of ions in the motor organ of *Samanea*. *Proc. Nat. Acad. Sci. USA* **78**, 2981–4.

Dainty, J. (1976). Water relations of plant cells. In *Encyclopedia of Plant Physiology*, New Series, vol. 2, Part A, eds U. Lüttge and M. G. Pitman, Springer-Verlag, Berlin, Heidelberg, New York, pp. 12–35.

Darwin, C. (1880). *The Power of Movement in Plants*. Assisted by F. Darwin, John Murray, London.

Di Palma, J. R., Mohl, R. and Best, W. (1961). Action potential and contraction of *Dionaea muscipula* (Venus' flytrap). *Science* **133**, 878–9.

Findlay, G. P. (1978). Movement of the column of *Stylidium crassifolium* as a function of temperature. *Aust. J. Plant Physiol.* **5**, 477–84.

Findlay, G. P. (1982). Generation of torque by the column of *Stylidium*. *Aust. J. Plant Physiol.* **9**, 271–86.

Findlay, G. P. and Findlay, N. (1975). Anatomy and movement of the column in *Stylidium*. *Aust. J. Plant Physiol.* **2**, 597–621.

Findlay, G. P. and Findlay, N. (1981). Respiration dependent movements of the column of *Stylidium graminifolium*. *Aust. J. Plant Physiol.* **8**, 45–56.

Findlay, G. P. and Hope, A. B. (1976). Electrical properties of plant cells. Methods and findings. In *Encyclopedia of Plant Physiology*, New Series, vol. 2, Part A, *Transport in Plants II*, eds U. Lüttge and M. G. Pitman, Springer-Verlag, Berlin, Heidelberg, New York, pp. 53–92.

Findlay, G. P. and Pallaghy, C. K. (1978). Potassium chloride in the motor tissue of *Stylidium*. *Aust. J. Plant Physiol.* **5**, 219–29.

Gad, J. (1880). Über die Bewegungserscheinungen an der Blüthe von *Stylidium adnatum* R. Br. *Bot. Ztg* **38**, 216–24.

Guggino, S. and Gutknecht, J. (1982). Turgor regulation in *Valonia macrophysa* following acute osmotic shock, *J. Memb. Biol.* **67**, 155–64.

Hill, B. S. and Findlay, G. P. (1981). The power of movement in plants: the role of osmotic machines. *Q. Rev. Biophys.* **14**, 173–222.

Hope, A. B. and Findlay, G. P. (1964). The action potential in *Chara*. *Plant Cell Physiol.* **5**, 377–9.

Iijima, T. and Sibaoka, T. (1981). Action potential in the trap-lobes of *Aldrovanda vesiculosa*. *Plant Cell Physiol.* **22**, 1595–1601.

Iijima, T. and Sibaoka, T. (1982). Propagation of action potential over the trap-lobes of *Aldrovanda vesiculosa*. *Plant Cell Physiol.* **23**, 679–88.

Jacobson, S. L. (1965). Receptor response in Venus' fly-trap. *J. Gen. Physiol.* **49**, 117–29.

Kamiya, N. and Tazawa, M. (1956). Studies of water permeability of a single plant cell by means of transcellular osmosis. *Protoplasma* **46**, 394–422.

MacRobbie, E. A. C. (1971). Phloem translocation. *Biol. Rev.* **46**, 429–81.

Racusen, R. H. and Galston, A. W. (1977). Electrical evidence for rhythmic changes in the cotransport of sucrose and hydrogen ions in *Samanea* pulvini. *Planta* **135**, 57–62.

Ricca, U. (1916). Soluzione d'un problema di Fisiologia Nuovo. *G. Bot. Ital.* **23**, 51–170.

Samejima, M. and Sibaoka, T. (1980). Changes in the extracellular ion concentration in the main pulvinus of *Mimosa pudica* during rapid movement and recovery. *Plant Cell Physiol.* **21**, 467–79.

Samejima, M. and Sibaoka, T. (1982). Membrane potentials and resistances of excitable cells in the petiole and main pulvinus of *Mimosa pudica*. *Plant Cell Physiol.* **23**, 459–65.

Satter, R. L. (1979). Leaf movements and tendril curling. In *Encyclopedia of Plant Physiology*, New Series, vol. 7, *Physiology of Movements*, eds H. Haupt and M. E. Feinleib. Springer-Verlag, Berlin, Heidelberg, New York, pp. 442–84.

Satter, R. L. and Galston, A. W. (1981). Mechanisms of control of leaf movements. *Ann. Rev. Plant Physiol.* **32**, 83–110.

Scott, B. I. H. and Gulline, H. F. (1972). Natural and forced circadian oscillations in the leaf of *Trifolium repens*. *Aust. J. Biol. Sci.* **25**, 61–76.

Scott, B. I. H., Gulline, H. F. and Robinson, G. R. (1977). Circadian electrochemical changes in the pulvinules of *Trifolium repens* L. *Aust. J. Plant Physiol.* **4**, 193–206.

Sibaoka, T. (1966). Action potentials in plant organs. *Symp. Soc. Exp. Biol.* **20**, 49–73.

Sibaoka, T. (1969). Physiology of rapid movement in higher plants. *Ann. Rev. Plant Physiol.* **20**, 165–84.

Spanswick, R. M. (1981). Electrogenic ion pumps. *Ann. Rev. Plant Physiol.* **32**, 267–89.

Stuhlman, O. and Darden, E. B. (1950). The action potentials obtained from Venus's-flytrap. *Science* **111**, 491–2.

Toriyama, H. (1955). The migration of potassium in primary pulvinus of *Mimosa pudica*. *Cytologia* **20**, 367–77.

Williams, S. E. and Bennett, A. B. (1982). Venus' fly-trap closure: an acid growth response. *Science* **218**, 1120–2.

Williams, S. E. and Pickard, B. G. (1972a). Properties of action potentials in *Drosera* tentacles. *Planta* **103**, 222–40.

Williams, S. E. and Pickard, B. G. (1972b). Receptor potentials and action potentials in *Drosera* tentacles. *Planta* **103**, 193–221.

Williams, S. E. and Spanswick, R. M. (1976). Propagation of the neuroid action potential of the carnivorous plant *Drosera*. *J. Comp. Physiol.* **108**, 211–23.

Circadian rhythms

10

T A Mansfield and P J Snaith

Introduction

The daily changes in orientation of the leaves and petals of many plants are so obvious to the casual observer that the causes of, and reasons for, these movements may well have been debated since pre-historic times. The subject was first mentioned in writings over 2000 years ago, but it was not until 1729 that a French scientist, De Mairan, performed an experiment which marked the beginning of one of the most fascinating and mysterious areas of study for plant physiologists. De Mairan showed that the movements were not dependent on the daily cycle of light and darkness even though they were normally synchronized with it. Because the movements continued even when plants were placed where they could not perceive daily changes in illumination, he suggested that they were controlled by an internal mechanism. Further studies, particularly those of Pfeffer, showed that the movements could persist for more than one day without known external influences. They were thus 'endogenous' rhythms (arising from within). Since these early studies many other rhythms have been investigated in plants and animals and not all are related to cycling on a daily basis. Some annual rhythms have, for example, been identified. Rhythms which are related to daily changes are now called 'circadian', and this chapter will be concerned only with *endogenous circadian rhythms* because these reflect phenomena in cells, tissues and organs which are fundamental to our understanding of the functioning of plants.

Occurrence of circadian rhythms

Many of the basic physiological and metabolic activities in higher plants are now known to be subject to circadian rhythms, and the following list is by no means exhaustive:

leaf and petal movements
photosynthesis
respiration
growth rate
stomatal movements
dark CO_2 fixation
root exudation
flowering.

Some of the rhythms have been found to possess features in common and a popular view has been that there is a basic oscillator or 'cellular clock' which provides the necessary control. The extent to which one rhythm determines another is often, however, difficult to ascertain because of the inter-relationships between processes in the cells. Rhythms of photosynthesis can illustrate the problem. Such rhythms have been observed in a wide range of organisms from single-celled dinoflagellates such as *Gonyaulax polyedra* up to the higher plants. Carbon dioxide fixation and the processes subsequent to it dominate much of the metabolism of green plant cells, and so it is inevitable that rhythms in many other activities must follow in the wake of a rhythm in photosynthesis. Synchronization is therefore no surprise and is not satisfactory evidence either for or against the existence of a basic oscillator. It is possible that several rhythms in a few fundamental processes could be the result of the operation of such an oscillator, and that many other overt rhythms are then caused indirectly. The answers to many such questions are still not known and the unraveling of cause–effect relationships is a demanding exercise for the physiologist who chooses to work on this subject.

Terminology

Figure 10.1 illustrates some of the simple terms used to describe rhythms. It depicts the activity of a

not seem to exert any major effect on the basic features of the rhythm. The *phase* is the state of the oscillation at any instant in time, and a *phase-shift* is said to have occurred if the oscillations change in relation to the time axis. Figure 10.1 shows the changes that can occur when a plant in an entrained rhythm is placed in constant conditions, in this case in continuous light. First, there is a cycle which is called the *transient* and which may be different in period from the constant period found in the eventual *free-running* cycles. When the rhythm becomes free-running the amplitude is not normally maintained, *i.e.*, a *damping* occurs.

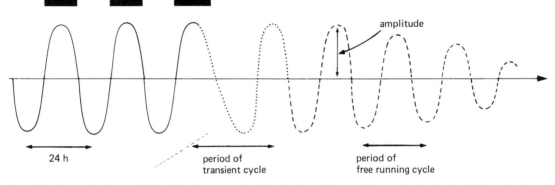

Fig. 10.1   The main characteristics of endogenous circadian rhythms. The entrained cycles (unbroken line) are in diurnal periods of light and darkness (black bars indicate darkness). The rhythm is also shown proceeding in continuous light, establishing its free-running form (dashed line) after a transient cycle (dotted line). The period of the free-running rhythm is not exactly 24 h, and the peaks and troughs do not coincide with changes in environmental factors. The phase may, however, depend on a change in external conditions, for example, the time of transfer to the constant environment. In this hypothetical case the phase of the free-running rhythm has been affected by the prolongation of the third period of darkness.

hypothetical process in a plant which passes through distinct phases that are normally associated with day and night. The first two cycles depicted are in a normal light–dark sequence, and here the rhythm is *entrained* because its features are obviously related to the events taking place in the photo-environment. The important feature of entrainment is that the *period* (the time for the completion of a cycle) does not deviate from the 24-h period of the natural environment. More eye-catching, but of little significance in relation to mechanisms behind rhythms, is the effect of the environment on *amplitude*. The influence that a factor such as light intensity may have on the amplitude of an entrained rhythm does

Endogenous and exogenous control of rhythms

At one time the development of this subject was threatened by the skepticism of a few physiologists who adamantly declared that the possibility of exogenous control had not been ruled out. The useful outcome of the controversy was the establishment of a set of formal rules or conditions that a rhythm should satisfy before it could be said to be endogenous. Pittendrigh (1954) proposed five conditions which met with general approval, and a useful, concise summary of these was later provided by Wilkins (1969). Three of the five are particularly important and if any one of these cannot be satisfied a circadian rhythm is unlikely to be truly endogenous. They are that:

(1) the rhythm should continue under environmental conditions which are kept as constant as is physically possible;

(2) the period of the free-running rhythm should not correspond exactly to known oscillations in the environment: this means that the period of a circadian rhythm should not be exactly 24 h;

(3) it should be possible to determine the phase by suitable pretreatments, and/or alter the phase of the free-running rhythm by abrupt changes in one or more of the environmental parameters normally held constant to satisfy condition (1).

If the third condition can be satisfied by a clear demonstration of phase-shift, there can be little remaining doubt about the endogenous nature of the control. Although environmental variables are not in phase with one another (for example, peaks of temperature are not coincident with those of light intensity), the ability to shift the phase of the rhythm so that its peaks and troughs appear at *any* selected time of the day must remove any suspicion of control from oscillating environmental factors.

Rhythms of CO_2 metabolism in succulents

The way in which experiments can be performed to establish the endogenous nature of, and the factors controlling, a particular overt rhythm can be illustrated by reference to the extensively investigated CO_2 metabolism of members of the *Crassulaceae*. These succulent plants possess a type of C_4 photosynthesis which involves a separation in time of the initial capturing of atmospheric CO_2 and its eventual use in the Calvin cycle. Carbon dioxide is taken in at night through open stomata and fixed first into oxaloacetic acid, from which other C_4 acids such as malic are formed and then stored in the cell vacuoles. During the day the stomata close and decarboxylation of the C_4 compounds occurs to release CO_2 for normal photosynthesis. This clear association of metabolic events with times of day produces diurnal patterns of a distinctive nature, for example, readily measurable changes in tissue acidity and the associated CO_2 exchange, and stomatal behavior which is the reverse of that found in mesophytes.

Continuous measurements of the CO_2 exchange of the leaves or shoots of succulents, or of the CO_2 concentration they maintain in a closed system, both under constant environmental conditions, have provided some of the most precise information about the ways in which endogenous circadian rhythms are controlled. Wilkins (1959) measured the CO_2 output from *Bryophyllum fedtschenkoi* into a flow of CO_2-free air and Jones (1973) was able to observe the rhythm by the alternative method of determining the CO_2 compensation point in a closed system.

Wilkins (1959) found that the rhythm of CO_2 output could persist for at least five days when leaves of *B. fedtschenkoi* were maintained in a flow of CO_2-free air throughout the period of measurement, though there was a damping of the amplitude with time (Fig. 10.2a). When, however, normal air containing $330\,\mu l\,l^{-1}\,CO_2$ replaced the CO_2-free air, the persistence of the rhythm was much reduced; indeed, under these conditions a free-running rhythm was not recognizable (Fig. 10.2b). The difference between Figs 10.2a and 10.2b demonstrates an important point which is sometimes overlooked. This is that while the persistence under constant conditions is of significance for physiological investigation (with rapid damping the basic characteristics of a rhythm cannot be explored), it must not be considered to be a measure of the physiological significance of the oscillations. The basic mechanisms within the cells were unlikely to have been changed by the different external conditions which altered the overt nature of the rhythm in Figs 10.2a and 10.2b. Under normal conditions free-running oscillations are never expressed as in Fig. 10.2a. Wilkins had a simple explanation of the lack of persistence of the rhythm in normal air, namely, the carboxylation centers became nearly saturated with CO_2 during the first cycle, so that periodicity in the activity of the CO_2 acceptor system was not evident. In the stream of CO_2-free air, however, such saturation was prevented and the carboxylation centers made repeated attempts at intervals to capture the CO_2 of respiration. The troughs in the rhythm therefore indicate the phase of carboxylation, and the peaks that of decarboxylation, in the normal metabolism of the succulent leaf.

A comparison of several succulents showed that a free-running rhythm like that in Fig. 10.2a was unusual, even when leaves were maintained in CO_2-free air. In *Crassula arborescens* and *Kalanchoë*

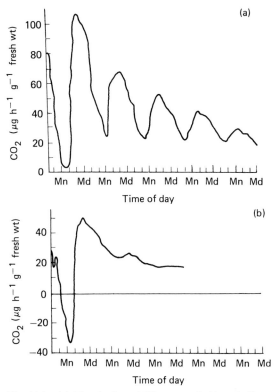

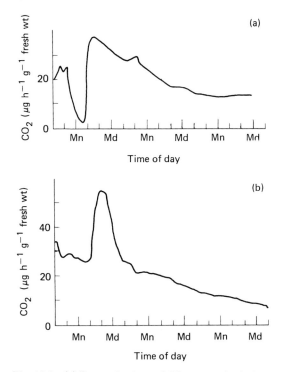

Fig. 10.2 (a) The rhythm of CO_2 output in *Bryophyllum fedtschenkoi* in darkness commencing at 4 p.m. Excised leaves were kept at 26°C in a stream of CO_2-free air. Illumination was given for 8 h prior to the measurements. Mn, midnight; md, midday. (b) As above, but in a stream of normal air. Negative values indicate CO_2 uptake. (From Wilkins, 1959.)

Fig. 10.3 (a) Determinations of CO_2 output in darkness from excised leaves of *Crassula arborescens* under conditions as in Fig. 10.2a. (b) As above, but for young leaves of *Kalanchoë blossfeldiana*. (From Wilkins, 1959.)

blossfeldiana, for example, there was one well-defined peak of CO_2 output (Figs 10.3a and b) and the behavior resembled that of *B.fedtschenkoi* in normal air (Fig. 10.2b). It is clear not only that a demonstrable free-running rhythm depends to a large extent on a correct choice of experimental conditions, but also that processes which are fundamentally similar may sometimes be overtly expressed as a rhythm in one plant species and not in another.

Control of phase

Changes in environmental factors have been shown to have a predominant role in determining the phase

of circadian rhythms, and in the CO_2 metabolism of succulents both light and temperature are effective. For the rhythm of CO_2 output in darkness, the time of commencement of darkness is important. Wilkins (1959) compared the phase of the rhythm in *B. fedtschenkoi* when darkness began at 4 p.m. (Fig. 10.2a) and when the photoperiod was extended for different lengths of time. An extension of 11 h resulted in a rhythm which was almost totally out of phase with that observed earlier, but with a longer extension of 22 h the rhythm progressed very nearly in its original phase. The period of the free-running rhythm calculated from several experiments was 22.4 ± 0.4 h and, consequently, the fact that an 11-h extension of the photoperiod placed the rhythm almost totally out of phase implied that it was the time of termination of light which was the critical factor. Although this was approximately true, the extent to which the photoperiod was lengthened did have a small effect on the time between the commencement of darkness and the first peak of CO_2 output (*i.e.*, the transient, as defined in Fig. 10.1). This peak occurred a little sooner after transfer to darkness as the photoperiod was progressively lengthened (Fig. 10.4) suggesting that there was some slight movement of the oscillating system during prolonged illumination.

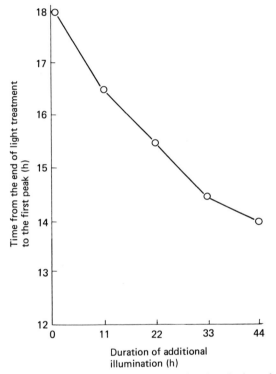

Fig. 10.4 Effect on the transient cycle of a rhythm of increased duration of light beyond the normal time of darkness. (Data of Wilkins, 1959, for the rhythm of CO_2 output in *Bryophyllum fedtschenkoi*.)

Later studies by Wilkins (1960) showed that a rhythm of CO_2 output could also be initiated by a sudden transfer from high to low light intensity, and that this rhythm persisted for several days, like the one in darkness. The fact that the same oscillating system could be induced to occur either in total darkness or in low-intensity light provided an opportunity to determine the response of different phases to light and darkness. In the case of the rhythm in darkness, 3-h or 6-h light interruptions given during a trough of CO_2 output caused the phase of the succeeding rhythm to be reset, *i.e.*, peaks and troughs of treated and control plants were out of phase with one another (Fig. 10.5a). A light interruption given at the peak of CO_2 output, however, had no effect on the subsequent course of the rhythm. In the case of the oscillations induced to occur in low-intensity light, the phase was reset when dark treatments were given at a peak, but the same interruption applied during a trough had no effect (Fig. 10.5a). It was thus clearly demonstrated that the sensitivity to both light and darkness is dependent on the phase of the rhythm at which they are applied.

These studies by Wilkins of the phase-control of the rhythm of CO_2 output demonstrate that the times of peaks and troughs can be predetermined by experimental manipulations. There is, therefore, no possibility that the rhythm is dependent on diurnal changes in unknown factors in the natural environment. This fact, coupled with the persistence of the free-running rhythm and the small deviation of its period from 24 h, fully satisfies the main criteria laid down by Pittendrigh (*see* page 202) for establishing that there is endogenous control.

The CO_2-output rhythm could be looked upon as one of alternating excitability to light and darkness (Fig. 10.5a) but such a description is inadvisable because there is evidence that other environmental factors can exert similar effects. For example, Wilkins (1962) found that a 10°C temperature rise for a few hours from a constant 26°C could cause a phase-shift in the free-running rhythm which was identical to that produced by light, and a sudden change in light intensity caused the phase of the rhythm of CO_2 compensation in *B. fedtschenkoi* to be reset (Fig. 10.5b). Attention should probably be focused, therefore, on changes in an otherwise uniform environment which are capable of producing phase-shift. The factors which are able to act in this way are, however, very specific and this is well illustrated by the action of light. Not all wavelengths are effective; in the case of the rhythm in *B. fedtschenkoi*, Wilkins (1973) found that only the red region of the spectrum was active (Fig. 10.6).

Entrainment

A circadian rhythm with a free-running period which is not exactly 24 h becomes entrained to one

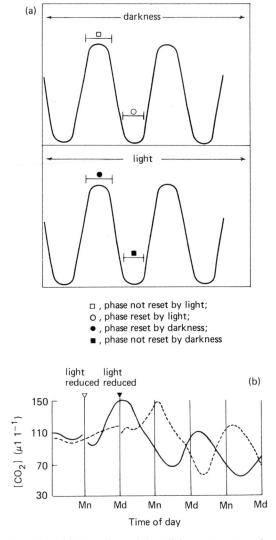

(a)

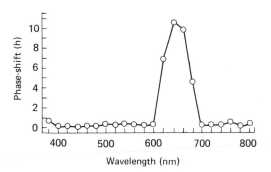

Fig. 10.6 Action spectrum for producing phase-shift of the rhythm of CO_2 output in *Bryophyllum fedtschenkoi*. Interruptions of darkness of 4 h duration were given between the peaks of the rhythm, with light of quantum fluence rate 8.9×10^{-13} mol m^{-2} s^{-1}. (From Wilkins, 1973.)

of 24 h in the natural environment. This is achieved by the phase-setting ability of factors such as light and temperature, and it is obvious that the precise form of an entrained rhythm will be different when the relative lengths of day and night change with time of year. The foregoing discussion has indicated that the effect of treatments which produce phase-shift are not identical throughout the complete cycle of the rhythm. It is possible to produce what is known as a *phase-response curve* by plotting the time of the treatment against the phase-shift produced. Several different rhythms have been studied in this way and the phase-response curves have been found to have the same general form (Fig. 10.7). Not only does the time of treatment affect

Fig. 10.5 (a) The effects of short light treatments on the phase of a rhythm in darkness, and of short dark treatments on the phase of a rhythm in light. (From Wilkins, 1960, and based on experimental data for *Bryophyllum fedtschenkoi*.) (b) Determination of the phase of the rhythm in CO_2 compensation in *Bryophyllum fedtschenkoi* by the timing of a reduction of light intensity (fluence rate). Measurements were made initially in a high fluence rate (95 J m^{-2} s^{-1}), which inhibited the rhythm, but which was then changed abruptly to 18 J m^{-2} s^{-1} at the times shown. $\triangledown$ refers to the continuous line and $\blacktriangledown$ to the broken line. (From Jones, 1973.)

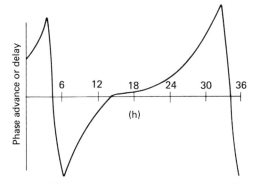

Fig. 10.7 Phase delay or advance relative to the time of an environmental perturbation given at different points along the abscissa.

the amount of phase-shift, but it also affects the direction, *i.e.*, the phase can both be delayed and advanced. The phase–response curves for light perturbations of darkness, or dark perturbations of light, for the same process in the same organism are found to be mirror images (Sweeney, 1969).

The form of the phase–response curve can be used in attempts to explain diurnal physiological or other changes during different lengths of day and night. It is important to remember, however, that the phase–response curve for the first cycle (the transient) can be different from that of the free-running cycles. The transient cycles may show a delay response like that of the free-running cycles, but they are rarely advanced (Zimmer, 1962; Aschoff, 1965). It is not known whether the overt expression of the rhythm during the transient response reflects what the basic oscillator is doing at this time, or whether, immediately after a treatment which brings about phase-shift, the basic oscillator changes into the phase which is eventually seen in the free-running rhythm. In this case, we could regard the transient as the time interval required for the cellular processes to become coupled to the new phase of the basic oscillator.

It is not difficult to envisage problems for the metabolic activities of a cell to adjust very rapidly to new signals from a basic oscillator. Consider the processes in Fig. 10.8, which are hypothetical but which do bear some resemblance to those of dark fixation in succulents. During darkness (and in what could be termed the 'night phase' of a rhythm under constant conditions) CO_2 is taken into an organelle (A) in the cytoplasm and fixed into a compound (X) which is then transported into the cell vacuole where

it is stored. During the day this store of material is drawn upon by the metabolic activities of another organelle, B. Now suppose that in the 'night phase' a product of the CO_2 fixation, X, is transported from A into the vacuole, and during the 'day phase' X is transported in the reverse direction across the membrane around the vacuole (the tonoplast) on its way to B. This sequence of events can occur both as an entrained rhythm in normal alternations of light and darkness and as a free-running rhythm under constant conditions. Let it now be supposed that a basic oscillator controlling the free-running rhythm is reset so that, at an instant in time, the cellular processes are called upon to rephase themselves completely so that night-phase activities occupy the time previously taken up by day-phase activities, and *vice versa*. A consideration of the metabolic state of the cell at different points in the cycle of events reveals that an instantaneous adjustment of phase may be an impossibility. At the beginning of the night phase, for example, there is no store of material in the vacuole and a switchover to the beginning of a period of 'day phase' activity at this point is not possible. Only after a lapse of time, during which there is a readjustment, can the metabolic sequences in the cell respond to the rephased signals from the basic oscillator.

Effects of light and temperature

Changes in light and temperature were mentioned in the previous section as agents concerned in the setting of the phase of rhythms in plants. It is clearly important to discover how such environmental factors are perceived in the cells and how a basic oscillator may be affected by changes which they induce. Other characteristics of the free-running rhythm apart from the phase can also be determined by changes in the environment, light and temperature again being of major importance.

Light requirements for phase shift and for initiation of rhythms

The action spectrum in Fig. 10.6 for the production of a phase shift in the CO_2 rhythm of *B. fedtschenkoi* resembles the absorption of phytochrome apart from the absence of a subsidiary peak in the blue region. Extensive investigations into the possible

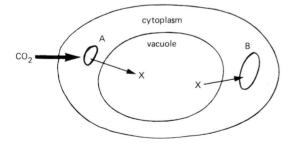

Fig. 10.8 Hypothetical processes taking place in a cell during different phases of a circadian rhythm. See text for explanation.

photoreceptors involved in determining leaf movement rhythms have been reviewed by Satter and Galston (1981). Phytochrome has been convincingly identified as one pigment involved. Lörcher (1958) found that plants of *Phaseolus vulgaris* which had been grown in the dark began to show rhythmic leaf movements if they were provided with red light between 600 and 700 nm, but simultaneous irradiation with far-red (700–800 nm) nullified the effect. Later studies (*e.g.*, Simon, Satter and Galston, 1976) have confirmed the importance of red light in the initiation or maintenance of circadian rhythms, implying that phytochrome in the P_{fr} form is the active agent in the cells. Five-minute pulses of red light could achieve phase-shift of the rhythm of leaf movement in *Samanea saman*, but there was no effect if each pulse was immediately followed by far-red (Simon *et al.*, 1976). There is thus the red/far-red reversibility expected of a phytochrome-dependent system.

Equally strong evidence has been obtained that other pigments as well as phytochrome act as photoreceptors for the production of phase-shifts. Different pigments may be involved in different tissues. For example, a phase-shift of the leaf movement rhythm in *Phaseolus multiflorus* was obtained by exposure of the lamina to far-red, and of the pulvinus to red light (Bünning and Moser, 1966). More recent work points to the involvement of a pigment absorbing blue and far-red light, but this is as yet unidentified. While short irradiations with red light could induce phase-shift in *Samanea*, this was not true of blue light. Even 30-min pulses of blue had little effect, but 2-h pulses could produce a characteristic phase-shift with advance or delay (*cf.* Fig. 10.7). The effects of these 2-h treatments with blue light were not reversed by subsequent exposure to far-red (Satter, Guggino, Lonergan and Galston, 1981).

Blue light is also effective in shifting the phase of circadian rhythms in organisms which do not contain phytochrome, such as *Drosophila* (Klemm and Ninnemann, 1976), and the most popular current view is that a flavin photoreceptor is involved. There are, however, other suggestions of the nature of the ubiquitous blue-light-absorbing pigment which controls many phenomena in plants and animals (Senger, 1980). The identification of any pigments which, in addition to phytochrome, are responsible for the control of phase-shift is of great importance, because this could be the key to understanding subcellular events which control the postulated basic oscillator.

The influence of light on overt free-running rhythms is very complex. It has often been reported that higher light intensities (>10% of full sunlight) cause rapid damping, but it is not possible to generalize because there are some remarkable examples of persistence even in strong light. For example, Hoshizaki and Hamner (1964) followed a rhythm of leaf movement in *Phaseolus vulgaris* for four weeks in light without any marked damping. The fact that there is often damping as light intensity is increased does, however, suggest that the light-dependence of overt rhythms is not primarily related to energy supply. This conclusion is also supported by the relatively low light requirement for the occurrence of some rhythms, which is near to or below the light compensation point for photosynthesis.

Temperature responses

The physiological and other processes in which rhythms are observed are usually greatly influenced by temperature, and this may mean that the amplitude of an overt rhythm is temperature-dependent. The periods of free-running rhythms are, however, remarkably unaffected by ambient temperature. As with other activities in the plant, the temperature coefficient (Q_{10}) is used to indicate the magnitude of temperature responses and in this case the Q_{10} is best defined as

$$Q_{10} = \frac{1/(\text{period at } t°C)}{1/[\text{period at } (t-10)°C]}$$

The reciprocal of the period is used because, if the processes behind oscillations occur faster at higher temperatures, the period will be shorter. A Q_{10} of unity would indicate temperature-independence, but it is perhaps more realistic to think in terms of temperature-compensation which would be the outcome of effects on a number of individual processes. There are unlikely to be single processes which are unaffected by a change in temperature.

The Q_{10} for many free-running rhythms has been found to be very close to unity; for example, the rhythm of leaf movement in darkness in *Phaseolus* (Leinweber, 1956) has periods of 28.3 h at 15°C and

28.0 h at 25°C ($Q_{10} = 1.01$) and that of CO_2 output in *Bryophyllum* (Wilkins, 1962) has periods of 23.9 h at 16°C and 22.4 h at 26°C ($Q_{10} = 1.06$). This degree of temperature compensation is, however, found only with the free-running rhythm and not with the transient, which is often strongly dependent on temperature. This fact supports the view (*see* p. 207 above) that the transient is a stage of readjustment which is governed by metabolic limitations in the cells.

The role of membranes

The considerable physiological and biochemical diversity of endogenous circadian rhythms has already been mentioned. Much of this diversity is known to exist at the cellular level; in other words, there are rhythms in many different processes within a single cell. The notion of a 'basic oscillator' or 'cellular clock' (see page 201) from which all the rhythms are derived has proved attractive because it seems logical for them to be coordinated from a central control point. As noted earlier, however, a rhythm in one major metabolic process inevitably leads to rhythms in others, and it is rarely possible to discover whether the oscillations of several related processes are separately derived from a cellular clock or whether some are secondarily derived.

Evidence of rhythmic changes in the properties of membranes which has come to light in recent years has been regarded as of major significance in relation to the control of circadian oscillations. The involvement of membranes would help to explain the occurrence of rhythms in many diverse processes. It is changes in inner compartment membranes which have attracted attention because no circadian rhythm has yet been reported in a prokaryotic organism.

Some of the recent evidence that has directed the attention of physiologists towards changes in the state of membranes has concerned the intimate connexion between circadian rhythms and the transformations of phytochrome between the P_r and P_{fr} states. It was noted above that phytochrome is implicated in phase-determination in leaf movement and other rhythms. There is evidence that, *in vivo*, the active form of phytochrome (P_{fr}) is membrane bound (*see* Chapter 16) and that a change in electrical potential takes place when P_r is trans-

formed to P_{fr} (Boisard, Marmé and Briggs, 1974; Mackenzie, Boisard and Briggs, 1974; Quail, Marmé and Schafer, 1973; Jaffe, 1968; Newman and Briggs, 1972). Relevant information has also accumulated from experiments involving the application of chemicals which are known to act at the membrane level and which have been shown to produce changes in circadian rhythms.

An observation of particular interest was made by Sweeney, Prézelin, Wong and Govindjee (1979), when examining the fluorescence of chlorophyll *a* in *Gonyaulax polyedra*, which displays distinct circadian variations when kept in constant environmental conditions (Table 10.1). The rhythm was no longer

Table 10.1 A circadian rhythm in the initial intensity of chlorophyll *a* fluorescence in *Gonyaulax polyedra* at the beginning of illumination. The values represent relative magnitudes and are compared with the circadian variation in rate of photosynthesis. (From Sweeney, Prézelin, Wong and Govindjee, 1979.)

Time of day	Rate of photosynthesis (μmol O_2 h^{-1} 10^6 cell^{-1})	Relative fluorescence
00.10	4.1	112
01.55	4.1	103
05.40	5.5	122
10.00	8.3	161
13.00	8.5	179
21.25	6.7	145

exhibited at the temperature of liquid nitrogen (77K (degree kelvin)), but the application of 3-(3,4-dichlorophenyl)-1,1-dimethylurea (DCMU) did not produce a similar response. DCMU inhibits the transfer of electrons from photosystem II to the electron-transport chain. Because of this, it blocks non-cyclic flow of electrons between the two photosystems without inhibiting cyclic electron flow within photosystem I. The continuation of circadian changes in fluorescence in spite of the action of DCMU shows that these changes are not dependent on electron flow between the photosystems. The loss of the rhythm at low temperature must, therefore, have a cause which is unrelated to such electron flow. Another factor behind the rhythmic changes could be alterations in the efficiency of energy spillover from photosystem II, which is highly fluorescent, to photosystem I, which exhibits substantially weaker fluorescence. However, lowering the temperature to 77K would not be expected to affect this spillover, which cannot therefore be the causative agent in the rhythm. An effect of the treatment at 77K would be to prevent changes in the fluidity of the membranes, and thus it was proposed that these might be the basis of the circadian oscillations.

There can be little doubt concerning the importance of ion balance in certain rhythms in plants, for example in petal and leaf movements and in stomata, where K^+ influx and efflux occur across the guard cell membranes. Electron microprobe analysis has demonstrated a clear correlation between K^+ and Cl^- distribution and the circadian leaf movements found in *Phaseolus coccineus* (Schrempf and Meyer, 1980). The K^+ uptake of *Lemna gibba* G3 (a long-day-flowering duckweed) followed a circadian rhythm which persisted for five days in continuous light (Kondo and Tsudzuki, 1978).

In both *Phaseolus* (Bünning and Moser, 1972) and *Gonyaulax* (Sweeney, 1974) the antibiotic valinomycin, which functions as an ionophore, has been successfully used to alter the phase of the rhythm. Valinomycin is thought to operate as an ion-carrier, complexing with ions such as K^+ (for which it is highly selective) and transporting them across the lipid phase of the membrane much faster than would be possible for the ions on their own (Clarkson, 1974). The production of phase-shift by valinomycin clearly implies that membrane properties are important for the rhythm, and it is interesting to note that the time of treatment with the ionophore is critical; indeed there is a similar phase–response relationship with valinomycin (with advance and delay) as with a physiological factor such as light (page 207).

Fusaric acid, a toxin produced by a fungus which causes wilt disease, is known to impair the semipermeability of the plasma membrane (Sundararajan, Subbaraj, Chandrashekaran and Shunmuzasundaram, 1979), producing a significant change in the ionic balance of infected cotton plants (Sandasivan, 1961). Because of this well-defined action, fusaric acid has been an interesting tool in experiments on circadian rhythms and its application has been found to produce phase-shifts in cotton, *Gossypium hirsutum*. Enzyme-induction studies demonstrated that this toxin altered the phospholipids of membranes,

and fluorescence techniques indicated conformational changes among the phospholipids within the membranes (Sundararajan *et al.*, 1979).

Other treatments which might be expected to influence membranes or ion gradients across them have also been reported to produce phase-shift. The simplest of these are wilting (which causes a physical stress on membranes) and addition of KCl in solution (Sweeney, 1979). The uncomplicated technique of placing a cut stem in various solutions and observing the effect on leaf movements has led to the discovery of compounds which can affect the rhythm. Exposures of this sort to ethanol for two hours were effective in producing phase-shift (Bünning and Baltes, 1962) and specific concentrations of ethanol, methanol and theophyllin (a stimulant found in tea) were able to increase the period length (Keller, 1960). Another stimulant, caffeine, was also able to bring about phase-shift (Mayer and Scherer, 1975). It would be interesting to know how researchers came to try out some of these compounds! A common feature they possess is action on membranes; for example, theophyllin may change the binding of Ca^{2+} and affect membrane function (Sweeney, 1979).

A further antibiotic whose effects on rhythms have received attention is cycloheximide (actidione), which is an inhibitor of protein synthesis. It suppresses or removes circadian oscillations in several organisms including *Euglena* (Feldman, 1967), *Neurospora* (Sargent, 1969), *Acetabularia* (Mergenhagen and Schweiger, 1975; Karakashian and Schweiger, 1976) and, more recently, it has been successfully employed in studies of the rhythm of CO_2 metabolism in *Bryophyllum*, the characteristics of which we have described earlier. These oscillations are the result of periodic changes in the dark fixation of CO_2 (Warren and Wilkins, 1961) and there is evidence that the interaction between the carboxylation enzyme, PEP-C (phosphoenolpyruvate carboxylase) and the end product of CO_2 fixation (malate) could provide a basis for the rhythm. Malate is a very effective inhibitor of PEP-C activity, and the only explanation of continued CO_2 fixation when there are high malate levels in the cells is a separation of the end product from the enzyme. Klüge and Osmond (1972) suggested that the malate must be actively transported from the cytoplasm (PEP-C is a cytoplasmic enzyme) across the tonoplast into the vacuole. Studies of isolated vacuoles have shown that they are the location of stored malate in leaf cells of *Bryophyllum* (Buser and Matile, 1977).

During the photoperiod in succulents the decarboxylation of malate occurs to provide the CO_2 necessary for photosynthesis to proceed behind closed stomata. At the end of the day, therefore, the endogenous malate levels are low and PEP-C activity is high, and during the first hours in darkness CO_2 fixation occurs rapidly (the initial trough in the rhythm of CO_2 output in Fig. 10.2a). The accumulation of malate eventually inhibits PEP-C, however, and the enzyme remains inactive until the malate is

relocated into the vacuole. Repetition of this sequence of events is suggested as the reason for the oscillations of CO_2 output which are the overt circadian rhythm.

When leaves were treated with cycloheximide the rhythm in *Bryophyllum* was no longer evident (Fig. 10.9). Experiments with $^{14}CO_2$ showed that the abolition of the rhythm was the result of inhibition of the dark fixation of CO_2, which did, of course, reduce malate production (Fig. 10.10). This was not achieved through an effect on the activity of PEP-C (Fig. 10.11), or a change in the rate of synthesis of the enzyme. The favored explanation was that cycloheximide brought about the release of accumulated malate from the vacuole into the cytoplasm. Oscillations in membrane potential and ion transport across membranes have been reported to be inhibited by treatment with cycloheximide (Läuchli, Lüttage and Pitman, 1973; Kelday and Bowling, 1975; Brinckmann and Lüttage, 1975). The nature

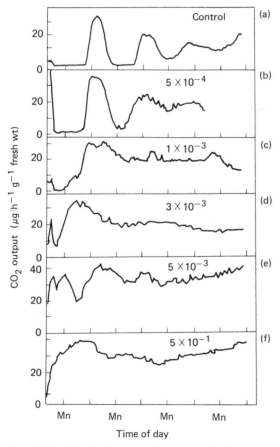

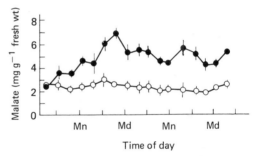

Fig. 10.10 Effect of cycloheximide on malate accumulation in leaves of *Bryophyllum fedtschenkoi* kept in darkness and normal air at constant temperature (17°C). ●, Control; ○, 5×10^{-1} mol m^{-3} cycloheximide. (From Bollig and Wilkins, 1979.)

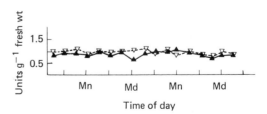

Fig. 10.11 Effect of cycloheximide on the activity of PEP carboxylase in leaves of *Bryophyllum fedtschenkoi* kept in darkness and normal air at constant temperature (17°C). ▽, Control; ▲, 5×10^{-1} mol m^{-3} cycloheximide. (From Bollig and Wilkins, 1979.)

Fig. 10.9 (a) Rhythm of CO_2 output in leaves of *Bryophyllum fedtschenkoi* as observed in controls. (b–f) Rhythm as affected by the different concentrations of cycloheximide indicated (mol m^{-3}). (From Bollig and Wilkins, 1979.)

of the action on the membrane is not understood, and it is possible that inhibition of protein synthesis could affect renewal both of structural components in the membrane and of ion carriers of a proteinaceous nature.

Bollig and Wilkins (1979) also found that the uncoupler of oxidative phosphorylation, dinitrophenol, could abolish the rhythm of CO_2 output in *Bryophyllum*. A reduction in the energy supply from mitochondria could directly affect the active transport of malate across the tonoplast. The end results of treatment with cycloheximide and dinitrophenol were similar, even though the two inhibitors are known to have different sites of action in the cells. It is not surprising that a rhythm whose phase-changes depend on membrane function and relocation of materials in the cells should be susceptible to metabolic inhibitors of many different kinds.

Models to explain circadian oscillations and photoperiodic time measurement

The capacity to distinguish relative lengths of day and night is found in many plants and animals (Chapter 18), and in recent years the topics of photoperiodism and circadian rhythms have overlapped considerably, particularly with regard to theoretical models put forward to explain observed phenomena.

It has previously been mentioned that there is evidence that the photoconversion of phytochrome between the P_r and P_{fr} states is involved in the phase control of circadian rhythms. At an early stage in the study of phytochrome-mediated responses it was proposed that interconversions between P_r and P_{fr} could provide plants with an 'hour-glass' type of time measurement. The physiologically active P_{fr}, which disappears slowly in uninterrupted darkness, was likened to the hour-glass sand (*see* Chapter 18). Circadian oscillations had no place in the simple idea that was advanced, namely that during a photoperiod with the balance of wavelengths found in daylight, phytochrome would be converted from P_r to P_{fr}, and the time-dependent disappearance of P_{fr} during the night would control cellular reactions which ultimately give rise to photoperiodic responses.

Later work demonstrated that the measurement

of time depended on something more than this kind of simple indication of the length of the dark period, for flowering and other photoperiodically-controlled responses were shown to be rhythmic in nature, with complex reactions to light at different phases (Claes and Lang, 1947; Hamner, 1960; Carpenter and Hamner, 1964; Takimoto and Hamner, 1965). The interaction between phytochrome and the endogenous circadian rhythm is now believed to be of central importance in photoperiodism in plants (*see* Chapter 18).

The original concepts concerning the involvement of rhythms in photoperiodism arose from the work of Bünning (1936, 1960, 1973). He suggested that the phases of the rhythm could be distinguished according to whether light is stimulatory (the *photophil* phase) or inhibitory (the *scotophil* phase). Strong evidence of these 'light-liking' and 'dark-liking' phases emerged from many experiments on different overt rhythms, including some on flowering (Carpenter and Hamner, 1964). More recent studies by Heide (1978), who examined the photoperiodic regulation of leaf regeneration in *Begonia*, indicated the existence of more than a single light-sensitive phase during the night, in support of conclusions by Zimmer (1976a,b). The response produced by a change in the light regime has been shown to depend directly on the particular state of the endogenous rhythm.

It is important to emphasize that in drawing particular attention to the role of phytochrome, we are discussing mechanisms in eukaryotic green plants from algae upwards in which this pigment system is found. In other organisms (*e.g.*, the fungi) different pigments may assume a similar role.

The coincidence of the action spectrum for producing phase-shift with the absorption spectrum of P_r, and the evidence of photoreversibility, are the indications of a role for phytochrome which have come from physiological measurements (*see* p. 207). Research into the mechanism by which phytochrome controls events in cells has recently pointed to the possibility that membrane-bound P_r and P_{fr} may undergo cyclic photoconversions and move to and fro across the membrane to achieve the flow of other molecules (Johnson and Tasker, 1979).

A model for circadian oscillations proposed by Njus, Sulzman and Hastings (1974) suggested that there are changes in both ion distribution and membrane permeability to the ions, and that feedback arrangements produce the self-sustained rhythm. Their model was able to accommodate responses of the rhythm to light and temperature. The photoreceptor was considered to be membrane-bound and to function as an 'ion gate', in a manner essentially similar to that later proposed for phytochrome by Johnson and Tasker (1979). The small effect of temperature on the period of a free-running rhythm was explained in terms of adaptation of membrane lipids, and phase change was supposed to occur through the action of the photoreceptor which could alter the precise positioning of transport proteins upon a change in irradiation.

Heide (1977) has produced a theoretical model for photoperiodic time measurement in higher plants which places phytochrome in a central role within a circadian timing system (Fig. 10.12). The basic assumption is made that the circadian clock involves structures which achieve membrane transport, and the compartmentalization of solutes, which interact to form a feedback loop. The model is dependent upon three other main assumptions: (1) that receptor sites containing phytochrome are able to undergo circadian alterations in configuration, due both to their own structural properties and to the rhythmic changes across the membrane (*e.g.*, potential differences, energy charge, pH etc.); (2) that P_{fr} is able to function as an allosteric effector to produce changes in its own receptor sites, bringing about rephasing and entrainment of the rhythm; (3) that the level of P_{fr} (relative to $P_{fr} + P_r$) is rhythmic and is utilized in maintaining the phase of the membrane rhythm. For each phase of the rhythm there is an 'optimum' percentage of the total phytochrome in the P_{fr} form.

The role of the pigment is therefore seen as providing a linkage between the membrane and changes occurring in the photo-environment. Heide extended his model to explain the photoperiodic basis of flower initiation; an example based on a hypothetical short-day plant is given in Fig. 10.13. Here the optimum level of P_{fr} (*see* (3) above) is represented by the continuous line and the actual levels in two different photoperiods, 10 h and 18 h, are shown as dashed and dotted lines respectively. The photophase (light-liking) is supposed to require a comparatively high P_{fr} level to maintain the running of the rhythm, but the converse is true in the scotophase (dark-liking). In a photoperiod of 10 h

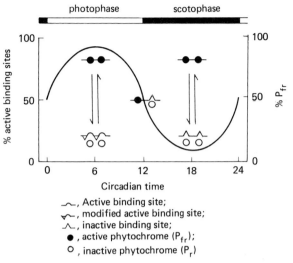

, Active binding site;
, modified active binding site;
, inactive binding site;
●, active phytochrome (P_{fr});
○, inactive phytochrome (P_r)

Fig. 10.12 Diagram based on Heide's (1977) suggestions of the oscillatory changes in phytochrome associated with structures which achieve membrane transport. The amounts of the two forms of phytochrome (P_r and P_{fr}) are supposed to change with circadian time. Membrane-bound P_{fr} is assumed to be physiologically active, and P_r inactive, and P_r is supposed to be able to modify active 'binding sites'. In addition to changes in P_r/P_{fr} in circadian time, the proportion of active binding sites also changes. Thus in the photophase P_{fr} and active binding sites normally predominate, and red light during this phase would reinforce (entrain) this state. Far-red light during the photophase would, however, convert active P_{fr} to inactive P_r, leading to modification of the binding sites. Structural features of the membrane on which transport of solutes depends are intimately connected with changes in the phytochrome-binding sites.

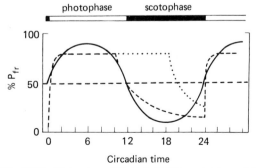

Fig. 10.13 Extension of Heide's (1977) model to explain photoperiodic time measurement in a short-day plant. See text for explanation.

the P_{fr} levels are assumed to stay sufficiently close to the optimum ones and this results in flower initiation. However, if a photoperiod is sufficiently long, the actual P_{fr} levels deviate considerably from the optimum, and the rhythm is forced to rephase. The rephasing is the cause of failure to flower, and only relative lengths of day and night which do not lead to rephasing induce flowering. Similar arguments may be applied to long-day plants, with the difference that in their case rephasing is a prerequisite for flowering (Fig. 10.14). The model has attractions

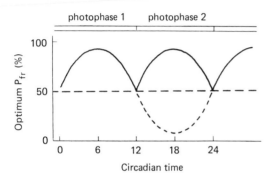

Fig. 10.14 Heide's (1977) model for long-day plants suggests that the critical photoperiod for flowering is the shortest photoperiod that is able to cause rephasing of the rhythm. In this diagram, it is supposed that light extending into the scotophase (*i.e.*, the second 12-h phase of a 24-h cycle) causes a rephasing so that two consecutive photophases occur (continuous line). Progress into the scotophase which would have occurred in darkness is shown as a broken line. The rephasing shown can also be achieved by a short interruption of the scotophase with red light.

insofar as it can explain some of the responses of short- and long-day plants to red and far-red light at different points in the diurnal cycle. Heide also considered the influence of temperature, which is known to be small on the period of free-running rhythms but may be considerable in the case of the transient cycle (page 207).

Photoperiodic responses are often temperature-dependent (*e.g.*, critical night length can increase with decreasing temperature) and such effects would be the result of the rates of phytochrome interconversions which are strongly affected by temperature. Aspects of the flowering response that appear to be temperature compensated would, on the other

hand, be those relating to the basic membrane rhythm.

In summary, the present view that endogenous timing is membrane-mediated is supported by the evidence of phytochrome involvement. The role of this pigment system must, however, be restricted to rhythms in phytochrome-containing organisms. Other pigments may replace phytochrome in other organisms, and it is possible that further pigments (such as the blue-absorbing flavin photoreceptor) may also have a role in higher plants. Rhythms which are insensitive to light must be phased in other ways, and thus we must regard phytochrome and other pigments as having a role in perception of the environment for the purpose of phase-control, and not as an essential part of the basic oscillator. However, further studies of the involvement of phytochrome, a pigment whose functions are being investigated in great detail (*see* Chapters 16 and 17), may well provide insight into the nature of the basic oscillator. It is hoped that such studies will stimulate further thought and research on a topic which has a bearing on many fields in plant physiology.

Consequences of rhythms for the research plant physiologist

Some understanding of the way in which various processes can be subject to endogenous circadian rhythms is essential for everyone engaged in research in whole plant physiology, and for many in cell physiology too. It is interesting to compare the situation of a research worker in chemistry with one in plant physiology. The chemist is able to take raw materials of known quality at any time of the day or night and can predict the kinetics of a particular process with reference only to the conditions established for the reaction. The physiologist who is concerned with rates of a process such as photosynthesis, on the other hand, cannot assume that the physical conditions prevailing at the time or the supply of raw material will exert the only, or even the main, control. During the past two decades the after-effects of various factors such as water stress have been increasingly well documented (*e.g.*, Mansfield, Wellburn and Moreira, 1978) and experimenters now usually pay careful attention to the pre-treatment of their material as a matter of course. Equally important, however, is the timing of experiments, particularly in relation to the previous light–dark regimes to which the plants have been exposed.

Much of the published work on circadian rhythms has been concerned with the characteristics of free-running oscillations. This has tended to distract attention from some of the practical consequences of changes in phase for various physiological processes. The magnitude of the differences that can occur can be illustrated by reference to the rate of stomatal opening in light. Stomata exhibit rhythms in both light and darkness, the rhythm in continuous darkness being manifest by a small degree of opening at the time when light would normally be experienced (Mansfield and Heath, 1963; Martin and Meidner, 1971). The course of the rhythm in darkness appears to have a profound effect on the reaction of the stomata to light, indeed it determines the rate of opening (Fig. 10.15a,b). This effect is of concern not only to the physiologist interested in stomata *per se*; control of the leaf's gas exchange by the stomata will mean that many other functions will also be affected. For example, the rate of CO_2 fixation will be restricted when stomata respond sluggishly to light, and many other aspects of metabolism will thereby be affected.

How may the physiologist design his experiments to guard against unwanted interference from rhythmic phenomena? The simplest precaution is to maintain experimental plants on light–dark cycles identical to those in which they have been grown. A common mistake is to take plants from a glasshouse where they have received natural light and put them in a growth room with a photoperiod which is out of phase with the natural day. A transient cycle (*see* Fig. 10.1) may then be experienced before the rhythm becomes entrained to the new regime. Measurements in the period immediately following transfer are therefore best avoided. The phase of the rhythm at the time of transfer is important and some problems may be avoided by consistency in this respect. Even when full regard has been paid to the phasing of the rhythm, the possibility of changes during a period of measurements must be considered. Figure 10.16 shows the rhythm of stomatal behavior in continuous light in *Tradescantia virginiana*, as observed by Martin and Meidner (1971), together with the behavior when a night period of 12 h was given at the accustomed time from 6 p.m. to 6 a.m. The stomatal behavior during the first photo-

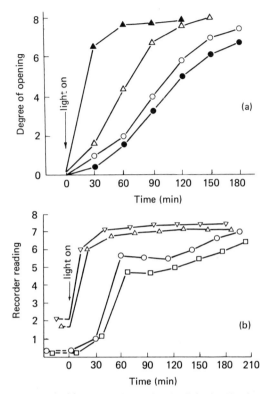

Fig. 10.15 (a) Stomatal opening in light in *Xanthium strumarium* after different lengths of night. ●, 3-h night; ▲, 9-h night; △, 17-h night; ○, 21-h night. (From Mansfield and Heath, 1963.) (b) Stomatal opening in light in *Tradescantia virginiana* after 3 h (○, □) and after 16 h in darkness (▽, △). (From Martin and Meidner, 1972.)

period clearly corresponds with the first cycle of the rhythm in continuous light. Towards the end of the photoperiod the stomata close appreciably even though the light intensity is maintained at a constant level. Suppose that the effect of CO_2 supply on photosynthesis is to be investigated and that, for convenience, CO_2 concentrations are increased step-wise during the day and that CO_2 uptake is measured at a fixed time after the establishment of each new CO_2 concentration. This may seem to be a perfectly reasonable experimental procedure, but only to someone unacquainted with the very large changes in stomatal aperture that can occur due to endogenous rhythmic factors.

It is possible to take advantage of our knowledge

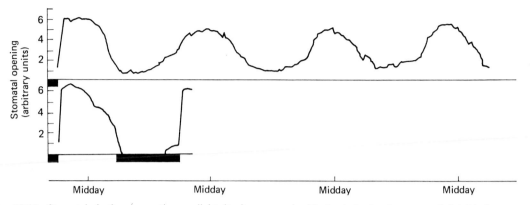

Fig. 10.16 Stomatal rhythm in continuous light (top) compared with the behavior in a normal light/dark sequence (bottom). Note that the closure during the first part of the first cycle in continuous light is matched by progressive closure during the photoperiod prior to darkness in the bottom figure. (From Martin and Meidner, 1971.)

of circadian rhythms by purposefully rephasing them to suit our convenience. The writers have done this themselves with plants to be used in undergraduate practical work which is timetabled for the afternoon in winter, when the normal time of dusk is about mid-way through the work period. Plants which are glasshouse-grown and are not rephased undergo processes such as photosynthesis very reluctantly, and the simple answer is to entrain them for a few days to more suitable cycles of light and darkness in a growth room. The students can then make their measurements on a winter afternoon on plants that are in the early morning phase of their rhythm.

References

Aschoff, J. (ed.) (1965). Response curves in circadian periodicity. In *Circadian Clocks*, Amsterdam, North-Holland.

Boisard, J., Marmé, D. and Briggs, W. R. (1974). *In vivo* properties of membrane-bound phytochrome. *Plant Physiol.* **54**, 272–6.

Bollig, I. C. and Wilkins, M. B. (1979). Inhibition of the circadian rhythm of CO_2 metabolism in *Bryophyllum* leaves by cycloheximide and dinitrophenol. *Planta* **145**, 105–12.

Brinckmann, E. and Lüttage, U. (1975). Inhibition of light-induced, transient membrane potential oscillations of *Oenothera* leaf cells by cycloheximide. *Experientia* **31**, 933–5.

Bünning, E. (1936). Die endonome Tagesrhythmik als Grundlage der photoperiodischen Reaction. *Ber. dtsch. Bot. Ges.* **54**, 590–607.

Bünning, E. (1960). Circadian rhythms and time measurement in photoperiodism. *Cold Spring Harbor Symp. Quant. Biol.* **25**, 249–56.

Bünning, E. (1973). *The Physiological Clock*, 3rd edn, Springer-Verlag, New York.

Bünning, E. and Baltes, J. (1962). Wirkung von Äthyl-alkohol auf die physiologische Uhr. *Naturwissenschaften* **49**, 19–20.

Bünning, E. and Moser, I. (1966). Response-kurven bei der circadianen Rhythmik von *Phaseolus*. *Planta* **69**, 101–10.

Bünning, E. and Moser, I. (1972). Influence of valino-mycin on circadian leaf movements of *Phaseolus*. *Proc. Nat. Acad. Sci. USA* **69**, 2732–3.

Buser, C. and Matile, P. (1977). Malic acid in vacuoles isolated from *Bryophyllum* leaf cells. *Z. Pflanzen Physiol.* **82**, 462–6.

Carpenter, B. H. and Hamner, K. C. (1964). The effect of dual perturbations on the rhythmic flowering response of Biloxi soybean. *Plant Physiol.* **39**, 884–9.

Claes, H. and Lang, A. (1947). Die Blütenbildung var *Hyoscyamus niger* in 48-stundigen Licht-Dunkel-Zyklen und in Zyklen mit aufgeteilten Lichtphasen. *Z. Naturforsch.* **2b**, 56–63.

Clarkson, D. T. (1974). *Ion Transport and Cell Structure in Plants*, London, McGraw-Hill.

De Mairan, M. (1729). Histoire de l'Académie Royal des Sciences, Paris.

Feldman, J. F. (1967). Lengthening the period of a biological clock in *Euglena* by cycloheximide, an inhibitor of protein synthesis. *Proc. Nat. Acad. Sci. USA* **57**, 1080–7.

Hamner, K. C. (1960). Photoperiodism and circadian rhythms. *Cold Spring Harbor Symp. Quant. Biol.* **25**, 269–77.

Heide, O. M. (1977) Photoperiodism in higher plants: an interaction of phytochrome and circadian rhythms. *Physiol. Plant.* **39**, 25–32.

Heide, O. M. (1978). Circadian rhythmicity in photoperiodic regulation of regeneration in *Begonia* leaves. *Physiol. Plant.* **43**, 266–70.

Hoshizaki, T. and Hamner, K. C. (1964). Circadian leaf movements: persistence in bean plants grown in continuous high intensity light. *Science* **144**, 1240–1.

Jaffe, M. J. (1968). Phytochrome-mediated bioelectric potentials in mung bean seedlings. *Science* **162**, 1016–17.

Johnson, C. B. and Tasker, R. (1979). A scheme to account quantitatively for the action of phytochrome in etiolated and light-grown plants. *Plant, Cell Environ.* **2**, 259–65.

Jones, M. B. (1973). Some observations on a circadian rhythm in carbon dioxide compensation in *Bryophyllum fedtschenkoi. Ann. Bot.* **37**, 1027–34.

Karakashian, M. W. and Schweiger, H. G. (1976). Evidence for a cycloheximide sensitive component in the biological clock of *Acetabularia. Exp. Cell. Res.* **98**, 303–12.

Kelday, C. S. and Bowling, D. J. F. (1975). The effect of cycloheximide on uptake and transport of ions by sunflower roots. *Ann. Bot.* **39**, 1023–7.

Keller, S. (1960). Über die wirkung chemischer Faktoren auf die tagesperiodischen Blattbewegungen von *Phaseolus multiflorus. Z. Bot.* **48**, 32–57.

Klemm, E. and Ninnemann, H. (1976). Detailed action spectra for the delay shift in pupae emergence of *Drosophila pseudoobscura. Photochem. Photobiol.* **24**, 369–71.

Klüge, M. and Osmond, C. B. (1972). Studies on phosphoenolpyruvate carboxylase and other enzymes of crassulacean acid metabolism of *Bryophyllum tubiflorum* and *Sedum praealtum. Z. Pflanzen Physiol.* **66**, 97–105.

Kondo, T. and Tsudzuki, T. (1978). Rhythm in potassium uptake by a duckweed, *Lemna gibba* G.3. *Plant Cell Physiol.* **19**, 1465–74.

Läuchli, A., Lüttage, U. and Pitman, M. G. (1973). Ion uptake and transport through barley seedlings: differential effect of cycloheximide. *Z. Naturforsch.* **28c**, 431–4.

Leinweber, F. J. (1956). Über die Temperaturabhängigkeit der Periodenlänge bei der endogenen Tagesrhythmik von *Phaseolus. Z. Bot.* **44**, 337–64.

Lörcher, L. (1958). Die wirkung verschiedener Lichtqualitäten auf die endogene Tagesrhythmik von *Phaseolus. Z. Bot.* **46**, 209–41.

Mackenzie, J. M., Boisard, J. and Briggs, W. R. (1974). The isolation and partial characterization of membrane vesicles containing phytochrome. *Plant Physiol.* **54**, 263–71.

Mansfield, T. A. and Heath, O. V. S. (1963). Studies in stomatal behaviour. IX. Photoperiodic effects on rhythmic phenomena in *Xanthium pennsylvanicum. J. Exp. Bot.* **14**, 334–52.

Mansfield, T. A., Wellburn, A. R. and Moreira, T. J. S. (1978). The role of abscisic acid and farnesol in the alleviation of water stress. *Phil. Trans. R. Soc. Lond.* **B284**, 471–82.

Martin, E. S. and Meidner, H. (1971). Endogenous stomatal movements in *Tradescantia virginiana. New Phytol.* **70**, 923–8.

Martin, E. S. and Meidner, H. (1972). The phase response of the dark stomatal rhythm in *Tradescantia virginiana* to light and dark treatments. *New Phytol.* **71**, 1045–54.

Mayer, W. and Scherer, I. (1975). Phase-shifting effects of caffeine on the circadian rhythm of *Phaseolus coccineus* L. *Z. Naturforsch.* **30c**, 855–6.

Mergenhagen, D. and Schweiger, H. G. (1975). The effect of different inhibitors of transcription and translation on the expression and control of circadian rhythm in individual cells of *Acetabularia. Exp. Cell Res.* **94**, 321–6.

Newman, I. A. and Briggs, W. R. (1972). Phytochrome-mediated electric potential changes in oat seedlings. *Plant Physiol.* **50**, 687–93.

Njus, D., Sulzman, F. M. and Hastings, J. W. (1974). Membrane model for the circadian clock. *Nature* **248**, 116–20.

Pittendrigh, C. S. (1954). On the temperature independence in the clock system controlling emergence time in *Drosophila. Proc. Nat. Acad. Sci. USA* **40**, 1018–29.

Quail, P. H., Marmé, D. and Schafer, E. (1973). Particle-bound phytochrome from maize and pumpkin. *Nature New Biol.* **245**, 189–91.

Sandasivan, T. S. (1961). Physiology of wilt disease. *Ann. Rev. Plant Physiol.* **12**, 449–68.

Sargent, M. L. (1969). Effects of four antibiotics on growth and periodicity of a rhythmic strain of *Neurospora. Neurosp. Newslett.* **15**, 17.

Satter, R. L. and Galston, A. W. (1981). Mechanisms of control of leaf movements. *Ann. Rev. Plant Physiol.* **32**, 83–110.

Satter, R. L., Guggino, S. E., Lonergan, T. A. and Galston, A. W. (1981). The effects of blue and far red light on rhythmic leaflet movements in *Samanea* and *Albizzia. Plant Physiol.* **67**, 965–8.

Schrempf, M. and Meyer, W.-E. (1980). Electron microprobe analysis of the circadian changes in potassium and chloride distribution in the laminar pulvinus of *Phaseolus coccineus. Z. Pflanzenphysiol.* **100** (3), 247–56.

Senger, H. (ed.) (1980). *The Blue Light Syndrome*, Berlin, Springer-Verlag.

Simon, E., Satter, R. L. and Galston, A. W. (1976). Circadian rhythmicity in excised *Samanea* pulvini. II. Resetting the clock by phytochrome conversion. *Plant Physiol.* **58**, 421–5.

Sundararajan, K. S., Subbaraj, R., Chandrashekaran, M. K. and Shunmuzasundaram, S. (1979). Influence of fusaric acid on circadian leaf movements of the cotton plant. *Gossypium hirsutum. Planta* **144**, 111–12.

Sweeney, B. M. (1969). *Rhythmic Phenomena in Plants.* London, Academic Press.

Sweeney, B. M. (1974). The potassium content of *Gonyaulax polyedra* and phase changes in the circadian rhythm of stimulated bioluminescence by short exposures to ethanol and valinomycin. *Plant Physiol.* **53**, 337–42.

Sweeney, B. M. (1979). Endogenous rhythms in the movements of plants. In *Physiology of Movements, Encyclopedia of Plant Physiology*, vol. 7, pp. 71–93.

Sweeney, B. M., Prézelin, B. A., Wong, D. and Govindjee (1979). *In vivo* chlorophyll *a* fluorescence transients and the circadian rhythm of photosynthesis in *Gonyaulax polyedra. Photochem. Photobiol.* **30**, 309–11.

Takimoto, A. and Hamner, K. C. (1965). Studies on red light interruption in relation to timing mechanisms involved in the photoperiodic response of *Pharbitis nil. Plant Physiol.* **40**, 852–4.

Warren, D. and Wilkins, M. B. (1961). An endogenous rhythm in the rate of dark-fixation of carbon dioxide in leaves of *Bryophyllum fedtschenkoi. Nature* **191**, 686–8.

Wilkins, M. B. (1959). An endogenous rhythm in the rate of carbon dioxide output of *Bryophyllum*. I. Some preliminary experiments. *J. Exp. Bot.* **10**, 377–90.

Wilkins, M. B. (1960). An endogenous rhythm in the rate of carbon dioxide output of *Bryophyllum*. II. The effects of light and darkness on the phase and period of the rhythm. *J. Exp. Bot.* **11**, 269–88.

Wilkins, M. B. (1962). An endogenous rhythm in the rate of carbon dioxide output of *Bryophyllum*. III. The effects of temperature on the phase and period of the rhythm. *Proc. R. Soc. Lond.* **B156**, 220–41.

Wilkins, M. B. (1969). Circadian rhythms in plants. In *Physiology of Plant Growth and Development*, ed. M. B. Wilkins, London, McGraw-Hill.

Wilkins, M. B. (1973). An endogenous circadian rhythm in the rate of carbon dioxide output of *Bryophyllum*. VI. Action spectrum for the induction of phase shifts by visible radiation. *J. Exp. Bot.* **24**, 488–96.

Zimmer, R. (1962). Phasenverschiebung und andere Störlichtwirkungen auf die endogen tagesperiodischen Blüttenblattbewegungen von *Kalanchoë blossfeldiana. Planta* **58**, 283–300.

Zimmer, K. (1976a). Die Wirkung von weissem Störlicht auf F_1-Nachkommen aus Kreuzungen von *Begonia socotrana* und *Begonia dregei. Gartenbauwissenschaft* **35**, 387–92.

Zimmer, K. (1976b). Weitere Untersuchungen an *Begonia socotrana. Gartenbauwissenschaft* **41**, 140–2.

Photosynthesis

11

M F Hipkins

Introduction

Light can affect plants in two distinct ways. Firstly, depending on its spectral quality or spatial direction, light can modify the growth pattern of a plant by acting on specific photoreceptors (*see* Chapters 7, 16, and 17), a relatively small amount of light activating a trigger or switch mechanism. Secondly, the energy of large amounts of sunlight can continually be transduced to chemical energy which is then used to maintain the organization and growth of the plant. The energy transduction reaction is called photosynthesis. It is central to life since it is the ultimate source of food for almost all organisms. The scale of photosynthesis is vast: on the Earth some 2×10^{11} tonnes of carbon are fixed each year (> 6000 tonnes s^{-1}).

An approximate equation for the overall process of photosynthesis is

$$CO_2 + 2H_2O \rightarrow (CH_2O) + H_2O + O_2$$
$$\Delta G = 4.8 \times 10^5 \, J \, mol^{-1} \quad (1)$$

where CH_2O represents carbohydrate, and the positive value for the Gibbs free energy, ΔG indicates that an energy input of at least 4.8×10^5 J is required for each mole of carbon that is fixed.

History of photosynthesis research

Research into photosynthesis began more than 200 years ago, when Priestley (1772) discovered that photosynthesis was the reverse of respiration. The discoveries of the next century indicated that water, carbon dioxide, light and the green parts of plants were all necessary for photosynthesis. The green pigment chlorophyll was isolated early in the nineteenth century, and later the hypothesis that photosynthesis led to the storage of chemical energy was proposed (*see* Rabinowitch and Govindjee, 1969). There followed careful work on the stoichiometry of gas exchange, and the finding that starch was the product of photosynthesis. Therefore, towards the latter half of the nineteenth century, photosynthesis could be described by the approximate Equation (1) given above. Then the elegant experiments of Englemann (1883) using oxygen-requiring motile bacteria and the green alga *Spirogyra* showed that the chloroplast was the seat of photosynthesis in the cell.

The variety of photosynthetic organisms

All higher plants and algae, together with certain species of bacteria, are capable of storing light energy *via* photosynthesis. The most primitive photosynthesizing cells are the photosynthetic bacteria, and their photosynthesis differs from that of other organisms because they are not able to evolve oxygen and because they contain the pigment bacteriochlorophyll, rather than chlorophyll as in green plants. They are, however, widely used in photosynthesis research because the cells are amenable to a range of biochemical and molecular biological treatments. Mainly for this reason, our knowledge of the photosynthetic processes in bacteria is more advanced than that of higher plants. The link between bacterial photosynthesis and green-plant photosynthesis is provided by the blue-green algae (cyanobacteria). These organisms have the characteristics of prokaryotic cells, but have the ability to evolve oxygen; their pigments are also more like those of green plants, since they contain chlorophyll.

Oxidation and reduction in photosynthesis

The approximate Equation (1) for photosynthesis does not indicate whether water or carbon dioxide is the source of the oxygen evolved in green plant photosynthesis. Three lines of evidence suggest that water is oxidized to give molecular oxygen.

van Niel's hypothesis van Niel (1941) studied the comparative biochemistry of photosynthesis in oxygen-evolving organisms and in the non-oxygen-evolving photosynthetic bacteria. The latter can be divided into three classes depending on their requirements for the assimilation of carbon dioxide. Two of these classes (the green- and the purple-sulfur bacteria) can use hydrogen sulfide as a substrate for growth, and produce sulfur as a by-product:

$$2H_2S + CO_2 \rightarrow CH_2O + H_2O + 2S \qquad (2)$$

The third class of bacteria (the purple non-sulfur bacteria) use organic substances like succinate as a substrate, giving fumarate as a by-product:

$$2COOHCH_2CH_2COOH + CO_2 \rightarrow CH_2O \\ + H_2O + 2COOHCHCHCOOH \qquad (3)$$

Taking Equations (2) and (3) with that for green plants

$$2H_2O + CO_2 \rightarrow CH_2O + H_2O + O_2$$

van Niel generalized the process of photosynthesis, and wrote a general equation:

$$2H_2A + CO_2 \rightarrow CH_2O + H_2O + 2A$$

where H_2A is an oxidizable substrate. van Niel's hypothesis has two important consequences: firstly, it gives a logical relationship between apparently diverse photosynthetic processes; secondly, it implies that water is the source of photosynthetic oxygen and that carbon dioxide is reduced to carbohydrate.

The Hill reaction A much more direct line of evidence came from work by Hill and Scarisbrick (1940) on isolated chloroplasts, which showed that the illuminated chloroplasts would evolve oxygen only in the presence of a reducible substance (an electron acceptor or 'Hill oxidant') such as potassium ferricyanide. The evolution of oxygen and the reduction of the electron acceptor have identical stoichiometries, and occur in the absence of fixation of carbon dioxide. The results of this experiment indicate that the evolution of oxygen involves the oxidation of water coupled to the reduction of another substance. In the leaf, this latter compound is carbon dioxide, but the isolation of chloroplasts frequently leads to the loss of the enzymes of the carbon dioxide fixation cycle and consequently oxygen cannot be evolved without an exogenous electron acceptor. It is only comparatively recently that techniques have been developed to isolate chloroplasts able to fix carbon dioxide at rates comparable to those found in leaves (Walker, 1971).

^{18}O studies An obvious experiment, in the light of Hill's findings, was to use the stable, heavy isotope of oxygen (^{18}O) to discriminate between water and carbon dioxide as the source of photosynthetic oxygen. When ^{18}O became available, such an experiment was performed by Ruben, Randall, Kamen and Hyde (1941), and the result suggested that water was the source of evolved oxygen. Such an interpretation is, however, not without problems because of the possibility of exchange reactions (Metzner, 1975).

Light and dark reactions

The results of the preceding section indicate clearly that during photosynthesis water is oxidized and carbon dioxide is reduced, but they give no indication of where in the overall process light energy intervenes to drive the reaction. However, it is possible to show that photosynthesis consists of a combination of light-requiring reactions (the 'light reactions') and non-light-requiring reactions (the 'dark reactions').

Blackman (1905) interpreted the hyperbolic light-intensity dependence of photosynthesis as evidence for a temperature-independent photochemical reaction coupled to a 'dark' reaction which is dependent both on temperature and on the level of carbon dioxide. Later, in the classic experiments of Emerson and Arnold (1932a,b) where the rate of photosynthesis was measured under flashing light, it was found that the yield of photosynthesis per flash depended not only on flash intensity, but also on the time interval between flashes. The time interval between flashes which gave the maximum yield varied with temperature, but was typically 100 ms, suggesting the existence of a relatively slow dark reaction which took about 100 ms to be complete.

The functional relationship between the 'light' and 'dark' reactions can be established by examining the requirements of the 'dark' reactions. As will be discussed in detail below, the 'dark' reactions comprise a complex cycle of enzyme-mediated reactions (the Calvin cycle) which catalyzes the reduction of carbon dioxide to sugar. As well as carbon dioxide the cycle requires reducing power in the form of reduced nicotinamide adenine dinucleotide phosphate (NADP) and chemical energy in the form of adenosine triphosphate (ATP). The reduced NADP (NADPH) and ATP are produced by the 'light' reactions. It is thus possible to divide a description of photosynthesis into those reactions associated with the Calvin cycle and the fixation of carbon dioxide, and those reactions (capture of light by pigments, electron transport, photophosphorylation) which are directly driven by light. This chapter will follow that pattern, describing the 'light' reactions first.

Structure and function of the photosynthetic apparatus

Figure 11.1 shows an electron micrograph of a mature chloroplast from a higher plant. The chloroplast is bounded by an outer double membrane or envelope, and contains a complex network of thylakoid membranes surrounded by a matrix called the stroma. The division of photosynthesis into 'light' and 'dark' reactions is reflected in this structure; the light-driven reactions all occur in or on the thylakoid membranes while the complex cycle of enzyme-mediated 'dark' reactions takes place in the stroma.

The pigments of photosynthesis

The first step in the 'light' reactions of photosynthesis is the absorption of light by pigment molecules. Pigments are molecules which absorb visible light strongly, and hence appear colored; photosynthetic pigments have extensive conjugated double bonds in the light-absorbing part of the molecule.

Types of pigments and their distribution

All oxygen-evolving organisms contain chlorophyll-a (chl-a) whose structure and absorption spectrum are shown in Figs 11.2 and 11.3. The molecule has two distinct parts: the 'head' group consists of a cyclic tetrapyrrole with a Mg ion complexed in the center, and has extensive conjugated double bonds. The 'tail' of the molecule is a phytol side-chain which has no role in the absorption of light. Higher plants and green algae also contain chlorophyll-b, in which the —CH_3 group on ring II of Chl-a is replaced by a —CHO group. Two other forms of chlorophyll (Chl-c and Chl-d) are found in diatoms and some algae (see Table 11.1). Blue-green algae lack Chl-b.

Photosynthetic bacteria contain pigments similar to chlorophyll, called bacteriochlorophylls (BChl). BChl has two fewer double bonds than Chl, and has an absorption spectrum in an acetone–methanol mixture which shows a major absorption band in the near-infra-red at 770 nm. The major type is BChl-a, but other types exist.

Photosynthetic organisms do not contain chlorophyll alone. All contain one or more carotenoids, which are linear molecules containing a number of conjugated double bonds. Figure 11.2 shows the structure of a typical carotenoid (β-carotene) and its absorption spectrum in organic solvent is shown in Fig. 11.3. The three peaks in the absorption spectrum are very characteristic. There is a wide variety of carotenoids, and they can be split into the purely hydrocarbon carotenes and the oxygenated carotenols. Their types and occurrence are listed in Table 11.1. Note that neither chlorophylls nor carotenoids are water-soluble.

The blue-green and red algae have distinctive colors which are due to the presence of bile pigments in their cells. Bile pigments have a linear tetrapyrrole as the chromophore, and in some cases a water-soluble pigment–protein complex can be extracted. The phycoerythrins absorb in the green and yellow regions of the spectrum and hence appear red; the phycocyanins appear blue. The unusual absorption characteristics of these pigments

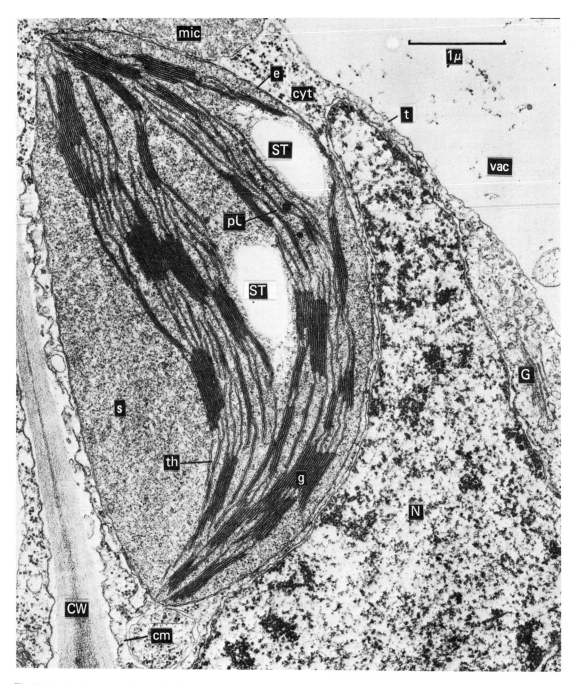

Fig. 11.1 An electron micrograph of a mature chloroplast in a spinach leaf mesophyll cell. The chloroplast is bounded by a double membrane (the envelope, e) which contains the stroma (s) and a complex network of thylakoid membranes (th) which are in places formed into stacks or grana (g). The chloroplast also contains starch grains (ST) and lipid droplets (plastoglobuli, pl). The micrograph also shows the cytoplasm (cyt), tonoplast (t), vacuole (vac), a mitochondrion (mic), a golgi body (G), together with part of the cell wall (cw), nucleus (N) and cell membrane (cm). Micrograph kindly provided by A. D. Greenwood, Imperial College.

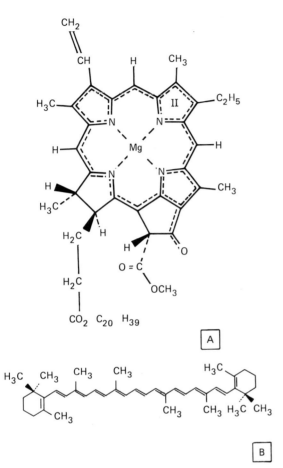

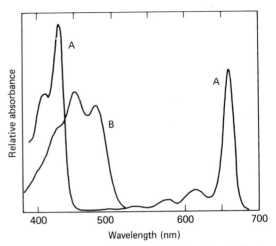

Fig. 11.3 The absorption spectra of (A) chlorophyll-*a* in acetone (redrawn from Zschiele and Comar, 1941) and (B) β-carotene in ether (redrawn from Zschiele, White, Beadle and Roach, 1942).

Fig. 11.2 The structures of chlorophyll and β-carotene. A, chlorophyll-*a*: the —CH₃ on ring II is replaced by —CHO in Chl-*b*. The conjugated double-bond system is represented by -----. B, β-carotene.

Table 11.1 Types and occurrence of photosynthetic pigments

Pigment	Occurrence
Chlorophyll *a*	All oxygen-evolving organisms
Chlorophyll *b*	Higher plants and green algae
Chlorophyll *c*	Diatoms and brown algae
Chlorophyll *d*	Red algae
α-carotene	Higher plants, most algae
β-carotene	Most plants, some algae
Luteol	Higher plants, green and red algae
Violaxanthol	Higher plants
Fucoxanthol	Brown algae, diatoms
Phycoerythrin	Red algae, some blue-green algae
Phycocyanin	Blue-green algae, some red algae
Allophycocyanin	Blue-green algae, red algae
Bacteriochlorophyll *a*	Purple and green bacteria
Bacteriochlorophyll *b*	Some purple bacteria

have been useful in determining the functional role of the photosynthetic pigments; the unicellular red alga *Porphyridium cruentum* has been a particularly favored organism.

Functional organization of pigments

With such a variety of types of pigments it is clearly important to know how each one plays its part in the capture of light energy. They are subject to a high degree of organization, and this contributes in large measure to the efficiency of the early steps of photosynthesis.

Light harvesting The experimental technique that led to the concept of light energy being transferred from one pigment to the next is called sensitized fluorescence. Fluorescence itself is an important property of a number of pigment molecules. If a pigment is dissolved in an organic solvent and then illuminated, the absorbed light energy has a limited number of routes *via* which it may be dissipated. One of these routes involves the re-emission of the light, normally at a wavelength somewhat longer than that of the exciting light, and the emitted light is called fluorescence (Clayton, 1970). The fluorescence emission spectrum is a characteristic feature of a pigment. Experiments by Dutton, Manning and Duggar (1943) using a diatom showed that light absorbed by the carotenoid fucoxanthin excited fluorescence from Chl-*a*. Similarly, in the early 1950s Duysens investigated the fluorescence properties of pigments in the unicellular red alga *Porphyridium* which, as Table 11.1 indicates, contains bile pigments. He found that if the algae were illuminated with monochromatic radiation absorbed mainly by phycoerythrin then the emitted fluorescence had a fluorescence spectrum characteristic of Chl-*a* (Duysens, 1951). Since Chl-*a* itself was not directly excited, this suggests that it must have been excited indirectly by excitation energy transfer from phycoerythrin. Working in this way with each pigment in turn, Duysens was able to draw up an excitation energy transfer map (Fig. 11.4) where

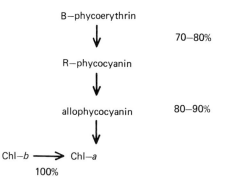

Fig. 11.4   The pathway of excitation transfer from the accessory pigments to Chl-*a* in the red alga *Porphyridium cruentum*. The percentages indicate the efficiency of the energy transfer.

light absorbed by the bile pigments is transferred to Chl-*a* through a kind of molecular bucket-brigade, and likewise Chl-*b* feeds light energy directly to Chl-*a*. Moreover, the efficiencies of energy transfer to Chl-*a* were established; they are generally rather high, but sometimes the transfer from carotenoids has an efficiency of only 20–50%.

In theory, the excitation energy transfer map of Fig. 11.4 should be open to confirmation by another type of experiment. Because of the time needed for excitation energy to transfer from one pigment to the next there will be a short, but finite, lag-time between excitation of B-phycoerythrin by light and the emission of fluorescence by pigments which are 'downstream' and accept the transferred excitation. Experiments designed to measure such lag-times were first performed with *Porphyridium* by Brody and Rabinowitch (1957), but the most striking evidence has come with the advent of mode-locked lasers, capable of producing a train of intense pulses of light, each pulse having a duration of a few picoseconds (one picosecond = 10^{-12} s; light travels approximately 0.3 mm in 1 ps). Using a mode-locked laser and a rapidly-responding photodetector, Porter and his colleagues (Porter, Tredwell, Searle and Barber, 1978) excited one of the bile pigments in *Porphyridium* (B-phycoerythrin) with a pulse of wavelength 530 nm and of 6 ps duration, and monitored the fluorescence emission from B-phycoerythrin itself at 576 nm, then from R-phycocyanin (640 nm), allophycocyanin (661 nm) and Chl-*a* (685 nm). The results are shown in Fig. 11.5; there is a finite lag of a few picoseconds between the emission peaks of the pigments. The result is a direct confirmation of Duysens' (1951) work. Intermolecular excitation energy transfer is an important part of the function of the photosynthetic pigments, enabling wavelengths of light which Chl cannot absorb to contribute to photosynthesis.

Prevention of the photodynamic effect   In organic solution chlorophyll is a labile pigment. If a bright green solution of chlorophyll in acetone is left in sunlight it is bleached quickly and irreversibly to a pale straw color. This reaction requires light and oxygen, both of which are available in the leaf, and yet chlorophyll *in vivo* is not decolorized. Why might this be?

The precise mechanism of the bleaching of chlorophyll (the 'photodynamic effect') is not altogether clear (Krinsky, 1978), but it appears that the over-excitation of chlorophyll molecules by light results in the formation of so-called Chl triplet states. A triplet state is an excited state in which the spin of the excited electron has been reversed: the state is stable for several microseconds and in a reaction with molecular oxygen can form either singlet oxy-

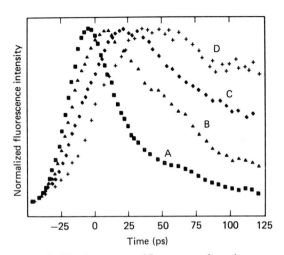

Fig. 11.5 The time-course of fluorescence from the accessory pigments and Chl-*a* in the red alga *Porphyridium cruentum*. The alga was excited with a pulse of light (duration 6 ps, wavelength 530 nm) which was absorbed by B-phycoerythrin. The fluorescences from B-phycoerythrin (576 nm, curve A), R-phycocyanin (640 nm, curve B), allophycocyanin (661 nm, curve C) and Chl-*a* (685 nm, curve D) were then detected. Because excitation energy is transferred from B-phycoerythrin through the other bile pigments to Chl-*a*, the peak of the fluorescence of the pigments which receive the excitation energy is progressively delayed. (Redrawn from Porter *et al.*, 1978.)

gen ($^1\Delta_g O_2$) or superoxide (O_2^-). Both of these forms of oxygen are very reactive, and lead to irreversible damage not only to photosynthetic pigments but also to nucleic acids and other compounds. In the leaf the formation of triplet Chl appears to be followed very rapidly by the quenching of the chlorophyll triplet due to transfer of excitation energy to neighboring carotenoid molecules. This transfer leads to the formation of a harmless carotenoid triplet, whose energy is dissipated to heat. Reactions of this type have been detected by monitoring the rapid transient absorbance changes due to carotenoid triplets in chloroplasts. The protective role of carotenoids is of vital importance: it may be the primary function of the pigments, with the light harvesting role being secondary (Krinsky, 1978). In the absence of carotenoids a photosynthetic organism is at a distinct disadvantage: a mutant strain (designated R26) of the purple non-sulfur bacterium *Rhodopseudo-*

monas sphaeroides which lacks colored carotenoids must be grown strictly anaerobically since it has no mechanism to deal with bacteriochlorophyll triplet states (Sistrom, Griffiths and Stanier, 1956).

The photosynthetic unit In the early part of this century there was much discussion about the part played by chlorophyll in photosynthesis. A major step forward was made by Emerson and Arnold (1932a,b) in the same series of experiments that led to the idea of 'light' and 'dark' reactions. Emerson and Arnold used neon tubes with a flash duration of about 10 μs to illuminate algal suspensions and measured manometrically the amount of oxygen evolved by a long sequence of flashes. They calculated the oxygen yield per flash and found that it was related to the energy of the flash up to a point, but that the oxygen yield saturated at relatively low flash energy. Taking the point at which saturation has just been reached, Emerson and Arnold analyzed the algae for their chlorophyll content, and calculated the number of chlorophyll molecules associated with the evolution of one molecule of oxygen. They found the number was about 2500, and that this number was fairly constant over a range of measurements.

Another ratio must be introduced here: the number of quanta that are required to evolve one molecule of oxygen. After fierce controversy in the 1930s and 1940s it is now generally agreed that at least 8 quanta are required (Emerson, 1958). Thus a fundamental association of $2500/8 \simeq 300$ chlorophyll molecules is concerned with the processing of each quantum of light. The group of about 300 chlorophyll molecules is termed a 'photosynthetic unit'; its significance is that of the 300 chlorophyll molecules, only one 'special' Chl is concerned with the processing of light energy (*see* below), while the remainder act to absorb and channel light energy to the special chlorophyll. In this way, the effective capture cross-section of the 'special' chlorophyll is increased some 300-fold.

The concept of the photosynthetic unit received further support from some calculations by Gaffron and Wohl (1936), who estimated the time taken for oxygen evolution to begin in dark-adapted cells exposed to weak light. They calculated a long lag-time, but experimentally found a much shorter one, and suggested that chlorophyll molecules were able in some way to pool their resources by exchanging excitation energy among themselves.

The mechanism of intermolecular energy transfer
The photosynthetic unit is a very important idea because it describes the high level of organization among photosynthetic pigments; however it is still unclear whether the photosynthetic unit is a statistical or a physical entity. Nevertheless, the occurrence of excitation energy transfer between different molecules, and between like molecules, is beyond doubt. The first physical description of such excitation transfers was given by Förster (1959) who suggested that the precise mechanism would depend on the distance between donor and acceptor molecules: if R is the intermolecular distance then for large values of R ($\geqslant 1.5$ nm), the efficiency of energy transfer would depend on R^{-6}, but for small values of R ($\leqslant 1.5$ nm) the donor and acceptor molecules can 'see' each other, and the efficiency of energy transfer would depend on R^{-3}. As well as depending on intermolecular distance, the efficiency of energy transfer is also a function (a) of the degree to which the fluorescence spectrum of the donor molecule overlaps with the absorption spectrum of the acceptor molecule (the 'overlap integral') and (b) of intermolecular orientation, so that energy transfer is more likely when the dipole moments of donor and acceptor molecules are aligned. This is like rotating a transistor radio to get a strong signal from the transmitter. Förster's theories have been reviewed and extended by Knox (1977), and some experimental studies have been reported. In physical systems of pigment mixtures, Förster's R^{-6} transfer mechanism appears to describe the intermolecular energy transfer well, but in biological systems the picture at present is not as clear.

Absorption spectra of the photosynthetic apparatus
In higher plants there appear to be only two chemically distinct forms of Chl, Chl-a and Chl-b. In the red region of the spectrum the only pigments that absorb are the chlorophylls, so that it might be expected that the absorption spectrum of the photosynthetic apparatus would be relatively simple. But in practice this is not the case: Fig. 11.6 shows the absorption spectrum of a sample of isolated chloroplasts, and this clearly differs from the absorption spectrum of Chl in organic solvent (Fig. 11.3). The difference can be understood by realizing that in the thylakoid membrane Chl does not exist in free solution, but as pigment–protein complexes. The physical environment of a pigment may affect its

absorption spectrum, and the absorption spectrum of Chl-a is shifted to the red to a greater or lesser extent by interaction with protein, giving rise to a number of Chl-a species (really Chl-a-protein complexes) with absorption maxima in the wavelength range 650 to 720 nm. An attempt at identification of

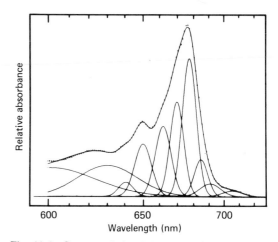

Fig. 11.6 Curve analysis of the absorption spectrum of spinach chloroplasts. A computer-assisted, least-squares analysis was used to fit a number of Gaussian components, each of which is taken as a Chl–protein complex, under the curve. The analysis demonstrates the existence of several distinct Chl–protein complexes in the chloroplasts. (Adapted from French *et al.*, 1972.)

the pigment–protein complexes can be made from the absorption spectrum by assuming that each complex has a single Gaussian (bell-shaped) absorption profile, and that the observed spectrum is the sum of a number of such profiles (French, Brown and Lawrence, 1972). Computer-assisted curve-fitting procedures can then be used to determine the number, wavelength maximum and width of each component (Fig. 11.6). The task is made easier if the sample is cooled to 77 K, since the components have reduced half-width; moreover, a fourth derivative spectrum obtained by differentiating the absorption spectrum four times with respect to wavelength can help determine the number of components. In a similar way the number of fluorescent Chl-a-protein components can be identified by careful study of the low-temperature fluorescence emission spectrum of chloroplasts (Papageorgiou, 1975).

The photosystems and electron transport

The action spectrum of photosynthesis

In the study of any photobiological process it is essential to determine the action spectrum. This spectrum shows the effectiveness or efficiency of various wavelengths of light (*i.e.*, the relative effectiveness of quanta of different energies) in exciting the process, and can indicate the pigments which are involved. That the action spectrum of photosynthesis is not identical to the absorption spectrum of chlorophyll further supports the idea that other pigments absorb light which is used in driving photosynthesis.

Careful studies on the action spectrum of photosynthesis were started by Emerson and his colleagues in the 1940s (*see* Emerson and Lewis, 1943) and continued for many years. When compared to absorption spectra the action spectra show two important features (Fig. 11.7). Firstly, at the short-wavelength end of the spectrum (*i.e.*, about 420–500 nm) some wavelengths of light are relatively inefficient at exciting photosynthesis; this corresponds to the rather poor efficiency of excitation energy transfer between carotenoid and chlorophyll

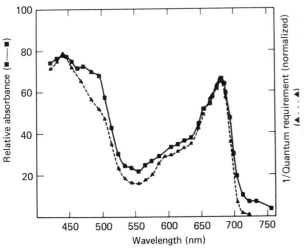

Fig. 11.7 The absorption spectrum and action spectrum of the green alga *Ulva*. The 'red drop' in the action spectrum is seen beyond 680 nm. (Redrawn from Haxo and Blincks, 1950.)

that has already been noted above. Secondly, at the long-wavelength end of the spectrum, light between 680 and 720 nm is also inefficient at exciting photosynthesis, although the absorption due to chlorophyll in this spectral region is still strong. This deficiency in the ability of red light to excite photosynthesis was termed the 'red drop', and might be interpreted as evidence for the existence of a long-wavelength-absorbing pool of Chl which, although it absorbs light, does not participate in photosynthesis. This interpretation is, however, inconsistent with the results of further experiments conducted by Emerson and his colleagues.

Emerson enhancement In continuing and extending studies on the action spectra of photosynthesis, Emerson, Chalmers and Cederstrand (1957) found that the low efficiency of far-red light in exciting photosynthesis could be increased by simultaneously illuminating the sample with a second beam of light of shorter wavelength, say, at 650 nm. Not only was the low efficiency of the far-red light made up to the expected level, but the yield of photosynthesis in the *combined* beams was greater than the sum of the yields of the beams applied separately. This effect is known as Emerson enhancement, and can be given a numerical value, E, by the equation

$$E = \frac{\Delta O_2 \text{ (combined)} - \Delta O_2 \text{ (short-wavelength alone)}}{\Delta O_2 \text{ (long-wavelength alone)}}$$

where ΔO_2 represents the rate of oxygen evolution.

Emerson enhancement is as much a photobiological phenomenon as photosynthesis itself, so an action spectrum was necessary to ascertain which pigments are involved. The results were rather complicated, but in red algae (which lack Chl-*b*) it appeared that the action spectrum for enhancement was the same as the absorption spectrum of the accessory pigments (Emerson and Rabinowitch, 1960). Thus Emerson proposed that there were two light-driven processes in photosynthesis, one sensitized by Chl-*a* and the other sensitized by the accessory pigments. It proved impossible to generalize this proposal: close examination of the action spectra of enhancement in green algae revealed that one of the species of chlorophyll-*a* itself was involved in sensitizing enhancement. This observation changed the picture: the idea of two light-driven reactions sensitized by different sets of pigments remained, but the sets of pigments were not divided as simply as originally envisioned by Emerson.

The two light-driven reactions were named by Duysens, Amesz and Kamp (1961) as Photosystem II (PSII) and Photosystem I (PSI). PSII is sensitized by several types of Chl-*a*-protein complexes and by Chl-*b* which is also present in a pigment–protein complex (*see* below). In red algae, light absorbed by the bile pigments is preferentially transferred to PSII. In all oxygen-evolving organisms the pigments sensitizing PSII absorb only up to about 690 nm. PSI is also sensitized by Chl-*a*-protein complexes, but in this case the maximum wavelength of absorption is about 720 nm. PSI is not strongly sensitized by Chl-*b*.

If the two reactions are consecutive then the overall rate of photosynthesis will be governed by the rate-limiting step. The explanation for the 'red drop' now becomes clear, since at exciting wavelengths above about 690 nm PSII ceases to function, and hence the overall rate of photosynthesis decreases even though PSI still absorbs light and tries to work. The 'red drop' is eliminated when the second beam of light excites PSII and speeds up the rate-limiting step; enhancement occurs because some of this second beam excites PSI as well.

Other evidence for two photosystems The concept of two photosystems each sensitized by a different set of pigment molecules helped to resolve a number of outstanding problems in the literature at that time. For example, several workers had studied the action spectrum of chlorophyll fluorescence *in vivo*, and found that in red algae light absorbed by the phycobilins was more efficient in exciting chlorophyll fluorescence than light absorbed by chlorophyll itself (Duysens, 1951). The data were now interpreted as suggesting that there were two 'pools' of Chl-*a*, one which was functionally associated with the phycobilins, and which had a high fluorescence yield, and the other which was not excited by the phycobilins, and which had a low fluorescence yield. In retrospect, it is seen that the Chl with a high fluorescence yield is associated with PSII, and that with the low fluorescence yield with PSI.

Another curious observation was called 'chromatic transients'. In experiments designed to adjust the fluence rates of two wavelengths of light to produce an equal rate of photosynthesis in each, Blincks (1957) noted a transient increase in the rate of photosynthesis when one beam was switched off and the other switched on, suggesting that enhancement as described above was occurring, but that the two wavelengths of light need not be applied simultaneously to the sample. Similarly, Myers and French (1960) noted that enhancement could be elicited by flashes of different wavelengths separated by a time interval of several seconds. Taken together, these observations suggest that the photosystem excited by the one wavelength has the ability to store some intermediate which is required by the second photosystem for it to function fully. As will be seen below, the intermediates are associated with oxidized or reduced components of an electron transport chain which connects the two photosystems.

The Z-scheme

The evidence for two light-driven photoprocesses was now clear, but the connexion between them was as yet undiscovered. In principle two methods of connexion were possible, either in series with the products of one reaction becoming the reactants of the second, or in parallel, where the two reactions proceed side by side. In 1960 Hill and Bendall considered the role that might be played by the cytochromes present in chloroplasts. Cytochromes are complexes of protein and heme: the prosthetic group of the molecule is the heme, and this consists of a cyclic tetrapyrrole with an iron atom complexed in the center. Cytochromes were well known to undergo oxidation and reduction changes which were associated with changes in the cytochrome absorption spectrum. Hill had found both a c-type cytochrome (Hill and Scarisbrick, 1951) and a b-type cytochrome (Hill, 1954), and suggested (Hill and Bendall, 1960) that the cytochromes might act as oxidizable and reducible intermediates connecting the two photoreactions in series *via* electron transport.

Absorption spectrophotometry of cytochromes The proposal of Hill and Bendall (1960) received support from the work of Duysens *et al*. (1961) who used a sensitive absorption–spectrophotometric technique to determine the redox state of the c-type cytochrome (cyt-*f*) in the red alga *Porphyridium*. The results of a typical experiment are illustrated in Fig. 11.8. The algae were illuminated with light at

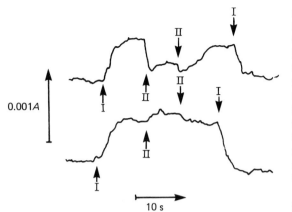

0.001A

← 10 s →

Fig. 11.8 The oxidation and reduction of cyt-f in *Porphyridium*. Algae were illuminated with light absorbed by PSI (680 nm; marked I on the figure) and by PSII (562 nm; II on the figure). An upward deflection is an increase in absorption at 420 nm, which corresponds to oxidation of the cytochrome. An upward arrow denotes switching the actinic light on, a downward arrow switching it off. Upper curve, control; lower curve, in the presence of DCMU, which blocks non-cyclic electron transfer between Q and PQ (see Fig. 11.9). (Redrawn from Duysens *et al.*, 1961.)

680 nm, which is absorbed mainly by the Chl molecules associated with PSI, and at 562 nm, which is absorbed mainly by bile pigments and transferred to PSII, while the redox state of the cytochrome was monitored by the relative absorption of a weak measuring beam at 420 nm. As Fig. 11.8 shows, wavelengths of light absorbed by PSI and PSII have antagonistic effects on cyt-f; light absorbed by system I tends to oxidize the cytochrome whereas light absorbed by PSII tends to reduce it. These observations are consistent with the series formulation or 'Z-scheme' of Hill and Bendall (1960) where the cytochrome is placed between the two photosystems. It is reduced when electrons are received from PSII and oxidized when they are passed on to PSI.

The electron-transport scheme in detail Since the Hill and Bendall scheme was proposed, a wide variety of experimental techniques has been used further to elucidate the way in which PSII and PSI are connected. The techniques have included: (1) absorption difference spectrophotometry (as described above for cyt-f) applied to cytochromes, quinones and the copper-containing protein plastocyanin (PC); (2) extraction and re-addition procedures, particularly for quinones (Trebst, 1963); (3) electron paramagnetic resonance (epr) spectroscopy, used for those compounds which do not show marked absorbance changes when oxidized or reduced (Bolton and Warden, 1976); (4) inhibitor studies (Izawa, 1977); and (5) studies on mutant algae and higher plants which show impaired electron transport (Levine, 1969). The results of all this research are shown in the detailed electron transport map of Fig. 11.9, where the route of electron transport from water to NADP is shown.

Non-cyclic electron transport Non-cyclic electron transport involves the flow of electrons from water, through PSII and PSI in series to the terminal electron acceptor NADP. The area of the electron transport chain between water and PSII is still in doubt, but Mn is probably involved, electron transport requires Cl^- and is inhibited by a number of treatments (*see* below). On the reducing side of PSII the first stable electron acceptor is Q, probably a quinone, which can apparently be detected through its absorbance at 320 nm (Van Gorkom, 1974). Recently it has become apparent that the pool of Q is composed of two components which can be distinguished on the basis of mid-point redox potential; the low potential component (Q_{lp}) has a mid-point redox potential of around -250 mV while the high potential component (Q_{hp}), the previously designated Q, has a mid-point potential of approximately 0 mV (Horton and Croze, 1979; Malkin, 1978). Q is a single electron acceptor; the next major component in the electron transport chain is plastoquinone (PQ) which accepts two electrons and at the same time takes up two protons. The reactions of plastoquinone *in vitro* have been studied recently by Rich (1981). In order for electron transport to proceed smoothly from Q to PQ there is an intermediate called R (Bouges-Bocquet, 1973), also called B, which acts as an electron buffer, accepting two electrons from Q before handing them on to PQ. Plastoquinone also has a very important role to play in photophosphorylation (*see* below). Experiments using a mutant of the duckweed *Lemna* (Malkin and Aparicio, 1975) have indicated that electrons are passed from plastoquinone to a non-heme iron–sulfur (Fe–S) protein which is similar to that found in complex 3 of animal mitochondria, the

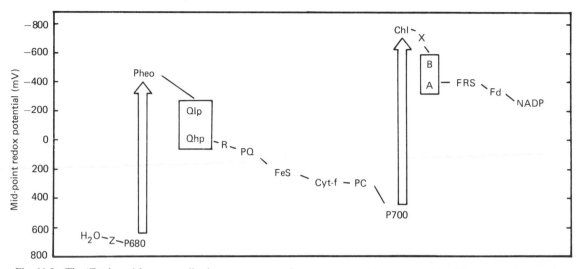

Fig. 11.9 The 'Z-scheme' for non-cyclic electron transport. Some components have been omitted from the reducing side of PSI for clarity. Where components are enclosed in boxes, their order is in doubt. Z, unknown intermediate; Pheo, pheophytin; Q, quinone; R, intermediate electron buffer; PQ, plastoquinone; FeS, 'Rieske' iron–sulfur center; cyt-f, cytochrome-f; PC, plastocyanin; X,A,B, bound iron–sulfur centers; FRS, ferredoxin-reducing substance; Fd, ferredoxin. The scale on the figure is mid-point redox potential, with strongly reducing couples at the top of the figure.

so-called Rieske center. The Rieske center is unusual for an iron–sulfur protein, since they generally have rather negative redox potentials. In contrast to cytochromes, the iron is held to the protein *via* cysteine–sulfur linkages; the sulfur is called labile, since it is released by mildly acidic conditions (Hall and Rao, 1977). The Rieske iron–sulfur center is oxidized by cytochrome f (Cyt-f) which is in its turn oxidized by plastocyanin (PC), a copper-containing protein (Katoh, 1977) which is the electron donor to P700. Plastocyanin appears not to be strictly necessary: in some algae non-cyclic electron transport proceeds unimpeded although PC is absent.

The reducing side of PSI is rich in iron–sulfur proteins: the electron acceptors X, A and B are all thought to be bound Fe–S centers. The electrons then pass *via* ferredoxin-reducing substance (FRS) to a soluble Fe–S protein, ferredoxin (Fd), which is now well studied (Hall and Rao, 1977). The terminal electron acceptor is NADP (Vishniac and Ochoa, 1951) which is reduced by ferredoxin *via* the ferredoxin-NADP reductase. The complete non-cyclic electron transport chain has been reviewed by Malkin (1982).

At first sight it might be thought that the Z-scheme implies a fixed relationship between one PSII and one PSI that are rigidly linked by the electron transport chain. This is not so: it must be realized that the thylakoid membrane is fluid, and the PQ pool can make communication between several PSII and PSI possible. Moreover, chloroplasts may not contain equal numbers of PSII and PSI centers.

Cyclic electron transport Electron transport does not only occur from water to NADP; there is a second 'mode' which involves neither the oxidation of water nor the reduction of NADP, and moreover only requires the excitation of photosystem I. The pathway is called cyclic electron transport (Fig. 11.10), and it was originally described by Arnon (*see* Arnon, 1977) in his studies on photosynthetic phosphorylation. Cyclic electron flow is not inhibited by 3(3,4-dichlorophenyl)-1,1-dimethylurea (DCMU), a powerful inhibitor of non-cyclic electron flow. In addition, it may involve a cytochrome (cyt-b_6) which does not appear to be involved in non-cyclic flow. *In vitro* cyclic electron flow normally requires the addition of cofactors to give appreciable rates.

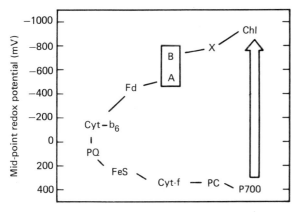

Fig. 11.10 A scheme for cyclic electron flow. Cyt-b_6, cytochrome-b_6; X, bound iron–sulfur center. Other symbols as in Fig. 11.9.

Partial reactions of electron transport

It has been noted above that the isolation of chloroplasts normally yields a preparation that requires an exogenous electron acceptor before electron transport and the evolution of oxygen can proceed. The Hill oxidant potassium ferricyanide is one of the best known, but a variety of electron acceptors can be used which, depending on their mid-point redox potentials, will accept at different sites on the Z-scheme. Similarly, if the oxidation of water is absent or inhibited, a suitable electron donor can be provided which will enable electron transport to proceed. In this way various sections of the electron transport chain ('partial reactions') can operate (Hauska and Trebst, 1977). For example, PSII alone mediates a partial reaction from water to the oxidized dye dichlorophenolindophenol (DCIP) which is reduced to its colorless form; PSI alone mediates electron transport from reduced DCIP to a PSI acceptor such as methyl viologen. While the evolution of oxygen is normally inhibited in the latter reaction, the oxygen electrode may still be used to detect electron transport since the reduction of methyl viologen leads to the uptake of oxygen in a process known as the Mehler reaction (Mehler, 1951).

Inhibitors of electron transport

A powerful technique in research into photosynthesis is the use of specific inhibitors of the electron transport chain (Izawa, 1977). The best known and most widely used is DCMU, which inhibits electron transport between Q and plastoquinone. Commercially DCMU is used as a herbicide (diuron), and a good deal of attention has been paid to the protein on the chloroplast membrane to which it binds (Mullet and Arntzen, 1981). The binding protein can be modified by the enzyme trypsin, which can then render electron transport DCMU-insensitive.

On the oxidizing side of PSII, electron transport can be inhibited by hydroxylamine, high concentrations of ammonia or methylamine, washing with high concentrations of the pH-buffer Tris, the uncoupler carbonyl cyanide 3-chlorophenyl hydrazone (CCCP), or treatment with a number of heavy metals. Between the two photosystems the plastoquinone analog 2,5-dibromo-3-methyl-6-isopropyl-p-benzoquinone (dibromothymoquinone or DBMIB) inhibits the oxidation of plastoquinone, while plastocyanin is inhibited by poly-L-lysine, KCN or mercury. The reducing side of PSI is inhibited by disalicylidenpropane diamine (DSPD) and its water-soluble analog sulfo-DSPD (Izawa, 1977).

Primary photoreactions

The reaction center, where the photoreactions take place, is clearly central to the energy transduction of photosynthesis since it mediates the first conversion of light energy to redox energy (Cogdell, 1983). It does this by stimulating the separation of charge, and creating a reduced electron acceptor and an oxidized electron donor. Briefly, all reaction centers contain one or two molecules of Chl that are found in a special environment which makes the Chl molecules able to use very rapidly the excitation energy for photochemical work before it can move to the next light-harvesting Chl molecule. The reaction center chlorophyll therefore constitutes a trap for excitation energy. When the energy is captured by the reaction center chlorophyll, it is used to put the chlorophyll into a state in which giving away an electron is a favored process; the structure of the center ensures that there is a suitable electron acceptor held very close to the reaction center chlorophyll, and which can take the electron, leaving the chlorophyll oxidized. The spatial arrangement of the reaction center is also organized so that there is an electron donor in close proximity which can give an electron to the chlorophyll and return it to its former state. Designating the electron acceptor by A, the donor by D and the reaction center chlorophyll by P, the sequence is then

$$DPA + \text{light energy} \rightarrow DP^*A \rightarrow DP^+A^- \rightarrow D^+PA^-$$

where P^* represents the excited state of the chlorophyll.

In purple bacteria, the ability of biochemists to isolate and purify reaction center complexes (*see* below), and the ability of physicists to build light sources and detectors for spectrophotometers with picosecond time resolution has led to a good understanding of the details of charge separation in these organisms (Parson and Cogdell, 1975; Cogdell, 1983). In green plants, however, the biochemistry is more difficult and consequently our knowledge is less advanced. Nevertheless, there is a good deal of evidence which gives clues to how the reaction centers work.

Photosystem I

Photosystem I is characterized by a small reversible light-induced bleaching (decrease in absorbance) at 700 nm, first observed by Kok (1961). The absorption difference spectrum has the characteristics of chlorophyll, and the pigment responsible for it was named P-700. The bleaching was associated with an electron paramagnetic resonance (epr) signal, which indicates the appearance of an unpaired electron, and suggests that the optical bleaching was associated with photo-oxidation. This result was confirmed by use of chemical oxidants. Careful study of the linewidth of the P700 epr signal (Signal 1) indicates that it is consistent with P700 being a pair of Chl-*a* molecules, which are so closely aligned in space that functionally they act as one single molecule. However this interpretation is not unequivocal.

The nature of the substance that accepts the electron when P700 is oxidized to P700$^+$ has been studied mainly by epr spectroscopy since optical spectroscopy has not proved a fruitful technique. The picture is rather complex, but it appears that the primary acceptor may also be a Chl molecule, and the first stable acceptor is a bound iron–sulfur center denoted X; the secondary acceptors appear to be two further iron–sulfur centers (A and B) close by (Malkin, 1982).

Photosystem II

In some senses the reactions of photosystem II are easier to study than those of photosystem I because a wider range of experimental techniques can be applied. But in spite of this apparent advantage some of the reactions of photosystem II, particularly those associated with the oxidation of water, remain enigmatic.

Like PSI, PSII exhibits a small, reversible light-induced bleaching of a chlorophyll molecule, but in this case the maximum of the bleaching is at 680 nm, and hence the reaction-center pigment of PSII is termed P680 (Witt, 1975). Several epr signals are associated with PSII; the majority appear to be related to the oxidizing side of the reaction center. The components of the reaction center, other than P680, have been difficult to identify, but it has been suggested that the primary electron acceptor is a pheophytin molecule (Fig. 11.9), that is, a Chl molecule whose central Mg has been replaced by two H$^+$ (Klimov, Klevanik, Shuvalov and Krasnovsky, 1977).

Two phenomena associated with PSII are the emission of fluorescence from the light-harvesting chlorophylls and the oxidation of water. These will be described in turn.

Chlorophyll fluorescence Chlorophyll fluorescence spectroscopy has considerably aided the study of PSII. It is very easy to measure, requiring only simple apparatus. Perhaps as a consequence of this simplicity the literature contains many reports on chlorophyll fluorescence: it is worth bearing in mind that even under the simplest circumstances a large number of factors come into play and influence the deceptively simple measurements.

It has been noted above that some of the energy absorbed by the chlorophyll molecules of PSII can be lost again as fluorescence. At room temperature in unfractionated chloroplasts, the emission is largely restricted to photosystem II; the findings of Duysens and Sweers (1963) suggested that the yield of fluorescence (equal to the intensity of fluorescence divided by the intensity of illumination) was under the control of the redox state of the electron acceptor Q. When Q was oxidized the fluorescence was quenched, but when Q was reduced the quenching was relieved. Indeed, Q stands for 'quencher'. The action of Q in quenching fluorescence can be readily understood in terms of the possible fates for excitation energy in photosystem II; these are (i) re-emission as fluorescence, (ii) radiationless decay to heat, (iii) transfer to the weakly fluorescent PSI, and (iv) use in photochemistry. Giving these processes the rate constants K_f, K_h, K_t and K_p respectively, then the yield of fluorescence φ_f is given by

$$\varphi_f = \frac{K_f}{K_f + K_h + K_t + K_p A}$$

where A is the fraction of photosystem II reaction centers that can accept an excitation, that is, centers where Q is oxidized. K_p is a relatively large rate constant when compared with the other rate constants, so φ_f is typically around 0.05. But if A decreases (more Q becomes reduced), φ_f will increase, because the denominator of the equation becomes smaller. Conversely, if A increases (more Q becomes oxidized), φ_f will decrease; hence fluorescence is an inverse measure of the yield of photochemistry. There are two extreme cases, when $A = 1$ and $A = 0$. These give

$$\varphi_o = \frac{K_f}{K_f + K_h + K_t + K_p} \quad \text{and} \quad \varphi_m = \frac{K_f}{K_f + K_h + K_t}$$

If chloroplasts are dark-adapted so that all Q becomes oxidized ($A = 1$) and then illuminated in the presence of DCMU (to prevent reduced Q becoming re-oxidized by PQ), a fluorescence induction from a low to a high level is seen. Qualitatively, the Duysens and Sweer's hypothesis accounts quite well for the low and high levels.

While chlorophyll fluorescence is strictly speaking a property of PSII, the phenomenon can give information on electron transport if measurements are made in the absence of DCMU. The chlorophyll fluorescence induction under such circumstances is called the Kautsky effect (Kautsky, Appel and Amann, 1960; Lavorel and Etienne, 1977). The redox state of Q certainly plays a role in determining the fluorescence yield, but in addition a more subtle factor comes into play. The absorption spectra of PSI and PSII are not identical, and consequently it is

possible to illuminate a photosynthetic system so that either PSI or PSII is over-excited, and the under-excited photosystem becomes the rate-limiting step overall. This is the origin of the 'red drop' described above. It now appears that the rate constant K_t for excitation energy transfer between PSII and PSI is not fixed, but can vary in order to equalize the excitation to the two photosystems in a process known as 'spillover'. It seems likely that the complex time course of the Kautsky effect is a mixture of spillover and quenching by Q, although there is now evidence (Walker, 1981) that even the carbon dioxide fixation cycle may have an influence.

Chlorophyll fluorescence should be distinguished from another PSII light emission process, called delayed light emission, which is also known as luminescence or delayed fluorescence. This emission, which has an emission spectrum similar to that of 'prompt' fluorescence, is only seen after the exciting light has been extinguished. It was first observed by Strehler and Arnold (1951), and is thought to arise as a back-reaction between metastable stored intermediates in the reaction center of PSII. The yield of emission is very low. In principle it offers a probe into the very heart of PSII, but the intensity of the emission can be modulated by a number of factors, and in practice the phenomenon is very complex (Malkin, 1977).

Oxygen evolution The evolution of oxygen from water is one of the most important reactions of photosynthesis, and yet it remains one of the least understood. A glance at the stoichiometry indicates that the oxidation of water is chemically not an easy task

$$2H_2O \rightarrow O_2 + 4H^+ + 4e^-$$

since four electrons must be withdrawn from two molecules of water before molecular oxygen can be evolved. PSII can only withdraw one electron at a time: there are then two possibilities, either (i) four PSII reaction centers co-operate and pool their oxidizing equivalents to oxidize two molecules of water, or (ii) each PSII reaction center works independently and has a means of storing oxidizing equivalents until four are collected and can be used together.

The first experimental data pertinent to distinguishing between these alternatives were obtained by Allen and Franck (1955) who found, using a sensitive phosphorescence method, that in fully dark-adapted algae no oxygen was evolved by illumination with a single flash. Similarly Joliot (1961) found that the evolution of oxygen in dark-adapted material required at least two photoacts for 'activation'. But a full answer to the question of co-operation or independence awaited a technical development, that of a rapidly-responding and sensitive oxygen electrode (Joliot and Joliot, 1968). The new electrode differed from the more usual Clarke-type (*see* Delieu and Walker, 1972) found in many laboratories, in that the photosynthetic material was placed directly on the platinum electrode, and was not separated from it by a thin membrane. Using a train of brief (10 µs) saturating flashes, the oxygen evolution elicited by each flash in the train could be observed. Figure 11.11 gives an example of the data obtained.

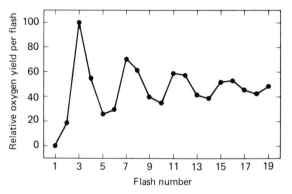

Fig. 11.11   The evolution of oxygen from dark-adapted *Chlorella* in response to a series of 5-µs flashes. The time interval between each flash is 500 ms. (Hipkins, unpublished.)

In this case the unicellular green alga *Chlorella* has been thoroughly dark-adapted before the experiment; there is no oxygen evolved on the first flash, little on the second, but a maximal amount on the third. The sequence continues with a period 4 in flash number, and gradually damps out until a steady-state level is seen on each flash. An interpretation of this phenomenon was offered by Kok, Forbush and McGloin (1970), and incorporated the concept of charge storage by each center. Each PSII reaction center is imagined to have a charge storage mechanism which can assume five states: S_0, S_1, S_2, S_3 and S_4 where the subscript denotes the oxidizing equivalents (power to extract electrons) stored.

Each absorbed photon transforms one state to the next with equal efficiency ($S_n \rightarrow S_{n+1}$); S_4 exists only transiently and reverts to S_0 after oxidizing water. The states are thus cycled round as shown in Fig. 11.12, which has become known as the 'Kok clock'. The scheme needs the additional hypothesis that in the dark S_0 and S_1 are stable, and that during dark adaption the higher S-states revert mainly to S_1 and partially to S_0. Taken thus far the scheme would predict that on illumination with a flash train, the resulting sequence would stay in perfect synchrony but, as Fig. 11.11 shows, this is not the case, and the sequence damps out. Kok *et al.* (1970) suggested that this was due to two 'errors' that might appear in

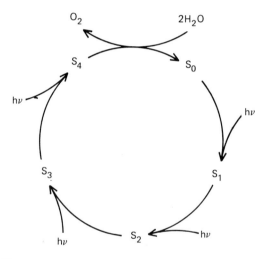

Fig. 11.12   The 'Kok-clock' hypothesis for the storage of oxidizing equivalents in photosystem II.

the transformation of one state to the next during a flash: a state might 'miss' during a flash, and stay in the same state ($S_n \rightarrow S_n$), or it might suffer a 'double hit' and be moved on two states ($S_n \rightarrow S_{n+2}$). There is some evidence for double hits, obtained from changing the duration of the flashes (Joliot, Joliot, Bouges and Barbieri, 1970), but misses have posed a serious problem of interpretation. Lavorel (1978) suggests that the independence of centers is not quite as strict as originally conceived by Kok *et al.* (1970), and that the oxygen-evolving apparatus consists of a fixed photoreaction which generates an oxidizing equivalent, and a mobile protein which stores the oxidizing equivalents. Misses occur when the mobile storage protein cannot collect an oxidizing equivalent from the fixed photoreaction center.

The phenomenology of flash-induced oxygen evolution says nothing about the biochemical nature of the intermediates involved. It has been known for some time that Mn is implicated in the evolution of oxygen (Radmer and Cheniae, 1977), although not all of the pool of chloroplast Mn is essential. Similarly, Cl^- (Izawa, Heath and Hind, 1969) and bicarbonate (Govindjee and van Rensen, 1978) are essential co-factors for oxygen evolution. But interest has centered on Mn, since the multiplicity of oxidation states it possesses makes it plausible that it is involved in charge storage. Some use has been made of the Mn epr signal, which shows six strong peaks when it is free in solution, but these peaks disappear when it is bound. Washing chloroplasts with high concentrations of the buffer Tris is well known to inhibit the evolution of oxygen; using an epr spectrometer, Blankenship and Sauer (1974) showed that the inhibition was correlated with the release of bound Mn. Similarly treatment of chloroplasts with hydroxylamine (NH_2OH) or mild heat, both of which inhibit oxygen evolution, are associated with the release of Mn (Radmer and Cheniae, 1977). This circumstantial evidence was very useful, and a further step forward was made by studying the amount of Mn that could be released by mild heat treatment after each flash in a series of flashes (Wydrzynski and Sauer, 1980). The amount of heat-releasable Mn shows oscillations with flash number of period 4, suggesting that it is closely associated with the oxygen-evolving mechanisms. Details of the roles Mn may play have been reviewed by Sauer (1980).

A further problem in the mechanism of oxygen evolution lies in the release of protons along with the oxidation of water. It might be thought that the protons would be released altogether at the same time as oxygen is released, but this is not the case. The sequence of proton release with respect to flash number is tricky to determine, but there is growing agreement on the sequence 1, 0, 1, 2 (Förster, Hong and Junge, 1981). The experimental findings thus show a complex pattern, indicating that the oxidation of water is a multi-step process.

Biochemical investigations

The investigations described in the foregoing sec-tions have resulted in a picture of the way that pigments and electron transport components operate together, and in particular suggest the existence in higher plants and algae of two photosystems each of which contains a reaction center and has a light-harvesting complex of molecules attached to it. It is a natural question to ask whether these functional units can be separated as physical units using biochemical techniques. There are two aims: firstly to confirm the description derived from other experiments, and secondly to make a simpler experimental material for the study of certain phenomena. Since oxygen-evolving organisms and photosynthetic bacteria show rather different characteristics, they will be treated separately.

Oxygen-evolving organisms Starting with isolated chloroplasts, a number of techniques may be used to derive preparations enriched in PSI or PSII: Boardman, Thorne and Anderson (1966) treated chloroplasts with the detergent digitonin and found that after differential centrifugation the lighter of the two fractions so obtained had enhanced PSI activity and depressed PSII activity when compared to the control, and was moreover depleted in Chl-b. The heavier fraction was enriched in PSII activity and Chl-b as compared to the control. Subsequently it was shown that the heavy fraction was composed mainly of granal stacks (see Fig. 11.1) and that the light fraction consisted primarily of stromal lamellae. Similar separation can be achieved using other detergents, for example Triton X-100, or by sonication or by passage of samples through the French pressure cell (see Brown, 1973). However, it can be seen that the sub-chloroplast particles obtained by these methods are not a pure preparation of one photosystem. Nevertheless, the separation provides confirmation that PSI and PSII exist.

Further purification must consist in removing the pigments and electron transport components that are extraneous to the reaction centers, and it is there that the problem lies. Many attempts have been made, but it has proved very difficult satisfactorily to fractionate the preparations further, mainly because of the instability of the extracts (Cogdell, 1983). Nevertheless, progress has been made towards purifying preparations of PSII (Hiller and Goodchild, 1981) and of PSI (Golbeck, 1980).

The protein composition of green plant chloroplasts has also attracted much interest (Thornber,

Markwell and Reinman, 1979; Anderson, 1980); thylakoid membranes are solubilized using the detergent sodium dodecyl sulfate (SDS) and the proteins separated using polyacrylamide gel electrophoresis. The results indicate that most (Anderson, 1980) if not all chlorophyll exists as pigment–protein complexes (Markwell, Thornber and Boggs, 1979) and that there are at least six major pigment–proteins, although these probably represent only three distinct chlorophyll–protein complexes (Hiller and Goodchild, 1981). These are: (i) a light-harvesting chlorophyll–protein complex which contains Chl-b, but has no light-induced oxidation-reduction activity associated with it (Bennett, 1979); (ii) a PSI complex, which contains P700, several forms of Chl-a (as few as 40 Chl molecules, which include long-wavelength-absorbing forms) and a number of polypeptides; and (iii) a fraction showing some PSII activity, but not the oxidation of water (Hiller and Goodchild, 1981). These three components which have presently been identified may well not describe the experimental data for long, since work is progressing rapidly in this field. Nevertheless, tentative models for the organization of two reaction center complexes and a light-harvesting complex have been proposed (Hiller and Goodchild, 1981), including the 'tripartite' model of Butler (1978). In addition, there have been attempts to correlate the very striking small and large particles which are apparent on freeze-fracture electron micrographs of thylakoid membranes with functional photosynthetic complexes (*see* Arntzen, 1978).

Photosynthetic bacteria In marked contrast to higher plants, the isolation of functional complexes from photosynthetic bacteria was achieved several years ago (Parson and Cogdell, 1975). The isolation procedure is normally to break the cells, treat the exposed membranes with detergent and then use protein separation techniques to isolate a specific complex. Photochemical reaction centers devoid of light-harvesting pigments have been obtained from a number of species. The reaction center from the purple non-sulfur bacterium *Rhodopseudomonas sphaeroides* contains 4 molecules of BChl and 2 of bacteriopheophytin (like BChl, but the Mg replaced by two H^+), one or two molecules of ubiquinone and one atom of ferrous iron, together with three polypeptides of apparent molecular weights in the region of 28, 24, and 21 kD (Cogdell, 1983). Reac-

tion centers of this type have been the subject of intense study, and the photochemical reactions occurring in them are now quite well understood (Parson and Cogdell, 1975). In addition, the reaction center polypeptides are now the subject of sequencing and crystallographic investigations (Cogdell, 1983). In agreement with the concept of reaction centers and light-harvesting pigments, there are also purely light-harvesting pigment–protein complexes, the study of which is only now beginning (Cogdell and Thornber, 1980). An atypical (water-soluble) bacterial light-harvesting pigment–protein complex has been crystallized and subjected to X-ray analysis (Fenna and Matthews, 1975). Seven BChl molecules are held inside a cylindrical protein, with the pigments being about 1.4 nm apart on average.

Photophosphorylation

ATP is the universal cellular energy carrier. It is formed from adenosine diphosphate (ADP) and inorganic phosphate (P_i) in oxidative and photosynthetic phosphorylation, and hydrolyzed with the consequent release of energy in a multitude of cellular reactions. The fixation of carbon dioxide within the chloroplast requires ATP; the experiments of Arnon, Allen and Whatley (1954) demonstrated that ATP could be generated by the chloroplast itself, and not, as had previously been proposed, by oxidative phosphorylation.

Hypotheses of phosphorylation

The energy required to make the phosphate bond between ADP and P_i comes from energy lost in electron transport, whether oxidative or photosynthetic. The energy is passed *via* an intermediate known as the 'high-energy state' and often denoted by a so-called 'squiggle' bond, '$\sim$': it is the nature of the high-energy state that distinguishes the proposed mechanisms of phosphorylation.

The chemical hypothesis This hypothesis, first proposed by Slater (1954) suggested that a covalent bond was formed between a component of the electron transport chain, A, and some other entity, sometimes written as X. At the start A is reduced by a reaction center, but on oxidation by the next member of the electron transport chain, B, the redox energy which is lost is captured in turning the normal bond to the '~' bond between A and X. The '~' bond is then transferred until it is the phosphate bond between ADP and P_i. In summary

$$AH_2 + X + B \rightarrow A \sim X + BH_2$$
$$A \sim X + P_i \rightarrow A + X \sim P_i$$
$$ADP + X \sim P_i \rightarrow ATP + X$$

This mechanism has been examined very closely since it does not predict transmembrane electro-chemical potential gradients, nor does it give a role for closed vesicles in phosphorylation (Jagendorf, 1975).

The chemiosmotic hypothesis Proposed by Mitchell (1966), this hypothesis suggests that the high-energy state consists of a proton-concentration gradient and an electric field across the functional membrane. It is general, and applies equally well to photosynthetic and oxidative phosphorylation both in prokaryotes and eukaryotes. There are five main postulates (*see* Fig. 11.13; Jagendorf, 1975; Nicholls, 1982) which are here applied to the chloroplast: (a) the reactions take place on the thylakoid membrane, which forms closed vesicles, and which is essentially

impermeable to protons; (b) electron transfer from water to NADP takes place *via* pure electron carriers (like cytochromes) and hydrogen atom (electron *plus* proton) carriers; these carriers are arranged vectorially in the membrane; (c) electron transfer is obligatorily linked to the pumping of hydrogen ions from the stroma into the intrathylakoid space, leading to the build-up of a transmembrane proton concentration gradient (ΔpH) and a transmembrane field ($\Delta\psi$); in the chloroplast the build-up of ΔpH is favored over that of $\Delta\psi$ by movement of counter-ions; (d) the combination of ΔpH and $\Delta\psi$ constitutes a store of energy (the high-energy state) and tends to expel protons from the intrathylakoid space; (e) there is a membrane-bound vectorial ATP-synthetase enzyme that takes the energy of the electrochemical potential gradient of protons and uses it to form ATP from ADP and P_i.

The conformational hypothesis The third type of hypothesis suggests that events which occur within the membrane itself are important, and that the transfer of energy from electron transport to ATP formation is *via* forced conformational changes in the membrane or in the enzymes concerned with phosphorylation (*see* Boyer, 1974).

The chemiosmotic hypothesis

One major advantage of the chemiosmotic hypothesis was that it was open to experimental testing in a number of ways. In general, it has received good support from experiment, but there remain a number of gray areas where the precise mechanism remains in doubt. One such area is the arrangement of the proton pump; another is the functioning of the ATP-synthetase.

The experimental evidence will be taken for each part of the hypothesis in turn.

The proton pump It is an essential part of the chemiosmotic hypothesis that electron transport from water to NADP and the translocation of protons from the stroma to the intrathylakoid space are obligatorily linked or coupled. That proton movements of this type take place in chloroplasts was demonstrated by Hind and Jagendorf (1963), where the loss of protons from an essentially unbuffered medium was detected by a pH-electrode in the

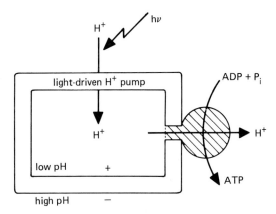

Fig. 11.13 Mitchell's chemiosmotic scheme for photophosphorylation in chloroplasts. The chloroplast thylakoid vesicle is shown with a membrane-bound proton pump (the electron transport chain) which causes the build-up of a transmembrane electrochemical potential gradient of protons. The protons leave the vesicle *via* the ATP-synthetase, losing energy and forming ATP as they do so.

medium. The increase in pH soon reached a steady state; the proton movements were reversed in the dark, and inhibited by inhibitors of electron transport, as well as uncouplers (Neumann and Jagendorf, 1964; *see* below). The relationship between the electron transport chain and the proton pump clearly depends on the spatial arrangement of the electron-transport components in the thylakoid membrane (Trebst, 1974). Evidence for the location of particular components can be grouped into three forms: (i) antibody agglutination or chemical labeling studies; (ii) the accessibility of components to exogenous electron donors or acceptors before and after disrupting the thylakoid membrane; and (iii) indirect studies. For example, interaction between the reaction center chlorophyll of PSI, P700, and exogenous plastocyanin occurs only after the thylakoid membrane has been disrupted, suggesting P700 is located at the interior of the thylakoid vesicle (Wood and Bendall, 1975). Similarly, antibodies raised to the protein plastocyanin cross-react only after disruption of the thylakoid, also suggesting that this component is at the interior. Using techniques of this type, the picture of vectorial electron transport illustrated in Fig. 11.14 was built up (Trebst,

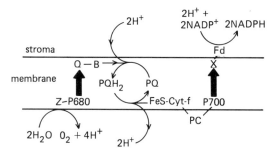

Fig. 11.14 The arrangement of electron-transport components in the thylakoid membrane that leads to the movement of protons from the stroma to the thylakoid space concomitant with the transport of electrons from water to NADP. The diagram does not include the possibility that a Q-cycle exists.

1974; Cox and Olsen, 1982). According to this scheme there are two modes for proton movements. In the first, plastoquinone, which when reduced is an electron-plus-proton carrier, is reduced at the stromal side of the membrane and becomes protonated there. The reduced entity (PQH$_2$) then dif-

fuses to the inside (Rich, 1981) of the membrane where it is oxidized, and the protons are released. This is a direct example of the obligatory coupling between electron transport and proton pumping. In the second mode, protons are taken up from the stroma as NADP is reduced, and released into the intrathylakoid space as water is oxidized, leading to an apparent movement of protons across the thylakoid membrane which is still strictly coupled to electron transport.

The ratio of protons translocated across the thylakoid membrane to electrons passed down the electron-transfer chain (the H$^+$/e ratio) has been the subject of a number of studies (Cox and Olsen, 1982). In most cases the value of H$^+$/e has been found to be 1, but there are a significant number of reports of H$^+$/e being 2. The experiments are rather difficult to perform and interpret; more data are needed before the value of the H$^+$/e ratio is finally settled.

It should not be thought that the scheme shown in Fig. 11.14 is the final word on electron transport and proton pumping. There has been considerable discussion (Velthuys, 1980; Malkin, 1982; Cox and Olsen, 1982) concerning the presence of a proton-motive quinone cycle (Q-cycle) in chloroplasts. Q-cycles are thought to exist in the electron-transport chains of both mitochondria and bacteria. Such a cycle (Mitchell, 1976) might be able to take account of hitherto unexplained features of the electrochromic bandshift (*see* below), light-induced redox changes of cyt-b_6, and the uncertain stoichiometry of protons pumped through the membrane to electrons transported down the chain.

The high-energy state In the chemiosmotic hypothesis the high-energy state is composed of the proton concentration gradient and the transmembrane field. Taken together, these two factors form the electrochemical potential gradient of protons, and comprise a store of energy which can be used to drive reactions in the energy-requiring ('uphill') direction. Thermodynamically the parameter used to describe such energy is Gibbs free energy, which is calculated from the difference in the electrochemical potentials on either side of the membrane, $\bar{\mu}_{H^+}$ and these are given by

$$\bar{\mu}_{H^+} = \bar{\mu}^*_{H^+} + RT \ln H^+ + FE$$

where $\bar{\mu}^*_{H^+}$ is a constant, R is the gas constant, T the absolute temperature, F the Faraday and E the electrical potential. Taking the difference in $\bar{\mu}_{H^+}$ between the stroma and intrathylakoid space yields

$$\Delta\bar{\mu}_{H^+} = 2.303\,RT\,\Delta pH + F\Delta\psi$$

where ΔpH is the pH gradient across the membrane, and $\Delta\psi$ is the electrical potential difference. $\Delta\bar{\mu}_{H^+}$ represents the maximum amount of free energy available to do work; the expression is usually modified by dividing through by F, and renaming $\bar{\mu}_{H^+}/F$ the proton motive force (pmf). Recognizing that $2.303\,RT/F$ is $59\,mV$ at $20°C$, then

$$pmf = 59\Delta pH + \Delta\psi \quad mV$$

The value of this expression is that it indicates the equal utility of ΔpH or $\Delta\psi$ in storing energy; thermodynamically the pmf can be all ΔpH and no $\Delta\psi$ or *vice versa*, or some of each. The estimation of ΔpH and $\Delta\psi$ can be quite tricky (Rottenberg, 1979), and usually entails following the distribution of a radioactively labeled permeant weak acid or amine to measure ΔpH, and a labeled permeant ion in the case of $\Delta\psi$. In this way, the pmf in chloroplasts was found to be mainly ΔpH, with the gradient attaining values of three pH units.

An experimental confirmation that a transmembrane pH gradient is the driving force for photophosphorylation in chloroplasts was provided by Jagendorf and Uribe (1966). These workers decided to eliminate the H^+-pumping part of the photophosphorylation system, and impose an artificial transmembrane pH gradient in the dark. They suspended chloroplasts in a medium of low pH, and after allowing them to equilibrate transferred them to a medium of high pH, containing ADP, P_i and Mg^{2+}. In this way a pH gradient of the correct polarity and size was imposed across the thylakoid membrane, and ATP was formed.

There is an important consequence of the build-up of a large pH gradient across the thylakoid membrane. If protons are to be pumped across the thylakoid membrane to achieve a concentration difference of around 1000-fold ($\Delta pH = 3$) then such an accumulation of positively charged ions will be accompanied by a membrane potential which is so large that it will oppose further proton influx. This will be prevented only if there is movement of counter-ions (anions inwards or cations outwards) to reduce the field to a small fraction of what it otherwise would be. Measurements of the fluxes of

H^+ and other ions have been made, either with ion-sensitive electrodes (Hind, Nakatani and Izawa, 1974), or by indirect methods (Barber, Mills and Nicolson, 1974) and the results indicate that the movement of Mg^{2+} from the intra-thylakoid space to the stroma and Cl^- in the reverse direction may be important counter-ion fluxes.

The measurement of $\Delta\psi$, at least at short times after flash excitation, can be studied very conveniently in photosynthetic organisms using spectrophotometry. The basis of the method is that under the influence of the transmembrane field the pigments embedded in the membrane undergo a small shift in their absorption spectra, known as the electrochromic bandshift (Junge, 1977). In chloroplasts, the maximum absorption is found at $515\,nm$ in the green region of the spectrum. On flash illumination the bandshift shows a very rapid onset ($\leqslant20\,ns$), concomitant with the separation of charge at the reaction centers, followed by a slow decay whose half-time (20–$200\,ms$) is related to the decay of the transmembrane field due to ion movement through the membrane. In steady-state illumination the absorption changes are much more complex because of light-induced scattering changes.

Careful study of the kinetics of the onset of the electrochromic bandshift has revealed that under some circumstances the initial rapid absorption increase is followed by subsequent slower increase (half-time 5–$20\,ms$) that may represent the electrogenic movement of charge across the membrane that occurs at sites other than the reaction centers (Crowther and Hind, 1980). Observations of this type have stimulated the interest in the existence of proton motive Q-cycles in chloroplasts that has been mentioned above.

The ATP-synthetase The vectorial enzyme which catalyzes the formation of ATP at the expense of the proton-motive force is complex: it consists of a membrane-bound proton-conducting channel termed CF_0 (coupling factor 0) and a more loosely attached catalytic section called CF_1 (coupling factor 1). If the CF_1 is removed from the membrane either by brief sonication or by incubation with ethylene diamine tetra-acetic acid (EDTA; chelates divalent cations) the phosphorylation activity of the chloroplasts is lost and the thylakoids become leaky to H^+, as judged by H^+ uptake (Avron, 1963). In electron micrographs of suitably stained material the CF_1

protein appears as a sphere of diameter about 9 nm; it has an apparent molecular weight of 325 000 daltons. After separation from the membrane CF_1 can be reconstituted and some Mg^{2+}-dependent phosphorylation activity is restored. CF_1 has latent ATP hydrolysis activity which can be activated with trypsin, and which is Ca^{2+}-dependent when the enzyme is removed from the membrane (Jagendorf, 1975; McCarty, 1977a; Nelson, 1982).

The question of how many protons efflux through the ATP-synthetase per ATP formed (the H^+/ATP ratio) can be approached in two ways. The first approach is thermodynamic, and seeks to equate the energy lost by protons as they pass through the ATP-synthetase to the energy required to phosphorylate ADP to ATP. Suitable calculations indicate that a H^+/ATP ratio of 3 is required to account for the synthesis of ATP found under circumstances when the pmf was known. Secondly, the approach can be experimental (Hauska and Trebst, 1977), when H^+/ATP ratios between 2 and 4 have been found.

CF_1 is composed of five subunits (Nelson, 1982) whose molecular weights and subunit functions are now becoming clear (Table 11.2). One technique to determine subunit function has been to isolate and purify each subunit and to raise antibodies to it, then to challenge the intact coupling factor with the antibodies and to see which functions are inhibited.

There has been a good deal of study on the binding of nucleotides to the CF_1 and on the influence of the high-energy state on such binding, as well as the concept of conformational changes generally. During catalysis conformational changes on the enzyme take place, detected by the energy-dependent inclusion of tritiated water into the enzyme (Ryrie and Jagendorf, 1971). The soluble CF_1 contains several nucleotide-binding sites: thoroughly washed CF_1 contains 2 ATP and 1 ADP per CF_1

which exchange only slowly in the dark, but exchange in the light is more rapid and uncoupler-sensitive (Harris and Slater, 1975).

An elegant reconstitution study was performed by Racker and Stockenius (1974) where a light-induced proton pump and the ATPase from beef heart were inserted into artificial phospholipid vesicles. The clever part of the experiment was the choice of the proton pump; rather than try to extract and reconstitute the complex proton pumps of the chloroplast or mitochondrion, Racker and Stockenius (1974) chose the simplest possible proton pump which is the 'purple patch' (a single protein called bacteriorhodopsin which pumps protons when illuminated) from a halophylic bacterium *Halobacterium halobium*. The purple patches and ATPase were reconstituted into the vesicles which were then illuminated, and ATP formation was found, suggesting that physical links between proton pump and ATPase were not necessary.

Uncouplers and ionophores A brief consideration of the diagrammatic chemiosmotic scheme (Fig. 11.13) shows that there are three separate places where photophosphorylation may be inhibited. The first place is the proton pump which, because it is obligatorily linked to electron transport, may be inhibited by the type of electron-transport inhibitor that has already been described. Secondly, inhibition can take place at the ATP synthesizing enzyme which can either be removed from the membrane by EDTA treatment or be blocked by substances called energy-transfer inhibitors (McCarty, 1977b). Lastly, the chemiosmotic hypothesis requires that the functional vesicle be impermeable to protons, but a number of substances can artificially increase the proton permeability and hence wastefully dissipate the pmf. Since this breaks the link between electron transport and phosphorylation, these compounds

Table 11.2 The subunits of the ATP-synthetase enzyme*.

Designation	Molecular weight	Possible stoichiometry	Suggested function
α	59 700	2	High affinity nucleotide-binding site; regulation
β	55 700	2	Low affinity nucleotide-binding site; active site
γ	36 800	1	Energy transduction from pmf to ATP
δ	19 400	1	Attachment of CF_1 to the membrane
ε	14 100	2	Inhibits ATP-hydrolase activity

*Adapted from Nelson, 1982.

are called 'uncouplers'. Classically, they are lipophilic weak acids which can traverse the membrane in either protonated or unprotonated form, and hence dissipate both the proton concentration gradient and the transmembrane field. Some examples are dinitrophenol (DNP), carbonyl cyanide-*p*-trifluoromethoxyphenylhydrazone (FCCP), and CCCP which has been mentioned above. There also exist ionophores which form lipophilic ionophore-ion complexes and hence render membranes permeable to specific ions, *e.g.*, valinomycin is a specific K^+ carrier; nigericin catalyzes the neutral exchange of K^+ for H^+.

The fixation of carbon dioxide

The key experimental tools to solve the problem of working out the biochemical pathway for the incorporation of carbon dioxide into sugar were firstly the availability of the stable radioactive isotope of carbon, ^{14}C, and secondly the ability to separate similar substances by paper chromatography. With these tools, Calvin and his colleagues working in Berkeley in the 1940s made great strides in the understanding

of the dark reactions of photosynthesis. The experiments consisted in illuminating algae in the presence of $H^{14}CO_3^-$ and after various times killing and extracting the algae in hot alcohol. The extract was then separated using two-dimensional chromatography, and radioactive compounds were identified by autoradiography (Robinson and Walker, 1981).

The Calvin cycle (C-3 cycle)

Calvin and his co-workers found a general pattern in their experiments (Fig. 11.15); the first radioactively labeled compound was phosphoglycerate (PGA) which contains 3 carbon atoms—hence the name 'C-3 cycle'. This was produced as a result of the carboxylation of the 5-carbon compound ribulose-1,5-bisphosphate (RuBP), and the reaction was catalyzed by the most abundant protein in the world (Ellis, 1979), ribulose-1,5-bisphosphate carboxylase (reaction I).

$$3\,RuBP + 3\,CO_2 + 3\,H_2O \rightarrow 6\,PGA$$

After carboxylation, PGA is phosphorylated to diphosphoglycerate (1,3-DPGA) at the expense of a molecule of ATP (reaction II).

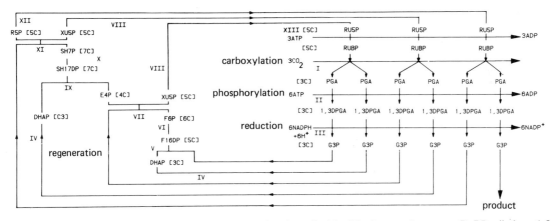

Fig. 11.15 The Calvin (C-3) cycle for the fixation of carbon dioxide. The intermediates are: RuBP, ribulose-1,5-bisphosphate; PGA, 3-phosphoglycerate; 1,3DPGA; 1,3-diphosphoglycerate, G3P, glyceraldehyde-3-phosphate; DHAP, dihydroxyacetone phosphate; F16DP, fructose-1,6-diphosphate; F6P, fructose-6-phosphate; Xu5P, xylulose-5-phosphate; Ru5P, ribulose-5-phosphate; E4P, erythrose-4-phosphate; SH17DP, sedoheptulose-1,7-diphosphate; SH7P, sedoheptulose-7-phosphate; R5P, ribose-5-phosphate. The enzymes are: I, ribulose-1,5-bisphosphate carboxylase or carboxydismutase; II, 3-phosphoglycerate kinase; III, NADP-glyceraldehyde-3-phosphate dehydrogenase; IV, triose phosphate isomerase; V, aldolase; VI, fructose-1,6-diphosphatase; VII, transketolase; VIII, ribulose-5-phosphate-3-epimerase; IX, aldolase; X, sedoheptulose-1,7-diphosphatase; XI, transketolase; XII, ribosephosphate isomerase; XIII, phosphoribulokinase.

$$6\,PGA + 6\,ATP \rightarrow 6\,(1,3\text{-}DPGA) + 6\,ADP$$

The 1,3-DPGA is then reduced with NADPH and a proton to glyceraldehyde-3-phosphate (G3P), which can be regarded as the product of the cycle (reaction III)

$$6\,(1,3\text{-}DPGA) + 6\,NADPH + 6\,H^+ \rightarrow 6\,G3P \\ + 6\,NADP^+ + 6\,P_i$$

Looked at in its simplest terms, the cycle involves four reactions: carboxylation of RuBP to 2 PGA, phosphorylation of PGA to 1,3-DPGA, reduction of 1,3-DPGA to G3P and lastly the regeneration of RuBP from G3P, a reaction which also calls for chemical energy derived from ATP (reactions IV to XIII). The net equation of the Calvin cycle can be written

$$3\,CO_2 + 9\,ATP + 6\,NADPH + 6\,H^+ \\ \rightarrow G3P\,(Product) + 9\,ADP + 6\,NADP^+ + 8\,P_i$$

Now that the Calvin cycle itself is well established, attention has more recently been centered on the regulation of the cycle (Robinson and Walker, 1981) and on the properties of its enzymes.

Regulation of the Calvin cycle has been discussed in detail by Edwards and Walker (1983). It is important because both oxidative and reductive carbohydrate metabolism occur in the same organelle. If the Calvin cycle were not de-activated in the dark there is the possibility that ATP would be wasted in 'futile cycles', one of which might be the continuous inter-conversion of fructose-6-phosphate and fructose-1,6-diphosphate catalyzed by the enzymes phosphofructokinase and fructose-1,6-diphosphatase. Each turn of the cycle would wastefully consume one molecule of ATP. The problem of such a 'futile cycle' could be resolved if one of the enzymes were regulated, so that both were not simultaneously catalytic (Robinson and Walker, 1981).

Activation of some enzymes of the Calvin cycle by light has been known for some time (Buchanan, 1980). In particular, the enzymes fructose-1,6-diphosphatase (reaction VI), sedoheptulose-1,7-diphosphatase (reaction X), NADP-glyceraldehyde-3-phosphate dehydrogenase (reaction III) and phosphoribulokinase (reaction XIII) are activated by light in chloroplasts in a DCMU-sensitive manner. This activation can also be achieved outside the chloroplast by reducing agents such as dithio-threitol, and has led to the suggestion that activation *in vivo* is due to reducing agents which are formed in the light as a result of non-cyclic electron transport. The hypothesis has two variations; the first is proposed by Buchanan and suggests that small soluble proteins, named thioredoxin and ferredoxin-in-thioredoxin reductase (FTR) participate. Thioredoxin is thought to be reduced by ferredoxin *via* FTR, and reduced thioredoxin activates the sensitive enzyme by reducing disulfide groups on the enzyme to sulfhydryl groups. The second variation has been proposed by Anderson and co-workers (*see* Robinson and Walker, 1981), and suggests the involvement of a membrane-bound component (the 'light effect mediator', LEM) which is reduced directly by the electron-transport chain, and then alters the conformation of the sensitive enzyme.

An alternative hypothesis for the regulation of the Calvin cycle is linked to the idea of proton movements from the stroma into the intra-thylakoid space being coupled with the movement of Mg^{2+} in the opposite direction to maintain electroneutrality. Since the chloroplast envelope is relatively impermeable to ions (Gimmler, Schäfer and Heber, 1975), the movements of H^+ and Mg^{2+} imply that their concentrations in the stroma will vary; both fructose-1,6-diphosphatase (reaction VI) and sedoheptulose-1,6-diphosphatase (reaction X) are sensitive to pH and Mg^{2+} concentration (Portis, Chon, Mosbach and Heldt, 1977) and it has been suggested that the movement of these ions in response to electron transport may switch on the Calvin cycle in the light. It has been argued, however, that light-induced changes in stromal Mg^{2+} levels have little significance for enzyme activation (Buchanan, 1980).

The C-4 cycle

The labeling pattern ^{14}C found with the Calvin cycle is not found in all plants. In some, the first radioactively labeled compound contains 4 carbon atoms (*e.g.*, malate, aspartate) and plants showing this character are termed 'C-4 plants' (Edwards and Huber, 1981). They have a distinguishing anatomical feature as well: the presence of a layer of bundle-sheath cells containing chloroplasts around the vascular tissue in the leaf. This feature is called Kranz anatomy (Kranz means 'wreath' in German), and is not seen in C-3 plants. Moreover, there are differences in chloroplast ultrastructure between the mesophyll and bundle-sheath cells; mesophyll cell chloroplasts have normal thylakoid membranes which are stacked into grana in some regions, but lack starch grains, whereas bundle-sheath cell chloroplasts are agranal but have starch grains. There are, however, subtle differences in the ultrastructure of the bundle-sheath cell chloroplasts between the different types of C-4 plants (*see* below).

The distinguishing biochemical feature of C-4 plants is the original carboxylation of CO_2. The CO_2 acceptor is the 3-carbon compound phosphoenolpyruvate (PEP), and the product is the 4-carbon compound oxaloacetate (OAA). The reaction occurs in the mesophyll cells. The fate of the OAA is of the same general pattern in all C-4 plants, but varies in the detail. Taking maize (*Zea mays*) as an example, the OAA is reduced to malate at the expense of reduced NADP; the malate so formed is then exported to the bundle-sheath cells where it is decarboxylated to pyruvate, and the CO_2 released is carboxylated by RuBP carboxylase into PGA. The pyruvate returns to the mesophyll cells where it is converted to PEP at the expense of ATP. This cycle is illustrated in Fig. 11.16.

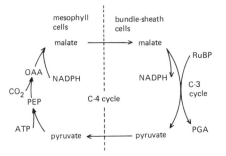

Fig. 11.16 The C-4 cycle for the fixation of carbon dioxide in a malate-forming species (e.g. maize). PEP, phosphoenolpyruvate; OAA, oxaloacetate.

It should be noted that the C-4 cycle does not achieve any *net* carboxylation; it merely supplies carbon dioxide to the C-3 cycle *via* a different route. It costs metabolic energy (extra ATP per CO_2), but since the K_m for CO_2 of PEP carboxylase is lower than that of RuBP carboxylase, the cycle can act as a CO_2-concentrating device.

C-4 plants can be divided into three types, depending on the details of the intermediates of the

cycle. The example given above, maize, is known as a malate former, since most of the radioactively-labeled carbon is found in malate, although some is found in aspartate. On the other hand there are two types of aspartate former, both of which form aspartate from OAA in mesophyll cells and then transfer it to the bundle-sheath cells, where the difference between the two types becomes apparent. In aspartate formers of type 1, aspartate is converted back to OAA, which is then decarboxylated to PEP using a molecule of ATP as it does so. The fate of the PEP is not clear: it may be returned to the mesophyll cells, or it may participate in the return of nitrogen to the mesophyll cells. Aspartate formers of type 2 transfer aspartate to the bundle-sheath cells as described above, but the formation of OAA is followed by that of malate, which is then decarboxylated to pyruvate. The pyruvate is transaminated to alanine, which is the route by which nitrogen returns to the mesophyll cells.

The three types of C-4 plants can equally well be named according to the predominant enzyme which catalyzes the decarboxylation of the C-4 compound in the bundle-sheath cells. Malate formers have the NADP-dependent malic enzyme (NADP-ME), aspartate formers of type 1 have phosphoenolpyruvate carboxykinase (PEP-CK) and aspartate formers of type 2 have the NAD-dependent malic enzyme (NAD-ME). Edwards and Huber (1981) have summarized the features of each type of C-4 plant; it should be realized that the relative amount of ^{14}C found in a C-4 product is to some extent dependent on environmental and other factors.

C-4 plants can be distinguished from C-3 plants by means other than the pattern of incorporation of $^{14}CO_2$, and the vascular anatomy. Due to the low K_m for CO_2 of PEP-carboxylase, and the low rate or absence of photorespiration (*see* below), C-4 plants when allowed to photosynthesize in a closed atmosphere can reduce the CO_2 concentration to a much lower level than can C-3 plants. The equilibrium level, reached after a prolonged period, is called the CO_2-compensation point. Alternatively, C-4 and C-3 plants can be distinguished by their different abilities to discriminate between the naturally-occurring ^{13}C and ^{12}C isotopes of carbon (O'Leary, 1981).

It has been suggested that there exist species which exhibit characteristics intermediate between C-3 and C-4 plants. Such observations, while at present few, raise the possibility that plants currently recognized as C-3 or C-4 may be extreme examples of a continuum of carbon-dioxide fixation types.

Crassulacean acid metabolism

Crassulacean acid metabolism (CAM) is an interesting variation of carbon metabolism that occurs in succulent plants that normally inhabit dry conditions (Osmond and Holtum, 1981). During the night the stomata are open, and the level of malic acid in the vacuole increases, whereas during the day the stomata close and the level of malic acid falls. Carbon dioxide uptake from the atmosphere occurs predominantly during darkness. CAM plants contain the enzymes of the C-4 cycle, and it appears that carboxylation of PEP to malate occurs during the night when the stomata are open, and a pool of malate is formed. During the day the stomata are closed to reduce water loss, and the malate is decarboxylated to provide a source of CO_2 for the Calvin cycle. Hence, what the C-4 plant separates in space, the CAM plant separates in time. CAM appears to be an adaptation to reduce water loss during the day in dry habitats, and yet to maintain photosynthesis. The CAM plant of greatest economic importance is the pineapple, *Ananus sativus*.

Photorespiration

RuBP carboxylase acts not only as the enzyme which carboxylates RuBP to form PGA. It also acts as an oxygenase, whereby at high oxygen concentrations RuBP and oxygen combine to yield phosphoglycollate and PGA (Lorimer and Andrews, 1981). Phosphoglycolic acid is then converted to glycollate, which is metabolized to glyoxylate with further oxygen uptake. Glyoxylate is then transaminated possibly by serine, but also by glutamate, to glycine in peroxisomes, and is then converted to serine and carbon dioxide by leaf mitochondria. Further metabolism of serine yields PGA, and a cycle of reactions, sometimes known as the C-2 cycle feeds PGA back into the Calvin cycle (Fig. 11.17). It is not yet known if photorespiration has a useful purpose (Heber and Krause, 1980); it appears to have very low activity or to be absent in C-4 plants. In C-3 plants it has been estimated that at least one sixth of the CO_2 fixed is released again under the

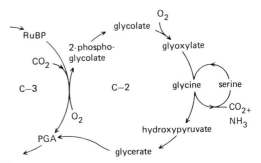

Fig. 11.17 A possible pathway for photorespiration *via* the C-2 cycle. (Serine may follow glycine in the cycle.) (Adapted from Lorimer and Andrews, 1981.)

relatively favorable conditions of 21% O_2 and 25°C (Chollet and Ogren, 1975).

Further reading

Clayton, R. K. (1980). *Photosynthesis: Physical Mechanisms and Chemical Patterns*, Cambridge, Cambridge University Press.

Clayton, R. K. (1970). *Light and Living Matter*, Parts I and II, London, McGraw-Hill.

Edwards, G. and Walker, D. A. (1983). *C-3, C-4: Mechanisms, and Cellular and Environmental Regulation, of Photosynthesis*, Oxford, Blackwell.

Gregory, R. P. F. (1977). *Biochemistry of Photosynthesis*, 2nd edn, New York, Wiley.

Haliwell, B. (1981). *Chloroplast Metabolism*, Oxford, Oxford University Press.

Nicholls, D. G. (1982). *Bioenergetics: An Introduction to the Chemiosmotic Theory*, New York and London, Academic Press.

Nobel, P. S. (1974). *Introduction to Biophysical Plant Physiology*, New York, Freeman.

References

Allen, F. L. and Franck, J. (1955). Photosynthetic evolution of oxygen by flashes of light. *Arch. Biochem. Biophys.* **58**, 124–43.

Anderson, J. M. (1980). Chlorophyll–protein complexes of higher plant thylakoids: distribution, stoichiometry and organisation in the photosynthetic unit. *FEBS Lett.* **117**, 327–31.

Arnon, D. I. (1977). Photosynthesis 1950–75: Changing concepts and perspectives. In *Encyclopaedia of Plant Physiology* (New Series), vol. 5, eds A. Trebst and M. Avron, pp. 1–56.

Arnon, D. I., Allen, M. B. and Whatley, F. R. (1954). Photosynthesis by isolated chloroplasts. *Nature,* **174,** 394–6.

Arntzen, C. J. (1978). Dynamic structural features of chloroplast lamellae. *Curr. Topics Bioenerg.* **8,** 112–60.

Avron, M. (1963). A coupling factor in photophosphorylation. *Biochim. Biophys. Acta* **77,** 699–702.

Barber, J., Mills, J. and Nicolson, J. (1974). Studies with cation-specific ionophores show that within the intact chloroplast Mg^{2+} acts as the main exchange cation for H^+ pumping. *FEBS Lett.* **49,** 106–10.

Bennett, J. (1979). The protein that harvests sunlight. *Trends Biochem. Sci.* **4,** 268–71.

Blackman, F. F. (1905). Optima and limiting factors. *Ann. Bot.* **19,** 281–95.

Blankenship, R. E. and Sauer, K. (1974). Manganese in photosynthetic oxygen evolution I. EPR study of the environment of manganese in Tris-washed chloroplasts. *Biochim. Biophys. Acta* **357,** 252–66.

Blincks, L. R. (1957). Chromatic transients in photosynthesis of red algae. In *Research in Photosynthesis,* eds H. R. Gaffron *et al.,* pp. 444–9, Interscience, New York.

Boardman, N. K., Thorne, S. W. and Anderson, J. (1966). Fluorescence properties of particles obtained by digitonin fractionation of spinach chloroplasts. *Proc. Natn. Acad. Sci. USA* **56,** 586–93.

Bolton, J. R. and Warden, J. T. (1976). Paramagnetic intermediates in photosynthesis. *Ann. Rev. Plant Physiol.* **27,** 375–83.

Bouges-Bocquet, B. (1973). Electron transfer between the two photosystems in spinach chloroplasts. *Biochim. Biophys. Acta* **314,** 250–6.

Boyer, P. D. (1974). Conformational coupling in biological energy transductions. In *Dynamics of Energy-Transducing Membranes,* eds L. Ernster, W. R. Estabrook and E. C. Slater, B.B.A. Library, vol. 13, pp. 289–301, Elsevier, Amsterdam.

Brody, S. S. and Rabinowitch, E. (1957). Excitation lifetime of photosynthetic pigments *in vitro* and *in vivo. Science* **125,** 555.

Brown, J. S. (1973). Separation of photosynthetic systems I and II. In *Photophysiology,* ed. A. C. Giese, vol. 8, pp. 97–112, Academic Press, New York.

Buchanan, B. B. (1980). Role of light in the regulation of chloroplast enzymes. *Ann. Rev. Plant Physiol.* **31,** 341–74.

Butler, W. L. (1978). Energy distribution in the photochemical apparatus of photosynthesis. *Ann. Rev. Plant Physiol.* **29,** 345–78.

Chollet, R. and Ogren, W. L. (1975). Regulation of photorespiration in C-3 and C-4 species. *Bot. Rev.* **41,** 137–79.

Clayton, R. K. (1970). *Light and Living Matter. Part I. The Physical Part,* McGraw-Hill, New York.

Cogdell, R. J. (1983). Photosynthetic reaction centres. *Ann. Rev. Plant Physiol.* **34,** 21–45.

Cogdell, R. J. and Thornber, J. P. (1980). Light-harvesting pigment–protein complexes of purple photosynthetic bacteria. *FEBS Lett.* **122,** 1–8.

Cox, R. P., and Olsen, L. F. (1982). The organisation of the electron transport chain in the thylakoid membrane. In *Electron Transport and Photophosphorylation, Topics in Photosynthesis,* vol. 4, ed. J. Barber, pp. 49–79, Elsevier, Amsterdam.

Crowther, D. and Hind, G. (1980). Partial characterisation of cyclic electron transport in chloroplasts. *Arch. Biochem. Biophys.* **204,** 568–77.

Delieu, T. and Walker, D. A. (1972). An improved cathode for the measurement of photosynthetic oxygen evolution by isolated chloroplasts. *New Phytol.* **71,** 201–25.

Dutton, H. J., Manning, W. M. and Duggar, B. M. (1943). Chlorophyll fluorescence and energy transfer in the diatom *Nitzschia closterium. J. Phys. Chem.* **47,** 308–13.

Duysens, L. N. M. (1951). Transfer of light energy within the pigment systems present in photosynthesising cells. *Nature* **168,** 548–50.

Duysens, L. N. M., Amesz, J. and Kamp, B. M. (1961). Two photochemical systems in photosynthesis. *Nature* **190,** 510–11.

Duysens, L. N. M. and Sweers, H. E. (1963). Mechanism of two photochemical reactions in algae as studied by means of fluorescence. In *Studies in Microalgae and Photosynthetic Bacteria,* ed. Jap. Soc. Pl. Physiol., pp. 353–72, Univ. of Tokyo Press, Tokyo.

Edwards, G. and Walker, D. A. (1983). *C-3, C-4: Mechanisms, and Cellular and Environmental Regulation, of Photosynthesis,* Blackwell, Oxford.

Edwards, G. E. and Huber, S. C. (1981). The C-4 Pathway. In *The Biochemistry of Plants,* vol. 8, eds M. D. Hatch and N. K. Boardman, pp. 237–81, Academic Press, New York.

Ellis, R. J. (1979). The most abundant protein in the world. *Trends Biochem. Sci.* **4,** 241–4.

Emerson, R. (1958). The quantum yield of photosynthesis. *Ann. Rev. Plant Physiol.* **9,** 1–24.

Emerson, R. and Arnold, W. A. (1932a). A separation of the reactions in photosynthesis by means of intermittent light. *J. Gen. Physiol.* **15,** 391–420.

Emerson, R. and Arnold, W. A. (1932b). The photochemical reaction in photosynthesis. *J. Gen. Physiol.* **16,** 191–205.

Emerson, R., Chalmers, R. and Cederstrand, C. (1957). Some factors influencing the long-wave limit of photosynthesis. *Proc. Natn. Acad. Sci. USA* **43,** 133–43.

Emerson, R. and Lewis, C. M. (1943). The dependence of the quantum yield of *Chlorella* photosynthesis on wavelength of light. *Am. J. Bot* **30,** 165–78.

Emerson, R. and Rabinowitch, E. (1960). Red drop and role of auxiliary pigments in photosynthesis. *Plant Physiol.* **35**, 477–85.

Englemann, T. W. (1883). Farbe und Assimilation. *Botan. Ztg* **41**, 1–13.

Fenna, R. E. and Matthews, B. W. (1975). Chlorophyll arrangement in a bacteriochlorophyll protein from *Chlorobium limicola*. *Nature* **258**, 573–7.

Förster, T. (1959). Transfer mechanisms of electronic excitation. *Discussions Faraday Soc.* **27**, 7–17.

Förster, V., Hong, Y-Q. and Junge, W. (1981). Electron transfer and proton pumping under excitation of dark-adapted chloroplasts with flashes of light. *Biochim. Biophys. Acta* **638**, 141–52.

French, C. S., Brown, J. S. and Lawrence, M. C. (1972). Four universal forms of chlorophyll-*a*. *Plant Physiol.* **49**, 421–9.

Gaffron, H. and Wohl, K. (1936). Zur Theorie der Assimilation. *Naturwissenschaften,* **24**, 81–90.

Gimmler, H., Schäfer, G. and Heber, U. (1975). Low permeability of the chloroplast envelope towards cations. In *Proc. 3rd Int. Cong. Photosynthesis*, vol. 2, ed. M. Avron, pp. 1381–92, Elsevier, Amsterdam.

Golbeck, J. H. (1980). Subchloroplast particle enriched in P700 and iron-sulphur protein. *Meth. Enzymol.* **69**, 129–41.

Govindjee and van Rensen, J. J. S. (1978). Bicarbonate effects on the electron flow in isolated broken chloroplasts. *Biochim. Biophys. Acta* **505**, 183–213.

Hall, D. O. and Rao, K. K. (1977). Ferredoxin. In *Encyclopaedia of Plant Physiology* (New Series) vol. 5, eds A. Trebst and M. Avron, pp. 206–16.

Harris, D. A. and Slater, E. C. (1975). Tightly-bound nucleotides of the energy-transducing ATPase of chloroplasts and their role in photophosphorylation. *Biochim. Biophys. Acta* **387**, 335–48.

Hauska, G. and Trebst, A. (1977). Proton translocation in chloroplasts. *Curr. Topics Bioenerg* **6**, 151–220.

Haxo, F. T., and Blincks, L. R. (1950). Photosynthetic action spectra of marine algae. *J. Gen. Physiol.* **33**, 389–422.

Heber, U. and Krause, G. H. (1980). What is the physiological role of photorespiration? *Trends Biochem. Sci.* **5**, 32–4.

Hill, R. (1954). The cytochrome-*b* component of chloroplasts. *Nature,* **174**, 501–3.

Hill, R. and Bendall, F. (1960). Function of the two cytochromes in chloroplasts: a working hypothesis. *Nature* **186**, 136–7.

Hill, R. and Scarisbrick, R. (1940). Production of oxygen by isolated chloroplasts. *Nature* **146**, 61–2.

Hill, R. and Scarisbrick, R. (1951). The haematin compounds of leaves. *New Phytol.* **50**, 98–111.

Hiller, R. G. and Goodchild, D. J. (1981). Thylakoid membrane and pigment organisation. In *The Biochemistry of Plants,* vol. 8, eds M. D. Hatch and N. K. Boardman, pp. 1–49, Academic Press, New York.

Hind, G. and Jagendorf, A. T. (1963). Separation of light and dark stages in photophosphorylation. *Proc. Natn. Acad. Sci. USA* **49**, 715–22.

Hind, G., Nakatani, H. Y. and Izawa, S. (1974). Light-dependent redistribution of ions in suspensions of chloroplast thylakoid membranes. *Proc. Natn. Acad. Sci. USA* **71**, 1484–8.

Horton, P. and Croze, E. (1979). Characterization of two quenchers of chlorophyll fluorescence with different midpoint oxidation-reduction potentials in chloroplasts. *Biochim. Biophys. Acta* **545**, 188–201.

Izawa, S. (1977). Inhibitors of electron transport. In *Encyclopaedia of Plant Physiology* (New Series) vol. 5, eds A. Trebst and M. Avron, pp. 266–82.

Izawa, S., Heath, R. L. and Hind, G. (1969). The role of chloride ion in photosynthesis III. The effect of artificial donors upon electron transport. *Biochim. Biophys. Acta* **180**, 388–98.

Jagendorf, A. T. (1975). Mechanism of photophosphorylation. In *Bioenergetics of Photosynthesis*, ed. Govindjee, pp. 413–92, Academic Press, New York.

Jagendorf, A. T. and Uribe, E. (1966). ATP-formation caused by acid-base transition of spinach chloroplasts. *Proc. Natn. Acad. Sci. USA* **55**, 170–7.

Joliot, P. (1961). Cinétique d'induction de la photosynthèse chez *Chlorella pyrenoidosa*. II Cinétique d'émission d'oxygène et fluorescence pendant la phase initiale d'illumination. *J. Chim. Phys.* **58**, 584–95.

Joliot, P. and Joliot, A. (1968). A polarographic method for detection of oxygen production and reduction of Hill reagents by isolated chloroplasts. *Biochim. Biophys. Acta* **153**, 625–34.

Joliot, P., Joliot, A., Bouges, B. and Barbieri, G. (1971). Studies of system II photocentres by comparative measurements of luminescence, fluorescence and oxygen emission. *Photochem. Photobiol.* **14**, 287–305.

Junge, W. (1977). Membrane potentials in photosynthesis. *Ann. Rev. Plant Physiol.* **28**, 503–36.

Katoh, S. (1977). Plastocyanin. In *Encyclopaedia of Plant Physiology* (New Series) vol. 5, eds A. Trebst and M. Avron, pp. 247–52.

Kautsky, H., Appel, W. and Amann, H. (1960). Chlorophyllfluorescenz und Kohlensäureassimilation XIII Mitteilung. Die Fluorescenzkurve und die Photochemie der Pflanze. *Biochem. Z.* **332**, 277–92.

Klimov, V. V., Klevanik, A. V., Shuvalov, V. A. and Krasnovsky, A. A. (1977). Reduction of pheophytin in the primary light reaction of photosystem II. *FEBS Lett.* **82**, 183–6.

Knox, R. S. (1977). Photosynthetic efficiency and exciton transfer and trapping. In *Primary Processes of Photosynthesis. Topics in Photosynthesis*, vol. 2, ed. J. Barber, pp. 55–97, Elsevier, Amsterdam.

Kok, B. (1961). Partial purification and determination of oxidation-reduction potential of the photosynthetic complex absorbing at 700nm. *Biochim. Biophys. Acta* **48,** 527–33.

Kok, B., Forbush, B. and McGloin, M. (1970). Co-operation of charges in photosynthetic O_2 evolution I: a linear four-step mechanism. *Photochem. Photobiol.* **11,** 457–75.

Krinsky, N. I. (1978). Non-photosynthetic functions of carotenoids. *Phil. Trans. R. Soc. Lond.* **B284,** 581–90.

Lavorel, J. (1978). On the origin of damping of the oxygen yield in sequences of flashes. In *Photosynthetic Oxygen Evolution,* ed. H. Metzner, pp. 249–68, Academic Press, London.

Lavorel, J. and Etienne, A-L. (1977). *In vivo* chlorophyll fluorescence. In *Primary Processes of Photosynthesis. Topics in Photosynthesis,* vol. 2, ed. J. Barber, pp. 203–68, Elsevier, Amsterdam.

Levine, R. P. (1969). The analysis of photosynthesis using mutant strains of algae and higher plants. *Ann. Rev. Plant Physiol.* **20,** 523–40.

Lorimer, G. H. and Andrews, T. J. (1981). The C-2 chemo- and photorespiratory carbon oxidation cycle. In *The Biochemistry of Plants,* vol. 8, eds M. D. Hatch and N. K. Boardman, pp. 329–74, Academic Press, New York.

McCarty, R. E. (1977a). The ATPase complex of chloroplasts and chromatophores. *Curr. Topics Bioenerg.* **7,** 245–78.

McCarty, R. E. (1977b). Energy transfer inhibitors of photophosphorylation in chloroplasts. In *Encyclopaedia of Plant Physiology* (New Series), vol. 5, eds A. Trebst and M. Avron, pp. 437–47.

Malkin, R. (1978). Oxidation-reduction potential dependence of the flash-induced 518nm absorbance change in chloroplasts. *FEBS Lett.* **87,** 329–33.

Malkin, R. (1982). Redox properties and functional aspects of electron carriers in chloroplast photosynthesis. In *Electron Transport and Photophosphorylation; Topics in Photosynthesis,* vol. 4, ed. J. Barber, pp. 1–47, Elsevier, Amsterdam.

Malkin, R. and Aparicio, P. J. (1975). Identification of a $g = 1.90$ high-potential iron–sulphur protein in chloroplasts. *Biochem. Biophys. Res. Commun.* **63,** 1157–60.

Malkin, S. (1977). Delayed luminescence. In *Primary Processes of Photosynthesis. Topics in Photosynthesis,* vol. 2, ed. J. Barber, pp. 349–431, Elsevier, Amsterdam.

Markwell, J. P., Thornber, J. P. and Boggs, R. T. (1979). Higher plant chloroplasts—evidence that all the chlorophyll exists as chlorophyll-protein complexes. *Proc. Natn. Acad. Sci. USA* **76,** 1233–5.

Mehler, A. H. (1951). Studies on reactions of illuminated chloroplasts. I: Mechanism of the reduction of oxygen and other Hill reagents. *Arch. Biochem. Biophys.* **33,** 65–77.

Metzner, H. (1975). Water decomposition in photosynthesis? A critical reconsideration. *J. Theor. Biol.* **51,** 201–31.

Mitchell, P. (1966). Chemiosmotic coupling in oxidative and photosynthetic phosphorylation. *Biol. Rev.* **41,** 445–502.

Mitchell, P. (1976). Possible molecular mechanisms of the proton motive function of cytochrome systems. *J. Theor. Biol.* **62,** 327–67.

Mullet, J. E. and Arntzen, C. J. (1981). Identification of a 32–34-kilodalton polypeptide as a herbicide receptor protein in photosystem II. *Biochim. Biophys. Acta* **635,** 236–48.

Myers, J. and French, C. S. (1960). Relationship between time course, chromatic transient and enhancement phenomena of photosynthesis. *Plant Physiol.* **35,** 963–9.

Nelson, N. (1982). Structure and function of the higher plant coupling factor. In *Electron Transport and Photophosphorylation, Topics in Photosynthesis,* vol. 4, ed. J. Barber, pp. 81–104, Elsevier, Amsterdam.

Neumann, J. and Jagendorf, A. T. (1964). Light-induced pH changes related to phosphorylation by chloroplasts. *Arch. Biochem. Biophys.* **107,** 109–19.

Nicholls, D. G. (1982). *Bioenergetics: An Introduction to the Chemiosmotic Theory,* Academic Press, London.

O'Leary, M. (1981). Carbon isotope fractionation in plants. *Phytochem* **20,** 553–67.

Osmond, C. B. and Holtum, J. A. M. (1981). Crassulacean acid metabolism. In *The Biochemistry of Plants,* vol. 8, eds M. D. Hatch and N. K. Boardman, pp. 283–328, Academic Press, New York.

Papageorgiou, G. (1975). Chlorophyll fluorescence. In *Bioenergetics of Photosynthesis,* ed. Govindjee, pp. 319–71, Academic Press, New York.

Parson, W. W. and Cogdell, R. J. (1975). The primary photochemical reaction of bacterial photosynthesis. *Biochim. Biophys. Acta* **416,** 105–49.

Porter, G., Tredwell, C. J., Searle, G. F. W. and Barber, J. (1978). Picosecond time-resolved energy transfer in *Porphyridium cruentum* Part I. The intact alga. *Biochim. Biophys. Acta* **501,** 232–45.

Portis, A. R., Chon, C. J., Mosbach, A. and Heldt, H. W. (1977). Fructose- and sedoheptulose-bisphosphatase. The sites of a possible control of CO_2 fixation by light-dependent changes of the stromal Mg^{2+} concentration. *Biochim. Biophys. Acta* **461,** 313–25.

Priestley, J. (1772). Observations on different kinds of air. *Phil. Trans. Roy. Soc. London* **62,** 147–264.

Rabinowitch, E. and Govindjee (1969). *Photosynthesis,* John Wiley, New York.

Racker, E. and Stockenius, W. (1974). Reconstitution of purple membrane vesicles catalysing light-driven proton uptake and adenosine triphosphate formation. *J. Biol. Chem.* **249,** 662–3.

Radmer, R. and Cheniae, G. (1977). Mechanisms of oxygen evolution. In *Primary Processes of Photosynthesis, Topics in Photosynthesis*, vol. 2, ed. J. Barber, pp. 303–48, Elsevier, Amsterdam.

Rich, P. R. (1981). Electron transfer reactions between quinols and quinones in aqueous and aprotic media. *Biochim. Biophys. Acta* **637**, 28–33.

Robinson, S. P. and Walker, D. A. (1981). Photosynthetic carbon reduction cycle. In *The Biochemistry of Plants*, vol. 8, eds M. D. Hatch and N. K. Boardman, pp. 193–236, Academic Press, New York.

Rottenberg, H. (1979). The measurement of membrane potential and ΔpH in cells, organelles and vesicles. *Meth. Enzymol.* **55**, 547–69.

Ruben, S., Randall, M., Kamen, M. and Hyde, J. L. (1941). Heavy oxygen (^{18}O) as a tracer in the study of photosynthesis. *J. Am. Chem. Soc.* **63**, 877–9.

Ryrie, I. J. and Jagendorf, A. T. (1971). An energy-linked conformational change in the coupling factor protein in chloroplasts. Studies with hydrogen exchange. *J. Biol. Chem.* **246**, 3771–4.

Sauer, K. (1980). A role for manganese in oxygen evolution in photosynthesis. *Acc. Chem. Res.* **13**, 249–56.

Sistrom, W. R., Griffiths, M. and Stanier, Y. (1956). The biology of a photosynthetic bacterium which lacks coloured carotenoids. *J. Cell. Comp. Physiol.* **48**, 473–515.

Slater, E. C. (1953). Mechanism of phosphorylation in the respiratory chain. *Nature,* **172**, 975–82.

Strehler, B. L. and Arnold, W. (1951). Light production by green plants. *J. Gen. Physiol.* **34**, 809–20.

Thornber, J. P., Markwell, J. P. and Reinman, S. (1979). Plant chlorophyll-protein complexes: recent advances. *Photochem. Photobiol.* **29**, 1205–16.

Trebst, A. (1963). The role of benzoquinones in the electron transport system. *Proc. Roy. Soc. London.* B **157**, 355–64.

Trebst, A. (1974). Energy conservation in photosynthetic electron transport of chloroplasts. *Ann. Rev. Plant Physiol.* **25**, 423–58.

van Gorkom, H. J. (1974). Identification of the reduced primary electron acceptor of photosystem II as a bound semiquinone anion. *Biochim. Biophys. Acta* **347**, 439–42.

van Niel, C. B. (1941). The bacterial photosyntheses and their importance for the general problems of photosynthesis. *Adv. Enzymol.* **1**, 263–328.

Velthuys, B. R. (1980). Mechanisms of electron flow in Photosystem II and toward Photosystem I. *Ann. Rev. Plant Physiol.* **31**, 545–67.

Vishniac, W. and Ochoa, S (1951). Photochemical reduction of pyridine nucleotides by spinach grana and coupled carbon dioxide fixation. *Nature* **167**, 768–70.

Walker, D. A. (1971). Chloroplasts (and grana): aqueous (including high carbon fixation ability). *Meth. Enzymol.* **23A**, 211–20.

Walker, D. A. (1981). Secondary fluorescence kinetics of spinach leaves in relation to the onset of photosynthetic carbon assimilation. *Planta,* **153**, 273–8.

Witt, H. T. (1975). Primary acts of energy conservation in the functional membrane of photosynthesis. In *Bioenergetics of Photosynthesis,* ed. Govindjee, pp. 493–554, Academic Press, New York.

Wood, P. M. and Bendall, D. S. (1975). The kinetics and specificity of electron transfer from cytochromes and copper proteins to P700. *Biochim. Biophys. Acta* **387**, 115–28.

Wydrzynski, T. and Sauer, K. (1980). Periodic changes in the oxidation state of manganese in photosynthetic oxygen evolution upon illumination with flashes. *Biochim. Biophys. Acta* **589**, 56–70.

Zscheile, F. P. and Comar, C. L. (1941). Influence of preparative procedure on the purity of chlorophyll components as shown by absorption spectra. *Bot. Gaz.* **102**, 463–81.

Zscheile, F. P., White, J. W., Beadle, B. W. and Roach, J. R. (1942). The preparation and absorption spectra of five pure carotenoid pigments. *Plant Physiol.* **17**, 331–46.

Nitrogen fixation

12

J I Sprent

Introduction

The vast majority of plants take up nitrogen in combined form, usually as nitrate or ammonia, via their root systems, sometimes assisted by mycorrhizal fungi. Small amounts of ammonia from animal excreta or other sources may volatilize from soil and be absorbed by the plant canopy. The ecological occurrence and significance of the various strategies of nitrogen acquisition are discussed by Guttschick (1981). Nitrate assimilation, which can occur in roots, leaves or other organs such as pods, may account for the incorporation into plant material of 2×10^4 Mt of N per year (Guerrero, Vega and Losada, 1981). This is approximately twice the annual estimate for biological nitrogen fixation. However, nitrogen fixation is the major route by which gaseous nitrogen is introduced into the ecosystem and the ability to carry out this process is confined to certain prokaryotic organisms (*see* list in Postgate, 1982). Some of these are free living, others occur in symbiotic partnership with eukaryotic organisms, generally plants, but including some animals such as certain termites (Prestwich and Bentley, 1981). In this chapter, the major types of nitrogen-fixing symbiosis involving vascular plants will be described briefly and then emphasis will be placed upon the physiology of legume root nodules.

Symbioses between nitrogen-fixing prokaryotes and vascular plants

These are listed in Table 12.1: further details and references can be found in Sprent (1979). The aquatic fern *Azolla* is particularly common in the humid tropics and is used as a source of nitrogen fertilizer in many rice plantations. Cycads have local significance, for example as an understory dominant in certain *Eucalyptus* forests in Australia (Grove, O'Connell and Malajczuk, 1980). *Gunnera* is the only angiosperm genus known to have a cyanobacterial endosymbiont. Its species range from small creeping plants to herbs with leaves 2 m high: all those examined are symbiotic and probably fix sufficient nitrogen for their needs. Both the filamentous *Frankia* and the unicellular *Rhizobium* invade roots and initiate swellings known as nodules. *Frankia* associates with a wide variety of genera, sometimes given the collective name 'actinorrhizal plants': Table 12.2 lists those genera where nitrogen fixation has been confirmed at the time of writing. With the exception of *Parasponia*, all the genera nodulated by *Rhizobium* are in the Leguminosae. The structures of actinorrhizal and leguminous nodules are compared in Table 12.3. The Leguminosae is a large family, usually divided into three subfamilies, the Caesalpiniodeae, Mimosoideae and Papilionoideae. For a current account of legume systematics, see Polhill and Raven (1981). Nodulation is least common in the Caesalpiniodeae and most common in the Papilionoideae, although only about half the genera in the world have been examined: a comprehensive record is given by Allen and Allen (1981).

Looser associations between angiosperms and nitrogen-fixing bacteria have been the subject of much research and publicity in recent years, particu-

Table 12.1   Symbioses between nitrogen fixing prokaryotes and vascular plants.

Prokaryotic symbiont			Vascular symbiont	
Class	Genus	Location	Class/order	Family/genus
Cyanobacteriales	*Anabaena*	Intercellular, in pockets	Pteridophyta	*Azolla*
	Nostoc?	Mainly intercellular in coralloid roots	Cycadales	All genera examined
	Nostoc	Intracellular, in petiole bases	Angiospermae	*Gunnera* spp.
Actinomycetales	*Frankia*	Intracellular, in root nodules	Angiospermae	Various non-leguminous genera
Eubacteriales	*Rhizobium*	Intracellular, in root nodules	Angiospermae	Ulmaceae, *Parasponia*
Eubacteriales	*Rhizobium*	Intracellular, in root nodules	Angiospermae	Leguminosae

larly in the case of certain grasses such as wheat, corn and sugar-cane. Although these may prove to be of considerable significance, especially in tropical areas, their physiology is poorly understood and they will not be considered further in this chapter. To obtain an idea of their potential the reader is referred to Boddey and Dobereiner (1982) and Brown (1982).

Table 12.2   Angiosperm genera reported to be nodulated by *Frankia* and to show nitrogenase activity: for most genera only a few species have been examined or found to be nodulated.

Family	Genus	Comments
Betulaceae	*Alnus*	Nodulation a generic attribute
	Arctostaphylos	Nodulates only in Alaska
Casuarinaceae	*Casuarina*	Nodulation rather variable
Coriariaceae	*Colletia*	One report on *C. paradoxa*
	Coriaria	Nodulation a generic attribute?
Datiscaceae	*Datisca*	One sp. each from Pakistan and California
Eleagnaceae	*Eleagnus*	⎫
	Hippophae	⎬ Nodulation a generic attribute?
	Shepherdia	⎭
Myricaceae	*Comptonia*	*C. peregrina* also known as *Myrica asplenifolia*
	Myrica	Nodulation a generic attribute?
Rhamnaceae	*Ceanothus*	
	Discaria	
	Trevoa	One report on *T. trinervis*
Rosaceae	*Cercocarpus*	
	Chaemaebatia	One report on *C. foliolosa*
	Cowania	One report on *C. mexicana*
	Dryas	Nodulates in Alaska and Canada, not in Europe or Japan
	Purshia	
	Rubus	

Table 12.3 Some characteristics of legume and non-legume root nodules.

Characteristic	Legume	Non-legume
Endophyte	*Rhizobium**	*Frankia*
Reaction to endophyte	Root hair curling and branching	Root hair curling and branching
Method of invasion	*Via* root hairs or epidermis[1]	*Via* root hairs[2]
Location of nodular vascular tissue	Cortical	Central*
Location of infected tissue	Central	Cortical*
Nitrogen-fixing structure	Bacteroid	Endophyte vesicle?[3]
Oxygen-transport pigment	Leghemoglobin*[4]	None confirmed or necessary?[5]

* Indicates feature to be found in *Parasponia* nodules.
(1) Depends on genus.
(2) In the few species which have been examined.
(3) *Casuarina* does not produce vesicles.
(4) Appleby, Tjepkema and Trinick (1983).
(5) Recent evidence suggests that some genera may contain a hemoglobin similar to leghemoglobin (Tjepkema, 1984).

Formation of legume root nodules

Infection

Not all legumes can be infected by all rhizobia. The degree of specificity varies widely. Systems which have been studied in sufficient detail fall into two broad categories, those from tropical/subtropical regions and those from temperate regions. Some of the more obvious differences in the rhizobia from these regions are given in Table 12.4. The magnitude of these differences has led to the proposal that the slow-growing types be placed in a new genus, *Bradyrhizobium* (Jordan, 1982).

Since soils frequently contain many different rhizobia, there must be means by which strains can recognize, and be recognized by, compatible legumes. The most widely-studied cases involve infection through root hairs. Unicellular hairs on underground portions of the epicotyl of *Vicia faba* are also capable of being infected and formation of functional nodules may follow (Fyson and Sprent, 1980). One of the factors involved in the recognition process appears to be a special type of protein (lectin) on the root surface which binds to surface

Table 12.4 Differences between certain rhizobia of tropical and temperate regions. *See* Sprent (1980, 1983) for further details and references.

	Tropical	Temperate
Examples of *Rhizobium* sp.	*japonicum*	*leguminosarum*
	Cowpea miscellany	*trifolii*
Examples of host	Soybean	Pea
	Cowpea	Clover
Characters of free-living form		
Growth rate on agar	Slow	Fast
Ability to take up and hydrolyze disaccharides	Present	Absent
Changes in symbiotic form (bacteroids)		
DNA content per cell	Increased	Little change
Size	Greatly increased, nearly filling the membrane envelope	Slightly increased
Shape	Pleomorphic	Swollen rods
Viability	Little or none	Some

polysaccharides on the bacterium. However, this is not sufficiently specific since, for example, pea roots bind large numbers of clover rhizobia. It is likely that there is a sequence of events leading to full recognition (Bauer, 1981).

Curling and sometimes branching of root hairs (Fig. 12.1A) have been known for many years to precede infection, but the way in which hair growth has been modified has not been understood. In soybean (*Glycine max*), infectible cells are those which are about to form hairs; the latter develop where the more rigid of two cell wall layers is absent. Bauer (1981) has put forward a model in which rhizobia modify the growth of the hair so that it becomes deformed and curls around the rhizobia. This implies that curling must occur close to the base

of the hair. This is not always observed to be the case (see Fig. 12.1A). A slightly different hypothesis was propounded by Hubbell (1981), in which the bacteria inhibit the growth of the cell wall at the point where they attach. This could occur at any point where a hair is still growing. At the point where curling is tightest, there is localized dissolution of the host cell wall (Callaham and Torrey, 1981). Wall-dissolving enzymes (such as pectinases, glycosyltransferases and polygalacturonases) may also act as lectins (Hubbell, 1981); some are sensitive to Ca^{2+}, consistent with the observations of Sethi and Reporter (1981) that where bacteria penetrate root hairs Ca^{2+} distribution is altered. In order for the cell wall to be penetrated, the correct amount of both enzyme and substrate must be

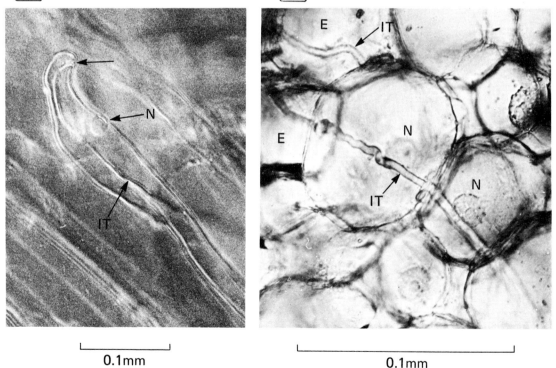

Fig. 12.1 Infection in roots of *Vicia faba*. A. curled root hair showing point of penetration by rhizobia (arrow) and infection thread (IT) passing down the hair. The host cell nucleus (N) is also shown; this is more usually found at the bases of hair cells with infection threads. B. infection threads (IT) passing through the outer layers of the root cortex. E, epidermal cell; N, host cell nucleus. Bar indicates 0.1 mm. (Photograph courtesy J. M. Sutherland.)

present, and the curling reaction may assist in the localized accumulation of reactive molecules. If the conditions are not quite correct, curling without infection may occur. Hubbell (1981) suggests that concentrations are very critical and that this is why only a very small proportion of root hairs becomes infected, as anyone who has searched for them will appreciate. For example Fyson (1981) examined 4051 hairs on the tap root of *Vicia faba* plants and found only 102 (2.5%) to be infected.

Once inside the wall of the root hair, dividing bacteria, surrounded by a matrix of bacterial origin, are kept in place by newly-formed plant cell wall material, forming an infection thread (Fig. 12.1A). The cell wall does not normally cover the tip of the thread; if it does, the thread aborts. Host cell nuclei appear to act in a controlling manner in thread growth, accompanying it as it passes down the root hair and across the cells of the root cortex (Fig. 12.1B).

Not all legumes are infected via root hairs. *Arachis hypogaea* (groundnut or peanut) and *Stylosanthes* species (a tropical forage legume), both of which belong to the same tribe within the Papilionoideae, are infected at the points where lateral roots emerge (Chandler, 1978; Chandler, Date and Roughley, 1982). Two related plants, *Aesche-*

nomene and *Sesbania*, can form nodules on aerial parts of their stems: these nodules are particularly profuse when plants are grown under waterlogged conditions (Dreyfus and Dommergues, 1981). It is likely that further exceptions to root hair infection will be found when more plants of the warmer regions of the world are examined. As Hubbell (1981) points out, tropical rhizobia tend to be more promiscuous than temperate species, in that one strain infects a wider variety of hosts: he suggests that this could result from less exacting requirements for root hair infection. A logical extension of this would be to obviate the need for root hairs. Tropical rhizobia are often also considered to be more primitive than temperate ones (Norris, 1965) and tropical legumes older (in the evolutionary sense) than temperate legumes (Polhill and Raven, 1981). It is thus possible that root-hair infection may be a comparatively recent evolutionary event.

Nodule development

There are two known basic types of developmental pattern, resulting in nodules with determinate and indeterminate growth, elegantly illustrated by Newcomb (1976) and Newcomb, Sippel and Peterson (1979) using soybean and pea respectively. Figure

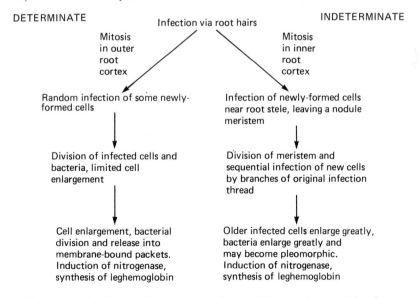

Fig. 12.2 Some differences in development between determinate and indeterminate nodules, for example soybean and pea, respectively.

12.2 outlines the major differences between the two and the final structure of the two types is illustrated in Figs 12.3 and 12.4. Some further structural differences are listed in Table 12.5.

It is generally assumed that nodule initiation and development are hormonally controlled, with growth substances being provided by both partners.

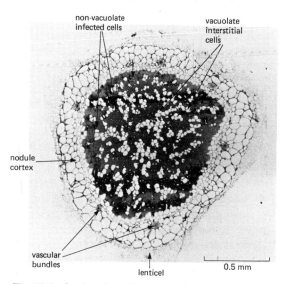

Fig. 12.3 Section through a typical determinate nodule, *Vigna radiata*. Note the absence of an apical meristem and the uniform nature of the central infected tissue. The lenticels and vascular bundles radiate from the point of attachment to the root and fuse at the distal end. Arrows indicate some air spaces. Bar indicates 0.5 mm. (Photograph courtesy J. M. Sutherland.)

The original indication of hormonal involvement came from the observation that cell divisions occurred in roots in advance of penetration by infection threads. Since then, auxins, gibberellins, cytokinins and abscisic acid have all been invoked, together with various unidentified 'factors'. The situation is almost as confused as that with respect to the hormonal control of flowering (Chapter 18). The exact relationships must vary with species to account for the wide variation in nodule shape (Corby, 1981; see also Fig. 12.5) and the fact that in some plants such as soybean, meristematic activity stops at an early stage in nodule development, whereas in others, such as pea, the nodule meristem persists beyond the onset of nitrogen fixing activity. The final shape of both the nodule and the active form of *Rhizobium*, called the bacteroid, are controlled, at least in part, by the host genotype. For example, Kidby and Goodchild (1966) produced effective (*i.e.*, nitrogen-fixing) nodules on *Lupinus luteus* and *Ornithopus sativus* with the same strain of *Rhizobium lupini*. Lupin nodules were of the typical collar-like shape commonly ascribed to but not always found in this genus, whereas those on *Ornithopus* were more clover-like (see Fig. 12.5). When *Rhizobium* strain 32H1 is inoculated on to peanut plants, bacteroids are almost spherical, but the same strain produces rod-shaped bacteroids in cowpea (*Vigna unguiculata* (Sen and Weaver, 1980)). Similarly, the rhizobial strain may modify plant morphology as shown by differences in internode length in the lima bean (*Phaseolus lunatus*) when inoculated with strains of *Rhizobium* whose bacteroids produce different quantities of gibberellins (Evensen and Blevins, 1981).

As a final example of the physiological interaction between host and endosymbiont, a recent study on *Medicago sativa* (lucerne, alfalfa) will be taken (Truchet, Michel and Denarié, 1980). Four rhizobial mutants, all lacking the ability to make the amino acid leucine, induced the formation of white, ineffective nodules (*i.e.*, not able to fix nitrogen) in which meristematic activity was limited and the nuclei remained small; bacteria were not released from the infection threads. When plants were given 0.075 mol m^{-3} or 0.2 mol m^{-3} leucine, apparently normal indeterminate nodules were formed, anatomically like the wild type and physiologically able to fix nitrogen. The rhizobia were released from their infection threads and differentiated into bacteriods: at the same time the nuclei of infected cells enlarged and appeared to become polyploid. This observation may resolve one problem which has been exercising nodule workers for many years, namely, whether the polyploidy regularly seen in infected cells of some nodules is a prerequisite for or a consequence of infection. In the case of lucerne, it appears to be a consequence, since release of bacteria from infection threads induces endoreduplication of DNA. As a result of their work, Truchet *et al.* made two proposals: (1) that bacteria can induce meristem formation by producing some agent that can traverse cell walls and plasmalemma, and (2) that differentiation of active nitrogen-fixing cells depends upon another agent which cannot pass from

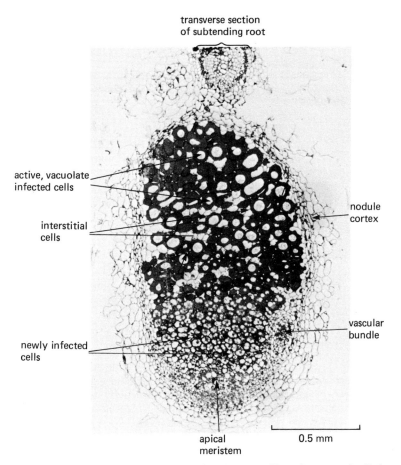

transverse section
of subtending root

active, vacuolate
infected cells

interstitial
cells

nodule
cortex

newly infected
cells

vascular
bundle

apical
meristem

0.5 mm

Fig. 12.4 Section through a typical indeterminate nodule, *Ononis repens*. Note the range of cells from meristematic to fully differentiated. Older nodules have senescent infected cells near to the subtending root. Arrows indicate some air spaces. Bar indicates 0.5 mm. (Photograph courtesy J. M. Sutherland.)

cell to cell and requires that bacteria first be released from their infection threads.

Mature nodule size and structure

The final size of nodules must depend on cell division and enlargement. In determinate nodules of the soybean type, cell numbers appear to be determined at an early stage in development (Bergersen, 1982). The following lines of evidence indicate that cell size is controlled independently: (1) nodules on plants of *Phaseolus vulgaris* (nodules similar to soybean) when transferred from low to higher light intensity increase in size due to cell enlargement (Antoniw, 1976); (2) in *Lotus pedunculatus* decreased nutrient supply increases the size but decreases the number of nodules, but only with certain strains of rhizobia (Pankhurst, 1981); (3) nodules formed on *Phaseolus vulgaris* at lower temperatures are larger than those formed at higher temperatures (Thomas and Sprent, 1984). In the last two examples it is not yet known whether nodule-size changes are due to cell-number changes or cell enlargement, but it is clear that, even though cell number is

Table 12.5 Differences between certain determinate and indeterminate nodules.

There are many indeterminate nodules of tropical legume tribes whose fine structure has not been studied. The tropical types described here are from the tribe Phaseoleae and serve to illustrate that there are wide sets of differences between groups; see also Sprent (1981).

	Determinate	Indeterminate
Examples	*Glycine, Vigna, Phaseolus, Desmodium Centrosema*	*Pisum, Vicia, Trifolium Medicago, Melilotus*
Geographical origin	Tropical, subtropical	Temperate
Major area for gaseous exchange	Lenticels formed from a cortical cambium	Intercellular spaces over most of surface
Meristem	Not persistent	Apical, persists for many weeks, may branch
Vascular supply	Fuses at distal end when growth ceases	Grows with nodule
Vascular transfer cells	Absent	Present
Infected cells	Virtually non-vacuolate, bacteroid envelopes throughout cell	Usually vacuolate, bacteroid envelopes in cell periphery
Interstitial cells in infected region	With prominent microsomes	Without prominent microsomes?
Principal product exported	Allantoin, allantoic acid	Asparagine, glutamine

determined at an early stage, adaptation to changing environment may occur *via* cell enlargement.

Final size in indeterminate nodules is governed by the longevity of the meristem, as well as size of individual cells. The developmental sequence here is quite different from that of soybean in that each infected cell is formed following an independent invasion by an infection thread; once invaded, the cells do not divide. Nodule growth may be terminated when all the cells of the meristem become invaded by infection threads. Thus we have a race between cell division and cell infection. Under certain conditions adverse to nitrogen fixation, such as low temperature (Fyson and Sprent, 1982) and salinity (Yousef and Sprent, 1983), *Vicia faba* nodules partially compensate by producing larger, rather than more, nodules. At low temperature, it is known that this is because meristematic activity is prolonged. Meristematic activity in lucerne is unaffected by 18 mol m^{-3} nitrate, whereas other aspects of nodule development are retarded (Truchet and Dazzo, 1982). Water stress reduces final cell size (Gallacher and Sprent, 1978); however, transfer from a stressed to a well-watered regime

results in both cell expansion and renewed meristematic activity, making the indeterminate type of nodule more flexible in its response to adverse conditions (Sprent, 1979). The pea and soybean types of nodule are probably only two of many. Preliminary observations on genera endemic to Western Australia (Pate and Sprent, unpublished), suggest that these, at least, may have a quite different outer structure to protect them in hot, dry periods. Peanut nodules have no interstitial cells. Doubtless many new variants will be discovered and, it is to be hoped, examined in parallel with their physiology (*see* Sprent, 1980). Growth of perennial legume nodules has been studied scarcely at all. In some plants, for example pigeon pea (*Cajanus cajan*), young nodules may be very similar to those of soybean, but after months or years develop into more massive lobed structures (*see* Fig. 12.5). How they do this is not clear.

Nodule senescence

Although it has been realized for many years that nodules have a limited life span, which if prolonged could improve total nitrogen fixation, few detailed

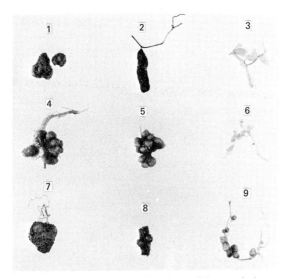

Fig. 12.5 Some examples of legume nodule morphology, natural size. (1) *Macrotyloma uniflorum*. Note the spherical and more elongated (older) forms. The ridges are lenticels. (2) *Acacia farnesiana*. Note the dark pigmentation. (3) *Coronilla varia*. (4) *Sesbania rostrata*. (5) *Pisum farnesiana*. Some of these indeterminate nodules are branched. (6) *Trifolium repens*. (7) *Abrus precatorius*. (8) *Desmodium dillenii*. (9) *Caragana arborescens*. Nodules from a preserved collection kindly donated to the author by H. D. L. Corby and photographed by C. Harris.

studies on senescence processes have been made. Sutton (1983) has reviewed the subject most recently. A change from pink to greenish-brown coloration is the first sign of senescence: in nodules of the soybean type this usually begins in the center and spreads rapidly throughout the infected region. Under some circumstances it may be reversible (Pfeiffer, Malik and Wagner, 1983). In the pea type, the oldest infected cells first become senescent and as nodules age the proportion of senescent to non-senescent tissue increases, although nodules may remain active for weeks, months or even years. The fine structure of senescent cells shows loss of membrane integrity and usually the dying cells become invaded by the free-living forms of the relevant rhizobia. As with other plant tissues, senescence may be advanced by adverse conditions (*see* Chapter 20). One commonly-held view is that nodule senescence is enhanced in grain legumes when there is competition for carbohydrate between growing pods

and nodules. Although nodule activity often declines during pod filling (*e.g.* Sprent, 1982) the decline is not necessarily due to carbohydrate shortage (Streeter, 1981).

Biochemistry of nitrogen fixation

The reduction of nitrogen to ammonia is catalyzed by an enzyme complex known as nitrogenase, the structure of which varies little among nitrogen-fixing organisms. In this section, only the major features of nitrogenase will be considered: more detail can be found in the volumes edited by Stewart and Gallon (1981) and Gibson and Newton (1981). Both of these books contain papers by Burris and his co-workers at Wisconsin and Fig. 12.6 gives an enlarged version of their proposed outline of nitrogenase action. Electrons are passed to the Fe protein part of the nitrogenase complex (also known as dinitrogenase reductase). This reduced Fe protein, joined to two MgATP molecules, transfers electrons singly to the Fe-Mo-protein part of the complex (dinitrogenase), releasing MgADP and Pi. To a site associated with molybdenum, electrons flow from the reduced Fe-Mo-protein to various substrates of which the natural ones are N_2 and H^+. Many other substrates can be reduced and these have proved useful in elucidating the mechanism of nitrogenase action. The most widely known is acetylene which diverts all of the available electrons towards production of ethylene and is the basis of the widely used acetylene reduction assay for nitrogenase activity. In natural conditions, *i.e.* in air, electrons are partitioned between the reduction of nitrogen and of protons, giving ammonia (probably immediately protonated to NH_4^+) and hydrogen gas. Note that if 2ATPs are used per e^- transferred, then 12 are required for the reduction of one N_2 and four for the production of one H_2. The subsequent account applies specifically to legume nodules.

Source of reducing power

This has been an enigma for many years until Carter, Rawlings, Orme-Johnson, Becker and Evans (1980) purified a ferredoxin from *Rhizobium japonicum* (which nodulates soybean) and showed that it could reduce the Fe protein. However, we still do not know how the ferredoxin is reduced. One

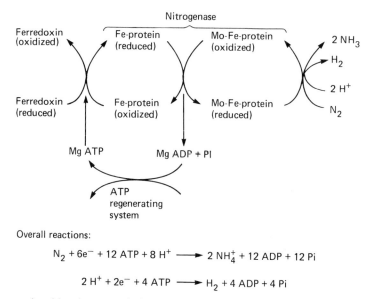

Fig. 12.6 Reactions catalyzed by nitrogenase in bacteroids of root nodules.

suggestion is that an intact bacterial membrane could aid a flow of electrons to ferredoxin if either (1) the membrane potential is sufficiently large, or (2) there is a proton-motive force sufficient to drive reversed electron transport (Haaker, Laane and Veeger, 1980). A combination of (1) and (2) is also possible. The reader is referred to the photosynthesis chapter (11) to see how membrane potential and proton motive force can be used to produce energy (ATP) and to drive electron flow. Although these processes are located in the bacteroids, the necessary carbon substrates come from the host cells.

The major C substrates found in nodules are carbohydrates, polyols (carbohydrates with alcohol side groups) and organic acids. Rhizobia may change their substrate requirement when they differentiate from free-living into bacteroid forms. For example, *Rhizobium leguminosarum* (which nodulates pea) has a glucose uptake system in the free-living state, but bacteroids isolated from pea nodules do not accumulate glucose: instead they have an active succinate-uptake system (Hudman and Glenn, 1980). [It is interesting to note that in the partially anaerobic conditions obtaining in pea nodules, succinate production is enhanced in host cells (de Vries, Veld and Kijne, 1980)]. Some of the variations in uptake properties may be related to

major changes in bacterial cell-wall structure (van Brussel, Planqué and Quispel, 1977). Alternatively, substrate availability may affect morphology: for example, succinate induces bacteroid-like morphology in free-living cultures of *R. trifolii* (Urban and Dazzo, 1982).

Substrate reduction

There is general agreement that the reduction of nitrogen is invariably accompanied by the reduction of protons, a minimum of one H_2 being produced per nitrogen reduced. Whether or not more than one H_2 is produced depends upon the relative proportions of the two subunits of the nitrogenase complex, the redox potential and probably other factors. Although taking place at the same general site on the enzyme, proton reduction is not inhibited by levels of carbon monoxide which inhibit nitrogen reduction. In the absence of nitrogen (when for example incubated in artificial air in which nitrogen is replaced by argon) all the electrons pass to protons. Obviously, the more hydrogen that a nodule produces, the more ATP energy appears to be wasted, in that it could have been more profitably used for nitrogen reduction.

Uptake hydrogenase

Not all nodules show net hydrogen evolution in air. Those that do not have rhizobia which possess an uptake hydrogenase (they are said to be Hup^+, rather than Hup^-). This enzyme oxidizes hydrogen, using gaseous oxygen, with the production of ATP. Estimates of ATP production vary, but assuming a $P/2e^-$ ratio of 3, the actual ATP saved in this way is 3/4 of that wasted in hydrogen production. However, each $2e^-$ used in the nitrogenase reaction might have been available for oxidative phosphorylation and therefore may be considered as equivalent to 3 ATP. Production of one H_2 thus uses the equivalent of 7 ATP (*see* Fig. 12.6). This gives a recovery rate by uptake hydrogenase of only 3/7. Is this significant? Pate, Atkins and Rainbird (1981) produced a matrix of values for mol ATP used per mol N_2 fixed, using a range of proportions of electrons allocated to H^+ and N_2 and a range of values for uptake hydrogenase activity. To take two examples:

(1) One H_2 produced for each N_2 reduced: total cost in ATP equivalents $7 + 21 = 28$. If all the H_2 is recycled we save only 3 of these, *i.e.*, 10.7%.

(2) Two H_2 produced for each N_2 reduced: total cost in ATP equivalents $14 + 21 = 35$. Recycling H_2 could save 6, *i.e.*, 17.1%.

Thus, the benefits of an efficient uptake hydrogenase are rather small if the normal output of H_2 is similar to that of N_2 reduced, but if for any reason more electrons are allocated to proton reduction, then hydrogen recycling becomes relatively more significant in energetic terms.

Many people believe that better crop yields can be obtained by using legumes inoculated with Hup^+ rhizobia (Zablotowicz, Russell and Evans, 1982, for soybean; de Jong, Brewin, Johnston and Phillips, 1982, for pea). Others have been unable to link the possession of Hup^+ with increased fixation of nitrogen (Gibson, Dreyfus, Lawn, Sprent and Turner, 1981). The situation is further complicated by the fact that the expression of Hup^+ may be suppressed in some unknown way by the host genotype. For example, *Rhizobium* strain CB756 is Hup^+; it is expressed in the nodules of *Vigna mungo* but not those of *Vigna radiata* (Gibson and Sprent, unpublished).

Are there any roles for Hup^+, other than saving ATP? Hydrogen is known to inhibit nitrogenase and

Dixon, Bluden and Searl (1981b) consider that H_2 accumulation in nodules may be sufficient to depress nitrogen fixation. Possession of Hup^+ could minimize this problem. A further possibility is that Hup^+ could act as an oxygen-scavenging system, preventing oxygen inactivation of nitrogenase (Dixon, 1972). Clearly, one of the ways in which we can assess these different possibilities is to try to understand how gases diffuse into and out of nodules.

Gaseous exchange in nodules

In most eukaryotic systems gaseous exchange is confined to uptake and evolution of oxygen or carbon dioxide. Nodules have the added problem of nitrogen and hydrogen. The reactions involved are summarized in Table 12.6; they are further complicated by the fact that bacteroids require oxygen for ATP production but that oxygen inactivates nitrogenase. As in all biological systems, water conservation must be considered concomitantly with gas exchange. Because nodules are found in soils varying from arid to waterlogged, the importance of water conservation may also vary. Generally nodules have a small surface-to-volume ratio which in itself tends to conserve water, spherical nodules of the soybean type being the most advantageous in this respect. Here, gaseous exchange with the soil atmosphere is confined to lenticels (*see* Pankhurst and Sprent, 1975a) which are particularly prominent in plants grown under waterlogged conditions and which collapse under water stress (Pankhurst and Sprent, 1975b; *see also* Ralston and Imsande, 1982). Bergersen and Goodchild (1973a) quantified the intercellular spaces in the various parts of soybean nodules, considering there to be a continuous system from central tissue to soil. However (*see* discussion in Bergersen, 1982), Bergersen has long considered there to be an oxygen limitation to soybean nodule nitrogen fixation. We return to the dilemma of admitting to nodules sufficient oxygen to support the respiratory processes of both symbionts without inactivating nitrogenase. There have been three approaches to the quantification of oxygen diffusion into nodules.

(1) Measurement of air spaces This incredibly tedious task has only been faced in detail for soybean nodules (Bergersen and Goodchild, 1973b). Serial

Table 12.6 Gases involved in nodule reactions. The significance of reactions marked* will be considered in later sections.

Gas	Reaction	Comments
Oxygen	Oxidative phosphorylation	Used in host cells and bacteroids via respiratory pathways Associated with Hup^+ in some strains of *Rhizobium*
	Urate oxidation*	Consumed: may regulate ureide synthesis*
Carbon dioxide	Respiratory decarboxylations, Ureide biosynthesis*	CO_2 produced Consumed in one stage, produced in another (Fig. 12.10)
	PEP† carboxylase	CO_2 incorporated into organic acids*
Nitrogen	Nitrogen reduction	Consumed
Hydrogen	Proton reduction	Produced
	Uptake hydrogenase	Consumed

† Phospho-enol pyruvate.

sections were made and photographed under the electron microscope; then the resulting micrographs were cut up into all the cellular and subcellular components and the pieces weighed (*see* Bergersen, 1982, for a discussion of both methods and results). Because soybean nodules have a prominent layer of sclereids in the outer cortex, sections tear readily and there was some doubt as to whether this layer was continuous (*i.e.*, without intercellular spaces) or not. Porosity (*i.e.*, the sum of all intercellular spaces) was measured indirectly by Pankhurst and Sprent (1975b) and found to vary with water status, being less in drought-stressed plants.

(2) Physiological measurements These began with Fraser (1942) who looked at penetration of the redox dye methylene blue into nodules. This dye is reduced to a colorless leukoform, but is readily re-oxidized by gaseous oxygen. Fraser concluded that free-oxygen levels were very low inside nodules, because of a restriction at the endodermis, a conclusion supported by the observations of Tjepkema and Yocum (1973) using an oxygen microelectrode. From measurements of uptake of oxygen from various external concentrations and at different levels of water stress, Pankhurst and Sprent (1975b)

also concluded that there is a diffusion barrier in soybean nodules.

(3) Theoretical studies Sinclair and Goudriaan (1981) set up a physical/mathematical model of a spherical (determinate) nodule and considered how simple diffusion in the gas phase may limit activity. They concluded that such nodules could only function at the observed rates if there was an adequate intercellular space system. However, a restriction to purely gaseous diffusion, equivalent to a liquid layer of about 45 μm in thickness, was necessary to maintain a low level of oxygen around the bacteroids.

The exact location of the hypothetical barrier to diffusion is still not known. Apart from the sclereid layer or endodermis, Sutherland and Sprent (unpublished) have observed very close packing in the cambial layer (equivalent to the phellogen) which gives rise to the lenticels. Collapse of lenticels (Ralston and Imsande, 1982), together with reduced porosity (Pankhurst and Sprent, 1975b) could further impede gaseous diffusion under water stress where the need to conserve water may override the need to maintain full metabolic activity. This would be equivalent to the closing of stomata on leaves, which reduces water loss but curtails photosynthesis.

Sheehy, Minchin and Witty (1983) have suggested a variable resistance to oxygen diffusion in some nodules.

Less evidence on gaseous diffusion is available for other species. Dixon *et al.* (1981b) examined lupins (*L. alba*) and peas using both a sectioning method and penetration of intercellular spaces by India ink in detergent solution. They concluded that the surface of these nodules is studded with intercellular spaces which are continuous with spaces in the outer cortex. Internal to this, any intercellular spaces form separate, discrete groups. Using appropriate differential equations they concluded that diffusion of gases into pea and lupin nodules was determined solely by the aqueous phase. Their concept of separate areas of infected tissue, each with its own discrete intercellular space system led to the idea of a build-up of hydrogen levels sufficient to inhibit nitrogenase. This concept has now been extended to soybean (Dixon, Berlier and Lespinat, 1981a), in apparent contradiction to the ideas of Sinclair and Goudriaan (1981). There is no obvious resolution of this conflict. One possible solution is that the size and internal structure of the nodules were different as a result of different environments during growth. This suggestion is supported by the fact that Dixon *et al.* (1981a) give respiratory quotients (CO_2 evolved/O_2 taken up) for soybean consistently less than unity whereas Pankhurst and Sprent (1975b) and Sprent and Gibson (unpublished) have found values consistently in excess of unity. In the latter cases the nodules were considered to be partially anaerobic, a suggestion backed by the stoichiometric production of ethanol under certain conditions (Sprent and Gallacher, 1976). Dixon *et al.* explained their low R.Q.s as resulting from partial oxidation of glucose and removal of products prior to decarboxylation steps, for the purpose of nitrogen assimilation (*see* below). No-one has considered PEP carboxylase, known to occur in nodules, as a complicating factor!

This leads us on to a further major point, namely that one cannot consider single gases in isolation. Obviously if both oxygen uptake and carbon dioxide evolution are measured, a better idea of respiratory processes is obtained than if only one of these is considered as is often the case. Even so we still have to consider the extent to which oxygen uptake may be involved in hydrogen oxidation (in *Hup*[+] systems) and CO_2 in carboxylation reactions. Both

gases may be involved in the ammonia-assimilatory machinery (*see* below).

All of the data discussed above have been obtained with plants from mesic environments. We know nothing of the gas-exchange problems of nodules native to arid areas or to waterlogged environments. The extreme case of the latter is the aquatic genus *Neptunia*, a native of the river Amazon which forms submerged nodules which are entirely covered with aerating tissue.

The role of leghemoglobin

The correlation between pink coloration in legume nodules and their effectiveness in nitrogen fixation has been known for many years and it was in 1939 that the coloration was shown to be due to hemoglobin (Kubo, 1939). Bergersen (1982) has written an entertaining history of the development of knowledge on this subject.

Leghemoglobin, as the pigment is now most commonly called, is a symbiotic compound *par excellence*, the two parts of the molecule, apoprotein and heme, being coded for by plant and rhizobial DNAs respectively. It appeared for some years that these two components formed the active complex in a region developmentally extracellular to both partners, namely within the peribacteroid membrane envelope (since this develops from invaginations of the host plasmalemma, its contents are really an extracellular compartment). Recent evidence indicates that this is not entirely true; it appears that most of the leghemoglobin is located in the plant cell cytoplasm but some, possibly of considerable physiological significance, bathes the bacteroids within their envelopes.

There is now overwhelming evidence that leghemoglobin acts, as does hemoglobin in blood, as an oxygen-carrying pigment. As already described, one of the nodule's chief dilemmas is to allow sufficient oxygen to reach the bacteroids for ATP synthesis without causing inactivation of nitrogenase. What is needed is a high *flux* of oxygen at a low *concentration*. The oxygen-binding properties of leghemoglobin achieve this.

There are several distinct protein moieties in leghemoglobins from the same and from different species; the proteins differ in amino-acid sequence and oxygen-binding properties, and each is coded by a different host gene. Within individual nodules, the ratio of different components varies with age (Uheda and Syono, 1982a), and it is suggested that this may result in an improvement in oxygen-carrying capacity as nodules age. It would be interesting to see whether or not these changes are coupled with the changes in resistance to gaseous diffusion outlined earlier. In another paper, Uheda and Syono (1982b) reported that they had tested leg-hemoglobins from both pea and soybean for their ability to transport oxygen to bacteroids. They do not say whether they did the reciprocal experiment, *i.e.*, soybean leghemoglobin with pea bacteroids and *vice versa*. In view of the differences in bacteroid structure between these two species (Table 12.4), and the fact that bacteroids in soybean occupy a smaller fraction of the total envelope volume than do those of pea, their oxygen-carrying requirements may well vary in degree. Further, the influence of the host, which codes for the variable protein part of the molecule, in determining bacteroid shape and possibly oxygen tolerance (Sen and Weaver, 1980, 1981), suggests that there may be quantitative differences in oxygen-carrying requirements in the liquid phase of nodules as well as in the gaseous phase. These suggestions are consistent with the co-evolution of nodules of different shapes and physiology (*e.g.*, most ureide exporting nodules are determinate) which, as far as the available evidence goes, parallels the evolution of specific leghemoglobins (Jing, Paau and Brill, 1982).

Leghemoglobin is only the most obvious of a number of nodule-specific molecules. Host genes are known to code for 18–20 proteins tentatively termed 'nodulins' which are only produced in effective nodules (Auger and Verma, 1981). Their role is, as yet, obscure, although one has been identified as uricase (*see* Fig. 12.10, p. 265) in soybean.

Ammonia-assimilating pathways

It is general dogma that ammonia must be assimilated quickly: in nitrogen-fixing systems there are two reasons for this, the first is that ammonia is toxic and the second that ammonia represses nitrogenase synthesis. Although there is some evidence that rhizobia can metabolize high concentrations of ammonia (Dilworth and Glenn, 1982), their host cells probably cannot. The synthesis of nitrogenase

in rhizobia may be rather more resistant to ammonia than in other organisms and in any case, at least in indeterminate nodules, there may be little concurrent, co-located nitrogenase synthesis and nitrogenase activity. There is considerable evidence that, as in leaves and roots, the major pathway of ammonia assimilation is into the amide group of glutamine, using the high affinity glutamine synthetase (K_m about 2×10^{-5} M) rather than the lower affinity glutamic dehydrogenase system (Fig. 12.7). The enzymes are located mainly in the host cells. In other words, once the bacteroids have reduced nitrogen gas, the ammonia is handed on to the host cell for assimilation—a good division of labor between the symbionts. Free-living rhizobia do not normally reduce nitrogen, although some may be induced to do so under special cultural conditions (*e.g.*, Pagan, Child, Scowcroft and Gibson, 1975). This is because the induction of nitrogenase genes is coupled to the repression of genes for ammonia assimilation.

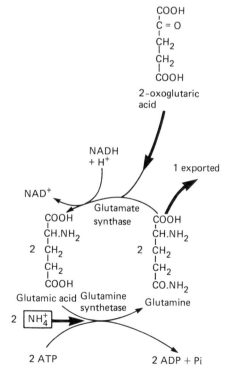

Fig. 12.7 Incorporation of ammonium into glutamine as in legume nodule host-cell cytoplasm.

Amide synthesis

Temperate legumes export from their nodules principally the amides glutamine and asparagine. The normal assimilatory pathway couples glutamine synthetase with glutamate synthase, as outlined in Fig. 12.7. The bold arrows indicate how net glutamine synthesis may occur at the expense of one molecule each of 2-oxoglutarate and reduced pyridine nucleotide and two of ATP. NADH + H[+] may be regarded as 3 ATP equivalents, since it might otherwise be used in oxidative phosphorylation. Synthesis of asparagine in lupin, and probably other legumes occurs as shown in Fig. 12.8 (Scott, Farnden and Robertson, 1976). Here there is effectively a net utilization of one molecule of oxaloacetate for the production of one molecule of asparagine, but with the use of 2 ATP equivalents (ATP → AMP + Pi being energetically equivalent to 2ATP → 2ADP). However, there is a net saving of one reduced carbon, i.e., $=CH_2$. The export product has a better C:N ratio (4:2) than glutamine (5:2). Being zwitterions, the amides carry little net charge at xylem pH and therefore do not need an accompanying counterion for transport.

Ureide synthesis

Recently it has been found that a number of legumes of tropical and subtropical origin, such as soybean, *Phaseolus*, and *Vigna*, export the ureides allantoin and allantoic acid (Fig. 12.9) (*see* reviews by Sprent, 1980, and Thomas and Schrader, 1981). Current evidence favors the following pathways for ureide

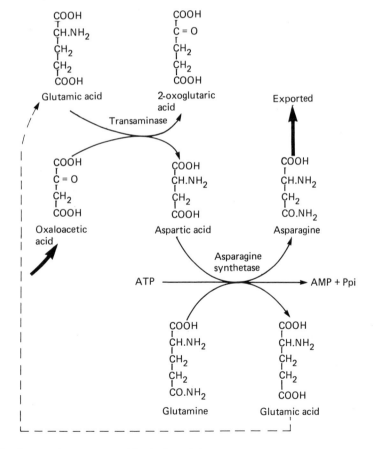

Fig. 12.8 Synthesis of asparagine as proposed for lupin nodules.

synthesis in host cells (*see* Fig. 12.10). The initial assimilation of ammonia is into glutamine and glutamate (Fig. 12.7), and then transamination reactions either directly or indirectly yield the other purine precursors glycine and aspartate. The process of purine synthesis appears to be broadly similar to that in other organisms. The likely origins of the various atoms of the heterocyclic ring are indicated in Fig. 12.9.

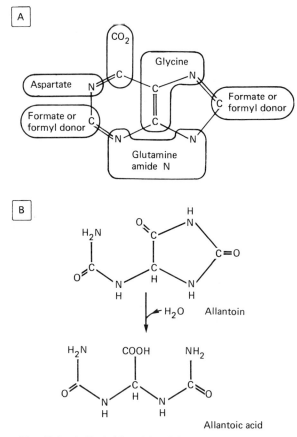

Fig. 12.9 A. Probable origin of C and N in skeleton of purine ring; B. Structure of the ureides allantoin and allantoic acid.

Whereas the amounts of ATP energy, or its equivalent, used in synthesizing amides are well established, there are several incompletely understood reactions in purine synthesis which may affect the total amount of ATP used. For example, the amide group of glutamine may be transferred to phosphoribosyl pyrophosphate (PRPP), made in the host cytosol from ribose-5, phosphate and ATP ($\rightarrow$AMP) forming phosphoribosylamine. This is the basic building block of the purine ring. The relevant enzyme is probably located in the host cell proplastids (Boland, Hanks, Reynolds, Blevins, Tolbert and Schubert, 1982). However, direct addition of ammonium to ribose phosphate, utilizing 1 ATP$\rightarrow$ADP cannot be ruled out (Susuki and Takahashi, 1977): this could save one of the two ATP equivalents used in making PRPP and the ATP required for incorporating the amide group into glutamine (R. J. Thomas, personal communication, *see also* Atkins, Ritchie, Rowe, McCairns and Sauer, 1982).

The reactions up to the formation of free purine, xanthine, are necessary for nodule growth as well as to produce export products. Subsequent reactions are specific to the formation of export products: this is indicated by the high correlation between the activities of the relevant enzymes and nitrogenase (Atkins *et al.*, 1982; Reynolds, Boland, Blevins, Schubert and Randall, 1982). NADH is produced during xanthine oxidation to uric acid (Fig. 12.10)—if used in oxidative phosphorylation, this could result in the synthesis of 3 ATP. The conversion of uric acid to allantoin is an interesting reaction for several reasons. First, it appears to be located in the microsomal fraction (probably peroxisomes) of the interstitial cells of the infected region of nodules (Newcomb and Tandon, 1981). Interstitial cells had previously been suggested to have a functional role, because of the large number of plasmodesmatal connexions between them and their neighboring infected cells (Sprent, 1971), but this is the first definite reaction to be assigned to them. Second, the level of activity may be controlled by oxygen because urate oxidase has a low affinity (high K_m) for oxygen (Rainbird and Atkins, 1981). Thus, if nodules are short of oxygen, ureide synthesis may be curtailed. Even under normal conditions, oxygen may be a major factor affecting ureide production.

The ureides allantoin and allantoic acid were assayed as a group by many of the earlier workers in the field, who tended to refer to the sum of the molecular species as 'allantoin'. Sprent (1980) pointed out that both allantoin and allantoic acid are relatively insoluble compounds and under cool conditions this low solubility could restrict export from nodules. The solubility of potassium allantoate is

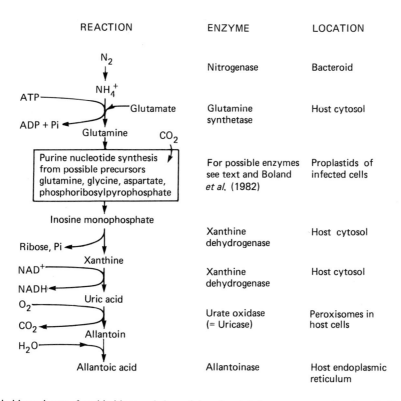

REACTION	ENZYME	LOCATION
N_2	Nitrogenase	Bacteroid
NH_4^+		
ATP → Glutamate	Glutamine synthetase	Host cytosol
ADP + Pi ← Glutamine		
Purine nucleotide synthesis from possible precursors glutamine, glycine, aspartate, phosphoribosylpyrophosphate	For possible enzymes see text and Boland *et al.* (1982)	Proplastids of infected cells
Inosine monophosphate	Xanthine dehydrogenase	Host cytosol
Ribose, Pi ← Xanthine	Xanthine dehydrogenase	Host cytosol
NAD⁺ → NADH ← Uric acid	Urate oxidase (= Uricase)	Peroxisomes in host cells
O_2 → CO_2 ← Allantoin		
H_2O → Allantoic acid	Allantoinase	Host endoplasmic reticulum

Fig. 12.10 Probable pathway of ureide biogenesis in nodules of certain legumes, generally of tropical/subtropical origin.

much higher and analyses of xylem sap suggest that this compound may account for up to half the ureide exported.

It is interesting to consider why some legumes export ureides and others amides. The most commonly attributed advantage to ureides is a very favorable C:N ratio, namely 4:4; they thus sequester a much smaller fraction of plant carbon in export products. However, analysis of the C:N ratio of xylem sap of soybeans gave a value in excess of 2, even when 83% of the N compounds were ureides (Israel and Jackson, 1982). This is largely because of high levels of malate in the sap. The significance of organic acids in sap will be discussed later, but it is pertinent to note that the C:N ratio of the nitrogenous export product alone may be misleading.

Synthesis of ureides involves far more steps than synthesis of amides and the energy cost of producing and maintaining the necessary enzymes is quite unknown. Attempts have been made to calculate the ATP balance of the corresponding reaction sequences and generally production of the two types of compound is thought to require approximately the same number of ATP equivalents. The metabolism of amides and ureides when they arrive at leaves or pods is quite different and will be considered later. Whether ureides have any competitive advantage in natural habitats (for example, xylem-sucking insects, as far as the author can ascertain, cannot live on ureides) remains to be seen.

Other compounds

Many nodulated non-leguminous plants such as *Alnus* spp. export citrulline which is often classified as a ureide; others, such as *Myrica* export amides (*see* Sprent, 1979). Nodules from plants of dry habitats contain a wide range of nitrogen compounds, some of which may be significant in osmotic adjustment (Rudelier, Goas and Larher, 1982).

Relationship with photosynthesis

Ultimately, the carbon compounds used by nodules derive from photosynthesis. The extent to which nodules are dependent upon newly-made photosynthate depends upon plant species, age and growing conditions. Generally, field-grown material is better buffered against short interruptions of photosynthate supply than laboratory-grown material. This is reflected, for example, in the less-pronounced diurnal rhythms of nitrogenase activity found in field-grown plants (*see* Table 9 in Minchin, Summerfield, Hadley, Roberts and Rawsthorne, 1981).

Several lines of evidence have indicated a longer term photosynthate limitation to nitrogen fixation. For example, in soybean, nitrogen fixation (acetylene reduction) is increased by supplementary lighting placed within the leaf canopy (Lawn and Brun, 1974) and by grafting two shoot systems on to one rootstock (Streeter, 1974). Thus it is possible that genotypes with higher rates of photosynthesis may be able to support higher rates of nitrogen fixation. In a study with field grown peas, Mahon (1982) looked at growth, carbon dioxide exchange rate per unit leaf area (CER) and acetylene-reducing activity in six genotypes previously selected for high or low CER. He concluded that nodule activity was more closely related to leaf area than to CO_2 exchange *per se*. Because the genotypes selected for low CER had a greater leaf area index (m^2 leaf per m^2 land surface) the nodule activity per unit area of land was greater in the plots containing plants with the lower CER. This somewhat surprising result indicates the complexity of the relationships between photosynthetic structures and nodules. As discussed in detail by Mahon (1982), these relationships will need to be understood much better before it can be decided which characters it is desirable to select in programs for breeding plants with better photosynthesis and nitrogen fixation.

Many nodules have been shown to possess PEP carboxylase activity (Lawrie and Wheeler, 1975; Christeller, Laing and Sutton, 1977), and $^{14}CO_2$ incorporated using this enzyme contributes significantly to the organic acids used as acceptors of fixed N (ammonium). It is thus often considered to effect a significant saving in photosynthate. Clearly it conserves carbon, but since this carbon is fully oxidized, it is not saving energy in terms of potential ATP production, which is associated with the oxidation of reduced carbon.

Obviously, any environmental factors which affect photosynthesis will ultimately affect nitrogen fixation, in addition to any effects they may have directly on nodule physiology. For further discussion of these factors, *see* Sprent, Minchin and Thomas (1983).

Effects of environment on nitrogen fixation

Factors affecting distribution of rhizobia in soil and on the processes of infection and nodulation will not be considered here: the reader is referred to Sprent *et al.* (1983) and the references therein. All the environmental factors listed in Table 12.7 also affect nodule growth, often differentially affecting the processes of cell division and enlargement as briefly considered in the section on formation of nodules. Extremes of soil pH and supplies of essential nutrients such as P, K, Mo and Fe affect both nodule development and physiology. In all species studied,

Table 12.7 Some effects of environment on the physiology of nitrogen fixing root nodules.

Factor	Effect
Temperature	N$_2$ fixation and/or assimilation Gaseous diffusion and solubility Respiratory activity
Water stress	Direct effects on nodules including reduced porosity affecting O$_2$ uptake Indirect effect of reduced supply of photosynthate ? effects on transport to and from nodules
Waterlogging	Generally by anoxia May get adaptation
Salinity	Reduced nitrogenase activity Inhibition of leghemoglobin synthesis
Combined nitrogen	Diversion of photosynthate to regions assimilating combined N Inhibition by nitrite formed from nitrate reduction in bacteroids

nodules with indeterminate growth have better recovery potential than those with determinate growth because they can resume meristematic activity and hence make new nitrogen-fixing tissue.

Temperature

Optimum temperature curves for nitrogen fixation, generally measured by the acetylene reduction assay, have been drawn for many species. Generally for any one host–rhizobial combination, maximum rates of activity are obtained over a temperature range of about 10°C. The actual values vary widely, usually around 15–25°C for temperate and 25–35°C for tropical species. The optima are broader than those found for single enzymes *in vitro*, reflecting the complexity of the nitrogen-fixing process. Table 12.7 lists three of the ways in which temperature may affect overall nodule activity. Not all of these will have identical responses: for example, a rise in temperature from 10° to 15°C will generally increase the rate of an enzyme catalyzed reaction, but decrease the solubility of oxygen. Under oxygen-limited conditions these two responses will act in opposition. It is obviously useful to the plant to be able to continue to fix nitrogen at approximately maximal rates over a wide temperature range, but there may be penalties to pay, in that the process may not use resources equally efficiently over the whole range. For example, the proportion of electrons used in proton reduction may vary with temperature (Fyson, 1981).

For the temperate forage legume *Trifolium repens* (white clover), Masterson and Murphy (1976) concluded that soil temperature was the environmental factor having the greatest effect on nitrogen fixation and nodule growth. However, in spite of the fact that the isolated nitrogenase complex may be unstable at temperatures near 0°C, measurable rates of acetylene reduction may be obtained in nodulated plants down to 0.5°C (Fyson, 1981). This means that active enzyme molecules may survive cold periods, so that as soil warms, nodule activity and growth may be resumed, sometimes even before measurable plant growth can be detected (Masterson and Murphy, 1976).

Water stress

Water stress has been reviewed by Sprent (1976). Its effects, like those of temperature, are complex. Under conditions of photosynthate limitation, nitrogen fixation may fall rapidly following stomatal closure, as found by Huang, Boyer and Vanderhoef (1975) with soybean plants. However, by maintaining a water supply to nodulated roots at the same time as exposing the shoot system to a very low relative humidity, Sprent and Gibson (unpublished) were able to reduce photosynthesis in soybean to a very low level without affecting nodule activity. Direct effects on the nitrogen-fixing process have been found in a number of species (*see* Sprent, 1976). The possibility that concentrations of nitrogen-fixing products may build up when transpiration is reduced may further complicate the issue (Minchin and Pate, 1974). Under field conditions water stress may seriously reduce nitrogen fixation and comparative studies are now being carried out to examine the sensitivity of different species and lines. For example, in *Medicago sativa*, Aparico-Téjo, Sanchez-Diaz and Peña (1980) found that, at leaf water potentials of −1.5 MPa, the cultivar Tierro de Campos reduced acetylene at about 50% of its maximal rate whereas the cultivar Aragon showed virtually no activity.

Waterlogging

In some plants, for example *Pisum sativum*, this may be a potentially more severe limitation to nitrogen fixation than water stress (Minchin and Pate, 1975), although the related species *Vicia faba* is more tolerant (Gallacher and Sprent, 1978). Intermittent waterlogging reduces the oxygen supply to the nodules and is therefore inhibitory to nitrogen fixation. Prolonged exposure to wet soils may result in the formation of nodules with more air spaces and thus improved tolerance to waterlogging.

Salinity

Considering the proportion of soils in the world which are saline, remarkably little effort has been expended on examining the effects of salinity on nitrogen fixation. Generally, nodulated crop plants do not like saline conditions, although at low salt levels (25 mol m^{-3} NaCl) they may show compensation by producing larger nodules (Yousef and Sprent, 1983). It is unfortunate that many legumes grown in salt-affected soils, for example *Vicia faba* in the Middle East, are not tolerant of salinity. *Glycine wightii*, a forage legume, shows some salt tolerance, although plants grown on combined nitrogen are more tolerant than those dependent on fixed nitrogen (Wilson, 1970).

Combined nitrogen

The reasons why combined nitrogen reduces nodulation as well as activity of preformed nodules have been studied extensively. There have been two schools of thought: (a) that photosynthate is diverted away from nodules towards areas where combined nitrogen is being assimilated, and (b) that in the case of nitrate, bacteroid nitrate reductase produces nitrite in concentrations high enough to inhibit nitrogenase activity. There is evidence for and against both of these hypotheses (Sprent *et al.* 1983). A further suggestion, made recently by Streeter (1983), is that nitrate reduction in the host cells of nodules may be the controlling reaction.

Comparison of nitrogen fixation and assimilation of combined nitrogen

When nitrogen-fixing plants have adequate mineral nitrogen available (nitrate, ammonium), the processes of nodule formation and nitrogen fixation are slowed down or completely stopped. The exception to this is that small doses of 'starter' nitrogen may promote nodulation and hence the amount of N fixed. Since plants apparently fix nitrogen only when they must, one intuitively thinks that the process must be more expensive in some way(s) than to take up and incorporate mineral nitrogen. Many comparisons between nitrogen-fixing and non-fixing plants of the same legume species have been made in an attempt to find out why the plants prefer combined nitrogen. These comparisons are usually centered on the carbon economy of the plants, balancing photosynthesis in the shoots against respiration of roots, shoots and nodules and considering the distribution of dry matter among various plant organs. Most of these comparisons have been made with crop plants which have been selected for yield when grown on plentiful supplies of nutrients, including combined nitrogen. It is thus possible that there has been unconscious selection for those plants which grow best on combined nitrogen. There are several physiological differences between plants grown on N_2, nitrate or ammonium fertilizers which may affect the final yield of a crop (Table 12.8).

Formation of specific organs

Production of nodules obviously requires expenditure of energy in terms of C and N compounds before any benefits can accrue and this may be a big disadvantage for a legume growing without any soil nitrogen. The problem is particularly acute in young seedlings and its magnitude almost certainly depends on the size of seed reserves and the rates of growth of different plant parts during the early

Table 12.8 Ways in which plants grown on different forms of N may have different energy requirements.

	N source		
	N_2	NO_3^-	NH_4^+
Formation of a specific organ	Nodule	None	None (mycorrhizas)
Uptake	Passive (diffusion)	Active (requires also ion exchange or uptake of counterion)	Active
Site of assimilation	Nodule	Root or shoot	Root
Requirement for organate in xylem	Yes	Yes, if reduced in root	Yes
pH regulation required for	Slight net H^+	Net OH^-	Net H^+
Form of N sent to shoots	Amides or ureides	Amides, NO_3^-	Amides

seedling stage. To take extremes, the relatively large-seeded species such as *Vicia faba* can nodulate in the dark. Small-seeded species, such as *Trifolium repens* (white clover), need newly-formed photosynthetic products. Under some conditions, generally those which favor high shoot-growth rates, a definite nitrogen hunger phase may be observed (*e.g.*, in soybeans). Plants literally go from pale to dark green overnight, when nitrogen fixation begins. These are conditions under which starter nitrogen may be particularly useful (*see* Sprent *et al.*, 1983).

Are there any morphological changes associated with the assimilation of mineral N? This question has rarely been directly addressed. There is some evidence that soil nitrate stimulates production of lateral roots—this would lead to increased area for ion uptake. On the other hand, nodulated plants of clover have larger root-to-shoot ratios than plants grown on nitrate (Ryle, Powell and Gordon, 1981). In some soils, mycorrhizas may be of great benefit in ammonium uptake (Raven, Smith and Smith, 1978).

Nitrogen uptake

The problem of gaseous exchange in nodules has already been dealt with. There is no evidence that N_2 gas is ever limiting (Sinclair and Goudriaan, 1981). Even with its low solubility, diffusion from an atmosphere containing nearly 80% N_2 to an enzyme in whose vicinity the partial pressure is low is likely to be more than adequate. Nitrate and ammonium uptake, on the other hand, frequently occur against a concentration gradient and involve the expenditure of energy (*see* Chapter 15). However, compared with the energy involved in making root nodules this can probably be disregarded. The problem of counterions will be discussed later.

Site of assimilation

Ammonium ions taken up from soil are assimilated in root cells by the same pathway that nodules use for newly-fixed N, except that the products exported are almost exclusively amides. Nitrate may be assimilated in roots, leaves or even pods. In roots, amides are the chief exported product, although small quantities of ureides may be produced in species such as *Phaseolus vulgaris* (Thomas, Feller and Erismann, 1980). When either nitrate or ammonia is assimilated in roots, unlike nodules, the amide balance is towards glutamine rather than asparagine.

Ion balance

When the supply of nitrogen leaving roots for shoots in xylem sap is in an organic form the plant (legume or otherwise) has a problem of ionic balance. Such plants generally take up more cations (K^+, Ca^{2+}, Mg^{2+} etc.) than anions ($H_2PO_4^-$, SO_4^{2-}, Cl^-). Since the nitrogenous export products (with the exception of allantoic acid) carry very little net charge at xylem pHs, the plant must synthesize anions to balance the excess cations in the sap. In those species which have been examined, strong organic acids are produced. These ionize into organate$^-$ and H^+. The latter may be disposed of into the soil, sometimes in exchange for the cations taken up, and the organate travels in the xylem. Israel and Jackson (1982) found that 70% of the excess cations taken up by nodulated soybean plants was balanced by malate. Plants which transport nitrate to the leaves for reduction do not have this problem: they may even have an excess of anions, although this is unlikely in natural conditions. They have the added potential advantage of using ATP produced by photophosphorylation as a source of energy for nitrate assimilation rather than respiratory ATP as in root nitrate assimilation.

pH regulation

As noted in Table 12.8, the assimilation of nitrogen in various forms to the level of amide results in the production of either excess H^+ or OH^-. The reasons for this are summarized in Table 12.9 which is based on the arguments of Raven and Smith (1976). In order to balance pH discrepancies, plants employ two basic strategies, (1) the biophysical and (2) the biochemical pH stats (Fig. 12.11). The former is applicable to aquatic plants (when water acts as the sink) and root- or nodule-based processes (where soil may be used as sink). The biochemical pH stat is also a possibility for root- and nodule-based processes, but it is the only one available when nitrate is reduced in leaves. Here excess OH^- is neutralized by H^+ from strong organic acids such as malate or oxalate, made from neutral precursors. Oxalate may be precipitated as its insoluble Ca^{2+} salt, whereas malate is usually transported in the

Table 12.9   Reasons why pH regulation is needed as a result of N assimilation (*see* Raven and Smith, 1976, for full details).

Basic facts:
(1) On average, plants have a net negative charge as a result of their cell wall and protein components.
(2) The net negative charge in cell proteins is <1 per N. Therefore, the manufacture of protein from inorganic N generates positive or negative charge as follows (expressed per N):

N source and charge		Charge produced in making protein	
		Theoretical	Observed
$0.5N_2$	0	$<1\ H^+$	0.1–0.2
NH_4^+	$1\ H^+$	$>1\ H^+$	about 1.1
NO_3^-	$1\ -ve$	$<1\ OH^-$	0.5–0.9

phloem (together with K^+) to roots, where it may be further processed so that, in effect, the OH^- is passed into the soil. Thus, soil pH may be altered according to the nitrogen nutrition of the plants growing on it. The buffering capacity of some soils is sufficient to ensure that large pH changes do not occur, but in others, especially those with low base-exchange capacity, the problem may be acute enough to affect availability of essential elements. For example, acidification of soil can reduce the availability of Mo, a constituent element of nitrogenase, nitrate reductase and xanthine dehydrogenase; it can also have the positive effect of leading to the release of insoluble phosphates. Legumes may acidify soil far in excess of the values predicted from the ammonium-assimilation processes, because of exchange of H^+ for essential cations (Griffiths, 1982). Although pH regulation may be costly, especially when organic acids are produced, in whole plant terms costs may not be very significant (Raven and Smith, 1976).

Fate of N assimilated in shoots

This is another area of uncertainty. Where glutamine is exported, the amide group may be transferred to oxoglutaric acid by glutamate synthase (Fig. 12.7). Transamination of the amino group of the two glutamates so produced can lead to the production of a variety of amino acids. There is no such reaction known for asparagine; the amide group is split off by asparaginase and reassimilated into glutamine and glutamate. The amine group of asparagine is amenable to transaminations (Miflin and Lea, 1976, have reviewed these processes). Incorporation of the two nitrogens of asparagine into protein thus requires one more ATP than incorporation of the two nitrogens of glutamine (compare Figs 12.7 and 12.8). In the case of ureides, there are various possible pathways (Thomas and Schrader, 1981), but the limited evidence available favors their breakdown into urea and glyoxylate, and the further breakdown of urea into carbon dioxide and ammonia. If this is the case, all the nitrogen in ureides has to be reassimilated in shoot tissues. In whole plants terms, the energy cost of these reactions should be considered in association with the cost of producing the relevant compounds in nodules. It may be no accident that ureide-exporting legumes come largely from tropical and subtropical areas where excess ATP from photophosphorylation may be available for the reassimilation of ammonium. It is also interesting to note that the same species, when supplied with nitrate send most of it to the leaves for assimilation, whereas many of the temperate legumes reduce nitrate principally in roots. Therefore, light energy should not be forgotten when considering the overall costs of nitrogen assimilation. This has now been done for nitrate (Raven, 1984).

Comparison of a legume with a cereal

An alternative way of comparing nitrogen fixing and non-fixing plants is to look at the energy inputs and outputs of a grain legume and a cereal at the farm level (Table 12.10). In large areas of the USA, alternate crops of soybean and maize (corn) are grown. The farms are large and highly mechanized, so that labor components of costing are small, but there is a large energy input in terms of fuel for machines, herbicides and insecticides, as well as seed and fertilizer. Both the inputs and outputs are much larger for maize than soybean, but the ratio of the two is in the same general range over the four States for which data are cited. However, maize is about 8.9% protein whereas soybean averages 34% protein; in terms of protein output per total input soybeans are way ahead of maize. When the total N in the harvested grain is compared with the N applied, again soybean (which relies largely on fixed

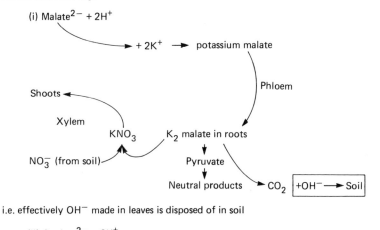

1. (a) Nodulated legumes and non-legumes

   (b) Plants assimilating NH_4^+
   (always in roots)

   $\boxed{H^+ \longrightarrow \text{soil} \\ \text{(pH falls)}}$

2. Plants assimilating NO_3^-

   (a) In roots: OH^- produced

   (i) $\boxed{OH^- \longrightarrow \text{soil (pH rises)}}$

   and/or (ii) OH^- neutralized by H^+ from strong acid made from neutral precursor (e.g. sucrose $\longrightarrow$ malic acid) in roots

   (b) In shoots, mainly leaves
   OH^- must be neutralized by H^+ from strong acid made from neutral precursor e.g.

   (i) $Malate^{2-} + 2H^+$

   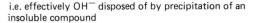

   $+ 2K^+ \longrightarrow$ potassium malate

   Shoots

   Xylem

   KNO_3 K_2 malate in roots

   NO_3^- (from soil)

   Phloem

   Pyruvate

   Neutral products $\longrightarrow CO_2$ $\boxed{+OH^- \longrightarrow \text{Soil}}$

   i.e. effectively OH^- made in leaves is disposed of in soil

   (ii) $Oxalate^{2-} + 2H^+$

   $\longrightarrow$ Vacuoles of leaf cells, where $+ Ca^{2+} \longrightarrow$
   crystals of calcium oxalate

   i.e. effectively OH^- disposed of by precipitation of an
   insoluble compound

Fig. 12.11 Examples of pH regulation. Disposal of ions into soil represents the biophysical pH stat, formation of organic acids the biochemical pH stat.

N) is superior. It is interesting to note that in Illinois, where no fertilizer N was given to soybean, the yield was highest, but in Georgia, where most fertilizer N was given, yield was lowest. This suggests that, given the right conditions, fixed N may be sufficient for soybeans. As far as corn is concerned, between 13 and 32% of the N applied was recovered in the crop. This represents an enormous waste of fertilizer N as well as possible trouble in terms of nitrate run-off into water supplies.

Is there a future for legumes?

Much has been made of the fact that grain legume yields have not increased nearly as much as cereal yields in recent years and upper limits to yields have been proposed, based on hypothetical 'self-destructive' tendencies (Sinclair and de Wit, 1975). Should we therefore continue to grow legumes? We must remember their high protein value (and oil content in some cases) and that their amino acid balance

Table 12.10   Energy inputs and outputs (MJ ha^{-1}) for maize (M) and soybean (S) in four of the United States of America. Data recalculated from Pimental (1980).

		Illinois	Ohio	Nebraska	Georgia
INPUT					
Seed, fertilizer, insecticide, herbicide	M	14 690	12 291	9 357	12 105
	S	4 199	4 481	4 076	8 320
Machinery, transport	M	10 955	8 291	11 396	12 303
	S	6 621	5 033	11 415	6 920
OUTPUT					
Grain	M	116 613	86 366	48 345	50 466
	S	43 749	33 552	37 186	27 999
Output/input	M	4.55	4.20	2.33	2.07
	S	4.46	3.53	2.40	1.84
Protein/input	M	0.465	0.429	0.238	0.211
	S	1.508	1.194	0.780	0.622
N in crop (kg ha^{-1})	M	44.6	33.0	18.5	21.7
	S	55.3	42.4	45.1	35.4
N added (kg ha^{-1})	M	140.2	112.1	84.1	136.8
	S	0.0	5.3	5.6	27.0

complements that of cereals. The traditional rice and bean diet of South Americans testifies to the healthiness of this mixture. Also with increasing costs, and problems of run-off of fertilizer from agricultural land into domestic water supplies, the growth of legumes should be encouraged (Stewart and Rosswall, 1982). Modern methods of genetic engineering and biotechnology are beginning to open up the possibility for rapid improvements of both partners in the legume symbiosis. We now know, for example, that the *nif* and *Hup* genes, together with factors leading to successful nodulation, are borne on plasmids (*i.e.*, extrachromosomal DNA). Techniques are being developed for crossing rhizobia so that potentially desirable traits (such as *Hup*$^+$) may be routinely incorporated into strains produced for soil inoculant programs. If we can modify legumes so that farmers can rely on them as much as they do on crops grown on fertilizer nitrogen, they are likely to be more widely grown in the developed world. The total N fixed by legumes in the field is difficult to assess accurately (LaRue and Patterson, 1981) and estimates vary from 10 to 600 kg ha^{-1} per year, depending on species and environmental conditions. These estimates are based on seasonal growth patterns. In humid tropical areas, where growth can continue all the year

round, values of 1 tonne are not impossible, provided that pests and diseases can be controlled, and plants can afford to devote to N fixation the necessary 15 or more tonnes of dry matter.

From a less sophisticated, but very practical point of view, examination of how legumes adapt to extreme environments could lead to an extension of those areas where these crops may be grown. An admirable start has been given by the US National Academy of Sciences in the production of their book *Tropical Legumes; Resources for the Future.*

Further reading

Allen, O. N. and Allen, E. K. (1981). *The Leguminosae*, Macmillan (Europe and British Commonwealth except Canada), and University of Wisconsin Press (USA, Canada and Japan).

Broughton, W. J. (ed.) (1982). *Nitrogen Fixation*, vol. 2, Blackwell, Oxford.

Gibson, A. H. and Newton, W. E. (eds) (1981). *Current Perspectives in Nitrogen Fixation*, Australian Academy of Sciences, Canberra.

Guttschick, V. P. (1981). Evolved strategies in nitrogen acquisition by plants. *The American Naturalist* **118**, 607–37.

Miflin, B. J. and Lea, P. J. (1976). The pathway of nitrogen assimilation in plants. *Phytochemistry* **15**, 873–85.

Newcomb, W. (1981). Nodule morphogenesis and differentiation. *Int. Rev. Cyt. Suppl* **13**, Academic Press, N.Y., pp. 247–97.

Polhill, R. M. and Raven, P. H. (eds) (1981). *Advances in Legume Systematics*, Royal Botanic Gardens, Kew.

Postgate, J. R. (1982). *The Fundamentals of Nitrogen Fixation*, Cambridge University Press.

Stewart, W. D. P. and Gallon, J. R. (eds) (1980). *Nitrogen Fixation*, Academic Press, London.

Stewart, W. D. P. and Rosswall, T. (eds) (1982). *The Nitrogen Cycle*, Royal Society, London.

Sprent, J. I. (1979). *The Biology of Nitrogen Fixing Organisms*, McGraw-Hill, UK.

Veeger, C. and Newton, W. E. (eds) (1984). *Advances in Nitrogen Fixation Research*, Nijhoff, Pudoc/Wageningen.

References

Antoniw, L. D. (1976). *Effect of Irradiance on Growth and Nitrogen Metabolism of* Phaseolus vulgaris *L.*, PhD thesis, Univ. Dundee, Scotland.

Aparico-Téjo, P. M., Sanchez-Diaz, M. F. and Peña, J. I. (1980). Nitrogen fixation, stomatal response and transpiration in *Medicago sativa, Trifolium repens* and *T. subterraneum* under water stress and recovery. *Physiol. Plant.* **48**, 1–4.

Appleby, C. A., Tjepkema, J. D. and Trinick, M. J. (1983). Hemoglobin in a nonleguminous plant, *Parasponia*: possible genetic origin and function in nitrogen fixation. *Science* **220**, 951–3.

Atkins, C. A., Ritchie, A., Rowe, P. B., McCairns, E. and Sauer, D. (1982). *De novo* purine synthesis in nitrogen-fixing nodules of cowpea (*Vigna unguiculata* [L.] Walp.) and soybean (*Glycine max* [L.] Merr.). *Plant Physiol.* **70**, 55–60.

Auger, S. and Verma, D. P. S. (1981). Induction and expression of nodule-specific host genes in effective and ineffective root nodules. *Biochemistry* **20**, 1300–6.

Bauer, W. D. (1981). Infection of legumes by rhizobia. *Ann. Rev. Plant Physiol.* **32**, 407–49.

Bergersen, F. J. (1982). *Root Nodules of Legumes: Structure and Functions*, Research Studies Press (a division of Wiley), Chichester.

Bergersen, F. J. and Goodchild, D. J. (1973a). Aeration pathways in soybean root nodules. *Aust. J. Biol. Sci.* **26**, 729–40.

Bergersen, F. J. and Goodchild, D. J. (1973b). Cellular location and concentration of leghaemoglobin in soybean root nodules. *Aust. J. Biol. Sci.* **26**, 741–56.

Boddey, R. M. and Dobereiner, J. (1982). Association of *Azospirillum* and other diazotrophs with tropical Gramineae. *Proc. 12th Int. Congr. Soil Sci., New Delhi*, 28–47.

Boland, M. J., Hanks, J. F., Reynolds, P. H. S., Blevins, D. G., Tolbert, N. E. and Schubert, K. R. (1982). Subcellular organization of ureide biogenesis from glycolytic intermediates and ammonium in nitrogen-fixing soybean nodules. *Planta* **155**, 45–51.

Brown, M. E. (1982). Nitrogen fixation by free-living bacteria associated with plants—fact or fiction? In *Bacteria and Plants*, eds M. E. Rhodes-Roberts and F. A. Skinner, Academic Press, London, pp. 25–41.

Brussel, A. N. N. van, Planqué, K. and Quispel, A. (1977). The wall of *Rhizobium leguminosarum* in bacteroid and free-living forms. *J. Gen. Microbiol.* **101**, 51–6.

Callaham, D. and Torrey, J. G. (1981). The structural basis for infection of root hairs of *Trifolium repens* by *Rhizobium. Can. J. Bot.* **59**, 1647–64.

Carter, K. R., Rawlings, J., Orme-Johnson, W. H., Becker, R. R. and Evans, H. J. (1980). Purification and characterization of a ferredoxin from *Rhizobium japonicum* bacteroids. *J. Biol. Chem.* **255**, 4213–33.

Chandler, M. R. (1978). Some observations on infection of *Arachis hypogaea* L. by *Rhizobium. J. Exp. Bot.* **29**, 749–55.

Chandler, M. R., Date, R. A. and Roughley, R. J. (1982). Infection and root-nodule development in *Stylosanthes* species by *Rhizobium. J. Exp. Bot.* **33**, 47–57.

Christeller, J. T., Laing, W. A. and Sutton, W. D. (1977). Carbon dioxide fixation by lupin root nodules. 1. Characterization, association with phosphoenolpyruvate carboxylase and correlation with nitrogen fixation during nodule development. *Plant Physiol.* **60**, 47–50.

Corby, H. D. L. (1981). The systematic value of leguminous root nodules. In *Advances in Legume Systematics*, eds R. M. Polhill and P. H. Raven, Royal Botanic Gardens, Kew, pp. 657–69.

Dilworth, M. J. and Glenn, A. R. (1982). Movement of ammonia in *Rhizobium leguminosarum. J. Gen. Microbiol.* **128**, 29–37.

Dixon, R. O. D. (1972). Hydrogenase in legume root nodule bacteroids; occurrence and properties. *Arch. Mikrobiol.* **85**, 103–201.

Dixon, R. O. D., Berlier, Y. M. and Lespinat, P. A. (1981a). Respiration and nitrogen fixation in nodulated roots of soybean and pea. *Plant Soil* **61**, 135–43.

Dixon, R. O. D., Blunden, E. A. G. and Searl, J. W. (1981b). Intercellular space and hydrogen diffusion in pea and lupin root nodules. *Plant Sci. Lett.* **23**, 109–16.

Dreyfus, B. L. and Dommergues, Y. R. (1981). Nitrogen-fixing nodules induced by *Rhizobium* on the stem of the tropical legume *Sesbania rostrata. FEMS Lett.* **10**, 313–17.

Evensen, K. B. and Blevins, D. G. (1981). Differences in endogenous levels of gibberellin-like substances in nodules of *Phaseolus lunatus* L. plants inoculated with two *Rhizobium* strains. *Plant Physiol.* **68**, 195–8.

Fraser, H. L. (1942). The occurrence of endodermis in leguminous root nodules and its effects upon nodule function. *Proc. Roy. Soc. Edinb.* B **61**, 328–43.

Fyson, A. (1981). *Effects of Low Temperature on the Development and Functioning of Root Nodules of* Vicia faba, PhD thesis, Univ. Dundee, Scotland.

Fyson A. and Sprent, J. I. (1980). A light and scanning electron microscope study of stem nodules in *Vicia faba* L. *J. Exp. Bot.* **31**, 1101–6.

Fyson, A. and Sprent, J. I. (1982). The development of primary root nodules on *Vicia faba* L. grown at two temperatures. *Ann. Bot.* **50**, 681–92.

Gallacher, A. E. and Sprent, J. I. (1978). The effects of different water regimes on the growth and nodule development of greenhouse grown *V. Faba. J. Exp. Bot.* **28**, 413–23.

Gibson, A. H., Dreyfus, B. L., Lawn, R. J., Sprent, J. I. and Turner, G. L. (1981) Host and environmental factors affecting hydrogen evolution and uptake. In *Current Perspectives in Nitrogen Fixation*, eds, A. H. Gibson and W. E. Newton, Australian Academy of Sciences, Canberra, p. 373.

Griffiths, H. (1982). *Interactions between Hydrogen Evolution, Nitrogen Fixation and Nitrogen Assimilation in* Lupinus albus *L.*, PhD thesis, Univ. of Dundee, Scotland.

Grove, T. S., O'Connell, A. M. and Malajczuk, N. (1980). Effects of fire on the growth, nutrient content and rate of nitrogen fixation of the cycad *Macrozamia riedlii. Aust. J. Bot.* **28**, 271–82.

Guerrero, M. G., Vega, J. M. and Losada, M. (1981). The assimilatory nitrate-reducing system and its regulation. *Ann. Rev. Plant Physiol.* **32**, 169–204.

Haaker, H., Laane, C. and Veeger, C. (1980). In *Nitrogen Fixation*, eds W. D. P. Stewart and J. R. Gallon, Academic Press, London, pp. 113–38.

Huang, C-Y., Boyer, J. S. and Vanderhoef, L. N. (1975). Limitations of acetylene reduction (nitrogen fixation) by photosynthesis in soybean having low water potentials. *Plant Physiol.* **56**, 228–32.

Hubbell, D. M. (1981). Legume infection by *Rhizobium*: a conceptual approach. *BioScience* **31**, 832–7.

Hudman, J. F. and Glenn, A. R. (1980). Glucose uptake by free living and bacteroid forms of *Rhizobium Leguminosarum. Arch. Microbiol.* **128**, 72–7.

Israel, D. W. and Jackson, W. A. (1982). Ion balance, uptake and transport processes in N_2-fixing and nitrate- and urea-dependent soybean plants. *Plant Physiol.* **69**, 171–8.

Jing, Y., Paau, A. S. and Brill, W. J. (1982). Leghemoglobin from alfalfa (*Medicago sativa* L vernal) root nodules. I. Purification and in vitro synthesis of five leghemoglobin components. *Plant Sci. Lett.* **25**, 119–32.

Jong, T. M. de, Brewin, N. J., Johnston, A. W. B. and Phillips, D. A. (1982). Improvement of symbiotic properties in *Rhizobium leguminosarum* by plasmid transfer. *J. Gen. Microbiol.* **128**, 1829–38.

Jordan, D. C. (1982). Transfer of *Rhizobium japonicum* Buchanan 1980 to *Bradyrhizobium* gen. nov., a genus of slow-growing, root nodule bacteria from leguminous plants. *Int. J. Syst. Bacteriol.* **32**, 136–9.

Kidby, D. K. and Goodchild, D. J. (1966). Host influence on the ultrastructure of root nodules of *Lupinus luteus* and *Ornithopus sativus. J. Gen. Microbiol.* **45**, 147–52.

Kubo, H. (1939). Uber Hamaprotein aus den Wurzelknollchen von Leguminosen. *Acta Phytochim. (Japan)* **11**, 195–200.

LaRue, T. A. and Patterson, T. G. (1981). How much nitrogen do legumes fix? *Adv. in Agron.* **34**, 15–38.

Lawn, R. J. and Brun, W. A. (1974). Symbiotic nitrogen fixation in soybeans I. Effect of photosynthetic source-sink manipulations. *Crop Sci.* **14**, 11–16.

Lawrie, A. C. and Wheeler, C. T. (1975). Nitrogen fixation in the root nodules of *Vicia faba* in relation to the assimilation of carbon. *New Phytol.* **74**, 437–45.

Mahon, J. D. (1982). Field evaluation of growth and nitrogen fixation in peas selected for high and low photosynthetic CO_2 exchange. *Can. J. Plant Sci.* **62**, 5–17.

Masterson, C. L. and Murphy, P. M. (1976). Application of the acetylene reduction technique to the study of nitrogen fixation by white clover in the field. In *Symbiotic Nitrogen Fixation in Plants*, ed. P. S. Nutman, Cambridge Univ. Press, Cambridge, pp. 299–316.

Minchin, F. R. and Pate, J. S. (1974). Diurnal functioning of the legume root nodule. *J. Exp. Bot.* **25**, 295–308.

Minchin, F. R. and Pate, J. S. (1975). Effect of water, aeration and salt regime on nitrogen fixation in a nodulated legume—definition of an optimum root environment. *J. Exp. Bot.* **26**, 60–9.

Minchin, F. R., Summerfield, R. J., Hadley, P., Roberts, E. H. and Rawsthorne, S. (1981). Carbon and nitrogen nutrition of nodulated roots of grain legumes. *Plant, Cell and Environ.* **4**, 5–26.

Newcomb, E. H. and Tandon, S. R. (1981). Uninfected cells of soybean root nodules: ultrastructure suggests a key role in ureide production. *Science* **212**, 1194–6.

Newcomb, W. (1976). A correlated light and electron microscope study of symbiotic growth and differentiation in *Pisum sativum* root nodules. *Can. J. Bot.* **54**, 2163–86.

Newcomb, W., Sippel, D. and Peterson, R. L. (1979). The early morphogenesis of *Glycine max* and *Pisum sativum* nodules. *Can. J. Bot.* **57**, 2603–16.

Norris, D. O. (1965). Legumes and the *Rhizobium* symbiosis. *Empire J. Exp. Agric.* **24**, 247–70.

Pagan, J. D., Child, J. J., Scowcroft, W. R. and Gibson, A. H. (1975). Nitrogen fixation by *Rhizobium* cultured on a defined medium. *Nature* **256**, 406–7.

Pankhurst, C. E. (1981). Effect of plant nutrient supply on nodule effectiveness and *Rhizobium* strain competition for nodulation of *Lotus pedunculatus*. *Plant Soil* **60**, 325–39.

Pankhurst, C. E. and Sprent, J. I. (1975a). Surface features of soybean nodules. *Protoplasma* **85**, 85–98.

Pankhurst, C. F. and Sprent, J. I. (1975b). Effects of water stress on the respiratory and nitrogen-fixing activity of soybean root nodules. *J. Exp. Bot.* **26**, 287–304.

Pate, J. S., Atkins, C. A. and Rainbird, R. M. (1981). Theoretical and experimental costing of nitrogen fixation and related processes in nodules of legumes. In *Current Perspectives in Nitrogen Fixation*, eds A. H. Gibson and W. E. Newton, Australian Academy of Sciences, Canberra, pp. 105–16.

Pimental, D. (ed.) (1980). *Handbook of Energy Utilization in Agriculture*. CRC Press Fla.

Prestwich, G. D. and Bentley, B. L. (1981). Nitrogen fixation by intact colonies of the termite *Nasutitermes corniger*. *Oecologia (Berl.)* **49**, 249–51.

Rainbird, R. M. and Atkins, C. A. (1981). Purification and some properties of urate oxidase from nitrogen-fixing nodules of cowpea. *Biochim. Biophys. Acta* **659**, 132–40.

Ralston, E. J. and Imsande, J. (1982). Entry of oxygen and nitrogen into intact soybean nodules. *J. Exp. Bot.* **33**, 208–14.

Raven, J. A. (1984). The role of membranes in pH regulation: implications for energetics and water use efficiency of higher plant growth with nitrate as nitrogen source. In *Ann. Proc. Phytochem. Soc. Europe*, eds A. Bondet *et al.*, Oxford University Press (in press).

Raven, J. A. and Smith, F. A. (1976). Nitrogen assimilation and transport in vascular land plants in relation to intracellular pH regulation. *New Phytol.* **76**, 415–31.

Raven, J. A., Smith, S. E. and Smith, F. A. (1978). Ammonium assimilation in Scottish plants: the role of mycorrhizas in ammonium uptake and metabolism and regulation of pH. *Trans. Bot. Soc. Edinb.* **43**, 27–35.

Reynolds, P. H. S., Boland, M. J., Blevins, D. G., Schubert, K. R. and Randall, D. D. (1982). Enzymes of amide and ureide biogenesis in developing soybean nodules. *Plant Physiol.* **69**, 1334–8.

Rudelier, D. de, Goas, G. and Larher, F. (1982). Onium compounds, amides and amino acid levels in nodules and other organs of nitrogen fixing plants. *Z. Pflanzenphysiol.* **105**, 417–26.

Ryle, G. J. A., Powell, C. E. and Gordon, A. J. (1981). Assimilate partitioning in red and white clover either dependent on N_2 fixing in root nodules or utilizing nitrate nitrogen. *Ann. Bot.* **47**, 515–23.

Scott, D. B., Farnden, K. J. F. and Robertson, J. G. (1976). Ammonium assimilation in lupin nodules. *Nature* **263**, 705–7.

Sen, W. and Weaver, R. W. (1980). Nitrogen fixing activity of rhizobial strain 32Hl in peanut and cowpea nodules. *Plant Sci. Lett.* **18**, 315–18.

Sen, D. and Weaver, R. W. (1981). Nitrogenase activity (C_2H_2) of isolated peanut and cowpea bacteroids under nitrogen, argon and helium atmospheres. *J. Exp. Bot.* **32**, 713–16.

Sethi, R. S. and Reporter, M. (1981). Calcium localization pattern in clover root hair cells associated with infection processes: studies with aureomycin. *Protoplasma* **105**, 321–5.

Sheehy, J. E., Minchin, F. R. and Witty, J. F. (1983). Biological control of the resistance to oxygen flux in nodules. *Ann. Bot.* **52**, 565–71.

Sinclair, T. R. and Goudriaan, J. (1981). Physiological and morphological constraints on transport in nodules. *Plant Physiol.* **67**, 143–5.

Sinclair, T. R. and Wit, C. T. de (1975). Comparative analysis of photosynthate and nitrogen requirements in the production of seeds by various crops. *Science* **189**, 565–7.

Sprent, J. I. (1971). The effects of water stress on nitrogen-fixing root nodules. I. Effects on the physiology of detached soybean nodules. *New Phytol.* **70**, 9–17.

Sprent, J. I. (1976). Water deficits and nitrogen fixing root nodules. In *Water Deficits and Plant Growth*, ed. T. T. Kozlowski, Academic Press, N.Y., pp. 291–315.

Sprent, J. I. (1980). Root nodule anatomy, type of export product and evolutionary origin in some Leguminosae. *Plant, Cell and Environ.* **3**, 35–43.

Sprent, J. I. (1981). Functional evolution in some Papilionoid root nodules. In *Advances in Legume Systematics*, eds R. M. Polhill and P. H. Raven, Royal Botanic Gardens, Kew, pp. 671–6.

Sprent, J. I. (1982). Nitrogen fixation by grain legumes in the U.K. *Phil. Trans. Roy. Soc. B.* **296**, 387–95.

Sprent, J. I. (1983). Agricultural and horticultural systems: implications for forestry. In *Nitrogen Fixation in Forestry*, eds J. G. Gordon and C. T. Wheeler, pp. 213–32.

Sprent, J. I. and Gallacher, A. E. (1976). Anaerobiosis in soybean root nodules under water stress. *Soil Biol. Biochem.* **8**, 317–20.

Sprent, J. I., Minchin, F. R. and Thomas, R. J. (1983). Environmental effects on the physiology of nodulation and nitrogen fixation. In *Temperate Legumes: Physiology, Genetics and Nodulation*, eds D. G. Jones and D. R. Davies, Pitman Books, London, pp. 269–318.

Streeter, J. G. (1974). Growth of two soybean shoots on a single root. *J. Exp. Bot.* **25**, 189–98.

Streeter, J. G. (1981). Seasonal distribution of carbohydrates in nodules and stem exudate from field grown soya bean plants. *Ann. Bot.* **48**, 441–50.

Streeter, J. G. (1982). Synthesis and accumulation of nitrite in soybean nodules supplied with nitrate. *Plant Physiol.* **69**, 1429–33.

Sutton, W. D. (1983). Nodule development and senescence. In *Nitrogen Fixation, Vol. 3, Legumes*, ed. W. J. Broughton, Clarendon Press, Oxford, pp. 144–212.

Suzuki, T. and Takahashi, E. (1977). Biosynthesis of purine nucleotides and methylated purines in higher plants. *Drug Metabolism Rev.* **6**, 213–42.

Thomas, R. J., Feller, U. and Erismann, K. H. (1980). Ureide metabolism in non-nodulated *Phaseolus vulgaris*. *J. Exp. Bot.* **31**, 409–17.

Thomas, R. J. and Schrader, L. E. (1981). Ureide metabolism in higher plants. *Phytochemistry* **20**, 361–71.

Thomas, R. J. and Sprent, J. I. (1984). The effects of temperature on vegetative and early reproductive growth of a cold tolerant and a cold sensitive line of *Phaseolus vulgaris* L. I. Nodulation, growth and partitioning of dry matter, carbon and nitrogen. *Ann. Bot.* **53** (in press).

Tjepkema, J. D. (1984). Physiology of actinorrhizas. In *Advances in Nitrogen Fixation Research*, eds C. Veeger and W. E. Newton, Nijhoff, Pudoc/Wageningen (in press).

Tjepkema, J. D. and Yocum, C. S. (1973). Respiration and oxygen transport in soybean nodules. *Planta* **115**, 59–72.

Truchet, G. L. and Dazzo, F. B. (1982). Morphogenesis of lucerne nodules incited by *Rhizobium meliloti* in the presence of combined nitrogen. *Planta* **154**, 352–60.

Truchet, G., Michel, M. and Denarié, J. (1980). Sequential analysis of the organogenesis of lucerne (*Medicago sativa*) root nodules using symbiotically defective mutants of *Rhizobium meliloti*. *Differentiation* **16**, 163–72.

Uheda, E. and Syono, K. (1982a). Physiological role of leghaemoglobin heterogeneity in pea root nodule development. *Plant Cell Physiol.* **23**, 75–84.

Uheda, E. and Syono, K. (1982b). Effects of leghaemoglobin components on nitrogen fixation and oxygen consumption. *Plant Cell Physiol.* **23**, 85–90.

Urban, J. E. and Dazzo, F. B. (1982). Succinate-inducing morphology of *Rhizobium trifolii* 0403 resembles that of bacteroids in clover nodules. *Appl. environ. Microbiol.* **44**, 219–26.

Verma, D. P. S. (1984). Leghemoglobins and nodulin genes: two major groups of host genes necessary for nitrogen fixation in legumes. In *Advances in Nitrogen Fixation Research*, eds C. Veeger and W. E. Newton, Nijhoff, Pudoc/Wageningen (in press).

Vries, G. E. de, Veld, 'T. and Kijne, J. W. (1980). Production of organic acids in *Pisum sativum* root nodules as a result of oxygen stress. *Plant Sci. Lett.* **20**, 115–23.

Wilson, J. R. (1970). Response to salinity in *Glycine* VI. Some effects of a range of short-term salt stresses on growth, nodulation and nitrogen fixation of *Glycine wightii* (formerly *javanica*). *Aust. J. agric. Res.* **21**, 571–82.

Yousef, A. N. and Sprent, J. I. (1983). Effects of NaCl on growth, nitrogen incorporation and chemical composition of inoculated and NH_4NO_3 fertilized *Vicia faba* (L.) plants. *J. exp. Bot.* **34**, 941–50.

Zablotowicz, R. M., Russell, S. A. and Evans, H. J. (1980). Effect of hydrogenase system in *Rhizobium japonicum* on the nitrogen fixation and growth of soybeans at different stages of development. *Agron. J.* **72**, 555–9.

Translocation of nutrients and hormones

13

M J Canny

The plant body, its traffics and pathways

All the cells of a plant have the same general needs: water, reduced carbon, reduced nitrogen, phosphate, potassium, and a diminishing indefinite list of other elements in lesser amounts, magnesium, calcium, sodium, ... All but the second and third of these are supplied from the soil to the roots. The reduced carbon is made in abundance where the cells are green—remote from the soil (*see* Chapter 11). The reduced nitrogen has to be made from oxdized nitrogen of the soil, either in the roots, or, in some plants, in the leaves (*see also* Chapter 12). A complex network of traffics is called for to bring the reduced carbon to the non-green parts, the soil-derived substances everywhere, and to circulate the nitrogen and protons to and from the sites of reduction.

To accomplish these traffics the cells have only two resources: the space within their cell membrane where life is, and the dead extra-cellular matrix outside the cell membrane. Because the living space of the cells is connected all over the plant body through plasmodesmata, surrounded in effect by a single cell membrane enveloping all cells and lining all the plasmodesmata, this space may be thought of as a continuum, the *symplast*. The cell wall matrix that lies outside it for a distance of one to several micrometers forms a second continuum of dead (though not inactive) matter, the *apoplast*. The boundary between the two, the cell membrane or *plasmalemma* defines by its activities and capacities the conditions of the life it contains. It is therefore as complicated as that life in its powers of selection,

synthesis, signalling, perception and response, and rich in molecules of great structural variety such as proteins, lipoproteins, glycoproteins and polysaccharides (*see* Chapter 15). The simplicity of its appearance in electron micrographs as two dark lines with a gap between will mislead no thoughtful physiologist.

These two spaces are extended in vascular plants into the two components of the vascular system, the phloem and the xylem. The apoplast continuum is connected by the death of tracheary elements to the low-resistance continuum of water contained in them, forming the pathway of water traffic that supplies transpiration (Chapter 14) and carries a large part of the traffic of inorganic material from the roots to the rest of the body (Chapter 15). The symplast continuum is connected *via* the plasmodesmata of companion cells and phloem parenchyma to the contents of the sieve elements, extending through the sieve pores into another low-resistance pathway inside the plasmalemma. Quite in what sense the low resistance of this pathway is to be understood lies at the heart of the dispute over how the traffic here is moved.

Long-distance transport in the symplast— translocation

A pre-scientific understanding, a folklore, of the importance of the inner bark to the life and vigor of woody plants grew from the thoughtful observations and manipulations of hundreds of generations of orchardists, vine growers and foresters. A general view of a sap rising from the roots, being enriched by the leaves and returning in the inner bark, sufficed for the successful managements of pruning, ringbarking, cincturing and grafting, and indeed suffices today for millions of successful gardeners unacquainted with plant physiology. The only obvious saps of plants, that which bleeds from cut xylem especially in spring, and the milky or resinous juices oozing from broken parts of plants of several groups, though they sustained this folklore with the evidence of nutritive fluids in the plant body, have in fact nothing to do with translocation. It was the German foresters of the mid-nineteenth century who found first the sieve tubes (Hartig, 1837), and later the obscure sap they contain. In 1860 Hartig demonstrated that careful shallow cuts into the phloem of some trees sometimes released small droplets of a quite different sap, rich in sugar, and known since then as phloem exudate. From this has grown the study of translocation.

The anatomists of the time made careful and still valid observations of the phloem tissue, its component parenchyma, sieve tubes and companion cells, their sizes, relative positions and connexions, while the physiologists set out theories with little supporting evidence. One study at this time was to have lasting influence. Büsgen in 1891, investigating the phenomenon of honeydew, the sticky rain that falls from the canopy of many broad-leaved trees in the north-temperate summer, found that it was the secretion from dorsal glands of aphids feeding on leaves and young stems. Following the course of the aphids' mouthparts into the plant, he found the stylets always ended in a sieve tube, giving the aphid access to the vast synthetic machinery, stores and transport channel of the plant body. An account of the importance of this, together with a copy of Büsgen's splendid plate of the feeding aphid, will be found in Esau (1961).

A general understanding was arrived at by the early years of the twentieth century that the sugar-rich sap flowed through the elongated sieve tubes from where carbon was abundant to where it was consumed or stored. No quantitative estimates of the magnitude of the traffic were made. In 1922, however, Dixon published data on the growth of potato tubers showing that if the material passed through the sieve tubes of the tuber stalk as flowing sap, and the speed of the sap was calculated from the equation:

mass transfer per unit area
$$= \text{concentration of sap} \times \text{speed} \quad (1),$$
the required speed was about 50 to 100 cm h^{-1}. This, said Dixon, was clearly impossible through such small tubes interrupted by sieve plates with very small holes in them. Doubts of this kind, though more precisely formulated, persist today (see Mechanism). This question sparked off vigorous controversy and experimentation to answer a question that had been taken as settled for sixty years: whether it was indeed only the phloem that carried the traffic or whether the xylem was also used. It stimulated the researches of Mason and Maskell into translocation in cotton, which remain one of the great classic studies in the history of plant physiology. They showed beyond doubt, and unrevised by later work, three facts:

(1) The transport tissue was the phloem, and especially that part of it richest in sieve tubes.
(2) The moving compound was sucrose.
(3) The rate of movement was proportional to the gradient of sucrose concentration in the phloem.

A summary of their approach, experiments and findings may be found in Canny (1973). The paradox of Dixon's 'impossibility' and these clear demonstrations that it was true stimulated forty years of search for a mechanism by which the paradox might be reconciled, seeking to discover whether, on the one hand, the difficulty lay in the assumption of simple mass flow through the tubes, whether a more complicated and 'vital' transport was operating, and whether any structures could be found within sieve tubes that might provide the machinery for such a transport, or, on the other hand, whether there were forces available that could drive flowing solution through the tubes and narrow pores.

Throughout these years it became clear only slowly to experimenters how fragile a tissue and system they were dealing with. Dissections and observations which scarcely affected the ordinary robust parts of the plant destroyed the translocation process and caused visible damage to sieve tubes. A cut into the phloem stopped translocation for many centimeters on either side immediately, and until new sieve tubes could be regenerated; the sieve tube contents surged under the release of pressure, piling up debris on the sieve plates. Removal of leaves or parts of shoots (formerly popular manipulations) immediately and drastically altered the patterns of movement from those in the intact plant. More gentle and subtle experiments had to be devised, and much of the early work discarded.

From about 1960, translocation research was strongly directed by attempts to reveal the structure of working sieve tubes. Questions of mechanism, it was felt, would be resolved by finding whether or not the tubes and pores were empty except for a flowing solution, or contained some organization of protoplasmic material that could operate an alternative mechanism. Also, the enormous promise of high-resolution images from the electron microscope was at hand.

By about 1975 these surges of enthusiasm had run their course without providing any certainties about either mechanism or fine structure. From then until the present, most workers accept the limitations of present techniques that block progress in our understanding of sieve tube functioning, and attention has turned to the events of loading solutes into the sieve tubes and their release from them.

Phloem transport

Methods of study

Autoradiography The most rewarding initial experiment in a study of translocation, yielding the most information with the least effort, is the gross autoradiograph. Carbon dioxide labeled with ^{14}C is presented to a selected source region (usually a leaf) in the light and assimilated into ^{14}C-metabolites. It joins the traffic of assimilated ^{12}C from the leaf and is moved by translocation to whatever sinks the leaf is currently supplying—expanding leaves, growing root tips or developing fruit. After a chosen time the plant is killed quickly by some means that prevents solute migration (freezing and drying or rapid drying with heat) and exposed to photographic film. An image is developed of the places to which the ^{14}C has moved and the pathways it has traced in getting there. The intensity of the image at each spot is a measure of the amount of translocation from the source to that spot. An immense amount can be learned quickly and economically about the rates, time scales and patterns of movement of a new translocation system from a series of such autoradiographs (Fig. 13.1). Since beta-particles from ^{14}C

have low energy, they are absorbed by quite thin layers of dry material, and uneven thicknesses of the ^{14}C-containing tissues will make precise comparisons of the intensities of the images of little value. With favorable (thin) tissues and due care, resolutions of the order of $100\,\mu$m can be achieved for vein networks in leaves. Transverse sections taken through stems, etc., and freeze-dried will reveal which parts of the phloem are carrying the label. No interference with the intact and functioning plant is involved except the brief enclosure of the source tissue in a small chamber.

Specific mass transfer Translocation results in the transport of dry weight matter away from sources or into sinks, and a convenient quantification of the process is the rate of movement of dry weight. Translocation into a growing fruit, for example, may be measured as the rate of increase of dry weight there. Since the channel is known to be the phloem, comparisons between fruits are conveniently expressed in terms of the cross-sectional area of phloem in the stalk. This quantity, termed the specific mass transfer (SMT, g h^{-1} cm$^{-2}_{ph}$), was the measure used by Mason and Maskell and many workers of that time for long-period averages of translocation. In a great many different species and plant organs it has been found to give values in quite a narrow range, 1 to 5 g h^{-1} cm$^{-2}_{ph}$ (see for example, Canny, 1975, Table 1). Refining the measure by attempting to express it per area of sieve tube (cm$^{-2}_{st}$) adds greatly to the labor and uncertainty because of the difficulties of knowing which cells of the phloem are sieve tubes. The area proportion of them has been variously estimated from one fifth to three quarters in different species.

Applying a similar measure to translocation out of leaves is more difficult because of the large relative concurrent gain in dry weight in the light (photosynthesis) and loss in the dark (respiration). The old method was to compare samples of attached and detached leaves, the difference being ascribed to translocation. But this and all other methods of isolating leaves, by heating or freezing a zone of the petiole, drastically interferes with photosynthesis by causing rapid stomatal closure, making comparisons with the attached leaves invalid.

Recently two other methods to measure translocation from leaves have been used. One is to monitor continuously for (say) a few hours the net

photograph of plant   autoradiograph

Fig. 13.1 Plant of *Epilobium* (left) which assimilated $^{14}CO_2$ in a single leaf (A) and translocated the products of photosynthesis for 2 h, and the autoradiograph (right) it formed on film. Note the strong accumulations of ^{14}C in the developing fruit (B). Scale, 2 cm.

photosynthetic gain per unit time by a leaf in a controlled environment, simultaneously measuring the change in dry weight per unit area by taking samples from similar leaves at the beginning and end of the period, and calling the difference translocation (Lush and Evans, 1974, and Table 13.1). The second is to feed $^{14}CO_2$ of known specific activity to an area of the source leaf, and monitor the arrival of the ^{14}C-translocate at a certain sink (the plant being pruned so that this is the only one operating). Rate of arrival of ^{14}C may then be converted to rate of export of carbon from the source (Geiger and Swanson, 1965; Geiger, Saunders and Cataldo, 1969). Results of both these measurements are often expressed per dm^2 of source leaf, making comparisons with other SMT values uncertain.

The congruence of the values of SMT which had been evident in the old measurements has been shattered by several recent estimates. The new methods, no longer averaging translocation over long periods, can define it at peak times and in local regions, and have produced some very high rates. Lush (1976) comparing photosynthesis and dry weight gain for leaves of the C-4 grass *Paspalum maximum*, records a peak rate at the base of the leaf of $52 \, g \, h^{-1} \, cm^{-2}_{ph}$. The highest estimate so far registered is for transport through one seminal root of a wheat plant when the other roots had been removed, where Passioura and Ashford (1974) found

^{14}C after pulse-labeling in a source leaf with an adjacent detector while the translocate is moved away. The curve of declining radioactivity with time is a double-exponential washout curve which can be interpreted as revealing two compartments in the leaf from which the assimilate is mobilized.

Speed of translocation In addition to the transfer of mass, translocation is associated with a speed of

Table 13.1 Measured values of specific mass transfer (SMT).

Authors	Plant/Organ	Method	SMT $(g \, h^{-1} \, cm^{-2}_{ph})$
Passioura and Ashford (1974)	wheat root	dry wt increment	180.0
Lush and Evans (1974)	Leaves of	PHS/leaf dry wt	
	Paspalum		12.7
	Cynodon		6.6
	Chloris		14.9
	Digitaria		4.7
	Lolium		4.4
Lush (1976)	*Paspalum*	PHS/leaf dry wt	51.8
	Lolium		11.9
Milburn and Zimmermann (1977)	*Cocos*	sap flow	3 to 32
Smith and Milburn (1980)	*Ricinus*	sap flow	18 to 252

a value of $180 \, g \, h^{-1} \, cm^{-2}_{st}$. They used this unit rather than measure the phloem area since there were only six conspicuous sieve tubes in the root whose luminal area could be found without difficulty. If Dixon had been unable to accept an SMT of $5 \, g \, h^{-1} \, cm^{-2}_{ph}$ into a potato, and mechanisms operating with that capacity have exercised the ingenuity of scores of phloem physiologists, this enormous value would enforce a complete revision of tentative theories. The reader may picture what it means: a gram of solid sugar passing through a square centimeter of sieve tube lumen every 20s. An assortment of these more recent estimates is assembled in Table 13.1.

Semi-quantitative measures Many investigators feel that the full estimate of a translocation rate as SMT is unnecessary for the problem they are studying, and are content to express in numerical terms (*e.g.*, as percentage total radioactivity) the kind of distribution of radioactive tracer that an autoradiograph shows. An example will be found in Fig. 13.3 (p. 291). Another technique is to monitor the loss of

movement through the phloem. In the simple interpretation of equation (1) this would be the speed of the solution in the sieve tubes. It has not proved an easy attribute to measure. Modern estimates rely mostly upon following the progress of radioactive tracers through the system. When ^{14}C is assimilated into a leaf and spreads through the phloem in petiole and stem, the concentration of ^{14}C-translocate as a function of distance is called a profile. At successive times it is found to have progressed further, and often also to have changed its shape, becoming wider and flatter as it moves. Because of the weak beta emissions from ^{14}C, profiles of it cannot be found easily in living plants and the measurement demands destructive sampling of a series of plants at different times, with the inevitable variability between the plants. Another isotope, ^{11}C, has much stronger positron emissions and can be detected outside the living plant with ease, making it a simple matter to measure profiles and their progress. Since it has a half-life of only 20 min, the experiment must be done next to the high-energy particle source that makes the isotope, but this drawback is balanced by the fact that the radioactivity vanishes after a few hours, and successive replicate experiments can be carried out on one plant. From a set of profiles progressing through a plant, estimates may be made of the speed of advance of a chosen part of the curve, generally the point of half peak-height. Some estimates of measured speeds are given in Table 13.2. The initial shape of the profile is probably generated by the time course of loading labeled assimilate into the leaf veins, and its subsequent changes of shape to exchanges between the phloem and surrounding tissues as it moves.

ing on *Tuberolachnus*, the willow aphid. More crude and uncertain than both of these methods is the employment of the massive bleeding that comes from the cut inflorescence stalks of many palms, and which is used in tropical countries as a source of sugar and beverages. An example of the flux from one such is given in Table 13.1.

Mobile forms of carbon

The simple answer found by Mason and Maskell, that the organic material of cotton plants moved as

Table 13.2   Measured values of speed of translocation.

Authors	Plant/Organ	Method	Speed (cm h^{-1})
Troughton and Currie (1977)	*Zea* leaf	profile advance	15 to 660
Thompson, Fensom, Anderson, Drouin and Leiper (1979)	*Helianthus* stem	profile advance	30 to 240
	Heracleum stem	profile advance	90 to 210
	Nymphoides stem	profile advance	30 to 156
	Ipomoea stem	profile advance	30 to 72
	Zea leaf	profile advance	282 to 600
	Triticum leaf	profile advance	168
	Fraxinus stem	profile advance	48
	Ulmus stem	profile advance	9.6 to 120
	Picea stem	profile advance	9.6 to 13.2
	Pinus stem	profile advance	6 to 48

Sap sampling In a few favorable plants the sap which bleeds from severed sieve tubes can be encouraged to flow for considerable periods, providing samples whose rate of volume flow and concentration of solutes can be measured. The simplest method is Hartig's, a razor cut into the sieve tubes, but the number of contributing sieve tubes is then uncertain and the sap is exposed to changes of composition in the wound. More precise is an ingenious use of the feeding aphid devised by Mittler. Severing the mouthparts leaves the stylets in place and from the canal formed by the inner pair, drops of exudate well up which have come from a single sieve tube. The operation is delicate, the result not predictable, and it is often much simpler to collect and measure the secreted honeydew from a live aphid. The carbohydrate components of the honeydew are little altered from the stylet sap, since the aphid has such an abundance of it and absorbs mainly the nitrogenous compounds. Peel and his colleagues have made the most use of aphid techniques, concentrat-

sucrose, was found to apply to most other plants, and is still a generalization of great validity, having suffered only minor modifications by the many studies made since. The techniques of exudate analysis, and ^{14}C labeling of the translocate, followed by chemical analysis of extracts of the phloem have made the identification of the mobile carbon much easier. Of the 500 species listed by Ziegler (1975) all contain sucrose, and about a third contain also the higher alpha-galactosides of sucrose, raffinose, stachyose or verbascose. A few contain one or more of the sugar alcohols, mannitol, sorbitol or dulcitol. Hexoses are not natural transport sugars. When glucose or fructose is found in extracts this is usually evidence of hydrolysis during extraction. Free sucrose is translocated, not a phosphorylated form, and this role as the transported form of carbon seems to be its principal one in the plant, notwithstanding its use for storage in a few species, notably sugar beet and sugar cane.

Properties of the translocation system

Concentrations and gradients of concentration The exudates from sieve tubes contain sucrose (or the allied sugars discussed above) at high concentrations, from 5 to 15 or 25% (w/v) or even higher, depending on where the sample comes from and on the physiological state of the plant. As a general rule the sieve tube sap is the most concentrated solution to be found in any space of the plant body, has the most negative osmotic potential (ψ_s), and will therefore plasmolyze all other cells. It is at its highest concentration in the minor veins of leaves where an osmotic potential (ψ_s) of 3.29 MPa was measured in sieve elements of *Beta* by Fellows and Geiger (1974). This corresponds to the ψ_s of 1 M sucrose. From this origin of translocation the concentration declines along the pathway towards the sinks, so that exudates taken from different heights of a tree show a progressive decline in concentration downwards. This gradient is fairly modest, approximately 0.01 to 0.02 M sucrose per meter in several species of tree (Canny, 1975). The gradient of sucrose concentration was the single internal parameter that Mason and Maskell found to be closely linked to translocation, indeed they showed that the SMT was proportional to sucrose concentration in the phloem over the range of their measurements. They pointed out that this relation is the same as Fick's law of diffusion, and used their data to estimate an 'apparent diffusion coefficient' of translocation whose value was $0.07 \, cm^2 \, s^{-1}$, or about 10 000 times the coefficient of diffusion of sucrose in water.

Pressures and gradients of pressure The high solute concentration of the sieve tube sap contained within the plasmalemma will generate a large osmotic potential, and hence a large turgor pressure in the sieve tubes. Mittler estimated this pressure from the rate of flow of the exudate from the severed aphid stylets and the known dimensions of the stylet canal as about 4 MPa. Other attempts to measure the pressure directly by inserting sharp hollow needles into phloem and attaching them to pressure gauges have been disappointing because of the very great difficulty of getting a tight seal. Hammel (1968) made a number of such measurements on the trunk of *Quercus rubra*. The values at two heights, 6.3 and 1.5 m, differ in their pooled averages by 0.149 MPa (significant at the 0.1% level). Gravity accounts for

0.051 MPa, leaving a turgor pressure gradient of 0.020 MPa m^{-1}. Milburn (1980) reviewed the available technologies and contributed a number of ingenious devices of his own invention. He used these to make direct measurements of the turgor pressures in *Ricinus* and showed that the estimates agree well with indirect measures made by the algebraic difference between the solute osmotic potential of the exuded sap and the water potential of the xylem measured by a pressure bomb (*see* Chapter 14). He assumes that the water potential of the phloem is in equilibrium with that of the adjacent xylem. Values are in the range 0.7 to 1.2 MPa. The direct and indirect estimates agree to within about 0.1 MPa.

If there is a gradient of solute concentration along the sieve tubes there will also be a gradient of turgor pressure, and if translocation results from a flow of solution as in equation (1) then this turgor gradient is one possible force to drive the flow. On this view the proportionality of SMT to sucrose concentration derives from a dependence of flow rate on pressure gradient. Hence the magnitude of the pressure gradient is an important datum for discussions about whether or not the measured rates of SMT can be achieved by solutions flowing at measured speeds, and whether large enough pressure gradients exist to produce these speeds through sieve tubes and sieve pores.

A sieve tube containing solution at a pressure of 4 or even 1.5 MPa is going to react violently to the release of this pressure when it is cut open. Here lies the reason for the extreme fragility of the sieve tubes and of the translocation system. Sieve tubes are part of the symplast; they have cell membranes and a few organelles, and their structure (whatever it is) is essential to their functioning. Rapid surges of their contents on pressure release disrupt their organization for many centimeters, piling up fragments on the sieve plates and plugging the sieve pores. Plants being liable to damage, some means of sealing off this surge is obviously an advantage. The opinion is widespread that this is the function of the special β-1,3 glucan polymer, callose, which is found on sieve plates, and lining sieve pores (Fig. 13.2), and in a few other places in plants. Many reports show that callose can form within seconds in the course of preparing tissues for microscopy, lining the sieve pores with constricting collars. With flow through a tube varying as (radius)^4, this is a

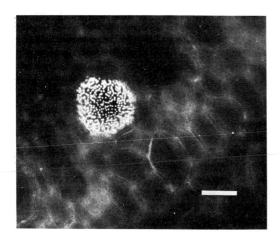

Fig. 13.2 Sieve plate in the phloem of the root of a field-grown maize plant; hand section stained with aniline blue and showing the fluorescent callose lining the sieve pores. Scale, 50 μm. (Preparation and photograph by courtesy of Dr M. E. McCully.)

most effective means of reducing flow and limiting the damage produced by pressure release. There is also evidence that callose from sieve plates may be remobilized and removed, especially in spring when some sieve tubes resume activity.

Relation between mass transfer and speed From what has been said about pressure gradients, the substantiation of the flux equation (1) is a first important test of the hypothesis of mass flow, while the second is the correspondence of measured speeds with pressure gradients necessary to produce them. The three quantities of equation (1) are all measurable, but because each is most easily measured in an individual translocation system and constitutes of itself a major research program, the simultaneous measurement of all three quantities remained for many years unaccomplished. A number of investigators measured two of them, assumed a reasonable value for the third, and showed that general agreement was likely. The one study that measured all three at once was that of Grange and Peel (1978). Working with rooted willow cuttings on which colonies of aphids were established, they measured the concentration of sucrose in the secreted honeydew, the speed by the time of passage of ^{14}C-assimilate between two colonies, and mass transfer by the constant-specific-activity technique,

equilibrating the assimilate–translocation system with $^{14}CO_2$ of constant concentration over a period, and measuring the specific activity of the honeydew as a sample of translocate. There is a broad general relation of the rather scattered points in their Fig. 5, a plot of the product: (speed × concentration) *vs.* SMT. 'There is a good correlation with $r = 0.76$, significant at the 1% level', falling slightly below the 45° slope expected for an equality. This is an encouraging agreement. Their SMT values span the range of 0.5 to 4.5 g h^{-1} cm^{-2}$_{ph}$. and their values of speed range from 3 to nearly 50 cm h^{-1}. These, until recently, were the accepted ranges for the two quantities, but now that SMT values approaching 200 g h^{-1} cm^{-2}$_{st}$ and speeds near to 600 cm h^{-1} have been recorded, it becomes pressing to extend the range of agreement for the flux equation into these regions.

Relation between pressure gradient and speed In the Poiseuille equation that describes mass flow through a tube, quantity transported varies as (radius)^4, speed as (radius)^2. A useful form is

$$\mathrm{d}p/\mathrm{d}l = (8\eta v \times 10^{-5})/r^2 \qquad (2)$$

where $\mathrm{d}p/\mathrm{d}l$ is the pressure gradient in MPa m^{-1}; η is the viscosity of the solution in poise; r is the radius of the tube in cm; and v is the speed in cm s^{-1}. The viscosity of 10% sucrose solution is 1.5×10^{-2} poise. Inserting in this a value for v of 100 cm h^{-1}, the corresponding pressure gradients needed to drive the flow through completely open tubes of various sizes are given in Table 13.3. It will be seen that tubes of the size of sieve tubes (say 10 μm radius) require quite small gradients of pressure to accommodate a speed of this magnitude, and that they may be neglected in favor of the sieve pores which will determine the pressure necessary. The third line of Table 13.3 lists some gradients required to force the same flow through tubes whose resistance is dominated by the pores in the sieve plates, for various radii, with a particular set of simple assumptions (20 plates per cm, each 1 μm thick, $\frac{1}{2}$ the plate occupied by pores). Some sieve pores, especially those in plants often favored for translocation studies like *Fraxinus* and *Cucurbita*, are quite large (2.5 to 5 μm). For such wide tubes and pores, flow is easy and seems an unavoidable result of small differences in internal pressure in the intact plant, or the opening of the tubes. However, though

Table 13.3 Pressure gradients (MPa m^{-1}) needed to force a flow of 100 cm h^{-1} through open tubes and pores of various radii.

Radius, r (μm)	20	10	5	2	1	0.5	0.2	0.1	0.05
$\Delta P/l$ (MPa/m) for continuous tube radius r	0.008	0.03	0.10	0.84	3.3	13	84	330	1300
$\Delta P/l$ (MPa m^{-1}) for tube interrupted by sieve plates whose pores are radius r^*			0.0004	0.003	0.013	0.04	0.33	1.35	5.40

*Assuming $\frac{1}{2}$ plate is pores, thickness 1 μm, 20 plates per cm.

protagonists of mass flow choose such values to demonstrate the feasibility of this as a general mechanism, most sieve pores are much smaller. Esau and Cheadle (1959), in a broad review of sieve pores of dicotyledons, found that the great majority had a radius close to 0.5 μm. For these the gradient needed in Table 13.3 is 0.04 MPa m^{-1}. The pores in the sieve cells of gymnosperms are smaller still (*see* for example, Murmanis and Evert, 1966), around 0.07 μm radius or less at the surface of the plate. Moreover, they have a cavity within the plate, the median nodule, filled with cytoplasmic material. The resistance of these can only be guessed at. Simple open-pore tubes of this radius in Table 13.3 would require a gradient of over 4 MPa m^{-1}. If the pores contain cytoplasmic material (*see* below) these gradients would be greatly increased.

We may now return to the gradients of pressure found in sieve tubes and compare them with these assessments of what might be needed. The gradient Hammel found in the red oak was 0.02 MPa m^{-1}, and is closely similar to the gradient that would be derived by converting solute concentration gradients into their pressure equivalents. Milburn (1974), working again with the bleeding sap from *Ricinus* showed that the gradient of concentration of solutes may in fact be unrelated to the gradient of pressure. The solute concentration gradient was often negligible along a meter or so of stem, in some cases even increasing downwards, while there were sudden changes at the two ends, near the root sinks, and particularly near the leaf sources. In a stem 87 cm long with a crown of leaves at the top, the osmotic potential of the exudate fell from 1.6 MPa at 70 cm to 1.2 MPa at 60 cm, stayed constant at this value down to 10 cm, and dropped just above the soil to about 0.6 MPa. He suggests that the pressure to drive the solution is generated only in the leaves,

that the solute potentials along the rest of the path need only be less, and are not part of the driving force. Thus, the turgor pressure near the source would impel through the sieve tubes a solution which might be depleted but whose concentration was of little relevance to movement. The gradients of concentration 'could be modified at will and made positive or negative by regulating the illumination and hence photosynthesis'. So it would be the pressure fall from source to sink (in this case 1 MPa in a meter) that should be considered, not the negligible gradient in between. In spite of the difficulties of measuring pressure in the sieve tube it appears that only by doing so can reliable estimates be obtained; the assumption that the local solute potential (except near the leaves) can be translated to a pressure may not be justified. This view of course conflicts with Milburn's (1980) assumptions previously discussed.

Leaves as sources: effects of light and sucrose With recent refinements in the rapidity of making translocation measurements it has become clear that short-term changes occur in response to the illumination of leaves. The measurement of speed by the progress of ^{11}C profiles has been particularly useful in this, repeated experiments on the same plant eliminating the variability inherent in older, destructive sampling methods. Troughton and Currie (1977) following profiles of ^{11}C-translocate in a single maize leaf, found a linear relation of the speed to light fluence rate (intensity) from 1.2 cm min^{-1} at zero light to 3.8 cm min^{-1} at 240 W m^{-2}. An obvious logical connexion of light with translocation would be *via* leaf photosynthesis, and Christy and Swanson (1976) using the ^{14}C steady-state labeling technique on sugar beet showed a linear relation of speed to leaf net photosynthesis (from

0.5 to 1.6 cm min^{-1} between 0 and 45 mg CO_2 dm^{-2} h^{-1}) as well as a general correspondence between rise and fall of SMT with increased or decreased photosynthesis.

The products of photosynthesis are transformed by leaf mesophyll cytoplasm to sucrose (Chapter 11) for transport out of the leaf, so the next logical relation to look for is between leaf sucrose concentration and translocation. This of course harks back to Mason and Maskell's original discovery, but has been made more precise with recent techniques. Sovonick, Geiger and Fellows (1974), using pruned sugar beet plants, showed that SMT out of the leaf was increased by feeding sucrose solutions to the leaf mesophyll through the abraded upper surface. It is therefore very likely that the effect of light operates through photosynthetic rate on the concentration of sucrose in the mesophyll and minor veins. Confirming evidence is included in the experiments of Troughton and Currie (1977), who state: 'High speeds of translocation (3 cm min^{-1}) were recorded in maize only when the sucrose concentration exceeded 7% (by dry weight). For sucrose concentrations between 1 and 7%, the speed for a normal maize leaf was 2 to 2.7 cm min^{-1} at high light, although this could be reduced to 1.5 to 2 cm min^{-1} for a leaf exposed to 48 h of darkness before the light treatment. This suggests that a prerequisite for high speeds may be a high sucrose concentration in the leaves.' Another indication that light probably acts through photosynthesis and sucrose concentration was that local shading of a portion of the leaf pathway did not affect the speed. The high sucrose concentration could act either by providing high sucrose potential in the leaf veins (osmotically), or by providing fuel for metabolic pumping machinery, or both.

Patterns of movement, 1, general rules

The plant achieves its form and life by delivering translocate to the right places at the right times. An understanding of translocation will include an understanding of how this is done. Although it is a question of agricultural importance, determining the proportion of plant found in the 'yield', whether leaf, fruit, tuber, etc., and has been the spur for much research, there are at present only elementary ideas about how it happens. Experiments extending the use of the gross autoradiograph show where current photosynthate from particular sources goes and roughly in what proportions. From many thousands of such experiments on many different species at different stages of their lives a few simple rules have become plain. Note, however, that nearly all such experiments have been conducted on small plants growing in culture, and that the rules for trees are practically unknown. The rules are as follows:

(1) There is no movement into mature leaves. This is shown in Fig. 13.1, and has been found to be almost universally true. Once a growing leaf has reached the stage of switching from import to export at about half size, it remains for the rest of its life an exporter only, and no amount of darkening or starvation makes it become a sink again. The one exception to this rule is when a local extraneous sink becomes established on the leaf, such as a feeding aphid or a fungal infection. These local sinks will then import labeled translocate from elsewhere, but the remainder of the leaf still does not. Thrower and Thrower (1980) discuss this rule and its exceptions.

(2) Upper leaves feed the larger proportion of their output to the stem apex, lower leaves to the roots. Leaves in between feed both, and the proportions of output of a leaf going in the two directions vary with the vigor of the sinks at apex and base.

(3) Removal of sources or sinks quickly alters the pattern of movement to compensate the loss. Thus the patterns just referred to are changed if (say) the apical leaves are removed; the lower leaves then send more of their output apically. It is this compensation which renders invalid much of the early work involving excision of plant parts. If plants are pruned to make a simplified experimental system, the pruning must be carried out well before the experiment starts, to allow the compensations to be completed.

(4) Translocate travels in straight lines along a narrow band of phloem, with very little sideways spread; this is revealed by autoradiographs of stem sections. Labeled translocate from one leaf, entering a single vascular strand in the stem, may be confined to that strand for hundreds of centimeters.

(5) Active sinks are fed by the nearest source. The substance of an apple comes from the crown of

leaves on its own short shoot; the glumes (and even awns where large) contribute a large part of the growing grass (cereal) grain, the rest comes from the flag leaf and reserves in the flowering stem; gourds grow spectacularly quickly only when subtended by a large and active leaf. When the nearest source is removed, Rule 3 comes into operation.

Patterns of movement, 2, effects of distance and sink size—'sink strength'

The final rule (5) above must be interpreted in terms of the plant's own internal geometry, its vascular connexions. A leaf is not always as 'near' a source as the node of its insertion, since many leaf traces do not join the stem vasculature at this node, but at the node below or even the further one. Hence translocate from a leaf travels down the stem for one or two nodes before entering the stem vasculature and traveling both apically and basally in the separate traces. Similarly, because of the operation of Rule 4, a leaf on another vertical line from a sink may not be accessible to the sink, however near it appears to be. In an attempt to amplify the operation of these five rules and to answer some of the practical questions about where the translocate in fact will go, Wareing and Patrick (1975) introduced the idea of 'sink strength' as a name for the observation that at a given stage of development some sinks dominate others, attracting the greater proportion of translocate from all sources. This was a blend of the size of the sink multiplied by its 'activity'. In an attempt to prevent the definition becoming circular, the 'activity' being an expression that this sink attracted the most translocate, they proposed that activity be measured by the potential of the isolated sink tissue to accumulate sugar from a culture solution. This raises many doubts: that the isolated tissue may behave quite differently from the intact tissue, that access of sugar *via* the apoplast may not represent the route in the plant (see below, Phloem unloading), that the tissues of the sink may be non-uniform in their activity. Sink size does, however, seem to be important. Peel and Ho (1970) showed that two aphid colonies of different sizes on a willow stem having equal access to ^{14}C-photosynthate, attracted ^{14}C-translocate in the proportion of their relative sizes. The experiments of Cook and Evans (1978) showed, similarly, that of two ears on separate

wheat tillers with equal access to ^{14}C-photosynthate in the flag leaf, the smaller ear (with grains removed) attracted less [14]C. They extended this observation to examine the interaction of the effects of size and distance, using now the two tiller ears and the flag leaf of one of them as the source (eliminating sources of the main plant), and showed that distance was more important than size, that the nearer ear gained most label, even when it was the smaller.

The form of the Poiseuille equation relating the *quantity* (M) of fluid carried through a tube by mass flow to the pressure gradient is

$$M = \frac{\pi r^4}{8\eta} \frac{\Delta p}{l} \times 10^{-3} \qquad (3)$$

where the units are as in equation (2), M is in grams of fluid of density 1, and l is in meters. It appears at first glance that the maximum SMT for a given pressure drop is inversely proportional to the distance (l). However, if the pressure driving the flow is generated by osmotic concentration gradients all along the tube, then p would also be a function of l, and the SMT would decrease with (distance)^2. This might explain the strong dependence of translocation on distance, though it raises awkward questions about translocation in tall trees. But if Milburn's findings for *Ricinus* exudates turn out to be generally applicable, that the pressure gradient is generated in the leaves and the concentration gradient along the tubes is largely irrelevant, then the naive view of the distance relation is correct and translocation over considerable distances is more easily understood.

Effects of temperature Translocation is irreversibly inactivated by temperatures above about 50°C. This was the classic test to distinguish movement in the phloem from movement in the xylem, and remains a useful and simple one. Heating a zone of an organ with steam or a hot electric wire leaves transpiration functioning but stops all translocation through the zone. Since sieve tubes are living and their subtle organization is necessary for functioning, this finding is not unexpected.

Low temperatures have been much used to test the involvement of metabolism along the path in generating the movement, that is, as a test to distinguish passive flow from active pumping. Local cooling of an organ to near 0°C almost always slows translocation (as SMT or speed) through it to a fraction of the value at room temperature. In one group of plants such as cucumber and tomato, the threshold for this inhibition is around 10°C; the inhibition persists for many hours if the cold block is maintained and recovers slowly (2–5 h) once it is removed. These are known as chilling-sensitive plants. Another group, chilling-insensitive, recover their translocation in speed and mass transfer after 60 to 90 min with the cold block still maintained, even near 0°C. Sugar beet, potato tubers and *Lolium* leaves are in this group. One explanation of these findings is that there may be two sorts of restriction of translocation in response to lowering temperature: first an increase in viscosity of the sieve tube sap which reduces the speed; and, at lower temperatures, an alteration of membrane structures which disorganizes the contents and causes plugging of the sieve pores. Chilling-sensitive plants go through the transition from the first to the second kind of damage around 10°C, while insensitive plants do not experience the more extreme damage above 0°C, being affected only by the viscosity change. For a review of the experiments and evidence, *see* Geiger and Sovonick (1975) and Fensom (1981).

Effects of inhibitors Another test for metabolic support along the channel of translocation, particularly for the use of respiratory energy in helping the movement, has been the response of translocation to respiratory inhibitors such as cyanide and dinitrophenol. It has been the general observation that cyanide applied locally to the phloem prevents translocation through the treated zone, and the inhibition is reversible on removal of the cyanide. Access of the inhibitor to the phloem is not easy in intact plants, so that it is usually necessary to expose a vascular bundle surgically. Care must be taken to distinguish responses at the site of application from responses at the source leaf where the inhibitor may be carried in the transpiration stream. Willenbrink (1978) demonstrated distinct and localized inhibition by CCCP, antimycin, atractylate and valinomycin.

Effects of water stress Since many crops spend a substantial proportion of their lives under a degree of water stress, it is of considerable interest to know

whether this affects their rate of translocation. The most comprehensive studies are still those of Wardlaw (1967, 1969) in which he measured speed of translocation into the developing ears of wheat and *Lolium* by following the progress of ^{14}C-translocate profiles. He concluded that the speed of translocation is little altered by quite severe water stress while the ear remains an active sink. Hoddinott, Ehret and Gorham (1979), in a study of the short-term effects of osmotic shock to the roots on photosynthesis and translocation in *Phaseolus*, found that photosynthesis was immediately reduced, while translocation stayed unchanged for several hours. Indeed translocation seems to be the plant function most resistant to water shortage, as one would perhaps expect for the living system operating at water potentials more negative than any of the others.

Effects of potassium After sucrose, potassium is the most constant and abundant solute present in phloem exudates. In addition, the general potassium status of a plant, as it has grown in soil either rich or poor in the element, has long been known to affect the growth of fruits and storage organs. All investigations into the interaction of potassium with translocation have shown that both general potassium abundance in the plant and local applications to leaf sources favor translocation however measured, but no clear reasons for this have been found. Mengel (1980) describes experiments on the bleeding sap of *Ricinus* in which those in high potassium solutions exuded sap at twice the rate of those in low potassium solutions, the saps of each having very similar concentrations of organic material. These effects are consistent with the hypothesis that K^+ ions are involved in the pumping process in leaves by which the sucrose is raised to a high potential (*see* below, Phloem loading).

Effects of hormones In the search for explanations of the rules of translocate movement and in attempts to redirect the partitioning of assimilate to other sinks, the various classes of plant hormones have been tested for their influence on the process. It has proved very difficult to sort out hormonal effects on translocation from hormonal effects on the metabolism of sink tissue which may then attract translocate by becoming a 'stronger' sink. It may be fairly said that in intact plants there is no evidence that the endogenous levels of hormones in the various tissues bear any relation to their capacity to attract translocated material. Some influence of exogenously supplied hormones has been reported in mutilated plants. Indoleacctic acid (IAA) applied to the cut stump of decapitated *Phaseolus* seedlings attracts to that region a small proportion of labeled photosynthate from the lower leaves, but this approach does not aid the understanding of partitioning in intact plants.

Phloem loading

All investigations of translocation lead back to the prime importance of the high sucrose concentration in the leaf phloem. From the mesophyll cell protoplasm where it is produced at quite modest concentration (say 10 to 50 mM), the sucrose must be concentrated 20 to 100 times as it moves to the sieve tubes of the leaf veins. This process, the origin of the sucrose gradient and the pressure gradient, and an active area of investigation in recent years, is known as phloem loading. The smallest veins of leaves seem to be where most of the loading occurs. In net-veined leaves they are the first to show accumulation of ^{14}C assimilate above the level in the surrounding mesophyll (Fondy and Geiger, 1977). In parallel-veined leaves, where small veins lie in groups between pairs of large veins connected sideways by very small and sparse transverse veins, the specialization of the small veins for loading, and the large ones for longitudinal movement, is especially clear (Lush, 1976; Altus and Canny, 1982).

In wheat leaves, the minor veins are concerned only with local longitudinal translocation, assembling the assimilate for transport sideways through the transverse veins to the major veins. In both types of leaf the two or three sieve tubes of the minor veins are very small, lying among companion cells and phloem parenchyma cells several times their diameter, a proportion which is, of course, reversed in phloem along the large leaf veins, petiole and stem.

The major event in translocation, the sucrose-concentration step, must happen somewhere along the route from mesophyll cytoplasm to sieve tube lumen. There must be some boundary at which energy is applied by a pumping mechanism to raise the sucrose to a high chemical potential, either a boundary between apoplast and symplast (plas-

malemma) or a boundary within the symplast (membrane-bounded space or specialized plasmodesmata). At present more is known about how the concentration step is accomplished than just where it happens or the path involved. No acceleration is necessary for the short journey of up to 100 μm between mesophyll cell and sieve tube. Diffusion over such distances is a very rapid process, producing equilibrium of concentration in a few milliseconds. There are two possible routes for the traffic, an apoplastic and a symplastic route, and both have their adherents.

The apoplastic route involves the release of sucrose through the plasmalemma of mesophyll cells into the free space (cell walls), its movement there by diffusion through all accessible areas of the free space, and concentration by a pump at the cell membrane surrounding the sieve tube/companion cell complex of the minor veins. That this route is used is indicated by two kinds of observations. First, that disks of leaf tissue floated on ^{14}C-sucrose solutions quickly accumulate ^{14}C in their vein networks by a pumping mechanism that has Michaelis kinetics and is specific for this sugar (Giaquinta, 1977). Second, that ^{14}C-sucrose can be recovered from cut or abraded pieces of leaf by washing out the free space, usually by centrifuging leaves injected with water or by trapping with a solution of ^{12}C sucrose. Geiger (1975) reviews the evidence up to that time. Both Geiger (1975) and Giaquinta (1980), the main protagonists of this view, state it in terms not of a general release into the mesophyll free space, but rather a very local release within the phloem and an immediate uptake of the released molecules by the companion cells and sieve elements. They now envisage the main part of the distance between mesophyll cytoplasm and sieve element being covered within the symplast, 'alhough', as Geiger says, 'experimental evidence is meager'.

The symplastic route would confine the sucrose to the cytoplasm of the chain of cells connecting the mesophyll cells *via* the bundle sheath cells and phloem parenchyma to the companion cells and sieve elements, traveling from cell to cell in plasmodesmata. Proponents of this view, for example Kuo, O'Brien and Canny (1974), have explored the plasmodesmatal connexions and assessed their adequacy to carry the flux. These authors studied wheat leaf veins, concentrating their attention on the inner tangential walls of the bundle sheath cells which all fluxes would have to cross, finding nearly all the plasmodesmata in the small veins, and there, in just two cells on each side of the bundle, opposite the phloem. In this route there must be a discontinuity to accomplish the jump in concentration, corresponding to the apoplastic stage and pumping, some intercellular space or a set of plasmodesmata. For a review of the potential of plasmodesmata for cell-to-cell transport, *see* Gunning and Robards (1976).

Madore and Webb (1981) criticize the use of leaf disks and abraded leaves, pointing out that leakage into the free space is to be expected after such treatments. Working with squash and taking precautions against damage and leakage, they analyzed the free-space solution and found that it contained all the sugars to be found in the metabolic space, but at 100-fold less concentration. On feeding $^{14}CO_2$ to the leaves, they found the same ^{14}C-sugars labeled in the free space as in the metabolic space but at 1000-fold lower concentration. There was no excess of the transport sugar (stachyose) in the free space. They find no evidence for a stage of release of the transport sugar to the free space, and suggest that '"loading" of transport sugars in leaves may very well commence by "loading" of a cytoplasmic compartment within the mesophyll cells'.

There is little disagreement about the probable mechanism by which the sucrose is accumulated across the boundary, wherever that boundary may be. All follow the bacterial and animal physiologists in adopting Mitchell's scheme for the translation of energy of ATP by vectorial ATPases in membranes coupled to the transport of protons across the membrane. For an explanation of this most important theory *see* Hinkle and McCarty (1978) and Nicholls (1982). The energy for accumulating sucrose against a gradient of chemical potential is generated by the extrusion of protons (H^+) from the space at the expense of ATP, and then as the protons return across the membrane, a 'symport' carrier system specific for sucrose accumulates sucrose molecules within the space. Such sucrose–proton co-transport pumps have been found in the cotyledons of *Ricinus* (Komor, Rotter, Waldhauser, Martin and Cho, 1980) which absorb sucrose from the endosperm during germination. When presented with sucrose (but no other metabolic sugars) they cause a transient alkalinization of the medium as the sucrose is taken up; uptake is stimulated by fusi-

coccin and inhibited by a number of substances including dinitrophenol (DNP), valinomycin and *p*-chloromercuribenzene sulfonic acid. These properties have been found in the loading of sucrose into minor veins. A summary of the experiments and evidence, with more details about the operation of the co-transport, may be found in Giaquinta (1980). Many of these electrogenic proton pumps do not simply push H^+ to one side of a membrane, but exchange it for K^+ moving in the opposite direction. There is some evidence that the sucrose pump of translocation is of this kind, which would explain the universal presence of K^+ in sieve tube exudates and the beneficial effects of K^+ on loading and translocation.

Phloem unloading

A simple means of supplying translocate to sinks at the downstream end would be to let it leak passively from the sieve tubes into the apoplast to be taken up by whatever cells are active in the neighborhood. This seems to happen in stems of sugar cane where Glasziou and Gayler (1972) have made extensive studies of the uptake of sucrose from the apoplast and storage in the vacuole. It is also the necessary route for supplying young embryos within their seed coats because, being another generation, they have no symplastic connexion with the parent plant. Patrick and McDonald (1981) have shown that ^{14}C-assimilates in *Phaseolus* spread through the seed coats and permeate the free space of the cotyledons while the embryo grows. This simple pathway of unloading is not found in all sinks. Dick and ap Rees (1975) found that in the growing tips of pea roots ^{14}C-assimilate arriving in the stele and entering the growing cortex was not in the apoplast but remained confined to the symplast.

Phloem transport of compounds other than sugars

From what has been said so far it will be plain that any substance which finds its way into the sieve tubes will probably be translocated with the sugar to currently-active sinks and at the same speed. It is easier to assess such transport by analyzing for the substances in the sap than by devising special experiments to measure their translocation, and probably just as reliable. The composition of most fruits is found to reflect quite closely the relative proportions of elements in the sap of the parent plant. A very comprehensive review of what has been found in phloem saps of many species will be found in Ziegler (1975).

Nitrogenous substances in the form of amino acids and amides are usually present at levels of 10 to 100 μM, aspartic and glutamic acids and their amides being very common. The amount varies with the state of the plant: at the stages of bud break, and of leaf senescence when the ground substance of leaf protoplasm is re-mobilized for return to the stem and storage, the nitrogenous component is especially rich. Nitrate is not usually found. Reduced nitrogen, formed from nitrate either in the roots or leaves, is the mobile form, and exogenously applied labeled amino acids have been shown to move at the same speeds as sugar but at only about 3% of the mass transfer.

Of the inorganic cations, potassium is the most abundant (1 to 2 mg ml^{-1}), magnesium, calcium and sodium at a tenth of this concentration or less. The transpiration stream delivers a small but long-continued load of these ions to the leaves, and they must either accumulate there or be re-exported in the phloem. All indications are that potassium is re-exported, but that calcium is not; it accumulates in the leaves and is shed with them. Magnesium, in contrast, appears to be translocated. The phloem immobility of calcium casts the responsibility for its transport back to the transpiration stream, and non-transpiring fruits like the peanut, and storage organs in the soil, need to absorb their calcium directly from the soil. The anions of the sap are principally phosphate, with occasionally chloride, and organic anions such as malate. The pH of the sap is high, 7.5 to 8.5.

Growth substances of all the major classes have been found in phloem sap, and exogenously applied IAA moves there at the same speed as the sugar. There is no evidence that this distribution of auxin or of the other growth substances is used by the plant as a hormone signaling system to activate any of its meristematic or elongating growth regions. It is ironic that the only hormone which is known to be translocated is the one whose chemical identity remains mysterious, the flowering hormone. Produced in leaves in response to inductive photoperiods, it travels at the speed of translocation to the buds and induces the transition to flowering there.

Its passage to buds can be blocked by leaving alternative, non-induced leaves subtending the buds (nearer sources) (Fig. 13.3). A useful review of this and other growth substance transport is that of King (1976).

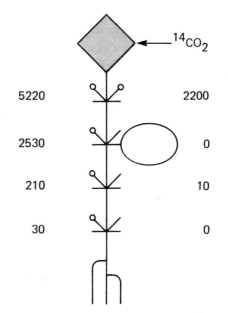

Fig. 13.3 The results of an experiment on the flowering of *Perilla*, a short-day plant. Induced (SD) leaves (rhombus) grafted at the apex of plants held in long days cause flowering in the axillary shoots of leaves that have been removed (circles). Axillary shoots do not produce flowers in the neighborhood of a remaining (LD) leaf. ^{14}C assimilated by the induced leaf is translocated to those buds which form flowers, but not to those buds which remain vegetative. The numbers indicate the relative ^{14}C content of the buds. (Redrawn after King and Zeevart (1973) by permission.)

Viruses move in the phloem. Transmission of many plant viruses is by aphids and other sap-feeding insects. Once injected they multiply and the particles may be found in sieve tubes. They move more sluggishly than the assimilate stream; a speed of 3.5 cm h^{-1} was measured by Helms and Wardlaw (1976) for tobacco mosaic virus at the same time as assimilate was moving at 60 cm h^{-1}. Esau (1961) gives a fascinating account of this three-way interaction of organisms.

Mechanisms of translocation—the search for a structural framework

Any explanation of how translocation works must be soundly based on knowledge of the internal organization of sieve tubes. In the absence of such knowledge all remains speculative, a balance of probabilities on the available conflicting evidence and personal prejudices about its interpretation. Even the least demanding of the hypotheses, Münch's proposal that the osmotic potential of the solutes generates a turgor pressure in sieve tubes of sources which drives a flow through the tubes and pores, requires an *absence* of structure within the tubes, and especially within the pores.

As revealed in the electron microscope by the best available techniques after either chemical fixation or rapid freezing and etching, the contents of sieve cells show two highly variable features: the presence or absence of microfilaments of protein (P-protein) in various configurations, and the filling or otherwise of the sieve pores with these filaments. Protagonists of mass flow are anxious to minimize the extent and pore-filling of these filaments; proponents of other hypotheses look to them to provide forces of motion either as some kind of contractile protein or as a charged matrix to facilitate electro-osmosis. Spanner (1978), after a careful review of all the published evidence, comes to the conclusion that the pores are more likely filled with filaments than not. The question of contractility (actin-like activity) of the filaments appears to have been finally settled in the negative. No positive actin reactions, such as binding with heavy meromyosin, have been detected, and Sabnis and Hart (1979) show that the P-protein is highly variable in its composition which bears no likeness to that of known contractile proteins.

However, it is quite certain that the images seen in the electron microscope are grave distortions of the living contents. In this chapter the special fragility of sieve tubes has constantly been stressed, and the difficulty of preserving them unchanged is to be expected. Images of robust plant cells are so full of interesting appearances that we often forget how gravely they are altered even by modern fixations. Mersey and McCully (1978) record a study of the course of fixation, monitoring changes as fixatives find their way into tomato hair cells. Even the best and most rapidly-penetrating fixatives took many minutes to produce a stable image while the proto-

plasm and membranes went through many transformations, distortions and vesiculations. A striking feature of the living protoplasm, its internal vacuolar network associated with microfibrils and streaming protoplasm, disappears entirely during fixation and has never been seen in the electron microscope. Sieve tubes are much less accessible to fixatives than hair cells, and, during the time that fixatives take to immobilize them, far more serious artefacts may be anticipated.

Rapid freezing seems to offer the chance of a quicker stabilization. Johnson (1978) has considered this question and concludes that the contents cannot be immobilized in less than $0.01 \, \text{s}$, and the time may be as long as $0.4 \, \text{s}$. During this time mass flow would move structures several micrometers, especially in the pores, where flow must be faster. At the scale of the organization of the P-protein microfilaments, even Brownian motion will move them in this time far enough to distort what must be a critical organization at the nanometer scale. He does not take into account the images published by Dempsey, Bullivant and Bieleski (1975) who used very cold metal to freeze celery phloem and showed that in the $12 \, \mu\text{m}$ nearest to the metal an organization of the filaments into thick strands could be revealed. Farther away, in regions of slower cooling, the usual artefact appearances were seen, the products of damage by movement or by the formation of ice crystals.

Observation of living, functioning sieve tubes is exceedingly difficult, again because of their fragility. A discussion of what has been observed and of the progressive breakdown of the organization with damage is given in Canny (1973). The techniques await fresh inventive genius.

With these uncertainties of the internal structure, models for translocation which demand a particular form of internal organization do not attract many adherents.

Electro-osmosis This mechanism in its simple original form pictured the sieve pores as the origin of the force for movement, not an obstruction. Filamentous material fixed in the pores bearing an electric ($-$) charge will exert a force on the solution around them if an electric potential is maintained across the sieve plate, causing flow through the plate. It was an elegant and satisfying idea, but so many modifications have had to be made to meet particular criticisms (Spanner, 1979) that its attractions have faded. The single most telling objection is that it could not transport ions with both positive and negative charges. The polarized potentials across the sieve plates have not been found, and the energy consumption would be very large.

Protoplasmic streaming variants Mechanisms based on the claims of Thaine to have seen intercellular strands of protoplasm moving through the sieve pores between elements were of several kinds, invoking peristaltic pumping in the strands and counter-current exchanges between strand and vacuole. These models are incapable of the speeds and SMTs in the higher ranges of Tables 13.1 and 13.2, and the membrane-bound strands have not been found. The strands seen by Dempsey *et al.* (1975) are of the right size but are not surrounded by membranes.

Contractile protein variants Contractile proteins would have been the source of streaming motion for the last-mentioned models, but various less definite proposals have been made suggesting the generation of movement by actin–myosin interactions during the period between the finding of P-protein filaments and the realization that they were not contractile proteins. Now that it seems certain that they are nothing of the kind, the heart has gone out of these proposals.

Pressure-driven flow The presence of some kind of pressure-driven flow in sieve tubes propelled by osmotic potentials generated in the leaves has been assumed in this chapter. In many plants it seems inescapable that it must occur in some parts of the phloem, where the pores are wide and open and pressure gradients are found. It is the most widely accepted hypothesis at present, though there are a number of reservations. In some plants, and perhaps in some parts of all plants, other mechanisms may well be operating. The gymnosperms, their extremely narrow pores and median nodules filled with protoplasmic elements, remain a barrier to universal acceptance for many. Also there are a number of situations in which sieve tubes appear to carry two substances in opposite directions simultaneously. The clearest demonstration of this for a single sieve tube is by Trip and Gorham (1968) who showed the presence of ^{14}C-assimilate and ^{3}H-glucose in a single sieve tube when they had arrived from opposite directions. In the minor veins of leaves also, and in the transverse veins of grass leaves, movement appears to go either way or both ways. It may well turn out that there is not a single universally-applicable mechanism of translocation. For a critique of this hypothesis see Fensom (1981), and of some others, Fensom (1975).

Polar transport of IAA

IAA shares with many other compounds the property of being translocated by the phloem once it finds its way in there. In a study to verify this Goldsmith, Cataldo, Karn, Brenneman and Trip (1974) supplied ^{14}C-IAA to leaves of *Coleus* through a cut flap of lamina, and showed that it traveled upwards and downwards in the stem at speeds of 16 to 20 cm h^{-1}, in all respects like ^{3}H-glucose fed simultaneously. They showed by autoradiography that it moved in the phloem and applied TIBA had no effect on the transport. In addition, IAA has a special transport system of its own in those organs much used for growth-substance studies, coleoptiles, hypocotyls and young roots, a transport system with quite different properties from translocation.

It is called *polar* transport because it is directed one way only, from apex to base in shoots. It differs from translocation in several ways:

(1) The speed is around 1 cm h^{-1}.
(2) It functions freely in short segments of tissue, 10 to 40 mm long.
(3) It is specific for auxin-like molecules.
(4) It will transport IAA against a gradient of concentration.
(5) It is blocked by TIBA.
(6) Feeding aphids do not have access to it.
(7) The rate declines as IAA content of the tissue falls.

A few experiments have been carried out on intact plants, applying labeled auxin to stem apices and young leaves, but the greater number of investigations have been made on short (10 to 20 mm) segments of stem, petiole, coleoptile or hypocotyl tissue between two blocks of agar. One, the donor block, is loaded with IAA (or ^{14}C-IAA) at the required concentration. The receiver block at the other end of the segment collects the IAA that arrives, and its contents may be estimated by bio-assay or radioactivity measurement.

Besides the seven properties listed above, the polar transport of IAA has been found to have many of the characteristics of diffusion, but the movement is in fact slower than diffusion of IAA through boiled segments (Chang and Jacobs, 1972). It may be thought of as an asymmetric diffusion with a greater coefficient basipetally than acropetally, using a limited volume of the tissue space. Vascular tissues are most active in the transport. Wangermann (1974) showed that interrupting the vascular strands of *Helianthus* stem segments prevented polar IAA transport, while interrupting the pith did not. Auto-radiographs showed ^{14}C-IAA most abundantly in the phloem region and rather less in the xylem. Plasmodesmata appear not to be the pathway between cells.

In roots, a similar polar transport system carries IAA in the same gravitational sense *towards* the apex, again predominantly in the stelar tissues.

Hypotheses to account for polar auxin transport have relied mostly on IAA-specific carrier mechanisms located in cell membranes at the basal end of each cell, secreting IAA into the free space to be picked up by the carrier in the plasmalemma of the next basal cell. Equilibration within each cell was believed to be by diffusion. A new hypothesis with more embracing explanations of the observed properties was suggested by Rubery and Sheldrake (1974) and Raven (1975). This, like the sucrose pumps, derives its energy from a proton extrusion pump, one in each cell, which makes the pH within high, and that of the free space low. As has been said, such proton pumps are found very widely in plant cells. The only other requirement for polar IAA transport is that each cell membrane should be slightly more permeable to the anion IAA$^-$ at the basal end. Because it is charged, the anion will have a much lower permeability through the plasma-lemma than the unionized acid. (Observations suggest a factor of 10^{-2} to 10^{-3}.) Because of the low pH outside, the IAAH favored by the equilibrium will enter the cell and there will be a total IAA (IAA$^-$ + IAAH) content within the cell greater than outside. With the slight polarity of permeability to IAA$^-$, more will leak out at the basal end, be turned to IAAH by the low pH and enter the next cell in the unionized form. Equilibration of concentrations within each cell is probably by diffusion in

the cytoplasm and vacuole. This hypothesis goes far to explain a great many of the observed properties of polar IAA transport, and focuses attention on the importance of controlling the pH of the transport-measuring systems by having buffers in the donor and receiver blocks. A review detailing the power and comprehensiveness of this simple mechanism will be found in Goldsmith (1977).

There is considerable doubt about the significance of this polar transport of IAA and its relevance to the polarity of differentiation of shoots or roots since Sheldrake (1974) showed that in inverted cuttings, which had lived for four months and produced roots from the morphologically upper end and shoots from the lower end, the original polarity of IAA transport still persisted.

Further reading

The literature sources where the current state of the subject may be explored are:
(a) several books and reviews by individuals presenting personal viewpoints and selections: Canny (1973), Crafts and Crisp (1971), MacRobbie (1971), Moorby (1981), Peel (1974), Wardlaw (1974);
(b) a number of conference reports and multiple-author volumes presenting a diversity of views: Aronoff *et al.* (1975), Wardlaw and Passioura (1976), Zimmermann and Milburn (1975); volume **93** (1) (1980) of *Ber. dtsch. bot. Ges.*; volume **58** (7) (1980) of *Can. J. Bot.*, pp. 745–832.

References

Altus, D. P. and Canny, M. J. (1982). Loading of assimilates in wheat leaves. I. The specialisation of vein types for separate activities. *Aust. J. Plant Physiol.* **9**, 571–81.

Aronoff, S., Dainty, J., Gorham, P. R., Srivastava, L. M. and Swanson, C. A. (1975). *Phloem Transport*, Plenum Press, New York.

Canny, M. J. (1973). *Phloem Translocation*, Cambridge University Press.

Canny, M. J. (1975). Mass transfer. In *Transport in Plants. I. Phloem Transport*, eds M. H. Zimmermann and J. A. Milburn, Springer, Berlin, pp. 139–53.

Chang, Y. P. and Jacobs, W. P. (1972). The contrast between active transport and diffusion of indole-3-acetic acid in *Coleus* petioles. *Plant Physiol.* **50**, 635–9.

Christy, A. L. and Swanson, C. A. (1976). Control of translocation by photosynthesis and carbohydrate concentrations of the source leaf. In *Transport and Transfer Processes in Plants*, eds I. F. Wardlaw and J. B. Passioura, Academic Press, New York, pp. 329–38.

Cook, M. G. and Evans, L. T. (1978). Effect of relative size and distance of competing sinks on the distribution of photosynthetic assimilates in wheat. *Aust. J. Plant Physiol.* **5**, 495–509.

Crafts, A. S. and Crisp, C. E. (1971). *Phloem Transport in Plants*, Freeman, San Francisco.

Dempsey, G. P., Bullivant, S. and Bieleski, R. L. (1975). The distribution of P-protein in mature sieve elements of celery. *Planta* **126**, 45–59.

Dick, P. S. and ap Rees, T. (1975). The pathway of sugar transport in the roots of *Pisum sativum. J. exp. Bot.* **26**, 305–14.

Esau, K. (1961). *Plants Viruses and Insects*, Cambridge, Mass.

Esau, K. and Cheadle, V. I. (1959). Size of pores and their contents in sieve elements of dicotyledons. *Proc. Nat. Acad. Sci. USA* **45**, 156–62.

Fellows, R. J. and Geiger, D. R. (1974). Structural and physiological changes in sugar beet leaves during sink to source conversion. *Plant Physiol.* **54**, 877–85.

Fensom, D. S. (1975). Other possible mechanisms. In *Transport in Plants. I. Phloem Transport*, eds M. H. Zimmermann and J. A. Milburn, Springer, Berlin, pp. 354–65.

Fensom, D. S. (1981). Problems arising from a Münch-type pressure flow mechanism of sugar transport in plants. *Can. J. Bot.* **59**, 425–32.

Fondy, B. R. and Geiger, D. R. (1977). Sugar selectivity and other characteristics of the phloem loading in *Beta vulgaris* L. *Plant Physiol.* **59**, 953–60.

Geiger, D. R. (1975). Phloem loading. In *Transport in Plants. I. Phloem Transport*, eds M. H. Zimmermann and J. A. Milburn, Springer, Berlin, pp. 395–431.

Geiger, D. R., Saunders, M. A. and Cataldo, D. A. (1969). Translocation and accumulation of translocate in the sugar beet petiole. *Plant Physiol.* **44**, 1657–65.

Geiger, D. R. and Sovonick, S. A. (1975). Effects of temperature, anoxia and other metabolic inhibitors on translocation. In *Transport in Plants I. Phloem Transport*, eds M. H. Zimmermann and J. A. Milburn, Springer, Berlin, pp. 256–86.

Geiger, D. R. and Swanson, C. A. (1965). Evaluation of selected parameters in a sugar beet system. *Plant Physiol.* **40**, 942–7.

Giaquinta, R. T. (1977). Phloem loading of sucrose. pH dependence and selectivity. *Plant Physiol.* **59**, 750–5.

Giaquinta, R. T. (1980). Mechanism and control of phloem loading of sucrose. *Ber. dtsch. bot. Ges.* **93**, 187–201.

Glasziou, K. T. and Gayler, K. R. (1972). Storage of sugars in stalks of sugar cane. *Bot. Rev.* **38**, 471–90.

Goldsmith, M. H. M. (1977). The polar transport of auxin. *Ann. Rev. Plant Physiol.* **28**, 439–78.

Goldsmith, M. H. M., Cataldo, D. A., Kam, J., Brenneman, T. and Trip, P. (1974). *Planta* **116**, 301–17.

Grange, R. I. and Peel, A. J. (1978). Evidence for solution flow in the phloem of willow. *Planta* **138**, 15–23.

Gunning, B. E. S. and Robards, A. W. (1976). Plasmodesmata and symplastic transport. In *Transport and Transfer Processes in Plants*, eds I. F. Wardlaw and J. B. Passioura, Academic Press, New York, pp. 15–41.

Hammel, H. T. (1968). Measurement of turgor pressure and its gradient in the phloem of oak. *Plant Physiol.* **43**, 1042–8.

Helms, K. and Wardlaw, I. F. (1970). Movement of viruses in plants: long distance movement of tobacco mosaic virus in *Nicotiana glutinosa*. In *Transport and Transfer Processes in Plants*, eds I. F. Wardlaw and J. B. Passioura, Academic Press, New York, pp. 283–93.

Hinkle, P. C. and McCarty, R. E. (1978). How cells make ATP. *Sci. Am.* **238** (3), 104–23.

Hoddinott, J., Ehret, D. L. and Gorham, P. R. (1979). Rapid influences of water stress on photosynthesis and translocation in *Phaseolus vulgaris. Can. J. Bot.* **57**, 768–76.

Johnson, R. P. C. (1978). The microscopy of P-protein filaments in freeze-etched sieve pores. *Planta* **143**, 191–205.

King, R. W. (1976). Implications for plant growth of the transport of regulatory compounds in phloem and xylem. In *Transport and Transfer Processes in Plants*, eds I. F. Wardlaw and J. B. Passioura, Academic Press, New York, pp. 415–31.

King, R. W. and Zeevart, J. A. D. (1973). Floral stimulus movement in *Perilla* and flower inhibition caused by non-induced leaves. *Plant Physiol.* **51**, 727–38.

Komor, E., Rotter, M., Waldhauser, J., Martin, E. and Cho, B. H. (1980). Sucrose proton symport for phloem loading in the *Ricinus* seedling. *Ber. dtsch. bot. Ges.* **93**, 211–19.

Kuo, J., O'Brien, T. P. and Canny, M. J. (1974). Pit-field distribution, plasmodesmatal frequency, and assimilate flux in the mestome sheath cells of wheat leaves. *Planta* **121**, 97–118.

Lush, W. M. (1976). Leaf structure and translocation of dry matter in a C3 and a C4 grass. *Planta* **130**, 235–44.

Lush, W. M. and Evans, L. T. (1974). Translocation of photosynthetic assimilate from grass leaves, as influenced by environment and species. *Aust. J. Plant Physiol.* **1**, 417–31.

MacRobbie, E. A. C. (1971). Phloem translocation. Facts and mechanisms: a comparative survey. *Biol. Rev.* **46**, 429–81.

Madore, M. and Webb, J. A. (1981). Leaf free space analysis and vein loading in *Cucurbita pepo. Can. J. Bot.* **59**, 2550–7.

Mengel, K. (1980). Effect of potassium on the assimilate conduction to storage tissue. *Ber. dtsch. bot. Ges.* **93**, 353–62.

Mersey, B. and McCully, M. E. (1978). Monitoring of the course of fixation in plant cells. *J. Microscopy* **114**, 49–76.

Milburn, J. A. (1974). Phloem transport in *Ricinus*. Concentration gradients between source and sink. *Planta* **117**, 303–19.

Milburn, J. A. (1980). The measurement of turgor pressure in sieve tubes. *Ber. dtsch. bot. Ges.* **93**, 153–66.

Milburn, J. A. and Zimmermann, M. H. (1977). Preliminary studies on sap flow in *Cocos nucifera* L. II Phloem transport. *New Phytol.* **79**, 543–58.

Moorby, J. (1981). *Transport Systems in Plants*, Longman, London.

Murmanis, L. and Evert, R. F. (1966). Some aspects of sieve cell ultrastructure in *Pinus strobus*. *Am. J. Bot.* **53**, 1065–78.

Nicholls, D. G. (1982). *Bioenergetics*, Academic Press, London.

Passioura, J. B. and Ashford, A. E. (1974). Rapid translocation in the phloem of wheat roots. *Aust. J. Plant Physiol.* **1**, 521–7.

Patrick, J. W. and McDonald, R. (1981). Pathway of carbon transport within developing ovules of *Phaseolus vulgaris* L. *Aust. J. Plant Physiol.* **7**, 671–84.

Peel, A. J. (1974). *Transport of Nutrients in Plants*, Butterworths, London.

Peel, A. J. and Ho, L. C. (1970). Colony size of *Tuberolachnus salignus* (Gmelin) in relation to mass transport of ^{14}C-labelled assimilates in willow. *Physiol. Plant.* **23**, 1033–8.

Raven, J. A. (1975). Transport of indoleacetic acid in plant cells in relation to pH and electrical gradients, and its significance for polar IAA transport. *New Phytol.* **74**, 163–72.

Rubery, P. H. and Sheldrake, A. R. (1974). Carrier-mediated auxin transport. *Planta* **118**, 101–21.

Sabnis, D. D. and Hart, J. W. (1979). Heterogeneity in phloem protein complements from different species. Consequences to hypotheses concerned with P-protein function. *Planta* **145**, 459–66.

Sheldrake, A. R. (1974). The polarity of auxin transport in inverted cuttings. *New Phytol.* **73**, 637–42.

Smith, J. A. C. and Milburn, J. A. (1980). Phloem transport, solute flux and the kinetics of sap exudation in *Ricinus communis* L. *Planta* **148**, 35–41.

Sovonick, S. A., Geiger, D. R. and Fellows, R. J. (1974). Evidence for active phloem loading in the minor veins of sugar beet. *Plant Physiol.* **54**, 886–91.

Spanner, D. C. (1978). Sieve-plate pores, open or occluded? A critical review. *Plant, Cell Environ.* **1**, 7–20.

Spanner, D. C. (1979). The electroosomotic theory of phloem transport: a final restatement. *Plant, Cell Environ.* **2**, 107–21.

Thompson, R. G., Fensom, D. S., Anderson, R. R., Drouin, R. and Leiper, W. (1979). Translocation of ^{11}C from leaves of *Helianthus, Heracleum, Nymphoides, Ipomoea, Tropaeolum, Zea, Fraxinus, Ulmus, Picea* and *Pinus*; comparative shapes and some fine structure profiles. *Can. J. Bot.* **57**, 845–63.

Thrower, Stella, L. and Thrower, L. B. (1980). Translocation into mature leaves—the pathway of assimilate movement. *New Phytol.* **86**, 145–54.

Trip, P. and Gorham, P. R. (1968). Bidirectional translocation of sugars in sieve tubes of squash plants. *Plant Physiol.* **43**, 877–82.

Troughton, J. H. and Currie, J. B. (1977). Relations between light level, sucrose concentration, and translocation of carbon 11 in *Zea mays* leaves. *Plant Physiol.* **59**, 808–20.

Wangermann, E. (1974). The pathway of transport of indolyl-acetic acid through internode segments. *New Phytol.* **73**, 623–36.

Wardlaw, I. F. (1967). The effect of water stress on translocation in relation to photosynthesis and growth. I. Effect during grain development in wheat. *Aust. J. Biol. Sci.* **20**, 25–39.

Wardlaw, I. F. (1969). The effect of water stress on translocation in relation to photosynthesis and growth. II. Effect during leaf development in *Lolium temulentum*. *Aust. J. Biol. Sci.* **22**, 1–16.

Wardlaw, I. F. (1974). Phloem transport: physical chemical or impossible. *Ann. Rev. Plant Physiol.* **25**, 515–39.

Wardlaw, I. F. and Passioura, J. B. (eds) (1976). *Transport and Transfer Processes in Plants*, Academic Press, New York.

Wareing, P. F. and Patrick, J. W. (1975). Source-sink relations and the partition of assimilates in the plant. In *Photosynthesis and Productivity in Different Environments*, ed. J. P. Cooper, Cambridge University Press, pp. 481–99.

Willenbrink, J. (1978). Localized inhibition of translocation of ^{14}C-assimilates in the phloem by valinomycin and other metabolic inhibitors. *Planta* **139**, 261–6.

Ziegler, H. (1975). Nature of transported substances. In *Transport in Plants. I. Phloem Transport*, eds M. H. Zimmermann and J. A. Milburn, Springer, Berlin, pp. 59–100.

Zimmermann, M. H. and Milburn, J. A. (1975). *Transport in Plants. I. Phloem Transport*, Springer, Berlin.

Water relations

14

D A Baker

Introduction

Over the past two decades our knowledge of plant–water relationships has progressed rapidly. This progress has resulted from the adoption of a thermodynamic approach towards the problems of water movement within plants and other biological systems, and to the development of increasingly sophisticated methods to determine the various physical parameters of such movement (*see* Dainty, 1976). As a consequence the student of plant–water relations meets concepts and analyses which are within the theory of irreversible thermodynamics. It is therefore essential that some basic knowledge of thermodynamics is obtained before attempting further to understand the fundamental theory of water movement. A text of thermodynamics suitable for biologists has been written by Spanner (1964) and an excellent account of the application of the theory of irreversible thermodynamics to water movement in plants is available (Dainty, 1963).

The present chapter presents a simplified account of water movement in plants which should provide readers with the background knowledge to delve into the current literature on the topic. The excellent text by Slatyer (1967) has not been superseded and should be used in conjunction with more recent accounts in House (1974), Nobel (1974), Meidner and Sheriff (1976) and Milburn (1979).

In studying water movement, the inextricable manner in which plant growth and development involve water should not be overlooked. All life is dependent on water and exists within a relatively narrow range of conditions determined by its physical properties. Water often comprises more than 70% of the fresh weight of plant tissues, in some cases exceeding 90%. As an essential constituent of cytoplasm, sometimes constituting 95% of the total weight, water is vital for the structural integrity of biological molecules participating either directly or indirectly in all metabolic activities of plants. Water also serves a vital function as a solvent for the various molecules transported within the plant and is involved in the development and maintenance of cell turgidity upon which growth and development of the organism depend.

Water movement

Continuity of liquid water exists from the soil through the plant to the liquid:gas interface at the evaporating surfaces within the leaf. Transport of water through this system involves liquid phase transport through the soil and plant, and gas phase transport from the liquid:gas interface to the bulk air, the driving forces for this transport differing at various stages in the system.

Water potential

Water movement occurs along gradients of decreasing total water potential, water potential being defined as the difference, in free energy per unit volume, between matrically-bound, pressurized, or osmotically-constrained water and pure water. The water potential of pure free water is for convenience designated as zero. When a solute is added to pure free water the water potential is lowered (becomes negative) as free energy is decreased, whereas when pure free water is compressed or heated the water potential is raised (becomes positive) as the free energy is increased.

Water potential (ψ, 'psi') can be determined from the relationship between the chemical potential of pure free water ($\mu_w°$) and that of water under the conditions mentioned above (μ_w) from the relationship

$$\psi = \frac{\mu_w - \mu_w°}{\overline{V}_w} = \frac{RT}{\overline{V}_w} \ln \frac{\theta}{\theta°} \qquad (1)$$

where R is the gas constant ($8.31\,\mathrm{J\,K^{-1}\,mol^{-1}}$), T is the absolute temperature (K), $\overline{V}_w$ is the partial molar volume of water in the system ($\mathrm{m^3\,mol^{-1}}$), θ and $\theta°$ are the water vapor density in the system and pure water respectively at temperature T. The dimensions of water potential are thus energy per unit mass or volume ($\mathrm{J\,kg^{-1}}$) and are equivalent to those of pressure ($\mathrm{kg\,m^{-1}\,s^{-2} = N\,m^{-2}}$) and water potential is usually expressed in units of pressure— pascals, bars or atmospheres. The conversion factor is approximately $100\,\mathrm{J\,kg^{-1}} = 1\,\mathrm{bar} = 10^5\,\mathrm{Pa} = 0.987$ at. Most modern works use the megapascal ($1\,\mathrm{MPa} = 10\,\mathrm{bar} = 9.87\,\mathrm{at}$) and that unit will be used here.

Water movement through the soil and in the plant often involves flow in response to pressure or gravitational gradients, in which case the movement is described as a mass flow, the liquid moving simultaneously in the same direction, as in the movement of water through saturated soils, xylem vessels and other tissues.

Diffusion

Diffusion, involving the random movements of individual molecules, is of immense significance in plant–water relations, with evaporation, a diffusional process, being the overall driving force for most water movement through the plant. The rate of diffusion of a substance can be calculated from Fick's law:

$$J = -D \frac{d\mu}{dx} \qquad (2)$$

where J is the quantity of the substance diffusing in unit time across unit area, $d\mu/dx$ the chemical potential gradient and D the coefficient of diffusion.

The ease with which a particular substance diffuses across a membrane may be represented by a permeability coefficient, k. For a simple non-electrolyte such as sugar the net inward flux across a membrane may be expressed as

$$J_s = k_s \Delta c_s \qquad (3)$$

where J_s is the flux of solute, i.e., the quantity passing unit area in unit time ($\mathrm{mol\,m^{-2}\,s^{-1}}$), k_s is the solute permeability coefficient ($\mathrm{m\,s^{-1}}$) and c_s the solute concentration difference ($\mathrm{mol\,m^{-3}}$). The above equation is analogous with Fick's equation and may be used to calculate k_s for a number of solutes, including isotopic water in water, k_d, from which the permeability coefficient for water, k_w, may be estimated. k_w is often referred to as the hydraulic conductance, L_p.

For diffusion of water across a membrane the corresponding equation is

$$J_v = L_p \Delta \psi \qquad (4)$$

where J_v is the flux of water, that is, the volume of water passing unit area in unit time ($\mathrm{m^3\,m^{-2}\,s^{-1}}$), L_p the hydraulic conductance ($\mathrm{m\,s^{-1}\,Pa^{-1}}$) and $\Delta\psi$ the difference in water potential across the membrane (Pa).

When simultaneous flow of water and solutes occurs, the above equations are inadequate, and when electrolytes are present the situation becomes increasingly complex (see Hall and Baker, 1977).

Osmosis

When a solution is separated from water by a membrane permeable only to the water (a semi-permeable membrane) there tends to be a net flux of water into the solution, as ψ is higher in the pure water than in the solution. This process is *osmosis* and the positive pressure applied to the solution necessary to prevent a net flux of water is the osmotic pressure (π). This movement of water across a membrane involves both diffusion and mass flow (Dainty, 1965). When a semi-permeable membrane separates pure water from a solution, only water will enter the pores, the solute being excluded. At the pore aperture diffusion of water molecules into the solution will occur, due to the difference in chemical potential, which creates a localized decrease in pressure resulting in a mass flow of water along the pore. However, plant cell membranes do not always behave in an ideal semi-permeable manner, and many solute molecules move across them.

The term 'osmotic pressure' may be misleading, as an unconfined solution has an osmotic pressure although no pressure in the literal sense is exerted. It is when confined by a membrane that a solution may exhibit a pressure. Solute potential (ψ_s) is thus a preferable term, being equal in magnitude but opposite in sign to π. Thus $\pi = -\psi_s$

ψ_s is one of the colligative or binding properties of a solution and is directly proportional to the number of solute molecules in a given volume. From the van't Hoff relationship

$$\psi_s = \frac{-nRT}{V} \qquad (5)$$

where n is the number of solute molecules in solution of volume V, R and T as before. From this relationship a molar solution of a non-electrolyte should have a ψ_s value of -2.27 MPa at $0°C$. Measured values are lower (more negative) than the theoretical due to solute hydration and the volume of solution occupied by the solute. The development of ψ_s in plant cells is of primary importance in determining their water relationships.

Electro-osmosis

Water can also be induced to move in response to an electrical potential gradient under certain conditions. When charged particles are fixed, as on the surface of a pore in a membrane, an electrical potential gradient will induce a water flow, such movement being termed electro-osmosis. Fixed negative charges on the pore surface will induce a positive polarity in adjacent water molecules which will then move towards a negative potential, while fixed positive charges will induce water movement in the opposite direction. Electro-osmosis has not been demonstrated and may not be very significant in the movement of water in plants because the low resistance of the normal pathways results in mass flow and diffusion predominating. It may develop during sap flow through porous systems such as cell walls and it has been proposed that electro-osmosis may contribute to transport in the phloem (Fensom, 1957; Spanner, 1958).

Water relations of cells and tissues

Mature plant cells are typically composed of a limiting cell wall enclosing a cytoplasmically-bounded vacuole. The cell walls are composed of a mesh of cellulose microfibrils $10-30$ nm in diameter with gaps $1-100$ nm between them. The wall may also contain pectic substances, proteins and a variety of other constituents with an abundance of hydroxyl and carboxyl groups which adsorb water by hydrogen bonding. Water is also held by surface tension in the inter-fibrillar spaces. Because of this attraction between water molecules and the matrix (cell walls), forces of considerable magnitude are necessary to extract water from such a system. The water potential is lowered by these matric or imbibitional forces, the term matric potential, ψ_m, being used in this situation. Up to 50% of the cell wall volume may be occupied by water held by matric forces. Similarly, within the soil much of the water movement is in response to $\Delta\psi_m$. The presence of proteins and other colloids within the cytoplasm results in ψ_m, in addition to the ψ_s due to dissolved solutes. Vacuolar sap is largely a true solution composed of about 2% solids and 98% water; thus water is retained in the vacuole primarily by osmotic forces, ψ_s, although when a high proportion of colloidal material is present, as in the leaves of some conifers, ψ_m also contributes significantly.

In entire cells and tissues the various forces outlined above interact continuously, yielding a general relationship between the amount of water present relative to that in a fully turgid tissue (termed the relative water content) and the ψ_w (Fig. 14.1). The shape of the curve indicates that a significant proportion of the water is retained even when severe water potential gradients ($\Delta\psi$) are imposed. This proportion is greater in more drought-tolerant plants than in the more mesophytic ones.

Idealized cell

A useful concept in plant-water relations is that of an idealized cell, possessing an elastic wall, of negligible volume, and a single semi-permeable membrane, of negligible thickness, binding a vacuole containing an aqueous solution. The presence of the cell wall restricts the degree of volume change by imposing a hydrostatic pressure on the vacuolar sap. A pressure potential ψ_p, often termed

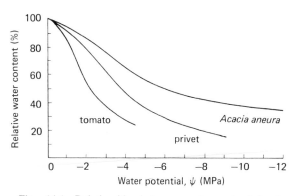

Fig. 14.1 Relationship between water potential, ψ (abscissa), and relative water content (ordinate) for mesophytic (tomato, privet) and xerophytic (*Acacia aneura*) tissue types (after Slatyer, 1960).

the turgor pressure P, is included as a component of the water potential. Thus

$$\psi_{cell} = \psi_p + \psi_s \qquad (6)$$

or

$$\psi_{cell} = P - \pi$$

When such a cell is placed in pure water, water will move into the cell until ψ_p is equal in magnitude to ψ_s and the total ψ of the cell is zero. This is the

condition of full turgor at which water absorption stops, where

$$-\psi_s = \psi_p \quad \text{or} \quad \pi = P \qquad \psi_{cell} = 0 \qquad (7)$$

Conversely, when the same cell is placed into a solution of low ψ_s, water will move out of the vacuole. At the point when the membrane ceases to press against the elastic cell wall, ψ_p becomes zero and ψ_s is then equal to ψ_{cell}. This is the point of incipient plasmolysis:

$$\psi_p = 0, \qquad \psi_{cell} = \psi_s \quad \text{or} \quad \psi_{cell} = -\pi \qquad (8)$$

The important concept implicit in the above relationship is that water movement is a response to ψ_{cell}, not just to ψ_s. At equilibrium, water movement into the cell is equal to water movement out, and the net flux is zero. The relationships between turgor pressure P, solute potential ψ_s and water potential ψ_{cell} are illustrated in Fig. 14.2.

The above discussion refers to an idealized plant cell and omits the contribution of the cell wall ψ_m and cytoplasmic ψ_m. A more accurate formulation for a typical plant cell should include ψ_m to give

$$\psi_{cell} = \psi_p + \psi_s + \psi_m \qquad (9)$$

ψ_p is usually positive or zero in individual cells but becomes negative in transpiring xylem tissues when tensions (negative pressures) arise. ψ_s and ψ_m are

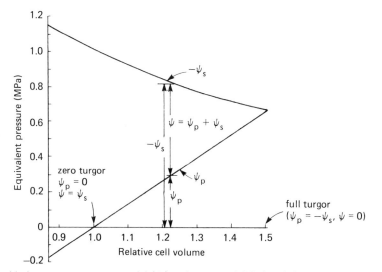

Fig. 14.2 Relationship between pressure potential (ψ_p), solute potential (ψ_s) and the resultant water potential (ψ) in an idealized (elastic) plant cell. The volume changes of the cell are subject to the cell wall elasticity (after Höfler, 1920).

always negative or zero. Total water potential, ψ_{cell}, is generally negative or zero, being positive in the sieve eiements and in the xylem when transpiration is zero (root pressure).

The flux of water across a plant cell membrane was given earlier as

$$J_v = L_p \Delta\psi \qquad (10)$$

which can be expanded for the three components of $\Delta\psi$ to give

$$J_v = L_p(\Delta\psi_p + \Delta\psi_s + \Delta\psi_m) \qquad (11)$$

although if water movement into a vacuole containing a true solution is being considered, $\Delta\psi_m$ can be omitted.

When the external solution is more concentrated than the vacuolar sap, water will continue to move out of the vacuole after incipient plasmolysis, causing separation of the membrane from the wall and allowing external solution to enter the space between them. This familiar laboratory phenomenon of plasmolysis seldom occurs under natural conditions unless the plants are suddenly inundated with a concentrated solution such as sea water. Plasmolyzed cells remain alive and may be deplasmolyzed by immersion in water, although prolonged plasmolysis is irreversible and results in death (Stadelmann, 1966, 1969).

Some of the water taken up by plant cells is retained in the cell wall and cytoplasm. Each of these cell components has different values of ψ_p, ψ_s and ψ_m, but when a cell is in equilibrium the ψ_w in each phase is the same and equal to the external ψ_w with which the cell is equilibrated. Withdrawal of water from the cell wall, as during evaporation, will lower ψ_w in this phase, creating a $\Delta\psi$ along which water moves from the cytoplasm and vacuole. Changes in the vacuolar volume as a result of absorption or loss of water are accompanied by changes in cytoplasmic volume, but such cytoplasmic changes are difficult to detect and measure.

Within structurally homogeneous groups of cells, such as parenchyma, individual cells may have quite different values of ψ_s, but when such a tissue is at equilibrium ψ_w has the same value for all cells.

Elasticity

The elasticity of the cell wall is a feature of plant cells which allows volume changes to occur over a range of hydrostatic pressures. These elastic properties of the cell wall are described by the volumetric elastic modulus, ε, which determines the slope of the pressure–volume curve in Fig. 14.2. This measure of the cell wall to resist pressure is defined by the following equation:

$$\varepsilon = \Delta\psi_p \frac{V}{\Delta V} \qquad (12)$$

where ε is the volumetric elastic modulus (Pa), $\Delta\psi_p$ the change in hydrostatic pressure, V the volume at full turgor and ΔV the volume change induced by hydrostatic pressure change. Large values of ε indicate a rigid, relatively inelastic wall, whereas small values indicate a very elastic wall.

ε can be estimated by applying small known hydrostatic pressures to the vacuole and measuring changes in cell volume. This technique has been successfully applied to giant algal cells and to some cells of higher plants (Zimmermann and Steudle, 1975). In situations where the hydrostatic pressure cannot be measured (i.e., where a probe cannot be inserted) ε must be estimated from ψ_{cell} and ψ_s as functions of V. Then

$$\varepsilon = (\Delta\psi_{cell} - \Delta\psi_s) \frac{V}{\Delta V} \qquad (13)$$

Such measurements and estimates indicate that ε increases with increasing cell turgor: when the cells are at or near full turgor, ε is in the order of $10{-}15$ MPa, but it becomes very small as the turgor approaches zero. Thus ε is an important parameter in cell water relations, controlling the manner in which ψ_{cell} alters as the cell volume changes (see Dainty, 1976; Zimmermann, 1978).

Determination of water status

The water status of plant cells and tissues can be determined by a variety of techniques. For a full review of methods see Barrs (1968), Boyer (1969) and Slavik (1974). ψ_{cell} and ψ_s may be measured directly by a number of methods, some of which are outlined below. ψ_m may also be measured directly, leaving ψ_p which is often obtained by difference when the other parameters have been obtained, but may be measured directly with a pressure probe (see Zimmermann, 1978).

Determination of water potential

ψ_w of a cell or tissue is the same as that of the liquid or vapor with which the tissue is in equilibrium. The ψ_w of an unknown tissue sample may therefore be determined by finding a solution in which the tissue neither gains nor loses volume or weight. The tissue is placed in a range of solutions, preferably of a non-permeating solute such as mannitol or polyethylene glycol (PEG), and changes in tissue volume or weight are measured after equilibration. When plotted against the ψ_s of the external solution, the intercept on the abscissa (x-axis) gives the value corresponding to the ψ_w of the tissue (Fig. 14.3).

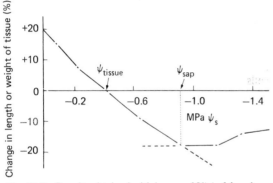

Fig. 14.3   Results obtained with leaves of *Vicia faba* when determining leaf water potential by floating leaf strips in a series of osmotic solutions, and measuring changes in the dimension of the strips under a dissecting microscope fitted with measuring graticule. The discontinuity of the curve −0.92 MPa indicates the solute potential of leaf cell sap. Other tissues can be used and changes in weight, instead of length, can be determined (after Kassam, 1972).

This method also gives a rough indication of ψ_s for the vacuolar sap, in that after incipient plasmolysis the tissue no longer loses weight or volume, the space between wall and protoplast being filled by the external solution.

An alternative approach is to measure changes in the external solution instead of changes in the tissue, the solution, which shows no change in its refractive index or density, being taken as equal to the tissue water potential. This is the basis of Shardakov's method where change of density of the bathing medium is detected by the rise or fall of an introduced colored droplet of the original medium (Knipling and Kramer, 1967).

The water vapor pressure of the air in equilibrium with a tissue in an enclosed chamber may be determined directly, using a thermocouple psychrometer or similar device, and the ψ_w deduced. Evaporation from a small water droplet on the wet bulb of a thermocouple causes a temperature depression which is proportional to the vapor pressure in the chamber. This temperature depression induces an electric current (the Seebeck effect) which is proportional to the temperature difference. Calibration is achieved by measuring the current when solutions of known ψ_w are in the chamber. Thermocouple psychrometers are also used extensively to measure ψ_w of soils. The two types of psychrometer most commonly used are developed from the proposals of Spanner (1951) and Richards and Ogata (1958), the former using the Peltier and the latter the Seebeck effect on small thermocouples.

The ψ_w of a leafy twig or shoot may be measured using a pressure bomb (Tyree and Hammel, 1972; Tyree, 1976). The pressure necessary to cause xylem sap just to exude from the cut surface will be equivalent to the ψ_w existing in the tissue prior to cutting. This value will be equal in magnitude but opposite in sign to the tension in the xylem when intact. Using this technique, ψ_w of branches taken at various heights on giant redwood trees has been determined, a marksman being employed to shoot off the branches which were then placed in a pressure bomb and the ψ_w measured (Scholander, Hammel, Bradstreet and Hemmingsen, 1965).

ψ_w may also be measured indirectly by establishing its relationship with relative water content (RWC) for a particular tissue. RWC is the amount of water present in a tissue compared with the amount present when the tissue is fully turgid. Disks of leaf tissue are weighed to give the fresh weight (W_f), floated on water for about one hour to eliminate any water deficit, reweighed to give the turgid weight (W_t) and finally oven-dried and weighed to give the dry weight (W_d). RWC (or relative turgidity) is then obtained from the ratio between the water content at sampling and the water content at full turgor, usually expressed as a percentage:

$$\text{relative water content} = \frac{W_f - W_d}{W_t - W_d} \times 100 \qquad (14)$$

Determination of solute potential

ψ_s for vacuolar sap may be determined on extracted

sap by measuring either its freezing point depression or its vapor pressure. With giant algal cells, pure sap samples may be obtained by micropipette; from higher plant cells, sap is extracted by freezing and thawing the tissue and then squeezing out the sap (*see* Crafts, Currier and Stocking, 1949). The depression of freezing point ΔT is related to ψ_s, ΔT of a molar solution of a non-electrolyte being 1.86°C, ψ_s of the same solution being -2.27 MPa. Thus, for an unknown solution,

$$\psi_s = \frac{-2.27}{1.86}\Delta T = -1.22\,\Delta T \qquad (15)$$

The vapor pressure of extracted sap can be measured with the thermocouple psychrometer (*see* above) although errors may arise due to contamination of the sap by other cell constituents.

An alternative procedure is to find the ψ_s of an external solution which causes incipient plasmolysis, at which point ψ_p is reduced to zero and ψ_s (medium) = ψ_s (tissue). In practice, thin strips of tissue are immersed in a series of solutions of graded ψ_s after which the strips are examined microscopically to determine the percentage of cells plasmolyzed. The results may be plotted graphically, the ψ_s corresponding to 50% plasmolysis being taken to equal the mean value of the vacuolar ψ_s at incipient plasmolysis. A correction is necessary so that the result may be related to the ψ_s of the sap in the fully turgid cell. This is because the vacuolar volume is less at incipient plasmolysis than at full turgor and

thus the value obtained must be multiplied by the ratio of the vacuolar volumes.
Thus

$$\psi_s \text{ (turgid)} = \psi_s \text{ (plasmolyzed)} \times \frac{V \text{ (plasmolyzed)}}{V \text{ (turgid)}}$$

$$(16)$$

The value of V (plasmolyzed)/V (turgid) depends on the volumetric elastic modulus (ε), that is, the cell wall elasticity of the tissue (Dainty, 1976). Adhesion between the cytoplasm and the cell wall, due in part to the presence of plasmodesmata, may lead to errors in this plasmometric method. Also if the osmotic solute can penetrate into the vacuole the apparent value of ψ_s will be greater (more negative). In this case a correction may be made, the ratio of the real to the apparent ψ_s being termed the reflection coefficient σ, where

$$\sigma = \frac{\psi_s \text{ (tissue)}}{\psi_s \text{ (medium)}} \qquad (17)$$

If the membrane is completely permeable to a solute, $\sigma = 0$; when it is completely impermeable, $\sigma = 1$. Thus $0 < \sigma < 1$, and for most biological membranes and small solutes $\sigma = 0.7–0.9$ (Slatyer, 1967). Some values of σ are presented in Table 14.1. Many measurements of ψ_s are made assuming $\sigma = 1$, sometimes leading to false conclusions regarding the water relations of a tissue. J_v is also influenced by σ,

Table 14.1   Reflection coefficients for some solutes of some giant algal cells (compiled by Dainty, 1976)

Solute	*N. flexilis*[a]	*N. translucens*[b]	*C. corallina*[b]	*V. utricularis*[c]
Sucrose	0.97	—	—	1
Glucose	0.96	—	—	0.95
Glycerol	0.80	—	—	0.81
Acetamide	0.91	—	—	0.79
Urea	0.91	1	1	—
Formamide	0.79	1	1	—
Ethylene glycol	0.94	1	1	—
Isopropanol	0.35	0.27	—	—
n-Propanol	0.17	0.16	0.22	—
Ethanol	0.34	0.29	0.27	—
Methanol	0.31	0.25	0.30	—

[a] Data on *N. flexilis* are from Steudle and Zimmermann (1974).
[b] Data on *N. translucens* and *C. corallina* are from Dainty and Ginzburg (1964).
[c] Data on *V. utricularis* are from Zimmermann and Steudle (1970).

and thus Equation (11) should be modified to give

$$J_v = L_p \left(\Delta\psi_p + \sigma\Delta\psi_s + \Delta\psi_m \right) \qquad (18)$$

Determination of matric potential

ψ_m usually makes quite a small contribution to cell–water relations and is often ignored. It may be measured by placing frozen and thawed tissue plus extruded sap in a psychrometer to obtain the combined value $\psi_s + \psi_m$, sometimes termed ψ_π, the osmotic potential. If a droplet of extracted sap is then used on the thermocouple, ψ_s will be eliminated, ψ_m only is measured, and ψ_s can be obtained by subtraction (Boyer, 1967).

ψ_p may be obtained by difference when the other components have been measured or may be measured directly with micromanometers (Steudle and Zimmermann, 1974).

Determination of L_p, the hydraulic conductance

The estimation of hydraulic conductance, L_p, and the reflection coefficient, σ, may be achieved in giant algal cells employing Equation (18) (without the $\Delta\psi_m$) by means of intracellular perfusion and transcellular osmosis, or from the kinetics of the shrinking and swelling of the cells. The perfusion technique, where vacuolar pressure is adjusted by an external regulatory system, has been extensively reviewed by Gutknecht et al. (1977) and by Gutknecht and Bisson (1977), who used this procedure on cells of Valonia ventricosa. The now classical method for measuring L_p is by the method of transcellular osmosis, first employed in a quantitative manner by Kamiya and Tagawa (1956). Intact, cylindrical giant internodal cells of giant Characeae are sealed into a double chamber with half the cell in each chamber. The two chambers are completely filled with water and equilibrated, then one of the chambers has the water replaced with a solution of sucrose (at a non-plasmolyzing concentration). The initial flow of water through the cell from the water compartment to the sucrose compartment is detected by a small potometer. From this initial flow rate, J_v, the areas of cell in the two chambers, A_1 and A_2, and $\Delta\psi_s$, the solute potential difference, L_p, can be calculated from the relationship

$$J_v = \frac{A_1 A_2}{A_1 + A_2} L_p \Delta\psi_s \qquad (19)$$

Values of L_p obtained by this technique for Characean cells are of the order of $10^{-6} \, \mathrm{m\,s^{-1}\,MPa^{-1}}$, indicating a high water permeability. However the situation is more complex in that the L_p for water entering the cell is greater than the L_p for water leaving the cell, i.e., there is a polarity of L_p (Dainty, 1976). This polarity is believed to be an intrinsic property of the plasma membrane (Kiyosawa and Tazawa, 1977). Kelly, Kohn and Dainty (1963) estimated L_p from measurements of the shrinking and swelling of cells of Nitella translucens using the relationship

$$t_c = \frac{V}{A L_p \left(\varepsilon - \psi_s \right)} \qquad (20)$$

where A and V are the area and volume of the cell, t_c is the swelling or shrinking time constant, ε and ψ_s as before. Similar values to those obtained by transcellular osmosis were obtained.

Recent developments with the pressure probe technique referred to above enable L_p to be calculated from the half-time of water exchange during the shrinking and swelling of giant algal cells in response to pressure (Zimmermann, 1978). Values obtained by this technique are again around $10^{-6} \, \mathrm{m\,s^{-1}\,MPa^{-1}}$. Interestingly, it was observed that L_p showed a marked dependence on ψ_p, the pressure potential, L_p, increasing as ψ_p decreased. No such pressure dependence was observed for the L_p values of bladder cells of Mesembryanthemum crystallinum, one of the earlier higher plant cells to have been investigated (Steudle et al., 1977) where a value of $2 \times 10^{-7} \, \mathrm{m\,s^{-1}\,MPa^{-1}}$ was obtained.

Subsequently the pressure probe technique has been applied to certain other individual cells of higher plants (Husken, Steudle and Zimmermann, 1978; Zimmermann, Husken and Schulze, 1980; Tomos, Steudle, Zimmermann and Schulze, 1981; Steudle, Zimmermann and Zillikens, 1982). The hydraulic conductance, L_p, is calculated from the half-time of the pressure potential relaxation process when the cell's pressure potential is altered with the probe (Equation (20)). The volumetric elastic modulus, ε, required for the calculation, is determined separately by measuring cell volume changes associated with pressure change (Equation (12)). L_p values determined by this technique show a marked turgor dependence. In Elodea densa L_p was $5.6 \times 10^{-8} \, \mathrm{m\,s^{-1}\,MPa^{-1}}$ for $\psi_p > 0.4 \, \mathrm{MPa}$ and

$14.1 \times 10^{-8}\,\mathrm{m\,s^{-1}\,MPa^{-1}}$ for $\psi_p < 0.4\,\mathrm{MPa}$ (Steudle et al., 1982).

The above L_p values are two orders of magnitude smaller than values estimated from the water permeability coefficient (P_d) measured for Elodea nutallii by Stout, Steponkus and Cotts (1977). Using a nuclear magnetic resonance (NMR) technique these authors obtained a value for P_d of $3 \times 10^{-4}\,\mathrm{m\,s^{-1}}$. P_d is related to L_p in the following manner (House, 1974):

$$L_p = P_d \frac{\overline{V}_w}{RT} \qquad (21)$$

The value of L_p from the above calculation is $2 \times 10^{-6}\,\mathrm{m\,s^{-1}\,MPa^{-1}}$.

The determination of accurate L_p values for higher plant cells is essential for a description of water transport within the plant (Lange and Lösch, 1979) and obviously the anomalies in the values determined by these different techniques must be resolved.

Active water transport

Equation (18) predicts that no net flow of water will occur when the driving forces are zero. However, there are a number of reports of a non-osmotic or active water transport occurring (see Tyree, 1973; Anderson, 1976). Root pressure exudation has been observed to continue from a cut root even when the bathing medium is isotonic with the xylem fluid. With an hydrostatic pressure gradient of zero, flow could only be explained by introducing an additional term, J_0, to Equation (18). As an active pump for water cannot exist for energetic reasons, it has been suggested that non-osmotic water flow is energized from the coupling of the water flow to a metabolically driven solute flow; this is termed the standing gradient osmotic flow hypothesis (Anderson et al., 1970; Ginzburg, 1971).

Water flow coupled directly to a metabolic reaction may be generated by the asymmetry of the cell membrane (see Chapter 15) and such flow has been demonstrated in an in vitro asymmetrical sandwich-membrane system (Meyer et al., 1974). The implications of such coupled flow are discussed in review articles by Stadelmann (1977) and Zimmermann (1978).

The significance of transpiration

In actively growing plants there is a continuous liquid water phase extending from the soil water to the liquid:gas interface within the leaves. Water is evaporating continuously from the surface of cells exposed to the air and this is replaced by absorption from the soil. Water can either move through the plant directly to the atmosphere or drain down to the water table (see Fig. 14.5 page 310). The amount of water moved through these pathways is dependent on $\Delta\psi$ and the relative magnitudes of the resistances to flow within the pathways. As with movement of water across a membrane, this can be expressed as

$$J_v = L_p \Delta\psi, \qquad L_p = 1/r \qquad (22)$$

where J_v is the flux of water, L_p the hydraulic conductance (the reciprocal of the resistance, r) and $\Delta\psi$ the water potential gradient, units as before.

The magnitude of transpiration is such that under favourable conditions an amount equivalent to the entire volume of water in a plant may be lost in the course of a single day. A single maize plant may absorb up to $250\,\mathrm{dm^3}$ of water in a growing season, though only a small fraction of this water is used in metabolic processes or retained in the plant, the remainder (about 98%) being transpired.

Leaves normally present a large surface area to the surrounding air to facilitate CO_2 assimilation. Simultaneous evaporation of water from the moist cell walls is inevitable and thus transpiration is a consequence of the structural organization of plants growing in air. Water lost in transpiration must be replaced by absorption from the soil and so the rate of transpiration is closely related to the availability of water in the environment. Plants growing in moist surroundings (mesophytes) have very little adaptation to reduce transpiration, whereas those that grow in dry situations (xerophytes) have various devices to restrict water loss. However, any structural feature which restricts water loss, such as reduced leaf surface area, inevitably reduces photosynthesis and hence the growth rate.

Although transpiration must be restricted as far as is compatible with an adequate growth rate, the water movement has several useful functions. The transpiration stream assists in the transfer of mineral salts and organic substances from the roots to the shoot. An appreciable amount of the energy absorbed by a leaf is utilized as latent heat in the evaporation of water and transpiration thus has a significant cooling effect. However, desert succulents, which are exposed to severe heating by solar radiation, have low transpiration rates and desiccation by excessive loss of water is generally a more potent danger to plants than overheating.

Water movement through the soil

Water enters the soil by natural precipitation or irrigation and is removed from the root zone by drainage, uptake by plant roots or evaporation. Infiltration of water into the soil is affected by the initial water content of the soil and its structure. Infiltration rate is markedly reduced by high initial water content due to both a decrease in $\Delta\psi$ within the soil and a swelling of the soil particles. Surface crusts, hardpan zones and other compacted conditions impede water entry into the soil and disruption of these zones is an important aspect of soil cultivation (see Kramer, 1969).

Within the soil the water moves downward under the influence of gravity, but some is retained within the surface region by matric forces, dependent on the degree of saturation of the soil. When the soil is not saturated with water and is not draining under gravity the major driving force for water movement is $\Delta\psi_m$ and thus for soils

$$J_v = L_p \Delta\psi_m \qquad (23)$$

If the water table is relatively near the surface ψ_m is obtained from the relationship

$$\psi_m = \rho_w g h \qquad (24)$$

where ρ_w is the density of water ($kg\,m^{-3}$), g the acceleration due to gravity ($m\,s^{-2}$) and h the height above the water table (m), thus giving ψ_m in pressure units ($kg\,m^{-1}s^{-2}$).

The presence of plant roots within the soil imposes an additional $\Delta\psi$ for water movement and profoundly affects the profile of soil water content. In order for water to move into the root there must be a $\Delta\psi$ from the soil to the root, $\psi_{soil} > \psi_{root}$. ψ_{soil} is affected by ψ_m and also by ψ_s, the solute potential of the soil solution (usually <-0.1 MPa, but important in saline soils where it may reach -5 MPa or lower). ψ_m is near to zero in water-saturated soils and decreases to -10 MPa or lower in very dry soils. As

ψ_{soil} decreases (becomes more negative), ψ_{root} must also decrease if the gradient is to be maintained. When $\psi_{soil} = \psi_{root}$, uptake of water ceases and the plant will dry out unless conditions are changed by additional infiltration of water.

Soil water content is usually expressed as a percentage of the soil dry weight. Soil which has drained of gravitational water and contains the maximum amount of water held by matric forces is said to be at field capacity. Clay soils with relatively small particles have a greater field capacity than sandy soils with larger particles and thus clays have a greater water storage capacity (Fig. 14.4). The permanent wilting percentage (PWP) is the soil water content which induces wilting of plants reversible only when more water is added to the soil. PWP values differ between different soils, but ψ_{soil} at PWP is usually about -1.5 MPa.

Water movement across the root

Except in submerged aquatics, the root is the major path of entry for water and minerals absorbed by higher plants, the zone of maximum water absorption being located some $20-200$ mm from the root tip behind the meristem and in front of the region of cutinization and suberization. This region is frequently characterized by the presence of root hairs which presumably serve to increase the area of contact between the root surface and the soil but are not essential for water absorption. Some roots, particularly the adventitious roots of bulbs, corms and rhizomes, do not have root hairs; plants grown in solution culture and aquatic plants also commonly lack root hairs.

As the root grows through the soil, the zone of rapid water uptake moves with it. The ability of the cells to absorb water partially diminishes with age and older parts of the root become suberized, while new rapidly-absorbing cells are differentiated from the meristem. Thus the root system of growing plants is continually exploring new regions of the soil, absorbing water and minerals from them. In trees and other woody perennials a large proportion of the root system becomes suberized and it is often assumed that absorption of water and salts does not take place through these regions. There is, however, an appreciable absorption of water through the suberized surface of older roots of trees through

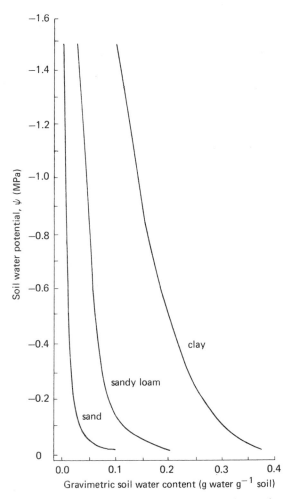

Fig. 14.4 Typical water characteristic curves for sand, sandy-loam and clay soils. For a soil water table at 3 m depth, soil water potential (ψ_{soil}) $\simeq 0$. Between this value and about -0.5 MPa to the limits of water extraction by roots (the permanent wilting point), around -1.5 MPa, lies the useful hydraulic capacitance or soil water reservoir which increases sand < loam < clay (after Slatyer and McIlroy, 1961).

cracks in the protective layer, and this is of particular importance in winter and during drought when young growing roots with root hairs may be lacking (Kramer and Bullock, 1966; Kramer, 1969).

Water moves centripetally across the root cortex to the stele, individual root cells competing with the stele for the available water. The initial movement of water is across the surface layer of the root with its associated root hairs and then through the cortex, a layer of some 5–15 large parenchyma cells with intercellular spaces.

There are three possible pathways for water movement across this region, through the cell walls, through the cytoplasm *via* plasmodesmata, and through the cytoplasm and vacuoles of the cortical cells. The relative flow through each pathway varies inversely with the resistance encountered. The path through the cell walls is considered to be the primary path for water movement across the cortex. This is sometimes referred to as the apoplast pathway, the non-living cell wall continuum external to the living cytoplasm forming the apoplast phase of the plant's structure (Läuchli, 1976). The cytoplasmic pathway is again through a three-dimensional network, the symplast. Resistance to water flow is higher for the symplast pathway due to the relatively high viscosity of cytoplasm, but cyclosis (cytoplasmic streaming) may assist the water movement (Tyree, 1970; Spanswick, 1976). The third pathway, involving passage across a number of membranes, layers of cytoplasm and vacuoles, also presents a higher resistance than does the apoplast and the volume of flow through this route will be proportionately restricted. However, these pathways are not mutually exclusive and there is appreciable transfer from one to another as water crosses the root cortex (Weatherley, 1970).

Whichever pathway is taken, water is ultimately diverted into the cytoplasm at the endodermis where further movement through the cell walls is prevented by a suberin-impregnated band, the Casparian strip, which seals the radial and transverse walls of the endodermal cells against penetration by water (Robards and Clarkson, 1976). In most roots, the walls of the endodermis are further thickened and suberized, with the exception of certain passage cells which are located opposite the xylem groups. Passage cells possess a Casparian strip and apoplastic continuity does not exist across the endodermis, which is therefore the major site of resistance to water movement across the root. Once across this barrier and within the stele, water encounters resistances similar to those of the cortex and will once again move primarily in the cell walls and enter the lumen of the xylem vessels and tracheids, the major water-conducting tissue in higher plants.

The driving force for water movement across the

root is the $\Delta\psi$ between the xylem sap of the root and the soil solution at the root surface. The gradient may be caused by the development of a negative pressure (tension) within the xylem due to the evaporation of water from the leaves, the process of transpiration. In the absence of transpiration, ion transport to the stele still occurs, causing an accumulation of ions within the stele which results in a lowering of ψ_s, the root then functioning as an osmometer. When ψ_{root} is lower than the ψ_{soil}, water moves into the root, resulting in a positive pressure in the xylem, as indicated by exudation from de-topped root systems (root pressure) and guttation from leaves under conditions of low transpiration (Stocking, 1956; Barrs, 1966).

Water movement through the vascular system

It is well established that water moves from the root to the leaf predominantly through the xylem. Evidence for this comes from the observation that water flow is not immediately impeded when a ring of tissues external to the xylem is removed, and that, conversely, when a section of xylem is removed leaving other tissues intact, water movement is impaired. When solutions labelled with water-soluble dyes, radioactive solutes or water containing ^{3}H or ^{18}O are supplied to plants, the label is rapidly detected in the xylem of the roots and stem, and particularly in the vessels and tracheids. In the giant redwoods of California, and the eucalypts of Australia, movement from roots to leaves can involve distances of over 100 m. Resistance to flow in the xylem is very low as there are no layers of relatively highly resistant cytoplasm to be traversed during movement through these conducting elements.

Water movement through the xylem is a bulk flow caused by $\Delta\psi_p$ between one end of the system and the other. A negative pressure is transmitted through the liquid continuity within the xylem to the root, the breaking of these liquid columns being prevented by cohesion between adjacent molecules and adhesion between water molecules and cell walls.

The cohesion theory has been criticized on the grounds that flow continues when large numbers of vessels are air-filled and also that deep overlapping incisions do not completely stop water flow. How-

ever, as only a relatively limited number of vessels is required to meet transpirational demand and as cohesion is effective laterally as well as vertically, liquid continuity can be maintained around cuts and air-filled vessels, although the resistance offered by this alternative pathway is greater.

Although movement of water takes place through non-living xylem vessels and tracheids, the presence of living cells is indirectly necessary as individual conducting elements function for only a relatively short time and their continued replacement is dependent on the activity of the adjoining meristematic cells of the cambium. In deciduous trees a new system of vessels is produced each year in association with each new crop of leaves and it is these vessels in the outer ring of xylem that supply the bulk of the water to the transpiring leaves.

The pressure gradient within actively transpiring trees range from 0.01 to 0.05 MPa m^{-1}, a normal gradient in trees well supplied with water being 0.015 MPa m^{-1}. Thus a tree of 100 m height would not require tensions lower than -1.5 MPa to move water up the xylem. Leaf cells commonly have ψ values of this order and it is possible that potentials of -10 MPa may develop in some species under conditions of severe water deficit (Milburn, 1979).

Theoretically it is estimated that a stretched water column would break at tensions between -100 and -200 MPa but in the living plant it is probable that cavitation occurs in most of the vessels before tensions of -20 MPa are reached, and will occur in the larger vessels at less negative values. The production of sound by cavitating water columns has been detected both in physical systems and in xylem vessels (Milburn and Johnson, 1966).

Flow through xylem vessels may be analyzed using the Hagen–Poiseuille equation which states that the volume flow of water through a conduit is proportional to the square of the radius

$$J_v = \frac{r^2}{8\eta}\Delta\psi_p \qquad (25)$$

where J_v is the volume flow of water (m^3 m^{-2} s^{-1}), η the viscosity (Pa s), r the radius, $\Delta\psi_p$ the hydrostatic pressure gradient (Pa). The hydraulic conductivity (L) and hydraulic conductance (L_p) may also be calculated from Equation (25) as

$$J_v A = \frac{LA\Delta\psi_p}{l} = L_p A\Delta\psi_p \qquad (26)$$

where A is the area and l the length through which flow is occurring. Hydraulic conductivity may also be determined experimentally by driving water through a length of wood with a known low pressure (Zimmermann and Brown, 1971).

Agreement with the Hagen–Poiseuille equation is found in vines, which have large open xylem vessels, ring porous species such as oak are close to theoretical values, while diffuse porous species such as beech are considerably lower than predicted values. Such deviations are not surprising when the presence of end walls in xylem vessels and tracheids is taken into account (Table 14.2).

Table 14.2 The efficiency of water conductivity in various plants expressed as the percentage ratio of the observed rates of water movement (v_0) and those predicted from the Hagen–Poiseuille relationship (v_t) (after Huber, 1956).

Plant	Conductive efficiency $v_0/v_t \times 100$
Vitis vinifera	100
Aristolochia sipho	100
Atragene alpina	100
Root wood of oak	84
Root wood of beech	37.5
Helianthus annuus	32
Rhododendron ferrugineum	20
Rhododendron hirsutum	13

Water movement through the leaf

Water moves into a leaf mainly through the leaf trace vascular bundles, the vessels of which form continuous channels from the root up the stem and through the petiole into the vascular strands of the leaf itself. From the vascular terminals water moves to the leaf parenchyma, from which evaporation and loss of water in the vapor phase takes place. There are two pathways for this outward movement. The first and major pathway involves movement of water through the leaf parenchyma to the liquid : air interfaces within the leaf and then, in the vapor phase, through the air-filled intercellular spaces of the leaf to the stomatal pores in the leaf epidermis. The second pathway, which is subsidiary to and runs parallel with the first, is through the parenchyma to the liquid : air interface at the outer surface of the

epidermal cells and then, in the vapor phase, through the cuticle to the outside air. Vapor-phase movement from the two pathways will be considered later; liquid phase movement will be discussed here.

In some instances, when transpiration is minimal, water in the liquid phase is transferred to the surface of the leaf and extruded as droplets; this phenomenon is known as guttation. Guttation occurs through hydathodes, when present, which are modified stomata separated from vascular terminals by a mass of thin-walled parenchyma called epithem. Guttation can also occur through ordinary stomata and lenticels, movement out of the plant taking the path of least resistance. It is brought about by the positive pressures which develop in the xylem as a result of root pressure. The volume of water exuded by a guttating plant varies from a few drops on a grass blade to the prodigious 200 ml recorded in one day from a single leaf of an Indian taro plant. The volume and composition of guttation fluid are extremely variable, ranging from almost pure water to a solution with ψ_s of -0.1 MPa or lower. The physiological significance of guttation is limited and generally the amount of water transported from the plant through this pathway is small in relation to losses *via* other routes. Guttation may, however, be important in the movement of material from roots to shoots in some submerged aquatic plants where transpiration cannot occur.

Transport of water through the leaf parenchyma occurs mainly in the cell walls. As with root cortical parenchyma, the cell wall pathway offers the least resistance to water movement, the ratio of water flow in the walls compared with that through the vacuole-to-vacuole pathway being about 50 : 1 in response to a given $\Delta\psi$ (*see* Weatherley, 1963, and Boyer, 1977).

Water movement through the whole system

Water transport through the whole system reflects a progressive drop in ψ from soil to atmosphere and encounters a number of resistances in each part of the system. By analogy with electrical current, flow through such a system should obey Ohm's law whereby

$$\text{flow} = \frac{\text{potential gradient}}{\text{resistance}}, \quad \left(J_v = \frac{\Delta\psi}{r} = L_p \Delta\psi \right) \quad (27)$$

This electrical analogy is used to describe water flow through the plant (Fig. 14.5). The path is visualized as a system of resistances arranged in series and in parallel. The soil has stores of water (represented by capacitors), from which water can either move through the plant directly to the atmosphere, or drain down to the water table (electrical earth). The flow of water along these paths is determined by the potential gradients and the relative magnitudes of the resistances encountered. Using the Ohm's law analogy, the following expression may be used for a steady rate of water flow, J_v, through the whole soil:plant:atmosphere system:

$$J_v = \frac{\Delta\psi}{r} = \frac{\psi_{soil} - \psi_{root}}{r_1} = \frac{\psi_{root} - \psi_{stem}}{r_2} = \frac{\psi_{stem} - \psi_{leaf}}{r_3}$$

$$= \frac{\psi_{leaf} - \psi_{air}}{r_4} = \frac{\psi_{soil} - \psi_{air}}{r_1 + r_2 + r_3 + r_4} \qquad (28)$$

where r_{1-4} are the resistances in the respective parts of the pathway (van den Honert, 1948).

Some approximate values for the water potential terms in Equation (28) are:

ψ_{soil}	-0.1 MPa
ψ_{root}	-1.0 MPa
ψ_{stem}	-1.5 MPa
ψ_{leaf}	-1.5 MPa
ψ_{air}	-100 MPa (at 50% relative humidity)

These values indicate that the leaf:air potential difference is by far the largest within the system, being greater, by more than an order of magnitude, than the combined potential differences between the other segments. It follows that the resistance of the leaf:air portion of the system must be proportionately high.

In a catenary process such as that proposed, the flow rate is controlled essentially by that part of the

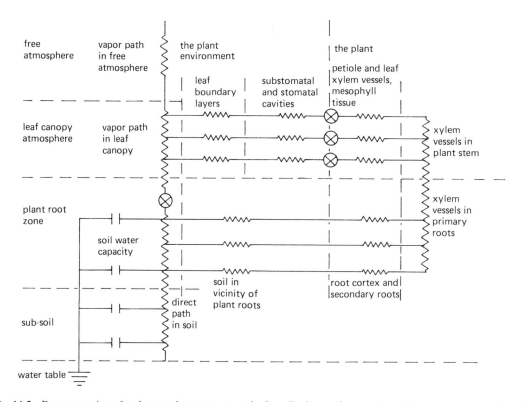

Fig. 14.5 Representation of pathways of water transport in the soil, plant and atmosphere. Site of phase change, liquid to vapor, are distinguished by the symbol ⊗ (after Cowan, 1965).

system with the highest resistance. An increase in any other resistance, provided that it is insufficient to have a significant effect on the total resistance, merely increases the pressure difference across that resistance and has little effect on the flow rate. If the resistance in any other part of the system rises appreciably, for example when seedlings are transplanted and soil:root contact is broken, the flow of water is reduced and the plant may wilt. The stomata which are located in the gaseous phase are the most effective regulators of water flow because they can significantly affect the high resistance of the gaseous phase.

The electrical analogy is not completely valid as, unlike the flow of electrons in an electrical circuit, water flow in a plant continues from any part of the system which is on the atmosphere side of a high-resistance barrier, and that part of the plant dries out. The presence of the stomata in the gaseous phase ensures that, when they are closed, water flow through the whole system is reduced, and that no part suffers from excessive desiccation. Although changes in root and stem resistance do not normally affect transpiration directly, they sometimes exert an indirect effect by causing the development of a water stress within the plant that results in stomatal closure. When stomata close in response to such a stimulus, the stress is decreased and in consequence the stomata reopen.

The relationship between transpiration rate and changes of resistance within the root may be illustrated by the following schematic relationship (Fig. 14.6). If a plant growing in water culture in the dark is suddenly illuminated, transpiration will increase to a steady level as the stomata open in the light. The absorption of water by the plant, lagging just behind transpiration, reaches a steady state somewhat later. Under these conditions a certain $\Delta\psi_w$ exists between the solution around the roots and the leaf. If at point A the conditions are suddenly changed and the root resistance is increased by, say, cooling the solution around the roots, the absorption rapidly drops but the transpiration remains unaltered, a new $\Delta\psi_w$ between leaf and solution being established as the absorption recovers. When the root resistance is further increased (B), absorption falls as before and a new gradient is established which cannot however sustain the original absorption rate and a partial stomatal closure causes a drop in transpiration to a lower steady state level,

paralleling the prevailing absorption rate. It is evident from Fig. 14.6 that whenever transpiration increases, the absorption of water lags behind for some time, a result of the resistance barrier within the root which restricts rapid water uptake.

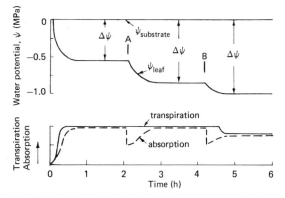

Fig. 14.6 Schematic relationship between transpiration rate, absorption rate, ψ_{leaf} and ($\psi_{substrate} - \psi_{leaf}$) when root resistance is increased in two stages, A and B, by, say, cooling the solution around the roots (after Slatyer, 1967).

The stomatal and cuticular pathways

Evaporation of water occurs from two sites within a leaf: directly from the outer epidermal cell walls through the cuticle into the external atmosphere; and from the walls of the palisade and mesophyll cells within the leaf into the intercellular space system, and then through the stomata to the air outside. These cuticular and stomatal pathways exist in parallel (Fig. 14.7) and the resistance to flow in

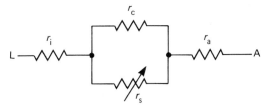

Fig. 14.7 Resistances encountered by a water molecule diffusing from a leaf cell (L) into the surrounding air (A). r_i is the resistance of the intercellular spaces, r_c the resistance of the cuticle, r_s the variable resistance of the stomata, and r_a the resistance of the boundary layer of unstirred air at the leaf surface through which water molecules must diffuse.

the vapor phase may therefore be generally represented as

$$\frac{1}{r_{leaf}} = \frac{1}{r_{cuticle}} + \frac{1}{r_{stomata}} \qquad (29)$$

Since the resistance of the cuticular pathway to diffusion of water vapor is usually very high, loss of water through the leaf epidermis is often small compared with the transpiration through open stomata. However, when stomata are closed, as under drought conditions, cuticular transpiration becomes very important in the water economy of many species. The rate of cuticular transpiration is determined by the thickness and age of the cuticle and its chemical composition. Because of differences in thickness and efficiency of the cuticular layer, transpiration through the cuticle varies widely. As leaf water content falls, transpiration through the cuticle decreases.

The resistance to water vapor flow through the stomatal pathway varies with the degree of opening of the stomata, as indicated by the varying resistance symbol in Fig. 14.7. The stomatal pathway resistance consists of a number of resistances in series which can be conveniently represented as

$$r_{stomata} = r_{cell\,walls} + r_{intercellular\,space} + r_{stomatal\,pore} \qquad (30)$$

Published values for $r_{stomata}$ vary with the species and the method of determination. For mesophytes, $r_{stomata}$ is around $150\,s\,m^{-1}$, whereas in conifers and xerophytes with sunken stomata the value may be as high as $3000\,s\,m^{-1}$ (Meidner and Mansfield, 1968). Externally to the leaf, the diffusive resistance to water vapor transfer in air, r_{air}, plays a critical part in the rate of water loss due to the presence of a boundary layer of unstirred air over the leaf surface. The relative vapor pressure of this boundary layer differs from that of the external bulk air and provides a considerable resistance to water vapor flow. The relationship of r_{air} to the diffusion coefficient of water vapor in air, D_w ($\approx 0.24 \times 10^{-4}\,m^2\,s^{-1}$) is expressed as $r_{air} = d/D_w$, where d is the thickness of the boundary layer (m). The unit of r_{air} is therefore $s\,m^{-1}$ and this same unit is used to express the other resistances in the vapor phase pathway. Calculated values of r_{air} range from about 10 to $300\,s\,m^{-1}$ for most leaves, larger values of r indicating a higher resistance (Slatyer, 1967).

The relative magnitude of the two major resistances r_{leaf} and r_{air} differ with the wind speed external to the plant. Under still air conditions, where r_{air} is comparatively high, there will be only limited stomatal control of transpiration until the stomatal aperture is nearly closed, whereas with moving air, when r_{air} is reduced, stomatal control will be effective throughout the entire range of stomatal apertures.

Factors affecting transpiration

Under most conditions, the efficiency of the absorbing surface of the root and the evaporating surface of the leaf will affect the transpiration rate. When water absorption lags behind transpiration, a water deficit will occur, ψ_{leaf} will fall and transpiration may be reduced by partial stomatal closure (Fig. 14.8).

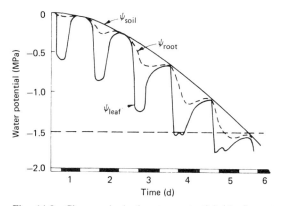

Fig. 14.8 Changes in leaf water potential (ψ_{leaf}), root surface water potential (ψ_{root}) and soil mass water potential (ψ_{soil}), as transpiration proceeds from a plant rooted in initially wet ($\psi_{root} \cong 0$) soil. The same evaporative conditions are considered to prevail each day. The horizontal dashed line indicates the value of ψ_{leaf} at which wilting occurs. Periods of light and dark are indicated by the bars on the abscissa (after Slatyer, 1967).

Because of this phenomenon high transpiration rates are favored by large root:shoot ratios which ensure that the transpiring plant is supplied with adequate water. A well-developed root system is also advantageous under conditions of low water supply; high root:shoot ratios are observed in xerophytes and halophytes (Troughton, 1974).

The magnitude of plant water loss is also related

to leaf area, although a simple proportionality does not exist. In general, plants with a large area of foliage transpire more rapidly than those with a smaller leaf area, but the rate is often lower per unit of leaf area (*see* Passioura, 1976). Removal of leaves from a plant by pruning increases the rate of transpiration from the remaining leaves; this is attributed to a reduced water stress resulting from the reduced overall water loss by the plant. When growing under dry conditions, some plants shed their leaves and in this way reduce transpiration (Ritchie, 1974). Many plants of arid habitats have a small leaf area, as a xeromorphic adaptation.

Considerable variation exists in the rate of transpiration of different plant species due to differences of leaf structure. Such features as the size, number, distribution and structure of stomata, composition and thickness of the cuticle, surface area exposed to intercellular spaces, and arrangement of the vascular tissues all exert marked effects on the rate of water loss. Leaves of the same species, or on the same plant, growing under different conditions of light intensity show differences in structure, such as the amount of photosynthetic tissue, thickness of cuticle, number of stomata, and other features which affect water loss. In addition to water loss from leaves, some transpiration takes place from young stems through cuticle and stomata, while water is lost from older stems through lenticels and the corky layers that surround them. Although the total amount of water lost by means of lenticular transpiration is low, winter transpiration from lenticels may cause water deficits and death by desiccation (Kozlowski, 1964).

Plants of arid habitats exhibit a number of xeromorphic features which reduce water loss. Of these, increased root development and reduced leaf area have already been mentioned above. In addition, the leaves of xerophytes frequently exhibit a thick cuticle, thick cell walls, well-developed palisade tissue and sunken stomata, and often have hairs or scales on the leaf surface. All of these features contrast strongly with the commoner mesophytic plants of less extreme habitats.

An important feature of xeromorphic leaves is that, because of their rigid structure, they do not wilt readily and are therefore less damaged when subjected to water deficit. Xeromorphic character assumes significance only under conditions of water stress, the transpiration of well-watered xeromorphic leaves being greater per unit of surface area than that of mesophytes, a reflexion of the greater ratio of internal to external surface area of xeromorphic leaves. These higher rates of transpiration are found only when the stomata are open and thus it is only the rate of stomatal transpiration which may be higher in xeromorphic species. When the stomata are closed and cuticular transpiration rates are compared, the water loss from mesophytes is often an order of magnitude higher than for xerophytes, $r_{cuticle}$ ranging from 2000 to $8000 \, s \, m^{-1}$ for mesophytes to $>20\,000 \, s \, m^{-1}$ for xerophytes (Holmgren, Jarvis and Jarvis, 1965).

In xerophytes the leaf area reduction is commonly associated with an increase in the surface area of the stem which may become the major photosynthetic as well as the main transpiring organ, as for instance in some cacti. In some species leafless stems may produce leaf-like appendages termed cladodes, although the benefit to the plant of this strategy is obscure. It may be that the lower stomatal frequency on the cladodes reduces the potential transpiration rate of such plants. Many xerophytes have succulent stems and leaves. One consequence of succulence is that the surface area of the plant is reduced relative to its volume and therefore water loss relative to total water content is reduced. Water stored in the large parenchyma cells of succulent tissues is gradually lost during periods of drought and this reservoir of water assists in the survival of these plants in arid habitats.

Any strategy devised for reducing water loss from plants also reduces the uptake of carbon dioxide and hence photosynthesis. Xerophytes are therefore usually slow-growing plants, but, as they occupy an ecological niche where mesophytes cannot compete, the slow growth rate is often not detrimental to their survival. Many of the stratagems adopted by xerophytes have been discussed by Kozlowski (1964), Slatyer (1964) and Morrow and Mooney (1974).

Halophytic plants often have xeromorphic characteristics such as well-developed root systems, succulence, reduced leaf area, sunken stomata and impermeable cuticles. Although there is often abundant water in their environment these plants suffer from 'physiological' drought because of the low ψ_s of the soil solution, which may reach $-5 \, MPa$ or lower (Flowers, 1975). However, the vacuolar sap of halophytes is more concentrated than in other plants and a gradient of water potential exists between plant and soil solution so that under normal conditions halophytes transpire quite rapidly. In areas such as salt marshes, the concentration of the soil solution may increase due to evaporation when the tide is out and the plants may suffer a temporary water deficit. Under these conditions the xeromorphic characters will exert an influence in reducing transpiration, while succulence will provide a reservoir of water, as with some of the xerophytes discussed above.

Transpiration is markedly affected by those environmental factors which influence physical evaporation, namely solar radiation, air temperature, humidity, air movement and water supply. These various factors fluctuate diurnally and seasonally and cause corresponding changes in transpiration rate. As may be expected, solar radiation shows the strongest correlation with transpiration rate as it provides the necessary energy for evaporation. However, light has a greater effect on transpiration than it has on evaporation as such, due to the opening and closing of the stomata in response to light and dark (*see* Chapter 10 and Willmer, 1983).

Transpiration is dependent on $\Delta\psi$ between the internal atmosphere of a leaf and the external atmosphere, and thus humidity, or the water content of air, has a marked effect on the transpiration rate. As the internal atmosphere of a leaf is saturated, or nearly so, $\Delta\psi$ is dependent on the external humidity, which is usually unsaturated. Low air water contents give a large $\Delta\psi$ and therefore cause high rates of transpiration, while high air water contents give low rates of transpiration. Air water content is often expressed in terms of vapor pressure (θ) which can be converted to water potential using the relationship

$$\psi = \frac{RT}{\bar{V}_w} \ln \frac{\theta}{\theta^\circ} \qquad (31)$$

Temperature exerts a profound effect on vapor pressure and will therefore affect transpiration by altering $\Delta\theta$. When the temperature of a leaf is higher than that of the air, $\Delta\theta$ is increased and the leaf will transpire even into water-saturated air as a $\Delta\psi$ will exist. For example, if leaf temperature is 30°C and external saturated air temperature is 20°C, then θ will be 4243 Pa for the leaf and 2336 Pa for the air, a vapor pressure gradient of some 1907 Pa. On the other hand, leaves sometimes become cooler than

the air and dew may be deposited on them under conditions of high atmospheric humidity. Temperature also influences stomatal movements and will therefore affect transpiration through its effect on stomatal resistance (Hall and Kaufmann, 1975).

The movement of air over the leaf surface tends to remove water vapor and, by decreasing the thickness of the boundary layer, reduces the value of r_{air}, thus increasing transpiration. Most of the increase occurs at relatively low wind velocities; with higher velocities the combined influence of cooling and stomatal closure results in a reduction in transpiration which may fall below the value obtained in still air.

Water availability has a profound influence on the rate of transpiration. As indicated previously, a decrease in soil water will reduce absorption and will cause a reduction of transpiration due to partial stomatal closure induced by water stress (*see* Fig. 14.8). When water is unavailable the plant suffers a severe water deficit and the stomata close completely, limiting transpiration to diffusion through the cuticle.

Measurement of transpiration

Several methods have been employed in the measurement of transpiration, usually involving the determination of either weight change or vapor loss from a plant or plant part. The amount of water transpired in a given time may be expressed as loss per plant or per leaf, fresh weight, dry weight or area.

One of the simplest ways of measuring transpiration is to weigh a plant and its container over intervals of time, evaporation from the container and soil being minimized by enclosing them in some waterproof material. Gain of weight due to photosynthesis and loss of weight due to respiration are inherent errors in this method but they are usually very small compared with transpirational losses and often may be neglected.

Transpiration of a single leaf or shoot may be measured by cutting off the plant part and determining the change in weight on a sensitive spring or torsion balance. Although this method gives useful information about the relative rates of transpiration of different plants, it is often a poor indication of normal transpiration of an intact plant.

Another method for measuring transpiration,

which can be applied either to an excised leaf or shoot or to a whole plant, is to use a potometer. The rate of water loss from the plant or plant part is determined indirectly by measuring the rate of absorption of water, assuming that absorption balances the water lost in transpiration—an assumption that is not always valid. Although the potometer is useful, particularly to demonstrate the effects of various environmental factors on transpiration, results obtained should be regarded as relative rather than as absolute. Water uptake by an excised leaf or cut shoot in a potometer will not necessarily be the same as that of the same material when on the plant, both water tensions in the xylem and the root resistance to water movement being eliminated in the excised material.

Transpiration may also be measured by enclosing a whole plant, or portion of a plant, in a chamber and determining the water content of the air passing through. The air water content may be determined using sensitive hygrometers of various kinds. This method allows for a more sensitive monitoring of transpiration than the above weighing methods, giving a response time of less than one minute. However, enclosure of the plant in a chamber creates abnormal conditions of light intensity, temperature and humidity, restricting the application of this particular technique.

A simple though somewhat inaccurate measure of transpiration is to measure the time taken for cobalt chloride paper to change from blue to pink, when clipped onto a transpiring leaf surface. Cobalt chloride paper may be made from pieces of absorbent paper soaked with slightly acid three per cent cobalt chloride solution and thoroughly dried. The dried paper is colored blue and changes to pink on absorbing moisture. As the portion of leaf beneath the paper is transpiring into a very dry atmosphere and the partial shading may induce stomatal closure, this method gives only a qualitative comparison of transpiration from different plants or different leaves on the same plant.

For plants in the field the total water loss including that evaporating from the soil, the evapo-transpiration, may be determined using a lysimeter. This is a large hydraulic or mechanical balance sunk into the ground. The amount of water supplied to the lysimeter can be compared with the quantity collected from it over a given period, enabling the loss by evapo-transpiration to be determined. Detailed accounts of the measurement of transpiration may be found in the texts of Kozlowski (1964), Slatyer (1967) and Kramer (1969).

Conclusions

The role of plant–water relations in the physiology of plant growth and development has been implicit rather than explicit in the preceding discourse. This is because the intricate relationships which exist between such parameters as cell wall elasticity, hydraulic conductance and cell turgor have only recently been partially resolved for single cells of giant algae. Attempts to extend such studies to the problems of water transport in higher plants are now being undertaken and as new sophisticated techniques are developed our understanding of these processes should progress.

The outstanding questions to be answered are those relating to the influence of water transport on the control of growth in plant cells. The effects of turgor on solute fluxes are well known (*see* Chapter 15; Zimmermann, 1977; Zimmermann and Steudle, 1978), with decreased turgor resulting in an increased net flux of solutes, such as potassium into a cell and increased turgor resulting in a decreased net solute influx. As cell turgor is determined by osmotically-driven water flow, the effect on solute flux will reflect the coupling of water and solute flow resulting in a regulatory system which may operate in plants in response to water and salt stress (Cram, 1976). This implies a turgor-sensing mechanism within the plant cell membranes which will transform a pressure signal into changes in the physicochemical properties of the cell membrane. Such changes as are observed are best explained in terms of the electro-mechanical model of the cell membrane which envisages the membrane as a capacitor filled with an elastic dielectricum (*see* Zimmermann, 1977). Such a membrane will be compressed by both electrical potential and pressure potential differences across the membrane. Compression of the cell membrane and interaction between the membrane and the cell wall may well affect the transport sites for ions within the membrane and explain the observed turgor-dependency of solute fluxes. Such a concept is consistent with the view that plant cells osmoregulate to achieve a particular pressure.

The effects of indoleacetic acid (IAA) on such

osmoregulation are significant in relation to turgor-mediated cell extension growth processes. IAA is known to increase the uptake of both solutes and water in plants (*see* Anderson, 1976). This may reflect the possibility that IAA causes a compression of certain regions of the cell membrane which will interact with ion transport. Certainly a similar effect of IAA and of turgor has been observed on both the electrical properties of the cell membrane and the ion transport of *Valonia* cells (*see* Zimmermann, 1977). Such an observation is consistent with a common molecular mechanism for IAA action and the pressure-sensing system. If such an IAA-stimulated ion flux is coupled with water transport the observed effect of IAA on water transport referred to above may be explained. Evidence in support of such hormone-mediated effects is limited, although the observations of Morré and Bracker (1976) that isolated plasma membranes of soybean were 10 to 15% thinner when treated with 10^{-6} M IAA is of interest in relation to the above predictions.

Although our knowledge within this area is still very restricted the concepts which may be reached, by extrapolation of results with giant algal cells, are profound in relation to plant growth processes. The observed pressure-dependence both of water transport and of active and passive ion transport, linked with the elastic properties of the cell wall, provides new insights into the control mechanisms of cell extension growth.

Further reading

Dainty, J. (1976). Water relations in plant cells. In *Encyclopedia of Plant Physiology*, N.S. vol. II, *Transport in Plants, Part A Cells*, eds U. Lüttge and M. G. Pitman, Springer-Verlag, Berlin, Heidelberg, New York, p. 12.

House, C. R. (1974). *Water Transport in Cells and Tissues*, Edward Arnold, London.

Kozlowski, T. T. (1968–1976). *Water Deficits and Plant Growth*. Vols I–IV, Academic Press, New York, London.

Kramer, P. J. (1969). *Plant and Soil Water Relationships: A Modern Synthesis*. McGraw-Hill, New York.

Lange, O. L. and Lösch, R. (1979). Plant water relations. In *Progress in Botany*, vol. 41, eds H. Ellenberg, K. Esser, K. Kubitzki, E. Schnepf and H. Zeigler, Springer, Berlin, Heidelberg, New York, p. 10.

Meidner, H. and Sheriff, D. W. (1976). *Water and Plants*, Blackie, Glasgow, London.

Milburn, J. A. (1979). *Water Flow in Plants*, Longman, London, New York.

Nobel, P. S. (1974). *Introduction to Physiological Plant Physiology*, Freeman, San Francisco.

Slatyer, R. O. (1967). *Plant–water Relationships*, Academic Press, New York, London.

Zimmermann, U. (1978). Physics of turgor- and osmoregulation. *Ann. Rev. Plant Physiol.* **29**, 121–48.

References

Anderson, W. P. (1976). Transport through roots. In *Encyclopedia of Plant Physiology*, N.S. vol. II, *Transport in Plants 2, Part B*, eds U. Lüttge and M. G. Pitman, Springer, Berlin, Heidelberg, New York, p. 129.

Anderson, W. P., Aikman, D. P. and Meiri, A. (1970). Excised root exudation—a standing-gradient osmotic flow. *Proc. Roy. Soc. Lond.* B **174**, 445–48.

Barrs, H. D. (1966). Root pressure and leaf water potential. *Science* **152**, 1266–8.

Barrs, H. D. (1968). Determination of water deficits in plant tissues. In *Water Deficits and Plant Growth*, vol. 1, ed. T. T. Kozlowski, Academic Press, New York, London, p. 235.

Boyer, J. S. (1967). Matric potentials of leaves. *Plant Physiol.* **42**, 213–17.

Boyer, J. S. (1969). Measurement of the water status of plants. *Ann. Rev. Plant Physiol.* **20**, 351–64.

Boyer, J. S. (1977). Regulation of water movement in whole plants. *Symp. Soc. Exp. Biol.* **31**, 455–70.

Cowan, I. R. (1965). Transport of water in the soil-plant-atmosphere system. *J. Appl. Ecol.* **2**, 221–39.

Crafts, A. S., Currier, H. B. and Stocking, C. R. (1949). *Water in the Physiology of Plants*, Chronica Botanica Company, Waltham, Mass.

Cram, W. J. (1976). Negative feedback regulation of transport in cells. The maintenance of turgor, volume and nutrient supply. In *Encyclopedia of Plant Physiology*, N.S. vol. II, *Transport in Plants*, Part A, *Cells*, eds U. Lüttge and M. G. Pitman, Springer-Verlag, Berlin, Heidelberg, New York, p. 284.

Dainty, J. (1963). Water relations of plant cells. *Adv. Bot. Res.* **1**, 279–326.

Dainty, J. (1965). Osmotic flow. *Symp. Soc. Exp. Biol.* **19**, 75–85.

Dainty, J. (1972). Plant cell-water relations: the elasticity of the cell wall. *Proc. Roy. Soc. Edinburgh* A **79**, 83–93.

Dainty, J. (1976). Water relations in plant cells. In *Encyclopedia of Plant Physiology*, N.S. vol. II, *Transport in Plants*, Part A, *Cells*, eds U. Lüttge and M. G. Pitman, Springer-Verlag, Berlin, Heidelberg, New York, p. 12.

Dainty, J. and Ginzburg, B. Z. (1964). The reflection coefficient of plant cell membranes for certain solutes. *Biochim. Biophys. Acta* **79**, 129–37.

Fensom, D. S. (1957). The bioelectric potentials of plants and their functional significance. I. An electrokinetic theory of transport. *Can. J. Bot.* **35**, 573–82.

Flowers, T. J. (1975). Halophytes. In *Ion Transport in Plant Cells and Tissues*, eds D. A. Baker and J. L. Hall, North-Holland, Amsterdam, Oxford, p. 309.

Ginzburg, H. (1971). Model for iso-osmotic flow in plant roots. *J. Theoret. Biol.* **32**, 147–58.

Gutknecht, J. and Bisson, M. A. (1977). Ion transport and osmotic regulation in giant algal cells. In *Water Relations in Membrane Transport in Animals and Plants*, eds A. M. Jungreis, T. Hodges, A. M. Kleinzeller and S. G. Schulz. Academic Press, New York, p. 3.

Gutknecht, J., Hastings, D. F. and Bisson, M. A. (1978). Ion transport and turgor pressure regulation in giant algal cells. In *Transport Across Biological Membranes*, vol. 3, eds G. Giebisch, D. C. Tosteson and H. H. Ussing, Springer, Berlin, Heidelberg, New York, p. 125.

Hall, A. E. and Kaufmann, M. R. (1975). Regulation of water transport in the soil-plant-atmosphere continuum. In *Perspectives of Biophysical Ecology*, eds D. M. Gates and R. B. Schmerl, Springer-Verlag, Berlin, p. 187.

Hall, J. L. and Baker, D. A. (1977). *Cell Membranes and Ion Transport*, Longman, London, New York.

Höfler, K. (1920). Ein Schema fur die osmotische Leistung der Pflanzenzelle. *Ber. dtsch. Bot. Ges.* **38**, 288–98.

Holmgren, P., Jarvis, P. G. and Jarvis, M. S. (1965). Resistance to carbon dioxide and water vapour transfer in leaves of different plant species. *Physiol. Plant.* **18**, 557–73.

Honert, T. H. van den (1948). Water transport in plants as a catenary process. *Discuss. Faraday Soc.* **3**, 146–53.

House, C. R. (1974). *Water Transport in Cells and Tissues*, Edward Arnold, London.

Huber, B. (1956). Die Transpiration von Sprossachsen und andaren nicht foliosen Organen. In *Encyclopedia of Plant Physiology*, vol. 3, ed. W. Ruhland, Springer-Verlag, Berlin, p. 427.

Husken, D., Steudle, E. and Zimmermann, U. (1978). Pressure probe technique for measuring water relations of cells in higher plants. *Plant Physiol.* **61**, 158–63.

Kamiya, N. and Tagawa, M. (1956). Studies of water permeability of a single plant cell by means of transcellular osmosis. *Protoplasma* **46**, 423–36.

Kassam, A. H. (1972). Determination of water potential and tissue characteristics of leaves of *Vicia faba* L. *Hort. Res.* **12**, 13–23.

Kelly, R. B., Kohn, P. G. and Dainty, J. (1963). Water relations of *Nitella translucens*. *Trans. Bot. Soc. Edinburgh*, **39**, 373–91.

Kiyosawa, K. and Tazawa, M. (1977). Hydraulic conductivity of tonoplast-free *Chara* cells. *J. Membrane Biol.* **37**, 157–66.

Knipling, E. B. and Kramer, P. J. (1967). Comparison of the dye method with the thermocouple psychrometer for measuring leaf water potentials. *Plant Physiol.* **42**, 1315–20.

Kozlowski, T. T. (1964). *Water Metabolism in Plants*, Harper and Row, New York.

Kramer, P. J. (1969). *Plant and Soil Water Relationships: A Modern Synthesis*. McGraw-Hill, New York.

Kramer, P. J. and Bullock, H. C. (1966). Seasonal variations in the proportions of suberized and unsuberized roots of trees in relation to the absorption of water. *Am. J. Bot.* **53**(2), 200–4.

Lange, O. L. and Lösch, R. (1979). Plant water relations. In *Progress in Botany* vol. 41, eds H. Ellenberg, K. Esser, K. Kubitzki, E. Schnepf, H. Zeigler, Springer, Berlin, Heidelberg, New York, p. 10.

Läuchli, A. (1976). Apoplasmic transport in tissues. In *Encyclopedia of Plant Physiology*, N.S. vol. 2, *Transport in Plants* 2, Part B, eds U. Lüttge and M. G. Pitman, Springer, Berlin, Heidelberg, New York, p. 3.

Meidner, H. and Mansfield, T. A. (1968). *Physiology of Stomata*, McGraw-Hill, London.

Meidner, H. and Sheriff, D. W. (1976). *Water and Plants*, Blackie, Glasgow, London.

Meyer, J., Sauer, F. and Woermann, D. (1974). Coupling of mass transfer and chemical reaction across an asymmetric sandwich membrane. In *Membrane Transport in Plants*, eds U. Zimmermann and J. Dainty, Springer Verlag, Berlin, Heidelberg, New York, p. 28.

Milburn, J. A. (1979). *Water Flow in Plants*. Longman, London, New York.

Milburn, J. A. and Johnson, R. P. C. (1966). The conduction of sap. II. Detection of vibrations produced by sap cavitation in *Ricinus* xylem. *Planta* **69**, 43–52.

Monteith, J. L. (1973). *Principles of Environmental Physics*, Edward Arnold, London.

Morré, D. J. and Bracker, E. (1976). Ultrastructural alteration of plant plasma membranes induced by auxin and calcium ions. *Plant Physiol.* **58**, 544–7.

Morrow, P. H. and Mooney, H. A. (1974). Drought adaptations in two Californian evergreen sclerophylls. *Oecologia* **15**, 205–22.

Nobel, P. S. (1974). *Introduction to Physiological Plant Physiology*, Freeman, San Francisco.

Passioura, J. B. (1976). The control of water movement through plants. In *Transport and Transfer Processes in Plants*, eds I. F. Wardlaw and J. B. Passioura, Academic Press, New York, San Francisco, London, p. 373.

Richards, L. A. and Ogata, G. (1958). Thermocouple for vapour pressure measurement in biological and soil systems at high humidity. *Science* **128**, 1089–90.

Ritchie, J. T. (1974). Atmospheric and soil water influences on the plant water balance. *Agric. Meth.* **14**, 183–98.

Robards, A. W. and Clarkson, D. T. (1976). The role of plasmodesmata in the transport of water and nutrients across roots. In *Intercellular Communications in Plants: Studies on Plasmodesmata*, eds B. E. S. Gunning, A. W. Robards, Springer, Berlin, Heidelberg, New York, p. 181.

Scholander, P. F., Hammel, H. T., Bradstreet, D. and Hemmingsen, E. A. (1965). Sap pressure in vascular plants. *Science* **148**, 339–46.

Slatyer, R. O. (1960). Aspects of the tissue water relationships of an important arid zone species (*Acacia aneura* F. Muell) in comparison with two mesophytes. *Bull. Res. Coun. Israel* **8D**, 159–68.

Slatyer, R. O. (1964). Efficiency of water utilization by arid zone vegetation. *Annals Arid Zone* **3**, 1–12.

Slatyer, R. O. (1967). *Plant–water Relationships*. Academic Press, New York, London.

Slatyer, R. O. and McIlroy, I. C. (1961). *Practical Microclimatology*, UNESCO, Paris.

Slavik, B. (1974). *Methods of Studying Plant Water Relations*, Academia, Prague, Chapman and Hall, London, Springer-Verlag, Heidelberg, New York.

Spanner, D. C. (1951). The Peltier effect and its use in the measurement of suction pressure. *J. Exp. Bot.* **11**, 145–68.

Spanner, D. C. (1958). The translocation of sugar in sieve tubes. *J. Exp. Bot.* **9**, 332–42.

Spanner, D. C. (1964). *Introduction to Thermodynamics*, Academic Press, London, New York.

Spanswick, R. M. (1976). Symplasmic transport in tissues. In *Encyclopedia of Plant Physiology*, N.S. vol. 2. *Transport in Plants* 2, Part B, eds U. Lüttge and M. G. Pitman, Springer, Berlin, Heidelberg, New York, p. 35.

Stadelmann, E. (1966). Evaluation of turgidity, plasmolysis and deplasmolysis of plant cells. In *Methods in Cell Physiology*, vol. 2, ed. D. M. Prescott, Academic Press, London, New York, p. 143.

Stadelmann, E. J. (1969). Permeability of the plant cell. *Ann. Rev. Plant Physiol.* **20**, 585–606.

Stadelmann, E. J. (1977). In *Regulation of Cell Membrane Activities in Plants*, eds E. Marrè and O. Ciferri, Elsevier, North-Holland, Amsterdam, p. 3.

Steudle, E. and Zimmermann, U. (1974). Determination of the hydraulic conductivity and of reflection coefficients in *Nitella flexilis* by means of direct cell-turgor pressure measurements. *Biochim. Biophys. Acta* **332**, 399–412.

Steudle, E., Zimmermann, U. and Lüttge, U. (1977). Effect of turgor pressure and cell size on the wall elasticity of plants cells. *Plant Physiol.* **59**, 285–9.

Steudle, E., Zimmermann, U. and Zillikens, J. (1982). Effect of cell turgor on hydraulic conductivity and elastic modulus of *Elodea* leaf cells. *Planta* **154**, 371–80.

Stocking, C. R. (1956). Guttation and bleeding. In *Encyclopedia of Plant Physiology*, vol. 3, ed. W. Ruhland, p. 489.

Stout, D. G., Steponkus, P. L. and Cotts, B. M. (1977). The diffusional water permeability of *Elodea* leaf cells measured by nuclear magnetic resonance. *Can. J. Bot.* **55**, 1623–31.

Tomos, A. D., Steudle, E., Zimmermann, U. and Schulze, E.-D. (1981). Water relations of leaf epidermal cells of *Tradescantia virginiana*. *Plant Physiol.* **68**, 1135–43.

Troughton, A. (1974). The growth and function of the root in relation to the shoot. In *Structure and Function of Primary Root Tissues*, ed. J. Kolek, Veda, Bratislava, p. 153.

Tyree, M. T. (1970). The symplast concept: a general theory of symplastic transport according to the thermodynamics of irreversible processes. *J. Theor. Biol.* **26**, 181–214.

Tyree, M. T. (1973). An alternative explanation for the apparently active water exudation in excised roots. *J. Exp. Bot.* **24**, 33–7.

Tyree, M. T. (1976). Negative turgor pressure in plant cells: fact or fallacy? *Can. J. Bot.* **54**, 2738–46.

Tyree, M. T. and Hammel, H. T. (1972). The measurement of turgor pressure and the water relations of plants by the pressure bomb technique. *J. Exp. Bot.* **24**, 267–82.

Weatherley, P. E. (1963). The pathway of water movement across the root cortex and leaf mesophyll of transpiring plants. In *The Water Relations of Plants*, eds A. J. Rutter and F. H. Whitehead, Blackwell Scientific, London, p. 86.

Weatherley, P. E. (1970). Some aspects of water relations. *Adv. Bot. Res.* **3**, 171–206.

Willmer, C. M. (1983). *Stomata*, Longman, London, New York.

Zimmermann, M. H. and Brown, C. L. (1971). *Trees—Structure and Function*, Springer-Verlag, New York.

Zimmermann, U. (1977). Cell turgor pressure regulation and turgor pressure-mediated transport processes. *Symp. Soc. Exp. Biol.* **31**, 117–54.

Zimmermann, U. (1978). Physics of turgor- and osmoregulation. *Ann. Rev. Plant Physiol.* **29**, 121–48.

Zimmermann, U., Husken, D. and Schulze, E.-D. (1980). Direct turgor pressure measurements in individual leaf cells of *Tradescantia virginiana*. *Planta* **149**, 445–53.

Zimmermann, U. and Steudle, E. (1970). Bistimmung von Reflexionkoeffizienten an der Membran der Alge *Valonia utricularis*. *Z. Naturforsch.* **25b**, 500–4.

Zimmermann, U. and Steudle, E. (1978). Physical aspects of water relations of plant cells. *Adv. Bot. Res.* **6**, 45–117.

Zimmermann, U. and Steudle, E. (1980). Fundamental water relation parameters. In *Plant membrane transport: current conceptual issues*, eds R. E. Spanswick, W. J. Lucas and J. Dainty, Elsevier/North-Holland Biomedical Press, Amsterdam, p. 113.

Ionic relations

15

D T Clarkson

Introduction

The absorption of ions by plant cells has been studied with great vigor over the last thirty years or so. Although the ionic relations of plant cells with their surroundings are dissimilar in some respects to those of animal cells, it occurred to some far-sighted workers that the electrophysiology worked out so elegantly for nerve axons might be applicable to plant cells. The problems arising from the small size of most plant cells could be avoided by using giant cenocytic cells from various algae. It is, perhaps, mainly for this reason that a great deal more is known about these rather unusual cells than about the more typical ones of higher plants (*cf*. Raven, 1976, with Pitman, 1976). There are several features which make ion transport in higher plants a more difficult but more intriguing matter for study. In land plants absorption of ions from the surroundings is confined largely to the cells of the root. It is through their activities that appropriate amounts of ions are supplied to support the growth of cells in the shoot. This implies effective communication between these two sets of cells, not only in the physical sense so that ions and water can be conducted efficiently from root to shoot, but also in a biochemical sense so that the activity of root cells is matched as far as possible to the requirements for nutrients in growth.

A note of candor seems appropriate before the mechanisms which bring about ion transport are discussed. Assertions that transport is mediated by carriers, pumps, porters, channels, etc., are made with such frequency and confidence that a spurious authority is sometimes lent to what is merely conjecture. With a few exceptions the exact nature of the molecules involved directly in transporting ions into plant cells is not known. Much of what is said is extrapolation of knowledge gained from animal cells and microbes. While it would be anticipated that certain details are different in plant cells, most people accept that there are general principles governing transport across lipid membranes of all types of organism.

Basic aspects of membrane structure

An appreciation of the chemical and physical properties of membranes is essential for the understanding of ion transport. The basic structure of the important molecules and the constraints which they impose on one another can quickly help to decide which explanations, or models, of transport processes are likely to be correct and those which are improbable. The literature of membranology is already vast and is growing rapidly. For economy of space, an arbitrary selection of statements about membranes is given in Table 15.1; the evidence for these statements can be found in the references cited, many of which are reviews. In the brief summary which follows only three matters will be discussed; the amphipathic nature of membrane constituents, the degree of ordering in membrane structure and the hypothetical arrangements of membrane proteins to provide ion-transporting channels.

Table 15.1 Some information on molecular aspects of membranes, its relevance to ion transport and sources of reference.

Information	Notes	Further reading
1. MEMBRANE = LIPID BILAYER + (INTRINSIC AND EXTRINSIC PROTEINS)	The bilayer acts as a barrier to general diffusion and provides the environment in which proteins and carriers operate	Singer (1974)
2. LIPID BILAYER—principal constituents are POLAR LIPIDS + STEROLS		
2.1 POLAR LIPIDS—mostly phospholipids in plasma membranes but some galacto lipid and possibly sulfolipid	The most abundant are phosphatidyl-choline (PC)—ethanolamine (PE)—glycerol (PG)—inositol (PI); these may have marked asymmetric distributions, *e.g.*, in some bacteria 70% of the PE is found on the cytoplasmic side of the cell membrane	Rothman and Lenard (1977) Op Den Kamp (1979)
—all have polar (hydrophilic) HEADGROUP + 2 (rarely 1) FATTY ACIDS with long hydrocarbon tails (hydrophobic)	Molecule interacts with water at one end and avoids contact with it at the other: amphipathic structure. Results in spontaneous formation of bilayer when lipid dispersed in water	Brockerhoff (1977)
2.1.1 HEADGROUPS		
—wide range of size and shape, especially if galacto lipids are present. They pack together to form a more or less flexible 'crust' over the surfaces of the bilayer	Packing of headgroups is disturbed by protrusion of intrinsic proteins (*see* entry 3.1) and perhaps by some plant growth regulators, e.g., gibberellins	Pauls *et al* (1982)
—have immobilized water layers associated with them	—PC and PE bind 10–11 water molecules firmly at each headgroup; stabilizes headgroup arrangement —Increase length of diffusion path across membrane	Schultz and Assunmaa (1970)
—may be interconnected by ionic bridges, especially if negatively charged	Ca^{2+} essential for maintenance of stability—in its absence membrane 'leaks' and membrane potential depolarizes	(see p. 326)
—may be associated specifically with intrinsic (and extrinsic?) proteins	Full enzymatic activity only when specific lipid present (*see* entry 3.1). May partly explain significance of asymmetric distribution of headgroups. Negatively charged species most likely to bind to proteins, *e.g.*, PE, PI and P-serine	Sandermann (1978)
—may carry net negative charge, *e.g.*, PG, PI, PE; no net charge (zwitterionic), *e.g.*, PC or very rarely net positive charge, *e.g.*, Lysl PG	A mosaic surface charge may determine prevailing ionic binding/transport. PC, which is zwitterionic involved exclusively in membrane stabilization—unlikely to engage in electrostatic bonding to membrane proteins	Brockerhoff (1977)
2.1.2 FATTY ACIDS		
—very many reported. The carboxyl groups associated with	In plants the most common acids are palmitic (16 carbon atoms: 0 double	

Information	Notes	Further reading
polar head in region known as the 'hydrogen belt'. Hydrocarbon tails mingle in fluid hydrocarbon core of bilayer	bonds), stearic (18:0), oleic (18:1), linoleic (18:2), linolenic (18:3)	
—hydrocarbon chains may be saturated (no double bonds) or unsaturated (with varying numbers of double bonds) —unsaturation occurs at the far ends of the chain (in the hydrophobic core) —thermal motion of unsaturated chains under greater fluidity increases with double bonds in chain	Double bonds act like an antifreeze in the hydrophobic core. At physiological temperatures the core remains predominantly in a liquid state. This means that intrinsic proteins may drift around (diffuse) in this layer, that hydrophobic ion carriers can diffuse across the membrane and that changes in protein configuration can occur with relative freedom. As the viscosity increases all these processes may slow down and they may cease if the core freezes	Shinitzky and Henkart (1979) Nobel (1974)
2.2 STEROLS —rigid plate-like molecules with small polar region and large hydrophobic region	—most common are β-sitosterol, campesterol, stigmasterol in higher plants; ergosterol common in fungi; cholesterol usually minor component ($<5\%$) in plant membranes	Grunwald (1975)
	—polar region associated with hydrogen belt: bulk of molecule in hydrophobic core	Brockerhoff (1977)
—inserted singly or in groups between hydrocarbon tails of fatty acids—not associated specifically with intrinsic protein	Tend to decrease the fluidity of the membrane at physiological temperatures but may prevent thermal transition by reducing the packing of hydrocarbon chains	Chapman (1973) Demel and De Kruyff (1976)
—more abundant in plasma membrane than elsewhere	—their role in structuring the hydrocarbon layer may be achieved by intrinsic proteins in other membranes	
—may increase in relative abundance as cells /organs mature and enter senescence	Probably due to turnover of phospholipids	
3. PROTEINS confer characteristic biochemical/ transport properties: may be associated with hydrophobic interior of membrane (INTRINSIC) or with membrane surface (EXTRINSIC)		
3.1 INTRINSIC PROTEINS —hydrophobic groups associated with hydrocarbon chains of fatty acids. Polar amino acids form bonds with polar headgroups or may protrude into cytoplasm or extracellular space.	Have amphipathic properties which allow them to 'bridge' the hydrophobic region of the bilayer—this property makes them potential channels for transport. Examples—ATPase proteins of various kinds, Pi/triose phosphate exchange	Sandermann (1978) Singer (1974) Hodges (1976)

Table 15.1 (*contd*)

Information	Notes	Further reading
	protein in plastid envelope, succinate dehydrogenase	
—can be visualized as spherical particles 6–9 nm in diameter on replicas of freeze-fractured membrane surfaces	Particles more abundant on the membrane face adjacent to the cytoplasm (p-face). Probably are aggregates of two or more protein subunits—may be stabilized by lipopolysaccharides	Branton and Deamer (1972)
—abundance greatest in membranes across which large fluxes of materials occur	—in mitochondrial and plastid membranes 80% of surface area occupied	Branton and Deamer (1972)
	—frequency lower in plasma membranes but may vary markedly between tissues in a single organ, *e.g.*, endodermis > cortex in plant roots	Robards, Newman and Clarkson (1980)
—may be free to diffuse laterally (*i.e.*, in the plane of the membrane)	Abundance may change at specific locations after some physical or hormonal signal—may be a factor in cell polarity. Can migrate into areas which remain fluid if membrane partly 'freezes'	Furtado, Williams, Brian and Quinn (1979)
—may exhibit strong lipid specificity	Probably dependent on association of polar amino acids with negatively charged lipid headgroups	Finean (1973)
—activity may be strongly influenced by fluidity of membrane lipids	As viscosity in hydrocarbon region increases, conformational changes become increasingly restricted	Sandermann (1978)
—hard to remove from associated lipids—detergents or organic solvents required	Enzymatic activity often lost or greatly reduced when lipid removed— conformation of the protein changes	Wheeler, Walker and Barker (1976)

3.2 EXTRINSIC PROTEINS
  —hydrophilic amino acid sequences dominant
  —readily released by mild osmotic shock
  —probably associated with polar ends of intrinsic proteins; linked by ionic bridges
	—unlikely to penetrate or move across membrane interior. Role restricted to binding ions which are subsequently transported across membrane by some other agency: an assembly point	Singer (1974)

—periplasmic ion-binding proteins probably of common occurrence	—binding proteins for sulfate and phosphate reported in bacteria, yeasts and algae; existence probable in roots of higher plants —loss of binding by osmotic shock leads to marked reduction in rate of ion transport	Pardee, Prestidge, Whipple and Dreyfus (1968) Jeanjean and Fournier (1979) Jeanjean, Bedu, Attia and Rocca-Serra (1982)

Amphipathic membrane constituents

When a molecule is sufficiently large for there to be appreciable separation of polar and non-polar re-
gions it is described as amphipathic. Polar groups on lipids and polar amino acid sequences on peptides make interactions with water molecules and with electrolytes; they are therefore hydrophilic. Non-

polar regions of these molecules are hydrophobic and will spontaneously assume configurations which minimize their contacts with water molecules. In Fig. 15.1 the structures of a phospholipid, phosphatidylethanolamine, and a common plant sterol, β-sitosterol, are illustrated. Both are amphipathic; the hydrophilic portion of the phospholipid is much larger than that of the sterol and may be subdivided into a charged region and a hydrogen belt (*see* Brockerhoff, 1977). When phospholipids are dispersed in water the molecules aggregate spontaneously to form the familiar bilayer structure which is the basic matrix of all biological membranes. The headgroups associate with one another

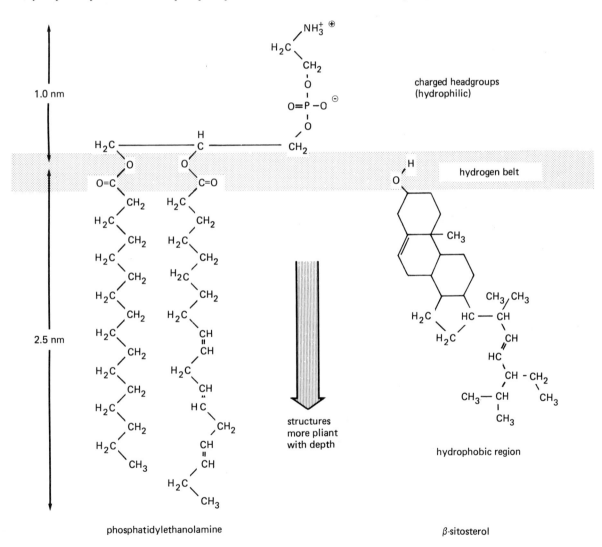

Fig. 15.1 Chemical structures of phosphatidyl ethanolamine and β-sitosterol, and their orientation in the lipid bilayer. The hydrocarbon chains are of unequal length in the phospholipid; that attached to the second carbon atom has three double bonds which are located in the distal part of the chain. The rigid ring structures of the sterol are toward the outside; the polar groups of the phospholipid form the headgroup which interacts with the aqueous external environment.

and with water molecules while the long hydrocarbon tails intermingle in a central hydrophobic belt.

Proteins are composed of sequences of amino acids which may be polar (charged), *e.g.*, glutamic acid and aspartic residues, or non-polar, *e.g.*, valine, leucine (uncharged, hydrophobic). These sequences in the peptide chain can fold independently of one another so that the hydrophobic ones are associated with the hydrocarbon tails of the lipids while the hydrophilic sequences interact with the headgroups, or protrude from the membrane surface into the cytoplasmic or extracellular surroundings (Fig. 15.2). Many membrane proteins span the hydrophobic belt and thus present opportunities for specific mechanisms (or channels) whereby electrolytes can cross the membrane without actually entering the hydrocarbon layer. This layer has a very low inherent conductance ($\equiv$ high resistance) to the permeation of ions. Except for a few unusual ions with high lipid solubility, *e.g.*, thiocyanate SCN^-, it is most unlikely that large quantities of ions can move across the hydrocarbon region of the membrane by simple diffusion—this process can occur but it is far too slow to account for the fluxes which are commonly observed. Polar channels provided by proteins might accommodate ion fluxes but there is also much speculation about the involvement of lipid-soluble molecules which can specifically bind an ion of a given type to form a complex ($\equiv$ carriers) which moves through the hydrocarbon belt. Being part of a lipophilic complex, the mobility of the ion is greatly increased. There are many substances of fungal and microbial origin which have this property—two examples are valinomycin which specifically carries K^+ and a dicarboxylic acid, known only as A23187, which carries Ca^{2+} and, less avidly, Mg^{2+}. These substances have antibiotic properties and it is not yet clear whether analogous substances are found normally in membranes of higher plants.

Order and fluidity in the membrane

The lipid bilayer may be likened to a semi-rigid crust of polar headgroups enclosing a fluid belt of hydrocarbon tails of long-chain fatty acids interspersed with sterols. Each biological species clearly sets a very high priority in maintaining the correct balance between order and disorder in these two layers. More is known about the regulation of fluidity in the hydrocarbon belt where the relative abundance of unsaturated chains ($\equiv$ those with double bonds) is continuously varied over the range of temperature in which a species can function. Such chains have a much higher thermal motion than saturated ones and melt at much lower tempera-

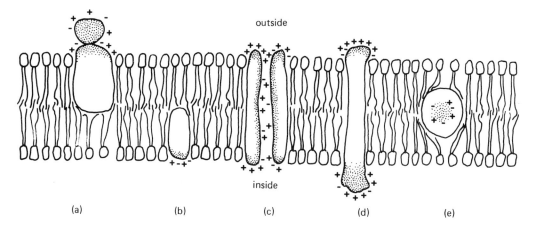

Fig. 15.2 Some of the many configurations of intrinsic membrane proteins. (a) Intrinsic protein protruding through headgroups on the outside of the membrane and bonded electrostatically to an extrinsic protein or receptor. (b) Polar amino acid sequences associated with the inner surface of membrane. (c) Two proteins forming a polar pore. (d) Protein with charged amino acid sequences at both ends. (e) Strongly hydrophobic protein, polar amino acid sequences, if any, folded into the interior of the molecule.

tures. They become much more abundant in cool conditions and decline in abundance in warm ones (Simon, 1974). This appears to maintain the fluidity of the membrane interior within acceptable limits. What is the significance of this? Let us anticipate some of the later discussion and say that ionic movements across membranes probably involve changes in the configuration or position of intrinsic membrane proteins. The hydrocarbon tails will clearly place some constraints on the configuration of these proteins—if they become very viscous or freeze into a gel state these constraints may become too great and the activity of the protein will be much inhibited. This can readily be shown in synthetic systems in which a transport ATPase is reconstituted into a bilayer membrane (Kimelberg and Papahad-

jopoulos, 1974). The activity of the enzyme is strongly dependent on the fluidity parameters of the lipids. Where transport occurs as a result of the movement of some complex between a lipid-soluble molecule and an ion, *i.e.*, a mobile carrier, the diffusion of the complex through the hydrocarbon belt will decrease exponentially as the viscosity increases.

The headgroups of polar lipids are more highly ordered than the hydrocarbon tails; ionic cross-linkages and hydrogen bonding with water molecules ensure this. The headgroups themselves differ markedly in size and shape, the galactosyl lipids being much bulkier than phospholipids (Fig. 15.3). The packing of the headgroup is, however, frequently disturbed by protrusions of intrinsic mem-

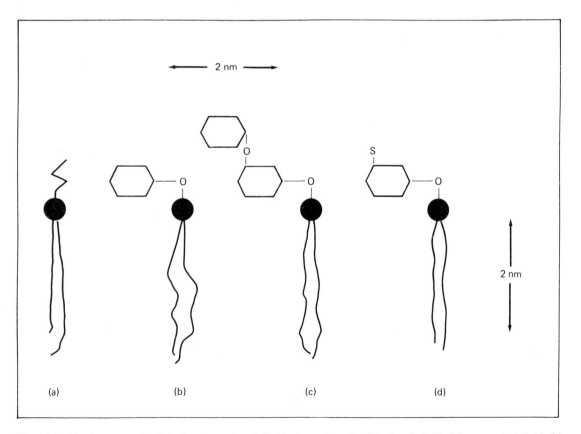

Fig. 15.3 Headgroups and 'tails' of various phospholipids drawn to scale: (a) phospholipid; (b) monogalactolipid; (c) digalactolipid; (d) sulfolipid. The opportunities for tight packing of headgroups are greater for phospholipids than for the other species.

brane proteins which can float around in the lipid (Fig. 15.2), pushing the headgroups aside as they do so (*see* Singer and Nicolson, 1972). Calcium ions seem to have a crucial role in preventing uncontrolled leakage of solutes across membranes, a role which is related to their capacity to form bridges between negatively-charged sites on adjacent headgroups. Stronger linkages, resulting in more rigid headgroup packing, are made by some other divalent, *e.g.*, Cd^{2+}, and trivalent, *e.g.*, La^{3+} and Al^{3+}, cations, but this is accompanied not only by a general decrease in permeability but by the inhibition of membrane-bound enzymes (*e.g.*, ATPase—*see* Caldwell and Haug, 1982). Conversely, ordering can be reduced by associations between headgroups and non-lipid molecules. Some physiologically active gibberellins reduce ordering of the headgroups and increase both ionic conductance and bilayer fluidity (Pauls, Chambers, Dumbroff and Thompson, 1982). It may be that less rigid packing of headgroups allows easier access of the ions to the proteins or other carriers which move them across the membrane.

Membrane proteins and ion transport

The way in which ion-transporting ATPase proteins are thought to work will be described in more detail in the next section; in this section the rather more general features of intrinsic and extrinsic proteins are discussed. In Fig. 15.4A, B two subunits of an intrinsic protein are shown—the polar amino acid sequences are folded so that there is a polar channel running through the core of the membrane. This arrangement is stabilized by hydrophobic bonding of the non-polar amino acid sequences with the hydrocarbon tails. A hydrated ion of the right radius can enter the pore and become specifically bound to some polar amino acid residue with the appropriate field strength (Fig. 15.4A). The binding of the ion, coupled with some metabolic process which puts energy into the system, induces some rearrangement of the subunits, closing the outside end of the pore and opening the cytoplasmic end; the ion is then released (Fig. 15.4B). This type of mechanism makes much more sense than models which usually show some large-scale movement or revolution of the transporter protein. Since the protein has polar groups at either end, any revolution of the molecule would involve dragging these groups through the

hydrophobic central belt; thermodynamically this would be a highly unfavorable process. The hydrophilic channel might also allow ion diffusion to occur without there being energy-dependent conformation changes. The channel would not have the 'constriction' shown in Fig. 15.4A and its dimensions would have to accommodate precisely any hydrated ion for which it showed selectivity. Of the physiologically important cations, K^+ has the smallest hydrated radius—a channel just large enough to allow the passage of K^+ would exclude the more bulky ions of Na^+, Mg^{2+} and Ca^{2+}.

A simple extension of the above model, Fig. 15.4C and D, accommodates the role of the extrinsic binding protein. Although these proteins may have specific binding sites they are strongly hydrophilic proteins which are, for thermodynamic reasons, most unlikely to diffuse across the membrane themselves. Their role must be envisaged in association with some intrinsic protein. An ion is bound to the extrinsic protein and the site (Fig. 15.4C) re-orients slightly as some biochemical activity in the intrinsic protein alters its configuration and opens up a channel along which the released ion diffuses across the membrane (Fig. 15.4D). The biochemical event most commonly associated with ion transport is the hydrolysis of ATP (*see* p. 332).

Driving forces on ions

An ion, like any other body, will move in a given direction only if driven by some force. Ion movements in and out of cells are of two kinds, those which occur spontaneously *down* a gradient of potential energy, *i.e.*, *diffusion*, and those which are driven *up* such a gradient by some process directly coupled to metabolism, *active transport*. The latter process does not occur spontaneously and is the equivalent of water flowing uphill; energy must be provided for, and work done by, some biochemical mechanism to bring it about. Formerly great emphasis has been given to distinguishing these two kinds of transport processes, but recently the distinction between them has become blurred by the recognition that the apparent active transport of one ion may be coupled to the diffusion of another: examples of this will be discussed later under *co-transport*.

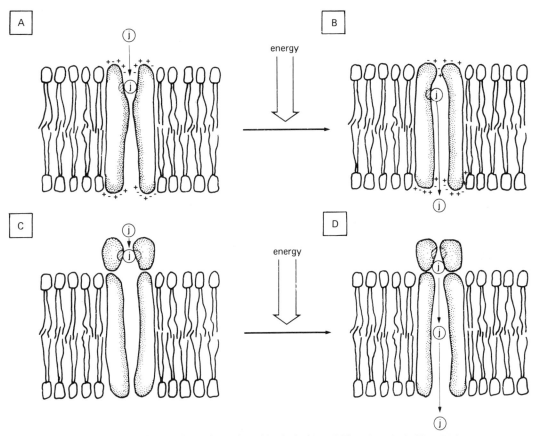

Fig. 15.4 Scheme to show how changes in configuration of intrinsic (A and B) and extrinsic (C and D) proteins can open up channels for the movement of ions. In A and B there is a specific binding site for the ion j on the intrinsic protein, in C and D this site is on the extrinsic protein. The binding of the ion and the input of free energy changes the position of the binding site with respect to the exterior and interior chemical environments; the ion is released when the field strength of the site changes.

Electrochemical potential gradients and diffusion

Because they carry electric charge, ions can be moved by electrical forces as well as by those dependent on concentration. Most cells are electrically negative relative to their surroundings and there will, therefore, be a tendency for positively charged ions to migrate into the cell. If the electrical gradient is strong enough it may allow an ion to *diffuse* from a dilute external concentration into the cell where its concentration is greater. This state of affairs is found quite commonly.

The electrochemical potential of an ion in a given system is made up of chemical and electrical terms:

$$\bar{\mu}_j = \bar{\mu}_j{}^* + RT \ln C_j + z_j F \psi \qquad (1)$$

where $\bar{\mu}_j{}^*$ is the electrochemical potential of the ion j in its standard state—a notional condition which need not concern us since in all subsequent steps it cancels out: R is the gas constant; T is the absolute temperature (K); $\ln C_j$ is the natural logarithm of the concentration of j in $mol\,m^{-3}$—strictly this should be modified by the activity coefficient which measures the effective concentration as a proportion of the total, but in most examples the solutions involved are sufficiently dilute for us to assume a value of unity for this coefficient; z_j is the algebraic valency of the ion j, e.g., +1 for a monovalent cation, −2 for a divalent anion, etc.: F is the Faraday which defines the amount of charge carried by 1 gram equivalent of an ion; ψ is the electric potential of the system which contains the ion—note that this is not a property of the ion j itself. The units of electrochemical potential are joules per mole, $J\,mol^{-1}$.

If an ion is present in two compartments which are not in equilibrium, the driving force on the ion across the intervening barrier is the gradient of electrochemical potential, $d\bar{\mu}_j$, which is obtained by differentiating Equation (1):

$$\frac{d\bar{\mu}_j}{dx} = -\frac{RT}{C_j}\cdot\frac{dC_j}{dx} - z_jF\cdot\frac{d\psi}{dx} \qquad (2)$$

The negative signs on the right-hand side of Equation (2) indicate that the driving force is exerted in a thermodynamically downhill direction. As an ion moves downhill it loses some of its potential energy, exactly how much being determined by the difference in the electrical potential between the two systems. Thus $\bar{\mu}^o_j - \bar{\mu}^i_j$ defines the decline in potential or free energy of 1 gram mol of the ion as it diffuses into the cell. If, on the other hand, j moves into the cell *up* a gradient of electrochemical potential (i.e., $\bar{\mu}^o_j$ greater than $\bar{\mu}^i_j$) then the difference in electrochemical potential defines the minimum amount of energy which must be supplied to 'drive' 1 gram mol up the gradient.

The direction of active transport

Derivations of Equation (1) can help in deciding whether the transport of an ion into a cell is likely to be uphill or downhill thermodynamically. To do this, a minimum of information about the system is required: the concentration of j, in the cell and surroundings and the electrical potential difference ($\psi^i - \psi^o$, usually abbreviated to E). The latter is measured across the cell membrane by inserting a glass microelectrode into the cytoplasm and placing another in the external solution; the two electrodes are connected via a salt bridge to an electrometer which registers the potential difference between them in millivolts. For a full description of the techniques involved in this essentially simple procedure *see* Clarkson (1974) or Lüttge and Higinbotham (1979).

First, consider a situation where the ion j is in equilibrium between the cell and the outside solution: it will have the same electrochemical potential both sides of the plasm membrane and there is no *net* movement of the ion in either direction, i.e.:

$$\bar{\mu}^i_j = \bar{\mu}^o_j \qquad (3)$$

Equation (3) can be expanded using Equation (1), and cancelling out $\bar{\mu}^*_j$ on either side:

$$RT\ln C^i_j + z_jF\psi^i = RT\ln C^o_j + z_jF\psi^o \qquad (4)$$

The electrical potential difference, E, measured experimentally is $\psi^i - \psi^o$. Gathering the electrical terms on the left-hand side the membrane potential difference (p.d.) can be defined as a function of concentration, e.g.,

$$E = \frac{RT}{z_jF}\ln\frac{C^o_j}{C^i_j} \qquad (5)$$

For a system at equilibrium, $\psi^i - \psi^o$ is given a special designation—the Nernst potential. We can see that the value of the Nernst potential is that which is needed to balance dissimilar concentrations of the ion j on opposite sides of the cell membrane. It is thus referred to as the Nernst potential of the ion j and abbreviated as E^N_j. There is a separate Nernst potential for each species of ion in the cell.

How effective the membrane potential is in maintaining ionic asymmetry across the membrane can be illustrated by a simple calculation. Assume that analysis has shown that the concentration of K^+ inside the cell is 100 times greater than that in the outside solution; the temperature of the system is 20°C (293K). Taking Equation (5) and giving numerical values to R and F (8.31 and 96.5 respectively), converting from natural logarithm to $\log_{10}(\times 2.303)$, then

$$E = \frac{58}{z_j}\log\frac{C^o_j}{C^i_j}\ \text{millivolts (mV)} \qquad (6)$$

Since K^+ is a monovalent cation, $z_j = 1$ and, given the measured ionic asymmetry,

$$E^N_{k^+} = 58\log 1/100$$
$$= 58 \times -2$$
$$= -116\,mV$$

Thus a membrane potential difference of $-116\,mV$ maintains a 100-fold concentration gradient at equilibrium. Values of the membrane potential are very commonly of this order and it need not be assumed, therefore, that because K^+ or other cations are more concentrated inside the cell they need to be associated directly with some energy-consuming mechanism in order to get there. The effectiveness of the membrane potential increases with the valency of the cation—if, for instance, $z_j = +2$, the 100-fold concentration difference of a gradient cation could be maintained at equilibrium by a potential of $-58\,\text{mV}$. Conversely, the existence of a negative membrane potential difference greatly *increases* the energy gradient up which anions must move if they enter the cell—a potential of $+116\,\text{mV}$ would be required to maintain a 100-fold concentration difference of, say, $H_2PO_4^-$ or Cl^-. Such potentials are never found and it must therefore be assumed that such ions are very far from being in equilibrium. Their movement into the cell is steeply uphill and energy must be expended to bring it about.

In practice, ions of either sign are rarely exactly in equilibrium between the cell and its surroundings—thus there is usually a difference in electrochemical potential difference for ions across the plasma-membrane:

$$\Delta\bar{\mu}_j = \bar{\mu}_j^i - \bar{\mu}_j^o = (RT \ln C_j^i + z_jF\psi^i)$$
$$- (RT \ln C_j^o + z_jF\psi^o) \quad (7)$$

$$= z_jF(\psi^i - \psi^o) - RT \ln \left(\frac{C_j^o}{C_j^i}\right) \quad (8)$$

From Equation (5),

$$z_jFE_j^N = RT \ln \left(\frac{C_j^o}{C_j^i}\right) \quad (9)$$

and substituting this in Equation (8) produces

$$\bar{\mu}_j^i - \bar{\mu}_j^o - z_jF(\psi^i - \psi^o) - z_jFE_j^N \quad (10)$$

$\psi^i - \psi^o$ is the measured potential E, thus

$$\bar{\mu}_j^i - \bar{\mu}_j^o = z_jF(E - E_j^N) \quad (11)$$

Thus the difference in electrochemical potential can be expressed as a millivoltage, ΔE_j. For a cation a positive value for ΔE_j shows that the measured potential is less than the Nernst potential so that there is likely to be some active component in its transport. If ΔE_j is negative the movement may be wholly passive. For anions the reverse of these statements applies. Table 15.2 shows the results of an analysis of ions in roots of oat seedlings in

Table 15.2 Comparison of observed[a] and equilibrium concentrations of ions in the roots of *Avena* (oat).

Ion(j)	External solution (C_j^o) (mol m^{-3})	Root tissue (C_j^i) Observed	Equilibrium prediction[b] (mol m^{-3} tissue water)	Direction of physical driving force
K^+	1.0	66.0	27.00	outward
Na^+	1.0	3.0	27.00	inward
Mg^{2+}	0.25	8.5	175.00	inward
Ca^{2+}	1.0	1.5	700.00	inward
NO_3^-	2.0	56.0	0.08	outward
$H_2PO_4^-$	1.0	17.0	0.04	outward
SO_4^{2-}	0.25	2.0	0.0004	outward
Cl^-	1.0	3.0	0.04	
Membrane potential		$E = -84\,\text{mV}$		

(a) These concentrations did not change appreciably with time and the roots were judged to be in a steady state with respect to the external solution. (Data from Higinbotham *et al.*, 1967.)

(b) Predicted values obtained by substituting the measured value of E (*i.e.*, $-84\,\text{mV}$) in the following equation

$$C_j^i = C_j^o[\exp(\pm z_jFE/RT)]$$

Where C_j^i is at equilibrium, E is equal to the Nernst potential (*see* p. 328).

relation to the electrical potential difference between the vacuole and the outside solution. It is evident that none of the ions is at its equilibrium concentration in this system. Potassium comes closest but the discrepancies between the predicted and observed values for the other ions are very great, especially for the anions which would tend to be excluded from the negatively-charged interior. On the basis of this table it would be concluded that the movement of the anions into the cell was steeply uphill and that of Na^+, Mg^{2+} and Ca^{2+} steeply downhill. In the latter case it is necessary to envisage some mechanism linked to metabolism which will transport these ions *out* of the cell to maintain the observed concentrations. There is evidence for the active extrusion of Na^+ (Jefferies, 1973) and Ca^{2+} (Macklon, 1975) but evidence for Mg^{2+} is not available.

The electrochemical state of the ions and the direction of physical driving forces recorded in Table 15.2 are representative of most land plants from non-saline habitats. Although in this and in other investigations K^+ is much closer to equilibrium than the other ions, there are numerous reports that K^+ transport into cells can be correlated with the *in vitro* activity of K^+-activated ATPase (Hodges, 1976). Where $[K^+]$ in the outside solution is less than 1 mM it seems probable that active transport will be required to maintain adequate concentrations in the cell; the soil solution of many natural habitats would contain substantially less than 1 mM K^+. It does, however, seem very peculiar that the K^+-stimulated ATPase activity found in plasma-membrane fractions should continue to rise when the external concentration is raised well above 1 mM K^+. In this range the electrochemical potential gradient, $\Delta\bar{\mu}_{K^+}$, is inwardly directed and cells often contain *less* than their equilibrium concentration of K^+. It seems possible that the increased ATPase activity may be concerned with K^+ efflux from the cytoplasm rather than with influx (*see* Etherton, 1963).

The cell and its surroundings should be in a steady state for the analysis of electrochemical potential gradients to be properly made—for this reason non-growing cells were selected for investigation in early research. If net flows of ions and water occur while measurements are being made, in theory at least, a rigorous analysis is much more complicated. For instance, there may be coupled flows of water and ions. Attempts to deal with this make use of the thermodynamics of irreversible processes, a matter well beyond the scope of this chapter and which remains of questionable practical value; a simple account can be found in Bowling (1976). For the majority of growing cells and those through which flows of water and ions occur, the calculation of Nernst potentials and their comparison with the measured value of E is a useful approximation of the electrochemical state of ions in the cell.

The electrical potential difference across membranes

With very rare exceptions, the electrical potential difference, E, across the plasma membrane is inside negative. This tells us that the cytoplasm has a small excess of (unpaired) negative charge relative to the external medium. The amount of charge which needs to be separated by the membrane to give rise to electrical potential differences of the order of -100 mV is minute, less than one unpaired anion in a million anions and cations (*see* Clarkson, 1974). How does this separation occur? Two major factors can contribute, one of which depends on the differing permeability of the membrane to anions and cations and the other on biochemical mechanisms which directly transfer an unpaired ion across the membrane.

It may be helpful to consider a very simple model (Fig. 15.5) in which a diffusion potential can develop because of differing permeability to anions and cations.

(a) The system consists of two compartments containing different concentrations of KCl and separated by a 'membrane' with high permeability to K^+ and very low permeability to Cl^-.

(b) Initially there will be a tendency for both ions to migrate from i towards o, but the rate at which K^+ can do so is much greater because of the properties of the membrane.

(c) For a very brief time i loses K^+ faster than Cl^-, a slight separation of charge occurs and the i face of the membrane becomes negatively charged—an electric potential difference has been generated.

(d) The negative p.d. slows down the movement of K^+ by creating an electrical driving force attracting it back towards i.

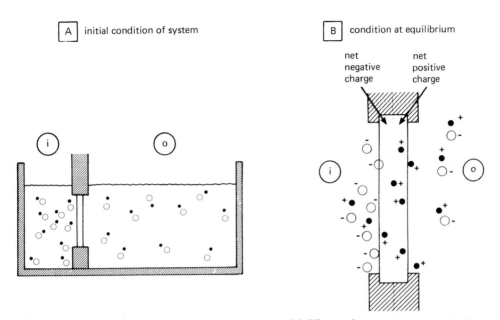

Fig. 15.5 Development of charge separation and an electrical potential difference between two compartments separated by a membrane selectively permeable to cations. For further explanation see text.

(e) If the membrane were totally impermeable to Cl⁻ an equilibrium would quickly be reached in which the higher [KCl] in i was balanced by the negative potential—*i.e.*, the Nernst potential for K⁺. Because there is a small finite permeability to Cl⁻, it too will migrate slowly from i to o. Since these movements result in the net loss of KCl from i the chemical driving force on K⁺ will diminish, the tendency to migrate from i to o will decline and the membrane potential will slowly decline.

Thus the potential developed in this simple system depends primarily on the differential permeability of the membrane to K⁺ and Cl⁻ and the difference in concentration of K⁺ on either side of the membrane.

The properties of this hypothetical membrane resemble those of the plasma membrane of plant cells in which the permeability coefficient for potassium, P_K, is at least one hundred times greater than P_{Cl}.

Each species of ion which can permeate the membrane will develop its diffusion potential just as in the example above. Its magnitude will be deter-

mined by the ratio of concentration of the ion and its permeability. The sum of all these diffusion potentials is calculated using the Goldman equation:

$$E = \frac{RT}{F} \ln \frac{P_K[K^+]^o + P_{cat}[cat]^o + P_{an}[an]^i}{P_K[K^+]^i + P_{cat}[cat]^i + P_{an}[an]^o} \quad (12)$$

where P_K, P_{cat}, P_{an} are membrane permeability coefficients for K⁺, and other cations and anions respectively. Notice that the ratio of the anion concentrations is inverted relative to the cationic terms. This is because an anion concentration gradient would reduce any negative potential set up by the asymmetric distribution of cations. The Goldman equation predicts the electrical potential across a membrane produced by the steady state concentrations of ions on either side of it and their permeabilities.

Although there are circumstances in which the actual value of the membrane potential corresponds quite closely to the value calculated from the Goldman equation, it is now widely recognized that a completely different process, *electrogenic* ion transport, can be an important contributor to the membrane potential (Spanswick, 1981).

Electrogenesis and membrane-bound ATPase

If an ion is transported in one direction across a membrane by some mechanism without being accompanied by an ion of opposite sign, charge is separated giving rise to an electrical potential difference. Such mechanisms are very commonly found and most of them share the common property that the ion is moved *up* a gradient of electrochemical potential. Many of them depend on ATP as an energy source which is hydrolyzed by intrinsic membrane ATPases (*see* Table 15.1 entry 3.1). In theory, any ion which could be selectively bound to the ATPase could be used for electrogenic ion transport, but in practice the choice is more restricted. A growing body of evidence supports the view that the proton, H^+, is the most common species used for this process in plants and bacteria (Spanswick, 1981); in animal cells there is greater emphasis on the electrogenic transport of Na^+.

Another simple model will be used to illustrate how an ATPase of this kind works (Fig. 15.6). First, it must be recalled that intrinsic proteins are amphipathic (Table 15.1 entry 3.1), have some of their polar groups associated with the membrane surface and may actually protrude into the cytoplasm and periplasm. These exposed regions of the protein may provide access to the membrane interior for transported ions. The model shows that on the cytoplasmic side of the ATPase there is a site which binds a Mg^{2+}-ATP complex (Fig. 15.6A). The terminal phosphate of ATP probably binds with the site since it remains attached to the protein after hydrolysis of the ATP has occurred. In different ATPase proteins this hydrolysis depends on ions bound to selective sites. It is not absolutely clear why this is necessary but many ATPases, even when largely separated from membrane lipids, are markedly activated by monovalent and divalent cations. Thus, there are ATPases specifically activated by Na^+ and K^+, by H^+, by Ca^{2+}, and one has been partially purified from the salt glands of *Limonium* which is activated by Cl^- (Hill and Hill, 1973). In general, ATPase preparations from plants are much less specific in their requirements for ions than those from animal cells (Hodges, 1976). It is possible that these ions induce subtle changes in the configuration of the protein when they are bound to it and that this exposes the active, catalytic site to its substrate, Mg-ATP. They are, therefore, co-factors in the enzymatic hydrolysis.

It is thought that the release of free energy during hydrolysis and the formation of the phosphoenzyme brings about a conformational change (Fig. 15.6B) which exposes the ion-binding site to the external environment, or at least to some aqueous environment which is in continuity with it. This might be in the core of the ATPase, the whole structure resembling a narrow channel crossing the hydrophobic region of the membrane. In this new configuration the field strength of the binding site may change and the ion which was bound from the cytoplasm is released (Fig. 15.6C). At this stage the electrogenic transport in which we are interested has occurred.

The ATPase is then thought to resume its initial configuration after the ion is released from the binding site, the phosphate is released from the active site into the cytoplasm and the enzyme is ready for another cycle.

In this model it would be expected that one co-factor ion or one charge would be separated for each ATP hydrolyzed—or $1\,\mu mol$ of charge, say H^+, for each μM ATP. The continuous operation of an ATPase would very quickly build up a very large and dangerous electrical potential across the cell membrane. Observation shows, however, that in a given set of conditions, the membrane potential remains constant, rarely exceeding $-150\,mV$ in higher plant cells. It must be concluded, therefore, that much of the potential generated by an electrogenic pump is dissipated in some way. At first sight this might appear to be a rather pointless use for valuable ATP but on closer inspection it is found that:

(1) there is a physiological requirement to remove excess H^+, Na^+ and Ca^{2+} from the cytoplasm so as to prevent damaging effects of pH, salt and cation bridge formation on a whole range of biochemical processes. The pH of the cytoplasm is generally closely controlled in the region of pH 7.0 to 8.0 even though processes such as nitrogen and sulfur metabolism may produce large amounts of H^+;

(2) there are fluxes of other ions which depend closely on the electrogenic transport.

Clearly a coupled movement of an *oppositely charged* ion in the *same direction* as the transported species, or a *similarly charged* one in the *opposite*

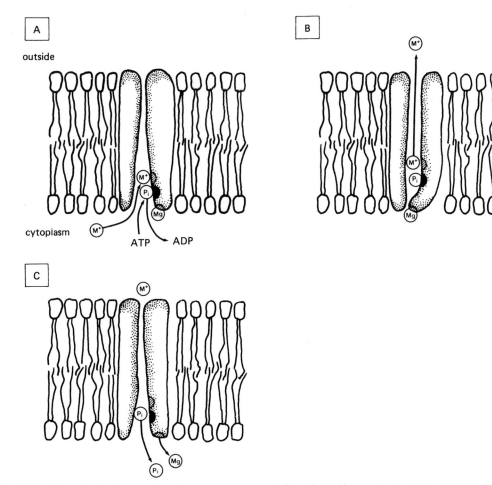

Fig. 15.6 A simplified scheme to show electrogenic transport of a cation, M$^+$, associated with the hydrolysis of ATP by a membrane-bound ATPase. For further explanation see text.

direction, will tend to short circuit the potential generated. Both of these processes are akin to the leakage of current from a cable. The rapidity with which potential is lost by 'leaks' can be seen by imposing a metabolic blockade on the ATP-generating systems of a cell. Fig. 15.7 shows how the addition of KCN very rapidly decreases the value of the negative membrane potential (depolarization) in pea epicotyl material; even more rapid depolarization from -170 mV to -80 mV occurs when cells of *Acetabularia* are placed in darkness. In the latter species there is an inwardly directed electrogenic transport of Cl$^-$ which depends heavily on photo-

phosphorylation for its ATP (Saddler, 1970). This rapid collapse of the membrane potential when energy metabolism is interrupted is one of the first signs looked for by the 'electrogenesis hunter'. It contrasts with a very slow depolarization of the diffusion potential caused by similar treatments (this is explained on p. 330 in the simple model of K$^+$ and Cl$^-$ diffusion). In Fig. 15.7 it is clear that within a few minutes the membrane potential has reached a new steady state. This value was found to correspond with the diffusion potential calculated from the Goldman equation (*see* Higinbotham, Graves and Davis, 1970).

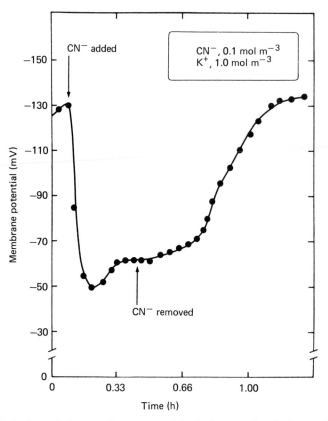

Fig. 15.7 Rapid depolarization of the membrane potential of *Pisum* epicotyl tissue after the application of $0.1 \, \text{mol m}^{-3}$ KCN to the bathing solution. Note that the effect is reversed when cyanide is washed out (based on Higinbotham *et al.*, 1970).

Where a positively charged ion such as H^+ is driven out of the cell, the negative membrane potential produced provides a driving force for the inward diffusion of a cation, say K^+. The exact nature of this coupling is far from clear, and to cover the confusion it is proposed that there is a separate category of membrane apparatus or channels which are called '*carriers*'. By implication, such carriers are not linked directly with energy metabolism and ATP hydrolysis, as in the ATPase proteins, but their existence is invoked to explain many phenomena.

Before discussing these carriers it should be pointed out that important ion-exchanges can be carried out by ATPase proteins themselves. The world's most studied membrane ATPase is the Na^+/K^+-activated ATPase in the red blood cell (*see*

Skou, 1965). The mechanism by which this transports Na^+ and K^+ is similar in outline to that illustrated in Fig. 15.6. The principal difference is that Na^+ is bound on the cytoplasmic side of the membrane and is exchanged for an approximately equivalent amount of K^+ from the outside during the conformational change of the protein. This K^+ is carried back into the cell and released into the cytoplasm. Both Na^+ and K^+ have moved uphill thermodynamically but there has been no charge separated—such an ATPase is sometimes described as 'electrically silent'. There is evidence that formally linked exchanges of ions through ATPases may occur across the plasma membrane of plant cells; coupled Na^+ and K^+ exchange has been reported in barley roots (Jeschke, 1970) and H^+/K^+ exchange occurs via an ATPase in yeast (Villalobo, 1982). Ion exchanges of this kind are 'useful' to the cell since the outgoing ion is frequently something which needs to be disposed of, while the incoming ion is needed either directly for biochemical purposes, or to provide the ionic asymmetry with which to maintain the diffusion potential across the cell membrane. At this point we have come a full circle back to p. 330 and Fig. 15.5. An electrically neutral K^+ transporter can generate the difference in $[K^+]$ between inside and outside which gives rise to a substantial part of the membrane potential.

The composite nature of the membrane potential

As a summary to the foregoing discussion it might be expected that in many circumstances both diffusion potentials and charge separation by electrogenic transport contribute to the observed membrane electrical potential difference. These contributions can be separated because metabolic blockade rapidly eliminates the latter, but affects the former component only slowly. Higinbotham et al. (1970) used potassium cyanide in various concentrations progressively to inhibit respiration and ATP synthesis. The fall in oxygen uptake in Fig. 15.8 was proportionally similar to the depolarization of the membrane potential in dilute solutions of KCN—at higher concentrations the membrane p.d. seemed to escape from respiratory control. The relatively constant value at high concentrations corresponded quite closely with the value of E predicted from the Goldman equation (see Equation (12)).

Proton-motive forces (PMF) across the plasma membrane

The pH of the cytoplasm is closely regulated so that the concentration of protons remains in the range

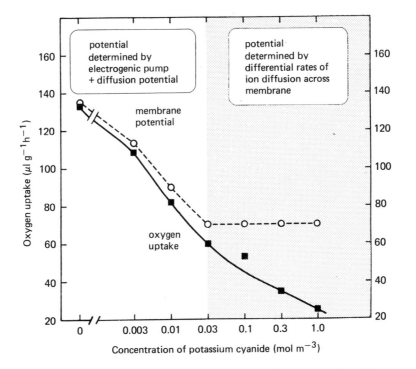

Fig. 15.8 The effect of potassium cyanide on the membrane potential and oxygen uptake in *Pisum* epicotyl tissue. The value of the membrane potential remains constant once the electrogenic component has been completely inhibited by cyanide (based on data in Higinbotham *et al.*, 1970).

10^{-8} to 10^{-7} M (Smith and Raven, 1979). Metabolic processes which generate additional H^+, *e.g.*, the assimilation of NH_4^+, or experimental perturbations designed to acidify the cytoplasm, activate proton extrusion from the cell (Lyalin and Ktitrova, 1976). This process is usually electrogenic and tends to polarize the membrane potential (Fig. 15.9). Conversely, the absorption of undissociated ammonia, NH_3, tends to alkalinize the cytoplasm. This depolarized the membrane potential in Lyalin and Ktitrova's experiments. These changes in membrane potential inevitably alter the driving forces acting on other ions moving across the cell membrane.

A simple calculation of the Nernst potential for H^+ in a hypothetical, but probable, situation shows how steep an uphill gradient may be traveled by a proton as it leaves the cell. Suppose that the concentrations of H^+ are 10^{-6} and 10^{-8} M in the outside solution and the cytoplasm respectively, and that the temperature is 293K (20°C): substituting these values in Equation (5)

$$E_{H+}^N = 58 \log \frac{10^{-6}}{10^{-8}}$$
$$= 58 \log 100$$
$$= 116 \text{ mV}$$

The membrane potential, E, of most cells is, say, -116 mV. In Equation (11) the electrochemical potential difference for H^+, $\Delta \bar{\mu}_{H+}$, between the inside and the outside solutions is given by $E - E_{H+}^N$,

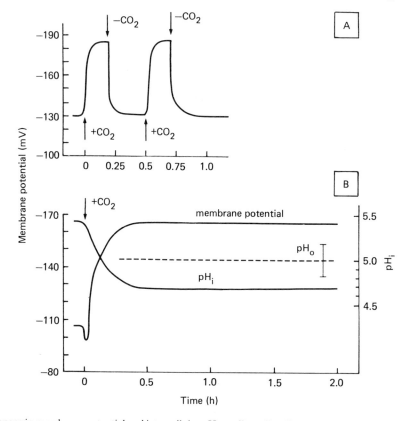

Fig. 15.9 Changes in membrane potential and intracellular pH_i as affected by dissolved CO_2. A. Results from the root hair of *Trianea bogotensis*. B. Simultaneous recording of membrane potential and internal pH (in vacuole) of *Nitella flexilis*. At the times indicated the bathing solution was either saturated with carbon dioxide at pH 5, or purged with air to remove CO_2. The cell interior is acidified as dissolved CO_2 dissociates and electrogenic proton extrusion increases in an attempt to maintain pH_i within acceptable limits. (Based on Fig. 3 of Lyalin and Ktitrova, 1976.)

that is, $-116 -(+116)\,mV = -232\,mV$. This is a very considerable force driving protons back into the cell interior from outside—it will become steeper as the surroundings become more acid. Even though the extrusion of protons involves a considerable cost in ATP, mechanisms have evolved which can make use of the membrane potential and the proton motive force thus developed to bring about the uphill transport of other much needed materials into the cell.

Assume that a cell contains $5\,mM\,NO_3^-$ in its cytoplasm and the external solution contains $0.1\,mM\,NO_3^-$, that the temperature is 20°C and the value of E is again $-116\,mV$. The Nernst potential for NO_3^- can be found by Equation (5) to be $+93\,mV$. The $\Delta\bar\mu_{NO_3^-}$ is, therefore, $-116 -(+93)\,mV = -209\,mV$. On this basis it appears that NO_3^- transport into the cell is steeply uphill and that some active transport mechanism, akin to the ATPase described earlier, may be operating on the NO_3^- ion. With the rare exception of the Cl^--stimulated ATPase from some plants of saline habitats, no anion-transporting ATPase has been detected. On the face of it, there would seem generally to be a lack of plausible biochemical mechanism to account for 'uphill' anion transport. Recently it has become clear that the driving force on H^+ diffusing into the cell—in the example above $\Delta\bar\mu_{H^+} = -232\,mV$—is frequently greater than the potential opposing the entry of other ions into the cell—coupling the two movements provides the energy necessary for their 'uphill' transport. Such a process is described as *co-transport* and in general terms it is envisaged as the formation of a ternary complex between one or more protons, the co-transported molecule or ion and some 'carrier' in the membrane. Protonation of the carrier favors the binding of the co-transported molecule; the complex re-orients or migrates in the membrane, exposing its binding sites to the cytoplasm where dissociation occurs.

Proton co-transport of sugars and ions

The most convincing evidence that such process can occur is from the movement of uncharged substrates rather than ions into cells. If a hexose sugar is added to the solution bathing cells of *Chlorella* (Komor and Tanner, 1976) or hyphae of *Neurospora* (Slayman and Slayman, 1974) a transient depolarization

of the membrane potential is observed (Fig. 15.10). It becomes less negative because of the inflow of protons accompanying the sugar. This inflow acidifies the cytoplasm and the proton extrusion mechanism is activated to keep cytoplasmic pH within prescribed limits. Since the transport is electrogenic the membrane potential approaches its original value. Closer study of such systems reveals that the rate of uptake of the hexose sugar depends strongly on $\Delta\bar\mu_{H^+}$ which can be manipulated by adjusting the pH of the external solution. The depolarization of E depends on the flux of hexose into the cell; E is not very sensitive to external pH in the absence of exogenous sugar. These observations lead to the conclusion that the membrane has rather low general permeability to protons and that their entry is through specific channels provided by the hexose carrier. Indeed, this appears to be a basic requirement for any system which is 'driven' by a PMF. If the membrane permeability is increased to H^+ these systems decline in effectiveness or cease. A number of substances can increase non-specific permeation of H^+ through membranes; they are usually molecules of high lipid solubility that bind protons reversibly and carry them in the direction of the gradient of $\Delta\bar\mu_{H^+}$ (examples are dinitrophenol (DNP), carbonyl cyanide *p*-trifluoromethoxy phenylhydrazone (FCCP)). In mitochondria and plastids these substances dissipate the PMF generated by electron transport and required for ATP synthesis, by allowing the diffusion of protons to occur generally over the membrane surface rather than restricting it to the specific channels associated with phosphorylation of ADP. They uncouple the processes and are thus known as *uncouplers*. They have much the same effects on PMF-dependent transport of materials across the plasma membranes of intact cells or in isolated membrane-bound vesicles.

The existence of proton co-transport systems for metabolites and anions is firmly established in the fungi (*see* Beever and Burns, 1980). Phosphate uptake has been shown to depend on the simultaneous intake of two protons. In the normal cell this process is obscured by the operation of the proton pump at the plasmalemma—in Fig. 15.11 the summary of the flows shows that inward and outward movements of H^+ are balanced, thus the pH of the cell and its surroundings are unchanged. It is only by measuring changes in pH that evidence of net proton movement can be inferred—tracers for

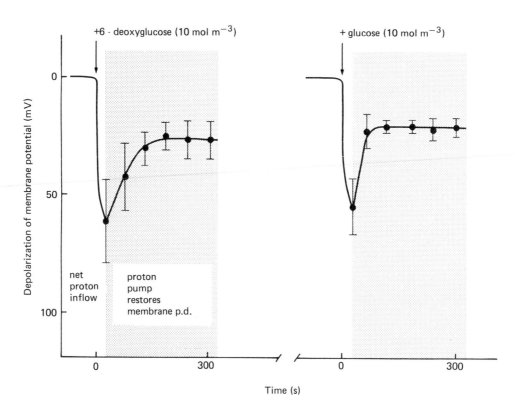

Fig. 15.10 Depolarization of the membrane potential of *Chorella vulgaris* when substrate-starved cells are treated with hexose sugars. The resting potential was about $-135\,mV$ before sugars were added; 6-deoxyglucose is not metabolized, therefore the partial recovery of the membrane p.d. is not due to substrate replenishment but depends on proton extrusion from the cell. (Adapted from Komor and Tanner, 1976.)

H^+ cannot be used because of exchange with water molecules. The activity of the ATPase in Fig. 15.11 can be inhibited by treating the cells with antimycin and by substrate starvation (Cockburn, Earnshaw and Eddy, 1975) which eliminate its ATP supply. In these circumstances there is an alkalinization of the external solution as P_i enters the cell, and K^+ moves *out* of the cell to balance the H^+ inflow. In such circumstances a direct relationship can be found between the initial pH of the external medium and the amount of P_i flowing into the cell.

Co-transport blurs the distinction between active transport and diffusion. Although anion transport into cells may be uphill it depends on the active transport of some other ion. Thus, the negative value of E, which appears at first to be such an obstacle to the entry of anions into cells, actually assists their entry because it provides a powerful driving force for the entry of H^+. In animal cells an inwardly directed gradient of electrochemical potential of Na^+ is used for co-transport.

The acceptance that this kind of transport can occur greatly simplifies the way in which the organization of transport across the plasma membrane is envisaged. The identity of the carriers themselves is completely unknown—it is not even certain that they are proteins—thus, much remains to be learned before it is established with certainty that this simple explanation is true rather than a hopeful delusion.

Compartments within the cell

A deliberately simplified view of the plant cell has been taken in the earlier discussion; the cell was regarded as little more than a mass of cytoplasm

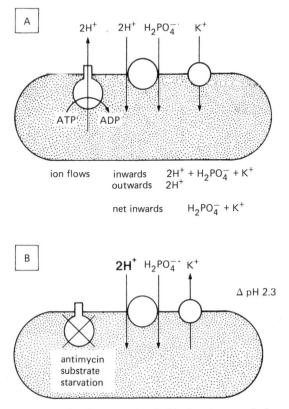

Fig. 15.11 Ion flows associated with phosphate uptake by fungal hyphae. A. Situation where the proton pump (ATPase) is operational; protons and phosphate enter *via* a common carrier. B. Proton pump inhibited; co-transport of protons and phosphate still occurs given a suitable pH across the membrane, but the direction of the K^+ flux changes to balance the H^+ inflow. (Compiled from Beever and Burns, 1980 and Cockburn *et al.*, 1975.)

surrounded by a plasma membrane. In fact the cell contains numerous membrane-bound compartments in which characteristic ionic compositions may be maintained. The principles governing transport across these membranes and the general biochemical features of the apparatus involved are the same as those that have already been considered. It is predictable that organelles, such as mitochondria and plastids, will have requirements for the transport of metabolites which differ from those of the plasma membrane and that fluxes of ions occur at much greater rates. The structural complexity of the organelle membranes reflects this

greater activity (*see* Chapter 11). The discussion of these matters is beyond the scope of this chapter but mention must be made of the vacuole and its membrane, the tonoplast.

In most higher-plant cells the vacuole occupies 85–90% of the cell volume once expansion has been completed. Gross ionic analysis of tissues largely reflects the contents of the vacuole (*e.g.*, Table 15.2), and when electrodes are inserted in such cells, the tip almost invariably impales both plasma membrane and tonoplast. This was also the case in the example in Table 15.2. It is only in unexpanded cells lacking a central vacuole and in unusual ones, like the distal region of root hairs having a relatively large cap of cytoplasm (Lyalin and Ktitrova, 1976), that separate electrical measurements can be made across the plasmalemma and across the tonoplast (*see* Clarkson, 1974). These operations are, of course, relatively straightforward in large algal cells where it has long been known that there is little electrical p.d. across the tonoplast (Raven, 1976); where potentials are detected, the vacuole is usually slightly positive relative to the cytoplasm. Measurements of electrical conductance (or resistance) show that the tonoplast has a 10-fold or greater electrolyte permeability than the plasma membrane. Such evidence as exists indicates that these properties are also found in the tonoplast of higher plant cells.

The greater permeability of the tonoplast means that perturbations which affect the activity or electrochemical state of ions in the cytoplasm are quickly reflected in changes within the vacuole, *e.g.*, in Fig. 15.9 the pH change was measured in the vacuolar sap while the p.d. was measured across the plasma membrane alone.

None of these findings implies that the vacuole has the same composition as the cytoplasm or that ions are necessarily at thermodynamic equilibrium across the tonoplast. The concentration of protons is usually 10 to 100 times that in the cytoplasm, *i.e.*, pH 5 to 6 rather than pH 7 to 8, and in some cases tens of thousands of times greater. In halophytes there may be similar differences of a similar magnitude in Na^+ and Cl^- concentration between cytoplasm and vacuole (Flowers, Troke and Yeo, 1977). It is also estimated that the activity of Ca^{2+} in the vacuole may be 10^4 times that of the cytoplasm (Clarkson and Hanson, 1980). When nutrient ions are supplied in excess of immediate metabolic requirements, they accumulate in the vacuole to a greater extent than in the cytoplasm, NO_3^- being a case in point.

The impression is, therefore, that the ionic composition of the vacuole may fluctuate since it acts as a store (or dump) which, by accumulating or releasing ions, can assist in maintaining ionic homeostasis of the cytoplasm. These flows of ions may, in some cases, require active metabolic support, *e.g.*, H^+ and Na^+ inflows. The great concentration differences suggest that passive permeability of the tonoplast to these ions should be quite low. The appreciable proton motive force or sodium gradient, $\Delta\bar{\mu}_{Na^+}$, which exists could, in theory, account for counter-movements of cations or coupled influx of anions (symport) into the cytoplasm; it is not yet clear whether these forces are harnessed in this way.

It has proved possible to isolate vacuoles in relatively large yields which are free of contamination by other membranes and organelles. Such isolated vacuoles appear to retain their functional integrity if kept in isotonic medium (Matile, 1978). Tonoplast from beet root (*Beta vulgaris*) possesses a number of enzymatic activities including ion-stimulated ATPase (Walker and Leigh, 1981). The prospects for reaching a detailed understanding of the working of tonoplast transport systems seem to be very bright; at present they are poorly understood although they are recognized as being of major importance in the ionic relations of cells.

Effect of ionic concentration on the rate of transport

The rate at which a given ion is absorbed by a cell can be determined by external factors of concentration and by internal regulatory processes (*see* p. 345). Great emphasis has been given to the description of ion transport in terms of kinetic parameters which measure the capacity and affinity of a transport system for the ion it carries (*see* Epstein, 1972). The weakness of such studies as an end in themselves is that the transport mechanism itself remains firmly enclosed in a 'black box' and its kinetic properties are not understood in molecular terms. Any system which has a finite capacity to process or handle some material will become saturated when the supply of material exceeds its maximum rate of working. Kinetics of the kind shown in Fig. 15.12

can be obtained from totally inanimate systems as well as for active transport processes.

Epstein and Hagen (1952) seem to have been the first to point out that transport of an ion into a plant cell may be analogous to the relationship between the binding of a substrate to an enzyme and the release of its products after catalysis. Consider a reaction

$$S + E \underset{K_2}{\overset{K_1}{\rightleftharpoons}} SE \overset{K_3}{\rightleftharpoons} E + P \qquad (13)$$

where S = substrate, E = enzyme and P = product. Two rate constants K_2 and K_3 govern the dissociation of SE while K_1 determines its rate of formation. The ratio of these constants describes the *affinity* of the enzyme for its substrate—it is known as the *Michaelis constant*, K_m. Thus

$$K_m = \frac{K_2 + K_3}{K_1} \qquad (14)$$

If K_m is a very small number, $K_1 \gg K_2 + K_3$ and the affinity between the enzyme and its substrate is high. The reverse is true as K_m approaches unity.

The maximum rate, V_{max}, of a reaction is related to the amount of enzyme present and the rate at which it turns over. Once the capacity has become saturated, the reaction rate becomes independent of the substrate concentration. Ion fluxes show exactly similar saturation kinetics (Fig. 15.12): the velocity at a given concentration, C_j, can be predicted from Equation (15) if V_{max} and K_m are known:

$$\varphi_j^{in} = \frac{V_{max} C_j}{K_m + C_j} \qquad (15)$$

More often than not V_{max} and K_m are being determined in an experiment where C_j is varied and φ_j^{in} is measured. A practical definition of K_m in terms of ionic concentration can be obtained by measuring the V_{max} at saturating concentrations of the ion j and estimating the concentration which produces half maximal velocity. In Fig. 15.12A two absorption isotherms are shown which have a similar 'saturation' or V_{max} while a third has a greater V_{max}. In practice it is tedious to gather enough data to draw curves like those in Fig. 15.12A. If the uptake rate and concentration are plotted as reciprocals, a straight line is frequently obtained (Fig. 15.12B). Extrapolation of this line gives an intercept at $1/V_{max}$;

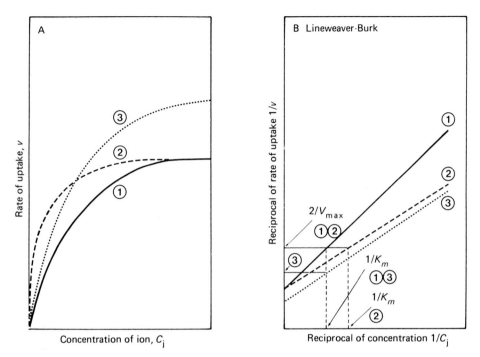

Fig. 15.12 The velocity of ion uptake *vs.* external concentration. A. Simple plots of results showing hyperbolas tending towards a maximum velocity. Curves 1 and 2 have a similar V_{max} but have different shapes; curves 1 and 3 differ in V_{max} but have the same shape. B. Double reciprocal plot (Lineweaver–Burk) of curves in A. Curves 1 and 3 have the same intercept on the $1/K_m$ axis; curves 1 and 2 have the same intercept on the $1/V_{max}$ axis.

the concentration at half maximal velocity, *i.e.*, $2/V_{max}$, gives the K_m. Note that the curves 1 and 3 in Fig. 15.12A are shown to have the same V_{max} in Fig. 15.12B.

The response of V_{max} and K_m of a transport process to physical and metabolic factors can undoubtedly provide some insight into the general nature of processes moving ions across membranes. The specificity of the mechanisms involved can be inferred from the effect of competing ions on V_{max}. Studies of this kind led Epstein (1966) to the conclusion that there were two sets of binding sites for potassium (labeled in his experiments by $^{86}Rb^+$), one specific for K^+, and another with affinity for Na^+ as well. At low concentrations of K^+, Na^+ had no effect on the φ_K^{in}, whereas, at higher concentrations of K^+, V_{max} was increasingly depressed by increasing $[Na^+]$. It was also found that at low $[K^+]$ the influx was indifferent to the rate of uptake of accompanying anions, whereas at high $[K^+]$ anions

had a great effect on the influx. When the uptake of K^+ is examined over a very wide range of concentration in the type of low-salt status seedlings that Epstein used, a Lineweaver–Burk plot of the results gives not one but at least two straight-line segments. Thus, potassium transport appeared to occur via two systems differing very markedly in V_{max} and K_m. The first, System 1, has limited capacity, a high affinity for K^+ and operates over the range of concentrations which might be expected for K^+ in soil solution around roots. The second, System 2, has a much larger capacity and a low affinity for K^+ and operates over the range of concentrations which one might expect to find in the xylem sap delivered to leaves when transpiration is low. Some critics of these ideas claim that the high concentrations necessary to get appreciable transport via System 2 are 'unphysiological'. This is a mistaken view because such concentrations are by no means uncommon in the soil solution of saline habitats and, more importantly, must be an everyday experience for the cells of leaf mesophyll.

A more significant criticism of Epstein's notion of dual uptake mechanisms is that the points may not fit a pair of superimposed rectangular hyperbolae in the way so commonly assumed. Exhaustive and painstaking research led Nissen (1974) to the view that the relationship between influx and concentration is much more complex with regular discontinuities showing up clearly on Lineweaver–Burk plots. Fig. 15.13 is taken from his study of SO_4^{2-} uptake by barley roots and shows three linear segments over the concentration range $10-250\,\mu M$. Nissen refers to these as 'phases'; the V_{max} and K_m of each increased regularly with sulfate concentration (Nissen, 1971). In an earlier paper, Epstein (1966) had illustrated similar behavior in the K^+ absorption isotherm at higher (System 2) concentration (Fig. 15.14). The important difference between the interpretations of these discontinuities offered by Nissen and Epstein is that the former believes that the changes in V_{max} and K_m over a very wide range of concentration reflect changes in the state of a *single* carrier system, whereas the latter has adhered to his view that there are two (or perhaps more?) quite separate systems. The changes of state envisaged by Nissen (1974) may be related to such factors as membrane surface charge, total electrolyte concentration, electrical potential difference across the plasma membrane and binding of the ion to various subunits of the carrier which may influence the accessibility of the carrier system to the ions it transports. Co-operative enzyme kinetics are well known and produce Lineweaver–Burk plots very similar in character to that in Fig. 15.13. They are explained by conformational changes in an enzyme when one or more of its subunits becomes linked to a ligand (or substrate). Such conformational changes have been discussed in relation to the operation of ATPase and will be introduced again in the next section where the regulation of ion transport by internal concentration is considered (see p. 343).

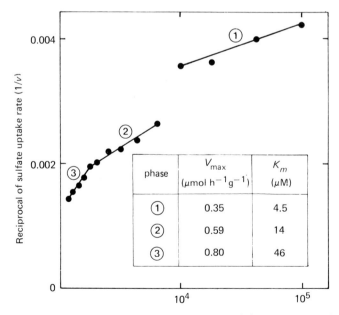

Fig. 15.13 Double reciprocal plots of sulfate absorption by barley roots. The plot can be resolved into three phases for which separate values for V_{max} and K_m can be found by extrapolation (as in Fig. 15.12B). As the concentrations of sulfate in the external medium increases the affinity of the carrier system for the ion (*i.e.* $1/K_m$) decreases. (From Nissen, 1971.)

In support of Epstein's view, there is compelling evidence from fungi that two transport systems differing in affinity (K_m) for $H_2PO_4^-$ exist side by side in the plasmalemma (*see* Beever and Burns, 1980). The wild type of *Neurospora crassa* has two transport systems with K_m values of about $2\,\mu M$ and $400–1000\,\mu M$ respectively; mutants, *nuc*-1 and *nuc*-2 have been isolated which cannot grow in media containing less than $300\,\mu M$ $H_2PO_4^-$. Kinetic analysis revealed that these mutants lack the high-affinity system entirely but have the one with low affinity (Lowendorf and Slayman, 1975). Kinetically and biochemically distinct sulfate permeases of high and low affinity have also been found in *Neurospora* (Marzluf, 1970).

Because low salt-status seedlings have been used so extensively in kinetic studies there have been many reports of very similar values for V_{max} and K_m for the transport of a given ion, particularly in the concentration range of 1 to $1000\,\mu M$. Unwittingly this has given rise to the belief that these parameters are fixed properties of the systems; this is another mistaken view since both parameters can be shown to vary in response to the plant's need for a given ion (*see* p. 345).

Regulation of ion transport

There are two levels at which the activity of transport systems can be regulated from within the cell: non-specifically, by variation in the energy supply (usually ATP) and specifically, by feedback between the system and the internal demand for the material it transports. The former is a blunt instrument and the latter a highly refined one with which to regulate the intake of nutrient ions into cells. The rate at which new cells are formed, either in a culture of

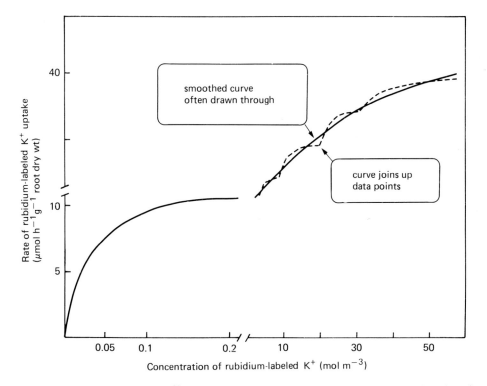

smoothed curve often drawn through

curve joins up data points

Rate of rubidium-labeled K^+ uptake ($\mu mol\ h^{-1}\,g^{-1}$ root dry wt)

Concentration of rubidium-labeled K^+ (mol m^{-3})

Fig. 15.14   Classical absorption isotherm of ^{86}Rb-labeled-K^+ by barley roots. At higher concentrations the points do not necessarily lie on a smooth curve but may show 'transitions' of the kind described by Nissen (1971). Each phase shows its own 'saturation' or V_{max}. (Based on Epstein, 1966.)

unicellular algae or in young trees, creates a demand for ions for biochemical and osmotic functions. This is self-evident but has been overlooked frequently in recent work. This may have been due to the fact that some of the most penetrating analyses of ionic relations have used non-growing plant cells such as the internodal cells of the charophyte *Nitella*. Such systems do not create any demand for net absorption of ions, indeed they were chosen partly with this in mind.

Growth rate and nutrient absorption

In the simplest growing systems, *e.g.*, *Chlorella* in synchronous culture, it can be shown that the rate of phosphate absorption per unit plasma membrane surface from a constant concentration of phosphate is related to the growth cycle of the cells, rising to a maximum where newly formed cells expand at the beginning of the photoperiod and slowing down again as they prepare for another cycle of division (Jeanjean, 1973). In a more complex experiment with *Lolium perenne* (ryegrass), diurnal variations in growth were matched by variations in the rate of nitrate absorption (Fig. 15.15) even though the concentration of nitrate remained constant in the flowing culture system used (Clement, Hopper, Jones and Leafe, 1978). These authors also showed that the rate of nitrate absorption was approximately constant over a 10^5-fold range of nitrate concentration and that the plants in each treatment grew at the same rate. It is evident, therefore, that varia-

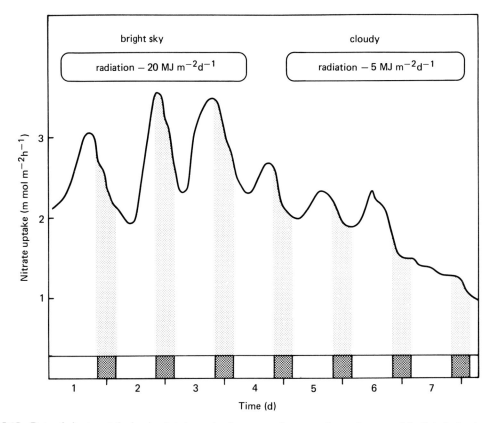

Fig. 15.15   Rate of nitrate uptake by simulated swards of rye grass plants growing under natural daylight in flowing culture solution where $[NO_3^-]$ held constant at approx. $7\,\mu M$. Note pronounced diurnal cycle in bright weather when plants grew rapidly and the fall in uptake rate during dull weather. (Adapted from Clement *et al.*, 1978.)

tions in uptake have little to do with the concentration of the nutrient in the external solution. The internal concentrations of ions within the cell or in a compartment within it seem more likely to regulate the rate of absorption.

Experimentally, these matters can be examined by altering the supply of the nutrient under consideration: it can be withheld or given in excess in the medium for a period and the subsequent response of the tissue can be observed by adding a labeled form of the ion. There are numerous reports of this kind of work (*see* Clarkson and Hanson, 1980) but the simplest for initial discussion are those where the ion is not metabolized after it is absorbed by the cells.

Regulation of K+ absorption by the cell [K+]

Figure 15.16A shows the uptake of labeled potassium (^{86}Rb) by barley roots from four different concentrations of KCl in relation to the internal concentration of K+ (Glass, 1976). The latter had been adjusted by holding the roots in the very high concentration (50 mol m^{-3}) of KCl for various times before the experiment began. In each case, increasing internal [K+] decreased the potassium influx into the root and the apparent V_{max} of the system. Analysis of these data indicated that there had also been a decrease in the affinity of the transport system for potassium, *i.e.*, the value of K_m had increased (Fig. 15.16B). This suggests that the properties of the carrier are modified in some way by the potassium concentration on the cytoplasmic side of the plasma membrane. In another paper (Glass and Dunlop, 1979) it is concluded that the decreased influx of K+ was not accounted for by depolarization of the membrane potential and a collapse of the driving force on the potassium ion. In explaining these results, and rather similar ones gathered quite

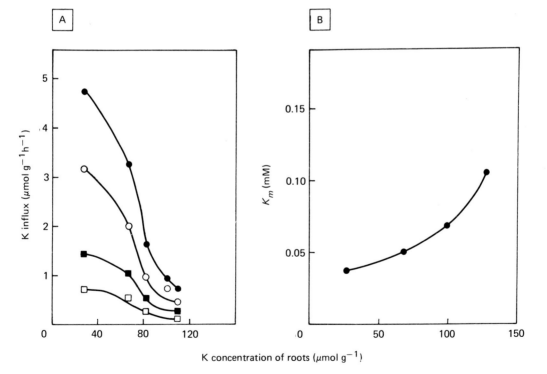

Fig. 15.16 Variation in K+ influx and K_m for K+ absorption with the internal K+-status of barley roots. A: relationships between influx from solutions containing KCl, 0.16 mol m^{-3} (solid circle), 0.08 mol m^{-3} (open circle), 0.04 mol m^{-3} (solid square) and 0.02 mol m^{-3} (open square), and [K+] in roots. B: relationship between K_m for K+ influx and [K+] in roots. (Taken from Glass, 1976.)

independently by Pettersson and Jensen (1978), it has been suggested that the configuration of the transport mechanism in the membrane can be modified by the presence of potassium ions bound to its inner surface and that the extent of binding depends on the activity of potassium ions in the cytoplasm. Glass (1976) illustrated his ideas about such allosteric regulation with Fig. 15.17. In assessing these ideas it should be said that there is ample precedence for the view that the conformation of macromolecules can be influenced by ions bound to them. This is the basis on which the ion-transporting activity of membrane ATPases are explained and it is clear that quite subtle changes in structure are all that is required to change both the binding site preference (field strength) and the direction from which such sites are accessible (*i.e.*, from the inside or outside). The evidence for allosteric regulation is not conclusive at present but it remains a reasonable hypothesis. It can hardly be expected that the matter will be resolved until the nature of the K^+ carrier has been described in molecular terms.

Effects of internal 'demand' on transport rate

In a growing organism there is usually a small surplus of nutrient ions which can be drawn upon should the external supplies fail for some reason. Growth dilutes the internal concentration in organisms 'living on their capital' in the absence of an external supply. This process cannot go on indefinitely, of course, and eventually the growth rate of the organism becomes limited by the nutrient(s) in short supply. The extent to which dilution can be tolerated before growth is affected differs with the nutrient, the extent to which it is stored in vacuoles and the ease with which it can be withdrawn from them and transferred (or translocated in higher plants) to the cytoplasmic locations where growth is occurring. While this kind of experiment can be objected to because the sudden cessation of nutrient supply does not occur very frequently in nature, there is evidence that, whether or not nutrient dilution (which we will call *nutrient-stress*) occurs rapidly or slowly, the effect is the same. It might also

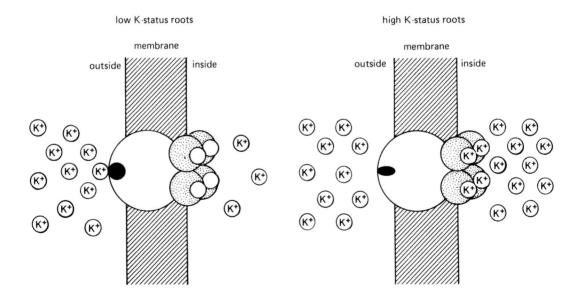

Fig. 15.17 A model to explain regulation of K^+ influx by internal $[K^+]$. When the binding sites on the 'inside' surface of the K^+-transport mechanism are empty, the field strength and/or accessibility of the binding site on the 'outside' surface is optimal. When the internal concentration increases the inner binding sites become progressively occupied, changing the conformation of the outer binding site and reducing its affinity for K^+; i.e., the K_m increases. (Adapted from Glass, 1976.)

be said that plants in natural vegetation are usually more likely to be in a state of incipient nutrient deficiency than burdened by a surfeit of nutrients; the same might not be true, however, for intensively grown crops.

When the supply of sulfate is withheld from the solution in which *Siratro* (*Macroptilium atropurpureum*—a legume closely related to the more familiar genus *Phaseolus*) is grown, the growth of the plant is unchanged for a period of time dependent on the growth rate set by other environmental factors. During this time the average concentration of both total sulfur and sulfate decreases in all tissues. The decrease is least in growing leaves due to redistribution from other parts of the plant (mainly from the roots). As soon as internal dilution begins, the potential rate for sulfate uptake by the roots increases and rises to maximum value (Fig. 15.18). If mildly S-stressed plants are returned to a medium containing sulfate, their internal sulfate concentration is rapidly restored to, or comes to exceed, that of the control and the rate of sulfate uptake falls back to the control level. The rate of sulfate uptake in the control plants is determined by the growth rate; in the stressed plants it rises as part of a strategy to maintain or restore the sulfate status of the plant and thus to maintain that growth rate.

Regulation in the above system might be achieved by a process directly controlled by the concentration of the nutrient in the cytoplasm or in some cytoplasmic pool. When growth dilutes this concentration to less than some limiting value the activity of carrier systems increases; the pool is thus topped up and represses the activity of the carrier. This repression may be similar to the process observed in the work by Glass and by Pettersson. In the *Siratro* experiment, interpretation is made somewhat easier by evidence that the 'pool', whatever its nature, probably contains sulfate and not some product of sulfate metabolism (*see also* Smith, 1980). Kinetic analysis indicates that derepression during stress is associated with an increased capacity of the sulfate transport system (reflected in very large increase in V_{max} (Table 15.3); the value of K_m is not significantly changed, however. During phosphate-stress in higher plants (Lefebvre and Glass, 1982; Lee, 1982) and in a number of phytoplanktonic algae (Gotham and Rhee, 1981) there were similarly large increases in V_{max} without a change in K_m. The simplest hypothesis is, then, that more carrier is made avail-

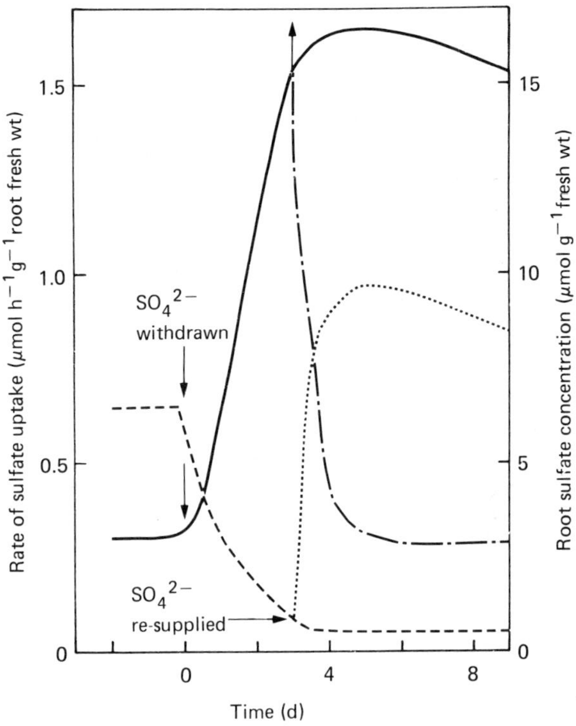

Fig. 15.18 Changes in the rate of sulfate uptake and sulfate concentration in roots of siratro (*Macroptilium atropurpureum*) during an interruption in the external sulfate supply. Solid line sulfate uptake by roots treated with 250 M sulfate after various times in medium lacking sulfate; -·—·—, uptake of sulfate by plants returned to sulfate-containing medium on day 3; ----, sulfate concentration in roots after various times in medium lacking sulfate; ·····, sulfate concentration in roots returned to sulfate containing medium on day 3. (Adapted from Clarkson *et al.*, 1983.)

able during stress rather than any modification of the property of the carrier such as its affinity for sulfate. It is possible that repression is achieved by a mechanism similar to that shown in Fig. 15.17, with the exception that a sulfate carrier is switched off for as long as a sulfate ion is bound to the allosteric site—its removal would make the carrier operational once more. Such a simple 'on–off' control would not be detectable by changes in K_m (Lee, 1982). Once again, in the absence of an exact mechanism for the sulfate permease it is rather idle to speculate further how regulation is achieved. If controls of this kind do not occur, however, one is left having to explain how the number of carriers can be increased or decreased five- to ten-fold over a 24-h period. These rates are very rapid in relation to general estimates of protein turnover (Davies, 1980) but the sulfate permease in *Neurospora* hyphae has been found to turnover rapidly, displaying a functional half-life of approximately 2 h (Marzluf, 1972).

where $V_{max(s)}$ is the maximum influx at a given internal concentration (s); max V_{max} is the maximum influx determined when the internal concentration was varied; b is the slope of relating V_{max} to the internal concentration. A similar relationship can be formulated for $K_{m(s)}$:

$$K_{m(s)} = \min K_m \cdot e^{-b'.s} \qquad (17)$$

Table 15.3 The effect of nutrient deprivation on kinetic parameters for sulfate and phosphate uptake.

Species	Nutritional status of plants	K_m (μm)	V_{max} (μmol h^{-1} g^{-1} root fresh wt)
Macroptilium[a]	High sulfate	5.0	0.34
atropurpureum	Low sulfate	8.0	1.95
Hordeum vulgare[b]	High sulfate	13.9	0.05
	Low sulfate	17.6	0.76
Hordeum vulgare[b]	High phosphate	6.6	0.26
	Low phosphate	4.9	0.48
Solanum tuberosum[c]	High phosphate	4.1	0.21
	Low phosphate	1.9	1.81

(a) From Clarkson, Smith and Vanden Berg (1983).
(b) From Lee (1982).
(c) From Cogliatti and Clarkson (unpublished).

Relationship between internal and external concentration of ions

In the original equation used by Epstein to relate influx and external concentration, V_{max} and K_m appear to be constants for a given species. This is not the case, as indicated above, but Siddiqi and Glass (1982) have attempted to incorporate the changes in the parameters which occur with internal concentration and demand into a modified form of Equation (15) (for convenience this is given again here).

$$\varphi_j^{in} = \frac{V_{max} C_j^o}{K_m + C_j^o} \qquad (15)$$

On the basis of earlier work by Glass and his colleagues both V_{max} and K_m were found to have an exponential relationship with the concentration of K^+ in the tissue. Thus,

$$V_{max(s)} = \max V_{max} \cdot e^{-b.s} \qquad (16)$$

Where $K_{m(s)}$ is the Michaelis constant at a given internal concentration (s); min K_m is the smallest value determined when the internal concentration was varied; b' is the slope relating K_m to the internal concentration.

The derived values for $V_{max(s)}$ and $K_{m(s)}$ can now be substituted in Equation (15) to give the influx at a given internal concentration:

$$\varphi_{j(s)}^{in} = V_{(s)} = \frac{\max V_{max} \cdot e^{-b.s} \cdot C_j}{\min K_m \cdot e^{-b'.s} + C_j} \qquad (18)$$

Siddiqi and Glass (1982) found a very good fit between predictions of this model and the observed rates of K^+ influx into barley roots when C_j^o and s were varied.

The general application of this model has yet to be examined. One problem which arises immediately is the appropriate measure of s when an ion which is metabolized or compartmented within the cell is being considered.

Communication about 'demand' between cells and organs

There is an added dimension in the response to nutrient deficiency in higher plants when comparison is made with simple algae like *Chlorella*. In the latter both nutrient dilution by growth and the response of the transport system occur within the same cell; in higher plants this is not always the case. Experiments where the root system of a plant is divided between solutions which contain or lack some nutrient show that the portion supplied with that nutrient takes it up more rapidly in response to deficiency elsewhere in the plant. This may occur even though gross chemical analysis reveals that the root treated with the nutrient is quite adequately supplied (*e.g.*, Drew and Saker, 1978). The activity of such a root seems to be governed by remotely-located demand. At first sight results such as those for treatment B in Table 15.4 seem to demolish the idea that the cytoplasm of the root cells has a 'pool'

Table 15.4 Effect of sulfur status on the uptake of sulfate by roots of *Siratro* divided between solutions containing or lacking external sulfate supply.

Sulfur status		Sulfate absorption by trained root
Trained root (μmol g^{-1} fresh wt) $\pm$ SEM	Main root	(μmol h^{-1} g^{-1} root fresh wt) $\pm$ SEM
A* 10 $\pm$ 1.0	10 $\pm$ 0.8	0.24 $\pm$ 0.04
B* 12 $\pm$ 0.6	5 $\pm$ 0.3	0.48 $\pm$ 0.04
C* 5 $\pm$ 0.1	4 $\pm$ 0.5	1.34 $\pm$ 0.16

*In A Both portions of root system grown in 250 μM SO$_4^{2-}$.
   B Trained root only in SO$_4^{2-}$.
   C Both portions of root system grown for 5 d without external SO$_4^{2-}$.

of the nutrient which regulates the transport system. Suppose, however, that the cytoplasmic pool of sulfate is quite small in relation to that in the vacuole

Table 15.5 Summary of views about the release of ions into the xylem.

Mechanism for ion release into xylem	Comment/References
(1) Leakage due to poor aeration of stelar tissues	Originally proposed by Crafts and Broyer (1938) and received some support from histochemical observations. Direct measurements of O$_2$ tension by Bowling (1973) show that stelar tissues in mature parts of root are well supplied with O$_2$.
(2) Release during xylem maturation	Proposed originally by Hylmo and supported later by Higinbotham and co-workers (*see* Lüttge and Higinbotham, 1979). As xylem cells die during differentiation, their ionic contents are discharged into the xylem vessels behind them. Probably does occur but only a minor fraction of ions loaded into xylem could be accounted for in this way. Also evident that ions can enter fully differentiated xylem at high rates (*see* Läuchli *et al.*, 1978).
(3) Passive release by diffusion from living cells in stele	The electrical potential of the xylem sap is less negative than that of the surrounding cells and both anions and cations are at a lower electrochemical potential in xylem sap than in, say, xylem parenchyma cells. Release is therefore 'downhill'. Dunlop and Bowling (1971). There are, however, clear indications that the process is under metabolic control (*see* Pitman, 1977).
(4) Metabolic regulation of release (active transport?)	Two observations seem to have given rise to this view; (a) that the cytoplasm of xylem parenchyma cells was unusually rich in organelles and vesicles, thus resembling a secretory tissue. (b) that release of ions into xylem can be inhibited independently of initial absorption into root cells. Transport to xylem seems more sensitive to the inhibition of protein synthesis and turnover than transport across the plasma membrane (*see* Pitman, 1977; Lüttge and Higinbotham, 1979, for thorough reviews). Does not necessarily imply that release in an active extrusion process (*i.e.*, uphill) but merely that the process is regulated by some protein activity—*see* the resistance 'R_2' in Fig. 15.19.

and that high demand elsewhere in the plant increases the rate of movement through the tissues of the root into the xylem. If the resistance of the pathway between the cytoplasmic pool and the xylem were lower than that between either the pool and the surroundings, or the pool and the vacuole, then flow out of the pool might exceed flow into it. A simple analogy of this situation is shown in Fig. 15.19: the nature of the 'resistances' referred to is not clear but they would represent a combination of physical resistances imposed by structure, *e.g.*, membrane reflexion coefficient, membrane electrical resistance, etc., and biochemical resistances related to the activity of membrane proteins. Such a model indicates a second site for the interaction of nutrient demand and ion fluxes, namely the cells which control radial movement of ions in the cortex and their release into the xylem (*see* Table 15.5). What sort of interaction could be envisaged? The model indicates that the value of resistance R_2 might be lowered if demand were high in the shoot or elsewhere in the root system, but this presupposes some method of communicating the information about demand between the shoot and the absorbing root. When communication of this kind is proposed in plant physiology, the response is usually arcane mumbling about hormonal messengers; the present consideration is no exception. To clarify matters, the following need to be established:

(i) the induction of a biochemical (hormonal) signal by the specific nutrient under consideration;

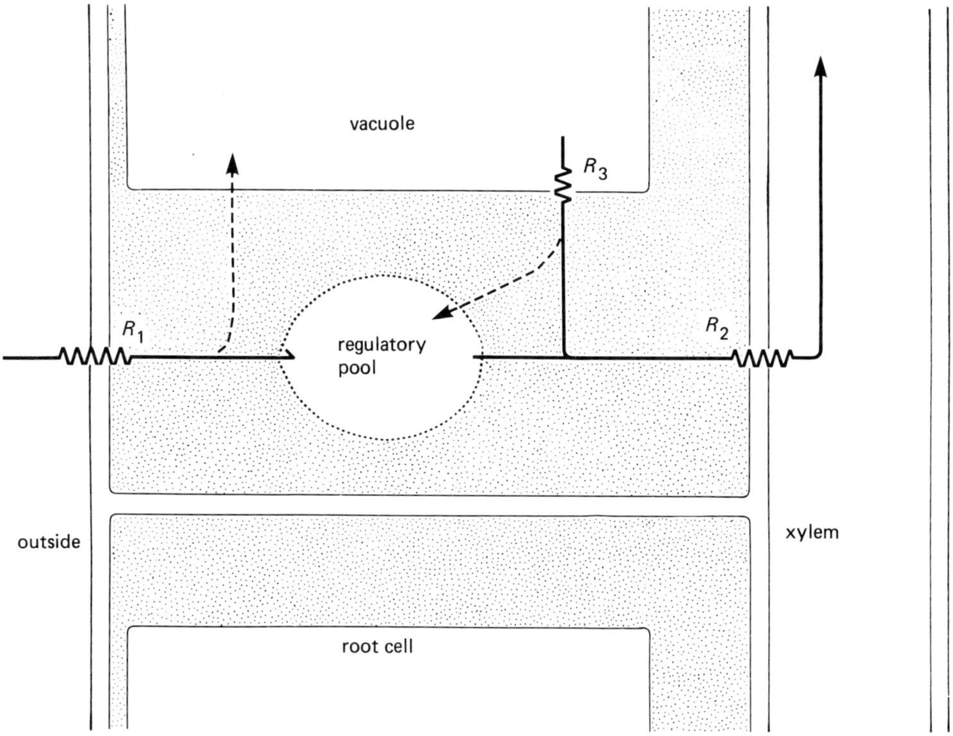

Fig. 15.19 A model showing notional 'resistances' to the flow of sulfate between the outside solution and xylem. R_1 is associated with the plasma membrane and can be modified by the $[SO_4^{2-}]$ in the regulatory pool; this is shown as a separate compartment, but could be the whole of the cytoplasm. R_2 is associated with the efflux of SO_4^{2-} from the symplast into the xylem. R_3 is associated with the tonoplast and influences the rate at which SO_4^{2-} can be removed from 'store'. A decrease in R_2 below that of R_3 and R_1 would dilute the regulatory pool which in turn would enhance absorption at the plasmalemma, effectively reducing R_1.

(ii) translocation of this message to the root (*via* the phloem);

(iii) the interaction between the message and some component of the resistance R_2 in the root— the proviso being that there must be a *specific* effect on R_2 for the ion in question and not on ion movement in general.

It is known that nutrient deficiencies may bring about major changes in the abscisic acid (Mizrahi and Richmond, 1972; Krauss, 1978) and gibberellin content of shoots and roots (*see* van Steveninck, 1976). Both groups of substances may have effects on membrane permeability by interacting with lipid headgroups (Lea and Collins, 1979; Wood, Paleg and Spotswood, 1974; Pauls *et al.*, 1982). The stumbling block at present is that these substances are likely to be non-specific in their effects on R_2 even if placed in a position to influence it.

Release of ions into the xylem of roots

Most of the ions absorbed by plant roots are readily transferred to the shoot *via* the xylem. Although radial movement of ions across the root may involve a combination of mass flow and diffusion, enough evidence exists to show that (a) the process can be precisely regulated, and (b) the final step involves efflux of the ions from living cells within the stele into the extracellular space represented by the xylem vessels. About this last step much uncertainty remains. A proper discussion of the conflicting viewpoints is not possible here and so the chapter ends, as it began, with a table which indicates the main hypothesis and some salient references (Table 15.5).

References

Beever, R. E. and Burns, D. J. W. (1980). Phosphorus uptake, storage and utilization by fungi, *Adv. Bot. Res.* **8**, 127–219.

Bowling, D. J. F. (1973). Measurement of a gradient of oxygen partial pressure across the intact root. *Planta* **111**, 323–8.

Bowling, D. J. F. (1976). *Uptake of Ions by Plant Roots*, Chapman and Hall, London, 212 pp.

Branton, D. and Deamer, D. W. (1972). *Membrane Structure, Protoplasmatologia* II E.1, Springer-Verlag, Vienna.

Brockerhoff, H. (1977). Molecular designs of membrane lipids. In *Macro and Multimolecular Systems*, ed. E. E. van Tamelen, Academic Press, N.Y., pp. 1–20.

Caldwell, C. R. and Haug, A. (1982). Divalent cation inhibition of barley root plasma membrane-bound Ca^{2+}-ATPase activity and its reversal of monovalent cations. *Physiol. Plant.* **54**, 112–188.

Chapman, D. (1973). Physical chemistry of phospholipids. In *Form and Function of Phospholipids*, eds G. B. Ansell, J. N. Hawthorne, R. M. C. Dawson, Elsevier, Amsterdam, pp. 117–42.

Clarkson, D. T. (1974). *Ion Transport and Cell Structure in Plants*, McGraw-Hill, London.

Clarkson, D. T. and Hanson, J. B. (1980). The mineral nutrition of higher plants. *Ann. Rev. Plant Physiol.* **31**, 239–98.

Clarkson, D. T., Smith, F. W. and Vanden Berg, P. J. (1983). Regulation of sulfate transport in a tropical legume, *Macroptilium atropurpureum. J. exp. Bot.* **34** (in press).

Clements, C. R., Hooper, M. J., Jones, L. H. T. and Leafe, E. L. (1978). The uptake of nitrate by *Lolium perenne* from flowing culture solution. II. Effect of light, defoliation and relationship to CO_2-flux. *J. exp. Bot.* **29**, 1173–83.

Cockburn, M., Earnshaw, P. and Eddy, A. A. (1975). The stoichiometry of the absorption of protons with phosphate and L-glutamate by yeasts of the genus *Saccharomyces. Biochem. J.* **146**, 705–12.

Crafts, A. S. and Broyer, T. C. (1938). Migration of salts and water into xylem of the roots of higher plants, *Am. J. Bot.* **25**, 529–35.

Davies, D. D. (1980). The measurement of protein turnover in plants. *Adv. Bot. Res.* **8**, 65–126.

Demel, R. A. and De Kruyff, B. (1976). The function of sterols in membranes. *Biochim. Biophys. Acta* **457**, 109–32.

Drew, M. C. and Saker, L. R. (1978). Nutrient supply and growth of the seminal root system of barley. III. Compensatory increases in growth of lateral roots and in rates of phosphate uptake, in response to a localized supply of phosphate. *J. exp. Bot.* **29**, 435–51.

Dunlop, J. and Bowling, D. J. F. (1971). The movement of ions to the xylem exudate of maize roots. III. The location of the electrical and electrochemical potential differences between the exudate and the medium. *J. exp. Bot.* **22**, 453–64.

Epstein, E. (1966). Dual pattern of ion absorption by plant cells and by plants. *Nature* **212**, 1324–7.

Epstein, E. (1972). *Mineral Nutrition of Plants: Principles and Perspectives*, Wiley, N.Y.

Epstein, E. and Hagen, C. E. (1952). A kinetic study of the absorption of alkali cations by barley roots, *Plant Physiol.* **27**, 457–74.

Etherton, B. (1963). Relationship of cell trans-membrane

electropotentials to potassium and sodium accumulation ratios in oat and pea seedlings. *Plant Physiol.* **38**, 581–5.

Finean, J. B. (1973). Phospholipids in biological membranes and the study of phospholipid-protein interactions. In *Form and Function of Phospholipids*, eds G. B. Ansell, J. N. Hawthorne and R. M. C. Dawson, Elsevier, Amsterdam, pp. 171–203.

Flowers, T. J., Troke, P. F. and Yeo, A. R. (1977). The mechanism of salt tolerance in halophytes. *Ann. Rev. Plant Physiol.* **28**, 89–121.

Furtado, D., Williams, W. P., Brian, A. P. R. and Quinn, P. J. (1979). Phase separation in membrane of *Anacystis nidulans* grown at different temperatures, *Biochim. Biophys. Acta* **555**, 352–7.

Glass, A. D. M. (1976). Regulation of potassium absorption in barley roots. An allosteric model. *Plant Physiol.* **58**, 33–7.

Glass, A. D. M. and Dunlop, J. (1979). The regulation of K^+ in excised barley roots. Relationships between K^+ influx and electrochemical potential differences. *Planta* **145**, 395–7.

Gotham, I. J. and Rhee, G-Y. (1981). Comparative kinetic studies of phosphate limited growth and phosphate uptake in phytoplankton in continuous culture, *J. Phycol.* **17**, 257–65.

Grunwald, C. (1975). Plant sterols. *Ann. Rev. Plant. Physiol.* **26**, 209–36.

Higinbotham, N., Graves, J. S. and Davis, R. F. (1970). Evidence for an electrogenic ion transport pump in cells of higher plants. *J. Membrane Biol.* **3**, 210–22.

Hill, B. S. and Hill, A. E. (1973). ATP-driven chloride pumping and ATPase activity in the *Limonium* salt gland. *J. Membrane Biol.* **12**, 145–8.

Hodges, T. K. (1976). ATPases associated with membranes of plant cells. In *Transport Plants II, Part A*, eds U. Lüttge and M. G. Pitman, Springer-Verlag, Berlin, pp. 260–83.

Jeanjean, R. (1973). Mechanisms d'absorption des ions phosphate par les Chlorelles: étude de l'absorption par des cultures synchrones. *Compt. Rend. Acad. Sci. Paris, D.* **277**, 193–5.

Jeanjean, R. and Fournier, N. (1979). Characterization and partial purification of phosphate binding proteins in *Candida tropicalis. FEBS Lett.* **105**, 163–6.

Jeanjean, R., Bedu, S., Attia, A. el F. and Rocca-Serra, J. (1982). Inorganic phosphate uptake of protoplasts and whole cells of yeast *Candida tropicalis*: absence of high affinity transport system in protoplasts, *Biochimie* **64**, 75–8.

Jefferies, R. L. (1973). The ionic relations of seedlings of the halophyte, *Triglochin maritima* L. In *Ion Transport in Plants*, ed. W. P. Anderson, Academic Press, London, pp. 297–321.

Jeschke, W. D. (1970). Evidence for a K^+ stimulated Na^+

efflux at the plasmalemma of barley root cells. *Planta* **94**, 240–5.

Kimelberg, H. K. and Papahadjopoulos, D. (1974). Effects of phospholipid acyl chain fluidity, phase transitions, and cholesterol on (Na^+ and K^+)-stimulated adenosine triphosphatase. *J. Biol. Chem.* **249**, 1071–80.

Komor, E. and Tanner, W. (1976). The determination of the membrane potential of *Chlorella vulgaris*. Evidence for electrogenic sugar transport. *Eur. J. Biochem.* **70**, 197–204.

Krauss, A. (1978). Tuberization and abscisic acid content of *Solanum tuberosum* as affected by nitrogen nutrition. *Potato Res.* **21**, 183–93.

Läuchli, A., Pitman, M. G., Lüttge, U., Kramer, D. and Ball, E. (1978). Are developing xylem vessels the site of ion exudation from root to shoot? *Plant Cell Environ.* **1**, 217–23.

Lea, E. J. A. and Collins, J. C. (1979). The effect of the plant hormone abscisic acid on lipid bilayer membranes. *New Phytol.* **82**, 11–18.

Lee, R. B. (1982). Selectivity and kinetics of ion uptake of barley plants following nutrient deficiency. *Ann. Bot.* **50**, 429–49.

Lefebvre, D. D. and Glass, A. D. M. (1982). Regulation of phosphate influx in barley roots: effects of orthophosphate deprivation and reduction of influx with provision of orthophosphate. *Physiol. Plant.* **54**, 199–206.

Lowendorf, H. S. and Slayman, C. W. (1975). Genetic regulation of phosphate transport system II in *Neurospora. Biochim. Biophys. Acta* **413**, 95–103.

Lüttge, U. and Higinbotham, N. (1979). *Transport in Plants*. Springer-Verlag, N.Y., 468 pp.

Lyalin, O. O. and Ktitrova, I. N. (1976). Experimental ways of shifting intracellular acidity and the effect of intracellular pH on the electrogenic hydrogen pump of the plant cell. *Soviet Plant Physiol.* **23**, 261–8.

Lyons, J. M., Raison, J. K. and Steponkus, P. L. (1979). The plant membrane response to low temperature; an overview. In *Low Temperature Stress in Crop Plants. The Role of the Membrane*, eds J. M. Lyons, D. Graham and J. K. Raison, Academic Press, N.Y., pp. 1–24.

Macklon, A. E. S. (1976). Cortical cell fluxes and transport to the stele in excised root segments of *Allium cepa* L. II. Calcium. *Planta* **122**, 131–41.

Marzluf, G. A. (1970). Genetic and biochemical studies of distinct sulfate permease species in different developmental stages of *Neurospora crassa. Arch. Biochem. Biophys.* **138**, 254–63.

Marzluf, G. A. (1972). Control of the synthesis, activity and turnover of enzymes of sulfur metabolism in *Neurospora crassa. Arch. Biochem. Biophys.* **150**, 714–24.

Matile, P. (1978). Biochemistry and function of vacuoles. *Ann. Rev. Plant. Physiol.* **29**, 193–213.

Mizrahi, Y. and Richmond, A. E. (1972). Abscisic acid in relation to mineral deprivation. *Plant Physiol.* **50**, 667–70.

Nissen, P. (1971). Uptake of sulfate by roots and leaf slices of barley: mediated by single, multiphasic mechanisms. *Physiol. Plant* **24**, 315–24.

Nissen, P. (1974). Uptake mechanisms: inorganic and organic. *Ann. Rev. Plant Physiol.* **25**, 53–79.

Nobel, P. S. (1974). Temperature dependence of the permeability of chloroplasts from chilling-sensitive and chilling-resistant plants. *Planta* **115**, 369–72.

Op Den Kamp, J. A. F. (1979). Lipid asymmetry in membranes. *Ann. Rev. Biochem.* **48**, 47–71.

Pardee, A. B., Prestidge, L. S., Whipple, M. B. and Dreyfus, M. B. (1968). A binding site for sulfate and its relation to sulfate transport into *Salmonella typhimurium*. *J. Biol. Chem.* **241**, 3962–9.

Pauls, K. P., Chambers, J. A., Dumbroff, E. B. and Thompson, J. E. (1982). Perturbation of phospholipid membranes by gibberellins. *New Phytol.* **91**, 1–17.

Pettersson, S. and Jensen, P. (1978). Allosteric and non-allosteric regulation of rubidium influx in barley roots. *Physiol. Plant.* **44**, 83–7.

Pitman, M. G. (1976). Ion intake by plant roots. In *Transport in Plants II, Part B*, eds. U. Lüttge and M. G. Pitman, Springer-Verlag, Berlin, pp. 129–88.

Pitman, M. G. (1977). Ion transport into the xylem. *Ann. Rev. Plant Physiol.* **28**, 71–88.

Raven, J. A. (1976). Transport in algal cells. In *Transport in Plants II, Part A*, eds U. Lüttge and M. G. Pitman, Springer-Verlag, Berlin, pp. 129–88.

Robards, A. W., Newman, T. M. and Clarkson, D. T. (1980). Demonstration of the distinctive nature of the plasma membrane of the endodermis in roots using freeze-fracture electron microscopy. In *Plant Membrane Transport: Current Conceptual Issues*, eds R. M. Spanswick, W. J. Lucas and J. Dainty, Elsevier/North-Holland, Amsterdam, pp. 395–6.

Rothman, J. E. and Lenard, J. (1977). Membrane asymmetry. *Science* **195**, 743–53.

Saddler, H. D. N. (1970). The ionic relations of *Acetabularia mediterranea*. *J. exp. Bot.* **21**, 345–59.

Sandermann, H. Jr. (1978). Regulation of membrane enzymes by lipids. *Biochim. Biophys. Acta* **515**, 209–37.

Schultz, R. D. and Assunmaa, S. K. (1970). Ordered water and the ultrastructure of the cellular plasma membrane. *Rec. Prog. Surface Sci.* **3**, 291–332.

Shinitzky, M. and Henkart, P. (1979). Fluidity of cell membranes—current concepts and trends. *Int. Rev. Cytol.* **60**, 121–47.

Siddiqi, M. Y. and Glass, A. D. M. (1982). Simultaneous consideration of tissue and substrate potassium concentration in K^+ uptake kinetics: a model. *Plant Physiol.* **69**, 283–5.

Simon, E. W. (1974). Phospholipids and plant membrane permeability. *New Phytol.* **73**, 377–420.

Singer, S. J. (1974). The molecular organization of membranes. *Ann. Rev. Biochem.* **43**, 805–33.

Singer, S. J. and Nicolson, G. L. (1972). The fluid mosaic model of the structure of cell membranes. *Science* **175**, 720–31.

Skou, J. C. (1965). Enzymatic basis for active transport of Na^+ and K^+ across cell membrane. *Physiol. Rev.* **45**, 596–617.

Slayman, C. L. and Slayman, C. W. (1974). Depolarization of the plasma membrane of *Neurospora* during active transport of glucose: evidence for a proton-dependent co-transport system. *Proc. Nat. Acad. Sci. USA* **71**, 1935–39.

Smith, F. A. and Raven, J. A. (1979). Intracellular pH and its regulation. *Ann. Rev. Plant Physiol.* **30**, 289–311.

Smith, I. K. (1980). Regulation of sulfate assimilation in tobacco cells. Effect of nitrogen and sulfur nutrition on sulfate permease and o-acetylserine sulfhydrase. *Plant Physiol.* **66**, 877–83.

Spanswick, R. M. (1981). Electrogenic ion pumps. *Ann. Rev. Plant Physiol.* **32**, 267–89.

Steveninck, R. F. M. van (1976). Effect of hormones and related substances on ion transport. In *Transport in Plants II, Part B*, eds U. Lüttge and M. G. Pitman, Springer-Verlag, Berlin, pp. 307–42.

Villalobo, A. (1982). Potassium transport coupled to ATP hydrolysis in reconstituted proteoliposomes of yeast plasma membrane ATPase. *J. Biol. Chem.* **257**, 1824–8.

Walker, R. R. and Leigh, R. A. (1981). Characterization of a salt-stimulated ATPase activity associated with vacuoles isolated from storage roots of red beet (*Beta vulgaris* L.). *Planta* **153**, 140–9.

Wheeler, K. P., Walker, J. A. and Barker, D. M. (1975). Lipid requirement of the membrane sodium-plus-potassium ion-dependent adenosine triphosphatase system. *Biochem. J.* **146**, 713–22.

Wood, A., Paleg, L. G. and Spotswood, T. M. (1974). Hormone-phospholipid interaction: a possible mechanism in the control of membrane permeability. *Aust. J. Plant Physiol.* **1**, 167–9.

Phytochrome

16

Pill-Soon Song

Introduction

The evolution and survival of virtually all living organisms entail interactions with radiant energy. Animal vision and plant photosynthesis are photobiological examples of such interactions, and these processes represent two of the better understood photophysiological processes of animals and plants, respectively. In addition to photosynthesis (*cf.* Chapter 11), plants exhibit a variety of responses to blue and/or red light, such as phototropism (*cf.* Chapter 7) and photomorphogenesis (described in the present and succeeding chapters). Over one hundred different responses of plants to light have been observed in the form of developmental and morphogenetic changes, as well as photoperiodical oscillation (circadian rhythm or, as it is more commonly called, the 'biological clock') of physiological activity modulated by red and far-red light. Thus, the discovery of phytochrome in the 1950s represents a milestone in the developing field of plant photophysiology, which is described in the next chapter.

Discovery

In 1952, the Beltsville research group of the USDA, headed by Borthwick and Hendricks, described important work which led to the discovery of phytochrome several years later. It was shown that red light elicits the germination of lettuce seeds, whereas far-red light reverses the process, *i.e.*, no germination occurs in the dark or in far-red light (Borthwick, Hendricks, Parker, Toole and Toole, 1952a). It was also shown that 1-min red-light irradiation followed immediately by 1-min far-red light inhibits germination, while reversing the order of light treatment, *i.e.*, far-red immediately followed by red elicits germination. In fact, as long as the last light treatment is red light, full germination may be observed.

A similar red and far-red reversibility was also demonstrated with floral initiation (Borthwick, Hendricks and Parker, 1952b). For example, 660-nm light inhibits flowering in short-day plants, whereas 740-nm light elicits flowering. Red irradiation immediately followed by far-red light negates the inhibition of flowering.

These observations indicate that the red-absorbing form of the photoreceptor is reversibly activated by light to the far-red-absorbing form; this physiologically active far-red-absorbing form triggers germination and flowering events in plants. The photoreceptor was subsequently named phytochrome after its discovery (Borthwick and Hendricks, 1960).

The first detection and isolation of the red/far-red reversible photoreceptor phytochrome were reported by the Beltsville group (Butler, Norris, Siegelman and Hendricks, 1959). Partial purification of phytochrome was also achieved at that time. Following the discovery of phytochrome, the next two decades witnessed rapid progress in the study of phytochrome-mediated responses in plants, as evidenced by the extensive phytochrome bibliography compiled by the Smithsonian group (Correll, Edwards and Shropshire, 1977).

Several personal accounts of the historical background behind the discovery of phytochrome provide interesting reading. Borthwick (1972) tells a fascinating story of the discovery of phytochrome in a phytochrome symposium volume (Mitrakos and Shropshire, 1972). Butler, the principal figure in the first detection and purification of phytochrome, reminisces in a personal story told at the Antwerpen symposium (Butler in De Greef, 1980). In another symposium volume (Smith, 1976), Briggs, whose work following the discovery of phytochrome in the late 1950s up to the present day has contributed significantly to the field, tells the history of the discovery and the impact of that discovery on plant photobiology (Briggs, 1976).

Phytochrome as a photoreceptor: general properties

In photomorphogenesis, the growth, development and differentiation processes of many plants are often regulated by light of red and far-red wavelengths. The photomorphogenetic responses of some plants are extremely sensitive to light, often responding to very low intensities (fluence rates) (for example, $500\,nW\,m^{-2}$; 3×10^8 photons $m^{-2}\,s^{-1}$), and sometimes responding to extremely low intensities ($100\,pW\,m^{-2}$; 6×10^4 photons $m^{-2}\,s^{-1}$). The scale of photosensitivity of plants is indicated by the fluence rate (intensity) of clear daylight, approximately $1\,kW\,m^{-2}$.

The photomorphogenesis of plants is manifested in truly diverse responses, ranging from red light-induced seed germination to the flowering of the morning glory. For the purpose of convenience, let us categorize these diverse photomorphogenetic responses into the following types: fast responses and slow responses.

Type I: Fast responses These are essentially energy-transducing responses. The Type I responses include those processes in which the quantum energy absorbed by the plant is transduced to another form of energy. Examples of this type include leaf movement in *Mimosa* and chloroplast movement in *Mougeotia* (*see* below). Other examples are membrane potential changes, surface potential changes and ion fluxes. These phenomena are relatively rapid, occurring on a time scale of seconds and minutes.

Type II: Slow responses The rates and activation of certain facets of growth and development, including enzyme induction and protein synthesis by plants, are switched on or modulated according to the spectral quality of the light environment (red *vs.* far-red). Examples of Type II responses include stem elongation and seed germination (*e.g.*, *see* the preceding discussion on lettuce seed germination). Type II responses occur on a time scale of hours and days. Other examples of Type II responses include hook opening, leaf expansion, flower initiation, and pigment biosynthesis (*e.g.*, chlorophyll and anthocyanin synthesis). Note that current knowledge does not exclude the possibility that a Type I response occurs completely independently of the mechanism that elicits a Type II response.

Just as rhodopsin acts as the photoreceptor for animal vision, the phytochrome molecule serves as the primary photoreceptor for the variety of morphogenetic responses summarized above. As will be described later, the phytochrome molecule can be thought of as a photochromic sensor that controls the photomorphogenetic machinery of plants (Shropshire, 1972, 1977).

Phytochrome exists in two forms, P_r and P_{fr}. The P_r form of phytochrome maximally absorbs red light of 660 nm and undergoes a phototransformation to the P_{fr} form. The latter maximally absorbs far-red light of 730 nm. The photochemical transformation of phytochrome, accompanied by a substantial spectral or color change, can be termed 'photochromism'. Thus, phytochrome is a photoreversible photochromic molecule:

$$P_r \underset{730\,nm}{\overset{660\,nm}{\rightleftharpoons}} P_{fr} \rightsquigarrow \text{physiological reactions}$$

where the photochemical and non-photochemical (thermal) processes are represented by straight and wavy arrows, respectively; this notation has been adopted throughout this chapter.

Since the original experiments of Borthwick *et al.* (1952a,b), the red/far-red modulation of phytochrome activity, occurring as in the above model of the phototransformation of phytochrome, has been demonstrated in a number of systems. A few examples of Type II responses suffice to illustrate this (examples of Type I responses can be found elsewhere: *e.g.*, Satter and Galston, 1976). Long before phytochrome was discovered, Flint and McAlister (1937) measured the

action spectrum of lettuce seed germination and
showed that red light ($\leqslant 600-680$ nm) promotes and
far-red light ($\leqslant 720-740$ nm) retards the germination of lettuce seed. Similar action spectra were
obtained for the seed germination of *Arabidopsis
thaliana* (Shropshire, Klein and Elstad, 1961). These
observations, along with those of Borthwick *et al.*
(1952a,b), clearly demonstrate that the photoreceptor phytochrome plays a primary role in seed germination through the phototransformation of P_r to
P_{fr}. Also, the rate of the biosynthesis of anthocyanin
pigments has been shown to be linearly proportional
to the concentration of the P_{fr} form in mustard
seedling (Drumm and Mohr, 1974). One of the
enzymes involved in the early phase of the biosynthesis of anthocyanins is phenylalanine ammonia lyase.
Synthesis of this enzyme is apparently induced by
red light (Oelze-Karow and Mohr, 1974) through
the so-called 'differential gene expression mechanism' (Mohr, 1966, 1982; Engelsma, 1967). On the
other hand, lipoxygenase synthesis is repressed by
P_{fr} (Oelze-Karow and Mohr, 1972; Schopfer, 1977;
see also Chapter 17 of this book).

The concept that P_{fr} is the physiologically active
form has recently been challenged: H. Smith (1981)
argues that plants monitor the changing spectral
photon distribution of natural radiation, which
changes the concentrations of P_{fr} relative to the total
phytochrome concentration, $(P_{tot}) = (P_r + P_{fr})$. In
other words, the P_{fr} form is not necessarily the active
form, but rather it is the ratio $(P_{fr})/(P_{tot})$ that
determines the photoresponse of plants. Thus, the
two measures of the red/far-red radiation absorbed
by plants, (P_{fr}) and $(P_{fr})/(P_{tot})$, can be compared
by spectrofluorometric and spectrophotometric
measurements of light intensity in a single beam,
absolute intensity mode or double beam, intensity
ratio mode, respectively. The prevailing evidence
favors the P_{fr} form being the active form in both
Type I and II responses, although alternative
mechanisms cannot be completely ruled out under
certain conditions (for example, in a natural radiation environment; Smith, H., 1978, 1981, and for
high irradiance reactions (HIR), *see* below and
Chapter 17).

Isolation, purification and assay

Subsequent to the original procedure for isolation
and purification of phytochrome (Butler *et al.*,

1959), three practical methods of isolation and
purification have been developed. These methods
include conventional chromatography, immuno-
affinity chromatography and hydrophobic affinity
chromatography; these methods have been critically
reviewed elsewhere (Pratt, 1982). Here, typical
procedures of each method are outlined briefly.

Conventional chromatographic method

There are several variants of the so-called 'conventional chromatographic procedure' originated by
Butler *et al.* (1959). The following procedure is
based on Rice, Briggs and Jackson-White (1973).

Etiolated rye seedling (or other plant sources)
↓
Crude extract (4°C)
↓
Brushite chromatography
↓
Ammonium sulfate fractionation
↓
DEAE-cellulose ion exchange chromatography
↓
Hydroxyapatite chromatography
↓
Gel filtration chromatography
| BioGel A1.5 m
↓ Sephadex G-200
Phytochrome, yield $\sim 5\%$

In this procedure, the hydroxyapatite chromatography may be omitted without a significant loss of
purity of the phytochrome. The purity criterion is
commonly assessed in terms of the specific absorbance ratio, $SAR = A_{660}/A_{280}$, where A_{660} and A_{280}
represent absorbance due to the P_r chromophore at
660 nm and due to the aromatic amino acid residues
of the phytochrome, respectively. Sodium dodecyl
sulfate–polyacrylamide gel electrophoresis has also
been used to establish the purity criteria. Rice *et al.*
(1973) obtained 1.2 mg of rye phytochrome with an

SAR value of 0.695 using the above procedure.

The SAR value of purified phytochrome depends on the plant source. In recent works, phytochromes of SAR greater than 0.8 have been obtained relatively routinely from oat seedling tissue and other plant materials.

As will be discussed in the following section, one of the major problems of phytochrome purification is the degradation of phytochrome by proteolysis. Virtually all isolation and purification procedures employed prior to 1973 resulted in proteolytically-degraded phytochrome, the predominant protein being a degraded phytochrome with a molecular weight of 60 000. Gardner, Pike, Rice and Briggs (1971) have demonstrated the degradation of 'large' phytochrome of molecular weight 120 000 to 'small' phytochrome of molecular weight 60 000 by proteases of endogenous origin.

The above procedure and its modified versions developed by several other workers minimize the proteolytic degradation of phytochrome, usually yielding 'large' phytochrome.

Immuno-affinity chromatographic method

This method, developed by Hunt and Pratt (1979), represents a major departure from conventional chromatographic method of purifying phytochrome. The following diagram outlines the immuno-affinity procedure.

Purification	*Immuno-column preparation*
Crude extract	Rabbit
↓	↓ phytochrome
Brushite chromatography	Antiphytochrome-IgG
↓	↓ agarose

Immobilized antiphytochrome-IgG-agarose column chromatography
↓
Elution of phytochrome
⎮ 3 M MgCl$_2$ and 1 M formic acid
Phytochrome
yield ~ 15%

The above procedure has been modified to achieve a greater yield of highly purified phytochrome (Cordonnier and Pratt, 1982).

Hydrophobic affinity chromatographic method

Based on the observation that phytochrome binds to agarose-immobilized Cibacron Blue dye, Smith and Daniels (1981) developed the Affi-Gel Blue chromatographic method, which yields highly purified phytochrome of SAR = 0.88–1.15 from rye seedlings. This procedure is outlined below.

Crude extract
↓
Brushite chromatography
↓
Ammonium sulfate fractionation
↓
Affi-Gel Blue (BioRad) column chromatography
↓
Elution with 10 mM flavin mononucleotide (FMN)
↓
Gel filtration chromatography
⎮ BioGel A1.5 m
↓
Phytochrome, yield ~ 10–25%

The immuno-affinity method has many advantages; its chief disadvantages seem to be the inconvenience of preparing the antiphytochrome column and its relative limitation for repeated usage. Of the three experimental methods outlined, the author's laboratory found the Affi-Gel chromatographic procedure to be the practical method of choice. A detailed, modified procedure is given elsewhere (Song, Kim and Hahn, 1981a).

In the modified Affi-Gel Blue chromatographic procedure, the eluting agent, FMN, must be carefully purified because flavin contaminants such as lumichrome are tightly retained in the purified phytochrome and may affect certain properties of 'large' phytochrome. However, this difficulty is readily rectified by using pure FMN and by avoiding photodegradation of the flavin by green safety light during phytochrome purification; flavins absorb green light, so the FMN solution should be shielded from light. There are also indications that the properties of native, proteolytically-undegraded phytochrome (molecular weight 124 000; *see* later) are not as affected by lumichrome as are the properties of 'large' phytochrome (molecular weight 118 000).

Figure 16.1 shows the absorption spectra of the P_r and P_{fr} forms of phytochrome purified by the Affi-Gel Blue chromatographic method; the absorption

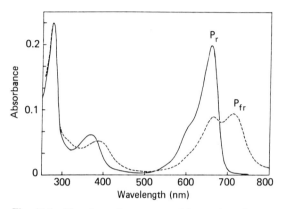

Fig. 16.1 The absorption spectra of large phytochrome isolated from oat seedlings; 0.1 M sodium phosphate buffer, pH 7.8, 50 mM KCl, and 0.1 mM EDTA. The phytochrome was purified in its P_r from by the Affi-Gel blue affinity chromatography method using lumichrome-free FMN as an eluant. The P_{fr} spectrum represents the composite absorbances due to a photostationary state mixture of P_r (19%) and P_{fr} (81%). (Redrawn from Song et al., 1981a.)

spectrum of P_{fr} represents the combined absorption spectra of a photostationary state mixture of ~81% P_{fr} and ~19% P_r. The P_r form shows absorption maxima at 378 and 664 nm, whereas the P_{fr} form has maxima at 392 and 719 nm.

Phytochrome is usually isolated and purified in its P_r form. However, the P_{fr} form behaves similarly on the chromatographic columns (including Brushite) employed for purification, and can, therefore, be isolated and purified from red-light-irradiated tissues or crude extracts (Briggs, Zolinger and Platz, 1968). However, the P_{fr} form is less recoverable from ammonium sulfate fractionation than is the P_r form. As will be discussed later, the so-called 'large' phytochrome of molecular weight 118 000 is not a native, undegraded phytochrome, due to a proteolytic cleavage of the 6000-dalton molecular mass peptide during the purification of phytochrome in the P_r form. This proteolysis can be minimized by isolating and purifying phytochrome in the P_{fr} form.

In addition to the problem of proteolysis, other difficulties such as (i) the small quantity of phytochrome present in plant tissues, (ii) thermal and photochemical instability, and (iii) the spectral perturbation or bleaching of the phytochrome absorption spectra by endogenous and exogenous substances, often pose obstacles to the successful purification of phytochrome. These difficulties are magnified in the case of phytochrome purification from green plants, which usually contain only a few percent of the phytochrome found in etiolated tissues, because of the chlorophylls (Pratt, 1982) and phytochrome killers (saponins) present in green plants (Yokota, Baba, Konomi, Shimazaki, Takahashi and Furuya, 1982).

Phytochrome assay

No specific biochemical assay of phytochrome in vitro is currently available in the literature. Instead, phytochrome is assayed spectrophotometrically both in vivo and in vitro. The spectrophotometric assay is based on the fact that the 660-nm absorbance of the P_r form of phytochrome decreases upon red-light-induced conversion to the P_{fr} form whose absorbance at 730 nm proportionately increases, whereas the 730-nm absorbance of the P_{fr} form decreases upon far-red-light-induced reversion to the P_r form (see Fig. 16.1). The amount of phytochrome being assayed is then proportional to such a 'double absorbance difference' $\Delta(\Delta A)$:

$$\Delta(\Delta A) = \Delta A_{fr}^{P_r} - \Delta A_r^{P_{fr}}$$
$$= (A_{660} - A_{730})_{fr}^{P_r} - (A_{660} - A_{730})_r^{P_{fr}}$$

where $\Delta A_{fr}^{P_r} = (A_{660} - A_{730})_{fr}^{P_r}$ represents the absorbance difference obtained after far-red irradiation of a P_r solution. $\Delta A_r^{P_{fr}} = (A_{660} - A_{730})_r^{P_{fr}}$ is obtained from the P_{fr} solution that results from red-irradiation of the first absorbance difference measurement ($\Delta A_{fr}^{P_r}$).

The absorbance difference $\Delta(\Delta A)$ is readily measured on a 'Ratiospect' photometer (Butler, Lane and Siegelman, 1963). There are several variations of the original design of the Ratiospect. An economical, simple light box consisting of a light source and red/far-red bandpass filters can be used with any spectrophotometer for spectrophotometric assay of phytochrome in crude extracts or in solution (Jung and Song, 1979).

The radioimmunoassay (RIA) of phytochrome has also been developed (Hunt and Pratt, 1979). The RIA procedure is outlined in Fig. 16.2. In most RIA, immunoprecipitates are collected and counted for radioactivity. For phytochrome RIA, the tritium-labeled P_r phytochrome in the supernatant can be counted, and from the difference between the supernatant count and the total added count, the phytochrome concentration can be determined. Thus, the less the tritiated immunoprecipitates, the less the difference between the free ^3H-P_r count and the total ^3H-P_r added count, and the more the amount of phytochrome in the sample being assayed.

The sensitivity of RIA for phytochrome is such that it can be applied in a range of 0–25 ng and detect phytochrome in a 1-μl sample of undiluted crude extract of etiolated oat seedling tissues. RIA can also be used to detect phytochrome in green plant tissues. Instead of tritium-labeled phytochrome, radio-iodinated phytochrome can be used with an overall increase in the RIA sensitivity. ^{125}I-Phytochrome can be prepared without loss of the spectral or photochemical integrity of the phytochrome, if the radioiodination is done carefully (Kim and Song, 1981).

Structure and properties of phytochrome

Chromophore

The action spectra of photomorphogenetic responses such as seed germination, stem elongation, hook opening, leaf expansion, anthocyanin and chlorophyll syntheses, and many more, show maxima in the red region (*ca.* 660 nm) (*see* Shropshire, 1977, for review). The 660-nm absorption responsible for the action-spectrum maximum in the red region is due to the electronic excitation of the tetrapyrrole chromophore bound to the P_r form of the chromoprotein phytochrome.

The chemical structure of the P_r form of phytochrome is now well established (Klein and Ruediger, 1978; Lagarias and Rapoport, 1980), as shown in Fig. 16.3. The tetrapyrrole chromophore is covalently linked to an apoprotein *via* a thioether bond to the vinyl group of ring A. Although it is possible that one of the propionic acid side chains is

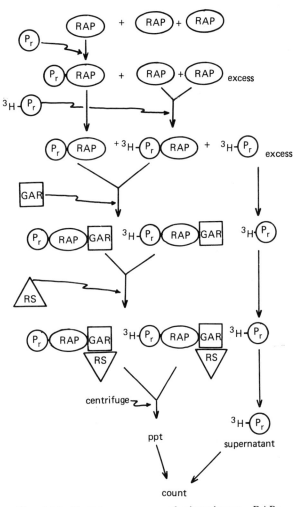

Fig. 16.2 Radioimmunoassay of phytochrome. RAP, rabbit antiphytochrome serum; ^3H-P_r, tritiated phytochrome (P_r form); GAR, goat antirabbit IgG; RS, non-immune rabbit serum. Either the immunoprecipitate (ppt) or supernatant is counted separately for radioactivity.

also covalently linked to the protein (Killilea, O'Carra and Murphy, 1980), no conclusive evidence for a second covalent linkage is available. The proteolytic peptide segments containing the chromophore possess no secondary covalent linkage other than the thioether bond to a cysteinyl residue (Lagarias and Rapoport, 1980). However, it cannot be ruled out that the second linkage, possibly an

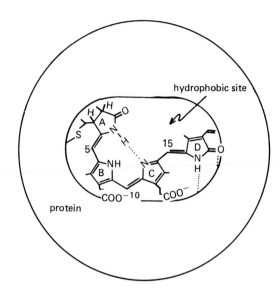

Fig. 16.3 The structure of phytochrome (P$_r$ form) covalently linked to apoprotein *via* a thioether bond (−S−). It is uncertain whether or not one of the propionic acid side-chains is also covalently linked. The chromophore-binding crevice is hydrophobic. There is an additional hydrophobic site on the protein moiety which binds flavin or Affi-Gel Blue dye (*cf.* text). The semi-extended conformation (rotation about the 14–15 single bond) is consistent with the absorption spectrum but it does not imply its actual conformation or configuration (redrawn from Song, 1983a).

ester or amide bond, was lost during the preparation of the chromopeptides.

The semi-extended (or semi-circular) conformation shown in Fig. 16.3 is consistent with the absorption spectrum of P$_r$ (Song, Chae and Gardner, 1979) (*see* following section). However, a proteolytically-produced chromopeptide is likely to assume a fully cyclic conformation, consistent with the strong near-UV absorbance (Thuemmler and Ruediger, 1983).

The cyclic P$_r$-chromopeptide (Fig. 16.4) can be photo-isomerized to a semi-extended chromopeptide in the presence of thiol, which acts as a nucleophilic catalyst. The semi-extended chromopeptide is spectroscopically similar to the P$_{fr}$-chromopeptide prepared from the P$_{fr}$ form of phytochrome. Does this mean that the P$_{fr}$ chromophore in native phytochrome is a geometric isomer of the

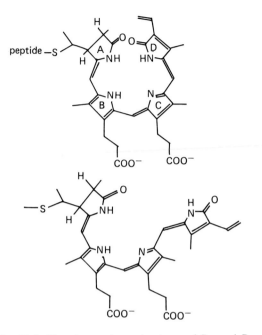

Fig. 16.4 The chromophore structures of P$_r$- and P$_{fr}$-chromopeptide. The P$_r$ and P$_{fr}$ chromophores are isomeric in configuration, *Z,Z,Z* and *Z,Z,E*, respectively.

P$_r$ chromophore with an identical chemical structure? This question cannot be answered definitively until the mechanism of the red shift of the visible absorption band (from about 660 to 730 nm) is known.

A number of proposals have been made regarding the chemical structure of the P$_{fr}$ chromophore. However, none of the proposed structures can be regarded as chemically proven. Whether or not the P$_r$ and P$_{fr}$ chromophores in native phytochrome represent geometric isomers (*cf.* Fig. 16.4) remains to be seen. In contrast, the absorption spectra of the P$_r$ and P$_{fr}$ proteins are consistent with the idea that the chromophore conformations/configurations of both the chromophores are very similar (Song *et al.*, 1979). One proposed structure of both forms of phytochrome is depicted in Fig. 16.5. The P$_{fr}$ chromophore results from the photo-induced addition of amino acid residue(s) to the ring A methene bridge (Klein, Grombein and Ruediger, 1977). The long wavelength absorption of P$_{fr}$ may then be attributed to the protein-stabilized, anionic chromophore.

However, the anionic, tripyrrolic structure for the P_{fr} chromophore does not satisfactorily account for the P_{fr} absorption spectrum because the observed spectral shape of the P_r and P_{fr} absorption is essentially conserved, suggesting that the π-electron network of both chromophores is also retained (Song

transitions (Q_x and Q_y) and the Soret band may include more than the two main contributions, B_x and B_y (Song, 1977, 1981; Song and Chae, 1979).

According to the selection rule, the visible band of a cyclic polyene is partially forbidden and weak, but the Soret band is allowed and much stronger

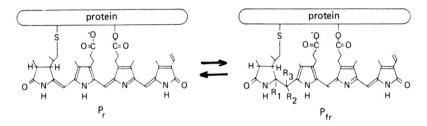

Fig. 16.5 A possible mechanism of the phototransformation of phytochrome based on the addition to the 4–5 double bond; residues $R_{1,2,3}$ are not specified.

and Chae, 1979) (*see* following section). It remains to be seen if the absorption spectrum of P_{fr} can be accommodated in terms of an anionic form of the isomeric type proposed for the P_{fr} chromopeptide (Fig. 16.4). Many other proposed P_{fr} structures have been reviewed elsewhere (Kendrick and Spruit, 1977; Pratt, 1978; Lagarias and Rapoport, 1980; Ruediger, 1980; Scheer, 1981).

The biosynthetic pathway of phytochrome has not been fully elucidated, although the early steps of biosynthesis are likely to follow the pathways leading to porphyrin biosynthesis. In etiolated pea seedlings, ^{14}C-δ-aminolevulinic acid is incorporated into phytochrome (Bonner, 1967). Most phytochrome is already present in dry seeds.

Spectroscopy

The absorption spectra of phytochrome (Fig. 16.1) are qualitatively similar to those of porphyrins and chlorophylls with their characteristic visible and Soret bands, except that the Soret band intensity is significantly weaker than the corresponding Soret band in porphyrins and chlorophylls. In cyclic polyene molecules such as porphyrins and chlorophylls, the two lowest electronic transitions of the $\pi \rightarrow \pi^*$ type are designated as $A \rightarrow Q_{x,y}$ and are responsible for the visible bands, whereas the higher-energy transitions are designated $A \rightarrow B_{x,y}$ correspond to the Soret band (Fig. 16.6). Note that the visible band is composed of two separate electronic

than the former. On the other hand, the reverse is true for an extended, linear molecule. The semi-extended conformation shown in Fig. 16.3 indicates that the visible and Soret bands possess comparable intensities, in terms of absolute intensity of oscillator strength, consistent with the observed absorption spectra (Fig. 16.1). Figure 16.6 shows the energy level diagrams for P_r and P_{fr}, in analogy to the chlorophyll system.

As mentioned earlier, the intensity ratios of the visible to near-UV (Soret) bands are strongly dependent on the shape of the π-electron conjugation network. Thus, the intensity ratio is significantly less and substantially greater than unity for cyclic and linear polyenes, respectively. In the case of both forms of phytochrome, the oscillator strength ratio, $f_{B_{x,y}}/f_{Q_{x,y}}$, is close to unity, which suggests unique chromophore configurations/conformations, neither fully cyclic nor fully extended linear molecular structures (Burke, Pratt and Moscowitz, 1972; Song et al., 1979; Song and Chae, 1979).

Fluorescence polarization of P_r yields an angle of about 50° between the polarization axes of the Q_y and $B_{x,y}$ bands, which agrees with the predicted angle of 53–63° between the Q_y and Soret band polarization axes for a semi-extended chromophore conformation (Song and Chae, 1979; Song, 1983a). In contrast, a cyclic conformation as in porphyrin (*cf.* Fig. 16.4) adequately accounts for the spectra of the synthetic P_r model and P_r-peptides in terms of the oscillator strength ratio and the polarization angle between the Q_y and $B_{x,y}$ transitions. The ratio $f_{B_{x,y}}/f_{Q_{x,y}}$ for a degraded phytochrome is 2.5, suggesting that the chromophore configuration/conformation is altered upon degradation (Burke *et al.*, 1972).

than $10^5 M^{-1} cm^{-1}$. Earlier, Briggs (private communication to Burke *et al.*, 1972) obtained a value of $1.1 \times 10^5 M^{-1} cm^{-1}$ for P_r, which agrees with the recently reported values in Table 16.1.

Until recently, the molar extinction coefficients of

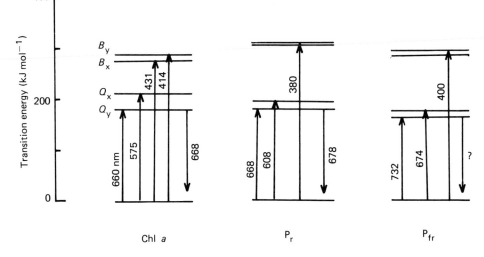

Fig. 16.6 The spectroscopic correlation between chlorophyll *a* (Chl *a*) and phytochrome. Note that the visible bands of phytochrome are correlated with the Q_x and Q_y bands of Chl *a*, whereas the near UV bands of phytochrome are derived from the Soret band of Chl *a* (B_x, B_y). The wavelength maxima for absorption and fluorescence are also indicated. (The wavelength maxima depend on buffer composition and plant source.)

The oscillator strength ratio for P_{fr} is also approximately unity (Burke *et al.*, 1972; Song *et al.*, 1979), indicating that the chromophore configuration/conformation has not been altered substantially in the $P_r \rightarrow P_{fr}$ phototransformation. However, precise structural changes accompanying the $P_r \rightarrow P_{fr}$ phototransformation (*see* above) are not fully known at present.

The molar extinction coefficients of the visible and Soret bands of phytochrome in the literature vary somewhat due mainly to uncertainties in the molecular weight of the protein and in the spectral resolution of the absorbance in the ultraviolet band, where both the aromatic amino acid residues and the chromophores absorb. It is generally assumed that the stoichiometry of the chromophore to the apoprotein unit is 1:1. Table 16.1 presents recent, representative data from the literature. From Table 16.1, it can be seen that a currently accepted value for the molar extinction coefficient of P_r is greater

phytochrome reported by Tobin and Briggs (1973) were widely used (Table 16.1). If the higher values are to be accepted, some of the phytochrome parameters, such as the photostationary equilibrium and the quantum yields of phototransformation [p. 369], must be recalculated.

Phytochrome (P_r) emits fluorescence from the lowest excited electronic state (Fig. 16.6). The fluorescence spectra and temperature-dependent quantum yield (0.04 at 14 K to about 10^{-4} at 298 K) of phytochrome (P_r form) have been reported (Song, Chae, Lightner, Briggs and Hopkins, 1973; Song, Chae and Briggs, 1975). The fluorescence lifetime ranges from ≤ 0.2 ns in glycerol–phosphate buffer (Song *et al.*, 1979) to 0.4 ns in Tris buffer (Hermann, Mueller, Schbert, Wabnitz and Wilhelmi, 1981). The fluorescence of P_r is substantially enhanced in D_2O over that in H_2O, provided that the P_r form has been photocycled through the P_{fr} form in D_2O (Sarkar and Song, 1981).

Table 16.1 The molar extinction coefficients (ε) for phytochrome.

Phytochrome	Apparent mol. wt	ε ($M^{-1} cm^{-1} \times 10^{-4}$ (λ in nm))	Reference
oat P_r	~ 60 000	7.6(664) 2.6(382) 8.2(280)	Anderson, Jenner and Mumford (1970)
oat P_{fr}	~ 60 000	4.6(724) 2.1(400) 8.2(280)	Anderson et al. (1970)
rye P_r	~ 120 000	7.0(665)	Tobin and Briggs (1973)
rye P_{fr}	~ 120 000	4.0(730)	Tobin and Briggs (1973)
oat P_r	~ 60 000	10.9(665)[a] 3.6(380)[a] 11.8(665)[b]	Brandlmeier, Scheer and Ruediger (1981)
oat P_r	~ 120 000	10.2(667)	Roux, McEntire and Brown (1982)
oat P_{fr}	~ 120 000	4.35(724)	Roux et al. (1982)

[a]Determined from denatured phytochrome.
[b]Determined from proteolysis.

Protein moiety

Molecular weight Siegelman and Firer (1964) reported a molecular weight of 90 000–150 000 for rye phytochrome. Subsequently, many different values for the molecular weight of phytochrome have appeared in the literature. Three factors seem to have contributed to this confusion: (1) different plant sources, (2) proteolytic degradation during isolation, and (3) the tendency of phytochrome to aggregate. Table 16.2 roughly summarizes the current status of the molecular weight data on oat and rye phytochromes, along with the proposed termi-

Table 16.2 Molecular weight data and terminology on oat and rye phytochromes.

Molecular Weight	Terminology	Reference
240 000	dimer	Briggs and Rice (1972)
124 000	intact	Vierstra and Quail (1982)
114 000–118 000	large	Gardner et al. (1971); Vierstra and Quail (1982)
		Rice and Briggs (1973)
		Roux, Lisansky and Stoker (1975)
		Smith and Correll (1975)
		Hunt and Pratt (1980)
60 000	small	Mumford and Jenner (1966)
		Gardner et al. (1971)
		Stoker, McEntire and Roux (1978)
		Kidd, Hunt, Boeshore and Pratt (1978)
42 000	42 K-fragment	Correll, Steers, Towe and Shropshire (1968)
20 000	20 K-fragment	
2000	chromopeptides	Fry and Mumford (1971)
		Lagarias and Rapoport (1980)
		Ruediger (1980)

nology designating different molecular weight chromoprotein species arising from proteolysis.

The term 'intact' phytochrome is preferred over the terms 'undegraded' and 'native' phytochromes, since the term 'undegraded' has been used erroneously in the literature to designate 'large' phytochrome and the term 'native' is reserved to refer to the denaturation of phytochromes of different proteolytic fragments including 'large' and 'small' phytochromes.

It appears that intact phytochrome has a molecu-

lar weight of 124 000 per monomer (Bolton and Quail, 1982). A 6000-dalton peptide fragment is readily lost from intact phytochrome, especially when the phytochrome is in the P_r form (Vierstra and Quail, 1982). Thus, the susceptibility of the P_r form to proteolysis accounts for the earlier observation that phytochrome isolated in the P_{fr} form exhibits a higher molecular weight than that isolated in the P_r form upon SDS polyacrylamide gel electrophoresis (Boeshore and Pratt, 1980).

Figure 16.7 shows the absorption spectra of intact

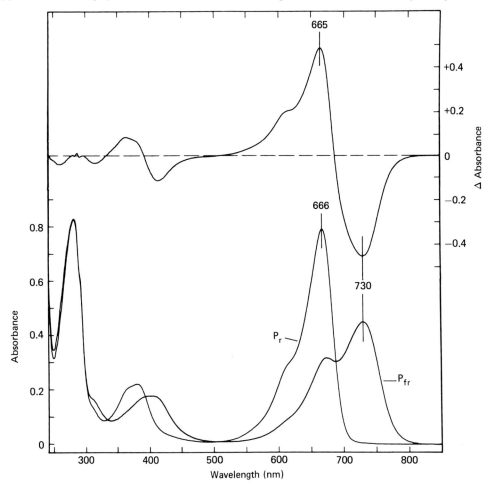

Fig. 16.7   Absorbance and difference spectra of 124 kdalton phytochrome ($A_{666}/A_{280} = 0.97$) purified by the Affi-Gel blue affinity procedure in 100 mM K-phosphate (pH 7.8, 4°C), 5 mM Na_4EDTA and 14 mM 2-mercaptoethanol. Absorbance spectra were measured at 3°C after saturating red (P_{fr}) and far-red (P_r) irradiation and the difference spectrum determined by subtracting the spectrum of P_{fr} from P_r. (Courtesy of R. D. Vierstra and P. H. Quail.)

phytochrome isolated from oat seedling. The absorption maxima are at 666 and 730 nm for the P_r and P_{fr} forms, respectively. The near-UV maxima are at 380 and 400 nm respectively.

The most significant spectral difference between intact and large phytochromes is the red shift of the visible absorption maximum in the P_{fr} form. This explains why the absorbance maximum at ≥ 730 nm for P_{fr} *in vivo* or crude extracts obtained after red irradiation of tissues is longer than that obtained for P_{fr} *in vitro* (Kendrick and Roth-Bejerano, 1978; Epel, 1981; Baron and Epel, 1982). This sort of absorption maximum difference *in vivo* and *in vitro* was already noted by Everett and Briggs (1970).

Amino acid composition and sequence Because of the variations in purity and proteolytic degradation of phytochrome, the amino acid compositions of phytochrome are not well established. Furthermore, the amino acid composition of intact phytochrome has not yet been reported. Table 16.3 lists typical amino acid compositions for large rye and oat phytochromes. Immunoaffinity-purified large oat

phytochrome of greater than 98% purity, which exists in solution as a dimer of its 118 000 molecular weight monomers, contains about 35% non-polar amino acid residues, with 115 carboxylic amino acids per monomer (Hunt and Pratt, 1980). The amino acid composition of large zucchini phytochrome is similar but not identical to that of oat phytochrome (Cordonnier and Pratt, 1982).

Although the amino acid sequence of phytochrome is not known, the primary structure of its chromopeptide segment is. For example, the following sequence has recently been elucidated (Lagarias and Rapoport, 1980):

Leu-Arg-Ala-Pro-His-Ser-Cys(-S-chromophore)-His-Leu-Gln-Tyr.

Protein structure The large phytochrome structure is composed of 20% α-helix, 30% β-pleated sheet, and 50% random coil conformations (Tobin and Briggs, 1973). The surface charges on large phytochrome are such that the isoelectric point (pI) ranges from 5.8 to 7.0 (Smith, W., 1981) depending

Table 16.3 Amino acid compositions of large phytochrome expressed as residues (rounded to the nearest integer) per $\sim 120\,000$-dalton subunit.

Amino acid	Rye phytochrome[a]	Oat phytochrome[b]	Oat phytochrome[c]
Lys	58	64	63
His	28	34	33
Arg	47	51	50
Asp	104	118	112
Thr	46	38	44
Ser	75	73	80
Glu	128	122	122
Pro	88	45	44
Gly	77	72	77
Ala	110	93	93
Cys	26	27	16
Val	89	79	81
Met	32	26	31
Ile	54	51	52
Leu	111	119	114
Tyr	23	23	21
Phe	43	45	43
Trp	—	8	9
Total residues	1139	1088	1085

[a]Data from Rice and Briggs (1973).
[b]Data from Hunt and Pratt (1980).
[c]Data from Roux *et al.* (1982).

on the source and proteolytic modification *in vitro*. The intact phytochrome monomer, however, is homogeneous by several criteria (Vierstra and Quail, 1982).

The partial specific volume of large (rye) phytochrome is $0.728 \, cm^3 \, g^{-1}$ (Rice and Briggs, 1973). Hydrodynamically, phytochrome behaves like an ellipsoidal molecule with an axial ratio of $10:1$ (Smith, W., 1975). It is likely that the long ellipsoid represents a dimeric form, rather than a monomeric molecule. The Stokes' radii of the P_r and P_{fr} forms of large (oat) phytochrome are 8.07 ± 0.1 and $8.27 \pm 0.4 \, nm$, respectively (Song, 1983a). These values, along with an analysis of the quasielastic scattering and rotational relaxation data (to be published) suggest that phytochrome at micromolar or lower concentrations exists as nearly spherical, globular proteins (Song, 1983a).

Immunochemically, phytochromes from different grasses are qualitatively identical, as mentioned earlier, although small and large phytochromes behave differently (Pratt, 1978). Phytochromes from dark- and light-grown pea shoots are also immunochemically and spectroscopically identical (Shimazarki, Moriyasu, Pratt and Furuya, 1981). Large and intact phytochromes are also immunochemically similar, although there are subtle differences in their spectroscopic behaviors (compare Figs 16.1 and 16.7; this is discussed in more detail later).

Molecular basis of the physiological activity— differences between P_r and P_{fr}

In order to describe adequately the differences between the P_r and P_{fr} forms of phytochrome, the awaited characterization of intact phytochrome is necessary. In the meantime, it is useful to discuss the available information on the differences between the two forms of large phytochrome, since the 6000-dalton molecular mass fragment is likely to be a local conformational modifier and/or supplement to the 118 000-dalton unit, rather than drastically altering the overall conformational and hydrodynamic properties of the latter. For example, the intact phytochrome protein exposes an additional hydrophobic surface upon $P_r \rightarrow P_{fr}$ phototransformation, similar to the case of large phytochrome, as monitored by the hydrophobic fluorescence probe, 8-anilinonaphthalene-1-sulfonate (ANS) (Song, 1983a). However, the chromophore of the intact P_{fr}

phytochrome is less accessible to tetranitromethane oxidation than that of the large P_{fr} phytochrome. This suggests a close interaction between the P_{fr} chromophore and the 6000-dalton peptide fragment and also is consistent with the fact that the wavelength maximum of absorbance for the intact P_{fr} phytochrome is red-shifted relative to that of large P_{fr} (*cf.* Fig. 16.7).

The hydrophobic surface on large phytochrome that becomes accessible to ANS upon a $P_r \rightarrow P_{fr}$ phototransformation is accompanied by additional hydrogen–tritium exchanges that seem to result from the 60 000-dalton chromophore-containing domain of the apoprotein (Hahn and Song, 1982). These observations suggest that the hydrophobicity of P_{fr} (Song *et al.*, 1979; Yamamoto, Smith and Furuya, 1980; Yamamoto and Smith, 1981a; Hahn and Song, 1981) is brought about by a chromophore phototransformation that involves a significant reorientation and exposure of the chromophore, along with a local conformational change of the chromopeptide and/or the chromophore binding crevice proper (*cf.* Fig. 16.3) that follows the chromophore movement (Sarkar and Song, 1982a; Song, 1983a,b). The fact that a considerable deuterium isotope effect on the fluorescence yield of P_r appears only after at least one cycling of phytochrome in D_2O, *i.e.*,

$$
\begin{array}{l}
\quad\quad\quad P_r \text{ (no H–D exchange)} \\
660 \, nm \downarrow \\
\quad\quad\quad P_{fr} \text{ (H–D exchange)} \\
730 \, nm \downarrow \\
\quad\quad\quad P_r \text{ (deuterated chromophore)}
\end{array}
$$

is consistent with the idea that the P_r chromophore is buried and becomes exposed in the P_{fr} form (Sarkar and Song, 1981). A similar effect of the photocycling of P_r in D_2O on the NMR spectrum has also been reported (Song, Sarkar, Tabba and Smith, 1982; Song, 1983b).

One model emerging from the above observation is that the $P_r \rightarrow P_{fr}$ phototransformation brings about a reorientation and exposure of the chromophore, with the result that part of the P_r chromophore crevice is more accessible to the medium and to chemical agents such as tetranitromethane, $KMnO_4$, and $NaBH_4$ in the P_{fr} form. To what extent does the chromophore movement entail conformational rearrangement of the apoprotein? This question cannot be answered quantitatively until a structure determination is made on as-yet unavailable phytochrome crystals. However, one could explain why a high viscosity medium slows down the $P_r \rightarrow P_{fr}$ phototransformation in terms of a possible hindrance of the chromophore reorientation by the viscous medium (Pratt and Butler, 1970).

Largely on the basis of the chemical modification of accessible amino acid residues (Hunt and Pratt, 1981), it has been suggested that a protein conformation change, even away from the chromophore-binding site, takes place in the $P_r \rightarrow P_{fr}$ phototransformation of phytochrome (Pratt, 1982). A conformational change in a remote area of the protein may occur as a result of a causative movement, either cooperative or non-cooperative, of the chromophore. Upon the $P_r \rightarrow P_{fr}$ phototransformation, an additional histidine residue becomes exposed as determined by the chemical accessibility method (Hunt and Pratt, 1981) and NMR spectroscopy (Song et al., 1982; Song, 1983a). The modification of the histidine residue does not abolish the photoreversibility of phytochrome. Significantly, a histidine residue is present in the chromopeptide segment of phytochrome (see page 365). It is thus possible that this histidine residue may become accessible to its modifying agent in aqueous medium as the chromophore rearranges around the binding crevice. The chemical modification of this histidine residue is not likely to affect drastically the photoreversibility of phytochrome, although it can affect the rate of photo- and dark-reversion of P_{fr} (see later). Whether the aforementioned exposure of additional histidine residue(s) results from (1) the chromopeptide, (2) the chromophore-binding site, or (3) catalytically important residues in the phototransformation mechanism must, therefore, be ascertained by a detailed kinetic analysis of the photo- and dark-reversion of P_{fr}.

Tetranitromethane specifically oxidizes the tetrapyrrole chromophore. Thus, it is not surprising that the P_{fr} form of large phytochrome, with its exposed chromophore, rapidly loses its far-red absorbance band. In fact, in the case of phytochrome, tetranitromethane is more suitable for the accessibility determination of the phytochrome than for tyrosyl residue tests. It was found that the P_{fr} chromophore of intact (oat) phytochrome is significantly shielded, compared with that of 'small' and 'large' phytochromes, suggesting that the 6000-dalton peptide chain 'protects' the P_{fr} chromophore (Song, 1983a).

Let us now summarize some of the available information on the molecular differences between the P_r and P_{fr} forms of phytochrome.

(i) The P_{fr} chromophore of large phytochrome is substantially more exposed than the P_r chromophore. Ionizable groups including histidine (Hunt and Pratt, 1981) and cysteine (Gardner, Thompson and Briggs, 1974) also become exposed. However, in intact phytochrome of molecular weight 124 000, the P_{fr} chromophore is shielded by a 6000-dalton peptide segment.

(ii) A hydrophobic surface is exposed upon $P_r \rightarrow P_{fr}$ phototransformation, as monitored by the effects of the hydrophobic fluorescence probe ANS on absorption, fluorescence and photo- and dark-reversion kinetics. The chromophore-related hydrophobic surface contains at least one tryptophan residue, and this tryptophan residue is apparently lacking in small phytochrome (Song et al., 1979; Sarkar and Song, 1982a). Thus, the hydrophobicity of the P_{fr} form of small phytochrome is minimized. In fact, the non-chromophore domain (about 60 000 molecular weight) of a tryptic digest of large phytochrome is more hydrophobic than the small phytochrome: the former is eluted from an Affi-Gel Blue column at lower ionic strength than is the latter (unpublished results).

It should also be noted that large phytochrome possesses an additional hydrophobic site that can be occupied by flavin and Affi-Gel Blue dyes. However, its location relative to the chromophore crevice is not established; it is probably also associated with the non-chromophore domain (Song, 1983a).

(iii) The extent of hydrogen-tritium exchange increases with the $P_r \rightarrow P_{fr}$ phototransformation in small and large phytochromes, suggesting that the P_{fr} form exposes additional peptide segments (Hahn and Song, 1982).

(iv) The secondary and tertiary structures of the apoprotein remain largely identical upon phototransformation of large phytochrome, as suggested by CD, NMR, and hydrodynamic data such as rotational relaxation times (Song, 1983b) and Stokes' radii.

(v) Energy transfer from the singlet-excited bound flavin to the large phytochrome chromophore occurs preferentially in the P_r form. Energy transfer from the triplet tryptophan residue at or near the chromophore binding crevice occurs in the P_r form, but not in the P_{fr} form (Sarkar and Song, 1982a,b).

These results are best explained on the basis of a change in the relative orientations of the energy donor (flavin or tryptophan) and the chromophore upon $P_r \rightarrow P_{fr}$ phototransformation.

(vi) The P_{fr} form of large phytochrome binds more tightly to hydrophobic sites of agarose-immobilized blue dextran (Smith, 1981), triterpenoid saponin ('P_{fr} killer') (Yokota, Baba, Konomi, Shimazaki, Takashi and Furuya, 1982), and cholesterol-enriched liposome than does the P_r form at high ionic strength (Kim and Song, 1981).

(vii) Only the P_{fr} chromophore is reversibly bleached by Cu^{2+}, Co^{2+}, and Zn^{2+} ions (Lisansky and Galston, 1974) and dehydration (Tobin, Briggs and Brown, 1973). In addition, urea, p-chloromercuribenzoate, glutaraldehyde, pronase, and trypsin all preferentially react with P_{fr} (*see* review by Briggs and Rice, 1972; Pratt, 1978). These results are consistent with the concept that the degree of P_{fr} chromophore exposure and/or segmental conformational changes with $P_r \rightarrow P_{fr}$ phototransformation.

We have already mentioned that the gross conformations of both the chromophore and apoprotein remain essentially identical for the P_r and P_{fr} forms of phytochrome. For example, several hydrodynamic parameters are the same for both forms (Briggs and Rice, 1972; Pratt, 1978).

Recently, Sundqvist and Björn (1983) affixed phytochrome to Sepharose beads via antiphytochrome immunoglobulin. From the dichroic photoconversion kinetics of the immobilized phytochrome, values of 32° or 148° were obtained for the change in direction of the Q_y-transition dipole upon phytochrome phototransformation, indicating that there is significant movement of the chromophore during the phototransformation. This result is qualitatively consistent with the chromophore reorientation/hydrophobic model of P_{fr}.

In summary, the picture that emerges from recent experiments is that the $P_r \rightarrow P_{fr}$ phototransformation brings about a certain degree of movement and exposure of the chromophore, resulting in the development of a hydrophobic surface on the P_{fr} protein. Figure 16.8 illustrates this molecular model for the physiologically active form of phytochrome. This model could explain the polarized light-induced chloroplast movement in *Mougeotia* (*see* later). What is the driving force behind the chro-

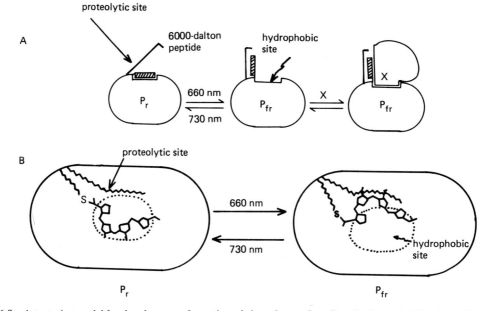

Fig. 16.8   A tentative model for the phototransformation of phytochrome from P_r to P_{fr} form. A: Side view with respect to the chromophore crevice; 'X' represents an as-yet unidentified membrane or receptor. B: Top view. Proteolytic digestion of P_r yields a 6000-dalton molecular mass peptide (arrow) (redrawn from Song, 1983a).

mophore movement? Is the alleged isomerization part of the chromophore movement? These questions remain unanswered.

Photochemistry of phytochrome

Photostationary state equilibrium

As can be seen from Figs 16.1 and 16.7, both the P_r and P_{fr} forms of phytochrome absorb in the red region, whereas far-red light at 730 nm is exclusively absorbed by the P_{fr} form. This means that monochromatic red irradiation of P_r cannot yield 100% P_{fr}, whereas $P_{fr} \rightarrow P_r$ photoreversion with monochromatic far-red light (730 nm) can be complete. Thus, one may expect the following photostationary state equilibrium after prolonged irradiation $(t = \infty)$.

$$P_r \underset{>0\%}{\xrightarrow[\substack{660\text{-nm irradiation} \\ (t=\infty)}]{}} P_{fr} {}_{<100\%} \tag{1}$$

$$P_{fr} \underset{\sim 0\%}{\xrightarrow[\substack{730\text{-nm irradiation} \\ (t=\infty)}]{}} P_r {}_{\sim 100\%} \tag{2}$$

For the forward reaction (1), the rate of P_{fr} formation at low concentrations with absorbance less than 0.1 is given by the expression

$$\left(\frac{1}{2.3}\right)\frac{d[P_{fr}]}{dt} = I_{660,abs} \cdot \Phi_{P_r} \cdot [P_r] - I_{660,abs} \cdot \Phi_{P_{fr}} \cdot [P_{fr}]$$
$$\simeq I^\circ_{660} \cdot E_{P_r,660} \cdot \Phi_{P_r} \cdot [P_r]$$
$$- I^\circ_{660} \cdot E_{P_{fr},660} \cdot \Phi_{P_{fr}} \cdot [P_{fr}] \tag{3}$$

where $I_{660,abs}$ and I°_{660} are absorbed and incident light intensities in $mol\,cm^{-1}\,s^{-1}$, respectively; $E_{P_r,660}$ and $E_{P_{fr},660}$ are the extinction coefficients in $cm^2\,mol^{-1}$ (to base 10; $E = 1000\varepsilon$, where ε is the more common form of the molar extinction coefficient in liters $mol^{-1}\,cm^{-1}$) of P_r and P_{fr} at 660 nm, respectively. At the photostationary state $(t = \infty)$, the rate of P_{fr} formation is approximately equal to zero. Thus,

$$\left(\frac{1}{2.3}\right)\frac{d[P_{fr}]}{dt} \simeq 0 \tag{4}$$

From Equations (3) and (4) and from $E = 1000\varepsilon$,

$$\varepsilon_{P_r,660} \cdot \Phi_{P_r} \cdot [P_r]_\infty = \varepsilon_{P_{fr},660} \cdot \Phi_{P_{fr}} \cdot [P_{fr}]_\infty \tag{5}$$

The photostationary state ratio of $[P_{fr}]_\infty$ to $[P_r]_\infty$ at 660 nm is then given by

$$\frac{[P_{fr}]_\infty}{[P_r]_\infty} = \frac{\varepsilon_{P_r,660}}{\varepsilon_{P_{fr},660}} \frac{\Phi_{P_r}}{\Phi_{P_{fr}}} \tag{6}$$

This ratio is 3.0 for large oat phytochrome (Pratt, 1978) and 4.26 for small phytochrome (Butler, Hendricks and Siegelman, 1964). Note that the photostationary ratio (Equation (6)) depends on the wavelength of irradiation which determines the ratio $\varepsilon_{P_r,\lambda}/\varepsilon_{P_{fr},\lambda}$ (Butler et al., 1964).

The photostationary state ratio is more commonly expressed as

$$\frac{[P_{fr}]_\infty}{[P_{tot}]} = \frac{[P_{fr}]_\infty}{([P_r]_\infty + [P_{fr}]_\infty)} \tag{7}$$

The plot of this ratio as a function of wavelength has been made by Hartmann (1966). This ratio (Equation (7)) at the red wavelength has been measured for large and small oat phytochromes, yielding 0.75 (Pratt, 1978) and 0.81 (Butler et al., 1964), respectively. Large rye phytochrome also yields the ratio of 0.81 (Gardner and Briggs, 1974), and this value has also been obtained for large oat phytochrome through a direct determination of $[P_{fr}]_\infty$ by selectively bleaching the absorption due to P_{fr} at 660 nm with tetranitromethane (unpublished). The photostationary state ratios of 0.84 and 0.80 have also been reported for rye and pea phytochrome by using an indirect method (Yamamoto and Smith, 1981b).

The quantum yield ratio $\Phi_{P_r}/\Phi_{P_{fr}}$, which is essentially independent of wavelength, has been determined to be 1.5 and 1.0 for small and large phytochromes from oat seedlings (Butler et al., 1964; Pratt, 1978). For large rye phytochrome the values are 1.4 ($\Phi_{P_r} = 0.28$, $\Phi_{P_{fr}} = 0.20$; Gardner and Briggs, 1974) and 1.53 (Yamamoto and Smith, 1981b). Pratt (1978) obtained a value for $\Phi_{P_r}/\Phi_{P_{fr}} = 0.17/0.17 = 1.0$, which differs from Gardner and Briggs' value of 1.4, by using the published data of Gardner and Briggs (1974) and by using $\varepsilon_{667} = 1.4 \times 10^5$ for a dimer of molecular weight 240 000 and $[P_{fr}]_\infty/[P_{tot}] = 0.75$ determined in the presence of EDTA. The lower value of $[P_{fr}]_\infty$ was largely attributed to the presence of EDTA. Since other workers have obtained photostationary state ratios greater than 0.75 and the accuracy of these ratios and quantum yields will eventually have to be re-evaluated, if the newly published molar extinction coefficients are correct (see above), the reported values of the quantum yield ratio for oat and rye phytochromes must also be re-examined. To illustrate, the previous values of $\Phi_{P_r} = 0.28$ and $\Phi_{P_{fr}} = 0.20$ (Gardner and Briggs, 1974) become 0.19 and 0.18 if the new molar extinction coefficients (Table 16.1) are used, assuming that the photostationary state ratio, $[P_{fr}]_\infty/[P_{tot}]$, is 0.81. Furthermore, all published data on the photostationary-state properties of phytochrome need to be re-examined in view of the evidence that the intact phytochrome exhibits a higher value of the photostationary state ratio than small or large phytochrome (compare Figs 16.1 and 16.7).

The photostationary state ratio, $[P_{fr}]_\infty/[P_{tot}]$, *in vivo* is much more uncertain than that *in vitro* because of the heterogeneity of phytochrome (*e.g.*, free *vs.* bound P_{fr}), the differential optical bias which changes as a function of wavelength (*e.g.*, screening pigments and scattering) and the effects of P_{fr}-bleaching substances (*e.g.*, phytochrome 'killers').

Photochemical transformation

The phototransformation of P_r to P_{fr} (Equation (1)) proceeds *via* several intermediates. Two different approaches have been employed to elucidate the kinetics of the phototransformation, namely, low-temperature spectroscopy and the flash photolysis technique.

From detailed low-temperature studies of small phytochrome (Cross, Linschitz, Kasche and Tenenbaum, 1968; Pratt and Butler, 1970), the following intermediate species have been resolved

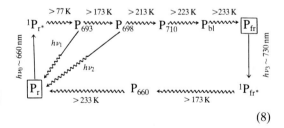

$$(8)$$

where $^1P_{r^*}$ and $^1P_{fr^*}$ are the excited singlet states (Q_y states) of P_r and P_{fr}, respectively. The number subscripts represent the absorption maxima of the intermediates, and P_{bl} stands for the bleached intermediate. Photochemical and thermal reactions are represented by straight and wavy arrows, respec-

tively. The combined, straight/wavy arrows represent both photochemical and thermal reaction paths. Note that the photoreversion of P_{fr} to P_r takes a different path than the $P_r \rightarrow P_{fr}$ photoconversion, and the two intermediates, P_{693} and P_{698} are photoreversible to P_r. Two additional intermediates were detected in a temperature-dependence study of the phototransformation of small phytochrome in 75% glycerol (Pratt and Butler, 1970).

In contrast to the relatively simple kinetic scheme obtainable from low-temperature studies, flash photolysis of small P_r (oat) phytochrome at ambient temperature resolves at least four intermediates involved in the phototransformation to P_{fr} *via* complex kinetic schemes, with either parallel or sequential pathways (Linschitz, Kasche, Butler and Siegelman, 1966).

In a recent laser flash-photolysis study of large phytochrome (oat at 275 K) the following kinetic scheme has been formulated as one possible pathway of the phototransformation (Cordonnier, Mathis and Pratt, 1981).

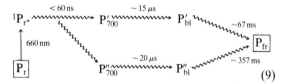

$$(9)$$

Although some of the intermediates detected at low temperatures are probably not resolved in a flash photolysis study, the parallel scheme proposed here for large phytochrome is essentially in agreement with earlier flash photolysis work on small phytochrome (Linschitz *et al.*, 1966) and large phytochrome (Shimazaki, Inoue, Yamamoto and Furuya, 1980). However, one major difference between the above scheme (9) for large phytochrome and the earlier data on small phytochrome is that large phytochrome exhibits a complex kinetic behavior with respect to the intermediate P_{700}, whereas small phytochrome shows a kinetic complexity with respect to the decay of P_{bl}. Whether this difference reflects the effect of apoprotein in small *vs.* large phytochrome on the kinetic mechanism or instrumental resolution and bias remains to be established. However, the parallel decays of P'_{700} and P''_{bl} are apparently not due to a proteolytic heterogeneity (Cordonnier *et al.*, 1981). Pea phytochrome behaves similarly, so at least two different plant materials, oat and pea, show the same kinetic mechanism for phytochrome phototransformation.

Briggs and Fork (1969a) described a general method for the detection of the intermediates involved in phytochrome phototransformation. Their basic idea is that, if phytochrome is irradiated with both red and far-red light, a cycling of the photochemical conversion/reversion should occur, and the steady state intermediates accumulated during cycling may then be detected by scanning with a monochromatic monitoring beam. A significant change in optical density monitored at 543 nm was readily observed under the cycling condition described. Kinetic analysis of the decay of this spectral change led to the conclusion that a parallel decay process of two distinct species occurs, on the basis that, if the actinic light is kept constant and the exposure times varied, the relative amounts of each component will vary, and each decay component will follow first-order kinetics. The intermediates monitored at 543 nm are probably due to the intermediate, P_{bl}, accumulated during cycling.

The phototransformation of phytochrome in vivo has been studied by low-temperature spectroscopy (Kendrick and Spruit, 1977, and references therein). The phototransformation pathway in vivo is similar to that in an in vitro solution:

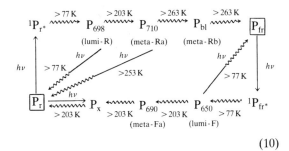

$$(10)$$

A comparison of the above scheme with the flash-photolysis kinetics shown in scheme (9) shows that the intermediates P_{700} and P_{bl} correspond to lumi-R (P_{698}) and meta-Rb, respectively.

The similarity of the kinetics of the $P_r \to P_{fr}$ phototransformation in vitro and in vivo has already been noted in a work on the parallel independent decay of the long-lived intermediates in oat coleoptile tissue (Briggs and Fork, 1969b). Whether or not a minimum of two phytochrome species exists and undergoes parallel kinetic processes must be evaluated

cautiously, since instrumental artefacts are possible when using a Ratiospect-type spectrometer (Everett, Briggs and Purves, 1970). However, the long-lived intermediates are accumulated more in vivo than in vitro. In solution, higher viscosity also enhances the accumulation of the intermediates. This may suggest that the rate-limiting formation of P_{fr} from P_{bl} involves a viscosity-dependent reorientation of the chromophore/peptide segments, as discussed earlier. This, however, does not rule out other explanations.

The chemical nature of P_{bl} is not known at present. However, it is likely that the P_{bl} intermediate has a more folded conformation or a more cyclic geometric isomer than the P_r form (Burke et al., 1972).

The early photoproduct $I_{698-700}$ or lumi-R may or may not be the primary photo-intermediate arising from the excited state of P_r. An intermediate has been observed at 685 nm with a lifetime of $70 \pm 15 \mu s$ after a 15-ns laser pulse excitation of small phytochrome (Braslavsky, Matthews, Herbert, de Kok, Spruit and Schaffner, 1980). Similarly, large pea phytochrome forms P_{692} upon a microsecond laser flash excitation (Shimazaki et al., 1980).

The difference in absorption spectrum observed with the above intermediates has been interpreted in terms of an isomerization, but it is more likely that the spectral change observed is due to a red shift of the visible and near-UV bands, relative to the absorption bands of P_r.

One possible mechanism for the primary photoprocess initiated from the excited state of P_r is an intramolecular proton transfer, as has been suggested on the basis of studies done at low temperature (Song, Sarkar, Kim and Poff, 1981b) and in heavy water (Sarkar and Song, 1981). Whatever the mechanism is, it should be noted that the primary photoreaction is fast and efficient, occurring in a subnanosecond time scale, in competition with fluorescence.

At present, the reaction mechanism of phototransformation largely remains to be elucidated. However, it can now be stated that the kinetics of the phototransformation process of phytochrome are complex and that they most likely depend on molecular species (e.g., small vs. large, large vs. intact, oat vs. pea, etc.) and to some extent on the environment (in vitro vs. in vivo). The fact that several intermediates, including the first intermediate produced at 77 K (Song *et al.*, 1981b), are reversibly photoreactive (Kendrick and Spruit, 1977) also contributes to the complexity of the phototransformation.

Dark reversion and destruction

Dark reversion and inverse dark reversion

The P_{fr} form of phytochrome in solution undergoes dark reversion to the P_r form. The rate of this thermal reaction is strongly dependent on several factors. The dark reversion is pH-dependent and specific-acid-catalyzed by deuteronium ions (Sarkar and Song, 1981). A variety of reducing agents greatly accelerates the dark reversion (Mumford and Jenner, 1971). The activation energy for the reaction is $102 \, \text{kJ mol}^{-1}$ (Anderson, Jenner and Mumford, 1969).

Small and large phytochromes (P_{fr}) undergo a rapid, biphasic dark reversion, whereas intact phytochrome shows virtually no reversion (J. C. Lagarias, private communication; Song, 1982a). The mechanism of dark reversion will not be known until the chemical structure of P_{fr} has been established. Interestingly, dark reversion is commonly observable in dicotyledons *in vivo*, but it is not readily observable in monocotyledons, wherein reversion is considered extremely slow if present at all (Briggs and Rice, 1972). It is possible that dark reversion occurs only in proteolytically-degraded phytochrome (even *in vivo*), since intact phytochrome shows very little dark reversion. The fact that *in vivo* reversion, when it occurs, is usually complete within an hour but is not always 100% complete may be explained in terms of two populations of phytochrome, namely, intact and proteolytically degraded phytochrome *in vivo*.

Dark reversion can be inhibited by hydrophobic substances such as ANS *in vitro* (Hahn and Song, 1981) and phytochrome killers *in vivo* (Shimazaki and Furuya, 1975). These inhibitors probably compete with the P_{fr} chromophore for a hydrophobic site on the protein.

Whether or not inverse dark reversion ($P_r \rightsquigarrow P_{fr}$) occurs at all *in vivo* is an interesting, open question. During seed imbibition, some inverse reversion has been observed, but it is more likely that the appearance of P_{fr} during imbibition is caused by hydration of the seed (Tobin, Briggs and Brown, 1973), the conversion of intermediates to P_{fr} (Kendrick and Spruit, 1974), and the shift of the equilibrium between the free and bleached P_{fr} forms (*cf.* Hahn and Song, 1981).

Kinetically, three mechanisms are possible: (i) heterogeneity of the phytochrome population, (ii) consecutive kinetics, $P_{fr} \rightsquigarrow P_{fr}' \rightsquigarrow P_r$, and (iii) an equilibrium between P_{fr} and P_{fr}', namely,

$$P_{fr} \underset{k_{-1}}{\overset{k_1}{\rightleftharpoons}} P_{fr}' \overset{k_2}{\rightsquigarrow} P_r \tag{11}$$

The last mechanism is consistent with the dark-reversion kinetics of large (oat) phytochrome. A recent analysis of the kinetics (Jung and Song, 1982) yielded $k_1 = 1.2 \times 10^{-4} \, \text{s}^{-1}$, $k_{-1} = 6.4 \times 10^{-4} \, \text{s}^{-1}$, and $k_2 = 1.1 \times 10^{-3} \, \text{s}^{-1}$.

Destruction

In contrast to dark reversion, a metabolic turnover of P_{fr} without regenerating P_r occurs *in vivo*. This irreversible disappearance of the P_{fr} form is referred to as destruction and forms the basis of a model for phytochrome action (Oelze-Karow, Schaefer and Mohr, 1976). Neglecting the dark reversion of P_{fr}, we write

$$P' \overset{k_s^\circ}{\rightsquigarrow} P_r \underset{k_2}{\overset{k_1}{\rightleftharpoons}} P_{fr} \overset{k_d}{\rightsquigarrow} P_{fr}' \tag{12}$$

where the destruction of P_{fr} (k_d) is first order and the synthesis of P_r (k_s°) is zero order under the steady state condition with respect to $[P_{fr}]$ under continuous irradiation. Thus, in the steady state, we have

$$\frac{d[P_{tot}]}{dt} = k_s^\circ - k_d[P_{fr}] \simeq 0 \tag{13}$$

The chemical mechanism of destruction is not well characterized, although it is likely to involve degradation of the protein moiety (Quail, Schaefer and Marmé, 1973; Pratt, Kidd and Coleman, 1974). However, if a spectrophotometric method is used to assay P_{fr}, it is not possible to discriminate between the metabolic turnover or destruction and the spectral bleaching of P_{fr} by hydrophobic substances including P_{fr} killers. Although it is generally assumed that destruction is specific for the P_{fr} form, the destruction of P_r cannot be neglected, especially if the dark reversion in dicotyledons reflects a proteolytic degradation of intact phytochrome (P_r) to large and small phytochromes *in vivo* and that only the intact phytochrome of molecular weight 124 000 leads to the physiologically active P_{fr} form.

Intracellular localization

In barley leaves, 75% of the total amount of phytochrome is compartmentalized in protoplasts, with the remaining 25% divided equally between the vascular and epidermal tissues (Duke, Netzler and McClure, 1981). Numerous attempts to identify an intracellular localization of phytochrome have been made without definitive evidence for or against specific localization (*see* review by Quail, 1982). In etiolated oat seedling cells, the P_r form of phytochrome is apparently diffusely distributed, whereas the P_{fr} form is present in a discretely-localized distribution pattern, according to immunocytochemical observations (Mackenzie, Briggs and Pratt, 1978). No specific interaction between phytochrome and any cellular organelle has been discerned in the coleoptile tips of *Avena sativa* L., *Zea mays* L., or *Triticum sativum* L. (Epel, Butler, Pratt and Tokuyasu, 1980).

In contrast to higher plants, there is a distinct possibility that many Type I responses in simpler organisms are mediated by membrane-bound phytochrome. For example, phytochrome in *Mougeotia* is membrane-bound. This contention is based on the fact that chloroplast movement in *Mougeotia* is polarotropically controlled. Thus, red light polarized perpendicular to the long axis of the cell is preferentially absorbed at the front and back sides of the cell, converting dichroically oriented phytochrome (P_r) to its P_{fr} form. As P_{fr} builds up at the front and back of the cell, chloroplast edges orient away from the higher P_{fr} gradient (profile position). On the other hand, red light polarized parallel to the long axis of the cell is not effective for chloroplast movement (Haupt, 1973). Similar polarotropic chloroplast movement is also operative in *Meso-*

taenium. When the *Mougeotia* cell (diameter about $20\,\mu m$) is irradiated with microbeams of $3-24\,\mu m$ diameters, different chloroplast patterns are observed in response to the resulting symmetric or asymmetric P_{fr} gradient, suggesting that phytochrome is not freely diffusible, translationally or rotationally.

Polarotropic observations and microbeam irradiations of the *Mougeotia* cell with red and far-red wavelength light have further shown that the dichroism of phytochrome most likely results from the transition moments of P_r and P_{fr} that are parallel and perpendicular to the cell surface, respectively. It is likely that phytochrome is bound to plasma membrane (plasmalemma), as is implicated from the above observations.

As mentioned earlier, *Mesotaenium* also exhibits red-light-induced chloroplast movement (Dorscheid and Wartenberg, 1966). It is also likely that phytochrome is anisotropically oriented in this organism. Other cells showing polarotropic responses mediated by phytochrome include single cell protonemata of *Adiantum capillus-veneris* L. (Kadota, Wada and Furuya, 1982) and *Dryopteris filix-mas* (Etzold, 1965). In these cells too, P_r and P_{fr} molecules are oriented parallel and perpendicular to the cell surface, respectively.

In view of the fact that phytochrome readily binds to membranous cellular fraction (such as mitochondria, endoplasmic reticulum, etioplast envelope, chloroplasts, plasmalemma, and nucleic acids) *in vitro* (*see* review on the 'pelletability' of P_{fr} by Pratt, 1978, and Quail, 1982) and to artificial membranes (liposomes), it is understandable that phytochrome molecules are dichroic in *Mougeotia* cells, although no direct evidence for the membrane binding to phytochrome *in vivo* is available at the present time.

Several Type I responses implicate some kind of interaction between phytochrome and the membrane. Examples of Type I responses that might involve membrane include the red-light-induced generation of bioelectric potentials (Newman and Briggs, 1972), the Tanada effect in mung bean root (Tanada, 1968), gibberellin efflux from plastids (Cooke, Saunders and Kendrick, 1975) and ion fluxes (*see* above; *see* review by Marmé, 1977).

In spite of the implied involvement of interactions between phytochrome and membrane, no direct elucidation of a physiologically specific, membrane-bound phytochrome has been achieved. Phytochrome, particularly P_r, is a soluble protein. However, phytochrome, especially P_{fr}, can bind and fuse to membrane, thus possibly modulating membrane activity. Whether or not this type of interaction between phytochrome and membrane is relevant to Type II responses remains to be seen. As mentioned earlier, the dichroism of phytochrome in the chloroplast movement (a Type I response) of *Mougeotia* strongly suggests that both the P_r and P_{fr} forms of phytochrome are fixed to the membrane so that the lateral and rotational diffusions of phytochrome in the membrane are restricted. Such an immobilized phytochrome can be envisioned if the phytochrome molecules are bound either to the membrane as a high-molecular-weight oligomer or to the phytochrome receptor protein in the membrane.

Mode of action

Earlier, we classified phytochrome-mediated photomorphogenetic changes kinetically into Type I and Type II responses. To aid the organization of this section, it is convenient to reclassify the responses in terms of their fluence–response relationship, namely, low-fluence-rate reactions (LFR or LIR, where I stands for intensity) and high-fluence-rate reactions (HFR or HIR) (Mohr, 1959).

LIR responses are wavelength-dependent; they exhibit characteristic induction with red light and reversion with far-red light, and they are fluence-dependent until a photostationary state equilibrium is reached. The Bunsen–Roscoe law of reciprocity, *i.e.*, a light-induced response is a function only of total fluence (fluence rate × irradiation time) and independent of the irradiation time or fluence rate, holds for LIR responses.

Etiolated seedlings (*e.g.*, mustard) do not fully transform into normal, green plants upon exposure to only short, low-fluence-rate red light. The seedlings develop fully only after a prolonged illumination, as in the natural radiation environment, *i.e.*, it is an HIR response. The action spectra of short and prolonged light effects on the growth and development of mustard seedlings are also different (Mohr, 1959).

HIR responses, after a photostationary state equilibrium of the phytochrome system is established, are still fluence-rate-dependent, but the reciprocity law does not hold. The action spectra of

HIR responses (*e.g.*, Hartmann, 1966) show maxima at about 710, 450, and 360 nm and differ in other ways from the action spectra of LIR responses.

A minimal model for the action of phytochrome is given by scheme (12). At steady state equilibrium under continuous light, the rate of synthesis and destruction are assumed to be equal. Thus, from Equation (13), we have

$$[P_{fr}]_\infty = k_s^\circ/k_d \qquad (14)$$

This means that $[P_{fr}]_\infty$ *in vivo* under continuous light is both wavelength- and fluence-rate-independent. This prediction has been confirmed in several cases, but it is probably not generally applicable to all LIR responses. For example, if the assumptions made about the zero order synthesis (k_s°) and the non-destructibility of P_r are relaxed, $[P_{fr}]_\infty$ will become both wavelength- and fluence-rate-dependent.

To account for the HIR responses (Schaefer, 1975), the following scheme has been postulated, where X is a phytochrome receptor/or membrane.

The processes designated by broken arrows do not operate under steady-state conditions. The above cyclic scheme accounts for both LIR and HIR responses. According to this scheme, $[P_{fr} \cdot X]$, accumulates in HIR responses, with approximate action maxima at 715, 410, and 310 nm (Schaefer, 1975). Although this model qualitatively explains the HIR action spectra with maxima at far-red and near-UV regions, it does not quantitatively accommodate the blue maximum, since phytochrome absorbs very little in the blue region (*cf.* Fig. 16.1). It is possible that a separate, blue photoreceptor participates under HIR conditions. However, no direct evidence for or against the participation of a blue photoreceptor has been described, although there are indirect lines of evidence that suggest a separate blue photoreceptor acting cooperatively with phytochrome. In this regard, it is interesting that flavin bound to phytochrome preferentially transfers its excitation energy to the P_r chromophore, thereby enhancing the $P_r \to P_{fr}$ phototransformation (Sarkar and Song, 1982c).

The molecular nature of phytochrome receptor 'X' is unknown. At present, one can only speculate as to what it is. In some Type I responses, 'X' may be the membrane itself or a membrane-bound receptor protein. In Type II responses, 'X' may be the membrane or a membrane-bound receptor, or a soluble receptor component of either macromolecular nature or a small secondary messenger molecule (*e.g.*, cAMP) (*see* Kendrick, 1983, for review).

One of the major difficulties of directly demonstrating the nature of 'X' *in vivo* is that its population may be very small and quickly saturated by phytochrome of high affinity, leaving behind excess phytochrome, which masks the detection of the phytochrome in $P_{fr} \cdot X$ complexes. Thus, very often, 1% or less of the total phytochrome converted to P_{fr} is enough to elicit responses in plants. At this time, whether or not 'X' is a membrane or a membrane-bound component is an open question.

To explain the biphasic fluence–germination response-curves of lettuce seed, a 'dichromophoric' model has been proposed (Van DerWoude, 1983). In this model, $P_{fr} \cdot P_{fr} \cdot X$ is assumed to be always active, whereas the activity of $P_r \cdot P_{fr} \cdot X$ depends on membrane properties near the receptor 'X' site. At the molecular level, this model is based on the assumption that the dimeric form of phytochrome is physiologically active. Since biphasic fluence–response phenomena are not unique to the phytochrome system, it is difficult to accept the dichromophoric model without validating the dimeric nature of physiologically active phytochrome.

As mentioned earlier, evidence favors the P_{fr} form as the active form (Schaefer, 1982), although the possibility that the photostationary state ratio $[P_{fr}]/[P_{tot}]$ acts as an apparent signal perceived by the plant under certain conditions, cannot be ruled out (Smith, H., 1981). For example, the HIR action spectrum for the inhibition of hypocotyl growth in lettuce can be explained in terms of the ratio $[P_{fr}]/[P_{tot}] = 0.03$. *i.e.*, a 3% $[P_{fr}]$ level is necessary for the HIR response (Hartmann, 1966). Does this mean that all the P_{fr} 'X' receptors are saturated at 3% $[P_{fr}]$, or does it mean that the lettuce plant responds to the specific photostationary state ratio? Regardless of whether plants perceive $[P_{fr}]$ or $[P_{fr}]/[P_{tot}]$, it is clear that P_{fr} plays a crucial role. Recall-

ing the previously mentioned analogy between a double-beam photometer and a single-beam fluorometer for light-intensity measurements, one would expect plants to employ a detection system that would measure $[P_{fr}]$, rather than the ratio, since the sensitivity of plants to red light is extremely high, as mentioned earlier.

The author's work described as part of this chapter has been supported by the Robert A. Welch Foundation (D-182) and the National Science Foundation (PCM81-19907).

Notes added in proof

In etiolated plant seedlings (*e.g.*, mustard), the $[P_{fr}]$ is definitely the quantity measured by the plant, rather than the ratio $[P_{fr}]/[P_{tot}]$ (Schmidt, R. and Mohr, H. (1982) *Plant Cell Environ.* **5**, 495–9; *cf.* p. 356).

Amino acid composition of intact phytochrome from oat has been reported (Vierstra, R. and Quail, P. H. (1983) *Biochemistry* **22**, 2498–505; *cf.* Table 16.3). The molar extinction coefficient of intact phytochrome (P_r) has also been determined ($1.21 \times 10^5 \, M^{-1} \, cm^{-1}$ at 668 nm; Cha, T.-A., Maki, A. H. and Lagarias, J. C. (1983) *Biochemistry* **22**, 2846–51; *cf.* Table 16.1).

Fluorescence lifetimes of large P_r phytochrome (oat) in tris-H_2O and tris-D_2O buffer, pH 7.56 and pD 7.6, respectively, were remeasured, yielding essentially exponential (greater than 90%) lifetimes of 0.14 and 0.54 nm at 292 K, respectively (Song, P.-S. (1983) *Laser Chem.*, in press; *cf.* p. 362).

References

Anderson, G. R., Jenner, E. L. and Mumford, F. E. (1969). Temperature and pH studies on phytochrome *in vitro. Biochemistry* **8**, 1182–7.

Anderson, G. R., Jenner, E. L. and Mumford, F. E. (1970). Optical rotatory dispersion and circular dichroism spectra of phytochrome. *Biochim. Biophys. Acta* **211**, 69–73.

Baron, O. and Epel, B. L. (1982). Studies on the capacity of Pr *in vitro* to photoconvert to the long-wavelength Pfr-form. A survey of ten plant species. *Photochem. Photobiol.* **36**, 79–82.

Boeshore, M. L. and Pratt, L. H. (1980). Phytochrome modification and light-enhanced-*in vivo*-induced phytochrome pelletability. *Plant Physiol.* **66**, 500–4.

Bolton, G. W. and Quail, P. H. (1982). Cell-free synthesis of phytochrome apoprotein. *Planta* **155**, 212–17.

Bonner, B. A. (1967). Incorporation of δ-aminolevulinic acid into the chromophore of phytochrome. *Plant Physiol.* **42**, Suppl. 11(A).

Borthwick, H. A. (1972). The biological significance of phytochrome. In *Phytochrome*, ed. K. Mitrakos and W. Shropshire, Jr., Academic Press, London and New York, pp. 27–44.

Borthwick, H. A. and Hendricks, S. B. (1960). Photoperiodism in plants. *Science* **132**, 1223–8.

Borthwick, H. A., Hendricks, S. B., Parker, M. W., Toole, E. H. and Toole, V. K. (1952a). A reversible photoreaction controlling seed germination. *Proc. Nat. Acad. Sci. USA* **38**, 662–6.

Borthwick, H. A., Hendricks, S. B. and Parker, M. W. (1952b). The reaction controlling floral initiation. *Proc. Nat. Acad. Sci. USA* **38**, 929–34.

Brandlmeier, T., Scheer, H. and Ruediger, W. (1981). Chromophore content and molar absorptivity of phytochrome in the Pr form. *Z. Naturforsch.* **36c**, 431–9.

Braslavsky, S. E., Matthews, J. I., Herbert, H. J., de Kok, J., Spruit, C. J. P. and Schaffner, K. (1980). Characterization of a microsecond intermediate in the laser flash photolysis of small phytochrome from oat. *Photochem. Photobiol.* **31**, 417–20.

Briggs, W. R. (1976). H. A. Borthwick and S. B. Hendricks—Pioneers of photomorphogenesis. In *Light and Plant Development*, ed. H. Smith, Butterworth, London and Boston, pp. 1–4.

Briggs, W. R. and Fork, D. C. (1969a). Long lived intermediates in phytochrome transformation. I. *In vitro* studies. *Plant Physiol.* **44**, 1081–8.

Briggs, W. R. and Fork, D. C. (1969b). Long lived intermediates in phytochrome transformation. II. *In vivo* studies. *Plant Physiol.* **44**, 1089–94.

Briggs, W. R. and Rice, H. V. (1972). Phytochrome: chemical and physical properties and mechanism of action. *Ann. Rev. Plant Physiol.* **23**, 293–334.

Briggs, W. R., Zolinger, W. D. and Platz, B. B. (1968). Some properties of phytochrome isolated from dark-grown oat seedlings (*Avena sativa* L.). *Plant Physiol.* **43**, 1239–43.

Burke, M. J., Pratt, D. C. and Moscowitz, A. (1972). Low-temperature absorption and circular dichroism studies of phytochrome. *Biochemistry* **11**, 4025–31.

Butler, W. L. (1980). Remembrances of phytochrome twenty years ago. In *Photoreceptors and Plant Development*, ed. J. De Greef, Antwerpen University Press, Antwerpen, Belgium, pp. 3–7.

Butler, W. L., Hendricks, S. B. and Siegelman, H. W. (1964). Action spectra of phytochrome *in vitro*. *Photochem. Photobiol.* **3**, 521–8.

Butler, W. L., Lane, H. C. and Siegelman, H. W. (1963). Nonphotochemical transformation of phytochrome *in vivo*. *Plant Physiol.* **38**, 514–9.

Butler, W. L., Norris, K. H., Siegelman, H. W. and Hendricks, S. B. (1959). Detection, assay, and preliminary purification of the pigment controlling photoresponsive development of plants. *Proc. Nat. Acad. Sci. USA* **45**, 1703–8.

Cooke, R. J., Saunders, P. F. and Kendrick, R. E. (1975). Red-light induced production of gibberelin-like substances in homogenates of etiolated wheat leaves and in suspensions of intact etioplasts. *Planta* **124**, 319–28.

Cordonnier, M.-M., Mathis, P. and Pratt, L. H. (1981). Phototransformation kinetics of undegraded oat and pea phytochrome initiated by laser flash excitation of the red-absorbing form. *Photochem. Photobiol.* **34**, 733–40.

Cordonnier, M.-M. and Pratt, L. H. (1982). Immunopurification and initial characterization of dicotyledonous phytochrome. *Plant Physiol.* **69**, 360–5.

Correll, D. L., Edwards, J. L. and Shropshire, Jr., W. (1977). *Phytochrome: a Bibliography with Author, Biological Materials, Taxonomic, and Subject Indexes of Publications Prior to 1975.* Smithsonian Institute Press, Rockville, MD, 411 pp.

Correll, D. L., Steers, Jr., E., Towe, K. M. and Shropshire, Jr., W. (1968). Phytochrome in etiolated annual rye. IV. Physical and chemical characterization of phytochrome. *Biochim. Biophys. Acta* **168**, 46–57.

Cross, D. R., Linschitz, H., Kasche, V. and Tenenbaum, J. (1968). Low-temperature studies on phytochrome: light and dark reactions in the red to far-red transformation and new intermediate forms of phytochrome. *Proc. Nat. Acad. Sci. USA* **61**, 1095-101.

Dorschied, T. and Wartenberg, A. (1966). Chlorophyll als Photoreceptor bei der Schwachlichtbewegung des *Mesotaenium* Chloroplasten. *Planta* **70**, 187–92.

Drumm, H. and Mohr, H. (1974). The dose response curve in phytochrome-mediated anthocyanin synthesis in mustard seedling. *Photochem. Photobiol.* **20**, 151–7.

Duke, S. O., Netzler, D. H. and McClure, J. W. (1981). Tissue localization of phytochrome in dark-grown barley leaves. *Phytochemistry* **20**, 2327–8.

Engelsma, G. (1967). Photoinduction of phenylalanine deaminase in gherkin seedlings. II. Effect of red and far-red light. *Planta* **77**, 49–57.

Epel, B. L. (1981). A partial characterization of the long wavelength 'activated' far-red absorbing form of phytochrome. *Planta* **151**, 1–5.

Epel, B. L., Butler, W. L., Pratt, L. H. and Tokuyasu, K. T. (1980). Immunofluorescence localization studies of the Pr and Pfr forms of phytochrome in the coleoptile tips of oats, corn and wheat. In *Photoreceptors and Plant Development*, ed. J. De Greef, Antwerpen University Press, Antwerpen, Belgium pp. 121–34.

Etzold, H. (1965). Der Polarotropismus und Phototropismus der Chloronemen von *Dryopteris filix-mas* (L.) Schott. *Planta* **64**, 254–80.

Everett, M. S. and Briggs, W. R. (1970). Some spectral properties of pea phytochrome *in vivo* and *in vitro*. *Plant Physiol.* **45**, 679–83.

Everett, M. S., Briggs, W. R. and Purves, W. K. (1970). Kinetics of phytochrome phototransformation: a re-examination. *Plant Physiol.* **45**, 805–6.

Flint, L. H. and McAlister, E. D. (1937). Wavelengths of radiation in the visible spectrum promoting the germination of light-sensitive lettuce seed. *Smithsonian Institute Miscellaneous Colloquia* **96**, 1–8.

Fry, K. T. and Mumford, F. E. (1971). Isolation and partial characterization of a chromophore-peptide fragment from pepsin digests of phytochrome. *Biochem. Biophys. Res. Commun.* **45**, 1466–73.

Gardner, G. and Briggs, W. R. (1974). Some properties of phototransformation of rye phytochrome *in vitro*. *Photochem. Photobiol.* **19**, 367–77.

Gardner, G., Pike, C. S., Rice, H. V. and Briggs, W. R. (1971). 'Disaggregation' of phytochrome *in vitro*—a consequence of proteolysis. *Plant Physiol.* **48**, 686–93.

Gardner, G., Thompson, W. F. and Briggs, W. R. (1974). Different reactivity of the red- and far-red absorbing forms of phytochrome to [^{14}C]N-ethyl maleimide. *Planta* **117**, 367–72.

Hahn, T.-R. and Song, P.-S. (1981). Hydrophobic properties of phytochrome as probed by 8-anilinonaphthalene-1-sulfonate fluorescence. *Biochemistry* **20**, 2602–9.

Hahn, T.-R. and Song, P.-S. (1982). Molecular topography of phytochrome as deduced from the tritium-exchange method. *Biochemistry* **21**, 1394–9.

Hartmann, K. M. (1966). A general hypothesis to interpret 'high energy phenomena' of photomorphogenesis on the basis of phytochrome. *Photochem. Photobiol.* **5**, 349–66.

Haupt, W. (1973). Role of light in chloroplast movement. *BioScience* **23**, 289–96.

Hermann, G., Mueller, E., Schbert, D., Wabnitz, H. and Wilhelmi, B. (1981). Fluorescence lifetime of the information processing plant pigment phytochrome. *Forschungsergebnisse* **7**, 1–7.

Hunt, R. E. and Pratt, L. H. (1979). Phytochrome immunoaffinity purification. *Plant Physiol.* **64**, 332–6.

Hunt, R. E. and Pratt, L. H. (1980). Partial characterization of undegraded oat phytochrome. *Biochemistry* **19**, 390–4.

Hunt, R. E. and Pratt, L. H. (1981). Physicochemical differences between the red and far-red absorbing forms of phytochrome. *Biochemistry* **20**, 941–5.

Jung, J. and Song, P.-S. (1979). A simple modification of the spectrophotometer for rapid phytochrome assay. *Photochem. Photobiol.* **29**, 419–21.

Jung, J. and Song, P.-S. (1982). Kinetic mechanism of dark-reversion of phytochrome. In *Proc. Int. Workshop on Photobiol.*, Cheju Island, May 26–29, 1982, p. 42.

Kadota, A., Wada, M. and Furuya, M. (1982). Phytochrome-mediated phototropism and different dichroic orientation of Pr and Pfr in protonemata of the fern *Adiantum capillus-veneris* L. *Photochem. Photobiol.* **35**, 533–6.

Kendrick, R. E. (1983). The physiology of phytochrome action. In *The Biology of Photoreceptors*, eds D. Cosens and D. V. Prue, Cambridge University Press, Cambridge (in press).

Kendrick, R. E. and Roth-Bejerano, N. (1978). Spectral characteristics of phytochrome *in-vivo* and *in-vitro*. *Planta* **142**, 225–8.

Kendrick, R. E. and Spruit, C. J. P. (1974). Inverse dark reversion of phytochrome: an explanation. *Planta* **120**, 265–72.

Kendrick, R. E. and Spruit, C. J. P. (1977). Phototransformations of phytochrome. *Photochem. Photobiol.* **26**, 201–14.

Kidd, G. H., Hunt, R. E., Boeshore, M. L. and Pratt, L. H. (1978). Asymmetry in the primary structure of undegraded phytochrome. *Nature* **276**, 733–5.

Killilea, S. D., O'Carra, P. and Murphy, R. F. (1980). Structures and apoprotein linkages of phycoerythrobilin and phycocyanobilin. *Biochem. J.* **187**, 311–20.

Kim, I.-S. and Song, P.-S. (1981). Binding of phytochrome to liposomes and protoplasts. *Biochemistry* **20**, 5482–9.

Klein, G., Grombein, S. and Ruediger, W. (1977). On the linkages between chromophore and protein in biliproteins, VI. Structure and protein linkage of the phytochrome chromophore. *Z. Physiol. Chem.* **358**, 1077–9.

Klein, G. and Ruediger, W. (1978). Uber die Bindungen zwischen Chromophor und Protein in Biliproteiden. V. Stereochemie von Modell–Imiden. *Liebigs Ann. Chem.* **2**, 267–79.

Lagarias, J. C. and Rapoport, H. (1980). Chromopeptides from phytochrome. The structure and linkage of the Pr form of the phytochrome chromophore. *J. Am. Chem. Soc.* **102**, 4821–8.

Linschitz, H., Kasche, V., Butler, W. L. and Siegelman, H. W. (1966). The kinetics of phytochrome conversion. *J. Biol. Chem.* **241**, 3395–403.

Lisansky, S. G. and Galston, A. W. (1974). Phytochrome stability *in vitro*. *Plant Physiol.* **53**, 352–9.

Mackenzie, Jr., J. M., Briggs, W. R. and Pratt, L. H. (1978). Intracellular phytochrome distribution as a function of its molecular form and of its destruction. *Am. J. Bot.* **65**, 671–8.

Marmé, D. (1977). Phytochrome: membranes as possible sites of primary action. *Ann. Rev. Plant Physiol.* **28**, 173–98.

Mitrakos, K. and Shropshire, Jr., W. (eds) (1972). *Phytochrome*, Academic Press, London and New York, 631 pp.

Mohr, H. (1959). Der Lighteinfluss auf des Wachstum der Keimblaetter bei *Sinapis alba* L. *Planta* **53**, 219–45.

Mohr, H. (1966). Differential gene activation as a mode of action of phytochrome. *Photochem. Photobiol.* **5**, 469–83.

Mohr, H. (1982). Phytochrome and gene expression. In *Trends in Photobiology*, eds C. Hélène, M. Charlier, Th. Montenay-Garestier and G. Laustriat, Plenum Press, New York, pp. 515–30.

Mumford, F. E. and Jenner, E. L. (1966). Purification and characterization of phytochrome from oat seedlings. *Biochemistry* **5**, 3657–62.

Mumford, F. E. and Jenner, E. L. (1971). Catalysis of the phytochrome dark reaction by reducing agents. *Biochemistry* **10**, 98–101.

Newman, I. A. and Briggs, W. R. (1972). Phytochrome-mediated electric potential changes in oat seedlings. *Plant Physiol.* **50**, 687–93.

Oelze-Karow, H. and Mohr, H. (1972). Repression of lipoxygenase synthesis in plant tissue through a threshold mechanism (cotyledons of mustard seedlings). *Proc. VI Int. Congr. Photobiol.* Abstr. No. 162(A).

Oelze-Karow, H. and Mohr, H. (1974). Interorgan correlation in a phytochrome-mediated response in the mustard seedling. *Photochem. Photobiol.* **20**, 127–31.

Oelze-Karow, H., Schaefer, E. and Mohr, H. (1976). On the physiological significance of dark reversion of phytochrome in the mustard seedling. *Photochem. Photobiol.* **23**, 55–60.

Pratt, L. H. (1978). Molecular properties of phytochrome. *Photochem. Photobiol.* **27**, 81–105.

Pratt, L. H. (1982). Phytochrome: the protein moiety. *Ann. Rev. Plant Physiol.* **33**, 557–82.

Pratt, L. H. and Butler, W. L. (1970). The temperature dependence of phytochrome transformation. *Photochem. Photobiol.* **11**, 361–9.

Pratt, L. H., Kidd, G. H. and Coleman, R. A. (1974). An immunochemical characterization of the phytochrome destruction reaction. *Biochim. Biophys. Acta* **365**, 93–107.

Quail, P. H. (1982). Intracellular localization of phytochrome. In *Trends in Photobiology*, eds C. Hélène, M. Charlier, Th. Montenay-Garestier and G. Laustriat, Plenum Press, New York, pp. 485–500.

Quail, P. H., Schaefer, E. and Marmé, D. (1973). Turnover of phytochrome in pumpkin cotyledons. *Plant Physiol.* **52**, 128–31.

Rice, H. V. and Briggs, W. R. (1973). Partial characterization of oat and rye phytochrome. *Plant Physiol.* **51**, 927–38.

Rice, H. V., Briggs, W. R. and Jackson-White, C. J. (1973). Purification of oat and rye phytochrome. *Plant Physiol.* **51**, 917–26.

Roux, S. J., Lisansky, S. G. and Stoker, B. M. (1975). Purification and partial carbohydrate analysis of phytochrome from *Avena sativa*. *Physiol. Plant.* **35**, 85–90.

Roux, S. J., McEntire, K. and Brown, W. E. (1982). Determination of extinction coefficients of oat phytochrome by quantitative amino acid analyses. *Photochem. Photobiol.* **35**, 537–43.

Ruediger, W. (1980). Phytochrome, a light receptor of plant photomorphogenesis. *Structure and Bonding* **40**, 101–41.

Sarkar, H. K. and Song, P.-S. (1981). Phototransformation and dark reversion of phytochrome in deuterium oxide. *Biochemistry* **20**, 4315–20.

Sarkar, H. K. and Song, P.-S. (1982a). Nature of phototransformation of phytochrome as probed by intrinsic tryptophan residues. *Biochemistry* **21**, 1967-72.

Sarkar, H. K. and Song, P.-S. (1982b). Blue light induced phototransformation of phytochrome in the presence of flavin. *Photochem. Photobiol.* **35**, 243–6.

Satter, R. L. and Galston, A. W. (1976). The physiological function of phytochrome. In *Chemistry and Biochemistry of Plant Pigments,* ed. T. W. Goodwin, Academic Press, New York, pp. 681–735.

Schaefer, E. (1975). A new approach to explain the 'high-irradiance responses' of photomorphogenesis on the basis of phytochrome. *J. Math. Biol.* **2**, 41–56.

Schaefer, E. (1982). Advances in photomorphogenesis. *Photochem. Photobiol.* **35**, 905–10.

Scheer, H. (1981). Biliproteins. *Angew. Chem. Int. Ed. (Engl)* **20**, 241–61.

Schopfer, P. (1977). Phytochrome control of enzymes. *Ann. Rev. Plant Physiol.* **28**, 223–52.

Shimazaki, Y. and Furuya, M. (1975). Isolation of a naturally occurring inhibitor for dark Pfr reversion from etiolated *Pisum* epicotyls. *Plant Cell Physiol.* **16**, 623–30.

Shimazaki, Y., Inoue, Y., Yamamoto, K. T. and Furuya, M. (1980). Phototransformation of the red-light-absorbing form of undegraded pea phytochrome by laser flash excitation. *Plant Cell Physiol.* **21**, 1619–25.

Shimazaki, Y., Moriyasu, Y., Pratt, L. H. and Furuya, M. (1981). Isolation of the red-light-absorbing form of phytochrome from light-grown pea shoots. *Plant Cell Physiol.* **22**, 1165–73.

Shropshire, Jr., W. (1972). Phytochrome, a photochromic sensor. In *Photophysiology*, vol. 8, ed. A. C. Giese, Academic Press, New York, pp. 33–72.

Shropshire, Jr., W. (1977). Photomorphogenesis. In *The Science of Photobiology,* ed. K. C. Smith, Plenum Press, New York, pp. 281–312.

Shropshire, Jr., W., Klein, W. H. and Elstad, V. B. (1961). Action spectra of photomorphogenic induction and photoinactivation of germination in *Arabidopsis thaliana*. *Plant Cell Physiol.* **2**, 63–9.

Siegelman, H. W. and Firer, E. M. (1964). Purification of phytochrome from oat seedlings. *Biochemistry* **3**, 418–23.

Smith, H. (ed.) (1976). *Light and Plant Development*. Butterworth, London, 516 pp.

Smith, H. (1978). Colour perception in plants. *Trends Biochem. Sci.* 204–6.

Smith, H. (1981). Evidence that P_{fr} is not the active form of phytochrome in light-grown maize. *Nature* **293**, 163–5.

Smith, Jr., W. O. (1975). *Purification and Physicochemical Studies of Phytochrome.* Ph.D. dissertation, University of Kentucky, Lexington, KY.

Smith, Jr., W. O. (1981). Characterization of the photoreceptor protein, phytochrome. *Photochem. Photobiol.* **33**, 961–4.

Smith, Jr., W. O. and Correll, D. L. (1975). Phytochrome: a re-examination of the quaternary structure. *Plant Physiol.* **56**, 340–3.

Smith, Jr., W. O. and Daniels, S. M. (1981). Purification of phytochrome by affinity chromatography on agarose-immobilized Cibacron blue 3GA. *Plant Physiol.* **68**, 443–6.

Song, P.-S. (1977). Physical methods and techniques: Part (iii) Ultraviolet and visible spectroscopy of bio-organic molecules. *Ann. Rep. (B), Chem. Soc., London,* pp. 18–40.

Song, P.-S. (1981). Electronic spectroscopy of photobiological receptors, *Can. J. Spectrosc.* **26**, 59–72.

Song, P.-S. (1983a). Protozoan and related photoreceptors: molecular aspects. *Ann. Rev. Biophys. Bioeng.* **12**, 35–68.

Song, P.-S. (1983b). The molecular basis of phytochrome (Pfr) and its interaction with model phytochrome receptors. In *The Biology of Photoreceptors,* eds D. Cosens and D. V. Prue, Cambridge University Press, Cambridge (in press).

Song, P.-S. and Chae, Q. (1979). The transformation of phytochrome to its physiologically active form. *Photochem. Photobiol.* **30**, 117–23.

Song, P.-S., Chae, Q. and Briggs, W. R. (1975). Temperature dependence of the fluorescence quantum yield of phytochrome. *Photochem. Photobiol.* **22**, 75–6.

Song, P.-S., Chae, Q. and Gardner, J. G. (1979). Spectroscopic properties and chromophore conformations of the photomorphogenic receptor: phytochrome. *Biochim. Biophys. Acta* **576**, 479–95.

Song, P.-S., Chae, Q., Lightner, D. A., Briggs, W. R. and Hopkins, D. (1973). Fluorescence characteristics of phytochrome and biliverdins. *J. Am. Chem. Soc.* **95**, 7892–4.

Song, P.-S., Kim, I.-S. and Hahn, T.-R. (1981a). Purification of phytochrome by Affi-gel blue chromatography; an effect of lumichrome on purified phytochrome. *Anal. Biochem.* **117**, 32–9.

Song, P.-S., Sarkar, H. K., Kim, I.-S. and Poff, K. L. (1981b). Primary photoprocesses of undegraded phytochrome excited with red and blue light at 77 K. *Biochim. Biophys. Acta* **635**, 369–82.

Song, P.-S., Sarkar, H. K., Tabba, H. and Smith, K. M. (1982). The phototransformation of phytochrome probed by 360 MHz proton NMR spectra. *Biochem. Biophys. Res. Commun.* **105**, 279–87.

Stoker, B. M., McEntire, K. and Roux, S. J. (1978). Identification of tryptic chromopeptides of phytochrome on sodium dodecyl sulfate gels. Implications for structure. *Photochem. Photobiol.* **27**, 597–602.

Sundqvist, C. and Björn, L. O. (1982). Light-induced linear dichroism in photoreversibly photochromic sensor pigments—II. Chromophore rotation in immobilized phytochrome. *Photochem. Photobiol.* **37**, 69–75.

Tanada, T. (1968). Substances essential for a red, far-red reversible attachment of mung bean root tips to glass. *Plant Physiol.* **43**, 2070–1.

Thuemmler, F. and Ruediger, W. (1983). Models for photoreversibility of phytochrome: Z,E isomerization of chromopeptides from phycocyanin and phytochrome. *Tetrahedron: Symp.* **39**, 1943–51.

Tobin, E. M. and Briggs, W. R. (1973). Studies on the protein conformation of phytochrome. *Photochem. Photobiol.* **18**, 487–95.

Tobin, E. M., Briggs, W. R. and Brown, P. K. (1973). The role of hydration in the phototransformation of phytochrome. *Photochem. Photobiol.* **18**, 497–503.

Van Der Woude, W. J. (1983). A dichromophoric mechanism of phytochrome action. Evidence from fluence-response studies of lettuce seed germination. *Proc. Nat. Acad. Sci. USA* (in press).

Vierstra, R. D. and Quail, P. H. (1982). Native phytochrome. Inhibition of proteolysis yields a homogeneous monomer of 124 kilodaltons from *Avena. Proc. Nat. Acad. Sci. USA* **79**, 5272–6.

Yamamoto, K. T. and Smith, Jr., W. O. (1981a). Alkyl and ω-amino alkyl agaroses as probes of light-induced changes in phytochrome from pea seedlings (*Pisum sativum* cv. Alaska). *Biochim. Biophys. Acta* **668**, 27–34.

Yamamoto, K. T. and Smith, Jr., W. O. (1981b). A re-evaluation of the mole fraction of Pfr at the red-light-induced photostationary state of undegraded rye phytochrome. *Plant Cell Physiol.* **22**, 1159–64.

Yamamoto, K. T., Smith, Jr., W. O. and Furuya, M. (1980). Photoreversible Ca^{2+}-dependent aggregation of purified phytochrome from etiolated pea and rye seedlings. *Photochem. Photobiol.* **32**, 233–9.

Yokota, T., Baba, J., Konomi, K., Shimazaki, Y., Takahashi, N. and Furuya, M. (1982). Identification of a triterpenoid saponin in etiolated pea shoots as phytochrome killer. *Plant Cell Physiol.* **23**, 265–71.

Photomorphogenesis

17

P Schopfer

Introduction

Everybody is familiar with the fact that the life of a green plant depends on light. It was only about 100 years ago that plant scientists recognized that this statement is true not only with respect to the conversion of light energy into organic matter during photosynthesis. In his famous textbook (*Vorlesungen über Pflanzen-Physiologie*, first edition 1882) the founder of modern experimental plant physiology, Julius von Sachs, described observations which are probably the first scientific documents of photomorphogenesis (Fig. 17.1). Sachs noted that darkened seedlings or parts of older plants developed an irregular, misshapen appearance, characterized by a thin, elongated stem and rudimentary, yellow leaves. He described this syndrome as an 'etiolation illness' which could be alleviated, but never cured, if another part of the plant was exposed to light.

Today photomorphogenesis is regarded in a slightly different way. The term 'photomorphogenesis' embraces all regulatory effects of light (visible and near-ultraviolet parts of the electromagnetic spectrum, *i.e.*, the range of about 300 to 800 nm) on the development of plants, independent of photosynthesis. Correspondingly, the developmental effects of complete darkness is designated (in contrast to development in light) as 'skotomorphogenesis' (Mohr, 1982) or 'etiolation'. Both photomorphogenesis and skotomorphogenesis can be regarded as specific adaptations of the autotrophic plant which is forced to cope with greatly variable light conditions in its natural environment in order to optimize the chances of survival.

Fig. 17.1 Experimental arrangement used by Julius von Sachs to demonstrate the etiolation syndrome of a partly darkened plant. On 25 July 1881, Sachs directed the apex of a healthy pumpkin plant (*A*) into a small hole (*d*) of a wood cabinet (*KK*). After five weeks the apex was again guided into the light (*c*). His drawing illustrates the situation on 7 October 1881, *i.e.*, after 74 days. Sachs was particularly impressed by the fact that the darkened plant part (*B*) developed, after artificial pollination, a fruit of 3 kg fresh weight containing 64 viable seeds. (After Sachs, 1887.)

The strict separation of photomorphogenesis and photosynthesis is more than a formal logical necessity. We know that the photoreceptors used in photomorphogenesis differ from the photoreceptors of photosynthesis in almost every respect. We recall that chlorophyll, phycobilin, or rhodopsin pigments are generally present in rather large amounts in the cell (*mass pigments*) and are densely packed in membranes specialized to convert light energy into chemical energy (*e.g.*, in the thylakoid membranes, or in the purple membrane of *Halobacteria*). In contrast, photomorphogenetic photoreceptors must be conceived as *sensory pigments*, the cellular level of which is generally extremely low and the function of which is not necessarily dependent on membranes. In fact, the low abundance of these photomorphogenetic pigments has prevented their physical detection until very recently and still greatly hampers their full chemical characterization, as well as the elucidation of their cellular localization and primary action mechanism (*see* Chapter 16). Basically, there are two different photoreceptor pigments in higher plants designed for sensing rather than converting light energy; these are known nowadays as *phytochrome* and 'cryptochrome'. Both pigments were detected indirectly by physiological means. However, while phytochrome was spectroscopically characterized and biochemically isolated shortly after its physiological detection, 'cryptochrome' has resisted physical and chemical identification so far, and remains a postulate derived from physiological experiments. Since the role of phytochrome in photomorphogenesis is much better established, this chapter will concentrate on the developmental functions of this pigment and refer only marginally to 'cryptochrome'. Furthermore, discussion will be restricted to photomorphogenesis of higher plants (spermatophytes), although photocontrol of developmental processes is also widespread in ferns, mosses and many algae. 'Cryptochrome' occurs in an even wider range of organisms, including also non-green plants such as bacteria and fungi. After a short recapitulation of some photoreceptor properties relevant for *in vivo* action, the multiplicity of photomorphogenetic responses will be examined. Finally, the mechanism of photomorphogenesis, that is, the signal transduction chains which connect phytochrome with the responding cell functions, will be discussed.

Phytochrome (a summary)

Basic properties

The biochemical and photochemical properties of phytochrome were dealt with in depth in the previous chapter and therefore only need to be reviewed very briefly. Phytochrome, a 124 000-dalton protein bearing an open-chain tetrapyrrole chromophore, represents a photochromic sensor pigment system. Excitation of the chromophoric group by saturating amounts of light leads to the establishment of photoequilibrium between the two interconvertible forms P_r and P_{fr} which is a function of wavelength in the whole range of visible and near ultraviolet radiation (Fig. 17.2):

$$\varphi_\lambda = \frac{[P_{fr}]_\lambda}{[P_{fr}] + [P_r]} = \frac{[P_{fr}]_\lambda}{[P_{tot}]}, \text{ where } P_{tot} = P_r + P_{fr}$$

(the brackets denote quantities). P_r (the physiologically inactive form) absorbs maximally at about 660 nm while P_{fr} (the physiologically active form) absorbs maximally at about 735 nm (*see* Chapter 16, Fig. 16.7). Thus, monochromatic light close to these two wavelengths (red and far-red light, respectively) can be conveniently used to drive the photoequilibrium of the phytochrome system either to a

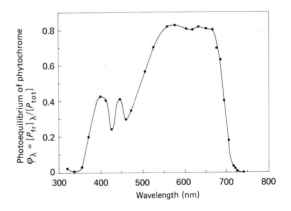

Fig. 17.2 Photoequilibrium (φ_λ) of the phytochrome system as a function of wavelength. Data from spectrophotometric measurements *in vivo* (hypocotyl hook segments from 3-day-old etiolated mustard seedlings, 25°C). The curve is normalized to $\varphi_{660} = 0.80$. (After Hanke, Hartmann and Mohr, 1969.)

relative maximum of P_{fr} ($\varphi_{660} \sim 0.8$; phytochrome system maximally active) or to a state with relative low amounts of P_{fr} (e.g., $\varphi_{730} \sim 0.01$, or $\varphi_{756} < 0.001$; phytochrome system moderately active or nearly inactive). This photoreversibility of the pigment is the rationale for the well-known red/far-red reversibility of phytochrome-mediated physiological responses. In fact, the induction of a response by a pulse of red light, the effect of which can be drastically reduced (or sometimes even abolished) by a subsequent far-red light pulse, provides a perfect operational criterion for the identification of phytochrome as the responsible photoreceptor pigment in a particular photoresponse (Fig. 17.3). However, close inspection of Fig. 17.2 reveals that the phytochrome system establishes a photoequilibrium with sizable amounts of P_{fr}, and is, therefore, more or less active, in the whole range of visible and near-ultraviolet radiation. For instance, the effect

Table 17.1   Induction and reversion of the anthocyanin response of mustard seedling cotyledons using red ($\varphi_{660} = 0.8$), medium far-red ($\varphi_{720} = 0.03$), and long-wavelength far-red ($\varphi_{756} < 0.001$) light pulses (5 min). Dark-grown seedlings were irradiated with one pulse (or two pulses) at 36 h after sowing; anthocyanin content was determined 24 h after the light treatment. (Unpublished data of R. Schmidt; see also Fig. 17.3.)

Light treatment	Relative amount of anthocyanin (%)
5 min 660 nm	100
5 min 720 nm	49
5 min 660 nm + 5 min 720 nm	51
5 min 756 nm	32
5 min 660 nm + 5 min 756 nm	32
5 min 720 nm + 5 min 756 nm	35
dark control	9

of far-red light at 720 nm can be reversed by far-red light of longer wavelengths (e.g., by 756 nm, Table 17.1). The significance of red and far-red light in experimental work with phytochrome results mainly from the unique opportunity of reversibly converting the phytochrome system by light pulses at moderate fluence rates between extreme states of activity.

Physiologically relevant dark reactions

The physiological effectiveness of the phytochrome system is further complicated by several dark reactions of P_r and P_{fr}. The most important dark reactions are: (1) the formation of P_r which seems to be related to a *de novo* synthesis of the protein moiety, and (2) the destruction of P_{fr}, which seems to include the degradation of the protein moiety. At least under certain conditions, there is evidence for several additional dark reactions, for instance a reversion of P_{fr} to P_r and a destruction of P_r. The physiological significance of these spectrophotometrically *in vivo*-detectable conversions is unclear; they are therefore not included in the kinetic minimal model of the phytochrome system

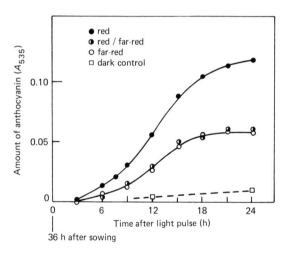

Fig. 17.3   Induction and reversion of a phytochrome-mediated photoresponse (accumulation of anthocyanin in mustard seedlings). Dark-grown seedlings were irradiated at time zero (36 h after sowing) with a 5-min pulse of red light ($\varphi_{660} = 0.8$), a 5-min pulse of far-red light ($\varphi_{720} = 0.025$), or a pulse of red light immediately followed by a pulse of far-red light. Far-red light fully reverses the effect of red light, i.e., the sequence red/far-red has the same effect as far-red alone. This demonstrates the operational criterion for the involvement of phytochrome. (After Lange, Shropshire and Mohr, 1971.)

shown in Fig. 17.4. This simple model allows the calculation of $[P_{fr}]_t$ (*i.e.*, the amount of P_{fr} at time t) from $[P_{tot}]_0$ and the spectrophotometrically measurable rate constants for P_r synthesis and P_{fr} destruction.

A system of light and dark reactions such as the one depicted in Fig. 17.4 tends to approach a

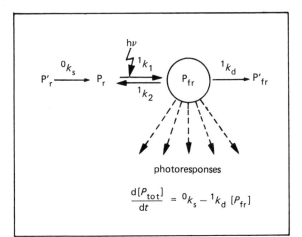

$$\frac{d[P_{tot}]}{dt} = {}^0k_s - {}^1k_d\,[P_{fr}]$$

Fig. 17.4 Linear kinetic model of the phytochrome system ($^1k_1$, $^1k_2$, first-order rate constants of the phototransformations; ^0k_s, zero-order rate constant of P_r synthesis from a precursor P'_r; ^1k_d, first-order rate constant of P_{fr} destruction to an inactive product P_{fr}'). This model has been verified experimentally by spectrophotometric measurements with mustard seedlings, according to which $^0k_s = 5 \times 10^{-7}[\Delta(\Delta A)]$ s^{-1}] and $^1k_d = 2.6 \times 10^{-4}$ [s^{-1}] at 25°C (Schäfer, 1976).

stationary state with constant levels of all intermediary components whenever the rate of P_{fr} destruction ($[P_{fr}] \cdot {}^1k_d$) approaches the rate of P_r synthesis (^0k_s). This takes place automatically in continuous light if the photon fluence rate is sufficient to establish the photoequilibrium $P_r \rightleftharpoons P_{fr}$ (Schäfer and Mohr, 1974). Thus, under these conditions the amount of P_{fr} depends solely on the ratio of the two dark reactions and is, therefore, wavelength-independent:

$$\frac{d[P_{tot}]}{dt} = 0, \quad [P_{fr}]_{\text{steady state}} = \frac{{}^0k_s}{{}^1k_d} = \text{const.}$$

This prediction has been verified spectrophotometrically in the hypocotyl and the cotyledons of the

mustard seedling (Schäfer, Schmidt and Mohr, 1973).

Effector/response relationships

The kinetic minimal model shown in Fig. 17.4 is sufficient to explain quantitatively, in terms of P_{fr}, the induction kinetics of certain physiological responses which are controlled by a critical P_{fr} threshold. The regulation of the enzyme lipoxygenase will be used to illustrate this threshold control. In the cotyledons of the young, dark-grown mustard seedling the activity of lipoxygenase accumulates rapidly. During the period of maximal increase (between about 36h and 48h after sowing) the enzyme activity rises by about 60%. This increase can be arrested instantaneously by the formation of a sufficient amount of P_{fr}. Sufficient means in this case $[P_{fr}]_t/[P_{tot}]_0$ exactly ≥ 0.125, *i.e.*, an absolute amount of P_{fr} equivalent to 1.25% or more of the amount of total phytochrome at time zero ($= 36$h after sowing in these experiments). This inhibition is fully reversible. After a period of light-induced repression, the lipoxygenase activity increases again at full speed when the P_{fr} level drops below the threshold of 1.25%.

Thus, the phytochrome system operates as a symmetric all-or-none switch, inhibiting enzyme formation completely at any P_{fr} level above the threshold and allowing enzyme formation with the dark-rate at any P_{fr} level below the threshold, no matter whether the threshold is surpassed in darkness (by P_{fr} destruction) or by adjustment of φ_λ with light. Within the limits of experimental error there is no hysteresis in this regulatory system. From the extensive work on the lipoxygenase response (*e.g.*, Mohr and Oelze-Karow, 1976) a basic set of experiments is shown in Fig. 17.5. It is evident from the time courses of enzyme activity that the duration of enzyme repression is related to the amount of P_{fr} produced by a saturating light pulse, *i.e.*, to the photoequilibrium adjusted by a particular wavelength up to $\varphi_{725} = 0.0125$. Using the photometrically determined rate constant of P_{fr} destruction ($^1k_d = 2.6 \times 10^{-4}$ s^{-1}, $\tau_{\frac{1}{2}} = \ln 2/^1k_d = 45$ min) it can be shown that the P_{fr} threshold of $[P_{fr}]_\lambda/[P_{tot}]_0 = 0.0125$ is passed at each wavelength precisely when the enzyme activity abruptly starts to rise again.

This type of response can obviously be used to

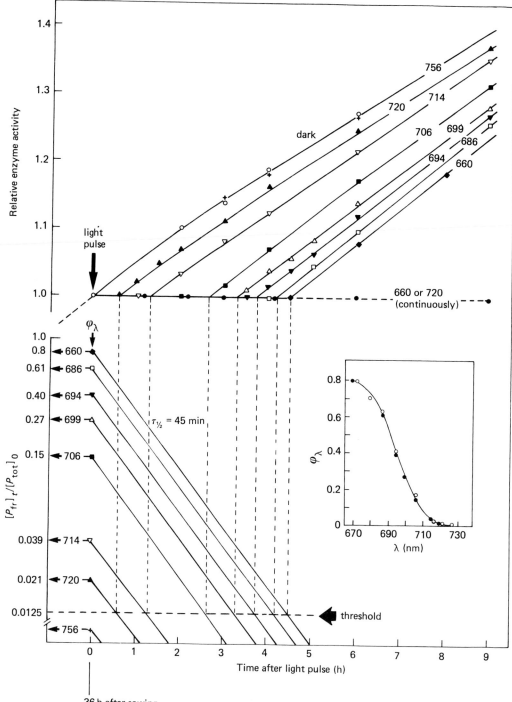

36 h after sowing

Fig. 17.5 Kinetics of the lipoxygenase response of mustard cotyledons. *Top*, kinetics of enzyme activity after saturating 5-min light pulses of various wavelengths (600 to 756 nm). The increase of the enzyme activity is immediately stopped by the light pulses, and again resumed with full speed after a time period depending on wavelength. *Bottom*, corresponding kinetics of P_{fr} destruction in darkness after establishment of φ_λ by a 5-min light pulse of the same group of wavelengths. At each wavelength, the same threshold value of $[P_{fr}]_t/[P_{tot}]_0 = 0.0125$ is reached whenever the response escapes from inhibition. *Inset*: Comparison of φ_λ values calculated from the lipoxygenase kinetics (●) and from spectrophotometric measurements (○, taken from Fig. 17.2). (After Mohr and Oelze-Karow, 1976.)

determine physiologically active levels of phytochrome *in vivo* and therefore provides a critical test for the physiological significance of spectrophotometric phytochrome measurements *in vivo*. A comparison of the numerical values for ^1k_d, ^0k_s, and φ_λ, determined either with *in vivo* spectrophotometry or with the lipoxygenase assay, are in amazing agreement. As an example, Fig. 17.5 (inset) demonstrates the perfect correspondence of φ_λ values between 660 nm and 725 nm obtained with these two methods.

A clear-cut quantitative relationship between the spectrophotometrically determined dynamics of the phytochrome system and the expression of a response such as the repression of lipoxygenase increase seems to be an exception rather than a rule. Although other threshold responses are known, the majority of phytochrome responses demonstrate a graded dependency on $[P_{fr}]$ down to very low P_{fr} levels and a complex φ_λ/effect function. Figure 17.6, for instance, shows the φ_λ dependency of the anthocyanin response induced with single saturating light pulses in the mustard seedling. This response turned out to be extremely sensitive to low levels of P_{fr} (down to $10^{-4}\%$, *see* inset of Fig. 17.6) but rather insensitive to changes between 5 and 80% P_{fr}. Unfortunately this is not the only difficulty with the anthocyanin response which may be representative for many—if not most—phytochrome responses. Figure 17.3 shows that after the start of P_{fr} formation the accumulation of anthocyanin cannot be observed until a lag-phase of 2–3 h has elapsed. Hence, the response first becomes detectable after a time which is equivalent to 3–4 half-lives of P_{fr}. There is obviously a large amount of hysteresis due to time-consuming steps within this signal response chain, which of course obliterates the $[P_{fr}]$/effect relationship on the time scale. Furthermore, phenomena such as the continuously changing responsiveness of the anthocyanin-forming cells (*see* Fig. 17.14) and the stimulation of the response through a previous light treatment (*see* Fig. 17.15) indicate that the signal transduction chain of the anthocyanin response—and probably of most other phytochrome responses—is considerably more sophisticated than the signal transduction chain of the lipoxygenase response. While there is no reason to doubt that a clear-cut $[P_{fr}]$/effect relationship does indeed exist in the anthocyanin and related responses, our present knowledge is far from sufficient

to present a comprehensive mathematical model that would allow the calculation of time functions or $[P_{fr}]$ functions of photomorphogenetic responses from the kinetic parameters of the phytochrome system.

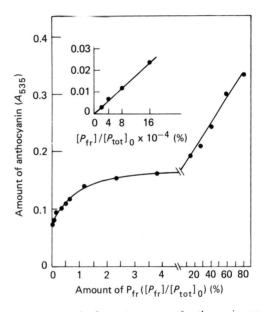

Fig. 17.6 The $[P_{fr}]$ response curve of anthocyanin accumulation in mustard seedlings. Saturating 5-min pulses with monochromatic light of various wavelengths (establishing various photoequilibria, φ_λ) were given 36 h after sowing. (P_{fr} levels below 1% were adjusted by defined amounts of red light.) Anthocyanin content was determined 24 h after the light treatment, when the accumulation had reached a constant value (*see* Fig. 17.3). The inset shows the range close to the origin. (After Drumm and Mohr, 1974.)

The high irradiance reaction

So far we have ignored one major complication of the phytochrome story which has confused workers almost from the beginning of research in photomorphogenesis. Theoretically, the simple phytochrome model of Fig. 17.4 can also be used to calculate P_{fr} levels for continuously irradiated plants, and, in fact, these calculations yield significant results in the case of the lipoxygenase response (Mohr and Oelze-Karow, 1976). However, in almost every other response studied so far there are significant discrepancies between calculated (and spectrophotometrically confirmed) P_{fr} levels and observed response levels in continuous light which are not resolved at present. Briefly, these complications result from the annoying observation that in continuous light most phytochrome responses depend strongly on photon fluence rate (light intensity) although the photoequilibrium is established after a few minutes and, therefore, the P_{fr} level apparently cannot increase further with increasing light absorption by P_i and P_{fr} (Fig. 17.7). Since this fluence rate dependency cannot be saturated even with very bright light, the

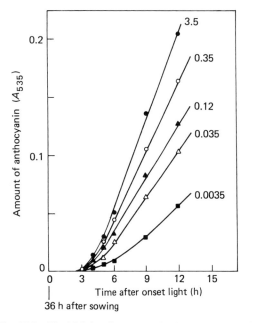

Fig. 17.7 The high irradiance reaction in the anthocyanin accumulation of mustard seedlings. Dark-grown seedlings were transferred to various energy fluence rates (shown at the right-hand side in $W\,m^{-2}$) of continuous far-red light at time zero (36 h after sowing). The highest fluence rate of the far-red source used in these experiments is sufficient to establish the photoequilibrium ($\varphi_{720} \sim 0.03$) within about 1 min. (After Lange *et al.*, 1971.)

term 'high irradiance reaction' has been coined for the underlying photoregulatory system. In addition to this strange fluence rate dependency, the action spectra determined under continuous light are generally characterized by a prominent peak in the

far-red region and smaller peaks in the blue and near-ultraviolet region (Fig. 17.8), although the linear phytochrome model (Fig. 17.4) predicts that $[P_{fr}]_\lambda = {}^0k_s/{}^1k_d = $ constant under these conditions. Evidently this simple kinetic model of the phytochrome system fails to explain the characteristic features of the high irradiance reaction which was, therefore, attributed to an unknown blue/far-red-absorbing pigment different from phytochrome by the early investigators of these phenomena.

Today this view has been replaced by the general understanding that the high irradiance reaction results from a specific mode of action of the phytochrome system, at least as far as the far-red action peak is concerned. This conclusion results from a

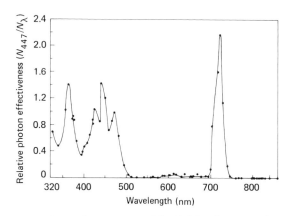

Fig. 17.8 Action spectrum of the high irradiance reaction of phytochrome mediating the inhibition of hypocotyl lengthening in seedlings of *Lactuca sativa*. (After Hartmann, 1967.)

brilliant analysis of the high irradiance response in the growth of the lettuce hypocotyl by Hartmann (1967). In essence, Hartmann showed that the effect of 720 nm light in the high irradiance response could be quantitatively mimicked by a mixture of 660 nm and 770 nm light which establishes the same φ_λ as 720 nm light, while both wavelengths are ineffective if given alone. Thus, the conclusion is unavoidable that the marked effect of continuous far-red light (around 720 nm) in photomorphogenesis must somehow be related to a specific state of the phytochrome system under continuous irradiation which appears to be characterized by an optimal action at apparently low P_{fr} levels (in the range of a few per cent of P_{tot}) and by a dependency on the total number of light quanta absorbed by both P_r and P_{fr} at photoequilibrium as defined in Fig. 17.4.

There have been various attempts to develop models of the phytochrome system which explain the basic features of the high irradiance reaction, a detailed discussion of which is beyond the scope of this chapter. It may suffice to mention that these models extend the minimal model of Fig. 17.4 by introducing at least one new element X which interacts with P_{fr} to form $P_{fr}X$. This complex, which is thought to be the true effector molecule of the phytochrome system, is formed from P_{fr} by a rapid dark reaction and establishes a separate photoconvertible phytochrome pool $P_rX \rightleftharpoons P_{fr}X$ which can be emptied only slowly back to the free phytochrome pool $P_r \rightleftharpoons P_{fr}$ via a slow dark reaction $P_rX \rightarrow P_r + X$. Such a cyclic extension of the phytochrome model would predict, for instance, that the physiologically effective $P_{fr}X$ pool increases at the expense of the P_{fr} pool with increasing fluence rate although the overall photoequilibrium $(P_r + P_rX) \rightleftharpoons (P_{fr} + P_{fr}X)$ remains constant.

The high irradiance reaction seems to be a characteristic feature of the phytochrome system of dark-grown plants which have accumulated relatively large levels of total phytochrome. Consequently, the high irradiance reaction is lost during prolonged light treatments which lead to a strong decrease of $[P_{tot}]$ due to P_{fr} destruction (*e.g.*, continuous red or white light). In plants raised in continuous natural white light the characteristics of the high irradiance reaction may be largely absent (*e.g.*, Jose and Vince-Prue, 1977; Beggs, Holmes, Jabben and Schäfer, 1980). However, for the researcher interested in the molecular mechanisms of photomorphogenesis, the high irradiance reaction with continuous far-red light provides a unique experimental tool to operate the phytochrome system under steady-state conditions with respect to the effector signal. Furthermore, continuous far-red light ($\lambda > 700$ nm) is inactive with respect to the photochemical protochlorophyll/chlorophyll transformation and, therefore, permits the study of photomorphogenesis in chlorophyll-deficient plants, avoiding interference with photosynthetic responses. These are the pragmatic reasons for the widespread use of continuous far-red light in photomorphogenetic research.

Table 17.2 Some typical blue-light-mediated photomorphogenetic responses in lower plants.

Response	Organism	Selected references*
Synthesis of carotenoids	bacteria (*Mycobacterium, Myxococcus, Flavobacterium*) and fungi (*e.g., Fusarium aquaeductuum, Neurospora crassa*)	Rau (1980)
Formation of conidia (sporulation)	lower fungi (*e.g., Trichoderma viride*)	Gressel and Galun (1970); Tan (1978)
Phototropism of sporangiophores	lower fungi (*e.g., Phycomyces blakesleeanus*)	Bergman *et al.* (1969); Delbrück, Katzir and Presti (1976)
Development of chloroplasts and photosynthetic capacity	*Euglena gracilis*	Chelm, Hallick and Gray (1979); Schiff (1980)
Biplanar growth of sporelings (prothallium formation)	fern gametophytes	Furuya (1978)

* *See* Senger, H. (1980) for a comprehensive survey of blue-light-mediated plant responses.

Cryptochrome

Basic properties

First of all it should be kept in mind that the term 'cryptochrome' describes a hypothetical photoregulatory pigment which has been characterized so far merely by its physiological action (Table 17.2). It may well turn out in the future that 'cryptochrome' represents a variety of photomorphogenetic pigments which have in common the ability to sense light in the blue-ultraviolet spectral range and which demonstrate a superficial similarity in their absorption spectra. For the time being the operational criterion for the involvement of 'cryptochrome' in eliciting a photoresponse is an action spectrum with a peak near 370 nm and three peaks or shoulders between 400 and 500 nm (Fig. 17.9; Mohr, 1982). In

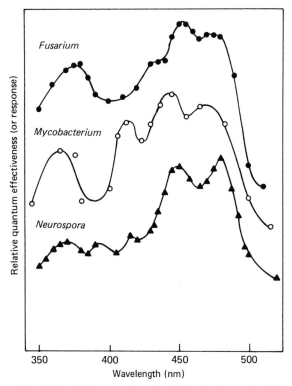

Fig. 17.9 Action spectra of photoinduced carotenoid biosynthesis in *Fusarium aquaeductuum*, *Mycobacterium* sp., and *Neurospora crassa*. (After various sources, redrawn from Rau, 1980.)

contrast to phytochrome the inductive effect of a blue-ultraviolet light treatment is not reversed by light of longer wavelengths. If the phytochrome system is also involved in a particular photoresponse, it is obviously difficult to distinguish clearly between the two photoreceptor systems since phytochrome also operates in the short wavelength region of the spectrum (*see* Figs 17.2, 17.8). However, there are many well-established photomorphogenetic blue-light responses also in phytochrome-containing higher plants which are clearly unrelated to phytochrome, such as, for instance, the well-known phototropic response of shoots or coleoptiles which is dealt with in Chapter 7.

Two types of plant pigment present themselves as candidates for the 'cryptochrome' photoreceptor, namely flavins and carotenoids. Unfortunately both groups of pigments show absorption spectra that are compatible with the action spectra of many of the blue-light photoresponses. Presently, flavins seem to be the favored candidates for some well-investigated blue-light responses, mainly because it has been shown that the effects of blue light can still be observed if the level of carotenoids is drastically reduced by mutation or other means.

Phytochrome/cryptochrome coactions

In higher plants containing phytochrome as well as some kind of 'cryptochrome', it is conceivable that both types of photoreceptor pigments must be activated in order to elicit a particular photomorphogenetic response. In fact, this situation seems to be realized in many plants, although the extent of interdependence between the two photoregulatory systems may vary greatly among different species. An extreme example is the following: the synthesis of anthocyanin in the mesocotyl of *Sorghum* seedlings is controlled by phytochrome. However, the phytochrome system can only act after a blue-ultraviolet light effect has occurred. On the other hand, the expression of the blue-ultraviolet effect is modified by phytochrome. It was concluded that there is an obligatory dependency ('obligatory sequential interaction') between the blue-ultraviolet light and light operating through phytochrome (Drumm and Mohr, 1978). It seems that somehow both pigment systems provide essential constituents of the biosynthetic chain leading to anthocyanin synthesis in this plant. Since the identification of these constituents has not been achieved so far, the neutral term *coaction* should be used to describe such an interdependence in photoreceptor action.

Phenomenology of photomorphogenesis

The adult plant

Photomorphogenesis is an essential property of practically all stages of the ontogeny of higher plants. Numerous morphological and physiological features that we are accustomed to regard as constitutive (genetically determined) in a normal green plant are in fact epigenetic adaptations to the light factor produced through photomorphogenesis. Some representative examples of such light-dependent responses, which are all mediated through

phytochrome, are summarized in Table 17.3. Moreover, a great number of developmental processes which are under photoperiodic control such as, for instance, flowering, fruit set, tuber initiation, tillering, or senescence could be included here since plants use the phytochrome system to measure the length of the daily photoperiod in order to synchronize their ontogenic cycle to the season of the year (*see* Chapter 18). Last, but not least, it should be mentioned that many plants produce photoblastic seeds, *i.e.*, dormant seeds which do not readily germinate unless they are treated with light. In most photoblastic seeds phytochrome can be demonstrated by red/far-red light pulse experiments to be the photoreceptor pigment responsible for induction of germination. In fact, the seeds (achenes) of *Lactuca sativa* have played a decisive role in the discovery of the phytochrome system (Borthwick, Hendricks, Toole and Toole, 1952). These pioneers

Table 17.3 Some typical examples of light-controlled properties (photoresponses) in green plants; in all cases phytochrome has been demonstrated to absorb the effective radiation.

Photoresponses	References
Inhibition of internode growth, promotion of leaf growth, increase of stomata number (leaf), increase of chlorophyll content (leaf)	Kasperbauer (1971); Kasperbauer and Peaslee (1973)
Leaf unrolling (grasses)	Wagné (1965)
Nyctinastic leaf movements	Fondeville, Borthwick and Hendricks (1966)
Leaf shape*	Sanchez (1971)
Lateral bud outgrowth (abolition of apical dominance)	Tucker (1977)
Induction of frost hardening	Williams, Pellett and Klein (1972)
Fruit ripening, including carotenoid accumulation (tomato)	Thomas and Jen (1975)
Parental control of seed germination	Shropshire (1972)
Accumulation of alkaloids and phenolics (tobacco)	Tso, Kasperbauer and Sorokin (1970)
Inhibition of chloroplast senescence including chlorophyll breakdown (*Marchantia* thallus, mustard cotyledons)	De Greef, Butler and Roth (1971); Biswal, Kasemir and Mohr (1982)

* *See* Fig. 17.10.

of phytochrome research used the photoblastic Grand Rapids variety of lettuce to demonstrate for the first time the repeated red/far-red photoreversibility of phytochrome action and to determine the first true action spectra of both phytochromes.

The vast amount of literature on the phenomenology of photomorphogenesis that has accumulated in the last three decades makes it quite clear that probably no higher plant can complete its life cycle without developmental control by light, mediated through the phytochrome system. It is important to realize, however, that the controlling effects of light are generally reversible within relatively short periods of time. For instance, careful measurements of stem extension growth using electronic position transducers have shown that the growth rate of this organ can be readapted to darkness within a few minutes after inactivation of the phytochrome system (*e.g.*, Morgan, O'Brien and Smith, 1980). Thus, at least during the period of rapid vegetative development, the plant can switch quickly from photomorphogenesis to skotomorphogenesis after cessation of the light stimulus. However, since the *products* of a previous period of development are in general rather stable, even a dramatic light-dependent modulation of the *rates* of morphogenetic processes will change only slowly the overall habit of a plant. At any rate, the natural light/dark cycle in principle leads to an intermediary state of photomorphogenesis and skotomorphogenesis. Plants growing in long photoperiods (or bright light) display predominantly features of photomorphogenesis while plants growing in short photoperiods (or in the shade of a dense canopy) display features of skotomorphogenesis. Obviously, photomorphogenesis and skotomorphogenesis are the extremes of a wide range of quantitative developmental adaptations to the light factor rather than all-or-none strategies.

The quantitative interplay of light and darkness in the natural environment can be demonstrated impressively in the leaf morphogenesis of the common dandelion. The characteristic dentate leaf contour of this plant is not, as one might assume, genetically determined but is a result of photomorphogenesis. In 10 h of light per day the expression of this taxonomically decisive feature requires the continued presence of P_{fr} for some time after the end of the photoperiod. If the phytochrome system is inactivated by a far-red light pulse before the onset of the daily 14 h dark period, the dandelion leaf de-

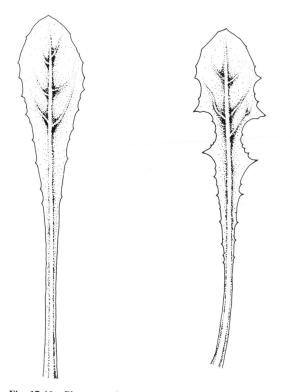

Fig. 17.10 Photomorphogenesis of the leaf of *Taraxacum officinale*. Genetically identical plants were grown in a photoperiod of 10 h white light/14 h darkness. *Left*, leaf of a plant which was irradiated for a few minutes with far-red light at the end of each photoperiod. *Right*, leaf of a control plant, which was transferred directly from white light (acting similarly to red light) to darkness. Plants which were treated with red light at the onset of the dark period (or with a sequence of far-red followed by red) developed leaves as shown on the right-hand side. (After Sanchez, 1971.)

velops only rudimentary signs of its specific leaf morphology (Fig, 17.10).

The seedling

A normal green plant is energetically dependent on light acting *via* photosynthesis. Therefore, experimental manipulation of photomorphogenesis is primarily limited to 'end-of-day' light treatments such as those used in Fig. 17.10 to adjust the effectiveness of the phytochrome system during the daily dark period. Because of the unavoidable interference with photosynthesis the adult green plant and its organs represent rather complex and experimentally cumbersome objects for the investigation of the mechanisms of photomorphogenesis. Mainly for such reasons the functional aspects of photomorphogenesis have been studied almost exclusively using young, dark-grown seedlings which are capable of rapid development on the basis of their own nutritional reserves. In fact, the few days following the rupturing of the seed coat are that period of plant ontogeny in which light exerts the most dramatic effect on development. From an ecological point of view, this is quite understandable. The ungerminated dicotyledonous seed contains an embryo, organized into embryonic organs (taproot, hypocotyl, cotyledons). The embryo is provided with a limited supply of organic nutrients which can support self-sustained growth of the seedling (*i.e.*, without access to light energy) but only for a few days. Obviously, this food store must be used most economically in order to reach the stage at which the young plant can depend on light as a source of metabolic energy. If this situation cannot be attained before exhaustion of the endogenous food reserves, the plant is irrevocably condemned to death. Consequently, it is most significant for the germinating embryo to concentrate all its developmental power on reaching the following goals: (1) pushing the cotyledons, which are designed as prospective light-harvesting organs, through the covering layers of soil; (2) building up

Table 17.4 The photomorphogenetic syndrome of the young dicotyledonous seedling (*e.g.*, from mustard); in all photomorphogenetic responses mentioned phytochrome has been shown to act as photoreceptor (*see* Mohr, 1972).

Skotomorphogenesis (growth in complete darkness)	Photomorphogenesis (growth in continuous white light)
(1) *Morphological features*	
cotyledons remain small and folded	cotyledons expand, unfold, and expose their surface to the light
hypocotyl forms apical hook (for penetration of the soil)	hypocotyl hook opens
hypocotyl lengthening reaches high rates (*e.g.*, $2\,mm\,h^{-1}$)	hypocotyl lengthening is reduced (*e.g.*, $0.1\,mm\,h^{-1}$)
(2) *Cellular features* (*cotyledon cells*)	
formation of etioplasts (without chlorophyll and other essentials of photosynthesis)	formation of functional chloroplasts (containing starch grains)
pre-existing stomatal meristems remain inactive	formation of functional stomata from pre-existing stomatal meristems
formation of glyoxysomes (involved in storage fat degradation)	formation of leaf peroxisomes (involved in storage fat degradation as well as photorespiration)
(3) *Physiological features* (*cotyledon cells*)	
no accumulation of screening pigments in the epidermis	accumulation of anthocyanin and other flavonoid pigments in the epidermis used for light protection
early peak in respiratory activity	later peak in respiratory activity
rapid export of organic materials	greatly reduced export of organic materials

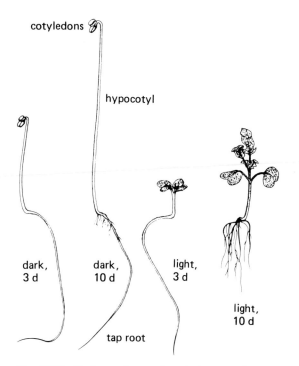

cotyledons

hypocotyl

dark, 3 d

dark, 10 d

light, 3 d

light, 10 d

tap root

Fig. 17.11 Skotomorphogenesis and photomorphogenesis of the mustard seedling. The seedlings were grown in complete darkness or white light for 3 or 10 days (25°C). (After Mohr, 1972, modified.)

the photosynthetic capacity of the cotyledons; and (3) exposing the cotyledons maximally to the sunlight. If we look at Fig. 17.11 and Table 17.4 with these goals in mind, the biological significance of skotomorphogenesis and photomorphogenesis of seedlings becomes immediately apparent. It should be noted at this point that all these developmental changes can occur in principle without further cell division. In the mustard seedling the DNA content of the cotyledons remains practically constant. In the hypocotyl the DNA content is increased by endopolyploidization; however, this takes place independently of light (Mösinger, Bolze and Schopfer, 1982).

In the following sections we shall deal with the question of how the switch from skotomorphogenesis to photomorphogenesis can be brought about on the cellular and molecular level. For the ex-

perimental foundation of general principles of photomorphogenesis we shall use, for didactic reasons, a single representative dicotyledonous plant system, the young seedling of mustard (*Sinapis alba* L., Fig. 17.11).

Cellular aspects of photomorphogenesis

Effector/responsiveness relationships

It was learned in the foregoing section that growth, differentiation, and the integration of these processes (= morphogenesis) in the multicellular plant are profoundly influenced by light acting through sensor pigments, of which phytochrome is the one most widely known. Before examining more closely the functional aspects of photomorphogenesis a few general facts need to be considered.

(1) Photomorphogenesis of a plant involves numerous individual photoresponses which are highly specific for a particular organ or tissue (Table 17.4).

(2) There is no evidence for a tissue specificity of phytochrome itself. All available data indicate that the phytochrome molecules in different parts of a plant are biochemically and spectroscopically identical.

(3) Although localization studies using immunological techniques to label phytochrome in thin sections of various plant tissues have shown that phytochrome is probably not evenly distributed within a plant (Pratt, Coleman and Mackenzie, 1976), the observed quantitative differences are clearly insufficient to explain the dramatic differences in the light responsiveness of plant tissues.

Obviously, phytochrome does not carry any information with respect to the specific effects of light within a particular cell. It must be concluded, therefore, that the specificity of phytochrome responses is determined exclusively by the responding cell or cell function. Furthermore, it must be expected that phytochrome responses depend on cell responsiveness not only qualitatively but also quantitatively. It is very unlikely that the cellular amount of P_{fr} is the decisive limiting factor determining the extent of a photoresponse under *all* circumstances. Rather it would be expected that the responsiveness of the cell with respect to P_{fr} can be at least as important as the P_{fr} level. It should be emphasized, at this point, that the term 'responsiveness' does not imply any particular biochemical mechanism. It is used here with a purely descriptive meaning to summarize all relevant properties of the black box on which P_{fr} acts. These could involve the amount of P_{fr} receptor molecules as well as the biosynthetic capacity of a metabolic pathway, to mention just two possibilities.

In summary, we have to be aware that photomorphogenesis is a complex product of a regulatory system in which P_{fr} is not the only variable parameter. In other words: photomorphogenesis results from an interaction of P_{fr} and certain cell functions, both of which can be used to regulate the final photoresponses. For an understanding of photomorphogenesis it is essential to investigate both of these aspects. Furthermore, we have to take into account that cell differentiation and morphogenesis can be described using either *time* or *space* as the parameter of the variable coordinate.

Spatial pattern of competence

Used in the spatial meaning, cell differentiation expresses itself in the appearance of various distinctly different cell types organized in a specific spatial pattern, notwithstanding the fact that all cells have inherited the same set of genes during cell division. An understanding of the basic mechanisms of spatial cell differentiation is still not in sight. However, it is known that this spatial pattern includes more than just those cell features which can be detected under the microscope. For instance, this pattern also involves instructions on how to respond to external developmental factors such as light in a tissue-specific manner. Mohr (1978) has coined the term 'pattern specification' (with regard to phytochrome) for this particular property of the differentiated cell, which is illustrated in Fig. 17.12. Two important conclusions can be drawn from an analysis of these simple drawings:

(1) The individual photoresponses are strictly confined to a particular cell type (or set of cell types) that is obviously determined for expressing a particular response. We use the term

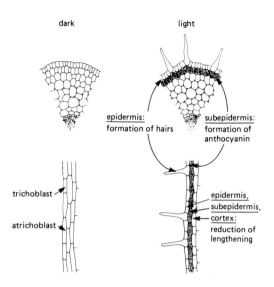

Fig. 17.12 Microscopic sections through the hypocotyl of dark-grown and light-grown mustard seedlings. *Top*, cross-sections; *bottom*, longitudinal sections through the outer cell layers. These drawings illustrate the spatial aspect of developmental responsiveness to phytochrome action (pattern specification). While particular epidermis cells (trichoblasts) are pre-programmed to form hairs (but no anthocyanin) under the influence of phytochrome, all cells of the subepidermis are able to accumulate anthocyanin. Irrespective of this specific response pattern, the cells of epidermis, subepidermis, and cortex respond to phytochrome by an inhibition of cell lengthening. (After Mohr, 1972).

'competence' for this state of responsiveness determination.

(2) Different spatial competence patterns can be superimposed without interfering with each other.

Thus, the basic spatial program of cell differentiation provides the cells with specific sets of competences to respond to the general developmental effector phytochrome. The ultimate response, *e.g.*, synthesis of anthocyanin in the subepidermal cell layer of the mustard hypocotyl, is brought about in a two-step process (Mohr, 1978):

(1) establishment of the cell-specific competence toward phytochrome (= *pattern specification*);
(2) expression of this pattern through the triggering action of phytochrome (= *pattern realization*).

It is the perfectly organized interplay of these two processes that is actually causing photomorphogenesis. This situation is, of course, not unique. The theoretical concept outlined above can be extended in principle to many other developmental effectors *e.g.*, to hormones.

Temporal pattern of competence

In principle, the theoretical framework describing photomorphogenesis in terms of space is equally applicable if we choose *time* as the developmental coordinate. Thus, each cell, in addition to having a specific position within the spatial pattern of competence, is further characterized by a specific *temporal* pattern of competence with respect to the expression of a particular photoresponse. Once again, the mustard seedling illustrates this point.

As in many other plants, the embryo of mustard is essentially insensitive to photomorphogenetically active light during the periods of histodifferentia-tion, seed maturation and early germination, although the phytochrome system operates perfectly under these conditions. It is not until some time after the young seedling has emerged from the seed coat that the responsiveness toward the phytochrome signal appears and the expression of photoresponses can be observed. However, the decisive point is that this takes place in a sequential rather than in a synchronized manner. As a first attempt to analyze the sequential pattern of competence, the different starting points of various photoresponses may be compounded. Figure 17.13 illustrates the starting points of some selected enzyme inductions of the mustard seedling cotyledons. There is an amazingly wide range of variation for the starting points of different photoresponses. Even closely related cell functions such as the activity of neighboring enzymes of a metabolic chain can display different starting points (*see* Table 17.5, p. 403).

Using appropriate irradiation programs it can be shown that the responsiveness toward phytochrome,

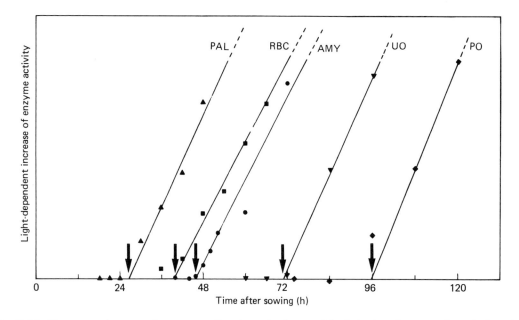

Fig. 17.13 The temporal pattern of competence ('starting points', *arrows*) for induction of enzymes through phytochrome in the cotyledons of mustard seedlings. The seedlings were irradiated with continuous red (or far-red) light from sowing (25°C). The data are compiled from Frosch, Drumm and Mohr (1977): PAL, phenylalanine ammonia-lyase; RBC, ribulose bisphosphate carboxylase; Sharma and Schopfer (1982): AMY, β-amylase; Hong and Schopfer (1981): UO, urate oxidase: Schopfer and Plachy (1973): PO, peroxidase. See Table 17.5 for starting points of further enzyme responses.

after surpassing the starting point, reaches an optimum and then declines to zero (Fig. 17.14). In the case of lipoxygenase repression it has been found that this enzyme expresses all-or-none kinetics (*see* Fig. 17.5) with respect to the competence toward phytochrome. It enters light control suddenly at 33 h and escapes from light control suddenly at 48 h after sowing (Mohr, 1978).

Thus it can be concluded from the data available that, in principle, each individual photoresponse is characterized by a specific temporal competence pattern. Photomorphogenesis of the multicellular

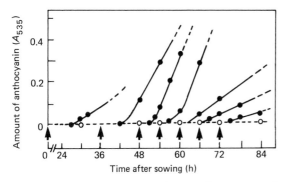

Fig. 17.14 The time course of responsiveness toward phytochrome in the anthocyanin response of mustard seedlings. Dark-grown seedlings were transferred to continuous far-red light at various times after sowing (*arrows*). ○, dark control; ●, far-red light. (After Lange, Bienger and Mohr, 1967; Steinitz, Drumm and Mohr, 1976.)

plant therefore results from the realization by light of a precise, pre-existent time schedule of responsiveness to phytochrome. It will shortly be realized that this general statement is too simple to provide a satisfactory description of the processes involved in eliciting most phytochrome-mediated photoresponses.

Potentiation of phytochrome action by phytochrome

There are good arguments for the notion that pattern specification as such is independent of light (Mohr, 1978). In certain responses it has been found, however, that a light pretreatment, *via* phytochrome, can dramatically influence the quantita-

tive effectiveness of a given amount of P_{fr} some hours later (Mohr, Drumm, Schmidt and Steinitz, 1979). An example of such a P_{fr}-mediated 'autosensitization' is illustrated in Fig. 17.15.

Using the anthocyanin response of mustard, Mohr *et al.* (1979) have shown that in this case an increase

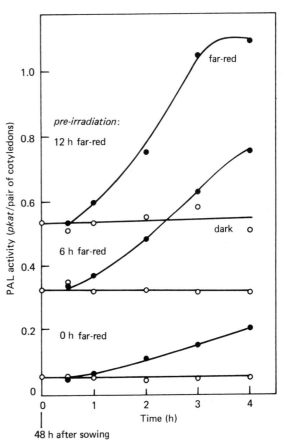

Fig. 17.15 The potentiation of phytochrome-mediated phenylalanine ammonia-lyase (PAL) accumulation by phytochrome (pre-irradiation with far-red light) in the cotyledons of mustard seedlings. The seedlings were pre-irradiated with 0, 6, 12 h of continuous far-red light followed by 4 h of darkness before measuring enzyme induction by far-red light (●) between 48 and 52 h after sowing (○, dark control). Although pre-irradiation reduces the amount of total phytochrome to some extent, this is greatly over-compensated for by a more than 5-fold stimulation of the effectiveness of the second light treatment. (After Acton, Fischer and Schopfer, 1980.)

as well as a reduction of phytochrome effectiveness can be produced by certain light pretreatments, even before the period of competence. This phenomenon of course greatly increases the difficulties in the interpretation of $[P_{fr}]$/response functions in continuous light. A great deal of the surprisingly strong effect of continuous irradiation (as compared with single light pulses producing even higher P_{fr} levels) may in fact be attributed to a light-mediated potentiation of responsiveness.

Mohr *et al.* (1979) have proposed that the P_{fr}-mediated change in responsiveness is caused by a change in the availability of P_{fr} receptor sites. It is, however, equally possible that the pre-irradiation mediates a change of some other metabolic factor which causes signal amplification or capacity changes at a later stage of the signal transduction chain.

Regulatory aspects of photomorphogenesis

Involvement of gene expression mechanisms

Formally, plant development can be regarded as a realization of temporal and spatial patterns which are laid down in the genetic information of DNA and finally become expressed through the catalytic action of enzymes. Factors such as light interfere with the realization of genetically determined patterns by selectively stimulating the expression of defined parts of the genetic information and inhibiting the expression of others. From the extensive work on enzyme photoresponses there is overwhelming evidence that regulation of enzyme activities is indeed an essential element of photomorphogenetic control. Meanwhile, there are more than 50 different enzymes that have been reported to be subject to phytochrome regulation (Schopfer, 1977) and this is probably only the tip of the iceberg. Thus, in a general sense, everybody agrees with the statement that photomorphogenesis ultimately results from the regulation of gene activity by photosensory pigments such as phytochrome. However, there remains the decisive problem that has been a matter of dispute for many years: at which stage of gene expression, detectable on the level of enzyme activities, does phytochrome interact with the flow of

genetic information? In a first approximation this problem can be rationalized by asking the following questions: does the phytochrome signal interfere with:

(1) the reading of the genetic information from the DNA (*transcription*), or
(2) the transformation of genetic information from the RNA code into the amino acid sequence of proteins (*translation*), or
(3) some *post-translational* process such as the activation or inactivation of existing enzymes?

The participation of gene activation and repression in phytochrome-mediated photomorphogenesis was suggested as early as 1964 by Mohr and collaborators (*e.g.*, Hock and Mohr, 1964; Mohr, 1972). However, mainly on the basis of experiments indicating rapid effects of phytochrome on ion transport processes and related phenomena (Marmé, 1977), there has been a strong tendency to deny the involvement of gene transcription and translation in photomorphogenetic control (*e.g.*, Smith, 1976; Smith, Billett and Giles, 1977). After many years of debate, the methodological progress in molecular biology has made a solution of this basic question feasible. There is now a solid body of experimental evidence that the regulation of mRNA formation and, concomitantly, the induction and repression of enzyme syntheses, are indeed essential links in the causal chain between phytochrome and developmental photoresponses.

In all cases of enzyme induction thoroughly investigated so far, the rise of enzyme activity has been shown to be related to induced *de novo* synthesis of the enzyme protein (Schopfer, 1977). This is true for short-lived enzymes such as phenylalanine ammonia-lyase (PAL, half-life about 3 h, Fig. 17.16) or physiologically stable enzymes such as β-amylase (half-life > 5 d, Fig. 17.17), both of which have been investigated intensively in the cotyledons of the mustard seedling (Tong and Schopfer, 1976; Sharma and Schopfer, 1982). As an example, Fig. 17.18 shows the kinetics of deuterium incorporation into β-amylase protein in seedlings grown on deuterium oxide. The labeling of an enzyme with deuterated amino acids synthesized in the presence of deuterium oxide can be conveniently determined from the position and profile shape of isopycnically banded enzyme after density gradient centrifugation (Fig. 17.19).

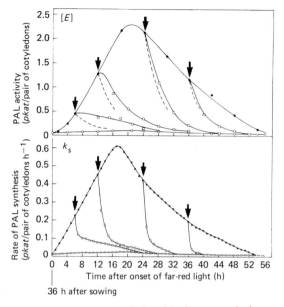

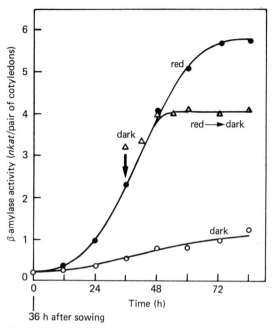

Fig. 17.16 Time course of phenylalanine ammonia-lyase (PAL) induction by phytochrome (continuous far-red light) in the cotyledons of mustard seedlings. *Top*, rise and fall of enzyme level, $[E]$, in continuous light (●) and darkness (○). At the time points indicated by *arrows*, the phytochrome signal was canceled by a 5-min pulse of 756 nm light ($\varphi_{756} < 0.001$) followed by darkness (□). From the decrease of $[E]$ in the virtual absence of enzyme synthesis (*e.g.*, after 36 h of light), a half-life of about 3 h (dashed lines) can be estimated. A similar half-life has been calculated from the kinetics of deuterium incorporation into PAL protein in light and darkness (Tong and Schopfer, 1976). *Bottom*, rise and fall of the rate of PAL synthesis, k_s, in the presence and absence of the phytochrome signal. The curves are calculated from the kinetics of $[E]$ (*top*) according to $k_s = d[E]/dt + k_d[E]$, where $k_d = \ln 2/\text{half-life} = 0.23\,\text{h}^{-1} = $ first-order rate constant for PAL degradation. k_s passes through a maximum about 6 h before $[E]$ reaches its highest value. After removal of P_{fr}, k_s drops with a half-life of less than 1 h. This provides an estimate for the lifetime of the longest-lived intermediate of the signal transduction chain, presumably the PAL-mRNA. The rise and fall kinetics of $[E]$ can be quantitatively explained by a temporally displaced rise and fall of k_s. The dynamics of k_s in continuous far-red light reflect the interaction of a virtually constant P_{fr} signal with a changing temporal pattern of competence. (After Acton et al., 1980.)

Fig. 17.17 Time course of β-amylase induction by phytochrome (continuous red light from sowing) in the cotyledons of mustard seedlings. Besides the different temporal pattern of competence, β-amylase induction differs from PAL induction in two decisive points: (1) The enzyme level continues to rise unchanged for about 12 h when the phytochrome signal is canceled by a 5-min pulse of 756 nm light followed by darkness. Hence, the longest-lived intermediate of the signal transduction chain must survive for this period of time. The implications of this result will be dealt with in a following section. (2) There is no indication of enzyme decay after cessation of phytochrome action in darkness. Labeling experiments with deuterium have shown that there is no detectable turnover of enzyme in light and darkness. (After Sharma and Schopfer, 1982.)

If photomorphogenesis involves a significant increase in the synthesis and translation of mRNAs, this should also be detectable on the level of ribosomal activity. Figure 17.20 shows that phytochrome indeed has a strong effect on the polysomal profiles obtained by density gradient centrifugation of the cytosolic and the membrane fraction of mustard cotyledon homogenates. Since both the fraction of polysomes (dimers up to about 10-mers) as well as the fraction of monomers and ribosomal subunits

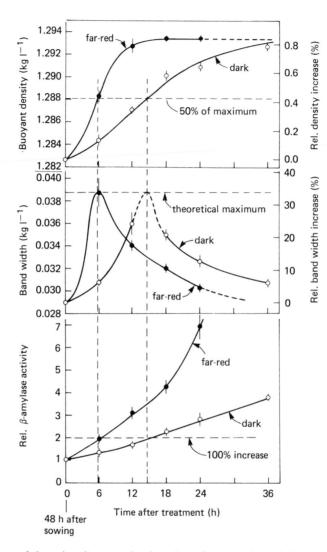

Fig. 17.18 Time course of deuterium incorporation into β-amylase protein in darkness and under the influence of phytochrome in the cotyledons of mustard seedlings. After 48 h growth in darkness on H_2O, the seedlings were transferred to 70% D_2O and either irradiated with continuous far-red light or kept further in darkness. Incorporation of deuterium (*via* deuterated amino acids) can be estimated, after isopycnic density gradient centrifugation, either from the band shift (*upper panel*) or from the transient band-width increase (*middle panel*) compared with the unlabeled control enzyme (*see* Fig. 17.19). Comparison of the enzyme activity kinetics (*lower panel*) with the overall density shift and band-width changes of the enzyme reveals a close correspondence of activity changes and appearance of density-labeled enzyme molecules. A doubling of enzyme activity (after about 6 and 15 h, respectively) coincides with a half-maximal band shift and a maximal band-width increase, *i.e.*, with the characteristics of a 1:1 mixture of pre-existing and newly synthesized enzyme molecules. (After Sharma and Schopfer, 1982.)

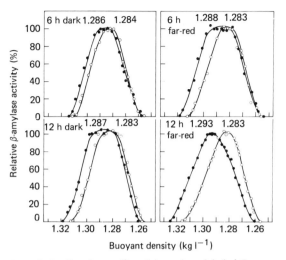

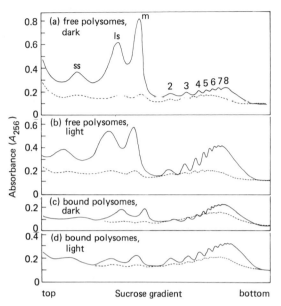

Fig. 17.19 Density profiles of deuterium-labeled β-amylase extracted from the cotyledons of mustard seedlings kept on D₂O in darkness (*left*) or continuous far-red light (*right*) for either 6 h (*top*) or 12 h (*bottom*). See Fig. 17.18 for further experimental details. The enzyme profiles were determined after fractionation on RbCl density gradients, spun to equilibrium at 176 000 × g and are normalized to the buoyant density of the marker enzyme β-galactosidase. The analysis of such profiles with respect to peak densities (indicated by numbers: kg l⁻¹) and band widths at half-peak height reveals the time course of appearance of newly synthesized β-amylase molecules on the background of pre-existing enzyme. (After Sharma and Schopfer, 1982.)

Fig. 17.20 Distribution of ribosomal material from dark-grown and light-grown mustard cotyledons on density gradients. The seedlings were either grown in darkness for 48 h or treated with 12 h of far-red light after 36 h growth in darkness. ER-bound polysomes were isolated from the membrane fraction using the detergent Triton X-100. Both ribosomal fractions were centrifuged on linear sucrose density gradients (10–60% w/w) for 4 h at 182 000 × g. The profiles were scanned photometrically. For comparison, the dashed lines indicate the profiles from 36-h-old dark-grown cotyledons. *ss*, small ribosomal subunit; *ls*, large ribosomal subunit; *m*, monomer; the numbers denote dimers, trimers, etc. (After Mösinger and Schopfer, 1983.)

can be quantitatively resolved, the ratio of active to inactive ribosomal material (with respect to protein synthesis) can be assessed from such gradient profiles. Figure 17.21 shows the kinetics of ribosomal activation in the cytosolic fraction by continuous far-red light. It is obvious that phytochrome can induce a very rapid (lag-phase < 30 min) and strong (doubling after about 1 h) increase of the relative amount of polysomes at the expense of inactive ribosomal material. The conclusion that the induced polysome formation is brought about by a rapid phytochrome-mediated increase in the supply of translatable mRNAs is supported by the fact that cordycepin, a potent inhibitor of mRNA synthesis, completely blocks the light effect on polysome formation. Moreover, labeling experiments with radioactive uridine have shown that the radioactivity

is preferentially incorporated into the polysome fraction rather than into the ribosomal monomers and subunits during the first 30 min of the light-mediated rise of polysomes and cannot, therefore, be attributed to newly synthesized ribosomal RNA (Mösinger and Schopfer, 1983). The drastic increase of polysomes shown in Fig. 17.21 indicates that the induction of new mRNAs in the light must quantitatively dominate the repression of dark-specific mRNAs, at least in an organ such as the cotyledon of the mustard seedling.

The next step toward the gene has been pioneered by Hahlbrock and collaborators who used parsley cell cultures to demonstrate the light-induced, co-

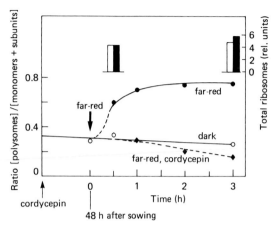

Fig. 17.21 Time course of ribosome activation by phytochrome (continuous far-red light) in the cotyledons of mustard seedlings. The ratio active *vs.* inactive ribosomal material, = amount of polysomes/amounts of monomers + subunits (cytosolic fraction), was determined from the respective peak areas of profiles such as shown in Fig. 17.20. The total amount of ribosomal material in darkness (*open bars*) and far-red light (*solid bars*) is hardly changed by light during the first 3 h of irradiation. Cordycepin $(0.5\,g\,l^{-1})$, an inhibitor of mRNA synthesis, abolishes the light effect (*dashed curve*). (After Mösinger and Schopfer, 1983.)

ordinated increase of the rates of PAL and chalcon synthase synthesis and the activity of the corresponding mRNAs in an *in vitro* translation assay (Schröder, Kreuzaler, Schäfer and Hahlbrock, 1979). Since then, similar results have been obtained for a variety of other phytochrome-controlled proteins, including the small subunit of ribulose bisphosphate carboxylase and the light-harvesting chlorophyll protein of the thylakoid membrane of barley and duckweed, respectively (Apel, 1979; Tobin, 1981). The synthesis of protochlorophyllide oxidoreductase in barley leaves is repressed by phytochrome. This effect can be attributed to a repression of translatable mRNA coding for the enzyme (Apel, 1981).

The demonstration of an increased level of mRNA activity, as revealed by *in vitro* translation, is not conclusive evidence for an increased gene transcription since there are possibly additional controlling steps in the processing and transport of mRNA which cannot be excluded by this assay. It is there-

fore of significance that Link (1982) demonstrated phytochrome control of the total amount of a mRNA sequence, coding for a 35 000-dalton plastid protein (of unknown function) in mustard cotyledons, using cloned DNA fragments as hybridization probes. However, it can be argued that even this experimental approach does not provide final proof for transcription control since control through mRNA degradation is still a theoretical possibility. This objection can only be ruled out by determining directly the synthesis of primary gene transcripts.

The synthesis of rRNA provides at present the only system in eukaryotic plants in which the *in vivo* transcription of genes of known function can be studied directly by measuring a primary transcript. The DNA sequences coding for the various RNAs of cytoplasmic ribosomes represent defined segments of a transcription unit (operon) producing a high-molecular-weight precursor rRNA which gives rise to the mature rRNAs by a sequence of specific cleavage steps. This precursor rRNA (2.5 Mdalton) can be regarded as a direct gene transcript. The high degree of amplification of the rRNA operon (in the range of 10^3-10^4 copies per nucleus) and the specific size of the rRNA precursor molecule greatly facilitate the detection and quantification of this particular gene product by electrophoretic techniques. It has previously been shown that the accumulation of cytoplasmic rRNAs in the cotyledons of mustard seedlings can be increased through phytochrome on the basis of a constant amount of DNA (Thien and Schopfer, 1975). Of course this effect, which is characterized by a relatively long lag-phase of 6 h, can hardly be related to specific photomorphogenetic responses. It may, however, be important for the adjustment of the capacity of protein synthesis during light-mediated development of the cotyledons. The unique properties of the rRNA precursor suggested the exploration of this system for a critical test of the gene regulation hypothesis. As expected, labeling experiments with radioactive uridine in the cotyledons of mustard seedlings demonstrated a rather rapid turnover of the rRNA precursor pool, leading to an adjustment of constant specific radioactivity about 60 min after application of the label. However, the investigation of rRNA precursor labeling before reaching the steady state provided clear evidence for a significantly enhanced rate of synthesis in far-red light-grown cotyledons as compared with the dark control. Figure 17.22 shows a typical experiment for a labeling period of 15 min, at the end of which the difference in rRNA precursor synthesis between dark-grown and light-grown cotyledons is optimally displayed by differential labeling. More detailed kinetics revealed a two-fold stimulation of the rate of rRNA precursor synthesis after 12 h of irradiation

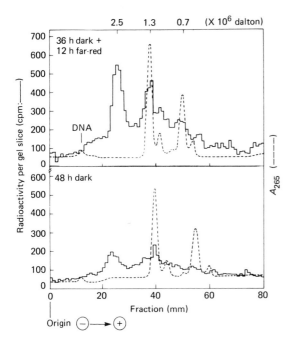

Fig. 17.22 Stimulation of rRNA precursor labeling by phytochrome (12 h continuous far-red light) in the cotyledons of 48-h-old mustard seedlings. After application of radioactive uridine for 15 min, total nucleic acids were extracted and subjected to polyacrylamide gel electrophoresis. The distribution of radioactivity and UV absorbance (marking the molecular weight of RNA species) are indicated by the *stepped* curves and the *dashed* curves, respectively. The amount of rRNA precursor (2.5 Mdalton) is too small to be detectable in the absorbance profiles. However, this fraction is predominantly labeled after the short period of uridine incorporation. Far-red light induces a strong increase in the amount of radioactive rRNA precursor while the uptake of labeled uridine and its incorporation into the UTP pool is the same as in the dark control. The peaks at 1.3 and 0.7 Mdalton represent the mature cytoplasmic rRNAs. (After Thien and Schopfer, 1982.)

with far-red light, exactly what is needed to account for the increased accumulation of mature cytoplasmic rRNA under these conditions (Thien and Schopfer, 1982). This result shows that the phytochrome effect on rRNA accumulation is in fact brought about by a change in the rate of transcription of the rRNA operon while the subsequent processing of the transcript is not influenced.

This brief survey of recent achievements in the molecular analysis of photomorphogenesis strongly supports the thesis that gene transcription indeed plays a causal role in the elicitation of developmental photoresponses. It is probably only a matter of time before the efforts presently under way in various laboratories applying cloning and hybridization techniques to genes of specific enzymes provide definite proof of the transcriptional control also in those cases where phytochrome regulation has so far been traced back only to the level of the amount or *in vitro* activity of the responsible mRNA. However, we should be aware that even a complete uncovering of the biochemical events elicited by the phytochrome signal at the gene level will hardly solve the mystery of how the striking *specificity* in time and space of phytochrome-mediated gene activation and repression can eventually be understood in molecular terms. This latter problem has not advanced beyond the level of description even during the 30 years of photomorphogenetic research.

Although the dominant role of transcriptionally mediated control in photomorphogenesis is hardly deniable, it would be premature to exclude post-transcriptional and post-translational mechanisms completely. An example warranting caution in this respect is the photoregulation of cell lengthening. It has been found, for instance, that the response of hypocotyl growth to light can be extremely fast (lag-phase of the order of seconds or minutes; *e.g.*, Cosgrove and Green, 1981). The low hysteresis reasonably excludes the causal involvement of the gene regulation pathway in this response. On the contrary, there is some evidence that control of cell lengthening by light (phytochrome or 'cryptochrome') involves a control of cell wall loosening through proton excretion (Pike and Richardson, 1977; Cosgrove and Green, 1981). Thus, the mechanism used in the photocontrol of cell growth by extension demonstrates striking similarities with the ion transport model currently discussed in auxin-mediated growth responses (*see* Chapter 1).

In this respect, growth by cell extension might be related more closely to those photoresponses in which the involvement of ion fluxes as the underlying cause of turgor-driven leaf movements and similar non-developmental processes is clearly established (*see* Marmé, 1977). The same may be true in seed germination which can be regarded as a special type of growth response of the young embryo at the transition from quiescence to active life (Nabors and Lang, 1971).

Hysteresis of the signal transduction chain of phytochrome

The tremendous progress in elucidation of the biochemical mechanism of single photoresponses on the enzyme level should not distract our attention from the fact that we still have no idea how the phytochrome signal is transduced from the effector molecule P_{fr} to the transcription of competent genes. Our knowledge is presently limited to indirect evidence, obtained by physiological means, which does, however, shed light on some very interesting features of this black box. For example, it might be asked whether the switching on (or off) of the signal transduction chain of a particular photoresponse is virtually free of hysteresis, in the sense that signal transduction is rapid (and rapidly reversible) and therefore obviously involves only short-lived elements, such as P_{fr} itself. Alternatively, a hysteretic control system, characterized by a certain degree of back-lash, would imply at least one rather stable element which can accumulate to some extent and support the terminal response even after cessation of the phytochrome signal. The term 'transmitter' has been used to signify such a stable intermediate of the signal transduction chain (Schopfer and Plachy, 1973).

The experimental approach for investigating the hysteresis of signal transduction chains is based on the fact that the initial phytochrome signal can be instantaneously canceled by irradiating the plant with long-wavelength far-red light, establishing a φ_λ close to zero. Measurement of the subsequent response kinetics provides information on the lifetime of the most stable intermediate. A comparison of the phytochrome switch-off kinetics shown in Figs 17.16 and 17.17 clearly indicates that PAL and β-amylase in mustard cotyledons display large dif-

ferences in this respect. From the after-effect of phytochrome, following the elimination of P_{fr}, we have to assume the involvement of a long-lived transmitter element in the signal chain leading to β-amylase production, but not in the signal chain leading to the production of the lyase. Detailed time-course studies of the far-red reversibility in responses such as β-amylase induction have led to the surprising result that transmitters can accumulate under the influence of phytochrome even before the cotyledons become competent actually to produce the terminal response. An example for this kind of response is illustrated in Fig. 17.23. The 'coupling point', operationally defined as the start of escape from full photoreversibility, marks the time-point when the photomorphogenetic signal no longer remains restricted to the photoconvertible P_{fr} molecules but is about to pass on to a metabolic intermediate which persists even if P_{fr} is removed by the reversing light treatment. In the case of urate oxidase of mustard cotyledons this event ('coupling') can take place about 26 h before the onset of the terminal response (starting point). Similarly, it can be concluded from the escape kinetics of Fig. 17.23 that the accumulation of the transmitter reaches completion about 20 h after the surpassing of the starting point. In the case of peroxidase induction in mustard cotyledons the competence for transmitter formation ceases even before the starting point, leading to the seemingly paradoxical situation that the response is insensitive to P_{fr} if the plants are treated with light only from the starting point onwards (Schopfer and Plachy, 1973).

In terms of the transmitter concept, the situation for urate oxidase induction (Fig. 17.23) can be described as follows. The cotyledons pass through a 40–50 h period of competence toward P_{fr} with respect to transmitter formation. Shifted by about 26 h, a period of competence toward the transmitter with respect to urate oxidase formation follows. Except for the deviating temporal patterns, very similar conclusions can be reached from investigation of numerous other enzyme responses of the mustard cotyledon (Table 17.5). It is immediately evident that, depending on the extent of overlap of the two competence periods, there may be a considerable time-gap between the initial action of P_{fr} and the photomorphogenetic expression of the P_{fr} signal. Thus, a response may have escaped phytochrome control, and may therefore have nothing to

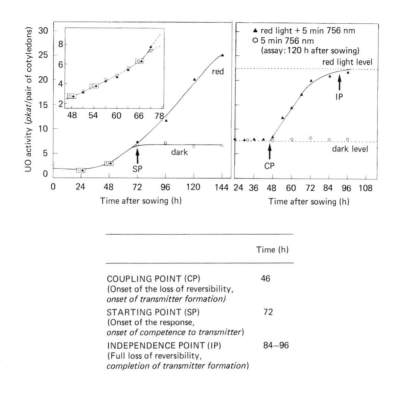

Fig. 17.23 Induction of urate oxidase (UO) by phytochrome in the cotyledons of mustard seedlings as revealed by the kinetics of response realization and reversibility escape. *Left*, response kinetics in continuous red light from the time of sowing. The starting point (*SP*) can be estimated at about 72 h after sowing. The insert shows the region close to the starting point. *Middle*, kinetics of the loss of photoreversibility ('escape kinetics', ▲). After irradiation with red light from the time of sowing, the seedlings were treated with a 5-min pulse of 756-nm light at the times indicated on the abscissa and kept in darkness until the end-point determination of enzyme activity (120 h after sowing). The dashed lines indicate the enzyme levels in continuous red light or complete darkness. The controls (dark-grown seedlings treated with a single 756-nm pulse, ○) indicate that the reversing light alone was ineffective. Thus, the take-off point of the escape kinetics at about 46 h indicates the onset of the loss of photoreversibility (= coupling point, *CP*). Full loss of reversibility is reached at about 84–96 h (independence point, *IP*). *Right*, survey of the three decisive time points and their interpretation in terms of the signal transduction chain of urate oxidase induction. The terms 'starting point' and 'coupling point' have been introduced by Steinitz *et al.* (1976) in a theoretical analysis of the primary process of the anthocyanin response in mustard. (After Hong and Schopfer, 1981.)

do with the actual state of the phytochrome system, at the time when it is actually observed.

The biochemical nature of transmitters is presently still unresolved. Likely candidates are, of course, mRNAs for particular proteins, at least in those responses that involve enzyme synthesis. In the case of the mustard peroxidase response, however, it was found that the enzyme protein can be deuterium-labeled if the label is applied only during the period of transmitter formation before reaching the starting point. Moreover, inhibitors of protein synthesis are effective in suppressing the response if applied before transmitter formation, whereas the rise of enzyme activity after the starting point proves to be independent of an inhibition of protein synthesis (Schopfer and Plachy, unpublished). These results

Table 17.5 Temporal specificity of coupling points (onset of escape from reversibility) and starting points (onset of induced activity increase) of some phytochrome-mediated enzyme responses in the cotyledons of mustard seedlings.

	Coupling point (h)	Starting point (h)	Time interval (h)	References
Ascorbate oxidase	14	27	13	Lercari, Drumm and Mohr (1980)
Ribulosebisphosphate carboxylase	15	42	27	Frosch *et al.* (1977)
Glutathione reductase	18	48	30	Mohr (1978)
Allantoinase	24	47	23	Hong and Schopfer (1981)
Phenylalanine ammonia-lyase	27	27	0	Frosch *et al.* (1977)
β-Amylase	35	46	11	Sharma and Schopfer (1982)
Glycolate oxidase	36	60–66	~27	Schopfer, unpublished
Hydroxypyruvate reductase	45	60	15	Schopfer, unpublished
Urate oxidase	46	72	26	Hong and Schopfer (1981)
Peroxidase	~60	~96	~36	Schopfer and Plachy (1973)

suggest that the transmitter may be identical to the inactive peroxidase protein which is made under the influence of phytochrome only before the starting point, but acquires enzymatic activity only afterwards.

Multiple action mechanisms of phytochrome?

In the early days of photomorphogenetic research it was generally assumed, at least implicitly, that P_{fr} participates in a single, general reaction mechanism which could in principle be discovered by analyzing any one of the many phytochrome effects. The increasing knowledge of the enormous heterogeneity of photoresponses even within the same organism, together with the apparent inability to elucidate the so-called 'primary reaction mechanism' of phytochrome, have prompted the serious questions of whether a uniform primary reaction $(P_{fr} + X \rightarrow P_{fr}X)$ and whether an, at least initially, uniform signal transduction mechanism of phytochrome really exist. Indeed, an alternative concept, advocated by Mohr (e.g., Drumm and Mohr, 1974; Frosch, Drumm and Mohr, 1977), proposes a multiplicity of primary reactions of P_{fr} $(P_{fr} + X_1 \rightarrow P_{fr}X_1, \ldots, P_{fr} + X_n \rightarrow P_{fr}X_n)$ to account for the conspicuous qualitative differences in the dynamics of individual photoresponses. Although a definite decision on this fundamental problem is not yet in sight, there is some relevant information from studying the hysteresis of signal transduction in

the presence of full competence (*i.e.*, after the starting point of a response).

First, it must be recognized that the high degree of specificity of individual photoresponses even within a single cell makes it necessary to assume that the signal transduction chains, leading, *e.g.*, to the level of enzyme synthesis, must separate somewhere in the pathway of signal transduction. This assumption is even stronger with regard to the dualism of non-developmental, rapid phytochrome responses (related to ion fluxes rather than to gene regulation; *see* Marmé, 1977) and the developmental phytochrome responses discussed in this chapter. However, the decisive question is: at which point of the signal transduction chain does specificity come into play?

It can be concluded that probably most features of response specificity, including the direction (promotion *vs.* suppression), the timing of coupling and starting points, or the quantitative differences in $[P_{fr}]$/response functions, can be attributed to the later steps of the signal transduction chain, *i.e.*, where the phytochrome signal interacts with cellular competences at or beyond the gene level. This 'coupling reaction' is operationally defined as the first step in creating a phytochrome signal which persists for some time even after removal of P_{fr} in a reversion experiment. The time for reaching this point can be as short as 1 min in the case of the flowering response of short-day plants (Frédéricq, 1964). In the mustard seedling, coupling times of about 2 min and more than 5 min have been reported for different responses occurring at the same time. Obviously, the phytochrome signal can be irreversibly transduced to the responding cell functions *via* different coupling reactions (Jabben and Mohr, 1975).

It has been stressed repeatedly that co-operative (all-or-none) responses such as the lipoxygenase repression (*see* Fig. 17.5) and non-co-operative (graded) responses such as anthocyanin accumulation (*see* Fig. 17.6) are difficult to reconcile even with a common primary reaction of P_{fr} (*e.g.*, Mohr and Oelze-Karow, 1976). This interpretation is true, however, only if one disregards the possibility that the initial steps of the signal transduction chain up to the coupling reaction are non-hysteretically synchronized with all changes of the active pool of P_{fr}.

In conclusion, there is unequivocal evidence for a multiplicity of signal transduction chains subsequent to the coupling reaction, which marks the irreversible manifestation of the phytochrome signal. This most probably takes place when the signal meets the level of cellular competence at the gene level. The steps prior to coupling, including the primary reaction of P_{fr}, may be mechanistically uniform if they are precisely linked to all changes of active P_{fr}. Regardless of the ultimate answer to this question we have to accept the fact that P_{fr} represents an unspecific triggering agent of an independently running pattern of competences which decides on the specific expression and integration of photoresponses. This unavoidably leads to the disillusioning conclusion that even a full uncovering of the primary reaction(s) of P_{fr} will hardly provide significant information on the basic problems of photomorphogenesis. With respect to disclosure of the mechanisms of pattern *specification*, the formulation of appropriate questions that could be experimentally explored is still awaited.

Further reading

Hendricks, S. B. and Borthwick, H. A. (1965). The physiological functions of phytochrome. In *Chemistry and Biochemistry of Plant Pigments*, ed. T. W. Goodwin, pp. 405–36, Academic Press, London.

Mohr, H. (1972). *Lectures on Photomorphogenesis*. Springer, Berlin, Heidelberg, New York.

Mohr, H. (1978). Pattern specification and realization in photomorphogenesis. *Bot. Mag. Tokyo Special Issue* **1**, 199–217.

Schopfer, P. (1977). Phytochrome control of enzymes. *Ann. Rev. Plant Physiol.* **28**, 223–52.

Senger, H., ed. (1980). *The Blue Light Syndrome*. Springer, Berlin, Heidelberg, New York.

Senger, H. and Briggs, W. R. (1981). The blue light receptor(s): primary reactions and subsequent metabolic changes. *Photochem. Photobiol. Rev.* **6**, 1–38.

Smith, H. ed. (1976). *Light and Plant Development*. Butterworth, London.

Shropshire, W. and Mohr, H., eds (1982). *Photomorphogenesis. Encyclopedia of Plant Physiology*, New Series, vol. 16, Springer, Berlin, Heidelberg, New York.

References

Acton, G. J., Fischer, W. and Schopfer, P. (1980). Lag-phase and rate of synthesis in phytochrome-mediated induction of phenylalanine ammonia-lyase in mustard (*Sinapis alba* L.) cotyledons. *Planta* **150**, 53–7.

Apel, K. (1979). Phytochrome-induced appearance of mRNA activity for the apoprotein of the light-harvesting chlorophyll *a/b* protein of barley (*Hordeum vulgare*). *Eur. J. Biochem.* **97**, 183–8.

Apel, K. (1981). The protochlorophyllide holochrome of barley (*Hordeum vulgare* L.). Phytochrome-induced decrease of translatable mRNA coding for the NADPH:protochlorphyllide oxidoreductase. *Eur. J. Biochem.* **120**, 89–93.

Beggs, C. J., Holmes, M. G., Jabben, M. and Schäfer, E. (1980). Action spectra for the inhibition of hypocotyl growth by continuous irradiation in light- and dark-grown *Sinapis alba* L. seedlings. *Plant Physiol.* **66**, 615–8.

Bergman, K. *et al.* (1969). *Phycomyces. Bacteriol. Rev.* **33**, 99–157.

Biswal, U. C., Kasemir, H. and Mohr, H. (1982). Phytochrome control of degreening of attached cotyledons and primary leaves of mustard (*Sinapis alba* L.) seedlings. *Photochem. Photobiol.* **35**, 237–41.

Borthwick, H. A., Hendricks, S. B., Parker, M. W., Toole, E. H. and Toole, V. K. (1952). A reversible photoreaction controlling seed germination. *Proc. Nat. Acad. Sci. USA* **38**, 662–6.

Chelm, B. K., Hallick, R. B. and Gray, P. W. (1979). Transcription program of the chloroplast genome of *Euglena gracilis* during chloroplast development. *Proc. Nat. Acad. Sci. USA* **76**, 2258–62.

Cosgrove, D. J. and Green, P. B. (1981). Rapid suppression of growth by blue light. Biophysical mechanism of action. *Plant Physiol.* **68**, 1447–53.

De Greef, J., Butler, W. L., Roth, T. F. and Frédéricq, H. (1971). Control of senescence in *Marchantia* by phytochrome. *Plant Physiol.* **48**, 407–12.

Delbrück, M., Katzir, A. and Presti, D. (1976). Responses of *Phycomyces* indicating optical excitation of the lowest triplet state of riboflavin. *Proc. Nat. Acad. Sci. USA* **73**, 1969–73.

Drumm, H. and Mohr, H. (1974). The dose response curve in phytochrome-mediated anthocyanin synthesis in the mustard seedling. *Photochem. Photobiol.* **20**, 151–7.

Drumm, H. and Mohr, H. (1978). The mode of interaction between blue (UV) light photoreceptor and phytochrome in anthocyanin formation of the *Sorghum* seedling. *Photochem. Photobiol.* **27**, 241–8.

Fondeville, J. C., Borthwick, H. A. and Hendricks, S. B. (1966). Leaflet movement of *Mimosa pudica* L. indicative of phytochrome action. *Planta* **69**, 357–64.

Frédéricq, H. (1964). Conditions determining effects of far-red and red irradiations on flowering response of *Pharbitis nil. Plant Physiol.* **39**, 812–16.

Frosch, S., Drumm, H. and Mohr, H. (1977). Regulation of enzyme levels by phytochrome in mustard cotyledons: multiple mechanisms? *Planta* **136**, 181–6.

Furuya, M. (1978). Photocontrol of developmental processes in fern gametophytes. *Bot. Mag. Tokyo Special Issue* **1**, 219–42.

Gressel, J. and Galun, E. (1970). Sporulation in 'Trichoderma': a model system with analogies to flowering. In *Cellular and Molecular Aspects of Floral Induction*, ed. G. Bernier, pp. 152–70, Longman, London.

Hanke, J., Hartmann, K. M. and Mohr, H. (1969). Die Wirkung von 'Störlicht' auf die Blütenbildung von *Sinapis alba* L. *Planta* **86**, 235–49.

Hartmann, K. M. (1967). Ein Wirkungsspektrum der Photomorphogenese unter Hochenergiebedingungen und seine Interpretation auf der Basis des Phytochroms (Hypokotylwachstumshemmung bei *Lactuca sativa* L.) *Z. Naturforsch.* **22b**, 1172–5.

Hock, B. and Mohr, H. (1964). Die Regulation der O_2-Aufnahme von Senfkeimlingen (*Sinapis alba* L.) durch Licht. *Planta* **61**, 209–28.

Hong, Y.-N. and Schopfer, P. (1981). Control by phytochrome of urate oxidate and allantoinase activities during peroxisome development in the cotyledons of mustard (*Sinapis alba* L.) seedlings. *Planta* **152**, 325–35.

Jabben, M. and Mohr, H. (1975). Stimulation of the Shibata shift by phytochrome in the cotyledons of the mustard seedlings *Sinapis alba* L. *Photochem. Photobiol.* **22**, 55–8.

Jose, A. M. and Vince-Prue, D. (1977). Light-induced changes in the photoresponses of plant stems; the loss of a high irradiance response to far-red light. *Planta* **135**, 95–100.

Kasperbauer, M. J. (1971). Spectral distribution of light in a tobacco canopy and effects of end-of-day light quality on growth and development. *Plant Physiol.* **47**, 775–8.

Kasperbauer, M. J. and Peaslee, D. E. (1973). Morphology and photosynthetic efficiency of tobacco leaves that received end-of-day red or far-red light during development. *Plant Physiol.* **52**, 440–2.

Lange, H., Bienger, I. and Mohr, H. (1967). Eine neue Beweisführung für die Hypothese einer differentiellen Genaktivierung durch Phytochrom 730. *Planta* **76**, 359–66.

Lange, H., Shropshire, W. and Mohr, H. (1971). An analysis of phytochrome-mediated anthocyanin synthesis. *Plant Physiol.* **47**, 649–55.

Lercari, B., Drumm, H. and Mohr, H. (1980). Phytochrome-mediated induction of ascorbate oxidase as affected by light treatments. In *Photoreceptors and Plant Development*, ed. J. DeGreef, pp. 317–28, Antwerpen University Press, Antwerp.

Link, G. (1982). Phytochrome control of plastid mRNA in mustard (*Sinapis alba* L.) *Planta* **154**, 81–6.

Marmé, D. (1977). Phytochrome: membranes as possible sites of primary action. *Ann. Rev. Plant Physiol.* **28**, 173–98.

Mohr, H. (1982). An introduction to photomorphogenesis for the general reader. In *Photomorphogenesis*, eds W. Shropshire and H. Mohr, *Encyclopedia of Plant Physiology*, New Series, vol. 16A. Springer, Berlin–Heidelberg–New York, pp. 24–38.

Mohr, H., Drumm, H., Schmidt, R. and Steinitz, B. (1979). The effect of light pretreatments on phytochrome-mediated induction of anthocyanin and phenylalanine ammonia-lyase. *Planta* **146**, 369–76.

Mohr, H. and Oelze-Karow, H. (1976). Phytochrome action as a threshold phenomenon. In *Light and Plant Development*, ed. H. Smith, pp. 257–84, Butterworth, London.

Morgan, D. C., O'Brien, T. and Smith, H. (1980). Rapid photomodulation of stem extension in light-grown *Sinapis alba* L. Studies on kinetics, site of perception and photoreceptor. *Planta* **150**, 95–101.

Mösinger, E., Bolze, K. and Schopfer, P. (1982). Evidence against the involvement of DNA synthesis in phytochrome-mediated photomorphogenesis. *Planta* **155**, 133–9.

Mösinger, E. and Schopfer, P. (1983). Polysome assembly and RNA synthesis during phytochrome-mediated photomorphogenesis in mustard (*Sinapis alba* L.) cotyledons. *Planta* **158**, 501–11.

Nabors, M. W. and Lang, A. (1971). The growth physics and water relations of red-light-induced germination in lettuce seeds. I. Embryos germinating in osmoticum. *Planta* **101**, 1–25.

Pike, C. S. and Richardson, A. E. (1977). Phytochrome-controlled hydrogen ion excretion by *Avena* coleoptiles. *Plant Physiol.* **59**, 615–17.

Pratt, L. H., Coleman, R. A. and Mackenzie, J. M. (1976). Immunological visualization of phytochrome. In

Light and Plant Development, ed. H. Smith, pp. 75–94, Butterworth, London.

Rau, W. (1980). Blue light-induced carotenoid biosynthesis in microorganisms. In *The Blue Light Syndrome*, ed. H. Senger, pp. 283–98, Springer, Berlin–Heidelberg–New York.

Sachs, J. (1887). *Vorlesungen über Pflanzen-Physiologie*. 2nd edn, Engelmann, Leipzig.

Sanchez, R. (1971). Phytochrome involvement in the control of leaf shape of *Taraxacum officinale* L. *Experientia* **27**, 1234–7.

Schäfer, E. (1976). The 'high irradiance reaction'. In *Light and Plant Development*, ed. H. Smith, pp. 45–59, Butterworth, London.

Schäfer, E. and Mohr, H. (1974). Irradiance dependency of the phytochrome system in cotyledons of mustard (*Sinapis alba* L.). *J. Math. Biol.* **1**, 9–15.

Schäfer, E., Schmidt, R. and Mohr, H. (1973). Comparative measurements of phytochrome in cotyledons and hypocotyl hook of mustard (*Sinapis alba* L.). *Photochem. Photobiol.* **18**, 331–4.

Schiff, J. A. (1980). Blue light and the photocontrol of chloroplast development in *Euglena*. In *The Blue Light Syndrome*, ed. H. Senger, pp. 495–511, Springer, Berlin–Heidelberg–New York.

Schopfer, P. and Plachy, C. (1973). Die organspezifische Photodetermination der Entwicklung von Peroxidaseaktivität im Senfkeimling (*Sinapis alba* L.) durch Phytochrom. *Z. Naturforsch.* **28c**, 296–301.

Schröder, J., Kreuzaler, R., Schäfer, E. and Hahlbrock, K. (1979). Concomitant induction of phenylalanine ammonia-lyase and flavanone synthase mRNAs in irradiated plant cells. *J. Biol. Chem.* **254**, 57-65.

Sharma, R. and Schopfer, P. (1982). Sequential control of phytochrome-mediated synthesis *de novo* of β-amylase in the cotyledons of mustard (*Sinapis alba* L.) seedlings. *Planta* **155**, 183–9.

Shropshire, W. (1972). Photoinduced parental control of seed germination and the spectral quality of solar radiation. *Solar Energy* **14**, 99-105.

Smith, H. (1976). Phytochrome-mediated assembly of polyribosomes in etiolated bean leaves. Evidence for post-transcriptional regulation of development. *Eur. J. Biochem.* **65**, 161–70.

Smith, H., Billett, E. E. and Giles, A. B. (1977). The photocontrol of gene expression in higher plants. In *Regulation of Enzyme Synthesis and Activity in Higher Plants*, ed. H. Smith, pp. 93–127, Academic Press, London.

Steinitz, B., Drumm, H. and Mohr, H. (1976). The appearance of competence for phytochrome-mediated anthocyanin synthesis in the cotyledons of *Sinapis alba* L. *Planta* **130**, 23–31.

Tan, K. K. (1978). Light-induced fungal development. In *The Filamentous Fungi*, eds J. E. Smith and D. R. Berry, vol. 3, pp. 334–57, Wiley, New York.

Thien, W. and Schopfer, P. (1975). Control by phytochrome of cytoplasmic and plastid rRNA accumulation in cotyledons of mustard seedlings in the absence of photosynthesis. *Plant Physiol.* **56**, 660–4.

Thien, W. and Schopfer, P. (1982). Control by phytochrome of cytoplasmic precursor rRNA synthesis in the cotyledons of mustard seedlings. *Plant Physiol.* **69**, 1156–60.

Thomas, R. L. and Jen, J. J. (1975). Phytochrome-mediated carotenoids biosynthesis in ripening tomatoes. *Plant Physiol.* **56**, 452–3.

Tobin, E. (1981). Phytochrome-mediated regulation of messenger RNAs for the small subunit of ribulose 1,5-bisphosphate carboxylase and the light-harvesting chlorophyll *a*/*b*-protein in *Lemna gibba*. *Plant Molec. Biol.* **1**, 35–51.

Tong, W.-F. and Schopfer, P. (1976). Phytochrome-mediated *de novo* synthesis of phenylalanine ammonia-lyase. An approach using pre-induced mustard seedlings. *Proc. Nat. Acad. Sci. USA* **73**, 4017–21.

Tso, T. C., Kasperbauer, M. J. and Sorokin, T. P. (1970). Effect of photoperiod and end-of-day light quality on alkaloids and phenolic compounds of tobacco. *Plant Physiol.* **45**, 330–3.

Tucker, D. J. (1977). The effects of far-red light on lateral bud outgrowth in decapitated tomato plants and the associated changes in the levels of auxin and abscisic acid. *Plant Sci. Lett.* **8**, 339–44.

Wagné, C. (1965). The distribution of the light effect from irradiated to non-irradiated parts of grass leaves. *Physiol. Plant.* **18**, 1001–6.

Williams, B. J., Pellett, N. E. and Klein, R. M. (1972). Phytochrome control of growth cessation and initiation of cold acclimation in selected woody plants. *Plant Physiol.* **50**, 262–5.

Juvenility, photoperiodism and vernalization

18

B Thomas and D Vince-Prue

Introduction

The development of most plants is seasonal, with flowering and other developmental processes occurring at particular times of the year. The most obvious explanation for such seasonal effects is that they are immediate responses to environmental conditions such as temperature, water supply or light intensity. In fact seasonal changes in plants are rarely controlled in this way, but rather are dependent on the use of certain characteristics of the environment as inductive signals to set in motion complex patterns of development.

Photoperiodism and vernalization are two major mechanisms underlying such seasonal responses and the words themselves indicate the characteristics of the inductive environmental signals. 'Photoperiodism' combines the Greek words for light and length of time. 'Vernalization' is a translation of *yarovizatzya* which combines the Russian root for spring with a suffix meaning 'to make', after the observation that exposure to winter low temperatures causes winter strains of cereals to behave like spring strains. As initial definitions, then, photoperiodism is a response to the length of day and vernalization is an effect on flowering brought about by exposure to cold. These two mechanisms and their combinations clearly enable plants to make seasonal adaptations. Such adaptations can be highly advantageous. Some examples include synchrony of reproduction within a population to increase outbreeding; synchrony of reproduction with favorable environments such as availability of water, favorable temperatures or a high daily light integral; avoidance of reproduction in unfavorable environments such as water stress or low temperature; and avoidance of the damaging effects of below-freezing temperatures by leaf abscission and the development of winter dormancy in response to autumnal short days.

With a mechanism sensitive to daylength, seasonal changes can be precisely controlled. The absolute daylength changes with latitude; it ranges from 6 to 19 h at 60°, from 9 to 15.5 h at 45° and from 10 to 14 h at 30°. The average rates of change for April and August are 40 min wk^{-1} at 60°, 25 min wk^{-1} at 45° and 12 min wk^{-1} at 30°. Thus a plant would need to discriminate only within these limits to time its response within a week. The greatest precision of timing is seen in the tropics where daylength changes are small; the flowering time of many tropical plants is, nevertheless, strongly influenced by daylength. Photoperiod also controls an enormous range of other developmental responses which include bud dormancy, the formation of storage organs, leaf development, stem elongation, and germination. Only flowering will be considered in this chapter but, with respect at least to the perception of daylength, the evidence points to a common mechanism.

Vernalization plays an additional role in seasonal flowering and is found in regions with winter temperatures that are unfavorable for growth. A requirement for exposure to cold for several weeks will prevent flowering until a winter has been experienced by the plant. Germination in late summer would then not lead to immediate flowering in favorable photoperiods, and vernalization becomes a device to ensure that the winter has passed before the onset of reproduction. Daylength alone is an ambiguous signal during spring and autumn and combination with a vernalization requirement is one way of ensuring that flowering occurs in spring and not at the onset of winter.

Juvenility is important in controlling when, in its life, a plant changes from vegetative to reproductive growth. Thus it appears to be a device to ensure that flowering does not occur until a plant is large enough to support the energetic demands of seed production.

Juvenility, vernalization and photoperiodism are, therefore, three important processes which determine when plants flower with respect to both ontogeny and season. Vernalization and photoperiodism also allow plants to occupy temporal ecological niches and are major determinants of the latitudinal distribution of both crop plants and wild species.

In this chapter we shall consider the perception of environmental signals, the physiology and biochemistry of vernalization and photoperiodism (with reference only to flowering) and the physiology of the transition from the juvenile to the adult phase of development. The initiation and development of flowers at receptive shoot meristems in response to vernalization or photoperiodic induction are not discussed.

Juvenility

Characteristics of the juvenile condition

Juvenility is the name usually given to the early phase of growth during which flowering cannot be induced by any treatment. In many plants the transition to reproduction occurs after the juvenile phase has been completed, without exposure to any particular stimulus; in others, appropriate environmental treatments are necessary. Sometimes the term 'ripe-to-flower' is used for plants which have completed the juvenile phase but have not yet experienced the correct conditions for flowering. The duration of the juvenile period varies widely. In most woody plants it lasts for several years and can be as long as 30–40 years in some forest trees. In contrast, the juvenile phase in herbaceous plants is usually quite short, rarely being more than a few

days or weeks in duration. In extreme cases, there is no juvenile phase since flower primordia are found in the seed.

The accepted criterion of the juvenile state is that the plant does not have the ability to form flowers. Associated with the transition from juvenile to mature with respect to the ability to form flowers, there are changes in physiological and morphological characters such as leaf shape and thickness, leaf retention, thorniness, phyllotaxis, pigment content and rooting capacity (Table 18.1). Morphological

Table 18.1   Juvenile and adult characters of English ivy (*Hedera helix* L.) (adapted from Wareing and Frydman, 1976).

Juvenile characters	Adult characters
Three- or five-lobed, palmate leaves	Entire, ovate leaves
Alternate phyllotaxy	Spiral phyllotaxy
Anthocyanin pigmentation of young leaves and stem	No anthocyanin pigmentation
Stems pubescent	Stems glabrous
Climbing and plagiotropic growth habit	Orthotropic growth habit
Shoots show unlimited growth and lack terminal buds	Shoots show limited growth terminated by buds with scales
Absence of flowering	Presence of flowers

differences between juvenile and adult are usually much less distinct in herbaceous plants, although they do occur. It seems that no particular morphological character is associated with flowering ability and the different characteristics do not necessarily develop simultaneously. In English ivy (*Hedera helix*), for example, reversion to the juvenile growth habit can be induced by the application of gibberellic acid (GA_3) but the various morphological characters are observed at different GA_3 doses; the ability to flower is lost at low doses ($\leqslant 1.0 \, \mu g$ per plant) and progressively higher doses are needed to induce aerial rootlet formation, anthocyanin, alternate phyllotaxy, and juvenile leaf shape (Rogler and Hackett, 1975a).

The transition from juvenile to adult appears to be associated with changes or events in the apex; the base of the plant may remain in the juvenile condition even after the transition to maturity has occurred at the apical meristem. Cuttings taken from the base of ivy develop into juvenile plants whereas those from the tip develop into mature plants; when taken from the transition zone between apex and base, shoots show a mixture of adult and juvenile characters but, during growth, become fully juvenile or mature (Wareing and Frydman, 1976). The basal parts of deciduous trees often retain their leaves during winter, which is a juvenile character. Similarly, shoots from the more basal parts of trees are often easier to root. The flowering behavior of grafts in which scions originated either from the top or base of a mature tree (Table 18.2) supports the concept that the basal part retains the juvenile condition.

Table 18.2   Comparison of grafts originating from different parts of an old, flowering specimen of *Betula verrucosa* (from Longman, 1976).

Origin of scions	Mean extension growth of main shoot for years 1 and 2 (m)	Flower initiation	
		Year 1	Year 2
Basal sprouts	1.13	none	none
Flowering crown	0.86	few	many

Once the phase change to maturity has occurred, the adult condition appears to be highly stable and reversion is infrequent, except by sexual or apomictic reproduction when the juvenile condition is regenerated (Hackett, 1980). Once attained, the adult condition appears to be propagated through all subsequent cell divisions. The relative stability of the two phases can be seen in the maintenance of the characteristic juvenile or adult growth habit when plants are propagated from cuttings taken from juvenile or adult parts of the tree.

The physiology of phase change in plants has often been divided into two interrelated problems: the cause of the transition from the juvenile to the adult phase of development, and the basis of the stable differences that are transmitted through cell divisions.

The attainment of maturity

The transition to maturity is not necessarily accompanied by flowering. Even a fully mature tree may not flower if it is growing very vigorously and irregular flowering is common in many trees (*e.g.*, biennial bearing in apples). Lack of flowering in itself, therefore, does not necessarily indicate the juvenile condition. Nevertheless, attempts to use vegetative characters as markers for phase change have not been successful since no single character has been shown invariably to be associated with the transition to flowering. For these reasons, the interpretation of experimental results is not always straightforward.

Size appears to be more important than chronological age in the transition to maturity and, in general, conditions that promote growth reduce the duration of the juvenile period. In contrast, treatments that slow growth and promote flowering in mature trees, such as dwarfing rootstocks and growth retardants, have little effect or may increase the length of the juvenile phase. The optimum conditions appear to be those which allow rapid growth to the minimum size appropriate to the species and then to expose plants to the appropriate flower-inducing treatment (Hackett, 1980).

While it is generally agreed that size is important, it is not yet clear what component of size is critical for the attainment of maturity. Two quite different views have been examined. One is that a plant of sufficient size transmits one or more signals to the apex. The second is that the apical meristem behaves independently and undergoes the phase transition at a particular time. There is some evidence for both views.

If maturity at the apex is determined only by signals received from a sufficiently large plant, grafting juvenile apices on to mature bearing trees should induce them to become mature. However, when Robinson and Wareing (1969) grafted juvenile or nearly mature apices of Japanese larch on to mature bearing trees, only one (out of 56 juvenile grafts) flowered in the first year, while 36 of the nearly mature grafts flowered. It was concluded that the juvenile-to-adult transition could not be determined by signals reaching the apex from the rest of the plant, but rather that the apex behaved more or less autonomously. However, other grafting experiments have produced less clear-cut results, or contradictory ones (Doorenbos, 1965). Few experiments have ensured that there is a flow of materials from the mature tree into the grafted part and it is evident that this is important. For example, 1–1½ year old seedlings of mango flowered when approach grafted to mature trees only when the receptor was defoliated and the mature tree was girdled below the graft union—treatments which would have increased the movement of assimilates into the receptor (Singh, 1959). These experiments indicated that juvenility is a property of the plant rather than the apex. A similar conclusion might be drawn from experiments with ivy, where grafting of mature scions on to juvenile stocks caused reversion to juvenility of the adult partner; in this case, too, the importance of treatments ensuring translocation from the presumptive donor to the receptor has been demonstrated (Clark and Hackett, 1980). However, in ivy, it has not been found possible to induce flowering by grafting to mature plants, even when the direction of translocation is manipulated. Thus both the apex itself and the supply of substances to it seem to be important for phase change in this plant. Conflicting results have been obtained with herbaceous plants. In some cases juvenility appears to be a property of the apex whereas, in others, it appears to be a property of the leaves.

Nutritional and hormonal factors The apex, of course, receives a supply of both nutritional and hormonal factors from the rest of the plant. None of these has yet been shown to be essential for, or inhibitory to, the transition from juvenile to adult, although factors such as low light and high temperature, which may reduce the supply of carbohydrate to the apex, cause rejuvenation or prolong the juvenile phase in several plants. There is, however, a considerable body of evidence, albeit circumstantial, which suggests that endogenous gibberellins may play a role in phase transition. The application of gibberellins has been shown to accelerate flowering in members of the *Taxodiaceae* and *Cupressaceae* and (with less polar gibberellins such as GA_7) in the *Pinaceae* (Pharis, Ross, Wample and Owens, 1976). In contrast, exogenously applied GA_3 rejuvenates ivy, *Acacia melanoxylon, Citrus* and *Prunus*. The application of Amo 1618, which reduced the amount of GA-like substances in the apex, was not, however, able to cause a phase change to maturity in ivy (Frydman and Wareing, 1974). However, ancymidol, another inhibitor of gibberellin biosynthesis, prevented the spontaneous reversion to the juvenile state in low light (Rogler and Hackett, 1975b). From these and other results,

such as those from the grafting experiments in ivy, it has been concluded that, although many factors cause reversion to the juvenile condition, only time and/or size can bring about the transition to maturity. Nevertheless, it is abundantly clear that precocious flowering can be induced in some cases by hormone applications, by gibberellins in certain conifers and by ethephon in mango (Chacko, Kohli, Dore Swamy and Randhawa, 1976). However, the gibberellin application did not cause a true phase change, for plants reverted to the non-flowering condition when the treatments were stopped. A true phase change results in a metastable adult condition which is perpetuated through subsequent mitoses.

The presence of roots close to the apex has sometimes been suggested to be important in maintaining juvenility. In ivy, the aerial adventitious roots produced at the nodes of juvenile plants have a high concentration of extractable gibberellins and removing the roots decreased the amount of GA-like substances in the shoot apices (Wareing and Frydman, 1976). Since exogenous GA_3 rejuvenates ivy, the possibility exists that a minimum distance from roots as a source of gibberellins may be necessary before the transition to maturity can occur in the shoot apical meristem. If such a simple hypothesis were true, grafting to adult plants should be more effective in inducing phase change than it appears to be. There are, nevertheless, a number of observations which indicate that roots, or substances from roots, are important; for example, the chilling and water-stressing of roots have both been reported to accelerate maturity.

Role of the apex The size of the apex itself has also been invoked as a possible cause of the transition to maturity since older apices are usually larger (perhaps as a consequence of a better nutritional supply from larger plants). In ivy, mature apices have a much larger meristematic area than juvenile ones. However, recent experiments with ivy plants that were induced to revert to the juvenile phase with GA_3 have shown that phyllotaxis changes to the juvenile configuration before there is any significant reduction in the area of the apical meristem (Hackett and Srinivasani, 1983). These results seem to exclude apex size, *per se*, as a determining factor in phase change, at least during rejuvenation.

A somewhat different concept is that the change to maturity is determined by events in the apex but that the apex 'age' is dependent in some manner on the number of cell divisions that have occurred since germination—a kind of programmed ageing. The evidence supporting this concept is fragmentary. Apical parts of *Ribes nigrum* were removed and re-rooted several times so that the plants never attained the minimum height or leaf number necessary for photoperiodic responsiveness (Robinson and Wareing, 1969). After three to four such decapitations, the rooted cuttings flowered in response to photoperiodic induction. This was interpreted as indicating that phase change is intrinsic to the apex and may be related to the number of cell divisions that have occurred there. However, the cumulative height and leaf number were greater than the minimum for intact seedlings.

The molecular basis of phase change is still unknown. In different experiments, results have indicated that influences from the rest of the plant are important in maintaining juvenility but the apical cells themselves appear to have some autonomy and may determine the transition to maturity. Perhaps both occur. One or more juvenile factors from the rest of the plant may have to be removed before phase change can occur but the transition itself may require some additional factor(s) intrinsic to the apex. Using hybridization-competition techniques, Rogler and Dahmus (1974) demonstrated qualitative differences in RNA between juvenile and adult apices of ivy, indicating that phase change may involve differences in the rate of transcription of specific genes. Some DNA sequences transcribed in the adult phase appeared to be inactive in the juvenile phase.

Maintenance of phase

Once the phase change to maturity has occurred at the apex, the new state is stable and is perpetuated through subsequent mitotic divisions. Hardly anything is known about the underlying physiology of such metastable states but there are parallels in other systems, such as in habituated callus. Tobacco pith explants, which normally require cytokinins for growth, can be habituated by, for example, growing them on cytokinin (Meins and Binns, 1979). When the hormone is removed, the habituated phenotype produces sufficient hormone to sustain its own growth and so a new 'state' of cells with different biochemical properties has been formed; this new state is metastable and persists through many cell generations. It is, however, reversible. Gradients in the ability to flower and in some vegetative characters on explants also indicate that juvenility *versus* maturity are conditions perpetuated within individual cells and that they persist after removal of cells from the parent plant. Similarly, stem calluses originating from juvenile and mature sections of the stem in ivy consistently develop different characteristics. 'Juvenile' callus produces shoots which can be detached and re-rooted to form new plants, whereas mature callus forms embryos. Despite the differences in their growth characteristics, the newly organized meristems are in both cases juvenile (Banks, 1979). In this context it is interesting to note that the cytokinin-habituated state in tobacco always reverts to the non-habituated state when adventitiously regenerated.

Photoperiodism

Characteristics of photoperiodic responses

The biological effects of daylength, now accepted quite readily as a major controlling factor in plant development, remained largely unsuspected until the beginning of this century. At about this time Hans Klebs and Julien Tournois proposed independently that the duration rather than the quantity of light in the daily cycle was important in regulating flowering in certain plants. However, it was Garner and Allard (1923) who established that response to daylength was a major controlling influence in flowering. They were led to their discoveries by observations of two species of plants being used in breeding programs at the time. It was noted that certain varieties of soya beans (*Glycine max.*), in particular a late maturing strain 'Biloxi', tended to flower at the same time irrespective of planting dates. This can be seen in Fig. 18.1, showing how the time taken to flower in four strains of soya bean changes with the date of sowing. Note that with 'Mandarin', an early maturing variety, the time of sowing had little effect on the time to flowering, whereas with 'Biloxi', the later in the summer the beans were sown the shorter the time to flowering. 'Peking' and 'Tokyo', two varieties which were intermediate as far as maturing was concerned,

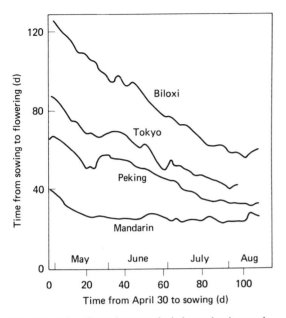

Fig. 18.1 The effect of progressively later planting on the length of the vegetative period before flowering in different varieties of soya beans. (Adapted from Garner and Allard 1920.)

showed responses intermediate between 'Mandarin' and 'Biloxi'.

The other lead was that a variety of tobacco, *Nicotiana tabacum* Maryland Mammoth, grew to a prodigious size out of doors in summer in Washington DC but failed to flower. On the other hand, when it was propagated vegetatively in the greenhouse during the winter even very small plants flowered rapidly. After eliminating temperature and light intensity as causal factors, Garner and Allard investigated the effect of daylength by transferring Biloxi soya beans and Maryland Mammoth tobacco to a darkened ventilated hut for part of the daily light period during the summer and comparing the response with controls growing in the open. In both species flowering was greatly accelerated by the artificially shortened days. They proposed that the answer to the two problems was the same: the varieties in question would flower only if the duration of the daily light period was sufficiently short—a condition met only in late September for the Biloxi soya beans (earlier for the other varieties of soya beans tested) and a little earlier for Maryland

Mammoth tobacco plants. Garner and Allard subsequently showed that not only flowering, but also branching, root growth, pigment formation, abscission, tuberization and dormancy were all under photoperiodic control. Of these, control of flowering, besides playing a central part in the discovery of photoperiodism, is still the most studied manifestation of the phenomenon in plants.

Some definitions The study of how daylength regulates flowering has, as in all specialist studies, evolved its own terminology. *Photoperiodism* may be defined descriptively as a response to the length of day which enables an organism to adapt to seasonal changes in its environment. A more useful operational definition has been given by Hillman (1969), who defined photoperiodism as a response to the *timing* of light and darkness. It is implicit in this definition that total light energy above a low level is relatively unimportant and so is the relative length of the light and dark period as a controlling factor. It is rather the timing of the exposures to light and darkness or, alternatively, the transitions between light and darkness which are crucial. The term *photoperiod* is often used for the daily light period and *skotoperiod* occasionally used for the period of darkness.

Photoperiodic response groups Plants in which flowering responds to daylength are usually classed in one of two groups.
(1) Short-day plants (SDP). These are plants which flower only in, or in which flowering is accelerated by, daylengths shorter than a particular duration known as the '*critical daylength*'.
(2) Long-day plants (LDP). These are plants which flower only in, or in which flowering is accelerated by, daylengths which exceed a particular duration, again the '*critical daylength*'. Plants in which flowering is not affected by daylength are known as day-neutral plants (DNP) or indifferent plants.

It is important to appreciate that the difference between SDP and LDP lies in the relationship between daylengths which cause flowering and the critical daylength, rather than in the value of the critical daylength itself. For example, *Xanthium strumarium*, an SDP, has a critical daylength of about $15\frac{1}{2}$ h, whereas the LDP *Hyoscyamus niger* has a critical daylength of 11 h. In terms of photoperiodism, then, a 'short-day' (SD) is less than the critical daylength and a 'long-day' (LD) is longer than the critical daylength and, in comparing different species, an SD for one may be longer than an LD for another. Under natural conditions the mechanism in LDP effectively delays flowering until the critical daylength of the plant in question is exceeded. It can therefore identify the lengthening days of spring and early summer and is associated with the summer flowering habit. Autumn flowering plants, on the other hand, are likely to be SDP as the days are progressively shortening at this time. Daylength alone is an ambiguous signal during the approximately equinoxial periods of spring and autumn, and, as mentioned earlier, vernalization is one way of ensuring that flowering occurs in the spring. An alternative strategy is used by plants with dual photoperiodic requirements. For example *Bryophyllum*, a long-short-day-plant (LSDP), requires long days followed by short days and would be expected to flower in the late summer or autumn but not in the spring. Conversely *Trifolium repens*, a short-long-day-plant (SLDP), flowers in the late spring or early summer. Two further variations on the basic mechanisms are provided by intermediate day plants which flower between narrow daylength limits (Lang, 1965), for example certain cultivars of sugar cane, and ambiphotoperiodic plants which flower in long or short days but not at intermediate daylengths, for example *Madia elegans*.

The capacity for diverse expression is a striking feature of photoperiodism and can be observed in species such as *Chenopodium rubrum* where strains from different latitudes show a wide range of photoperiodic requirements (Cumming, 1969). Such variations, however, do not alter the fact that the plants must be capable of measuring daylength even though the length of the day to which they respond may vary in the different strains.

Although flowering is dependent upon changes at the apex, it is the leaves that are the sensing organs for daylength. It follows, therefore, that the *perception* of daylength is a separate process from the transition to flowering at the apex, or *evocation*, and that the *transmission of a floral stimulus* is required. Perception, transmission and floral evocation can be regarded as three independent components in a chain of events which lead to flowering. This is not the whole story, however, as a number of plants need to experience favorable photoperiods for only a few days: they will then flower even if subsequent-

ly maintained in an unfavorable daylength. Here, favorable photoperiods cause a change of state in the leaf which is called *photoperiodic induction* and is regarded as the primary effect of a favorable daylength in the control of flowering.

Measurement of flowering It is not easy to carry out experiments on the photoperiodic control of flowering. As we have seen, plants usually have a juvenile stage and so have to reach an appropriate stage of development before they are photoperiodically sensitive. The time between experimental treatments and the appearance of flowers can be considerable and plants have to be maintained in a uniform environment during this period. Such experiments inevitably involve a great deal of labor and expense. It is not surprising, therefore, that the tendency has been to seek and use the least-complicated flowering systems for experimental purposes. The simplest systems are those in which the plants show a qualitative response to daylength, in which the experimental plants are small and where only a few cycles or preferably a single cycle at the appropriate daylength causes flowering. Plants in which these criteria are satisfied are relatively few and the bulk of research on the photoperiodic mechanisms has tended to concentrate on them.

The biochemical changes in the leaf which are caused by induction are not known and it is impossible to distinguish between induced and non-induced leaves except for the fact that the former cause flowering. The necessity to use some floral index as an assay for induction poses additional problems. The quantitative relationship between the degree of induction and the floral response may be adequately represented by the percentage of treated plants which flower, as used by Cumming (1969) in *Chenopodium rubrum*. Large numbers of plants are, however, required to obtain accurate estimates of flowering in this way and many other workers choose to use some other index of induction which may be a function of floral morphology or inflorescence structure. For example, in *Kalanchoë blossfeldiana* short days cause an increase in the order of branching, resulting in a logarithmic relationship between the number of short days and the number of flower buds (Fig. 18.2a). In *Glycine*, on the other hand, there is a linear relationship between the number of nodes bearing flower buds and the number of SD between the second and seventh cycles (Fig. 18.2b). The switch from vegetative to floral development in the terminal bud of *Pharbitis* when placed in SD limits the potential number of buds available to form flowers, resulting in the saturation of the response, in some cases after only one SD (Fig. 18.2a). In some species the stage of floral development rather than the number of flowers is the criterion employed. In *Xanthium*, arbitrary

be controlled closely. A more subtle problem is that floral development may itself be photoperiodically sensitive. In *Pharbitis nil*, for example, if plants are given relatively few SD cycles and then transferred to LD, many flower buds fail to develop and eventually abort, especially at high temperatures (Takimoto, 1969). To minimize such effects it is desirable to determine as early as possible, whether flower buds have formed, usually by dissection. Early

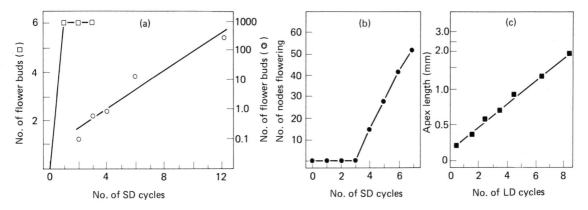

Fig. 18.2   Relationships between flowering response and number of inductive photoperiodic cycles in: (a) *Pharbitis nil* (cv. Violet) □, *Kalanchoë blossfeldiana* ○; (b) *Glycine max* ●; and (c) *Lolium temulentum* ■. (Adapted from Imamura, Muramatsu, Kitajo and Takimoto (1966) – *Pharbitis*; Schwabe (1956) – *Kalanchoe*; Blaney and Hamner (1957) – *Glycine*; and Evans (1960) – *Lolium*.)

stages in the transition from vegetative to floral are used to denote a scale of 1 to 6 (Salisbury, 1969). Using this system the mean floral stage is a linear function of time following induction which, in turn, means that the floral stage at some arbitrary time following induction can be used as a measure of the rate of floral development. Another morphological indicator of floral stage is employed for some graminaceous plants where apex length can be related to the degree of induction, such as in the LDP *Lolium temulentum* (Fig. 18.2c).

Whatever index is used to measure flowering, a common feature is that a period of time between the treatment and the assay is necessary for development at the apex to proceed to a point where flowers can be identified. For this reason floral assays are all prone to interference by environmental factors which affect flower bud development. Temperature and total light energy may be important and have to

assays have the added bonus of decreasing the length of experiments, allowing a more productive use of time and costly facilities.

Despite the formidable obstacles to an empirical approach to the study of photoperiodic induction it is nevertheless clear that good quantitative estimates can be obtained by using flowering indices. It is also clear that although floral evocation at a given apex is a qualitative change in development, induction is often quantitatively related to some aspect of the photoperiodic stimulus such as the number of inductive cycles. These relationships have been investigated extensively and have provided some understanding of the physiological events leading to induction.

Perception and induction   It is widely accepted that the leaves are the site of daylength perception even though the response is expressed at the apex. This was originally established by experiments in which leaves and apices on the same plant were treated simultaneously with different photoperiods. Under these conditions the flowering response was determined by the photoperiod presented to the leaves while the treatment of the site of the response had little effect. Knott (1934), for example, demonstrated that exposure of only the leaves of the LDP *Spinacia oleracea* to long photoperiods caused the initiation of floral primordia at the terminal growing point.

Confirmation of the special role of leaves is readily obtained by grafting those which have received inductive treatments to receptor plants, which subsequently flower although maintained in non-inductive conditions throughout the experiment. The most spectacular of these results have been observed in *Perilla* where Zeevaart (1958) has demonstrated that induced leaves can be grafted and regrafted sequentially to a number of receptor plants and retain their capacity to cause flowering.

Aerial tissues other than leaves are usually unable to respond to daylength in the absence of leaves, as can be demonstrated with excised tissues of *Streptocarpus nobilis* growing *in vitro* (Handro, 1977). In these experiments, leaf disks generated neoformed flowers directly in inductive conditions but stem and petiole explants first produced vegetative buds and hence leaves followed by subsequent flowering of the new shoots. Roots and stem tissue do not appear to be involved in the perception of daylength although in some cases roots are required for the expression of the response. The stage of development of a leaf is an important factor in its ability to perceive and respond to daylength. Such responses are generally associated with young expanding leaves although exceptions to this are not uncommon. There are some reports that vegetative buds in which the leaves are not yet expanded may also be sites of perception and, in the case of young *Pharbitis nil* seedlings, the responses of such buds may override the signals from expanded leaves or cotyledons (Gressel, Zilberstein, Porath and Arzee, 1980).

Leaves appear to be autonomous as far as transition to the induced state is concerned; for example, leaves of *Perilla* can be induced after being detached from the parent plants and prior to grafting to the receptor plants which subsequently flower (Zeevaart, 1969). Perception and induction are therefore closely linked within the leaf and largely independent of the final response at the apex. This allows consideration of the processes involved with perception and induction separately from those concerned with translating induction in the leaf into flowering at the apex.

Induction is a somewhat variable condition assuming different characteristics in different species. Two contrasting types of behavior are shown by the SDP *Perilla* and *Xanthium* where the nature of the inductive state has been considered in some detail. In *Perilla* induction is restricted entirely to leaves which have received inductive photoperiods (Zeevaart, 1969). In *Xanthium*, on the other hand, grafting a shoot induced in SD to a plant maintained in LD not only causes flowering but also leads to induction in the leaves of receptor plants which have received only LD. In *Xanthium*, Hamner and Bonner (1938) have shown that induced leaves are effective in causing flowering only when they are relatively young, their effectiveness declining with age. Once leaves are induced, however, the plants will continue to flower in LD as the young leaves are continually being induced by the older leaves. Self-perpetuation is not a characteristic of induction in *Perilla* but here individual leaves remain induced for several months and well into senescence.

In many plants there is a requirement for a number of favorable cycles before the leaves are induced. In this type of plant it is sometimes possible to show that non-inductive cycles interpolated into a series of inductive cycles do not alter the total number of inductive cycles needed to cause flowering. 'Fractional induction', as this is called, can often be shown to persist over several weeks of non-inductive cycles, as for example in *Plantago lanceolata*, a LDP (Hillman, 1969). Even in plants which are induced with a single or few inductive cycles a fractional or 'steady-state induction' can be observed. *Xanthium* will flower and subsequently fruit after only one SD but the rate of flower and fruit development is much slower than if a number of inductive cycles is given before transferring plants to LD. Although induction represents a qualitative change within the leaf there is clearly a quantitative dimension to this process. How the quantitative aspects of induction originate, however, is unlikely to be determined until the biochemical changes which constitute the change to the inductive state are identified.

Responses to light and darkness

The dark period Under natural conditions day and night lengths are mutually dependent, together forming a 24-h cycle of light and darkness. Under these conditions daylength, nightlength and the relative lengths of the light and dark periods are linked in a strict relationship. Any of these could therefore provide the required information for a functional photoperiodic mechanism. A valid question for early workers was, therefore, whether any of these features of the daily light/dark cycle was alone the controlling factor in the photoperiodic regulation of floral induction. In order to answer this question it was necessary to vary independently the length of light and dark periods, at the same time generating cycles of different lengths. In this way Hamner and Bonner (1938) were able to show almost half a century ago that photoperiodic timekeeping in the SDP *Xanthium* is essentially a question of dark period measurement. In this plant flowering occurred only if the dark period was in excess of 8.5 h irrespective of the relative duration of light and darkness in the experimental cycle or the length of the cycle itself. Cycles of 4 h light and 8 h dark did not cause flowering even though the light period was considerably less than the critical daylength of 15.5 h, whereas flowering occurred in 16 h light and 32 h dark, where the duration of the light period exceeded the critical daylength. Experiments of this sort established that neither the length of the light period nor the relative length of the light and dark periods was the controlling factor. The decisive importance of the length of the dark period has since been confirmed by a number of investigators in a range of species, although cycle length may also be important in some cases (*see* Fig. 18.4, p. 420). It has been found that in plants in which the duration of darkness is decisive, the dark period must exceed a certain minimal length to allow flowering in SDP or be shorter than a critical value to allow flowering in LDP. In normal daily cycles of light and darkness this results in an apparent critical daylength which, if exceeded, allows flowering in LDP and prevents it in SDP, although it is, in fact, the critical night length (CNL) which is important. This is not to say that the daily light period is inert or that all active processes concerned with photoperiodism proceed in darkness and are merely prevented by light, as we shall shortly see. Nevertheless, under natural conditions the time from dusk to dawn is the prime determinant of induction whereas the effects of light can usually be interpreted as interactions with this fundamental component.

An important feature underlining the importance of the dark period is that it can be rendered ineffective by an interruption with a short light treatment or 'night break'. Exposures of only a few minutes will usually prevent flowering in SDP such as *Xanthium*, *Glycine* or *Pharbitis*, if given in the middle of an inductive dark period. Conversely, the induction of LDP such as *Hordeum* and *Hyoscyamus* can be brought about by short night breaks. However, one difference between SDP and LDP is that many of the latter show little response to night-break treatments. This may indicate that LDP and SDP do not simply represent mirror images of a common mechanism and this question will be discussed again later.

Early experiments using long nights of 14–16 h duration showed that night breaks were most effective if given near the middle of the dark period. Night breaks were thought to act by merely splitting the long night into two short and hence ineffective dark periods. This explanation has since been shown to be inadequate, as in *Pharbitis* for example, where Takimoto and Hamner (1964) have shown that a night break given after 8 h of a 48-h dark period effectively prevents induction even though the subsequent 40-h dark period would normally be inductive. In such cases the night-break response is a transient period of light sensitivity related in time to the beginning rather than the end of the dark period.

The considerable effect of small amounts of light given as night breaks has proved to be a useful tool in the study of photoperiodic perception. Action spectra can be constructed with little concern for the non-photoperiodic (*e.g.*, photosynthetic) effects of the light as the added energy is negligible compared with that used in the main light periods. The most noticeable feature of night-break action spectra is the similarity between those for the prevention of flowering in SDP such as *Glycine* and *Xanthium* and the promotion of flowering in LDP such as *Hordeum vulgare* and *Hyoscyamus niger*, indicating that the same pigment acts as photoreceptor in the two classes of plants (Borthwick, Hendricks and Parker, 1948). Night-break action spectra show a maximum effect at wavelengths between 600 and

660 nm, little effect above 720 nm, and a relatively small response in the blue part of the spectrum. Such spectra are consistent with light absorption by the red-absorbing form of phytochrome (P_r) (*see* Chapters 16 and 17) and the participation of this pigment has been confirmed by the observation that the effect of a red night break can be reversed by subsequent far-red light (*e.g.*, Downs, 1956). To date there is no evidence that any pigment other than phytochrome is involved in the action of low-energy night breaks.

A further indication of the importance of the dark period is the often relatively greater effect of temperature here as compared with the light period. For example, Hamner and Bonner (1938) found that a long dark period ceased to be inductive in *Xanthium* if the temperature remained at 4°C whereas there was little effect of varying the day temperature between 4°C and 40°C during the inductive cycles. Low night temperatures have also been observed to prevent the inhibitory effect of long dark periods and promote flowering in some LDP such as *Hyoscyamus* kept in SD (Lang, 1965).

The light period The capacity to respond to inductive or inhibitory dark periods may be modified or in some cases eliminated by appropriate manipulations of the accompanying photoperiod. The duration, irradiance and spectral distribution of light presented during the photoperiod can each modify the flowering response to some extent.

It is possible to demonstrate that, without a preceding photoperiod, a dark period is not effective for induction in SDP, nor for preventing induction in LDP. Thus, in the single-cycle SDP *Xanthium*, an inductive dark period is only effective if preceded by a minimum of 3 to 5 h of light (Salisbury, 1965) and, in the LDP *Hyoscyamus*, two consecutive short days are more inhibitory to flowering than 48 h darkness (Joustra, 1970). In single-cycle SDP there is apparently no maximum length of photoperiod beyond which a subsequent dark period ceases to induce flowering; *Pharbitis nil*, for example, can be placed in continuous light for several days and still respond to a single inductive dark period. In SDP which require several inductive cycles this is apparently not so. *Glycine max* shows a sharply defined CNL of 10 h which is independent of photoperiod lengths in experiments using seven inductive cycles. If the photoperiod accompanying a 16-h

(*i.e.*, inductive) dark period is varied, flowering occurs only in plants which have received between 4 and 18 h light in each cycle (Fig. 18.3). Hence, although length of the dark period is decisive in regulating flowering, its effect is strongly modified by the accompanying light period. An analogous situation can exist in LDP where, in *Silene armeria* for example, long dark periods are not inhibitory when coupled with photoperiods in excess of 14 h (Takimoto, 1955).

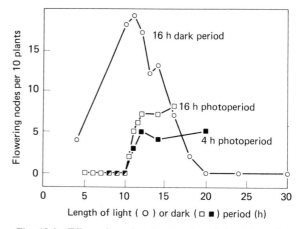

Fig. 18.3 Effect of varying the duration of the light and dark period on the flowering response of *Glycine max*. Plants received seven photoinductive cycles in which the length of only the light period accompanying a 16-h dark period or the dark period accompanying a 16-h or 4-h light period was varied. (Data of Hamner, 1940.)

The irradiance (fluence rate) of the light during the photoperiod can be important in so far as there appears to be a minimum requirement before the photoperiod functions as such. This may be very low, as in *Xanthium*, where Salisbury (1965) was able to produce a strong promoting effect with as little as ~0.02 W m^{-2}. Above the threshold values, increasing the irradiance had relatively little effect and in the *Xanthium* experiments there was no intensity dependence over the range ~20–55 W m^{-2}.

It has been suggested on several occasions that the requirement for light indicates a requirement for photosynthesis or photosynthetic products. This might explain why many SDP do not flower when photoperiods are very short or at low irradiance. Support for this suggestion comes from observations that a light requirement may be partially substituted for by applying sucrose to the leaves in several SDP and that the effectiveness of the light period may be prevented if CO_2 is withheld.

It would be surprising if there were no interaction between photosynthesis and flowering because of the fundamental role of photosynthesis in the plant. The question is whether photosynthetic products are directly involved in the photoperiodic control of induction. In *Chrysanthemum*, for example, Vince (1960) has shown that the irradiance of light given as part of inductive SD cycles is far more important than the previous photosynthetic history of the plant, indicating that products of the immediate light reactions during the photoperiodic cycles are more important than stored photosynthetic products. There are, however, some indications that such effects, while pointing to a strong interaction between photosynthesis and photoperiodic induction, are not essential features of the mechanism. In *Kalanchoë* very brief daily exposures to high-intensity light were found by Schwabe (1959) to allow induction although photosynthetically insignificant. Furthermore, in *Pharbitis nil*, Ogawa and King (1979) have demonstrated that dark-grown seedlings will flower if given five minutes of red light and simultaneously sprayed with benzyladenine, followed by an inductive dark period. In this latter case the plants do not have chlorophyll during the inductive process and hence photosynthetic effects are eliminated. Even here, however, there is an absolute requirement for light: controls with no red light never flower.

As mentioned earlier, the effectiveness of night breaks varies with wavelength, red being the most potent spectral region, which is consistent with the action of phytochrome. The importance of light quality during the main light period is more difficult to assess because prolonged light treatments with restricted spectral composition may result in photosynthetic differences and yield limited information as to the photoreceptors involved. The observation that an extension of a short main light period with low-irradiance light is effective in converting it to an LD as far as the plant is concerned has enabled the problems to be partially overcome. In such experiments red light is usually the most effective spectral region for SDP (*see later* for LDP) with blue and far-red being relatively without effect in most cases. It seems likely that the photoreceptor for continuous light is also phytochrome although the difficulties in interpreting action spectra, and the inability to test for red/far-red reversibility under such conditions, prevent definitive identification of the pigment concerned.

Photoperiodic mechanisms

The flowering of plants only in certain daylengths is the manifestation of precise and sensitive biochemical mechanisms which must not only be capable of distinguishing between light and darkness, but also be able to measure the duration of the dark phase in the daily cycle. These internal mechanisms must include at least a photoreceptor and a timekeeper coupled in some way to the control of the transition to the induced state which, in turn, determines whether flowering can take place. Most of the experimental work concerned with elucidating these internal mechanisms has been carried out in single-cycle SDP and the following consideration of photoperiodic mechanisms is based mainly on this work. Some alternative or additional mechanisms required to explain the responses of many LDP will also be discussed.

The measurement of time The observation that the length of the dark period, rather than the comparative length of light and darkness, is the decisive factor in induction must mean that the plant is capable of absolute rather than relative time measurement (*see also* Chapter 21). Two different types of mechanism have been postulated as the basis for the operation of the clock involved in the photoperiodic control of flowering.

The first hypothesis is that timing is a consequence of a series of unidirectional biochemical reactions beginning at the start of the dark period, each in turn proceeding to completion and leading, if not interrupted by light, to induction. The analogy between accumulation of the products of these reactions and the passage of sand through an hour-glass has led to the term 'hour-glass timing' for this type of mechanism.

The alternative hypothesis, which is supported by the bulk of available evidence, is that timing is related to the progress of a circadian rhythm in responsivity to light. Circadian rhythms are ubiquitous in nature and are often displayed as overt cycles of activity which, under certain constant environmental conditions, show a periodicity of approximately 24 h. Characteristic features and the basis of such rhythms are dealt with in Chapter 21 and here only the evidence for their involvement in photoperiodism will be discussed.

It is well established that the photoperiodic control of flowering is associated with rhythmic changes in the response to light and this can be demonstrated in a number of ways. For instance, if the SDP *Chenopodium rubrum* is treated by interrupting a long dark period with short night breaks at various times, a rhythm is observed in the flowering response (Fig. 18.4). Similar observations have been made on other SDP such as *Glycine max* and the LDP *Lolium temulentum*. *Xanthium*, on the other hand, is an exception as light interruptions have no effect if given between 24 and 48 h in a 48-h dark period (Salisbury and Ross, 1969). In this instance the response is characteristic of an hour-glass. Recent evidence indicates that the two patterns of

response need not necessarily be caused by separate timing mechanisms (Fig. 18.5). When dark-grown seedlings of *Pharbitis nil* are exposed to 24 h white light followed by 48 h in darkness at 25°C, there is a single maximum in the inhibitory response to a red-light interruption of the dark period as in *Xanthium*. If, on the other hand, the 24-h photoperiod is replaced with 5 min of red light, a rhythmic response to a second red pulse is obtained to give the type of pattern found in *Chenopodium*. The rhythm is still visible but reduced in the second circadian cycle if an intermediate photoperiod length of 6 h is used. As the timing mechanism is almost certainly the same in these treatments the apparent 'hour glass' must be caused by a rapid damping of the rhythmic flowering response following the long photoperiod.

The rhythmic nature of the flowering response can also be demonstrated by so-called resonance experi-

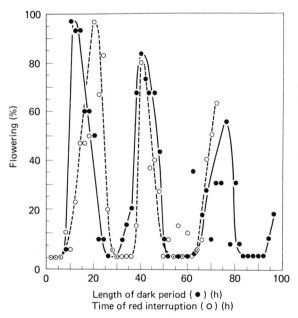

Fig. 18.4 Rhythm in light sensitivity in *Chenopodium rubrum*. Plants were given a dark period varying from 3 to 96 h or a 72-h dark period interrupted by 4 min of red light at different times. Plants were maintained in continuous incandescent light (~47 W m^{-2}) before and after the experimental dark period. (From Cumming, Hendricks and Borthwick, 1965.)

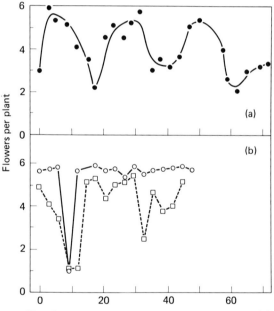

Fig. 18.5 Patterns of response to a red night-break in *Pharbitis nil* following a photoperiod of 5 min red light and simultaneous treatment with benzyladenine (●), 6 h white light (□) or 24 h white light (○). Length of the dark period was either 72 h (a) or 48 h (b). (Adapted from King, Schäfer, Thomas and Vince-Prue (1982) and authors' own data (unpublished).)

ments. In these, the flowering response changes rhythmically with increasing duration of the dark period which follows a short photoperiod. In Fig. 18.4 this rhythm in response to dark (or cycle) length in *Chenopodium* is plotted together with the rhythm in response to light interruptions of a long dark period. The two rhythms are similar, especially with respect to their amplitudes, although the peaks do not coincide during the first cycle. This may reflect differences in the stability of the rhythm under the different experimental protocols. Resonance experiments have produced similar results in both SDP and LDP, with some evidence that the phase of the rhythm is offset by 12 h in the latter. Other plants including *Xanthium* and *Kalanchoë* show no such changes in the magnitude of the flowering response with different cycle lengths although, as stated previously, this need not imply differences in the timing mechanism.

The phase of circadian rhythms is often found to be labile in that it can be advanced or delayed by light in a precise way depending upon the phase of the rhythm when light is presented. Therefore, if circadian rhythms are involved in the regulation of induction, light might be expected to have a dual action in photoperiodism. The first we have already seen in the rhythm of floral responsivity to short light treatments in *Chenopodium*. The second, phase-shifting effect can also be demonstrated in this species (Fig. 18.6). Phase shifting has also been shown in the floral rhythm in dark-grown seedlings of SDP *Pharbitis* and in the LDP *Hordeum vulgare*. A phase–response curve has so far only been constructed for *Chenopodium rubrum* and its characteristics are similar to those which have been obtained for many overt rhythms (King and Cumming, 1972a).

A feature which is peculiar to some circadian rhythms is that they can be entrained to skeleton photoperiods beginning and ending with short light pulses. Once again this can also be shown with flowering responses. Some work by the late William S. Hillman (1964) on this problem has provided some of the best evidence for the involvement of circadian rhythms in photoperiodism. Hillman's results obtained from a long series of complex experiments on the entrainment of the flowering response in the SDP *Lemna paucicostata* turned out to be in close agreement with the predictions of a theoretical model derived independently by Pittendrigh (1966)

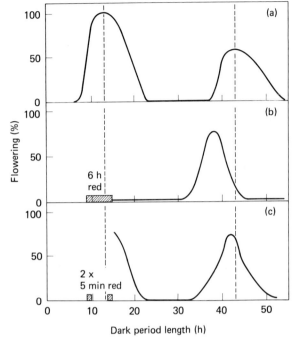

Fig. 18.6 Rephasing of a flowering rhythm by light in *Chenopodium rubrum*. Plants were grown in continuous light interrupted by a single dark period of various durations. The control rhythm (a) is advanced by 6 h red (b) given after 9 h of darkness but a 6-h skeleton beginning and ending with 5 min red (c) did not rephase the rhythm. (Adapted from King and Cumming, 1972b.)

from the characteristics of the overt circadian rhythm in pupal emergence in *Drosophila*. The correspondence of the *Lemna* results with a model based on an indisputable circadian rhythm is unlikely to be merely fortuitous.

If a rhythm in sensitivity to light is involved in night-length measurement, the phase of the rhythm must be established at the transition from light to darkness. The way in which this might be achieved in SDP is indicated in Fig. 18.7 which shows how the time of maximum night-break sensitivity in *Xanthium* is related to the length of the preceding photoperiod. When this is < 5 h maximum sensitivity is at a fixed time in relation to the beginning of the photoperiod, but when the photoperiod is > 5 h night-break sensitivity is delayed and remains constant at 8.5 h after the end of the photoperiod. Similar results have been obtained with *Pharbitis*

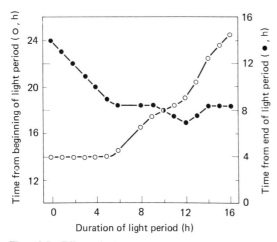

Fig. 18.7 Effect of photoperiod length on the time of maximum night-break sensitivity in *Xanthium strumarium*. Plants were given a phasing dark period of 7.5 h followed by a light period of various durations; after this an inductive test dark period of 16 h was given, interrupted by a short night break at various times. The time of maximum night-break sensitivity is plotted from the beginning (○) or the end (●) of the intervening light period. (Adapted from Papenfuss and Salisbury, 1967.)

and *Chenopodium* and indicate that the rhythm whose phase was set initially by the beginning of the photoperiod is subsequently suspended in continuous light and released at the light/dark transition. In this way the time of sensitivity to light in the subsequent dark period is established to form the basis of dark timing. This type of scheme is called *external coincidence* and is based on the interaction of light with a single circadian rhythm of sensitivity to light. An alternative scheme called *internal coincidence*, which ascribes photoperiodic responses to the interaction of two rhythms whose phases are set by 'light-on' and 'light-off', is not supported by the results obtained with single-cycle SDP where night-break timing is determined solely by the transition to darkness. They may, however, be important in other types of plant.

The action of phytochrome Apart from the clock, the other major component of the photoperiodic mechanism is the photoreceptor, which, as discussed earlier, appears to be phytochrome. The photoperiodic control of induction is, however, a consequence of the timing of the light and dark treatments

and hence the interaction of phytochrome with the clock must be a crucial feature in the mechanism. An attractive hypothesis in which phytochrome served as both timer and photoreceptor was advanced by Hendricks (1960). He proposed that CNL was determined by the time taken for the far-red absorbing form of phytochrome (P_{fr}) present at the end of the photoperiod to fall in darkness below a threshold level needed to prevent induction. This model, based on observations of P_{fr} disappearance by decay and reversion in dark-grown seedlings, represents an hour-glass form of time measurement. It is difficult, however, to reconcile this with the rhythmic responses which have already been mentioned. An even more compelling argument against this idea is the observation that in many SDP neither of the indices of dark timing, *i.e.*, the time of maximum night-break sensitivity and CNL, is greatly affected by photochemically lowering P_{fr} levels at the end of the photoperiod (*see* Fig. 18.8). In order to disprove entirely Hendricks' hypothesis it would be necessary to measure directly the rate of P_{fr} loss in darkness in photoperiodically-sensitive tissues. This has proved to be extremely difficult as the presence of chlorophyll effectively prevents the spectrophotometric assay of P_r, P_{fr} or total phytochrome (P_{tot}). Attempts have been made to follow changes in P_{fr}/P_{tot} ratios with an indirect physiological method known as the 'null' technique. This

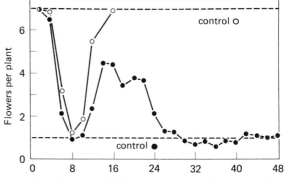

Fig. 18.8 Effect of 5 min far-red given immediately prior to a 48-h dark period on the patterns of response to a 5-min red night break in *Pharbitis nil*. Far-red treated, ●; far-red omitted, ○. Flowering levels in controls which received no red-light interruptions are shown. (Adapted from Takimoto and Hamner, 1965.)

method assumes that, if the P_{fr}/P_{tot} ratio established by a mixed light source is the same as is present in the tissues, no effect on the flowering response will be obtained. Experiments which have attempted to follow changes in the P_{fr}/P_{tot} ratio in an inductive dark period by this technique have, however, produced results which are often contradictory and difficult to interpret. They certainly do not fit in easily with the P_{fr} loss hypothesis. Recently, spectrophotometric measurements of phytochrome have been made in photoperiodically-sensitive plants which are free of chlorophyll, being either dark-grown or photo-bleached through the action of herbicides or low temperatures. In no case, however, has it been possible to show that changes in the form or content of spectrophotometrically assayable phytochrome are related to the time of night-break sensitivity or CNL (King, Vince-Prue and Quail, 1978).

Although the role of phytochrome in photoperiodism does not appear to be concerned directly with the measurement of time, there must at least be an intimate connexion between phytochrome and the clock, because the action of light in photoperiodism is very much dependent upon the time at which it is presented. For example, the rhythmic sensitivity to light interruptions in a dark period suggests that the system may respond differently to phytochrome at different times. Alternatively it is possible that the clock acts to change the properties of phytochrome in the dark period so that the rhythmic effect of light is due to changes in the ability of the photoreceptor to function as a transducer. This latter proposition has recently been examined for dark-grown seedlings of *Pharbitis nil* in which King, Schäfer, Thomas and Vince-Prue (1982) showed that rhythmic changes in sensitivity to a red-light pulse were accompanied by apparent variations in the quantum yield for photoconversion from $P_r \rightarrow P_{fr}$ and $P_{fr} \rightarrow P_r$. Other features of the photoreceptor such as total amount, relative amounts of P_r and P_{fr}, or spectral properties did not change. The changes in quantum yield were, however, quite small whereas the rhythmic response in *Pharbitis* is accompanied by changes of up to two orders of magnitude in the sensitivity of the response to light. Such changes in sensitivity appear to be common in photoperiodic plants, having been first documented in detail by Könitz (1958) working with *Chenopodium amaranticolor*. They are probably much too great to be accounted for solely in changes of photoreceptor properties and are

more likely to be based on a changing responsivity to the photoexcited photoreceptor. It is, however, difficult at present to rule out entirely changes in phytochrome properties as a basis for the rhythm in light sensitivity and more work is required to resolve this point.

A specific function of the photoreceptor must be to sense the light-to-dark transition at the end of the photoperiod. Red light is the most effective in preventing dark-time measurement, suggesting the involvement of phytochrome. Most available physiological evidence indicates that a decrease in irradiance values below a threshold level constitutes the light-off signal for photoperiodic timing. If we are to account for this solely in terms of P_{fr} which is supposedly the active form of phytochrome, it is necessary to propose that P_{fr} is lost rapidly through non-photochemical reactions which become significant at low irradiances. One problem with this hypothesis is evidence that in light-grown plants at least some P_{fr} persists and remains active in the dark. This can clearly be seen in many growth experiments where photochemically removing P_{fr} many hours after the end of the daily light period causes appreciable internode elongation. In some instances the flowering response is diminished if P_{fr} is photochemically removed at the beginning of the inductive dark period (*see later* in this section). The paradox that a reduction of P_{fr} is necessary for the initiation of dark timing while P_{fr} retention increases flowering may indicate the presence of two pools of phytochrome with different kinetic properties. Alternatively, a continuous generation of 'new' P_{fr} in the light may be required to prevent dark timing. The crucial experiments to discriminate between these possibilities remain to be performed.

Phytochrome not only functions as the photoreceptor which interacts with the clock in photoperiodism but is the major photoreceptor for the multiplicity of developmental responses, known collectively as photomorphogenesis, which plants show to light (*see* Chapter 17). In experiments on floral induction it is possible to distinguish the photoperiodic effects of phytochrome, which are concerned primarily with the timing of sensitivity to light, from developmental or photomorphogenetic effects of phytochrome. A good example of this is shown in *Pharbitis nil*, where far-red at the end of the photoperiod decreases the flowering response but does not affect night-break timing (Fig. 18.8). A requirement for P_{fr} during the dark period has been shown in several SDP. This is particularly evident after short photoperiods, which suggests that light in the photoperiod may partially substitute for P_{fr} in the dark period. Under certain conditions the requirement for P_{fr}, as shown by the inhibitory effect of far-red pulses, persists well into the dark period beyond the time of night-break sensitivity. Under these circumstances both red and far-red inhibit flowering and so red/far-red reversibility of the night-break response cannot be demonstrated. Once again there is a paradox, for P_{fr} appears simultaneously to inhibit and promote flowering at certain times in the dark period. The complex nature of the plant's changing sensitivity to P_{fr} in the dark is probably the main reason why attempts to measure P_{fr}/P_{tot} ratios from null responses, which assume only one action of P_{fr}, have not been successful.

Measurement of the dark period Most experiments on the mechanism of the photoperiodic clock have been concerned with the characteristics of the circadian rhythm in sensitivity to light pulses in long dark periods. Under natural conditions, however, the length of the dark period is the factor which allows

or prevents induction. An important question is, therefore, whether the CNL is under the control of the same timing mechanisms as night-break timing. Consider the situation in *Xanthium* and *Pharbitis* as illustrated in Fig. 18.9. In both species the time of maximum night-break sensitivity is not temperature-sensitive and, in *Xanthium*, this also applies to CNL. In *Pharbitis*, on the other hand, CNL does appear to be temperature-sensitive, implying a separate timing mechanism. In the *Xanthium* experiments a two-phase curve is observed; the first is relatively temperature-insensitive and indicates a common value for CNL, whereas the second phase is much more temperature-dependent. These results indicate that the response to increasing night length in *Xanthium* has two components. The first, or temperature-insensitive, component is concerned with timing and the second, temperature-sensitive component reminiscent of an hour-glass determines the magnitude of the flowering response and may possibly be related to the formation of flower-promoting hormones.

The *Pharbitis* data, on the other hand, suggest that the entire *measurable* flowering response to increasing night length is strongly modified by temperature and is equivalent to the second *Xanth-*

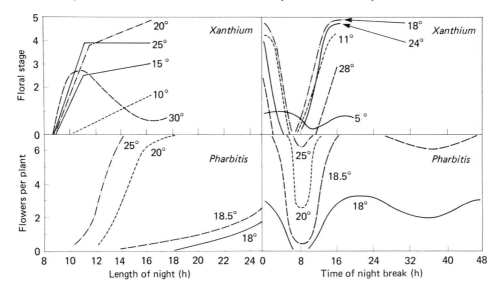

Fig. 18.9 Effects of temperature (°C) on CNL and night-break sensitivity in *Pharbitis nil* and *Xanthium strumarium*. Figures on the left show flowering response as a function of night length. Figures on the right show flowering responses of plants treated with brief intervals of light in a long dark period. (Adapted from Salisbury and Ross, 1969.)

ium component which is concerned with the magnitude, rather than the timing, of the response. The temperature-dependent component of the CNL may, therefore, merely represent the time required to achieve a measurable flowering response after a common temperature-insensitive timing point. The correspondence between CNL and the time of night-break sensitivity in some species and the observation that CNL is never shorter than the night-break time makes it likely that the timing component of both is the same circadian rhythm. Support for this idea can be gained from the rhythmic response to dark period length in *Chenopodium*, which was discussed earlier (Fig. 18.4).

To summarize photoperiodic induction in single-cycle SDP, it can be said that the phase of a circadian timer is set at the transition to darkness (at dusk) by the action of the photoperiod. During darkness the circadian timer establishes phases of sensitivity to light and the photoperiodic response of the plant depends on the phase of the rhythm at which the next exposure to light (dawn) occurs.

This type of external coincidence mechanism was originally proposed by Bünning, who suggested that the dark period is composed of alternating phases which require either light (photophase) or darkness (skotophase). An alternative version of this model is that a P_{fr}-operated switch, which prevents the initiation of processes leading to induction, can be activated only at a certain point in the rhythm known as the *photoinducible phase*. In this version the inhibitory effect of light at times following the photoinducible phase must be through a direct inhibition of the inductive process.

Photoperiodic induction in LDP The scheme outlined for SDP should account for the responses of LDP if the response groups are mirror images of a common mechanism. Thus, in LDP, the photoperiod would set the phase of the rhythm and flowering would occur only if light were given at the time of the photoinducible phase. In practice many LDP show a quantitative relationship between the duration or irradiance of the night break and the magnitude of the flowering response. Such plants show a negligible response to a brief night break and a strong flowering response to a long daily exposure to light. The distinction between plants in which flowering is controlled primarily by the dark processes which can be prevented by short night breaks

and those in which the light reactions predominate has been made by referring to them as *dark-dominant* or *light-dominant* response types. These correspond largely but not entirely with the designation of species as SDP or LDP.

Although light-dominant plants show a quantitative response to light, there is considerable variation in detail regarding the relationship between the response and how the light is presented. In some cases, such as in *Brassica campestris*, the response is a function of the light integral which may indicate a role for photosynthesis in setting the level of flowering (Friend, 1969). In others, such as *Lolium*, the duration of the photoperiod is important and here continuous light may be replaced by intermittent light pulses required perhaps to maintain a minimum level of P_{fr} (Evans, 1969).

Responses of light-dominant plants with respect to light quality follow a clearly discernible pattern (Table 18.3). Action spectra for prolonged photoperiods commonly show maxima near 710–720 nm

Table 18.3 Differences between the responses of dark-dominant and light-dominant species.

Photoperiod treatment	Dark dominant	Light dominant
SD + short night-break	LDR	SDR
LD (white light + far-red)	LDR	LDR
LD (white light only)	LDR	SDR or weak LDR
Red extension before main light	LDR	LDR
Red extension after main light	LDR	SDR or weak LDR

LDR: Plants respond as if receiving a long day.
SDR: Plants respond as if receiving a short day.

as opposed to nearer 660 nm when short night breaks are employed. A peak in the blue part of the spectrum is also sometimes obtained. The long wavelength peak is similar to that obtained in action spectra for prolonged illumination in dark-grown seedlings. This peak is characteristic of the phytochrome-mediated high-irradiance response which, in etiolated tissue, appears to deviate from the P_r absorbance spectrum as a consequence of P_{fr} instability (*see* Chapter 17). Recent work has, however, suggested that phytochrome (P_{fr}) is much more stable in light-grown plants, in which case the action spectrum maximum for a P_{fr}-mediated response should be closer to 660 nm than 730 nm. In *Hyoscyamus*, far-red and blue maxima were still obtained when the plants were simultaneously irradiated with red light (Schneider, Borthwick and Hendricks, 1967). This is extremely difficult to explain on the basis of the known properties of phytochrome and may suggest the involvement of another, as yet undiscovered, pigment. In *Lolium*, on the other hand, the peak moved to longer wavelengths when a background of red light was used, indicating that only phytochrome is involved in this plant (Blondon and Jaques, 1970).

Light-dominant plants frequently show a much greater flowering response when far-red is added to red or white light during the photoperiod. The promotion by mixtures of red and far-red over red (or white) alone is particularly interesting as the effect of far-red varies greatly depending on when in the photoperiod it is presented (Table 18.3). This can also be seen in experiments where a short period of far-red is added to either red (*Lolium*) or white (*Hordeum*) light at different times during a long photoperiod (Fig. 18.10). In both species the changing sensitivity to far-red follows a rhythmic pattern which persists in continuous light. This is in marked contrast to the situation in dark-dominant SDP, where the rhythm is apparently suspended after a relatively short period in continuous light. If this interpretation is correct, this represents a major difference between light-dominant and dark-dominant plants. In general, the responses of light-dominant plants are less well understood than those of dark-dominant plants, especially with regard to the action and even identity of the photoreceptor(s).

Vernalization

Characteristics of the response

Not all workers agree on the precise definition of vernalization but, in our opinion, the term is best restricted to the specific promotion of flowering by a cold treatment given to the imbibed seed or young plant. Dormancy breaking in certain seeds and buds also requires exposure to low temperature but these

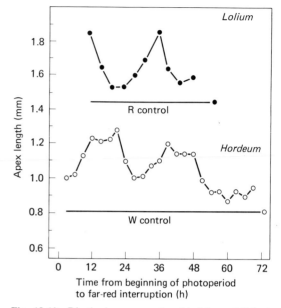

Fig. 18.10 Rhythmic response to added far-red light in the LDP *Lolium temulentum* Ba 3081 and *Hordeum vulgare* Wintex. *Lolium* plants were treated with 4 cycles consisting of 8 h sunlight and 40 h red light. *Hordeum* received a 72-h photoperiod in white fluorescent light. Far-red, 4 h in *Lolium* and 6 h in *Hordeum*, was added at various times. (Data from Vince-Prue, 1975, for *Lolium*, and Deitzer, Hayes and Jabben, 1979, for *Hordeum*.)

responses are excluded because they do not relate to flowering. There are direct effects of temperature on floral initiation in some plants but these can be distinguished from vernalization, which is an inductive phenomenon. After vernalization is complete, floral initials are not yet present and they differentiate only later when the plant is returned to higher temperatures and, in many cases, to particular photoperiodic regimes. In Brussels sprouts, *Brassica oleracea gemmifera*, and other plants with a similar direct response to low temperature, floral initials differentiate during the cold, or cool treatment. The biochemical changes that occur in response to cold are not necessarily different in the two response types.

The requirement for vernalization is most commonly, but not only, found in LDP (Ketellapper, 1966). These may be annuals (*e.g.*, winter cereals, pea), biennials (*e.g.*, carrot, sugar beet) or perennials (*e.g.*, *Lolium perenne*). Without exposure to an adequate period of cold, these plants show delayed flowering, or no flowering, and often exhibit a substantial reduction in internode length to grow as rosettes. The effect of low temperature increases with the duration of exposure until the response is saturated at a duration which varies widely with species and cultivar (<10 to >100 days). The vernalization effect also becomes progressively more stable and de-vernalization (*see later*) becomes more difficult (Table 18.4). The effective temperature range is from just below freezing to about 10°C,

Table 18.4 The effect of duration of exposure to low temperatures on stabilization of vernalization in winter rye (cv. Petkus) (adapted from Purvis and Gregory, 1952).

No. of weeks of cold treatment	2	3	4	5	6	8
Percentage of plants remaining vernalized after a de-vernalizing exposure to high temperature (3 d at 35°C)	0	42	44	75	84	97

with a broad optimum usually between 1° and 7°C (Ketellapper, 1966). Even at sub-optimal low temperatures, increasing the duration of exposure will, in most cases, effect complete vernalization.

The situation in cereals has been analyzed in a long series of papers by Gregory, Purvis and their co-workers, who worked mainly with winter and spring strains of Petkus rye (Purvis, 1961). In un-vernalized winter rye, ears are formed after about 22 leaves in continuous light. Following vernalization and in spring rye, only six or seven leaves are formed below the ear. In short photoperiods, flowers fail to emerge from the leaf sheath in both strains. In these winter annuals, the cold requirement can be satisfied by treating imbibed or germinating seeds. A water content of about 40–50% is sufficient and seeds can be vernalized on the parent plant before they have dried down.

Although some biennial plants can be vernalized as seeds (*e.g.*, beet), most of them must reach a certain size before the cold treatment is effective. This juvenile period persisted for about 10 days in *Hyoscyamus niger* and the response to low temperature subsequently increased up to about 30 days (Sarkar, 1958). In *Lunaria biennis*, a 12-week low-

temperature treatment was not effective unless the plants were more than seven weeks old when the treatment began (Wellensiek, 1958). The duration of the juvenile phase may be important in biennial root crops; a long juvenile phase can allow earlier sowing and so enable growers to take advantage of a longer growing season without incurring the crop losses which would result from flowering in the first year.

Cold-requiring perennials pose a special problem. As with juvenility, the vernalization requirement is re-established during sexual reproduction so that seeds or seedlings must again be exposed to low temperature before flowering can occur. The association of the perennial habit with a vernalization requirement at first sight appears contradictory because the vernalized state is transmitted through all subsequent mitotic divisions. All vernalized buds should, therefore, produce flowers and the plants would be monocarpic. This does not occur for several reasons.

In *Chrysanthemum morifolium*, the perennial habit is assured because some of the vernalized buds become de-vernalized during the summer. After chilling, all buds appear to be vernalized but, at the end of the summer, the perennating shoots which arise at the base of the plant are non-vernalized. A different strategy is seen in *Geum urbanum* where plants are perennial because chilling normally confers flowering capacity only to axillary buds at a certain stage in their development (Chouard, 1960). A third strategy occurs in some perennial grasses where the vernalization effect is not perpetuated indefinitely through cell divisions (Purvis, 1961). Tillers which develop late in the summer are not vernalized and overwinter without flowering.

De-vernalization

De-vernalization has already been mentioned as being important in maintaining the perennial habit in chrysanthemum. The possibility of de-vernalization, or loss of the vernalized condition, seems to be widespread and is probably ecologically significant. The most usual de-vernalizing agent is high temperature (~30°C) but de-vernalization by SD has also been reported (Wellensiek, 1965). High temperature is usually most effective immediately following a vernalizing treatment, especially when the latter is sub-optimal. In rye, complete loss of the vernalization effect occurred only after a short cold treatment and stability increased with increasing duration of low temperature, the vernalized condition becoming completely stable after about 8 weeks cold (Table 18.4).

Partial or sub-optimal low-temperature treatments can sometimes be stabilized by maintaining seeds or plants at a neutral temperature of about 12–15°C for a few days before exposing them to the de-vernalizing temperature. There are, however, many degrees of response ranging from no stabilization to no high-temperature de-vernalization. There are also complex interactions with light which, when given during the chilling treatment itself, appears to have a stabilizing effect in some plants. Wild lettuce, Lactuca serriola, can be de-vernalized only before germination and in the absence of light (Marks and Prince, 1979). Stabilization in this plant is thus associated with germination and de-vernalization of seeds may be important in maintaining a seed stock in the soil. Stabilization in Arabidopsis also seems to be associated with a period of growth at normal temperatures and here, too, light during the chilling treatment appears to have a stabilizing effect (Napp-Zinn, 1960).

Although high temperatures may lead to de-vernalization in the early stages of chilling, the vernalized condition is extremely stable once it is fully established and normally persists throughout the vegetative phase until the plants finally flower. In general it appears that all cell-division products of vernalized cells are in the vernalized condition. The vernalized state thus seems to have a self-perpetuating character; in this it resembles the metastable juvenile and adult states and poses the same kinds of problem with respect to mechanism.

Relationships between photoperiodism and vernalization

One of the interesting aspects of vernalization concerns the multiplicity of its interactions with photoperiodism.

A cold requirement for flowering is often linked with a requirement for a particular photoperiodic condition. The most common combination is the need for vernalization and LD which would lead to flowering in late spring or early summer, but a cold requirement can also be accompanied by a short-day photoperiodic response (Ketellapper, 1966).

There are also cases where vernalization can substitute for, or modify, a photoperiodic response, either reducing the critical daylength or causing plants to become indifferent to daylength. In certain strains of Spinacia, which normally require photoperiods of longer than 14 h, vernalization reduced the critical daylength to about 10 h (Vlitos and Meudt, 1955).

SD treatment can, in some cases, substitute either partly or entirely for cold. All plants where SD can substitute for low temperature are LDP; thus they could be classified either as SLDP or as cold-requiring LDP. The SD or low-temperature treatment must precede the LD exposure, the reverse order being ineffective. A detailed study of Campanula medium by Wellensiek (1960) has revealed several differences between the two alternative pre-long-day treatments. Fewer long photoperiods were necessary for floral initiation after SD than after cold and plants became sensitive to SD one month sooner than to low temperature. Consequently, SD and low temperatures appear to be different mechanisms leading to the same result. In Petkus rye, as with Campanula, SD seems to be an alternative to vernalization in hastening the flowering response to subsequent LD (Purvis, 1961). In rye, however, SD are much less effective than cold. A similar SD effect has been demonstrated in a range of wheat cultivars, in oat, and in a cold-requiring strain of Lolium temulentum. The possible substitution of SD for low temperature seems not uncommon and may thus have implications for the mechanism of vernalization. In this context, it is interesting to note that SD can also abolish the requirement for cold in Trifolium repens (a New Zealand strain), a plant in which the low-temperature treatment leads directly to floral initiation with no inductive (vernalization) effect (Thomas, 1979). As with vernalization, the SD effect is inductive and must be followed by long photoperiods if flowering is to occur. There is no photoperiodic effect when plants initiate flowers during a cold treatment.

Although SD cannot substitute for a cold treatment in some other LDP, they may promote flowering when given during or before an exposure to low temperature. Examples of this type of response are found in some perennial grasses and also in table beet. It is also seen in celery (Apium graveolens) where LD during vernalization delayed flowering but LD after vernalization promoted flowering (Table 18.5).

Table 18.5 Effect of daylength during and after vernalization on flowering in celery, *Apium graveolens* L. cv. Florida (adapted from Pressman and Negbi, 1980).

	Percentage of plants flowering
Daylength during vernalization	
Short days (9.75–12.5 h)	87
Long days (16 h)	37
Daylength after vernalization	
Short days (8 h)	0
Long days (16 h)	73

The nature of vernalization

Site of perception of low temperature A first step in studying the mechanism of vernalization is to locate the site of vernalization in the plant. Localized cooling treatments applied to biennial and perennial plants cause flowering when the stem apex alone is chilled; this effect appears to be largely independent of the temperature experienced by the rest of the plant. Excised shoot tips have also been successfully vernalized and, where seed vernalization is possible, fragments of embryos consisting essentially of the shoot tip are sensitive to a chilling treatment (Purvis, 1940).

All dividing shoot cells, including those in the leaves, may be potential sites of vernalization. When treated at 5°C, excised leaves from unvernalized plants of *Lunaria annua* regenerated plants in the vernalized condition (Wellensiek, 1964). However, when the lower five centimeters of leaf were removed after the low temperature, the regenerated plants were in the non-vernalized state; as this region is the main location of mitotic activity, these results point to the need for dividing cells. However, vernalization is possible in winter rye and in *Cheiranthus allionii* under conditions where mitosis is not occurring and so cell division, *per se*, seems not to be essential. Vernalization does, however, seem to occur only in meristematic zones of shoots and may be restricted to only a very small region of the shoot. In *Geum urbanum*, for example, Chouard (1960) has shown that young undifferentiated buds are insensitive to cold; sensitivity develops as they differentiate and buds finally become irreversibly determined as vegetative buds if they have not been vernalized in time. A study of the accompanying cell elongation, which is also a response to a vernalization treatment in *Chrysanthemum morifolium*, showed that only cells within 0.3 mm of the shoot apex were able to elongate in response to chilling. Cells at a greater distance from the apex were still capable of elongation (*e.g.*, in response to GA₃) but did not respond to low temperature (Mitra and Vince, 1962).

Metabolic changes during vernalization The characteristic feature of vernalization is that the optimum temperature lies between about 1° and 7°C (Fig. 18.11). Since vernalization requires both carbohydrate as an energy source and oxygen, it is unlikely to be simply the absence of an inhibitory reaction at low temperature. Some time ago Salisbury showed that a broad optimum-temperature curve of the type actually seen in many vernalization responses could be generated by assuming that vernalization consisted of an inhibitory and a promoting process with very different Q_{10} values (Fig. 18.11). However, this remains a purely theoretical concept.

The possibility of de-vernalization many weeks after an exposure to low temperature suggests that the process of vernalization probably does not result directly in the production of a transmissible stimulus, but rather leads to the development of a localized vernalized state in cells which are at a certain stage of development when they are exposed to cold. The vernalized condition is a property of all daughter cells resulting from divisions of vernalized cells, unless de-vernalization occurs. The nature of the changes which lead to the vernalized state has not yet been resolved. New proteins appeared in winter wheat during a cold treatment and, after vernalization, the protein pattern in embryos of winter wheat resembled those in a spring cultivar (Teroaka, 1972). The new proteins appeared only when embryos were maintained on a sucrose medium, which is essential for the vernalization process in excised embryos. The use of inhibitors gave further evidence that vernalization is essential for the appearance of the new proteins; these were produced in the presence of bromouracil, which allowed vernalization, but not in the presence of azaguanine, which did not. The use of inhibitory analogs indicated that vernalization (of leaf tissue) in *Streptocarpus wendlandii* involved the synthesis of a new messenger RNA (Hess, 1959).

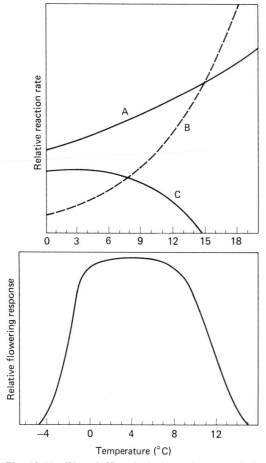

Fig. 18.11 (Upper) Hypothetical reaction rates during vernalization. Curves show reaction rates as a function of temperature for a promoting reaction with $Q_{10} = 1.5$ (A) and an inhibiting reaction with $Q_{10} = 4.0$ (B). The response at any temperature would then be a function of the relative rates of the two reactions (curve C). Compare the shape of curve C with the curve in the lower figure which shows vernalization in Petkus winter rye as a function of temperature. (Adapted from Salisbury, 1963.)

Intermediate physiology of flowering

Translocation of the floral stimulus

As we have already seen, in photoperiodism the barriers to flowering in non-inductive conditions

reside primarily in the leaf. Chailakhyan (1936) was the first to appreciate the significance of this observation with regard to the post-inductional events leading to flowering. He reasoned that flowering at the apex could take place only in response to a transmissible stimulus originating in the leaf, and for this stimulus he proposed the term *florigen*.

Leaving aside the nature of the floral stimulus for the moment, it is possible to ask how the stimulus is transmitted from the leaf to the apex. Most accounts of this process conclude somewhat guardedly that the stimulus moves in the phloem with assimilate although the evidence for this is not entirely conclusive. It is true that treatments which restrict phloem transport such as localized heat, cold treatments, girdling (which removes a ring of tissue external to the xylem) or narcotic treatments with chemicals such as chloroform, prevent the movement of the floral stimulus. All these treatments would, however, be expected to limit any form of symplastic movement and indicate only that the stimulus does not move in the xylem with the transpiration stream. Evidence that grafted leaves can induce flowering in receptor plants only when functional phloem connexions are formed is matched by reports that intergeneric grafts between *Xanthium* and *Silene* result in flowering in the absence of such connexions (Wellensiek, 1970). Even in this case though, some live tissue connexions are necessary.

The best evidence that the floral stimulus moves in the phloem comes from experiments showing good correlations between effects on translocation and flowering. For example, Chailakhyan and Butenko (1957) using $^{14}CO_2$ showed that a high flowering response in *Perilla* was associated with the presence in the bud of a large amount of label from an induced leaf and low flowering was associated with label in the bud from a non-induced leaf. Using a slightly different approach, Carr (1957) showed in *Xanthium* that the stimulus appeared to move out of the leaf much faster in high-intensity light than in darkness and that sucrose applied to the leaf could replace the effect of light. Movement against the expected flow of assimilate can, however, also be demonstrated, as young predominantly importing leaves are often highly active in causing flowering. Movement of the floral stimulus out of the leaf can be demonstrated in darkness in *Xanthium* (*see* Fig. 18.12), and in *Pharbitis* it has been shown to occur in

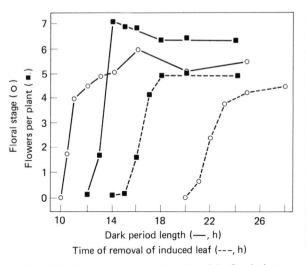

Fig. 18.12 Induction and translocation of the floral stimulus in *Xanthium strumarium* (○) and *Pharbitis nil* (■). Continuous lines indicate the effect of increasing dark-period duration on the flowering response. Dashed lines indicate the effect of removing the induced cotyledons (*Pharbitis*) after various lengths of dark period and returning the plants to light, or of removing the induced leaf (*Xanthium*) at various times, the plants remaining in darkness. In both cases flowering is prevented if the induced organ is removed before the translocation of the floral stimulus has occurred. (Adapted from Vince-Prue, 1975.)

the absence of assimilate movement. Simultaneous studies of the rate of assimilate movement and the movement of the floral stimulus have been carried out in only a few cases. In the LDP *Lolium temulentum* the translocation rate for the floral stimulus was estimated at 10–24 mm h^{-1} as compared with approximately 1000 mm h^{-1} for sucrose movement (Evans and Wardlaw, 1966). In *Pharbitis nil*, on the other hand, the rates were similar at 240–370 mm h^{-1} for the floral stimulus and 330–370 mm h^{-1} for labeled assimilate (King, Evans and Wardlaw, 1968). The floral stimulus, therefore, appears to travel from the induced leaf to the apex by a symplastic route which is the phloem in most but perhaps not all instances.

The nature of the floral stimulus

The studies concerned with the movement of the floral stimulus from the leaf to the apex show clearly

that a relatively slow-moving chemical rather than an electrical message passes between the two. The generation of a chemical message at one site to act remotely in a different type of tissue conforms entirely to the classical idea of hormonal regulation. There are two questions to be asked regarding the hormonal control of flowering and they are relatively simple. Is flowering regulated only by a floral promoter or inhibitor, or is the balance of promoters and inhibitors important? Secondly, are the hormones specifically concerned with flowering (*florigens* and *antiflorigens*) or are they the growth substances concerned with other aspects of plant growth and development (auxin, gibberellins, cytokinins, etc.)? The answers to the questions are, however, extremely difficult to obtain and can be dealt with only briefly here.

Promoters and inhibitors Much of the evidence for a flower-promoting substance has already been given within this chapter. The observations include the repeated transfer of the floral stimulus in grafting experiments, the cumulative nature of photoperiodic induction as seen in fractional and steady-state induction, and the requirement for only small amounts of induced leaf tissue, compared with the total leaf mass of the plant, for flowering to occur. There is, however, a substantial amount of evidence that leaves in non-inductive cycles may not be just neutral but may exert a positive inhibiting effect on flowering. Some plants, for example, will flower in non-inductive cycles if the leaves are removed and these include both LDP, such as *Hyoscyamus niger* (Lang and Melchers, 1943) and SDP, such as *Fragaria × ananassa* (Thompson and Guttridge, 1960). In *Hyoscyamus*, grafting back even a single leaf restored the requirement for LD for flowering. These results are extremely difficult to interpret other than through the action of an inhibitor produced by the leaves in non-inductive daylengths.

A second kind of experiment involves intercalating unfavorable daylengths into a series of favorable daylengths. It has already been shown that similar experiments were used to demonstrate fractional induction where the effects of favorable daylengths are remembered during the unfavorable ones. This does not seem to be the case, however, when only a few consecutive inductive cycles are given, for under these conditions the effect of the inductive cycles is annulled by the intercalated non-inductive cycles.

In the SDP *Kalanchoë blossfeldiana*, for example, Schwabe (1956) showed that the effects of SD treatments were reduced by intercalated LD. In these experiments, a single LD interposed between two sets of six SD reduced flowering to about 50% of the controls, which were given 12 consecutive SD. In experiments where single LD were interposed between different numbers of consecutive SD, it was found that one LD anulled the effect of 1.5–2 SD. These results are extremely difficult to explain in terms of lack of floral promoter and suggest the action of a floral inhibitor.

A cogent argument for the existence of a chemical which was produced in non-induced leaves and which actively inhibited flowering would be the demonstration of the transport of such an inhibition. In fact, this has been done by Guttridge (1959) with *Fragaria* × *ananassa*, the garden strawberry. In this species SD cause flowering and LD result in vegetative growth and stolon production. When parents are grown in LD and daughters attached to them by runners are maintained in SD, flower production is greatly restricted in the latter, especially under conditions which maximize the flow of assimilate from parent to daughter. These experiments have produced no evidence for a transportable floral stimulus, merely for a floral inhibitor which simultaneously promotes vegetative growth.

Evidence for inhibition under non-inductive daylengths by a flowering inhibitor is therefore very strong. Another type of inhibition can also be identified in some instances. In experiments where a few leaves on a plant are maintained in favorable daylengths while others are kept in unfavorable ones, the resultant flowering should then be the net result of the promotory and inhibitory substances. In such experiments unfavorable photoperiods given to leaves located between the leaves receiving favorable photoperiods and the apex are much more effective than if the treatments are reversed. These experiments have usually been interpreted in terms of the non-induced leaves acting as sinks for the floral stimulus moving with assimilates, although, as already mentioned, the pathway by which the floral stimulus is transmitted is not entirely clear.

Specific flowering hormones The production of flowers in photoperiodically-regulated plants is clearly under the control of transmissible regulators. In some species either a promoter or an inhibitor is the primary regulator while in others the balance between promoters and inhibitors is important. When trying to identify the regulators concerned experimenters have, however, experienced considerable difficulty. Evidence from experiments based on interspecific and intergeneric grafts can, in many instances, be taken as support for the existence of a specific flowering hormone common to or physiologically equivalent in a wide range of plants. The flowering stimulus, or *florigen*, will move from LDP to SDP as shown by Okuda (1953) with *Rudbeckia bicolor* grafted to *Xanthium canadense* and, conversely, from SDP to LDP as in *Nicotiana tobacum* grafted to *Hyoscyamus niger* (Lang and Melchers, 1948). In these experiments the transmitted stimulus is produced only in conditions which would normally result in induction in the donor plants. If we take the second example quoted above as a typical one, *Hyoscyamus*, the LDP will flower in short days if the grafted *Nicotiana* portion is also in short days but not if the *Nicotiana* is maintained under long days. The fact that the transmissible stimulus is equivalent in both LDP and SDP gives strong support to the *florigen* concept. On the other hand, grafting experiments are only possible, with a few isolated exceptions, between plants which are closely related, where it might be expected that the floral stimulus would in any case be similar. Obviously the conclusive demonstration of a universal flowering hormone would be its chemical identification in a wide range of species. Unfortunately it is here that the florigen concept has encountered most difficulty as attempts to isolate and identify the floral stimulus have been largely fruitless.

The reasons why the floral stimulus has not been identified may include lability of the compound, difficulty in finding a suitable assay, the problems of re-introducing compounds being investigated, or even the possibility that *florigen* is a mixture of different compounds. A common approach to this question has been to obtain alcoholic extracts of induced leaf tissue, concentrate and re-apply the extract to leaf or stem tissue in non-inductive conditions. The most promising investigation of this kind was by Lincoln and Cunningham (1964), who obtained an extract from *Xanthium* leaves which caused some flower initiation in *Xanthium* plants kept in LD. Preliminary purification indicated that the active substance was acidic in nature and for this reason it was called 'florigenic acid'. The effect of florigenic acid on flower initiation in *Xanthium* was found to be enhanced by GA$_3$ although even this combination resulted only in incomplete male flower development and absence of female flowers. Attempts at further purification have, unfortunatcly, resulted in a loss of activity from the extract.

The other approach has been to attempt to tap the flow of assimilate through the phloem either by bleeding tissue or by the use of aphids. In one such study, salicylic acid from *Xanthium* phloem sap and other benzoic acid derivatives not found in plant extracts have been shown to cause flowering in both LD and SD species of *Lemna*. This is one of the properties which we might expect of a true *florigen*. Unfortunately, these compounds do not cause flowering in *Xanthium*, the plant from which they were originally extracted (Cleland, 1974) and cannot therefore be the universal flowering stimulus.

The position with regard to the possibility of a specific flowering inhibitor (*anti-florigen*) is even less clear than with *florigen*. There is evidence from grafting experiments of an inhibitor which is transmissible between plants of different response types, but attempts to isolate and characterize such an inhibitor have been no more successful than attempts to extract and identify *florigen*.

The vernalization stimulus

One of the still unresolved questions about low-temperature vernalization is whether, like photoperiodic treatments, it leads to the production of a transmissible hormone. As with photoperiodism, the main approach has been grafting experiments, using non-cold-requiring or vernalized plants as donors and cold-requiring, non-vernalized plants as receptors. Both successes and failures have been reported (Lang, 1965). Transmission of a flowering stimulus has been observed between different species and genera and from both non-cold-requiring and vernalized donors. Failures have been reported in SD cultivars of *Chrysanthemum morifolium* (Schwabe, 1954) and LD species of *Oeonothera* (Chouard, 1960).

There are many parallels with the transmission of the stimulus resulting from photoperiodic induction. Some tissue union between grafted plants is essential, and treatments which favor the movement from donor to receptor also favor the transmission of the

flowering stimulus from plant to plant. The key question is whether the vernalization treatment leads directly to the production of a transmissible hormone which is different from the *florigen* thought to be produced in favorable daylengths. Alternatively, vernalization could lead to a change in the condition of the cells such that they are able to produce *florigen* when subsequently returned to favorable conditions. Most of the grafting studies are consistent with the second hypothesis, for the donor plants have usually been in conditions in which they flower; in these circumstances the vernalization effect could have simply been by-passed by the transmission of *florigen* from the donor plant. The very localized elongation response to low temperature in an SD cultivar of *Chrysanthemum* (where there is also no graft transmission of a floral stimulus from vernalized plants) supports the assumption that vernalization may not lead directly to the production of a transmissible stimulus. However, when *Nicotiana tabacum* cv Maryland Mammoth (a SDP) was used as a donor, it was able to cause flowering in non-vernalized biennial *Hyoscyamus niger* when the graft partners were maintained in LD and the Maryland Mammoth remained vegetative (Lang, 1965). The results point to the production of a distinct vernalization stimulus at least in this case. The results are thus in conflict and the possibility remains that vernalization does not always operate precisely in the same way.

If the exposure to a cold treatment results in the production of a transmissible hormone, it should be possible to extract and identify it. So far, however, the results have been equivocal.

Growth substances and flowering

Photoperiodism The failure so far to identify *florigen* forces us to consider the possibility that the transmissible stimulus in the photoperiodic control of flowering is one or more of the naturally-occurring growth substances. However, although auxins, ethylene, cytokinins and abscisic acid can be shown to have some effect on flowering in isolated instances, it is only the gibberellins (GAs) which appear to be involved in regulation by daylength in a range of species. Applied GAs cause flowering in non-inductive conditions in both LDP and SDP although much more commonly in the former than the latter. Applied GAs have also been found to inhibit flowering in a number of species, including a number of SDP and a few LDP.

GAs are extremely effective in causing flowering in rosette plants where flowering is usually accompanied by stem extension. Although the extension response is clearly under the control of endogenous GAs this is not so evident with respect to the formation of flowers. Cleland and Zeevaart (1970) have shown that in *Silene*, AMO 1618, an inhibitor of GA synthesis, prevented stem growth but not flowering in LD. Also, in some species the pattern of differentiation at the apex is different when flowering is caused by inductive daylengths or GA application.

GAs may also be involved in flowering in SDP such as *Pharbitis* where flowering can be suppressed by inhibitors of GA synthesis or action. In this species, however, applied GAs could not replace the inductive effect of a long night (Zeevaart, 1970). It appears from these studies that GAs are not *florigen* but rather that GAs are required for the formation or action of the flowering hormone. Chailakhyan has suggested that the flowering stimulus is composed of GAs and another class of hormone as yet unidentified which he has called *anthesins*. In his view, GAs would be limiting in LDP and *anthesins* in SDP (Chailakhyan, 1975). Unfortunately this hypothesis does not seem to account for the responses of all photoperiodic groups (Vince-Prue, 1983a) and *anthesins*, like *florigen*, still remain a concept rather than a physical reality.

Levels of endogenous GAs are found to be higher in LD than SD and this seems to be independent of the photoperiodic response group of the plants in question. More significant than photoperiodic effects on total GAs may be those on GA metabolism. Grigorieva, Kucherov, Lozhnikova and Chailakhyan (1971) showed that three GA-like substances could be extracted from the LDP *Nicotiana sylvestris* grown in LD, which caused flowering in the LDP *Rudbeckia*. In SD the leaves produced GAs which were ineffective in flowering. A different situation has been reported by Proebsting and Heftman (1980) where ^{3}(H)GA_9 was metabolized by the LDP *Pisum sativum* to more polar GAs in SD. They have suggested that polar gibberellins produced in SD inhibit flowering and their formation is blocked in LD. The recent advent of techniques such as gas chromatography, mass spectrometry and high-performance liquid chromatography has made it possible to quantitate individual GAs in different daylengths. Metzger and Zeevaart (1980, 1982) have used this approach in studying the interconversions of GAs in *Spinacia oleracea*, an LDP. Here, LD are associated with a decline in GA_{19} and increases in GA_{20} and GA_{29}. It appears that the increased levels of GA_{20} are accompanied by increases in both its rate of synthesis from GA_{19} and its conversion to GA_{29}. The presence of this large, rapidly-turning-over pool of GA_{20} is concerned with stem extension rather than flowering but such studies clearly indicate the pathway for future studies of endogenous GAs in the control of flowering by daylength.

Vernalization The application of GA_3 also enables many cold-requiring plants to flower without low temperature, although many are unresponsive (Table 18.6, Fig. 18.13).

Table 18.6 Responses to GA_3 treatment of non-vernalized plants grown in long days (adapted from Table 7.5, Vince-Prue, 1975).

Plants elongate and flower
  Daucus carota
  Hyoscyamus niger
  Myosotis alpestris

Plants elongate without flowering
  Beta vulgaris
  Oenothera biennis
  Scrophularia vernalis (small elongation response
                  see Fig. 18.1)

No response to GA_3
  Anagallis tenella (caulescent plant)
  Eryngium variifolium
  Euphorbia lathyris (caulescent plant)

The responsiveness to GA_3 suggests a possible role for gibberellin in the vernalization process, as well as in photoperiodism. However, no cause-and-effect relationship has yet been established and the precise role of endogenous gibberellins is unknown. The association of gibberellin with vernalization seems to be restricted essentially to rosette plants which are either DNP or LDP, and GA treatment does not seem to be effective in replacing a cold requirement in either caulescent plants or in SDP. However, not all rosette LDPs flower in response to GA application. Many elongate without flowering but some do not respond at all. Gibberellins may be

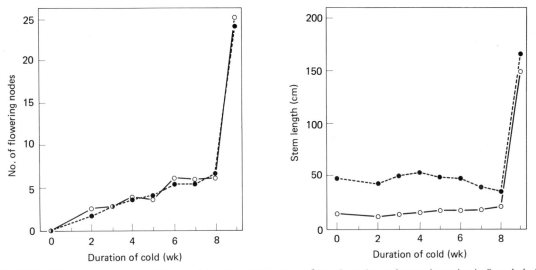

Fig. 18.13   Effect of vernalization (○) and gibberellic acid (2 × 1 μg, ●) on flowering and stem elongation in *Scrophularia vernalis*. (After Bismuth and Larrieu, 1978.)

effective only when plants have already been partially thermo-induced (as with cereals) or at temperatures close to those effective in vernalization (as in *Beta*). Although many plants can be vernalized as imbibed seeds, gibberellin application is usually not effective at this time, although it may cause flowering when applied at a later stage of growth.

There seems little doubt that the gibberellin content of plants can increase following a vernalization treatment as in LD. However, as with photoperiodism, there are problems of interpretation because the low-temperature treatment often has two effects: the induction of stem elongation and the initiation of floral primordia (Fig. 18.12). Where both responses can be effected by gibberellin application, stem elongation precedes initiation whereas, after vernalization, both responses occur more or less together. It has been shown that gibberellin application elicits only stem elongation in a number of cold-requiring rosette plants and that it is never an effective substitute for cold in caulescent plants. The possibility exists, therefore, that gibberellins are associated with the elongation response to cold rather than with flowering, a situation again paralleled in photoperiodism.

Nevertheless, gibberellin application can substitute for a cold treatment to induce both elongation and flowering in some plants and, moreover, can be shown to lead to the production of a graft-transmissible floral stimulus. Thus cold may remove the block to flowering by bringing about an alteration in gibberellin metabolism, at least in certain cases. It is important, too, to recognize that some gibberellins do not affect plant height and that these may also increase following a low-temperature treatment. In the caulescent plant, pea (*Pisum sativum* cv. Atsumikinusaya), seed vernalization accelerated flowering by two to four nodes; low temperature did not affect plant height but increased the total gibberellin content four-fold, apparently without causing the formation of different gibberellins (Suge, 1980). GA-like activity was present in diffusates from vernalized seeds; as determined by bioassays, these GAs were without an OH group on the C-3 position of the ent-gibberellane ring (perhaps GA_{20}). Application of GA_3 or GA_7 (with an OH group on the ring) increased stem height and delayed flowering in pea. Consequently, even in caulescent plants, specific endogenous gibberellins may play a role in vernalization.

Further reading

Brady, J. (ed.) (1982). *Biological Timekeeping*. Society for Experimental Biology Seminar, Series 14, Cambridge University Press, Cambridge, UK.

Champagnat, P. and Jaques, R. (eds) (1979). *La Physiologie de la Floraison*, C.N.R.S., Paris.

Doorenbos, J. (1965). Juvenile and adult phases in woody plants. In *Encyclopedia of Plant Physiology* XII (1) ed. W. Ruhland, Springer-Verlag, Berlin, pp. 1222–35.

Hackett, W. P. (1980). Control of phase change in woody plants. In *Control of Shoot Growth in Trees*, I.U.F.R.O., Fredericton, New Brunswick, USA, pp. 257–72.

Lang, A. (1965). Physiology of flower initiation. In *Encyclopedia of Plant Physiology* **15–1**, ed. W. Ruhland, Springer-Verlag, Berlin, pp. 1380–1536.

Purvis, O. N. (1961). The physiological analysis of vernalization. In *Encyclopedia of Plant Physiology*, **16**, ed. W. Ruhland, Springer-Verlag, Berlin, pp. 76–122.

Vince-Prue, D. (1975). *Photoperiodism in Plants*, McGraw Hill & Co., London.

Vince-Prue, D. (1983a). Photoperiod. In *Encyclopedia of Plant Physiology* (New Series): *Hormonal Regulation of Development*, eds R. P. Pharis and D. M. Reid, Springer-Verlag, Berlin, In Press.

Vince-Prue, D. (1983b). Photomorphogenesis and flowering. In *Encyclopedia of Plant Physiology* (New Series), vol. 16b, eds W. Shropshire Jr and H. Mohr, Springer-Verlag, Berlin, pp. 457–90.

Zimmermann, R. H. (1972). Juvenility and flowering in woody plants. *Hort. Sci.* **7**, 447–55.

References

Banks, M. S. (1979). Plant regeneration from callus from two growth phases of English ivy, *Hedera helix* L. *Z. Pflanzenphysiol.* **92**, 349–53.

Bismuth, F. and Larrieu, C. (1978). Etude de l'influence de l'acide gibbérellique (GA_3) et de l'acide *N*-dimethylaminosuccinamique (B_9) sur la croissance et la floraison de la *Scrophularia vernalis* L. *Biol. Plant.* **20**, 335–43.

Blaney, L. T. and Hamner, K. C. (1957). Interrelations among effects of temperature, photoperiod, and dark period on floral initiation of Biloxi soybean. *Bot. Gaz.* **119**, 10–11.

Blondon, F. and Jaques, R. (1970). Action de la lumière sur l'initiation florale du *Lolium temulentum* L.: spectre d'action et role du phytochrome. *C. R. Acad. Sci. Sér. D* **270**, 947–50.

Borthwick, H. A., Hendricks, S. B. and Parker, M. W. (1948). Action spectrum for the photoperiodic control of floral initiation of a long-day plant, Wintex Barley (*Hordeum vulgare*). *Bot. Gaz.* **110**, 103–18.

Carr, D. J. (1957). On the nature of photoperiodic induction. IV. Preliminary experiments on the effect of light following the inductive long dark period in *Xanthium pensylvanicum*. *Physiol. Plant.* **10**, 249–65.

Chacko, E. K., Kohli, R. R., Dore Swamy, R. and Randhawa, G. S. (1976). Growth regulators and flowering in juvenile mango (*Mangifera indica* L.) seedlings. *Acta Hort.* **56**, 173–81.

Chailakhyan, M. Kh. (1936). On the hormonal theory of plant development. *Dokl. Acad. Sci. USSR* **12**, 443–7.

Chailakhyan, M. Kh. (1975). Substances of plant flowering. *Biol. Plant.* **71**, 1–11.

Chailakhyan, M. Kh. and Butenko, R. G. (1957). Movement of assimilates of leaves to shoots under different photoperiodic conditions of leaves. *Fiziol. Rast.* **4**, 450–62.

Chouard, P. (1960). Vernalization and its relations to dormancy. *Ann. Rev. Plant. Physiol.* **11**, 191–238.

Clark, J. R. and Hackett, W. P. (1980). Assimilate translocation in juvenile-adult grafts of *Hedera helix*. *J. Am. Soc. Hort. Sci.* **105**, 727–9.

Cleland, C. F. (1974). Isolation of flower-inducing and flower-inhibiting factors from aphid honeydew. *Plant Physiol.* **54**, 899–903.

Cleland, C. F. and Zeevaart, J. A. D. (1970). Gibberellins in relation to flowering and stem elongation in the long-day plant *Silene armeria*. *Plant Physiol.* **46**, 392–400.

Cumming, B. G. (1969). *Chenopodium rubrum* L. and related species. In *The Induction of Flowering*, ed. L. T. Evans, Macmillan, Melbourne, London, pp. 156–85.

Cumming, B. G., Hendricks, S. B. and Borthwick, H. A. (1965). Rhythmic flowering responses and phytochrome changes in a selection of *Chenopodium rubrum*. *Can. J. Bot.* **43**, 825–53.

Deitzer, G. F., Hayes, R. and Jabben, M. (1979). Kinetics and time dependence on the effect of far red light on the photoperiodic induction of flowering in Wintex barley. *Plant Physiol.* **64**, 1015–21.

Downs, R. J. (1956). Photoreversibility of flower initiation. *Plant Physiol.* **31**, 279–84.

Evans, L. T. (1960). Inflorescence initiation in *Lolium temulentum* L. I. Effect of plant age and leaf area on sensitivity to photoperiodic induction. *Aust. J. Biol. Sci.* **13**, 123–31.

Evans, L. T. and Wardlaw, I. F. (1966). Independent translocation of ^{14}C-labelled assimilates and of the floral stimulus in *Lolium temulentum*. *Planta* **68**, 310–26.

Evans, L. T. (1969). *Lolium temulentum* L. In *The Induction of Flowering*, ed. L. T. Evans, Macmillan, Melbourne, London, pp. 328–49.

Friend, D. J. C. (1969). *Brassica campestris* L. In *The Induction of Flowering*, ed. L. T. Evans, Macmillan, Melbourne, London, pp. 364–75.

Frydman, V. M. and Wareing, P. F. (1974). Phase change in *Hedera helix* L. III. The effects of gibberellins, abscisic acid and growth retardants on juvenile and adult ivy. *J. Exp. Bot.* **25**, 420–9.

Garner, W. W. and Allard, H. A. (1920). Effect of the relative length of the day and night and other factors of the environment on growth and reproduction in plants. *J. Agric. Res.* **18**, 553–603.

Garner, W. W. and Allard, H. A. (1923). Further studies in photoperiodism, the response of the plant to the relative length of day and night. *J. Agric. Res.* **23**, 871–920.

Gressel, J., Zilberstein, A., Porath, D. and Arzee, T. (1980). Demonstration with fibre illumination that *Pharbitis* plumules also perceive flowering photoinduction. In *Photoreceptors and Plant Development,* ed. J. de Greef, Antwerpen University Press, Antwerp, pp. 525–30.

Grigorieva, N. Ya., Kucherov, V. F., Lozhnikova, V. N. and Chailakhyan, M. Kh. (1971). Endogenous gibberellins and gibberellin-like substances in long-day and short-day species of tobacco plants: a possible correlation with photoperiodic response. *Phytochemistry* **10**, 509–17.

Guttridge, C. G. (1959). Further evidence for a growth-promoting and flower-inhibiting hormone in strawberry. *Ann. Bot. N.S.* **23**, 612–21.

Hackett, W. P. and Srinivasani, C. (1983). *Hedera helix* and *Hedera canariensis*. In *CRP Handbook of Flowering,* ed. A. Halevy, in press.

Hamner, K. C. (1940). Interrelation of light and darkness in photoperiodic induction. *Bot. Gaz.* **101**, 658–87.

Hamner, K. C. and Bonner, J. (1938). Photoperiodism in relation to hormones as factors in floral initiation. *Bot. Gaz.* **100**, 388–431.

Handro, W. (1977). Photoperiodic induction of flowering on different explanted tissues from *Streptocarpus nobilis* cultured *in vitro. Bol. Botan. Univ. Sao Paulo* **5**, 21–6.

Hendricks, S. B. (1960). Rate of change of phytochrome as an essential factor determining photoperiodism in plants. *Cold Spring Harb. Symp. Quant. Biol.* **25**, 245–8.

Hess, D. (1959). Die selektive Blockierung eines an der Blühinduktion beteiligten Ribonucleinsäure-Eiweiss-Systems durch 2-Thiourasil (Untersuchungen an *Streptocarpus wendlandii*). *Planta* **54**, 74–94.

Hillman, W. S. (1964). Endogenous circadian rhythms and the response of *Lemna perpusilla* to skeleton photoperiods. *Am. Nat.* **98**, 323–8.

Hillman, W. S. (1969). Photoperiodism and vernalization. In *The Physiology of Plant Growth and Development,* ed. M. B. Wilkins, McGraw-Hill, London, pp. 557–601.

Imamura, S., Muramatsu, K., Kitajo, S. I. and Takimoto, A. (1966). Varietal differences in photoperiodic behaviour of *Pharbitis nil* Chois. *Bot. Mag. Tokyo* **79**, 714–21.

Joustra, M. K. (1970). Flower initiation in *Hyoscyamus niger* L. as influenced by widely divergent daylengths in different light qualities. *Meded. LandbHoogesch. Wageningen,* **70–19**, 78 pp.

Ketellapper, H. J. (1966). Vernalization requirement for flowering: Magnoliaphytes. In *Environmental Biology,* eds P. L. Altman and D. S. Dittmer, Fed. Am. Socs Exp. Biol. Bethesda, pp. 897–902.

King, R. W. and Cumming, B. G. (1972a). Rhythms as photoperiodic timers in the control of flowering in *Chenopodium rubrum* L. *Planta* **103**, 281–301.

King, R. W. and Cumming, B. G. (1972b). The role of phytochrome in photoperiodic time measurement and its relation to rhythmic timekeeping in the control of flowering in *Chenopodium rubrum. Planta* **108**, 39–57.

King, R. W., Evans, L. T. and Wardlaw, I. F. (1968). Translocation of the floral stimulus in *Pharbitis nil* in relation to that of assimilates. *Z. Pflanzenphysiol.* **59**, 377–85.

King, R. W., Schäfer, E., Thomas, B. and Vince-Prue, D. (1982). Photoperiodism and rhythmic responses to light. *Plant, Cell Environ.* **5**, 395–404.

King, R. W., Vince-Prue, D. and Quail, P. H. (1978). Light requirement, phytochrome and photoperiodic induction of flowering of *Pharbitis nil* Chois. III. *Planta* **141**, 15–22.

Knott, J. E. (1934). Effect of a localized photoperiod on spinach. *Proc. Am. Soc. Hort. Sci.* **31**, 152–4.

Könitz, W. (1958). Blühemmung bei Kurztagpflanzen durch hellrot und dunkelrotlicht in der photo- und der skotophilen phase. *Planta* **51**, 1–29.

Lang, A. and Melchers, G. (1943). Die photoperiodische Reaktion von *Hyoscyamus niger. Planta* **33**, 653–702.

Lang, A. and Melchers, G. (1948). Auslösung der Blutenbildung bei Langtagpflanzen unter Kurzbedingungen durch Aufpropfung von Kurztagpflanzen. *Z. Naturforsch.* **3b**, 108–11.

Lincoln, R. G. and Cunningham, A. (1964). Evidence for a florigenic acid. *Nature* **202**, 559–61.

Longman, K. A. (1976). Some experimental approaches to the problem of phase change in forest trees. *Acta Hort.* **56**, 81–90.

Marks, M. K. and Prince, S. D. (1979). Induction of flowering in wild lettuce (*Lactuca serriola* L.) II. Devernalization. *New Phytol.* **82**, 357–63.

Meins, F. Jr. and Binns, A. (1979). Cell determination in plant development. *BioScience* **29**, 221–5.

Metzger, J. D. and Zeevaart, J. A. D. (1980). Effect of photoperiod on the levels of endogenous gibberellins in spinach as measured by combined gas chromatography–selected ion current monitoring. *Plant Physiol.* **66**, 844–6.

Metzger, J. D. and Zeevaart, J. A. D. (1982). Photoperiodic control of gibberellin metabolism in spinach. *Plant Physiol.* **69**, 287–91.

Mitra, S. N. and Vince, D. (1962). A study of internode elongation in rosetted plants of *Chrysanthemum. Advances in Horticultural Science* (Proc. 15th Int. Hort. Congr. 1958) **2**, 384–90.

Napp-Zinn, K. (1960). Vernalization, licht und alter bei *Arabidopsis thaliana*. I. Licht und Dunkelheit während Kälte-und Wärmebehandlung. *Planta* 54, 409–44.

Ogawa, Y. and King, R. W. (1979). Establishment of photoperiodic sensitivity by benzyladenine and a brief red irradiation on dark grown seedlings of *Pharbitis nil* Chois. *Plant Cell Physiol.* 20, 119–22.

Okuda, M. (1953). Flower formation of *Xanthium canadense* under long-day conditions induced by grafting with long day plants. *Bot. Mag. Tokyo* 66, 247–55.

Papenfuss, H. D. and Salisbury, F. B. (1967). Properties of clock resetting in flowering of *Xanthium. Plant Physiol.* 42, 1562–8.

Pharis, R. P., Ross, S. D., Wample, R. L. and Owens, J. N. (1976). Promotion of flowering in conifers of the *Pinaceae* by certain of the gibberellins. *Acta Hort.* 56, 155–62.

Pittendrigh, C. S. (1966). The circadian oscillation in *Drosophila pseudo-obscura* pupae: a model for the photoperiodic clock. *Z. Pflanzenphysiol.* 54, 275–307.

Pressman, E. and Negbi, M. (1980). The effect of day-length on the response of celery to vernalization. *J. Exp. Bot.* 31, 1291–6.

Proebsting, W. and Heftman, E. (1980). The relationship of (^{3}H) GA$_9$ metabolism to photoperiod-induced flowering in *Pisum sativum* L. *Z. PflanzenPhysiol.* 98, 305–9.

Purvis, O. N. (1940). Vernalization of fragments of embryo tissue. *Nature* 145, 462.

Purvis, O. N. and Gregory, F. G. (1952). Studies in vernalization of cereals. XII. The reversibility by high temperature of the vernalized condition in Petkus winter rye. *Ann. Bot.* 1, 569–92.

Robinson, L. W. and Wareing, P. F. (1969). Experiments on the juvenile-adult phase change in some woody species. *New Phytol.* 68, 67–78.

Rogler, C. E. and Dahmus, M. E. (1974). Gibberellic acid induced phase change in *Hedera helix* as studied by deoxyribonucleic acid–ribonucleic acid hybridization. *Plant Physiol.* 54, 88–94.

Rogler, C. E. and Hackett, W. P. (1975a). Phase change in *Hedera helix*: induction of the mature to juvenile phase change by GA$_3$. *Physiol. Plant.* 34, 141–7.

Rogler, C. E. and Hackett, W. P. (1975b). Phase change in *Hedera helix*: stabilisation of the mature form with abscisic acid and growth retardants. *Physiol. Plant.* 34, 148–52.

Salisbury, F. B. (1963). *The Flowering Process*. Pergamon Press, London, Oxford.

Salisbury, F. B. (1965). Time measurement and the light period in flowering. *Planta* 66, 1–26.

Salisbury, F. B. (1969). *Xanthium strumarium* L. In *The Induction of Flowering*, ed. L. T. Evans, Macmillan, Melbourne, pp. 14–61.

Salisbury, F. B. and Ross, C. (1969). *Plant Physiology*, Wadsworth Publishing Co. Inc., Belmont, CA, USA, p. 608.

Sarkar, S. (1958). Versuche zur Physiologie der Vernalization. *Biol. Zbl.* 77, 1–49.

Schneider, M. J., Borthwick, H. A. and Hendricks, S. B. (1967). Effects of radiation on flowering of *Hyoscyamus niger. Am. J. Bot.* 54, 1241–9.

Schwabe, W. W. (1954). Factors controlling flowering in the chrysanthemum. IV. The site of vernalization and translocation of the stimulus. *J. Exp. Bot.* 5, 389–400.

Schwabe, W. W. (1956). Evidence for a flowering inhibitor produced in long days in *Kalanchoe blossfeldiana*. *Ann. Bot.* 77, 1–14.

Schwabe, W. W. (1959). Studies of long-day inhibition in short-day plants. *J. Exp. Bot.* 10, 317–29.

Singh, L. B. (1959). Movement of flowering substances into the mango leaves (*Mangifera indica* L.). *Hort. Advances* 3, 20–8.

Suge, H. (1980). Vernalization and gibberellins in pea. *J. Jap. Soc. Hort. Sci.* 49, 203–10.

Takimoto, A. (1955). Flowering response to various combinations of light and dark periods in *Silene armeria*. *Bot. Mag. Tokyo* 68, 308–14.

Takimoto, A. (1969). *Pharbitis nil* Chois. In *The Induction of Flowering*, ed. L. T. Evans, Macmillan, Melbourne, London, pp. 90–115.

Takimoto, A. and Hamner, K. C. (1964). Effect of temperature and pre-conditioning on photoperiodic response of *Pharbitis nil. Plant Physiol.* 39, 1024–30.

Takimoto, A. and Hamner, K. C. (1965). Effect of far-red light and its interaction with red light in the photoperiodic response in *Pharbitis nil. Plant Physiol.* 40, 859–64.

Teroaka, H. (1972). Proteins of wheat embryos in the period of vernalization. *Plant Cell Physiol.* 8, 87–95.

Thomas, R. G. (1979). Inflorescence initiation in *Trifolium repens* L.: influence of natural photoperiods and temperatures. *New Zealand J. Bot.* 17, 287–99.

Thompson, P. A. and Guttridge, C. G. (1960). The role of leaves as inhibitors of flower induction in strawberry. *Ann. Bot.* 24, 482–90.

Vince, D. (1960). Low temperature effects on the flowering of *Chrysanthemum morifolium* Ramat. *J. Hort. Sci.* 35, 161–75.

Vlitos, A. J. and Meudt, W. (1955). Interactions between vernalization and photoperiod in spinach. *Contrib. Boyce Thompson Inst. Plant Res.* 18, 159–66.

Wareing, P. F. and Frydman, V. M. (1976). General aspects of phase change with special reference to *Hedera helix* L. *Acta Hort.* 56, 57–68.

Wellensiek, S. J. (1958). Vernalization and age in *Lunaria biennis*. *Proc. K. ned. Akad. Wet.* C61, 552–60.

Wellensiek, S. J. (1960). Stem elongation and flower initiation. *Proc. K. ned. Akad. Wet.* C63, 159–66.

Wellensiek, S. J. (1964). Dividing cells as a pre-requisite for vernalization. *Plant Physiol.* **39**, 832–5.

Wellensiek, S. J. (1965). Recent developments in vernalization. *Acta Bot. Neerl.* **14**, 308–14.

Wellensiek, S. J. (1970). The floral hormones in *Silene armeria* L. and *Xanthium strumarium* L. *Z. Pflanzenphysiol.* **63**, 25–30.

Zeevaart, J. A. D. (1958). Flower formation as studied by grafting. *Meded. LandbHoogesch. Wageningen* **58**, 1–88.

Zeevaart, J. A. D. (1969). *Perilla*. In *The Induction of Flowering.* ed. L. T. Evans, MacMillan, Melbourne, London, pp. 116–55.

Zeevaart, J. A. D. (1970). Gibberellins and flower formation. In *Cellular and Molecular Aspects of Floral Induction*, ed. G. Bernier, Longman Group Ltd, London, pp. 335–44.

Germination and dormancy

19

A M M Berrie

Introduction

A seed is the ripened product of an ovule, and at maturity contains an embryo that has been derived either sexually or asexually. Most work on seed germination is carried out on units of dissemination whether they be seeds proper or single-seeded fruits, *e.g.*, lettuce (a cypsella), wheat (a caryopsis) and sycamore (a samara), three species favored by seed physiologists; or even aggregates of fruits, which may consist of two as in cocklebur and winter wild oat, or many as in wild beet. The word seed will be used for all disseminules mentioned in this text.

A true seed is morphologically complex, containing tissues of different genetic composition. The embryo, in most cases, is diploid and, except for apomictic species, unlike the diploid state of the maternal tissues which are to be found in the coverings—the testa and pericarp. The endosperm, a nutritive tissue which may not persist to maturity, is triploid. Incompatibility may exist amongst these tissues and sometimes a hybrid embryo cannot survive because it is incompatible with a particular endosperm, not because of genetic problems in the diploid embryo.

The covering layers are usually considered to be dead, but they should not be thought of as inert, since they can influence the behavior of the seed at germination.

Some seeds have in addition a perisperm which is derived from the nucellus. Like the endosperm this is nutritive. Those seeds which retain the endosperm to maturity are termed endospermous while in non-endospermous seeds the endosperm is utilized com-pletely during maturation. In many species the residual endosperm consists of no more than two or three layers of cells which are not part of a food reserve. They become a regulatory tissue, *e.g.*, in lettuce.

Non-endospermous seeds store their food reserves in the cotyledons. Dure (1977) considers that this pattern may not be biochemically the most efficient since the seed first mobilizes the reserves of the endosperm and deposits them as food materials in the cotyledons, then on germination hydrolyzes and exports the cotyledonary reserves to the embryonic axis. The endospermous types keep the endosperm reserves *in situ* as storage substance except for small amounts of soluble sugars and fats in the axis (scutellum in grass). Storage in cotyledons may be physiologically more efficient because transport to the axis is easier than from an endosperm during the early stages of germination, as there are no vascular connexions between the genetically different tissues. In the endospermous forms translocation over much of the distance involved is by diffusion, as is communication from cell to cell. In non-endospermous species with food reserves in cotyledons there is likely to be more rapid propagation of stimuli and translocation of solutes.

Figure 19.1 illustrates the main morphological features of three typical seed types.

Definition and measurement of germination

A seed is said to have germinated when the radicle bursts through the outer coverings. Cell division and

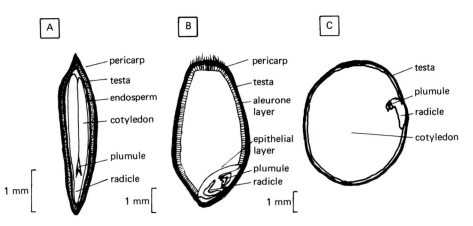

Fig. 19.1 Longitudinal sections of A lettuce, B wheat, and C pea 'seeds'.

elongation start at about this time and many seed physiologists take these events as terminating germination (Toole, Hendricks, Borthwick and Toole, 1956). The first phase, however, represents only part of what happens in seed germination and it is usual to include the processes which take place to ensure the establishment of the seedling.

In this account, phase I germination refers to the period up until the radicle emerges and phase II the subsequent events relating to seedling growth, dependent on seed reserves. These two phases can be separated physiologically because during phase I a seed can be desiccated and survive, though slightly changed, but if dehydrated in phase II it is killed (Berrie and Drennan, 1971; Sen and Osborne, 1974).

Germination is a quantile event and as such the percentage germination is distributed according to the binomial frequency function. This means that with small samples, and probabilities of germination removed from the extremes, i.e. <0.8 and >0.2, the usual sampling methods do not always give a good estimate of the true percentage germination of the seeds under test (Fig. 19.2). It can be seen that for a probability of germination of 0.5 (50%) and a sample size of 25 the 95% probability level extends from about 30 to 70% germination, a wide range indeed. Even when 100 seeds are tested, the 95% limits are 40 and 60% germination.

The rate of appearance of radicles is also important and various methods have been used to esti-

mate this parameter of the population (Orchard, 1977; Timson, 1965). The appearance of the radicles can be considered to be normally distributed in time and this allows the use of probit analysis to find the mean germination time and the rate of germination.

Table 19.1 A comparison of the use of germination percentage, and the parameters of the germinating population, in assessing the effects of treatments on the germination of lettuce seeds (from Berrie and Taylor, 1981).

Temp. of germination and treatment	Average percentage germination	Mean time to germinate (min)	Slope of probit line
27.5°C: red light (4 h from imbibition)	84*	1112**	6.33
27.5°C: 10^{-3} GA$_3$	78	1477	10.17

* Not significantly different.
** Significantly different at $p < 0.01$.

With lettuce seeds, which give the same percentage germination when treated with light or GA, it can be shown that the resultants are different populations and the conclusion that light induces the formation of GA is invalid (Berrie and Taylor, 1981); *see* Table 19.1.

(a)

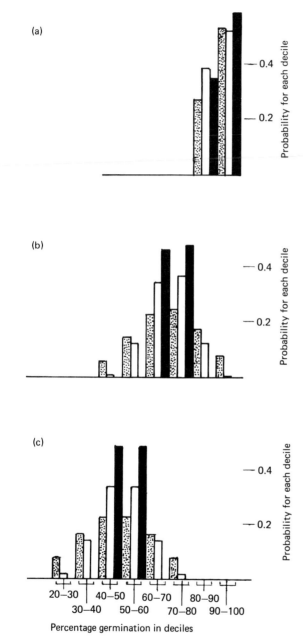

(b)

(c)

Percentage germination in deciles

Fig. 19.2 Distribution of germination by deciles for probabilities of germination of 0.9 (a), 0.7 (b) and 0.5 (c) for sample sizes of 10 seeds (stippled), 25 seeds (open) and 100 seeds (solid).

Seeds, water and solutes

Water content and viability

Seeds at maturity usually have a moisture content of less than 20% but a few species have seeds which cannot tolerate desiccation. In these, if the seed moisture content falls to less than 30% they are killed, so for the species to survive the moist seed must be shed on to moist soil. Such seed are usually found in tropical species, e.g., areca palm (Becwar, Stanwood and Roos, 1982; Harrington, 1974).

The majority of seeds are desiccation-tolerant and in dry storage their moisture content may fall to as low as 4%, though the actual value depends largely on the ambient humidity. Viability is lost with time and the rate of viability loss depends on the temperature and humidity of storage. Many empirical studies have been carried out on seed longevity and a set of charts, called nomographs, has been produced which allows the prediction of longevity of a particular species when stored in a given environment (Roberts, 1972). Storage in the absence of oxygen prolongs viability.

However, under natural conditions ripe seed are shed onto soil and are not likely to remain at a low moisture content. The soil contains large quantities of seed, not all viable; it has been estimated that average arable soils in the UK have about 2000 seeds m^{-2} while weedy soils may have as many as 55 000 seeds m^{-2} (Fryer and Makepeace, 1977).

The loss of viability has been attributed to protein denaturation, but old seed is a rich source of mutants. An alternative explanation is that nucleic acid is subject to fragmentation, and mutations occur so that eventually lethal mutants accumulate. An increase in DNase activity in dry stored rye has been reported, though RNase activity does not change. The DNase activity increase is due to loss of an inhibitor (Roberts and Osborne, 1973; Cheah and Osborne, 1978).

In many seeds, viability is retained under moist storage and it has been suggested that repair systems can operate only at normal water contents, unlike the destructive processes which can take place in dry seeds (Villiers, 1974). This repair process may be related to the turnover metabolism which is seen in dormant seeds (p. 452).

Water uptake

The 'dry' seed has a very low water potential (Chapter 14), perhaps as low as -100 MPa or 10^5 J kg^{-1}. Since most of this is due to matric potential it is difficult to obtain good estimates of the seed's water potential. Most values have been obtained by equilibrating seeds with soils of known water potential. There is every reason to believe that germination phase I is determined by the value of the matric potential rather than the osmotic potential because this phase is prevented to a greater extent than phase II by solutions which result in a 50% decrease in seedling growth (Hegarty and Ross, 1980/81). Germination is not possible in most species unless the water potential of the seed is greater than -1.5 MPa (Kaufmann and Ross, 1970).

When dry seeds are exposed to liquid water there is an initial rapid increase in moisture content which is followed by a slower uptake until the water content reaches about 60% of the initial dry weight. This plateau, of greater or lesser duration, is succeeded by a second increase in uptake (Fig. 19.3). This pattern is associated with the satisfaction of the matric potential—to hydrate macromolecules—and then the osmotic potential—to generate vacuolar sap and finally to permit growth in the axis. The overall pattern is a biphasic uptake but the plateau may be thought distinct so that there are three physiological phases, for it could be that during the plateau period specific germination events are

occurring (Bewley and Black, 1978; Shull, 1920), but the plateau may be virtually non-existent (*see* Fig. 19.3).

Heat-killed seeds show the same type of initial uptake but the second phase does not take place so it is assumed that water is absorbed by a purely physical process. However, the rate of water uptake is temperature-dependent and it has been thought that part of the process is physiological. Recently it has been shown that the temperature-dependent uptake can be correlated with the lowering of viscosity that takes place as the temperature rises and there is therefore no reason to suppose that the initial water uptake is other than physical (Murphy and Noland, 1982).

Since water may only be taken up at one point on the surface, *e.g.*, the micropyle, the movement of water through a large seed may not be homogeneous as wet tissues and dry tissues behave differently. The loss of solutes may be determined by the way in which cell constituents hydrate; a large seed may leach relatively more solute than a small seed and so be exposed to a greater chance of damage (Waggoner and Parlange, 1976).

Solute loss

During the early stages of water uptake, seeds lose solutes to the surrounding medium. The greater rate of loss corresponds to the very early stages while the seed has a water content of less than about 25% and there is no specificity of leached solutes. It has been concluded that the loss is passive (Simon, 1974; Simon and Raja Harun, 1972).

Phospholipids with a water content of less than 20% are claimed to adopt a hexagonal configuration so that a membrane in this state is not lamellar but possesses pores (Luzzati and Husson, 1962). Solutes could leak through these channels, passively, until the lamellar bilayer was established when normal physiological control systems would take over. This is an attractive hypothesis but when plant phospholipids from seed of *Lotus corniculatus* are hydrated to various degrees, to contain between 5 and 40% water, X-ray diffraction analysis shows all levels of hydration to exist in the lamellar configuration (McKersie and Stinson, 1980).

Fine-structure studies have shown that the bounding membranes of nuclei, mitochondria and proplastids of dry seeds appeared to have normal

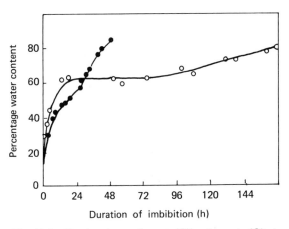

Fig. 19.3 Uptake of water by oats (●) and tomato (○) at 20° and 24°C respectively. The biphasic pattern is just observable in oat (from Berrie and Drennan, 1971).

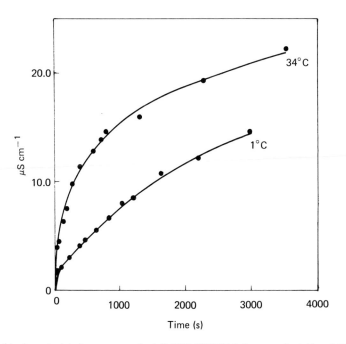

Fig. 19.4 Leaching of ionic materials from pre-washed (0.04% EDTA) lettuce seeds at 1° and 34°C. One gram of seeds were placed in 100 ml distilled water.

morphology, but that the internal membranes of these organelles were abnormal and the plasmalemma might not have been entire. There are difficulties associated with the preparation of dry seed tissue for examination so some of the features seen may have been artefacts. Within 20 min of the onset of imbibition all the membranes showed normal morphology (Chabot and Leopold, 1982).

Leaching is temperature-dependent (Fig. 19.4). Some species which are low-temperature sensitive leach more solute at temperatures just above freezing than at 20°C but most species show increased leaching with rising temperature, often exhibiting a sharp transition between 30° and 32°C (Hendricks and Taylorson, 1976, 1979; Leopold, 1980). Temperature-induced dormancy may be correlated with this transition and will be considered later (p. 446).

Seed dormancy

Plants which grow in regions with pronounced seasons have developed mechanisms that ensure surviv-

al during the unfavorable periods. Growth is prevented by low temperatures or water stress, often accompanied by supra-optimal temperatures. The survival mechanisms which allow persistence from year to year are short-term, but in many instances plant structures such as seeds, storage organs and buds can survive for decades or even centuries. Long-term survival is found in plants that appear first after the catastrophic destruction of existing communities, as happens after fire.

Dormancy, or the ability to retain viability while having restricted metabolic activity, and no observable growth, has evolved in both seeds and buds, but because of the ease with which seeds can be used for experimental purposes much more is known about dormancy in these organs. Much of the knowledge gained from observations on seeds may apply to buds, since the seed may be likened to a bud not attached to a mother plant.

Dormancy must be lost at a time when conditions are suitable for seedling growth, in the spring, and in a seed which is suitably placed in the soil for the seedling to be established before its food reserves are exhausted. There must then be sensors, or perhaps only one sensor, that detect conditions of the environment associated with the changes of the season, and the seed's position in the soil.

Physiologically a seed provided with adequate water, sufficient oxygen for normal aerobic metabolism, and surrounding temperature within physiological limits, but which does not germinate is termed dormant. A seed held in conditions not satisfying the above will not germinate and is said to be quiescent, *e.g.*, dry seed in a jar. Many seeds exhibit dormancy immediately on harvest and these are said to possess innate dormancy; others have dormancy induced by adverse environmental conditions such as high temperature (Borthwick and Robbins, 1928), reduced oxygen tension (Edwards, 1969) or water stress (Khan, 1960). Some authors consider innate dormancy as primary dormancy and induced dormancy as secondary dormancy since some seeds may show both types, the innate dormancy being broken but subsequently the seed being made dormant again.

Dormancy may be thought of as rest and deep dormancy would be the mid-rest period. At this stage dormancy can be broken only by very specific environmental conditions such as fluctuating temperature or light treatments or both. Away from the mid-rest period the environment has less influence on the seed (or bud) behavior, and eventually, as the organ moves out of rest, germination or growth will take place without the need for special environmental conditions (Vegis, 1964).

If innately dormant seeds are kept in dry storage, *i.e.*, quiescent, the degree of dormancy usually declines and eventually the seeds are non-dormant (Fig. 19.5). Exceptionally the opposite is the case and a seed which is not innately dormant becomes dormant on dry storage, *e.g.*, hazel (Jarvis, 1975). The loss of dormancy on dry storage has been called after-ripening but it is better to restrict this term to seeds which require to overcome an immature state, physiological or morphological, that exists when the seeds are shed at harvest. This after-ripening takes place only in moist seed, and seeds which require it, if forced to germinate without after-ripening, will produce abnormal seedlings, usually dwarfed, *e.g.*, peach, and many members of the Ranunculaceae.

Loss of dormancy in dry storage is temperature-dependent; the higher the temperature, the more rapidly is dormancy lost. If seeds are kept dry at

$-15°C$, or below, dormancy persists for long periods and viability is also kept. Seeds are not the only organs in which temperature determines the rate of loss of dormancy. Potato varieties vary considerably in sprouting during storage. The cultivar Majestic remains in rest for 28 weeks at 4.4°C but Craig's Defiance sprouts in eight weeks. At

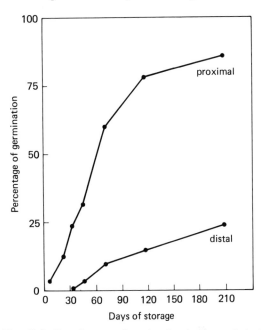

Fig. 19.5 Post harvest after-ripening in *Avena ludoviciana* held at room temperature. The proximal grain is inherently less dormant than the distal (data from Morgan, 1968).

22.5°C these varieties sprouted after eight and three weeks respectively (Burton, 1963). The water content of potato tubers is sufficient to sustain metabolism so it is easy to consider that loss of dormancy is associated with biochemical changes, but with dry seed no sort of metabolism seems to be taking place, and it is difficult to accept that dormancy wears off because of metabolic changes. Very little respiratory CO_2 is produced by dry seeds.

Most seeds other than crop species are dormant, to a degree, at harvest but some can germinate immediately, losing viability quickly, sometimes within a week. This is true of *Salix*, the willows. In a few species the seeds germinate on the mother plant, *e.g.*, mangroves. These seeds produce a heavy radicle and fall like a dart to embed in the mud of the estuarine reaches. This behavior may have arisen because phase I germination in mangrove species is more susceptible to the saline conditions of the mud than phase II (Mayer and Polyjakoff-Mayber, 1982).

Seed behavior is obviously closely associated with the ecology of the species. The conditions which break dormancy can often be deduced from the climatic conditions that prevail in a species' area of distribution and its phenology.

Types of dormancy

The factors or treatments which break dormancy are used to describe the type of dormancy exhibited by a seed. This leads to a catalog without offering an explanation of dormancy except in the type that will be discussed first, namely 'hard' seed.

Hard seed

If a seed is exposed to standard germination conditions (p. 445) and does not germinate because it is quiescent then it is hard, not dormant. Hardness is associated with the impermeability (partial or complete) of the outer covering to water, water-'hard' seeds, and to the metabolically important gases CO_2 and O_2, gas-'hard' seeds.

The former are easily seen because they do not swell but the latter cannot be recognized because the swollen seed does not show obvious indication of its state. Hardness is broken if the coverings are ruptured to allow the influx of fluid, but since puncturing may initiate events other than producing a physical opening, gas-hard seeds are difficult to determine.

Members of the Fabaceae (and the other legumes) are almost universally water-hard. The condition is also found in the Lamiaceae and Malvaceae. The waxy testas are wholly impermeable and water can enter only at the micropyle or the micropylar region. Any natural opening is blocked by a plug which is displaced by some factor of the environment such as fire. In *Albizia lophantha*, boiling the seed for two minutes displaces the plug and allows the seed to imbibe water at 25°C but if the opening is sealed with Araldite the hard condition is retained (Dell, 1980). Under natural condi-

tions unplugging takes place after a forest fire but the testa may be ruptured by scarification from soil movement or microbial action over a period of time.

Legume seeds are long-lived; an outstanding example of this was seen in World War II when seeds of herbarium specimens in the British Museum which were damaged by fire were water germinated. The oldest seed to germinate was that of *Albizia julibrissin* (147 years old, Anon, 1942).

Seeds which are gas-hard are probably widespread. If the testa is impermeable to CO_2 then respiratory CO_2 is kept within the seed to induce CO_2 narcosis. This prevents further development and the seed will not germinate. White mustard (*Sinapis alba*) and charlock (*Sinapsis arvensis*) are gas-hard but in both cases if the seed is dried and rewetted the testa becomes permeable. In the former, CO_2 seems to be the gas with restricted flux, but in the latter, it is O_2, lack of which allows the build-up of an inhibitor (Edwards, 1969; Kidd and West, 1917).

Many dormant seeds can be made to germinate by increasing the partial pressure of oxygen (pO_2), and this finding has led to the conclusion that the testa is only sparingly permeable to O_2 under ambient atmospheric conditions, thus not allowing an adequate entry of this gas to meet the needs of the embryo. The best example of this is the upper seed of cocklebur which is much more dormant than the lower one, but will germinate at high pO_2 levels. It has been found that oxygen is required to degrade an inhibitor the oxidation of which releases the seed from dormancy (Wareing and Foda, 1957; Porter and Wareing, 1974).

Temperature-sensitive seed

Fluctuating temperatures usually promote germination. Temperatures in nature follow both seasonal and diurnal cycles and also in the soil the amplitude and character of the cycle varies according to the depth (Russell, 1973). Seasonal variations are based on long periods, of the order of weeks, within one range of temperatures, *e.g.*, in winter, to be followed by another long period at higher temperature. The diurnal variations found in the uppermost few centimeters of soil, where seeds will germinate, are extreme, from as low as 5°C to as high as 35°C for the main 24-h cycle. During the day, short-term fluctuations of about 30 min duration can occur depending on the input of solar radiation and the wetness of the soil.

In many species dormancy is broken by exposing the moist seed to low temperature, between 0°C and 10°C, for anything from 7 to 180 days. This is termed stratification, and may be compared to vernalization with respect to flower induction. There are extensive lists of the optimum temperatures and times of stratification for a number of species but it appears that the average is 5°C for 100 days (Crocker and Barton, 1953). Species which show this behavior are wild oat (*Avena fatua*), birch (*Betula pubescens*) and ash (*Fraxinus* spp.). While kept at stratifying temperatures the seed may remain quiescent, and germination can only take place when the seed is exposed to higher temperatures. Since there is a time/temperature relationship in stratification, this method of breaking dormancy also senses the passing of the adverse season. Temperatures below freezing do not result in stratification.

A dormant seed which germinates when exposed to diurnal fluctuations is responding to a system which senses its position in the soil profile. It is found that limits of the cycle are separated by about 15°C and the most favorable temperatures are around 20° and 35°C, though other patterns may prevail and this response is not confined to temperate species (Attims and Côme, 1978; Totterdell and Roberts, 1980).

An example which shows the combined effects of stratification and germination in diurnal fluctuation of temperature is given in Table 19.2 and it can be seen that the effects of these treatments are quantitative rather than qualitative.

Table 19.2 Percentage germination of *Polygonum persicaria* stratified at 4°C for the number of days stated, and afterwards exposed to various temperature regimes (data of S. Mensah).

Duration of stratification (days)	Temperature regime		
	15°C constant	35°C constant	(8h at 15°C: 16h at 35°C)
0	0	20	28
14	40	45	90
28	55	67	98

High temperatures may remove dormancy, *e.g.*, in birch (Black and Wareing, 1955), but more often high temperatures induce dormancy (thermodormancy) which is often broken by an exposure to light (Reynolds and Thompson, 1971).

Light and seed germination

Seeds which respond to light are termed photoblastic. Non-responding seeds are non-photoblastic. Light may induce or promote germination, these seeds being positively photoblastic, or light may prevent or retard germination, in which case the seeds are negatively photoblastic.

Two types of positively photoblastic seeds occur. In the first a single brief exposure to red light (2 mmol m^{-2} at 660 nm) induces germination, and such seeds can be termed photosensitive. The second type of seed will only germinate when given protracted, repeated exposures to light; these are called photoperiodic. Both types are under phytochrome control. Photosensitivity in seeds is often induced and in lettuce the inductive agent may be temperature, water-stress or exposure to chemicals (Borthwick and Robbins, 1928; Khan, 1960; Berrie, Parker, Knights and Hendrie, 1968). Some seeds, *e.g.*, tomato, may be made photosensitive by exposure to far-red radiation (720 nm) (Mancinelli, Borthwick and Hendricks, 1966). In all cases photosensitivity is lessened, or even removed, by germinating the seed at low temperature (Berrie, 1966; Thompson, 1974).

Photoblastism in most cultivars of lettuce develops during imbibition at 30°C and the seed reaches a maximum sensitivity to light some 4–8 h from imbibition. Gradually the sensitivity declines. If not induced to germinate the seed may enter a stage of secondary dormancy which is difficult to break. The breaking of dormancy is fluence-dependent and if induced to germinate by red light the promotion can be reversed by exposure to far-red. Again this is fluence-dependent, and more far-red radiation is required than red. The interval between the red and far-red treatments affects the far-red response, and if four or more hours elapse the exposure to far-red is without effect, except at high doses (Bewley, Negbi and Black, 1968). Figure 19.6 illustrates these features in a strain of Grand Rapids in which thermodormancy is induced at 28°C.

Thermodormancy can be broken by treating seeds

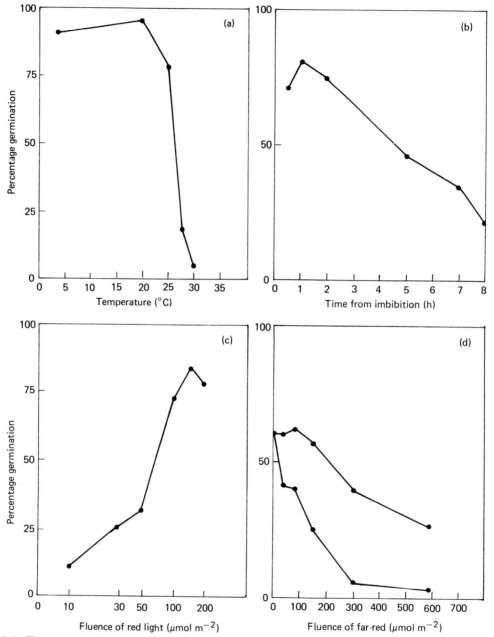

Fig. 19.6 The response of lettuce seed c.v. Grand Rapids germination to temperature and light (a). The induction of thermodormancy by high temperature (b). Sensitivity to red light from time of imbibition (660 nm: 100 μmol m^{-2}) (c). The dependence of germination on fluence of radiation of 660 nm (d). Photoreversal of red-induced germination (100 μmol m^{-2}) by far-red (730 nm) radiation. Lower curve: far-red given immediately after red; upper curve: far-red given 4 h after red irradiation (from Berrie, 1966, and unpublished).

with high doses of gibberellins (GA) (Table 19.2) but not all GAs are equally effective (*see* p. 452). Other growth regulators, *e.g.*, ethylene, kinetin and certain inorganic substances can also substitute for light (Hendricks and Taylorson, 1972; Ketring, 1977; Miller, 1956, 1958). The seeds, however, are responsive to a number of substances, and to incorporate each and every promoter and inhibitor into a general scheme is fraught with difficulty. One hypothesis is that light promotes the synthesis of GA which then stimulates germination, but this is not tenable because light-induced and GA-induced populations are different (Table 19.1). It has been shown that light acts within five minutes, which is too short a time for the *de novo* production of GA, and also that light and GA act synergistically (Bewley *et al.*, 1968). None of these findings would be expected if light induced the production of GA. Nor does it appear that light treatment results in the production of the other promoters.

Photoperiodism

Many early workers (*e.g.*, Gassner, 1915a,b) had demonstrated that some seeds germinated better in the light but it was not until the early 1950s that the term 'photoperiodism' was used with respect to seeds (Isikawa, 1954). This usage implies that seed germination may be like flower induction, but in seeds the photoperiodic behavior is not fixed. *Eragrostis ferruginea* appears to behave as a short-day

seed because germination is inhibited if a light flash is given in the dark period. However, both red and far-red radiation are effective, and this is unlike flowering behavior in a SD plant. Phytochrome is involved, since far-red-induced inhibition is reversed by a red irradiation immediately afterwards (Fujii, 1962).

In most cases increasing the exposure to light enhances germination, and irregular or intermittent irradiations are also effective, *e.g.*, *Epilobium cephalostigma* (Isikawa, 1962).

The interaction between photoperiod and temperature is complex and very specific. In birch chilled (stratified) seed are not photoblastic but unstratified seed, if germinated above 20°C, respond to the same extent regardless of photoperiod, although long days are more effective below 20°C. In 'artificial' days of more than 24 h duration but with a standard light period, germination is reduced as the 'night' increases in length (Black and Wareing, 1955). With *Chenopodium botrys* short days enhance germination in the temperature range 10° to 20°C but long days are optimal at 25° to 35°C. This species also responds according to the red/far-red light ratio, germinating only when there is sufficient red light. Thus in the field germination will not occur under a leaf canopy (Cumming, 1963). A similar situation exists in *Rumex* spp. (Totterdell and Roberts, 1980).

Many classifications of the main types of dormant seeds have been made, the most recent recognizing 15 forms of dormancy based mainly on the nature of

Table 19.3 A classification of seed dormancy, simplified (based on Nickolaeva, 1977).

	Exogenous dormancy (*i.e.*, not associated with the embryo)	Endogenous dormancy (associated with the embryo)		
Class	A	B(morphological)	B–C (Combined) BC (Complex)	C(physiological)
Sub-class	A_{ph}(water hard) A_{ch}(inhibitors in seed coverings) A_{mech}(restraint of embryo expansion)	B(immature embryos)	B–C_2 B–C_3	C_1(weak) C_2(moderate) C_3(strong)

e.g., *Avena fatua*, spring wild oat is C_2 (moderate physiological dormancy) but could be considered as A_{ch} because of the presence of inhibitors in the husks.
Anemone spp. are B since the embryos are immature when the seeds are shed.

the stratification needed and how the stratified seeds respond. The main categories are: A, not associated with the embryo; B, morphological; and C, physiological dormancy of the embryo. An embryo may possess both B and C types of dormancy (Nikolaeva, 1977). A simplified form of this classification is given in Table 19.3.

Epicotyl and double dormancy

These are species in which the different parts of the axis have separate dormancy characteristics. One class has non-dormant radicles and dormant plumules (epicotyl dormancy), and the other has radicles the dormancy of which is broken after one stratification but the plumules only 'germinate' after a second stratification (double dormancy). These are the so-called 'two-year' seeds, e.g., tree paeony and wake-robin.

Artificial breaking of dormancy

Many dormant seeds can be induced to germinate by puncturing or removing the coverings or excising the embryos (Hart and Berrie, 1966; Roberts and Smith, 1977). Mechanical damage to seeds might occur naturally, e.g., in soils subject to frost heave or after passage through an animal's digestive tract, but it is unlikely that germination in the field is solely dependent on either of these processes.

Surgical treatment of seeds may allow a hard condition to be removed, a mechanical ('strait-jacket') restraint released, an inhibitor removed, or metabolic activity initiated. In the first and second instances the treatment effect is self-explanatory but in seeds with 'strait-jackets', e.g., coconut palm, there is often a weak region through which the radicle grows. Some seeds, e.g., *Alisma plantago, Rubus fruticosa* and *Rosa* spp., may exhibit mechanical restraint of the embryo and chemical scarification with sulfuric acid will remove the 'strait-jacket'; often, however, stratification is sufficient (Crocker and Davis, 1914; Rose, 1919; Blundell and Jackson, 1971).

Naturally-occurring germination inhibitors are widespread and are often found in seeds, particularly in their coverings. The inhibitors may be specific, involved in growth regulation, e.g., ABA, or non-specific, e.g., phenolic acids. Some inhibitors have a restricted distribution, being found in a few families,

and are often very specific affecting only germination, e.g., coumarins (Murray, Mendez and Brown, 1982) and phthalides (Moewus and Schader, 1951). The Apiaceae is a family rich in these compounds and also terpenoid inhibitors. The phenolic compounds are water-soluble and are lost on leaching so the seed is released from auto-inhibition after enough phenolic has been removed by adequate rainfall. The other compounds are volatile and may be lost as vapor, and again auto-inhibition declines with time. ABA, on the other hand, may be involved intimately in the control system (*see* p. 454) (Sondheimer, Tzou and Galson, 1968).

Wounded seeds show increased dehydrogenase activity especially at the site of the wound. By initiating these redox reactions the whole metabolism of the seed may be invoked and germination may result. This seems to be so in wild oats (Hay, 1962). Such a finding provides circumstantial evidence that the control mechanism of germination is based on respiratory systems.

Germination can also be induced by treating seeds with chemicals which seem not to be growth regulators or not to promote metabolic activity, e.g., nitrate, nitrite, cyanide, azide, hydroxylamine, thiourea (Hendricks and Taylorson, 1972). The growth promoters GA, cytokinins, and ethylene usually induce germination of dormant seeds. The breaking of dormancy by chemicals will be considered later.

Inception of germination

During phase I the seed changes from being biologically inert to becoming a fully integrated and actively metabolizing axis. Since dormant seeds do not attain the state of active axis extension, a comparison of dormant and non-dormant batches of the same species, or closely related species, e.g., common oat and wild oat, might be thought to lead to the discovery of the 'trigger' for germination. The essential elements of this trigger may be 'pre-' or 'pro-' formed, like a zymogen, in the dry seed to be activated on hydration; or they may be created *de novo*.

The provision of respirable substrate was thought to be essential and this would come from the carbohydrate or fat reserves of the seed. However, the amylolytic and lipolytic enzymes only begin to be abundant during phase II and are associated with mobilization of food reserves to the extending axis. In oats (Fig. 19.7) only β-amylase is present in the dry grain and α-amylase appears only after the radicle has ruptured the coverings. In dormant samples α-amylase is not produced. A similar situation is present in pea with three different amylolytic enzymes (Shain and Mayer, 1968) and with lipase in wheat (Tavenar and Laidman, 1972). The trigger is not the provision of low molecular weight respirable substrate. Any soluble substrate needed is present in the dry seed usually as sucrose or raffinose.

Isolated wheat embryos (Marcus, Feeley and Volcani, 1966) and bean axes (Gillard and Walton, 1973) begin the synthesis of protein very shortly after the start of imbibition. In wheat, protein precursor ([^{14}C]leucine) began to be incorporated within 10 min, and by 20 min there was evidence of rapid protein synthesis. All types of protein were being produced within one hour. Bean axes were similar and it was shown that there was no difference between the proteins synthesized at one and five hours from imbibition, so the synthesis of a specific trigger protein may be ruled out.

This very early protein synthesis suggests that much of the machinery for protein synthesis is present in the dry seed. A brei of dry wheat embryos could support protein synthesis if ATP/GTP, mRNA, and ribosomes were added. If the dry seed contained functional mRNA and ribosomes then early protein synthesis can be explained.

Cordycepin does not prevent the early protein synthesis (Gillard and Walton, 1973; Spiegel and Marcus, 1975). Since this antibiotic prevents mRNA synthesis and also polyadenylation, this finding is good evidence that the seed has preformed mRNA and that polyribosomes are produced directly on hydration (Dure, 1979). Water uptake primes the system and polyribosome and endoplasmic reticulum formation must be rapid. Visual evidence of this should be obtainable and soya bean cotyledon tissue has been examined at the fine-structure level. After imbibition for 20 min at 25°C, abundant ER with patterned arrays of polyribosomes were observed. There was no evidence of ER in dry tissue (Webster and Leopold, 1977).

Since dormant seeds do not germinate when imbibed, they might be expected to show a different pattern from that described above for non-dormant seeds. However, when dormant and non-dormant

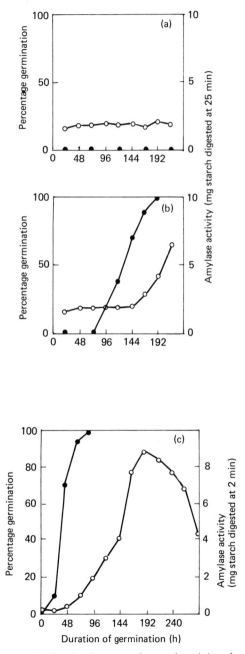

Fig. 19.7 The development of α-amylase ($\circ$) and germination ($\bullet$) in seeds of: (a) dormant *Avena ludoviciana*, (b) non-dormant *A. ludoviciana*, and (c) *A. sativa* (Drennan and Berrie, 1962).

wild oats were supplied with [^{14}C]leucine as precursor, the amount of label incorporated into protein was the same in both samples (Chen and Varner, 1970). This finding was confirmed independently and in addition, by looking at specific membrane proteins, it was shown that both dormant and non-dormant samples turn-over, rather than accumulate, protein. Only when the non-dormant embryos began to expand did their incorporation exceed that of the dormant (Fig. 19.8) (Cuming and Osborne,

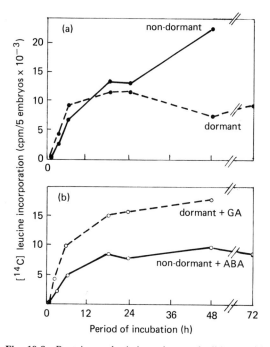

Fig. 19.8 Protein synthesis in embryos of wild oats. (a) Non-dormant and dormant seeds kept in water to retain their physiological status. (b) The dormant seeds induced to germinate by treatment with 10^{-4} M GA and the non-dormant seeds treated with 10^{-4} M ABA to prevent germination. (From Osborne and Cuming, 1979.)

1978a,b; Osborne and Cuming, 1979). These authors also showed that phospholipid behaved like protein. The dicotyledon *Xanthium pennsylvanicum* also showed little difference between protein synthesis in dormant and non-dormant samples prior to axis extension in the latter (Satoh and Esashi, 1979).

Protein synthesis alone cannot be the trigger for germination since dormant and non-dormant seeds

behave in the same way. DNA synthesis does not begin until long after imbibition and protein and RNA synthesis (Chen and Osborne, 1970). It has been suggested that the early events seen in both types of seeds are repair mechanisms to make good biochemical lesions that arise in storage (Berjack and Villiers, 1972; Villiers, 1974). The unique feature, if such exists, which characterizes phase I germination has yet to be found.

Germination and growth regulators

While both plants and seeds respond to applications of both naturally-occurring and synthetic growth regulators, there is a tendency among some physiologists to dispute that plant hormones have any regulatory function (Bewley, 1979; Trewavas, 1982). Certainly it is not always possible to correlate the physiological status of a seed with its hormone content, *e.g.*, gibberellins (Black, 1980/81) and ABA (Walton, 1980/81). The great danger is in the interpretation of results from experiments using very high, pharmacological doses of regulator; the findings cannot be disregarded but should be treated with caution.

Gibberellins

Gibberellin, at high concentration, can break the dormancy of positively photoblastic seed, *e.g.*, *Lactuca scariola* (Lona, 1956), negatively photoblastic seed, *e.g.*, *Phacelia tanacetifolia* (Chen and Thimann, 1964) and non-photoblastic seed which require stratification, *e.g.*, wild oat (Black and Naylor, 1959). It is also effective in breaking stress-imposed dormancy (Khan, 1960). The gibberellins are not equally effective; GA$_4$ is most active and able to break dormancy when applied at 10^{-4} M. In some species, *e.g.*, 'ever-bearing' strawberry, it was the only GA able to break dormancy, GAs 1, 3, 5, 7 and 9 being ineffective (Thompson, 1968). Though needed in large amounts, on the limits of physiological concentrations, GA might appear to be a universal dormancy-breaking agent and this led to the belief that other dormancy-breaking treatments induced the production of GA, but this view as already stated, seems to be invalid.

Lettuce seeds, dormant and non-dormant, synthesize protein as described for wild oat, but there are conflicting reports of the effects of GA on this process. For example, no difference, in [^{14}C]leucine incorporation into protein, between treated and untreated seeds, was observed by Bewley and Black (1972) whereas GA-treated seeds were found to incorporate more leucine than the untreated ones by Fountain and Bewley (1976). The latter authors claim that the methods used to extract protein can determine the results. GA increases the percentage polyribosomes after three hours but there is no change in mRNA (Frankland, Jarvis and Cherry, 1971). In all these cases, however, the effect of GA may not be causal since inducement to germinate results in the events occurring on the same time-scale as germination.

During seed maturation GA content rises and then declines. Immature seeds may not show dormancy and supplying developing heads of oats with GA reduces the amount of dormancy (Black and Naylor, 1959; Morgan and Berrie, 1970). It is possible that endogenous GA does control dormancy but there is no conclusive proof that GA initiates germination.

Extracts of hazel seeds during stratification showed an increase in GA-like substances in bioassays (Frankland and Wareing, 1962; Ross and Bradbeer, 1971). Bioassays, however, do not discriminate amongst the GAs and any increase in activity could have been due to hydrolysis of bound GA or interconversion among GAs to give larger amounts of those GAs most effective in the bioassay (Barendse, Kende and Lang, 1968; Hiraga, Yokota, Muorfushi and Takahashi, 1972; Reeve and Crozier, 1975). The changes in bioassay response reflect changes in the stratifying seed but their relation to germination is not clear. Hazel seeds treated with GA synthesize more protein and RNA (Jarvis, Frankland and Cherry, 1968), though the methods used are open to doubt (Bewley, 1979). Not all cold-requiring seeds show this increase in GA-like activity during stratification (Webb, Van Staden and Wareing, 1973a,b).

If GA is incorporated into liposomes their permeability towards glucose is altered. Phospholipid bilayers can also incorporate GA with altered transition temperatures (Wood and Paleg, 1974; Wood, Paleg and Spotswood, 1974). Berrie (unpublished) tried to obtain ESR spectra of bilayers incorporating spin-labeled gibberellin but the spectra could not be interpreted. GA may act by affecting membranes and though there was no effect on the amount of ER in GA-treated aleurone cells its physical properties were altered (Chrispeels and Jones, 1980/81).

Cytokinins

Kinetin can induce germination in photosensitive lettuce seeds and, like GA, a high concentration is needed (Miller, 1956, 1958; Thomas, 1977). Celery seed potentiated to germinate with sub-threshold amounts of GA can also be made to germinate by treatment with cytokinin (Biddington, Thomas and Dearman, 1980).

Different cytokinins differ in their effectiveness and the synthetic compounds are more effective than the natural. This may be due to their being absorbed more readily, or not being degraded by the enzymes involved in cytokinin metabolism (Biddington and Thomas, 1976). The synthetic cytokinins are inhibitory at high concentrations.

Increases in the amounts of extractable cytokinin have been found after stratification and light treatment in cold- and light-sensitive seeds (van Staden and Wareing, 1972; van Staden, Webb and Wareing, 1972; van Staden, 1973).

The naturally-occurring cytokinins are found in tRNAs located near the anti-codon (Kende, 1971). This finding would imply that if cytokinins were present in insufficient amounts then tRNAs would also be deficient and protein synthesis depressed. The cytokinins are very effective promoters of germination and since early protein synthesis is a necessary event, exogenously applied cytokinin could enhance protein synthesis, allowing phase I to be completed.

Ethylene

Like GA and cytokinin, ethylene promotes seed germination. The early work involving ethylene and other low molecular weight unsaturated hydrocarbons has been reviewed by Ketring (1977).

Ethylene is produced by germinating seed and the amounts appear related to the number of individual seeds germinating. Lettuce seed kept in various conditions—darkness, light or low temperature—each produced the same amount per germinating seed. It is easy to correlate the percentage germination with the amount of ethylene but it is clear that the production of ethylene is an effect of germination and not its cause (Abeles and Lonski, 1969). However, dormant seed of peanut can be made to germinate if treated with ethylene (Ketring and Morgan, 1970).

Recently it has been shown that ethylene can substitute for far-red radiation in inhibiting red-promoted germination of *Potentilla norvegica*, sequential treatments of red light and ethylene being the same as red and far-red (Suzuki and Taylorson, 1981).

Since ethylene is preferentially soluble in lipids it will partition into phospholipid membranes and might, like GA, alter the physical properties of the bilayer. Phytochrome could be subject to ethylene action by there being local, or general, perturbations in membrane architecture which might alter the configuration of phytochrome.

Abscisic acid

ABA is relatively easy to analyze in plant tissues and there is a large quantity of information on the endogenous amounts in different organs at different times during development. Unfortunately, there is not a clear-cut association between the physiological state of the seed and ABA content though some generalizations are possible (Walton, 1980/81).

Dry dormant seeds usually have relatively high concentrations of ABA compared with non-dormant samples, but in some cases there is little difference, *e.g.*, between non-dormant *Avena sativa* and dormant *A. fatua* (Berrie, Buller, Don and Parker, 1979). At germination there is either little difference in the amounts present in dormant and non-dormant seeds, or the capacity to germinate cannot be correlated with the ABA content, *e.g.*, in lettuce (Braun and Khan, 1975; Berrie and Robertson, 1976). Some seeds which require stratification lose ABA during the low-temperature treatment (Webb *et al.*, 1973a), but others such as ash (Sondheimer, Galson, Tinelli and Walton, 1974) and hazel (Williams, Ross and Bradbeer, 1973) do not show the decline even though only stratified seed germinates. A difficulty surrounding ABA inhibition of seed germination is the role of temperature. It is probable that ABA is more effective at high temperatures than low, and stratification reduces the effectiveness of an inhibitor, thus allowing phase I to be completed at a slow rate (Orlandini and

Bulard, 1977). No change in endogenous content is needed for a change in seed behavior.

Another problem regarding ABA, and all other hormones, is its distribution throughout the seed. ABA will be more effective in the axis than in the seed coverings. A redistribution from the axis to the coat without a change in the total amount would permit a seed to germinate, or if *vice versa* make it dormant, *e.g.*, apple (Rudnicki, 1973) and ash (Sondheimer *et al.*, 1968).

ABA inhibits certain enzymes associated with germination, *e.g.*, endo-β-mannanase in lettuce, but the reduced enzyme activity may be due to the reduced germination and thus not the primary action of ABA (Halmer, Bewley and Thorpe, 1976). ABA also affects RNA metabolism in non-dormant bean embryos and suppresses growth, but at the concentrations used, up to 10^{-4} M, did not inhibit growth completely (Walbot, Clutter and Sussex, 1975).

Other naturally-occurring growth regulators

IAA seems not to be involved in seed germination. Its role is probably associated with extension growth in the axis. Some attempts have been made to determine the amounts present in lettuce (Robertson, Hillman and Berrie, 1976). The amounts of IAA present in germinating and non-germinating batches were not significantly different, and in every instance where promoted seeds were treated with IAA germination was reduced. Those authors conclude by saying '. . . we question if IAA has any regulatory role in the germination of Grand Rapids lettuce seed'.

Table 19.4 Induction of germination in photosensitive lettuce seeds (cv. Grand Rapids) by gibberellic acid. Seeds kept in darkness at 28°C.

Conc. of GA$_3$ (mmol m^{-3})	0	50	100	500	1000
Percentage germination	3	37	59	86	92

Mention has already been made of phenolics, coumarin and phthalides. Coumarin and many of its derivatives can depress germination of wheat and lettuce (Mayer and Evenari, 1952) but when substituted at the 6 and/or 7 position with oxygenated functional groups the effect is reduced or removed. Disubstitution completely removes any biological activity possessed by the coumarin moiety. Berrie *et al.* (1968) showed that photoblasty in lettuce seeds was induced by the coumarins according to the same rules that applied to reduction of wheat germination. It was suggested that the lactone ring of the coumarin could occupy an active site that was normally occupied by a GA molecule through its lactone bridge, since there was a marked interaction between the coumarins and GA.

Much time has been expended in examining coumarin's effect on lettuce (*see* Mayer and Poljakoff-Mayber, 1982, for a summary) and even though these components occur widely in plants (Murray *et al.*, 1982), it is thought that they do not control germination.

Recently Berrie, Don, Buller, Alam and Parker (1975) proposed that certain volatile fatty acids may have a regulatory role in seed germination. They showed that these acids were present in a wide range of species and that the amounts could be correlated with the degree of dormancy shown by the organ. One acid in particular, nonanoic, when applied exogenously could influence thermodormancy in lettuce seed, lowering the temperature at which the seed required a light treatment to germinate. The acids could also affect the barley aleurone α-amylase system (Buller, Reid and Parker, 1975). Tuning (1981) considers that these substances are the only ones which can be correlated with the post-harvest dormancy found in barley.

However, from a concentration standpoint the effect of these acids is very abrupt, and though this can explain a threshold or switching type of response there is a distinct possibility that high concentrations are lethal.

The site of action of these acids is probably the phospholipid membrane. Liposome and spin-labeled studies indicate that if the acid is incorporated into the bilayers the temperature of the phase transition is reduced.

Interactions among growth regulators

The concept that germination is controled by interaction among growth regulators is based on the responses of seeds to mixtures of substances applied together, or given in sequence. Nearly always, the inhibitory effect of ABA can be overcome by cytokinin but not by GA, but GA will overcome the dormancy imposed by coumarin (Table 19.5). Physiological response to substances applied simul-

Table 19.5 Interaction between GA_3 and coumarin on the germination of lettuce seeds. Seeds kept at 20°C in darkness (from Berrie *et al.*, 1968).

Coumarin (mmol m^{-3})	Percentage germination in the presence of GA_3 (mmol m^{-3})			
	0	290	870	2900
0	86	83	83	83
750	66	69	71	71
2250	8	24	31	50
7500	4	7	1	7

taneously may not reflect a physiological interaction but rather chemical reactions between the agents, different rates of uptake, or different rates of inactivation and metabolism. Sometimes, as with ABA and cytokinin, where there appear to be independent effects on RNA, the observed interaction can be explained. Mostly the explanation of an interaction is left open, although a general scheme has been put forward by Khan and Waters (1969), Table 19.6. It was proposed that GA was essential

Table 19.6 The dependence of seed status on hormone combinations (after Khan and Waters, 1969).

	Hormone combinations (+ present, − absent)							
	1	2	3	4	5	6	7	8
Gibberellin	+	+	+	+	−	−	−	−
Cytokinin	+	+	−	−	+	+	−	−
Inhibitor (? ABA)	+	−	+	−	+	−	+	−
Seed status	G	G	D	G	D	D	D	D

G, germinates; D, dormant.

for germination, that cytokinin permitted germination and that an inhibitor fine-tuned the system but was ineffective in the presence of both GA and cytokinin. Any seed not possessing enough GA or cytokinin, even though it possessed inhibitor, would germinate if provided with these stimulators. As presented, the response is qualitative but in reality the responses would be quantitative and depend on the relative endogenous amounts of the regulators, so affording a wide range of interactions. Interactions have also been found among the regulators mentioned and ethylene, carbon dioxide, red light and far-red, and water-stress (Keys, Smith, Kumamoto and Lyon, 1975; Schonbeck and Egley, 1980).

More information may be gained by applying the regulators in sequence, and after the seed has imbibed, because during early imbibition amphipathic substances would be taken up passively and incorporated into membranes. The membrane function may be impaired, or enhanced, accidentally, since the normal active site of the test material may be elsewhere in the cell. This approach has been carried out on *Chenopodium album* where it is possible to recognize four morphological stages which probably correspond to distinct physiological phases. Stage 0—no visible morphological change; 1, splitting of the outer testa near the radicle; 2, extension of the radicle but clothed by the inner testa; 3, protrusion of the radicle. The seed can be dehydrated up to stage 2 so this corresponds to phase I germination (Cumming, 1963). Stage 3 is promoted by light. Regulators can be applied at these specific physiological times. ABA prevents the light-mediated radicle growth but does not affect the earlier stages. GA_{4+7} induces germination, *i.e.*, can promote activity up to stage 2, but GA_3 cannot. Red light acts synergistically with GA_{4+7} in promoting germination and ethylene could also promote at this stage. The ABA inhibition of radicle extension could be overcome by KIN, zeatin, GA_3, GA_{4+7} and ethylene. Two sites of hormone action were assumed: one at induction of germination, the other at embryo extension, with ABA involved only at the second (Karssen, 1976a,b). So far interaction studies have not fulfilled the promise expected. They have not provided an explanation of the control of germination.

Respiratory systems and germination

Oxidative processes can be linked to germination. Oxygen is consumed in the destruction of inhibitors in birch, *Xanthium* and *Sinapis*. The increase in dehydrogenase activity on puncturing seeds implies an increase in redox activity.

Since increasing respiratory activities are associated with germination there could be a correlation between any treatment that affects germination and at the same time affects respiration. Nitrate has long been known to enhance germination and it has been proposed that it, together with nitrite and hydroxylamine, may act as an alternative electron acceptor,

though hydroxylamine could act as an inhibitor of electron transport. This could alter the electron economy of metabolism which in turn affects respiration.

Roberts (1964) found that dormancy in rice could be broken by treating seeds with certain respiratory inhibitors. Among the best were cyanide and azide. Of some 22 respiratory inhibitors tested, only those which prevented electron flow in the terminal oxidase system were effective. Inhibitors of the EMP pathway or the Krebs' cycle did not break dormancy. Roberts concluded that the key enzyme involved was cytochrome oxidase: when it was inhibited, dormancy was broken.

Plant mitochondria can exhibit cyanide-resistant respiration (for reviews *see* Lambers, 1982; Laties, 1982) and in such cases electrons are diverted from the terminal oxidase pathway to an alternative path. The branching point is ubiquinone and by using aromatic hydroxamates, such as salicylhydroxamic acid (SHAM), the alternative path can be blocked. Seeds do possess cyanide-resistant mitochondria (Burgillo and Nicolás, 1977) so the effects of SHAM and cyanide or azide should be examined concurrently in any study of cyanide-promoted germination. SHAM should inhibit CN^- or N_3^- promotion.

SHAM has been found not to inhibit the germination of non-dormant wild oat (either inherently non-dormant or GA-induced) but it did inhibit azide-promoted germination of dormant lines. However, azide promoted germination was prevented by treating seeds with CCC which might be expected to inhibit GA synthesis. This may mean that in wild oats there are two non-related events induced by azide, namely promotion of alternative-path respiration and induction of GA synthesis, which seems a prerequisite in this species for dormant grain to germinate (Simpson, 1965; Upadhyaya, Naylor and Simpson, 1982). This study does not resolve the problem though the authors conclude that the alternative path is important in dormancy breaking.

While the alternative path allows electron flow to continue in the absence of cytochrome activity, it does not result in phosphorylation. Thus, there will be a depletion of the energy status of the system.

The involvement of a 'wasteful' metabolic pathway should not be thought a misdirection of the activities of a seed. During the early stages of germination the need for ATP seems to be met, but it is possible that the need for carbon skeletons required for early protein synthesis is not. The Krebs' cycle will produce these skeletons and, if provided in the required amounts, the seed will experience an excess electron flow. An overflow mechanism is needed and this is provided by the alternative path. Another possibility is that the seed needs oxygen for non-respiratory processes, *e.g.*, the oxidation of inhibitors. Since the affinity for oxygen of the alternative path is some two orders of magnitude less than that of the terminal oxidase pathway, oxygen consumption associated with electron flow will be reduced and more made available for other oxygen-consuming processes.

Roberts (1969) proposed the promotional effect of the respiratory inhibitors was due to the oxidative pentose phosphate pathway being brought into play and that this in some way initiated germination.

An increase in the enzymes of the PP pathway would be expected during the breaking of dormancy, especially glucose-6-phosphate dehydrogenase (G-6PDH) the first, and 6-phosphogluconic acid dehydrogenase (6PGDH) the second, enzymes of the pathway. In wild oats kept under two different dormancy-breaking regimes, the amount of G-6PDH increased three-fold under the regime which broke dormancy faster but five-fold in the other. 6PGDH activity did not follow any regular pattern (Adkins and Ross, 1981). These authors are of the opinion that previously reported increases in these enzymes are due to loss of dormancy and not the cause of breaking dormancy.

The inhibition of catalase activity has also been proposed as the mechanism which invokes the PP pathway. Both nitrite and hydroxylamine inhibit this enzyme but thiourea is more effective (Hendricks and Taylorson, 1975). Peroxidase activity is not affected and they considered that the hydrogen peroxide spared was used in the oxidation of phenols to quinones, so affording entry to NADPH oxidation/reduction, the oxidized NADPH acting as an oxidant for the PP pathway. Hydrogen peroxide can break dormancy in some cases.

Thiourea, nitrite and hydroxylamine were claimed to act by complexing with heme proteins of which catalase is one, but then so are the peroxidases. Catalase inhibition is not necessary to permit dormant cocklebur seed to germinate. However, by presoaking seed for seven days then treating with nitrite, cyanide, azide or thiourea, germination

could be induced though catalase activities were high. Non-soaked seed germinated and had reduced catalase activity. Nitrate does not induce germination in this species but there is a reduction in catalase activity after treatment with this ion (Eksashi, Sakai, Ushizawa and Tazaki, 1979). It would seem that this hypothesis does not apply generally and is another element in the germination paradox.

Energy charge and phosphorus metabolism

ATP is required by seeds during the early stages of germination and in wheat axes there is an immediate rise on imbibition of about 500% to 800 nmol per g of tissue (Obendorf and Marcus, 1974). Energy charges of between 0.61 and 0.85 were found between 1 and 16.5 h after imbibition implying that more of the adenosine phosphate is present as ATP. Growth is not possible until the energy charge exceeds 0.8 so that during phase I energy charge is built up to exceed this value. Between 0.8 and 0.9 there is a rapid switching between ATP-producing and ATP-consuming systems. Since the charge is low in seeds during early germination and a major ATP-producing system, oxidative phosphorylation, may not be operating, the PP pathway is brought into play to provide the necessary ATP. Other phosphorylating systems may be induced but have not been examined.

The dry seed contains five classes of phosphorus-containing compounds: (i) inorganic phosphate, $H_2PO_3^-$; (ii) storage compounds mainly phytin; (iii) constitutive substances, *e.g.*, phospholipids, nucleic acids etc.; (iv) metabolic intermediates, *e.g.*, sugar phosphates, NAD, etc.; and (v) 'energy-rich' compounds, *e.g.*, ATP. Changes among these would be expected but in the only analysis, in oat, which has attempted to prepare a phosphorus balance sheet during the course of germination, the critical early stages are not represented and classes (iii), (iv), and (v) were only semi-quantitatively estimated, if attempted at all (Hall and Hodges, 1966). The amount of inorganic phosphorus was 500 nmol per dry grain (approx. 20 mg) but this increased at the second day as a result of phytin hydrolysis. There was little mobilization from this pool of inorganic P to organic compounds though there was an almost stoichiometric transfer of phosphorus from the constitutive compounds of the storage tissues to the embryonic region. The inorganic pool went on increasing and was probably expended when active photosynthesis began.

Phytin hydrolysis begins eight hours after radicle emergence in lettuce and it is clear that release of inorganic P is a phase II event in this species. The timetable of events during the course of lettuce seed germination illustrates very well that the onset of hydrolytic enzyme activity is associated with phase II, Fig. 19.9 (Bewley and Halmer, 1980/81).

However, the phase changes may be associated with other features and the most readily accepted is permeability. The germination of a number of species over a temperature range was compared with amino acid leakage. In two cases, *Amaranthus albus* and *Abutilon theophrasti*, leakage was virtually the same at temperatures between 15° and 40°C. These species germinated readily at high temperature. In all of the other eight species tested, leakage of amino acid above 28°–32°C increased and germination reduced dramatically. If seeds were exposed to 20°C, where there was little leakage, and were then

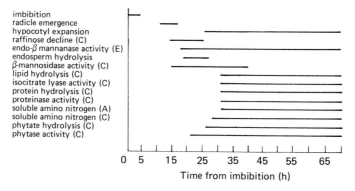

Fig. 19.9 The initiation of the main events associated with germination of lettuce. The duration of the event is indicated by the length of the bar: A, axis; C, cotyledon; E, embryo. (From Bewley and Halmer, 1980/81.)

Membranes and germination

It has already been mentioned that leaching of solutes from seeds has been correlated with membrane architecture. Because of the physical properties of phospholipid bilayers a number of phase states are possible over the range of temperatures that affects seed germination. At high temperatures the acyl chains are thermodynamically liquid and at low temperatures they are solid (frozen: crystalline). In the middle range the bilayer exists in a mesomorphic or liquid-crystal state. Transition temperatures between these states can be found. Differential thermal analysis (DTA), which can measure latent-heat changes as sensible heat changes, can pick up the transition as the bilayer is exposed to continuously-changing temperature. Although it is possible to use powdered seed for DTA the results are not very meaningful.

transferred to 30°C or 35°C, germination could be induced (Hendricks and Taylorson, 1976).

The change in leakage was correlated with a phase change in the membrane. Some analyses of the fatty acid composition of the phospholipids were made which showed a higher percentage of the C18 unsaturated fatty acids particularly C18:2, linoleic acid. Surprisingly *Amaranthus albus* had a very high percentage of linolenic acid (C18:3) which would be expected to result in a very fluid phospholipid system. However, the physical characteristics of a membrane cannot be deduced from the bulk fatty acid composition.

Wood and Paleg (1974) and Wood *et al.* (1974) looked at the physical properties of liposomes and tried to correlate these with the physiological properties of membranes. Liposomes made from soybean lecithin and certain additives have a transition temperature of 27°C, but the transition temperature can be altered by varying the amount and type of sterol incorporated into the bilayer. The ionic environment and the addition of GA to the bilayer will also alter the transition temperature. GA lowered it and also increased glucose leakage from the liposome to the medium.

Membrane fragments from lettuce and early wintercress (*Barbarea verna*) were labeled with fluorescent probes and then subjected to increasing temperature. The amount of fluorescence that could be measured relative to temperature change showed a transition between 28° and 30°C, which agrees with the transition found from amino acid leakage (Hendricks and Taylorson, 1979).

Hendricks and Taylorson (1967) were the first to suggest that phytochrome could be involved in membrane-associated reactions. Marmé (1977) reviews the available evidence for such an involvement and lists activities which would be related to membranes. Phytochrome could play a role in changing both membrane properties and membrane reactions (*see also* Chapters 16 and 17). The rapid response of photosensitive seeds, and the rapidity of certain phytochrome reactions that are membrane associated, strengthen the view that the membrane may be the locus of the primary germination stimulus.

Phase II events

Unlike phase I where the only morphological change is an increase in volume, phase II involves a marked change in the size of the embryonic axis, enlargement of cotyledons in species with epigeal germination and the beginnings of development. In parallel with the obvious morphological changes, there are also cellular and subcellular changes, especially in storage and regulatory tissues (endosperm, epithelial layer, cotyledons). All the changes are associated with conversion, mobilization, translocation and redistribution of the seed's storage materials.

Most of the observations have been made on economically important species like barley or those that are popular experimentally, such as lettuce.

Briggs (1978) has given an account of malt production and describes the nature of the chemical changes that take place in germinating barley. It is interesting to note that the amount of substance lost by respiration is only 5–7% of the weight of the dry

grain, and the residue of unused material, husks, etc., about another 10–12%. Since the embryo is about 4% of the grain weight this means that during phase II there is a redistribution of some 70% of the grain weight from storage tissue to the young seedling. The aleurone, most of which disappears, is relatively quite massive, making up about 10% of the weight of the grain.

Mobilization of the endospermic reserves cannot take place without the embryo (and scutellum). Embryo-less grains cannot accomplish the dissolution of starch or the production of diastatic activity (*i.e.*, amylolysis and amylase production).

Early work in the nineteenth century (Brown and Morris, 1890) showed that isolated germs (embryo + scutellum) could release enzymes. A very complex mixture of enzymes is secreted from the embryo to the medium in which they cultured (Briggs, 1962a). The amounts of hydrolytic enzyme that were secreted were insufficient to explain the rate of breakdown of endosperm storage material found in entire grain. The embryo was found to control the hydrolytic capacity of the aleurone and could influence the functioning of that tissue.

Yomo (1960) and Paleg (1960a,b, 1961) were the first to prove that the embryo region secreted gibberellin(s) to the aleurone and that this layer responded by developing increased α-amylase and ribonuclease activities (Chrispeels and Varner, 1967a). α-Amylase production induced in the aleurone by GA_3 is more sensitive than GA_3-induced germination of dormant seeds. The former is maximal at 10^{-7} M, and even 10^{-11} M will elicit some α-amylase synthesis, but dormancy can only be broken when the seeds are treated with GA_3 at concentrations of the order of 10^{-3} M. GA and ABA are mutually interactive in relation to enzyme production by barley aleurone; this finding contrasts with their dormancy-breaking and germination-inhibiting properties where GA does not usually overcome ABA inhibition (Chrispeels and Varner, 1967b).

By using protein and RNA synthesis inhibitors, Chrispeels and Varner (1967b) showed that GA promotion of enzyme activity was dependent on RNA and protein synthesis. They also found that GA had to be present continuously for continued enzyme synthesis and secretion; thus it was not simply an inductive trigger.

The only other cereal studied in any great detail is

rice. Miyata, Okamoto, Watanabe and Kazawa (1981) and Okamoto and Akazawa (1979) are of the opinion that the epithelial layer of the scutellum is the major source of α-amylase and it is this α-amylase which is the main amylolytic enzyme in the degradation of endosperm starch. They have shown conclusively that α-amylase is synthesized by the epithelium and that the synthesis involves a glycosylation prior to secretion.

Though, in barley, most attention has been paid to the aleurone (*see* Chapter 2) it should not be thought that the epithelium is of little consequence. Briggs (1962b) presented evidence that the scutellum is the source of hydrolytic enzymes. Brown and Morris (1890) showed that starch dissolution begins at the epithelial interface with the endosperm and it is only after some time that activity adjacent to the aleurone is observed. Maltsters are aware that the most rapid starch breakdown occurs in grains with large wrap-around scutella, but the geometry may relate as much to GA diffusion as the surface area of the epithelium and any α-amylase it produces. In maize α-amylase is found in the scutellum with β-amylase in the endosperm.

Recently it has been suggested that osmotic potential controls the production of hydrolytic enzymes. As the solute concentration increases there is feed-back to repress the synthesis of enzyme (Trevawas, 1982).

In other seeds where the major storage substance is starch, α-amylase production is induced by GA produced from the growing axis. Peas produce phosphorylase as well as amylase and its production is stimulated by GA. But there are conflicting reports regarding GA's regulatory role in legumes.

Where fat is the main storage substance, lipase activity increases during the course of germination and the metabolism of fat dissolution is well documented. The control mechanism(s) are not well known. GA does not induce dramatic increases in lipase activity. There is a small increase but it is insufficient to warrant the conclusion that GA is the major factor in inducing the increases associated with the germination of cotton (Black and Altschul, 1965). In castor bean endosperm GA was found not to affect lipase activity at all.

Many legumes contain a residual endosperm with a well developed aleurone layer, *e.g.*, soybean, guar, clover, fenugreek. This tissue, like the endosperm of lettuce, is not a major storage tissue but it does contain a considerable quantity of carbohydrate, usually a galactomannan. Eventually the galactomannan is hydrolyzed and the products are transported to the growing seedling (Reid and Meier, 1972). Though utilizable carbohydrate is produced, the main function of this tissue is to prevent severe water-stress developing in the embryo during germination. The galactomannans are gums and can absorb considerable amounts of water. The hydrated galactomannan acts as a buffer between the hydrated embryo and the soil and permits germination at soil water potentials which might be harmful to the expanding embryo.

Bud dormancy

Temperate herbaceous and woody perennials are exposed to low temperatures during winter; the temperatures they encounter in nature are lethal to actively-growing tissues. These plants have developed a capacity for their buds to enter a dormant state. In arid regions dormancy is associated with dry periods not low temperatures.

Dormancy of buds is defined as cessation of observable growth. In most trees growth declines after mid-summer and before leaf-fall, if the tree is deciduous, so that chronologically dormancy and leaf-fall are not related. With herbaceous perennials the aerial stems are reproductive and die before the onset of winter, basal axillary buds becoming dormant to ensure survival of the plant. The terminal and lateral buds do not usually survive but in some species, *e.g.*, raspberry, the aerial shoot is biennial, being vegetative in its first year and reproductive in the second. The aerial vegetative shoot becomes dormant and survives. In this species successive crops of shoots arise principally from adventitious buds.

Unlike seeds, buds are relatively well hydrated, at just over 40% moisture content in the winter months. Prior to reaching this stable winter value the summer moisture content was almost 60% and at bud-break it rose to over 70% (sugar maple, Dumbroff, Cohen and Webb, 1979). So while the winter bud is less hydrated than the summer and spring buds it is by no means desiccated.

Buds encounter a different set of environmental conditions from seeds. After entering the winter period in a dormant hydrated state they are exposed to natural conditions which prevent physiological activity. Quiescence can be maintained. In a climate with fluctuating temperatures, a period with temperatures within physiological limits may allow bud-break and then the bud is subject to winter-kill on the return of low temperatures. Regulation of this type of behavior can only be achieved if the breaking of dormancy is time/temperature dependent, and it is usually found that this depends on the locality in which the plant grows. For sugar maple, in Waterloo, Ontario (Canada), 2000 h exposure below 5°C is necessary to break dormancy, and in normal years this is attained by mid February or March. Ecotypic variation and noticeable effects of environmental factors on the character of dormancy have been observed (Downs and Bevington, 1981).

It is generally agreed that the primary factor for inducing dormancy in buds is daylength (Kramer, 1936; Downs and Borthwick, 1956; Wareing, 1949). Growth takes place in lengthening and long days but as the days shorten dormancy is induced. Other factors such as nutrition, water status, temperature and irradiance can modify the time of onset of dormancy (Perry, 1971).

Downs and Bevington (1981) looked at the dormancy of paper birch seedlings from different localities. Seedlings of Alaskan origin (64°–67°N) and from continental USA (44°–48°N) were exposed to different photoperiods. Under a 14-h day, only Alaskan material was dormant. Under a 9-h day all material was dormant if the night temperature was low (day temperature 30°C, night temperature 14°C). At a day/night temperature regime of 30°/26°C only 40% of the Alaskan material, and none of the other, were dormant. The sensing mechanism, therefore, seems to be detecting conditions suitable for growth. This is also the case with nutrients and water (Perry, 1971). Excess nitrogenous fertilizer can promote growth to such an extent that dormancy is not induced before lethal frosts occur and the plants are killed. This phenomenon is well known to farmers and growers who must ensure that their plants are in the correct physiological state to tolerate winter conditions.

Eagles and Wareing (1963, 1964) showed that an extract from birch kept under short-day conditions could suppress growth in birch seedlings kept under a photoperiod of 14.5 h. The seedlings, on extended treatment, produced buds with all the characteristics of dormant winter buds. They called the active

substance 'dormin', which was later found to be the same as Abscisin II, and was eventually renamed abscisic acid. Inhibitor content varied according to the daylength under which the birch seedling was growing, and as few as two short days could increase the quantity substantially. There have been many attempts to correlate ABA content and bud (stem) dormancy.

In willow, there is a seasonal change in the content of ABA in the leaves and stem and also in the sap (Alvim, Thomas and Saunders, 1978). The largest amounts were present in mid-summer but they declined in late summer and early autumn. There were larger amounts in trees given extended daylengths, although these trees continued extension growth (i.e., they did not enter the dormant phase) for about a month longer than the control group. It was concluded that ABA does not play a role in dormancy. The same conclusion was reached by Dumbroff et al. (1979) with regard to ABA and the induction of dormancy in the sugar maple. In this case free ABA had reached very low levels at bud-break, with concomitant increase in bound ABA at this time. There was no algebraic relationship between the sums of free and bound ABA over the season, so that the total ABA amounts were not the same. Birch seems to behave in the same way (Harrison and Saunders, 1975).

As the buds break there is a rise in the amount of cytokinin activity that can be found in extracts (Taylor and Dumbroff, 1975). This inverse relationship between ABA and cytokinin seems to be widespread. Possibly the function of ABA is to maintain rest; it may play a role in the winter behavior of trees but appears not to be the agent which induces dormancy.

Dormancy in buds of stems and tubers can be broken by treatment with ethylene or ethylene-generating chemicals (Abeles, 1973). The other growth-promoting compounds vary in their capacity to effect bud break. It is extremely difficult to apply solutions to a tightly closed bud covered by waxy bud scales, so the variety of the responses which have been noted probably relates to varying degrees of penetration by the regulators. Any attempt to inject the test solution results in wounding with the possible production of ethylene. Growth promoters such as GA and cytokinin accelerate the growth of the bud when dormancy is broken, and shoot length is often used as an assay for the breaking of dormancy. There is an inherent fallacy in experiments based on this measure (extent of shoot growth). A bud is either dormant—no measurable growth—or non-dormant—growing, however slowly.

Perennial herbaceous plants of the temperate zone show the same pattern of behavior just described for woody species. In climatic zones where seasonality is associated with aridity, plants enter the dormant phase in the dry period. Grasses rather than woody plants have been examined for drought-induced dormancy. Ecotypic variation is most marked and a species which has a range extending from the sub-arctic to the southern region of a Mediterranean-type climatic zone may exhibit both types of dormancy. A good example is found in the tall fescue (Morgan, 1964). In North Africa the growing season for this grass is from September to May and because of this the local ecotype grows under short-days rather than long-days. British races behave in the opposite fashion. If African ecotypes are grown in Britain they do not enter dormancy in response to the shortening days of autumn and early winter, and may be growing actively when the temperature reaches − 3°C which kills them. Although this is detrimental for species survival, growing these North African types is valuable agronomically since their use can extend the grazing season.

The agent responsible for the induction of summer dormancy in these grasses is not known and it is possible that the non-growth in summer is more akin to quiescence than to dormancy.

Concluding remarks

It could be said that if dormancy did not occur it would have to be invented. Since plants exist in a fixed location, they must have a mechanism to enable them to survive adverse conditions. One way of surviving is to reduce to insignificance metabolic activity; in seeds this is either achieved by desiccation or by maintaining a turn-over condition in the moist state. With buds, the low metabolic activity is induced when the plant is exposed to the shortening days of late summer and autumn. Bud growth ceases before leaf-fall and the dormant bud then senses low winter temperatures, and is able to sum the temperature deficits until the total reaches a particular value. When this value is reached the bud is out

of rest and will grow when climatic conditions are suitable for growth.

While there is much documentation of the types of dormancy and many examples, there is little that permits a statement about the nature of the molecular trigger that breaks dormancy, and allows metabolism to attain the rates associated with vigorously-growing axes. There is still much to be done in this area of plant physiology which allows the techniques of molecular biology to be practiced quite readily and within a reasonable time-scale.

Further reading

Bewley, J. D. and Black, M. (1978). *Physiology and Biochemistry of Seeds in Relation to Germination*, vol. 1, Springer-Verlag, Berlin, p. 306.

Bewley, J. D. and Black, M. (1982). *Physiology and Biochemistry of Seeds in Relation to Germination*, vol. 2, Springer-Verlag, Berlin, p. 375.

Khan, A. A. (ed.) (1982). *The Physiology and Biochemistry of Seed Development, Dormancy and Germination*, Elsevier Biomedical Press, Amsterdam, p. 560.

Laidman, D. L. and Wyn Jones, R. G. (eds) (1979). *Recent Advances in the Biochemistry of Cereals*, Academic Press, London, p. 391.

Mayer, A. M. (ed.) (1980/81). *Control Mechanisms in Seed Germination*, The Weizmann Science Press of Israel, Jerusalem, p. 322.

Mayer, A. M. and Poljakoff-Mayber, A. (1982). *The Germination of Seeds*, 3rd edn, Pergamon Press, Oxford, p. 211.

Roberts, E. H. (ed.) (1972). *Viability of Seeds*, Chapman and Hall Ltd., London, p. 448.

Rubenstein, I., Phillips, R. L., Green, C. E. and Gengenbach, B. G. (1979). *The Plant Seed: Development, Preservation and Germination*, Academic Press, New York, p. 266.

Taylorson, R. B. and Hendricks, S. B. (1977). Dormancy in seeds. *Ann. Rev. Plant Physiol.* **28**, 331–54.

Woolhouse, H. W. (1969). *Dormancy and Survival, Symp. Soc. Exp. Biol.* **23**, C.U.P., p. 598.

References

Abeles, F. B. (1973). *Ethylene in Plant Biology*, Academic Press, New York, p. 302.

Abeles, F. B. and Lonski, J. (1969). Stimulation of lettuce seed germination by ethylene. *Plant Physiol.* **44**, 277–80.

Adkins, S. W. and Ross, J. D. (1981). Studies in wild oat seed dormancy. II. Activities of pentose phosphate pathway dehydrogenases. *Plant Physiol.* **68**, 15–17.

Alvim, R., Thomas, S. and Saunders, P. F. (1978). Seasonal variation in the hormone content of willow. II. Effect of photoperiod on growth and abscisic acid content of trees under field conditions. *Plant Physiol.* **62**, 779–80.

Anon (1942). Recent work on germination—Report. *Nature,* **149**, 658–9.

Attims, Y. and Côme, D. (1978). Seed dormancy of a tropical plant *Oldenlandia corymbosa* Rubiaceae. Selection of two lines of plant. *C.R. Acad. Sci., Paris* **268D**, 1669–72.

Barendse, G. W. M., Kende, H. and Lang, A. (1968). Fate of radioactive gibberellin A_1 in maturing and germinating seeds of peas and Japanese Morning Glory. *Plant Physiol.* **43**, 815–22.

Becwar, M. C., Stanwood, P. C. and Roos, E. E. (1982). Dehydration effects on imbibitional leakage from desiccation-sensitive seeds. *Plant Physiol.* **69**, 1132–5.

Berjak, P. and Villiers, T. A. (1972). Ageing in plant embryos. II. Age induced damage and its repair during early germination. *New Phytol.* **71**, 135–44.

Berrie, A. M. M. (1966). The effect of temperature and light on the germination of lettuce seeds. *Physiol. Plant.* **19**, 429–36.

Berrie, A. M. M. and Drennan, D. S. H. (1971). The effect of hydration–dehydration on seed germination. *New Phytol.* **70**, 135–42.

Berrie, A. M. M. and Robertson, J. (1976). Abscisic acid as an endogenous component in lettuce fruits, *Lactuca sativa* L. cv. Grand Rapids. Does it control thermodormancy? *Planta,* **131**, 211–15.

Berrie, A. M. M. and Taylor, G. C. D. (1981). The use of population parameters in the analysis of germination of lettuce seed. *Physiol. Plant.* **51**, 229–33.

Berrie, A. M. M., Buller, D., Don, R. and Parker, W. (1979). Possible role of volatile fatty acids and abscisic acid in the dormancy of oats. *Plant Physiol.* **63**, 758–64.

Berrie, A. M. M., Don, R., Buller, D., Alam, M. and Parker, W. (1975). The occurrence and function of short chain length fatty acids in plants. *Plant Sci. Lett.* **6**, 163–73.

Berrie, A. M. M., Parker, W., Knights, B. A. and Hendrie, M. R. (1968). Studies on lettuce seed germination I. Coumarin induced dormancy. *Phytochemistry* **7**, 567–73.

Bewley, J. D. (1979). Dormancy breaking by hormones and other chemicals—action at the molecular level. In *The Plant Seed: Development, Preservation, and Germination,* eds J. Rubenstein, R. L. Phillips, C. E. Green and B. G. Gengenbach, Academic Press, New York.

Bewley, J. D. and Black, M. (1972). Protein synthesis during gibberellin-induced germination of lettuce seed. *Can. J. Bot.* **50**, 53–9.

Bewley, J. D. and Black, M. (1978). *Physiology and Biochemistry of Seeds in Relation to Germination,* vol. 1, Springer-Verlag, Berlin, p. 306.

Bewley, D. and Halmer, P. (1980/81). Embryo–endosperm interactions in the hydrolysis of lettuce seed reserves. *Israel J. Bot.* **29**, 118–32.

Bewley, J. D., Negbi, M. and Black, M. (1968). Immediate phytochrome action in lettuce seeds and its interaction with gibberellins and other germination promoters. *Planta* **78**, 351–7.

Biddington, N. L. and Thomas, T. H. (1976). Influence of different cytokinins on the germination of lettuce (*Lactuca sativa*) and celery (*Apium graveolens*) seeds. *Physiol. Plant.* **37**, 12–16.

Biddington, N. L., Thomas, T. H. and Dearman, A. S. (1980). The effect of temperature on the germination-promotion activities of cytokinin and gibberellin applied to celery seeds (*Apium graveolens*). *Physiol. Plant.* **49**, 68–70.

Black, H. S. and Altschul, A. M. (1965). Gibberellin acid-induced lipase and α-amylase formation and their inhibition by aflatoxin. *Biochem. Biophys. Res. Commun.* **19**, 661–4.

Black, M. (1980/81). The role of endogenous hormones in germination and dormancy. *Israel J. Bot.* **29**, 181–92.

Black, M. and Naylor, J. M. (1959). Prevention of the onset of seed dormancy by gibberellic acid. *Nature* **184**, 468–9.

Black, M. and Wareing, P. F. (1955). Growth studies in woody species. VII. Photoperiodic control of germination in *Betula pubescens* Ehrh. *Physiol. Plant.* **8**, 300–16.

Blundell, J. B. and Jackson, G. A. D. (1971). Rose seed germination in relation to stock production. *Natn. Rose Soc. Rose Ann.* 1971, 129–35.

Borthwick, H. A. and Robbins, W. W. (1929). Lettuce seed and its germination. *Hilgardia* **3**, 275–304.

Braun, J. W. and Khan, A. A. (1975). Endogenous abscisic acid levels in germinating and non-germinating lettuce seeds. *Plant Physiol.* **56**, 731–3.

Briggs, D. E. (1962a). Development of enzymes by barley embryos *in vitro. J. Inst. of Brew.* **68**, 470–5.

Briggs, D. E. (1962b). Biochemistry of barley germination. Action of gibberellic acid on barley endosperm. *J. Inst. Brew.* **69**, 13–19.

Briggs, D. E. (1978). *Barley,* Chapman and Hall Ltd., London, p. 612.

Brown, H. T. and Morris, G. H. (1890). XXX. Researches on the germination of some of the Gramineae. Part I. *J. Chem. Soc.* **57**, 458–528.

Buller, D. C., Reid, J. S. G. and Parker, W. (1976). Short chain fatty acids as inhibitors of gibberellin-induced amylase in barley endosperm. *Nature* **260**, 169–97.

Burguillo, P. de la F. and Nicolás, G. (1977). Appearance of an alternate pathway cyanide-resistant during germination of seeds of *Cicer arietinum*. *Plant Physiol.* **60**, 524–7.

Burton, W. G. (1963). Concepts and mechanism of dormancy. In *The Growth of the Potato,* eds J. D. Ivins and F. L. Milthorpe, p. 328, Butterworths, London.

Chabot, J. F. and Leopold, A. C. (1982). Ultrastructural changes of membranes with hydration in soybean seeds. *Am. J. Bot.* **69**, 623–33.

Cheah, K. S. E. and Osborne, D. J. (1978). DNA lesions occur with loss of viability in embryos of ageing rye seed. *Nature* **272**, 593–9.

Chen, D. and Osborne, D. J. (1970). Hormones in the translational control of early germination in wheat embryos. *Nature* **226**, 1157–60.

Chen, S. S. C. and Thimann, K. V. (1964). Studies on the germination of light-inhibited seeds of *Phacelia tanacetifolia*. *Israel J. Bot.* **13**, 57–73.

Chen, S. S. C. and Varner, J. E. (1970). Respiration and protein synthesis in dormant and after-ripened seeds of *Avena fatua*. *Plant Physiol.* **46**, 108–12.

Chrispeels, M. J. and Jones, R. L. (1980/81). The role of the endoplasmic reticulum in the mobilization of reserve macromolecules during seedling growth. *Israel J. Bot.* **29**, 225–45.

Chrispeels, M. J. and Varner, J. E. (1967a). Gibberellic acid-enhanced synthesis and release of α-amylase and ribonuclease by isolated barley aleurone layers. *Plant Physiol.* **42**, 398–406.

Chrispeels, M. J. and Varner, J. E. (1967b). Hormonal control of enzyme synthesis: on the mode of action of gibberellic acid and abscissin in aleurone layers of barley. *Plant Physiol.* **42**, 1008–16.

Crocker, W. and Barton, L. V. (1953). *Physiology of Seeds: An Introduction to the Experimental Study of Seed and Germination Problems,* Chronica Botanica Co., Waltham Mass., p. 267.

Crocker, W. and Davis, W. E. (1914). Delayed germination in seed of *Alisma plantago*. *Bot. Gaz.* **58**, 285–321.

Cuming, A. C. and Osborne, D. J. (1978a). Membrane turnover in imbibed and dormant embryos of the wild oat (*Avena fatua* L.). I. Protein turnover and membrane replacement. *Planta* **139**, 209–17.

Cuming, A. C. and Osborne, D. J. (1978b). Membrane turnover in imbibed and dormant embryos of the wild oat (*Avena fatua* L.). II. Phospholipid turnover and membrane replacement. *Planta* **139**, 219–26.

Cumming, B. G. (1963). The dependence of germination on photoperiod, light-quality, and temperature in *Chenopodium* spp. *Can. J. Bot.* **41**, 1211–33.

Dell, B. (1980). Structure and function of the strophiolar plug in seeds of *Albizia lophantha*. *Am. J. Bot.* **67**, 556–63.

Downs, R. J. and Bevington, J. M. (1981). Effect of temperature and photoperiod on growth and dormancy of *Betula papyrifera*. *Am. J. Bot.* **68**, 795–800.

Downs, R. J. and Borthwick, H. A. (1956). Effect of photoperiod on growth of trees. *Bot. Gaz.* **117**, 310–26.

Drennan, D. S. and Berrie, A. M. M. (1962). Physiological studies of germination in the genus *Avena*. I. The development of amylase activity. *New Phytol.* **61**, 1–9.

Dumbroff, E. B., Cohen, D. B. and Webb, D. P. (1979). Seasonal levels of abscisic acid in buds and stems of *Acer saccharum*. *Physiol. Plant.* **45**, 211–14.

Dure, L. S. (1977). Stored messenger ribonucleic acid and seed germination. In *The Physiology and Biochemistry of Seed Dormancy and Germination*, ed. A. A. Khan, North-Holland Publishing Co., Amsterdam.

Dure, L. S. (1979). Role of stored messenger RNA in late embryo development and germination. In *The Plant Seed Development, Preservation, and Germination,* eds I. Rubenstein, R. L. Phillips, C. E. Green and B. G. Gengenbach, Academic Press, New York.

Eagles, C. F. and Wareing, P. F. (1963). Experimental induction of dormancy in *Betula pubescens*. *Nature* **199**, 874–5.

Eagles, C. F. and Wareing, P. F. (1964). The role of growth substances in the regulation of bud dormancy. *Physiol. Plant* **17**, 697–709.

Edwards, M. M. (1969). Dormancy in seeds of Charlock. IV. Interrelationships of growth, oxygen supply and concentration of inhibitor. *J. exp. Bot.* **20**, 876–94.

Esashi, Y., Sakai, Y., Ushizawa, R. and Tazaki, S. (1979). Catalase is not involved in control of germination of cocklebur seeds. *Aust. J. Plant Physiol.* **6**, 425–9.

Fountain, D. W. and Bewley, J. D. (1976). Modulation of pre-germination protein synthesis by gibberellic acid, abscisic acid, and cytokinin. *Plant Physiol.* **58**, 530–6.

Frankland, B. and Wareing, P. F. (1962). Changes in endogenous gibberellins in relation to chilling of dormant seeds. *Nature* **194**, 313–14.

Frankland, B., Jarvis, B. C. and Cherry, J. H. (1971). RNA synthesis and the germination of light-sensitive lettuce seeds. *Planta* **97**, 39–49.

Fryer, J. D. and Makepeace, R. J. (1977). *Weed Control Handbook Vol. 1. Principles Including Plant Growth Regulators*. Blackwell Scientific Publications, Oxford, p. 510.

Fujii, T. (1962). Studies of photoperiodic responses involved in the germination of *Eragrostis* seeds. *Bot. Mag.* (Tokyo) **75**, 56–62.

Gassner, G. (1915a). Einige neue Falle von Keimungsauslosender Wirkung der Stickstoffoerbindungen (auf lichtemp findliche Samen). *Ber. deutsch. Bot. Ges.* **33**, 217–32.

Gassner, G. (1915b). Uber die Kiemungsauslosende Wirkung der Stickstoffsalze auf lichtempfindliche Samen. *Jahrb. wiss. Bot.* **55**, 259–342.

Gillard, D. F. and Walton, D. C. (1973). Germination of

Phaseolus vulgaris IV. Patterns of protein synthesis in excised axes. *Plant Physiol.* **51**, 1147–9.

Hall, J. R. and Hodges, T. K. (1966). Phosphorus metabolism of germinating oat seeds. *Plant Physiol.* **41**, 1459–64.

Halmer, P., Dewley, J. D. and Thorpe, T. A. (1976). An enzyme to degrade lettuce endosperm cell walls. Appearance of a mannanase following phytochrome- and gibberellin-induced germination. *Planta* **130**, 189–96.

Harrington, J. F. (1974). Seed storage and longevity. In *Seed Biology*, vol. 3, ed. T. T. Kozlowski, Academic Press, New York.

Harrison, M. A. and Saunders, P. F. (1975). The abscisic acid content of dormant birch buds. *Planta* **123**, 291–8.

Hart, J. W. and Berrie, A. M. M. (1966). The germination of *Avena fatua* under different gaseous environments. *Physiol. Plant.* **19**, 1020–5.

Hay, J. R. (1962). Experiments on the mechanisms of induced dormancy in wild oats, *Avena fatua* L. *Can. J. Bot.* **40**, 191–202.

Hegarty, T. W. and Ross, H. A. (1980/81). Investigations of control mechanisms of germination under water stress. *Israel J. Bot.* **29**, 83–92.

Hendricks, S. B. and Taylorson, R. B. (1967). The function of phytochrome in the regulation of plant growth. *Proc. Nat. Acad. Sci. USA* **58**, 2125–30.

Hendricks, S. B. and Taylorson, R. B. (1972). Promotion of seed germination by nitrates and cyanide. *Nature* **237**, 167–70.

Hendricks, S. B. and Taylorson, R. B. (1975). Breaking of seed dormancy by catalase inhibition. *Proc. Nat. Acad. Sci. USA* **72**, 306–9.

Hendricks, S. B. and Taylorson, R. B. (1976). Variation in germination and amino acid leakage of seeds with temperature related to membrane phase change. *Plant Physiol.* **58**, 7–11.

Hendricks, S. B. and Taylorson, R. B. (1979). Dependence on thermal responses of seeds on membrane transitions. *Proc. Nat. Acad. Sci. USA* **76**, 778–81.

Hiraga, K., Yokota, T., Murofushi, N. and Takahashi, N. (1972). Isolation and characterisation of a free gibberellin and glucosyl esters of gibberellins in mature seeds of *Phaseolus vulgaris*. *Agric. Biol. Chem.* **36**, 345–7.

Isikawa, S. (1954). Light sensitivity against the germination. I. 'Photoperiodism' of seeds. *Bot. Mag* (Tokyo) **67**, 37–42.

Isikawa, S. (1957). Interaction of light and temperature in the germination of *Nigella* seeds. *Bot. Mag.* (Tokyo) **70**, 264–75.

Isikawa, S. (1962). Light sensitivity against the germination. III. Studies on the various partial processes in light sensitive seeds. *Jap. J. Bot.* **18**, 105–32.

Jarvis, B. C. (1975). The role of seed parts in the induction of dormancy of hazel (*Corylus avellana* L.). *New Phytol.* **75**, 491–4.

Jarvis, B. C., Frankland, B. and Cherry, J. H. (1968). Increased DNA template and RNA polymerase associated with the breaking of seed dormancy. *Plant Physiol.* **43**, 1734–6.

Karssen, C. M. (1976a). Uptake and effect of abscisic acid during induction and progress of growth in seeds of *Chenopodium album*. *Physiol. Plant.* **36**, 259–63.

Karssen, C. M. (1976b). Two sites of hormonal action during germination of *Chenopodium album* seeds. *Physiol. Plant.* **36**, 264–70.

Kaufmann, M. R. and Ross, K. J. (1970). Water potential, temperature, and kinetin effects on seed germination in soil and solute systems. *Am. J. Bot.* **57**, 413–19.

Kende, H. (1971). The cytokinins. *Int. Rev. Cytol.* **31**, 301–38.

Ketring, D. L. (1977). Ethylene in seed germination. In *The Physiology and Biochemistry of Seed Dormancy and Germination*, ed. A. A. Khan, North-Holland Publishing Co., Amsterdam.

Ketring, D. L. and Morgan, P. W. (1970). Physiology of oil seeds I. Regulation of dormancy in Virginia–type peanut seeds. *Plant Physiol.* **45**, 268–73.

Keys, R. D., Smith, O. E., Kumamoto, J. and Lyon, J. L. (1975). Effect of gibberellic acid, kinetin, and ethylene plus carbon dioxide on the thermodormancy of lettuce seed (*Lactuca sativa* L. cv. Mesa 659). *Plant Physiol.* **56**, 826–9.

Khan, A. A. (1960). An analysis of dark-osmotic inhibition of germination of lettuce seeds. *Plant Physiol* **35**, 1–7.

Khan, A. A. and Waters, E. C. Jr. (1969). On the hormonal control of post-harvest dormancy and germination in barley seeds. *Life Sci.* **8**, 729–36.

Kidd, F. and West, C. (1917). The controlling influence of carbon dioxide. Part IV. On the production of secondary dormancy in seeds of *Brassica alba* following treatment with carbon dioxide and the relation of this phenomenon to the question of stimuli of growth processes. *Ann. Bot.* **31**, 456–87.

Kramer, P. J. (1936). The effect of variation in length of day on the growth and dormancy of trees. *Plant Physiol.* **11**, 127–37.

Lambers, H. (1982). Cyanide resistant respiration: a non-phosphorylating electron transport pathway acting as an energy overflow. *Physiol. Plant* **55**, 478–85.

Laties, G. G. (1982). The cyanide-resistant, alternative path in higher plant respiration. *Ann. Rev. Plant Physiol.* **33**, 519–55.

Leopold, A. C. (1980). Temperature effects on soybean imbibition and leakage. *Plant Physiol.* **65**, 1096–8.

Lona, F. (1956). L'acide gibberellico determina la germinazion dei semi di *Lactuca scariola* in fase di scoto inibizione. *L'ateneo Parmense* **27**, 641–4.

Luzzati, V. and Husson, F. (1962). The structure of liquid-crystalline phases of the lipid-water systems. *J. Cell Biol.* **12**, 207–19.

McKersie, B. D. and Stinson, R. H. (1980). Effect of dehydration on leakage and membrane structure in *Lotus corniculatus* L. seeds. *Plant Physiol.* **66**, 316–20.

Mancinelli, A. L., Borthwick, H. A. and Hendricks, S. B. (1966). Phytochrome action in tomato seed-germination. *Bot. Gaz.* **127**, 1–5.

Marcus, A., Feeley, J. and Volcani, T. (1966). Protein synthesis in imbibed seeds. III. Kinetics of amino acid incorporation, ribosome activation and polysome formation. *Plant Physiol.* **41**, 1167–72.

Marmé, D. (1977). Phytochrome: membranes as sites of primary action. *Ann. Rev. Plant Physiol.* **28**, 173–98.

Mayer, A. M. and Evenari, M. (1952). The relation between the structure of coumarin and its derivatives, and their activity as germination inhibitors. *J. exp. Bot.* **3**, 246–52.

Mayer, A. M. and Poljakoff-Mayber, A. (1982). *The Germination of Seeds*, 3rd edn, Pergamon Press, Oxford, p. 211.

Miller, C. O. (1956). Similarity of some kinetin and red light effects. *Plant Phys.* **31**, 318–19.

Miller, C. O. (1958). The relationship of the kinetin and red light promotions of lettuce seed germination. *Plant Phys.* **33**, 115–17.

Miyata, S., Okamoto, K., Watanabe, A. and Kazawa, T. A. (1981). Enzymic mechanism of starch breakdown in germinating rice seeds. 10. *In vivo* and *in vitro* synthesis of α-amylase in rice seed scutellum. *Plant Physiol.* **68**, 1314–18.

Moewus, F. and Schader, E. (1951). Uber die keimungs- und wachstuff hemmende Wirking einiger Phthalide. *Ber. deutsch. Bot. Ges.* **64**, 124–9.

Morgan, D. G. (1964). The eco-physiology of Mediterranean and north temperate varieties of tall fescue. *Outlook in Agriculture* **4**, 171–6.

Morgan, S. F. (1968). *Physiological Aspects of Seed Dormancy in* Avena ludoviciana *Dur.*, Ph.D. thesis, Univ. of Glasgow, Scotland.

Morgan, S. F. and Berrie, A. M. M. (1970). Development of dormancy during seed maturation in *Avena ludoviciana* Winter wild oat. *Nature* **228**, 1225.

Murphy, J. B. and Noland, T. L. (1982). Temperature effects on seed imbibition and leakage mediated by viscosity and membranes. *Plant Physiol.* **69**, 428–31.

Murray, R. D. H., Mendez, J. and Brown, S. A. (1982). *The Natural Coumarins. Occurrence, Chemistry and Biochemistry*, John Wiley and Sons Ltd., Chichester, p. 702.

Nikolaeva, M. G. (1977). Factors controlling the seed dormancy pattern. In *The Physiology and Biochemistry of Seed Dormancy and Germination*, ed. A. A. Khan, North-Holland Publishing Co., Amsterdam.

Obendorf, R. L. and Marcus, A. (1974). Rapid increase in adenosine 5′-triphosphate during early wheat embryo germination. *Plant Physiol.* **53**, 779–81.

Okamoto, K. and Akazawa, T. (1979). Enzymic mechanism of starch breakdown in germinating rice seeds 7. Amylase formation in the epithelium. *Plant Physiol.* **63**, 336–40.

Orchard, T. J. (1977). Estimating the parameters of plant seedling emergence. *Seed Sci. Technol.* **5**, 61–9.

Orlandini, M. and Bulard, C. (1977). Acide abscissique et germination des akènes de *Lactuca sativa* L. cv. Attraktion a l'obscurité à différentes températures. *Z. Planzenphysiol.* **85**, 77–81.

Osborne, D. J. and Cuming, A. C. (1979). Membrane protein and phospholipid turnover in imbibed dormant embryos of wild oat. In *Recent Advances in the Biochemistry of Cereals*, eds D. L. Laidman and R. G. Wyn Jones, Academic Press, London.

Paleg, L. G. (1960a). Physiological effects of gibberellic acid I. On carbohydrate metabolism and amylase activity of barley endosperm. *Plant Physiol.* **35**, 293–9.

Paleg, L. G. (1960b). Physiological effects of gibberellic acid II. On starch hydrolyzing enzymes of barley endosperm. *Plant Physiol.* **35**, 902–6.

Paleg, L. G. (1961). Physiological effects of gibberellic acid III. Observations on its mode of action on barley endosperm. *Plant Physiol.* **36**, 829–37.

Perry, T. O. (1971). Dormancy of trees in winter. *Science.* **171**, 29–36.

Porter, N. G. and Wareing, P. F. (1974). The role of the oxygen permeability of the seed coat in the dormancy of seed of *Xanthium pennsylvanicum* Wallr. *J. exp. Bot.* **25**, 583–97.

Reeve, D. R. and Crozier, A. (1975). Gibberellin bioassays. In *Gibberellins and Plant Growth*, ed. H. N. Krishnamoorthy, John Wiley & Sons Ltd., New Delhi.

Reid, J. S. G. and Meier, H. (1972). The function of the aleurone layer during galactomannan mobilisation in germinating seeds of Fenugreek (Trigonella *foenumgraecum* L.) Crimson clover (*Trifolium incarnatum* L.) and Lucerne (*Medicago sativa* L.): a correlative biochemical and ultrastructural study. *Planta* **106**, 44–60.

Reynolds, T. and Thompson, P. A. (1971). Characterisation of the high temperature inhibition of germination of lettuce (*Lactuca sativa*). *Physiol. Plant.* **24**, 544–7.

Roberts, B. E. and Osborne, D. J. (1973). Protein synthesis and loss of viability in Rye embryos. The lability of transferase enzymes during senescence. *Biochem. J.* **135**, 405–10.

Roberts, E. H. (1964). A survey of the effects of chemical treatments on dormancy in rice seed. *Physiol. Plant.* **17**, 30–43.

Roberts, E. H. (1969). Seed dormancy and oxidation processes. In *Dormancy and Survival*, ed. H. W. Woolhouse, C.U.P. Cambridge. *Symp. Soc. exp. Biol.* 23.

Roberts, E. H. (1972). Storage environment and the control of viability. In *Viability of Seeds,* ed. E. H. Roberts, Chapman & Hall Ltd., London.

Roberts, E. H. and Smith, R. D. (1977). In *The Physiology and Biochemistry of Seed Dormancy and Germination,* ed. A. A. Khan, North-Holland Publishing Co., Amsterdam.

Robertson, J., Hillman, J. R. and Berrie, A. M. M. (1976). The involvement of indole acetic acid in the thermodormancy of lettuce fruits, *Lactuca sativa* cv. Grand Rapids. *Planta* **131,** 309–13.

Rose, R. C. (1919). After-ripening and germination of seeds of *Tilia, Sambucus* and *Rubus. Bot. Gaz.* **67,** 281–308.

Ross, J. D. and Bradbeer, J. W. (1971). Studies in seed dormancy. V. The content of endogenous gibberellins in seeds of *Corylus avellana* L. *Planta* **100,** 288–302.

Rudnicki, R. (1973). Accumulation of abscisic acid in maturing apple seeds. *Proc. Res. Inst. Pom. Sk. Pol. Ser. E.,* 284–90.

Russell, E. W. (1973). *Soil Conditions and Plant Growth,* 10th edn, Longmans, London, p. 849.

Satoh, S. and Esashi, Y. (1979). Protein synthesis in dormant and non-dormant cocklebur seed segments. *Physiol. Plant.* **47,** 229–34.

Schonbeck, M. W. and Egley, G. H. (1980). Effects of temperature, water potential, and light on germination responses of redroot pigweed seeds to ethylene. *Plant Physiol.* **65,** 1149–54.

Sen, S. and Osborne, D. J. (1974). Germination of rye embryos following hydration–dehydration treatments: enhancement of protein and RNA synthesis and earlier induction of DNA replication. *J. exp. Bot.* **25,** 1010–19.

Shain, Y. and Mayer, A. M. (1968). Activation of enzymes during germination: amylopectin-1, 6-glucosidase in peas. *Physiol. Plant.* **21,** 765–76.

Shull, C. A. (1920). Temperature and rate of moisture intake of seeds. *Bot. Gaz.* **69,** 361–90.

Simon, E. W. (1974). Phospholipids and plant membrane permeability. *New Phytol.* **73,** 377–420.

Simon, E. W. and Raja Harun, R. M. (1972). Leakage during seed imbibition. *J. exp. Bot.* **23,** 1076–85.

Simpson, G. M. (1965). Dormancy studies in seed of *Avena fatua* 4. The role of gibberellins in embryo dormancy. *Can. J. Bot.* **43,** 793–816.

Sondheimer, E., Tzou, D. S. and Galson, E. C. (1968). Abscisic acid levels and seed dormancy. *Plant Physiol.* **43,** 1443–7.

Sondheimer, E., Galson, E. C., Tinelli, E. and Walton, D. C. (1974). The metabolism of hormones during seed germination and dormancy. IV. The metabolism of (S)-2-^{14}C-abscisic acid in ash seed. *Plant Physiol.* **54,** 803–9.

Spiegel, S. and Marcus, A. (1975). Polyribosome formation in early wheat embryo germination independent of either transcription or polyadenylation. *Nature* **256,** 228–30.

van Staden, J. (1973). Changes in endogenous cytokinins of lettuce seed during germination. *Physiol. Plant.* **28,** 108–11.

van Staden, J. and Wareing, P. F. (1972). The effect of light on endogenous cytokinin levels in seeds of *Rumex obtusifolus. Planta* **104,** 126–33.

van Staden, J., Webb, D. D. and Wareing, P. F. (1972). The effect of stratification on endogenous cytokinin levels in seeds of *Acer saccharum. Planta* **104,** 110–14.

Suzuki, S. and Taylorson, R. B. (1981). Ethylene inhibition of phytochrome-induced germination in *Potentilla norvegica* L. seeds. *Plant Physiol.* **68,** 1385–8.

Tavenar, R. J. and Laidman, D. L. (1972). The induction of lipase activity in the germinating wheat grain. *Phytochem.* **11,** 989–97.

Taylor, J. S. and Dumbroff, E. B. (1975). Bud, root, and growth regulator activity in *Acer saccharinum* during the dormant season. *Can. J. Bot.* **53,** 321–31.

Thomas, T. H. (1977). Cytokinins, cytokinin-active compounds and seed germination. In *The Physiology and Biochemistry of Seed Dormancy and Germination,* ed A. A. Khan, North-Holland Publishing Company, Amsterdam.

Thompson, P. A. (1968). The effect of some promoters and inhibitors on the light controlled germination of strawberry seeds: *Fragaria vesca semperflorens* Erh. *Physiol. Plant.* **211,** 833–41.

Thompson, P. A. (1974). Effects of fluctuating temperatures on germination. *J. exp. Bot.* **25,** 164–75.

Timson, I. (1965). New method of recording germination data. *Nature* **207,** 216–17.

Toole, E. H., Hendricks, S. B., Borthwick, H. A. and Toole, V. K. (1956). Physiology of seed germination. *Ann. Rev. Plant Physiol.* **7,** 299–324.

Totterdell, S. and Roberts, E. H. (1980). Characteristics of alternating temperatures which stimulate loss of dormancy in seeds of *Rumex obtusifolius* L. and *Rumex crispus* L. *Plant, Cell and Environ.* **3,** 3–12.

Trewavas, A. J. (1982). Growth substance sensitivity: the limiting factor in plant development. *Physiol. Plant.* **55,** 60–72.

Tuning, B. (1981). Groeiregulatoren in gerst in verband met keimvertraging. *Brouweerii-mouteri* **14,** 13–15.

Upadhyaya, M. K., Naylor, J. M. and Simpson, G. M. (1982). The physiological basis of seed dormancy in *Avena fatua* L. I. Action of respiratory inhibitors sodium azide and salicylhydroxamic acid. *Physiol. Plant.* **54,** 419–24.

Vegis, A. (1964). Dormancy in higher plants. *Ann. Rev. Plant Physiol.* **15,** 185–224.

Villiers, T. A. (1974). Seed aging: chromosome stability and extended viability of seeds stored fully imbibed. *Plant Physiol.* **53,** 875–8.

Waggoner, P. E. and Parlange, J. Y. (1976). Water uptake and water diffusivity of seeds. *Plant Physiol.* **57**, 153–6.

Walbot, V., Clutter, M. and Sussex, I. (1975). Effects of abscisic acid on germinating bean axes. *Plant Physiol.* **56**, 570–4.

Walton, D. C. (1980/81). Does ABA play a role in seed germination? *Israel J. Bot.* **29**, 168–80.

Wareing, P. F. (1949). Photoperiodism in woody species. *Forestry* **22**, 211–21.

Wareing, P. F. and Foda, H. A. (1957). Growth inhibitors and dormancy in *Xanthium* seed. *Physiol. Plant.* **10**, 266–80.

Webb, D. P., van Staden, J. and Wareing, P. F. (1973a). Seed dormancy in *Acer*: changes in endogenous cytokins, gibberellins and germination inhibitors during the breaking of dormancy in *Acer saccharum* Marsh. *J. exp. Bot.* **24**, 105–16.

Webb, D. P., van Staden, J. and Wareing, P. F. (1973b). Seed dormancy in *Acer*. Changes in endogenous germination inhibitors, cytokinins, and gibberellins during the breaking of dormancy in *Acer pseudoplatanus* L. *J. Exp. Bot.* **24**, 741–50.

Webster, B. and Leopold, C. A. (1977). The ultrastructure of dry and imbibed cotyledons of soybean. *Am. J. Bot.* **64**, 1286–93.

Williams, P. M., Ross, J. D. and Bradbeer, J. W. (1973). Studies in seed dormancy. VII. The abscisic acid content of the seeds and fruits of *Corylus avellana*. *Planta* **100**, 303–10.

Wood, A. and Paleg, L. G. (1974). Alteration of liposomal membrane fluidity by gibberellic acid. *Aust. J. Plant Physiol.* **1**, 31–40.

Wood, A., Paleg, L. G. and Spotswood, J. M. (1974). Hormone–phospholipid interaction: a possible hormonal mechanism of action in the control of membrane permeability. *Aust. J. Plant Physiol.* **1**, 167–9.

Yomo, H. (1960). Studies on the α-amylase activity substance. IV. On the amylase activating action of gibberellin. *Hakko Kyokaishi* **18**, 600–2, cited in *Chemical Abstracts* **55**, 26145 (1961).

Senescence and abscission 20

R Sexton and H W Woolhouse

Introduction

Defining senescence and abscission

Senescence may be simply defined as those changes which lead sooner or later to the death of an organism or some part of it. Another term which often creeps into the literature of this subject is ageing. Medawar (1957) clarified the relationship of the two terms by defining ageing as all those changes which occur in time without reference to death as a consequence; indeed its use need not be confined to living organisms.

The term abscission is used to describe the processes involved in the shedding of plant structures, characterized by the degradation of cell walls at the point of weakening. Present evidence suggests that cells surrounding the fracture line produce and secrete cell wall degrading enzymes which hydrolyze the central region of the wall, allowing the cells to separate and fracture to occur. In contrast, the less common process of 'mechanical tearing' involves the generation of large forces which tear apart an inherently weak band of cells. The cells along the fracture line thus play a passive role in this process, the flaking of bark providing a familiar example.

In this chapter senescence and abscission will be discussed as distinct phenomena but, as will be seen, there are interesting relationships between them in that they are frequently associated, as in the autumnal senescence and shedding of leaves.

Senescence

Distribution of senescence processes in the life cycle of plants

Senescence processes are encountered in all plants and at all stages of the life cycle (Woolhouse, 1978). In this chapter only senescence phenomena in the flowering plants will be considered. Figure 20.1 is a schematic diagram of the life cycle of a flowering plant in which senescence processes associated with successive stages of the cycle are listed.

A particular feature of this scheme is that the events which are listed are integral components of the developmental program of the plant, with the notable exception of those that occur within seeds.

It is convenient, in considering the multiplicity of senescence processes which contribute to the events which comprise the life cycle of a plant, to adopt this scheme of the cycle to provide a framework for discussion. A disadvantage in this approach is the risk of describing a wide range of apparently disparate phenomena in which elements of common ground become obscured. To obviate this difficulty, it may be useful at the outset to emphasize that although it is a commonly held view that senescence processes represent a descent into chaos, in terms of cellular and metabolic organization, they are in fact tightly controlled processes in which the sequence of events is usually highly ordered until the terminal stages are under way. The senescence of leaves in many species illustrates this point. The breakdown

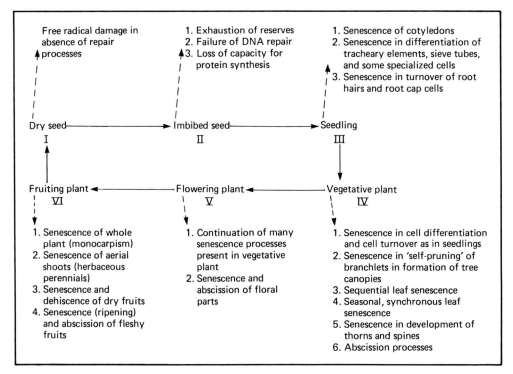

Fig. 20.1 Summary of the distribution of senescence processes in the life cycle of seed plants. Most of the phenomena listed, except those which occur within the seed, are integral parts of the developmental program of the plant.

of chlorophyll and dissolution of the thylakoid membranes may proceed almost to completion, yet the chloroplast envelope remains intact and the mitochondria and plasma membrane show little sign of disruption (Woolhouse and Jenkins, 1983). Thus, there occurs an orderly dismantling of the components of the leaf cells, with the retention to the last of those components which are needed for export of the contents to other parts of the plant.

Control processes in plant senescence

The schematic representation of a plant in Fig. 20.2 shows the principal organs with a listing against each of the factors which experimental evidence indicates may be involved in regulating the particular senescence processes they undergo (Wang and Woolhouse, 1982). It is immediately evident that at all stages a wide range of growth substances appears to be involved. Beyond this level of regulation involv-

ing growth substances it is difficult to make many generalizations about the control of senescence because species differ greatly in their behavior.

Natural selection and the distribution of senescence phenomena

The senescence strategies that have evolved in different species are very varied. For instance, among the evergreens there are species that retain their leaves for a single year, others which retain them for two or three years, while in the bristle cone pine (*Pinus aristata*) the needles are retained in a functional state for up to thirty years. Similarly, in many orchids the petals may be retained for months until the flower is pollinated, when they senesce and shrivel within a day (Arditti, 1979). On the other hand, in species such as morning glory (*Ipomea caerulea*) the flowers fade and then collapse within a day of bud opening, as their contents break down

whole plant
(monocarpism)
fruits, roots, daylength
auxins, ABA, cytokinins

apex
fruits, daylength
gibberellins

flower
pollination
ABA, cytokinins
ethylene

fruit
(including
senescence
of seeds)
all hormones

stem
(including senescence of
vascular tissues)
auxins, cytokinins
sugars

leaf
(including senescence
of abscission zone)
fruit, root, light
all hormones

cotyledon
axis, shoot, light,
fruit, root, all hormones

root
(including senescence of
vascular tissues, root cap and
hairs)
auxins, cytokinins

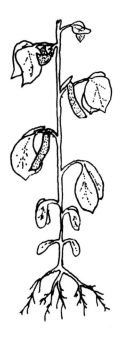

Fig. 20.2 Organs of a plant in which some or all the tissues undergo senescence during an annual cycle of growth. Some of the factors implicated in the regulation of the senescence processes are shown against each organ.

and are withdrawn to the developing gynecium. In foxgloves (*Digitalis purpurea*) and many other species the petals do not undergo any obvious senescence but are shed from the plant a few hours after pollination.

It is pertinent to enquire why there should have arisen this great diversity in the timing and pattern of senescence events. Though experimental verification is often difficult, it seems highly probable that natural selection based on factors in the habitat, which may often be quite subtle, has played a major part. Thus in the case of leaves it is found that plants of open meadows characteristically shed their

lower leaves in rapid succession as new ones are developed above to compete with neighboring plants for light; this is a process of sequential senescence (Woolhouse, 1967, 1974). Plants of similar life-form in the herb layer of woodland, on the other hand, where light is usually limiting, are better able to sustain their carbon balance by less profligate behavior and retain a single set of leaves for a whole season, or even for several years in some instances. It has often been noted that the evergreen habit prevails in nutrient-deficient habitats. Under these circumstances the plants often carry out a highly efficient process of nutrient cycling, producing a next generation of leaves and transferring all except the most immobile nutrients from the old to the new, before the exhausted ones are finally shed.

Other factors are believed to have influenced the evolution of the senescence programs of petals. For instance it has been suggested that the corolla of the foxglove is shed rapidly after pollination but before senescence, to prevent the diversion of scarce insect pollinators from the unfertilized flowers on the spike. In some desert succulents where the buds open at dusk the flowers senesce at dawn, presumably to restrict water loss during the day.

Metabolic regulation in senescence

Regarding senescence in most of its manifestations as a facet of the developmental program, distinguished from other developmental events only in that it leads to death of the plant or some part thereof, introduces by implication the hypothesis that it is ultimately a particular pattern of expression of gene activity that determines the course of events. Such regulation of gene expression may operate at several levels, as for example transcription of the DNA, translational control of the gene transcripts or by some form of regulation of the gene products, most notably the enzymes, once formed (see Fig. 20.7).

When viewed in this way senescence is seen to be what one author has referred to as a 'death program' or 'death clock' in which cells become committed to the expression of certain genes the action of which leads the cell to terminate its own existence. The study of senescence at a fundamental level thus becomes a matter of how the death sentence is encoded, how the sentence is passed and the manner of execution. There is, of course, no *a priori* reason to suppose that, even if senescence phenomena can all be reduced to such a set of conceptual essentials, the nature of the sentence, the manner of its expression or the mode of its execution will be the same in all cases. It will also be evident that the depth at which a discussion of these matters can be pursued must depend ultimately on the state of knowledge concerning such questions as the manner in which plant growth regulators function in the control of gene expression in plants, the organization of the controlling elements in the regulation of gene expression in plants, and the nature of the key events in the senescence process at the biochemical level. These are all areas of conspicuous ignorance

at the present time so that at the very outset not much can be hoped for at this level of our enquiry. Two examples serve to illustrate the nature of these difficulties: these are the problem of chlorophyll degradation, and the regulation of the synthesis and degradation of macromolecules which are essential to the functioning of the cells.

The problem of chlorophyll degradation One of the cardinal events by which senescence in leaves is generally recognized is the loss of chlorophyll, revealed in the change from green to yellow coloration. It must be emphasized that the timing of chlorophyll degradation in a senescing leaf may not always be correlated with other aspects of its functioning, not even with its photosynthetic capacity (Woolhouse, 1967; Hardwick, Wood and Woolhouse, 1968), and indeed mutants are known in which senescence and death occur without the chlorophyll ever being lost at all (Thomas and Stoddart, 1975, 1980; Thomas, 1977). Notwithstanding the fact that loss of chlorophyll may not be inseparably linked to the overall process of senescence, it is clearly such a general concomitant of the process that a knowledge of the pathway of chlorophyll degradation should contribute to our understanding of the senescence syndrome. Recent work with *Phaseolus vulgaris* and *Hordeum vulgare* (Maunders, Brown and Woolhouse, 1982), indicates that the degradation pathway of chlorophyll may involve an hydroxylation at C_{10} to yield chlorophyll a-1 (Fig. 20.3).

The observation that the chl. a-1 content rises, passes through a maximum and then decreases (Fig. 20.4), shows that it is not a stable degradation product but is an intermediate which is itself degraded. It has not yet proved possible to detect chl. a-1 in senescing leaves attached to the plant, but it seems possible that this may be because its further degradation proceeds so much faster in these circumstances that no accumulation takes place. It would not be surprising if hydroxylation was the first step in the chlorophyll breakdown pathway, since similar reactions initiate the catabolism of many cyclic molecules, including heme (Brown and Troxler, 1982). The subsequent steps in the degradation of the chlorophyll are unknown at present.

The synthesis and degradation (turnover) of cellular constituents Evidence has accumulated over the past

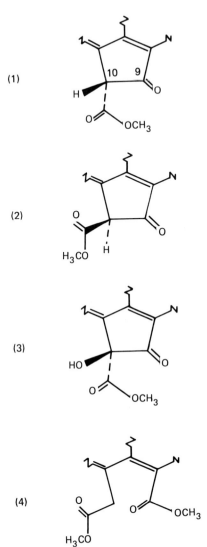

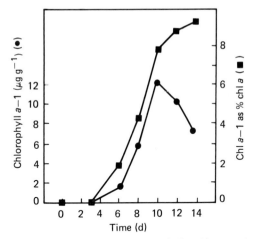

Fig. 20.4 The appearance of chlorophyll *a-1* in senescing leaf tissue of *Phaseolus vulgaris*. Chlorophyll *a-1* levels following excision (at day zero) are expressed in μg g^{-1} leaf tissue (●) and as a percentage of the total 'chlorophyll *a*-type 'pigments' (■). Chlorophyll *a* declined from 156 μg g^{-1} tissue at day zero to 80.2 μg g^{-1} tissue at day 14.

factor in senescence. In the nineteenth century it was first suggested that the proteins in plant cells are in a dynamic state, being continually broken down and resynthesized (*see* Davies, 1980). Similar ideas concerning RNA emerged later. From the notion of the dynamic state of the proteins in the living cell there developed the concept of a protein cycle (Fig. 20.5) (Mothes, 1933; Gregory and Sen, 1937), which

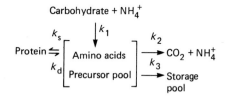

Fig. 20.5 A simple scheme to illustrate the concept of a protein cycle. k denotes rate constant: k_s for synthesis of protein, k_d for degradation of protein, k_1 and k_2 for synthesis and degradation of amino acids and k_3 for transfer to a storage pool. (After Davies, 1980.)

Fig. 20.3 Partial structures of chlorophyll derivatives. The region of the isocyclic ring is shown. (1) Chlorophylls *a* and *b*, (2) Chlorophylls *a'* and *b'*, (3) Chlorophyll *a-1*, (4) Chlorophyllin *a* (methyl ester form).

fifty years that in many species there is a progressive decline in the amount of protein and RNA in the leaves in the course of their maturation and senescence (Michael, 1936; Wollgiehn, 1967; Hardwick and Woolhouse, 1967). The loss of these essential cellular constituents is considered a potential causal

envisaged a continuous breakdown of proteins to their constituent amino acids. These then entered a common pool from which they could be withdrawn for use in the re-synthesis of new proteins as required.

When the net loss of protein from senescing tissues is considered in the context of this protein cycle, it is evident that it must arise from an imbalance between the rate of protein synthesis and the rate of degradation (Woolhouse, 1982a). Such an imbalance might be caused by a slowing of the rate of protein synthesis or an increase in the rate of degradation, or both (Fig. 20.6). This formulation of the problem leads to a number of specific questions.

for unresolved mixtures of proteins can reveal relatively little. A protein carrying out an essential function in the cell (as opposed to one which is involved in some facet of secondary metabolism), which is synthesized at one stage of development but not renewed thereafter, could, if subject to gradual inactivation or breakdown, become a factor in the declining functional capacity of a senescing cell. It therefore becomes necessary to purify to

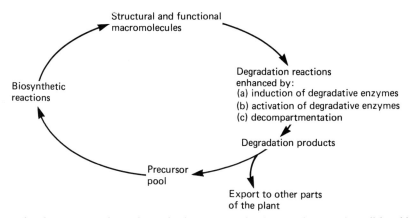

Fig. 20.6 Scheme for the turnover of proteins and other macromolecular constituents of a cell in which net loss of a constituent, leading to senescence, may arise from suppression of the biosynthetic system or enhanced activity of enzymes on the catabolic side of the cycle. Some breakdown products are shown as being lost to the system by translocation while some are re-cycled through the biosynthetic pathways.

Firstly, there are the practical problems of how to measure the relative rates of protein synthesis and degradation, or protein turnover as the process is now called. Secondly, when the measurement of turnover has been achieved, one may ask what these processes of synthesis and degradation involve and how the rates at which they take place are regulated.

Despite the availability of several powerful techniques there remain great practical obstacles to the accurate measurement of rates of protein turnover in plants (*see* Davies, 1980). In leaves, published values of general protein turnover are relatively few but suggest a half life of 3–5 days. Such measurements represent an average value for the leaf protein as a whole and may involve some proteins which are turning over rapidly and others which are turning over slowly or not at all. If the regulation of protein turnover in relation to the onset of senescence is to be understood it is necessary to measure turnover of individual proteins, since average values

homogeneity specific proteins so that their rates of turnover may be measured.

In considering which enzymes to focus on for measurements of turnover, it would seem logical to study either those which are nearest to the threshold level at which they are required in the economy of the cell or those which occupy key positions in the regulation of metabolism. Unfortunately most of the regulatory enzymes have so far proved difficult to purify and little is known concerning their turnover. For practical reasons the leaf enzyme which has received most attention is RuBP carboxylase, although whether this enzyme is rate-limiting to photosynthesis in senescing leaves is problematical (Woolhouse and Jenkins, 1982). Turnover studies on RuBP carboxylase have been carried out on leaves of cucumber, *Perilla frutescens* (Woolhouse, 1967), barley (Peterson, Kleinkopf and Huffaker, 1973; Peterson and Huffaker, 1975), maize (Simpson, Cooke and Davies, 1981) and *Phaseolus vulgaris*

(Barrett and Woolhouse, 1981). In *Perilla*, barley and cucumber, RuBP carboxylase is synthesized during development of the leaves but formation of the enzyme ceases when expansion is completed; thereafter the amount and activity of the enzyme declines gradually in the mature leaf. In contrast, RuBP carboxylase continues to be synthesized in the mature maize leaf, with degradation proceeding faster than synthesis in the latter stages (Simpson *et al.*, 1981).

A particular feature of RuBP carboxylase is that the large subunit of the enzyme is encoded in the chloroplast and synthesized on 70s ribosomes within the organelle, while the small subunit is encoded in the nucleus and synthesized on 80s ribosomes in the cytoplasm before entering the chloroplast. In many species both photosynthesis and the RNA and protein synthesis taking place in the chloroplast begin to slow down before there is a loss of organization and metabolic functions in the rest of the cell (Butler and Simon, 1972). The question therefore arises as to whether in the case of RuBP carboxylase there may be a loss of co-ordination in the synthesis of the two subunits of the enzyme. This problem has been examined in some detail in wheat (Brady, 1981; Spiers and Brady, 1981) which, like maize, synthesizes the enzyme in the mature leaf. At the time that the wheat leaf approaches completion of expansion the rate of synthesis of the carboxylase is similar to the mean rate for all proteins, but after the completion of leaf expansion the rate of synthesis falls sharply relative to that of other proteins. The decline in rate of synthesis was similar for the large and the small subunits of the carboxylase. Using wheat germ and *E. coli* systems for *in vitro* measurements of the amounts of translatable mRNA from leaves of different ages it was shown that there was a co-ordinated decline in the amounts of message for the two subunits of the carboxylase. The amounts of message for this enzyme declined more sharply after the completion of leaf expansion than did the other mRNAs. Thus in the case of the carboxylase enzyme in leaves of wheat, the evidence points to a decline in the production of the mRNAs for the enzyme as an underlying factor in the progressive loss of the enzyme; that is to say, in terms of Fig. 20.6 it is the arrest on the synthetic side of the protein cycle that is critical, rather than an enhanced rate of degradation.

Control of protein synthesis

It seems to be a general rule that in prokaryotic cells and in the organelles of eukaryotes, the mRNAs are relatively short-lived and gene expression is controlled primarily at the level of transcription. Conversely the cytoplasmic mRNAs of eukaryotes are longer-lived and control of protein synthesis occurs at the levels of both transcription and translation (Ochoa and De Haro, 1979). As a result, regulation of protein synthesis in the organelles and in the cytoplasm are likely to be different and are dealt with separately.

(a) *Chloroplasts* Most of the evidence for changes in the biosynthetic side of protein turnover in relation to leaf senescence comes from studies of the photosynthetic apparatus.

In leaves of *Perilla*, cucumber and *Phaseolus* there is a loss of capacity for chloroplast protein synthesis in the course of leaf senescence, although cytoplasmic protein synthesis is maintained (Callow, Callow and Woolhouse, 1972). Declining chloroplastic protein synthesis is associated with a loss of polysomes and the cessation of chloroplast RNA synthesis (Callow and Woolhouse, 1973).

The arrest of protein synthesis in the chloroplasts of *P. vulgaris* at the time that leaf expansion is completed, has been traced to the disappearance of chloroplast RNApolymerase activity (Ness and Woolhouse, 1980a,b). The RNApolymerase found in chloroplasts is encoded in the nucleus and synthesized on cytoplasmic ribosomes (Bunger and Feierabend, 1980) which raises the possibility that the control of this key enzyme may be one of the avenues by which the nucleus holds sway over the life-span of the chloroplast population. Further evidence of an overriding role for the nucleus in the control of senescence in leaf cells comes from the work of Yoshida (1961). Leaves of *Elodea densa* were plasmolyzed in calcium chloride solution, causing some of the cells to separate into two parts, one with and one without a nucleus. The leaves were then cultured over a period of five days during which the chloroplasts of the enucleate protoplasts remained green, while chloroplasts in the protoplasts containing nuclei disintegrated and were broken down.

(b) *Mitochondria* In the leaves of many plants the rate of respiration is maintained and may actually increase during yellowing and senescence (Arney, 1947; Eberhardt, 1955; James, 1953; Hardwick *et al.*, 1968). This finding is compatible with electron-microscopic studies which show that the mitochondria of yellowing leaves retain their structural integrity even at an advanced stage of senescence when the chloroplasts are showing extensive degradation and the rate of photosynthesis is declining (Butler and Simon, 1972).

It is reasonable to suppose that the sustained mitochondrial activity in senescing leaves serves to fuel vein-loading as the breakdown-products of the cells are remobilized and also the synthesis of lipids, amides (Yemm, 1950) and specific enzymes (Farkas and Stahmann, 1966; Sacher and Davies, 1974) takes place. A discussion of how the shutdown of organelles occurs at different times involves details of molecular biology which are beyond the scope of this chapter, but are examined by Woolhouse and Jenkins (1983).

(c) *Cytoplasm* There is substantial evidence that although the total protein content of leaves declines in the course of senescence (Woolhouse, 1967), there is nonetheless an active synthesis of proteins on 80s cytoplasmic ribosomes throughout this period (Callow *et al.*, 1972). The RNA content of senescing leaves also declines (Osborne, 1967, 1982; Dyer and Osborne, 1971; Pearson, Thomas and Thomas, 1978; Callow and Woolhouse, 1973), as does the rate of RNA synthesis (Wollgiehn, 1967; Callow *et al.*, 1972). Generally speaking, inhibitors of RNA synthesis, such as actinomycin D, are without effect on leaf senescence (Von Abrams, 1974; Thomas and Stoddart, 1977); this has been interpreted as suggesting that senescence is not controlled at the transcriptional level (Thomas and Stoddart, 1980). It may be that leaf senescence involves relatively long-lived mRNA for proteins associated with senescence which are controlled at the level of processing or translation of the mRNAs. A possibility of a processing control is suggested by the finding that treatment with cordycepin, an inhibitor of polyadenylation of RNA, delayed senescence in leaf disks of *Nicotiana* (Takegami and Yoshida, 1975).

Inhibitors of the 80s ribosome cycle in eukaryotes are frequently found to be active inhibitors of leaf senescence, which has been widely interpreted as evidence of control of senescence at the level of

mRNA translation. The most frequently used inhibitors in this type of work have been cycloheximide (Knypl and Mazurczyk, 1972; Makovetzki and Goldschmidt, 1976; Thomas, 1975) and MDMP 2-(4-methyl-2,6-dinitroanilino)-*N*-methyl-propionamide (Thomas, 1976).

Control of protein degradation

As has already been noted, a net loss of protein from senescing leaves may be achieved simply by shutting down the biosynthetic side of the turnover process, operating through the type of control mechanism described in the preceding section. Attention must now turn to the contribution of increased rates of degradation to the net loss of protein and to how this side of the protein cycle changes and is controlled in the course of senescence.

Table 20.1 lists proteases and other hydrolytic enzymes which have been reported to increase in

Table 20.1 Hydrolytic enzymes reported to show increased activity in senescing leaves. For details of the references, *see* Woolhouse (1982a).

1. PROTEASES: Tobacco, mung bean, pea, wheat, maize, barley, oats
2. RIBONUCLEASES: *Rhoeo discolor, Lolium temulentum*
3. ACID INVERTASE: *Lolium temulentum*
4. CHLOROPHYLLASE: Barley, oats
5. ESTERASE: *Festuca pratensis*
6. ACID PHOSPHATASE: *Perilla frutescens*
7. β 1–3 GLUCAN HYDROLASE: *Nicotiana glutinosa*

activity in senescing leaves. It is pertinent to ask whether these increases in hydrolase activity are in fact necessary for senescence to take place. In many cases the amount of protease activity detectable before the onset of senescence is more than adequate for the observed rate of breakdown of protein and the activity during senescence is increased by a factor of only two or three (Waters, Peoples, Simpson and Dalling, 1980); a change of this magnitude may be of much less significance than a change of compartmentation bringing enzymes and substrates into contact. Evidence which casts further doubt on the significance of measured activities of acid proteases comes from the work of Van Loon, Haverkort and Lockhurst (1978). Excised oat leaves floated on water in darkness showed the normal yellowing and increase of protease activity; leaves placed in an upright position with their bases in water, on the other hand, did not show the increased protease activity, but breakdown of protein occurred to the same extent in both sets of leaves.

The many sites at which there could, in principle, be regulation of levels and activity of proteolytic enzymes are summarized in Fig. 20.7. For convenience the regulatory controls proposed in Fig. 20.7 may be summarized in three groups: (i) *de novo* synthesis of proteolytic enzymes, (ii) activation and *in situ* regulation, and (iii) compartmentation. For detailed discussion of the evidence relating to each of these stages, see Woolhouse (1982a) and Woolhouse and Jenkins (1983).

Growth regulators and the control of leaf senescence

The preceding sections considered the problem of how intervention in the turnover of constituents of a tissue may be modified in a manner which leads it to senesce, but ignored the broader question of what overall controls are operating, whereby the sequence of metabolic changes is regulated. The circumstances in which leaves may come to senesce vary greatly (Simon, 1967). Excision of a leaf from the parent plant is the most certain way of accelerating its senescence, but opinions vary as to whether the changes which take place in an excised leaf represent merely a speeded-up version of the events

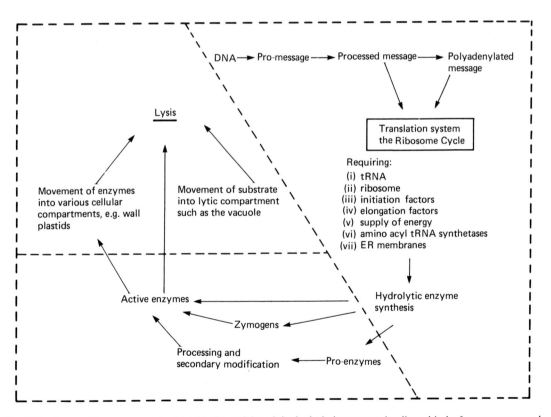

Fig. 20.7 Some of the many points at which the activity of the hydrolytic enzymes implicated in leaf senescence may be controlled. Each step or factor is potentially regulatory; those in the right-hand third are concerned with *de novo* synthesis, those in the bottom third are reactions involving *in situ* regulation by processing etc. The upper left-hand third indicates enzyme regulation by compartmentation and regulation at the substrate level.

in a naturally senescing leaf; they probably do not (Woolhouse and Jenkins, 1983). Detailed discussion of the ways in which environmental and developmental-correlative factors such as the formation of fruits or new adjacent leaves may influence leaf senescence can be found in several recent accounts (Woolhouse, 1974, 1982b; Noodén and Leopold, 1978; Thomas and Stoddart, 1980). Our present purpose is to examine the mass of fragmentary evidence from experiments on a wide range of species and ask whether any general principles underly the types of control mechanism involved.

Chibnall (1939) found that senescence of excised leaves of *Phaseolus vulgaris* could be arrested if roots developed on the cut petioles: he inferred that a substance originating in the roots may be necessary to maintain the viability of the leaves. Following this observation, much evidence has accumulated that growth substances originating in the roots, and indeed in other parts of the plant, may influence leaf senescence (Osborne, 1967; Woolhouse, 1974; Thomas and Stoddart, 1980). It would, however, be an exaggeration to claim that there is any clear understanding at the present time of how these substances act.

The experimental approach most frequently used to study the effects of growth substances on leaf senescence has been to place excised leaves with their cut petioles in solutions containing the growth substances, or to float excised leaf disks on the appropriate solutions and observe the rate of senescence, usually in terms of loss of chlorophyll and protein. The results obtained vary greatly between species. In excised leaves of *Xanthium* (Richmond and Lang, 1957) and many other species, cytokinins delay senescence. Indolyl acetic acid is needed to delay the senescence of leaves of *Prunus* species (Osborne, 1967). Gibberellins will delay the senescence of excised leaves of *Taraxacum officinale* but the cytokinins are without effect (Whyte and Luckwill, 1966). In *Rumex* species and *Tropaeolum majus*, cytokinins or gibberellins will delay the senescence of excised leaves over similar ranges of concentration (Beevers, 1968). The reason for this diversity of responses is not known. In the case of *Tropaeolum* it was found that the gibberellin content of the leaves decreased as they senesced. It is also known that in many species there is a continuous supply of cytokinins passing from the roots to the leaves in the transpiration stream, and that

cytokinins may be metabolized to other products at relatively high rates in leaf tissues. These observations indicate the possibility that the effects of a given growth substance in delaying senescence, when applied to an excised leaf of a particular species, may be dependent on the rate at which the endogenous growth substances are turned over. If the turnover of the growth substance is such that it falls below the level necessary to sustain metabolism, senescence will ensue and it is that particular growth substance which has the senescence-retarding effect when applied exogenously (Woolhouse, 1967, 1978).

Exogenously applied abscisic acid (ABA) accelerates senescence in leaves of many species (Paranjothy and Wareing, 1971). Colquhoun and Hillman (1975) failed to find any correlation between ABA content and the stage of senescence in leaves of *Phaseolus vulgaris*. It is not of course to be expected that any such correlation between total content of ABA and degree of senescence should of necessity exist. It could be serving a trigger function in the early stages of senescence and thereafter changing in amount in a manner unrelated to the progress of the senescence syndrome.

Excised leaves of oat incubated in darkness show a five-fold increase in ABA content which may be involved in promoting the rapid senescence which is observed (Gepstein and Thimann, 1980). The aleurone layer shares many attributes of a senescing tissue and it is noteworthy that ABA exerts an action in this system, which raises interesting possibilities for further work with leaves (Mozer, 1980).

Although the effect of cytokinins in delaying leaf senescence is so well known as even to form the basis of a bioassay, the mode of action of this family of hormones is still obscure. Studies using cultured cells of *Glycine max*, which are stimulated to growth by cytokinins, suggest a translational level of control. Cytokinin stimulated recruitment of monosomes into polysomes in cultured cells of *Glycine* within 15 min of application, apparently by activating existing mRNAs to a form amenable to translation (Tepfer and Foskett, 1978). The biochemistry of this effect is not yet understood. A similar mechanism could account for the effect of cytokinins in stimulating protein synthesis and thereby delaying senescence in leaf tissues. Another suggested mechanism of cytokinin action in stimulating protein synthesis may involve activation of the protein kinases, although a direct effect of cytokinins on phosphorylation of proteins seems unlikely (Ralph and Wojcik, 1981); an action involving initial effects of cytokinin on the intracellular distribution of ions, which in turn alters the pattern of protein synthesis, seems more probable.

The relationship of ethylene to the process of foliar senescence is of particular interest and suggests some intriguing avenues for further research. It has been suggested that ethylene may be without effect on the rate of senescence in some species (Thimann, 1978); other evidence, however, is positive. Ethylene induces senescence in leaves of tobacco and is produced by senescing leaves of this species. Excised leaves of evergreens such as *Ilex aquifolium* (holly) and *Hedera helix* (ivy), may be maintained in darkness for many months without undergoing senescence but can be induced to senesce rapidly by exposure to ethylene. It has also been found that treatment of leaves of these species with ethylene will promote senescence but if the gas is then withdrawn while senescence is yet incomplete, the process is arrested (Woolhouse, unpublished).

It has recently been shown that the pathway of ethylene biosynthesis from methionine involves *S*-adenosyl methionine (SAM) and 1-aminocyclopropane-1-carboxylic acid (ACC) as intermediates with a recycling of the methylthiol group as a unit into methionine (Adams and Yang, 1977). Using excised leaves of tobacco it was shown that there is a rapid rise in ethylene production prior to the final rapid phase of senescence (Aharoni, Lieberman and Sisler, 1979a). Treatment of attached leaves of *P. vulgaris* with 1 mM ACC led to ethylene production and rapid yellowing and senescence.

Auxin treatment of many plant tissues leads to an enhanced production of ethylene (Abeles, 1973) which has been shown to result from increased activity of ACC synthase, the enzyme which converts SAM to ACC (Adams and Yang, 1979). Aminoethoxyvinyl glycine (AVG), which inhibits ACC synthase and hence ethylene synthesis, and Co^{2+} which inhibits the conversion of ACC to ethylene, delay the senescence of excised leaves. Silver ions and high CO_2 concentrations, which inhibit ethylene action, give rise to similar effects (Aharoni and Lieberman, 1979; Aharoni, Anderson and Lieberman, 1979b). It is of interest to note that

molecular oxygen is also needed for the conversion of ACC to ethylene, as it is also for ethylene action (Beyer, 1976) (*see* Chapter 5). These results suggest explanations for the delay of leaf senescence under anaerobic conditions.

Monocarpic senescence

Some plants die at the time at which their fruits mature. The ripening of a field of corn is perhaps the most common manifestation of this phenomenon. Species which show this pattern of behavior are referred to as monocarpic to distinguish them from polycarpic species, which may go through many cycles of growth and reproduction. Examination of the detailed behavior of some species reveals that this monocarpic/polycarpic distinction may not be entirely clear-cut; for example, in some species it can be influenced by the nutrient supply and climatic conditions at the time of fruiting (Woolhouse, 1982b).

Monocarpic species such as cereals are characterized by having all of the vegetative apices transformed into determinate reproductive structures. Other monocarpic species such as *Pisum sativum* and *Glycine max* L. Merr. are indeterminate although the relatively abrupt cessation of growth which is observed in some genotypes of these species has led to the incorrect use of the term 'determinate' to describe their growth response (Wang and Woolhouse, 1982).

The close correlation between seed-filling and senescence in monocarpic plants invited the suggestion that death was a consequence of exhaustion as the vital resources of the plant were diverted to maximize the production of seed. Unfortunately the exhaustion or 'nutrient diversion' hypothesis, as it is sometimes called, has proved very difficult to assess experimentally and there is still no clear view of how significant it is. The long-standing observation favoring the nutrient diversion hypothesis is that in many monocarpic species removal of the flowers or developing fruits will delay senescence and thereby prolong the life of the plant (Molisch, 1938). The problem with this sort of experiment is that the removal of all flowers or fruits from a plant as they are formed may upset it in far more ways than simply removing a potential source or sink of nutrients; they may be major sources of growth substances, for instance.

One of the earliest expressions of doubt concerning the nutrient demand hypothesis was that of Murneek (1926) on the grounds that fruit-induced effects on growth were seen even in the presence of excess nutrients. There are other facts which are difficult to equate with this hypothesis. Thus in spinach, which is diecious and monocarpic, both male and female plants show similar patterns of senescence even though the nutrient demand is much higher in the female, seed-bearing, plant. Both sexes in spinach may be induced to 'bolt' by the application of gibberellic acid; in these circumstances there is a substantial diversion of nutrient to the massive inflorescence structure which develops, but flowers are not formed and senescence does not occur. A converse example, in which the nutrient sink can be removed without prevention of senescence, is in *Xanthium*, in which the disbudding of plants induced to flower fails to prevent senescence (Krizek, McIlrath and Bergara, 1966).

Before proceeding to alternative theories concerning the causes of monocarpic senescence, a word of caution is needed. Monocarpic senescence is found in many different families of plants, including monocotyledons and dicotyledons, and in many cases there is overwhelming evidence that it is an evolutionarily-derived condition. That is to say, the monocarpic species appears to have evolved from polycarpic progenitors; *Zea mays* which derives from *Z. diploperennis* is an example of this situation. The point to be emphasized here, however, is that if the monocarpic habit has such diverse evolutionary origins, we may not be justified in seeking a common mechanism for its regulation.

There is general agreement among physiologists that growth substances are implicated in the regulation of monocarpic senescence, although their mode of action is obscure. McCollum (1934) suggested that a hormone produced by young developing fruits might be involved in the diversion of nutrients to the fruits. Wareing and Seth (1967) removed the ovules from attached pods of *Phaseolus vulgaris*. Replacement of the ovules by auxin pastes diverted nutrients to the site of application. In the same group of experiments these authors showed that the level of cytokinins bioassayed from leaves was lower in plants with fruits than in leaves of plants from which the fruits had been removed.

Much of the more recent work on the involvement of growth substances in the regulation of plant senescence has been carried out with soybeans by Nooden and his co-workers. In soybean cv. Anoka, senescence begins when the majority of the fruits are filling rapidly (Lindoo and Nooden, 1977); removal of the fruits up to a stage just prior to ripening of the pods prevented senescence. Up to 50% of the pods could be removed from a plant without greatly affecting the rate of senescence. By removal of the stem apex, Y-shaped plants were developed and fruits were removed from one of the branches. Leaves on the depodded branch remained green while those on the fruiting branch senesced. There was some incipient senescence of leaves in the lower reaches of the depodded branch; this was attributed to the leakage across the depodded branch, of a senescence-inducing factor produced by the fruits, but there are other possible explanations. The effect of removal of fruits on the pattern of stomatal opening on the leaves of that shoot were not investigated. There is a strong possibility of such effects, in which case the relative proportions of the transpiration stream entering the two shoots would be altered, with a consequent imbalance in the amounts of nutrients and growth substances they received. When soybean plants were trimmed to a single leaf and one cluster of pods at a distance of three internodes above or below the leaf, the leaf senesced more rapidly when it was below the pods (Nooden, Rupp and Derman, 1978). This effect was interpreted as further evidence for a senescence-inducing factor coming from the fruits with a greater potential for movement in a downward direction. Again, however, the experiment is not as simple as it looks. The relative ages of the leaves above and below the pods are very different, and this alone will mean that they do not have the same propensity for senescence. In addition, it is well known that stomatal conductance declines as a function of leaf age (Woolhouse and Jenkins, 1983). If such a difference existed between the leaves above or below the pods, there would again be a difference in amount of water and dissolved substances passing through these leaves in the two treatments. Another difficulty raised by these experiments is that the movement of mineral nutrients to the pods, which normally takes place *via* the leaves along with photosynthate (Hocking and Pate, 1977), may well be differentially constrained in the two cases because of the difference in the pattern of vascular connexions when the leaf is above or below the pods.

Surgical severance of the phloem below the pods diminished the senescence of the leaves below (Murray and Noodén, 1981), a finding which was offered as evidence for a senescence factor moving downward from the fruits in the phloem. On the other hand, cutting of the phloem in the petioles of the leaves below the pods did not prevent senescence, suggesting that the hypothetical senescence factor was moving into the leaves in the transpiration stream. The best that can be said for these various experiments involving surgical treatment of monocarpic plants, is that they yield a rich variety of interesting results. It is the general case, however, that the treatments are much more complicated than they often seem at first sight, with the result that the experiments raise many questions but afford few, if any, answers. The general lack of progress in the direct search for extractable senescence-inducing substances from monocarpic plants reinforces the doubts arising from the surgical experiments (Wang and Woolhouse, 1982).

Abscisic acid (ABA) sprayed on to the leaves of fruiting soybean plants accelerated their senescence but ABA did not enhance the senescence of depodded plants. Thus, although ABA is known to increase in maturing fruits it does not seem a likely candidate for a senescence-inducing factor. Indeed experiments with ^{14}C-ABA indicate that it is carried preferentially from the leaves to the fruits, rather than the reverse (Noodén and Obermeyer, 1981).

If a general criticism were to be made of experiments on monocarpic plants to date, it is that they have not sufficiently emphasized the interplay of different parts of the whole plant and in particular the profound changes in the growth and metabolic activity in the roots of many species, as their fruits develop (Woolhouse, 1982b). A general theory to explain how such a system of interactions between organs might arise is shown in Fig. 20.8. This scheme envisages that the vegetative plant produces growth substances such as IAA and cytokinins from the actively-growing shoot and root apices. These substances, by their action, perpetuate the vegetative condition; however, as the apex switches to differentiating flowers, this feedback loop is broken. The leaves all mature and the supply of IAA from the shoot apex is substantially reduced. This in turn diminishes the supply of cytokinins (CK) from the roots as their growth becomes less active. The general decline in regulators that promote the

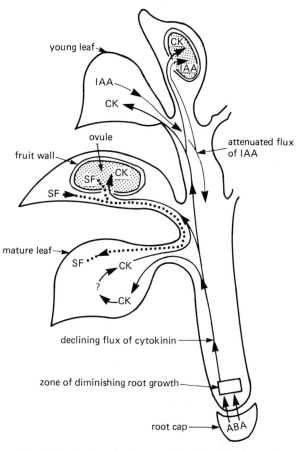

Fig. 20.8 An hypothesis concerning the rôle of growth substances in the regulation of plant senescence. The scheme envisages a control loop in the vegetative plant which is broken in two ways: (a) by elimination of the sources of IAA as the vegetative shoot apices switch to differentiating as flowers, and the young leaves mature, and (b) a hypothetical senescence factor (SF) which may be produced by the developing fruits in some species and promotes senescence in the immediately adjacent tissues. (CK = cytokinins.) (For detailed discussion of this scheme see Woolhouse 1982b.)

vegetative condition is accompanied by an increase in regulators from the developing fruit. These will include factors such as that associated with the expanding pods (*see above*) and ethylene which promote senescence. They will also further attenuate the supply of regulators that promote the vegetative state thus precipitating rapid senescence. It must be emphasized that this is only a hypothesis; this whole area is, however, a very fertile one for further research.

Abscission

Abscission zones

The more conspicuous examples of abscission such as the shedding of fruits, leaves, bud scales, floral structures and branchlets will be familiar, but it is important to realize that virtually any aerial part of a plant can be shed in this way. Structures may range in size from the complete shoot system of tumbleweeds down to the hairs shed from developing leaves. Fracture is caused by wall breakdown in 'separation layers'. Plants do not have the ability to produce separation layers anywhere; they seem genetically limited to specific locations called abscission zones and are characteristic of a given species. In some developmental mutants the ability to abscise is apparently lost, for instance the tomato varieties 'jointless' and 'lateral suppressor' have no floral abscission zone.

It was established by the nineteenth century anatomists that, although abscission was restricted to predictable positions, there was no obvious distinct class of cells associated with cell separation. The 'abscission zone' or general region through which fracture will occur contains the same cell classes as adjacent tissues, though it can be identified by the series of more subtle characteristics described below (Webster, 1973). It must be emphasized, however, that wall breakdown is usually confined to a 'separation layer' one to three cells wide in a five- to fifty-cell-wide 'abscission zone'.

In general, cells of the abscission zone are smaller than their counterparts in adjacent tissues. This results from lack of cell enlargement and more persistent meristematic activity. It is possible that these extra divisions are essential to program the cells to respond to the abscission signal. Osborne and Sargent (1976) have described how the ability of young leaves to abscise develops as these cells become distinguishable. Earlier Gawadi and Avery (1950), after studying the formation of abscission zones, advanced the view that the term was a misnomer as the tissue had little to do with abscis-

sion and was more concerned with scar-tissue formation. However, the frequent absence of lignified structural elements such as fibers and sclereids from the zone and their replacement by increased development of collenchyma does seem related to cell separation. Lignified walls are extremely resistant to enzymatic hydrolysis while collenchyma walls are readily degraded. The degree to which these features develop in the abscission zone varies; in delicate herbaceous tissues such as petal bases the zone may be scarcely discernible.

The time course of abscission

Experimental abscission is usually induced by excising the abscission zone together with small amounts of adjacent tissue and enclosing the 'explant' in a humid container. Excision is necessary to release the abscission zone from the inhibitory influence of the subtending organ. In modern work 1–50 ppm ethylene is added to accelerate and synchronize the process. Few botanists appreciate what a dynamic process abscission can be. Geranium buds will shed their petals within three hours of exposure to ethylene and the whole flower bud of tobacco or tomato is shed after 4–4.5 h exposure to the gas (Lieberman, Valdovinos and Jensen, 1982). The favored experimental materials take rather longer, bean leaf abscission, for instance, taking 36–48 h. It is possible to follow the kinetics of the weakening process by measuring the tensile strength or 'break strength' of the abscission zone. Usually break strength starts to decline after a lag phase which is of 2.5 h duration in tobacco flowers and 20 h in bean (Fig. 20.11).

Cell biology of abscission

One of the first detectable changes during the lag phase is the accumulation of cytoplasm and organelles in the abscission zone cells. Associated with this change are increases in the rate of respiration (Marynick, 1977) and incorporation of precursors into both RNA and protein (Abeles, 1968). Respiratory inhibitors or inhibitors of RNA and protein synthesis (Abeles, 1968) added during this period markedly inhibit the weakening process. These results indicate that essential species of RNA and protein are formed early in the lag period. Recently *in vitro* translation has been used to show that new species of mRNA appear in the abscission zone during the first six hours of the lag and other species are differentially amplified (Kelly, Trewavas, Sexton, Durbin and Lewis, 1982).

It is worth noting that all these results have been obtained with the slower leaf and fruit systems. The one floral abscission process studied appeared insensitive to actinomycin D but was inhibited by cycloheximide (Henry, Valdovinos and Jensen, 1974). It is conceivable, therefore, that the progress of abscission is suspended at a later stage in floral zones.

Toward the end of the lag phase it becomes possible to distinguish histochemically those cells that will break down their walls. They stain intensely for respiratory enzymes (Poovaiah and Rasmussen, 1974), have conspicuous nuclei and nucleoli and can be shown by autoradiography to be the site of highest rates of RNA and protein synthesis (Stösser, 1971). All these characteristics are features of intense metabolic activity.

As the break strength of the abscission zone starts to decline, wall breakdown becomes apparent in the 'separation layer'. At the light-microscopic level this is detectable as wall swelling and loss of pectin staining (Webster, 1973). At the ultrastructural level the process is characterized by a swelling of the middle lamella and adjacent areas of the primary wall. Eventually this central wall area fenestrates and disappears (Valdovinos and Jensen, 1968).

Virtually all microscopic work has concentrated on changes in the cortical parenchyma but the fracture plane must also pass through all the other tissues of the organ including the stele. It has been established that all living classes of cell along the fracture line degrade their walls (Sexton, 1976). Of the cells examined, transfer cells and collenchyma showed most evidence of degradation and epidermal cells least. Usually wall breakdown is confined to two rows of cells with maximum breakdown in the common wall, but separation layers may be more extensive.

The cytoplasm of separation zone cells is quite normal when wall breakdown starts and is not in a state of lysis as often envisaged (Fig. 20.10). Particularly noticeable is the preponderance of Golgi and rough endoplasmic reticulum (Valdovinos, Jensen and Sicko, 1972; Sexton, Jamieson and Allan, 1977). These membrane systems and their associated vesicles have been implicated in the secretion of wall-degrading enzymes. Various tests of plasma

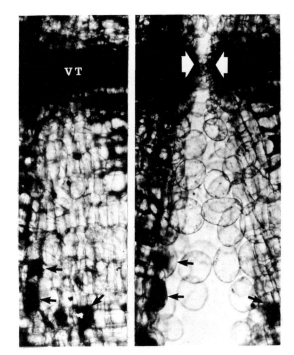

Fig. 20.9 A longitudinal living section through an abscission zone before (left) and after (right) being induced to abscise between two glass slides. Cell wall degradation has resulted in the cells of the separation layer becoming 'unglued' and expanding osmotically to form spheres. The resulting increase in cell volume widens the separation layer and in doing so stretches (white arrows) and eventually snaps the vessels of the vascular trace (VT) that bridge it. The black arrows indicate the position of three reference cells. (From Sexton and Redshaw, 1981.)

membrane permeability have shown that the separation zone cells retain their selective permeability; indeed the maintenance of turgor may be important in the mechanics of separation (see below). Cellular collapse may occur, but only as cell separation is completed.

Fracture usually follows the line of the middle lamella, leaving intact cells on the fracture surfaces (Sexton, 1976; Fig. 20.9). Where large forces develop across the separation layer, for instance when a heavy fruit is shed, cell rupture may also occur. Scanning electron-microscope studies of the fractured surfaces have led to the conclusion that xylem vessels, tracheids and cuticle are stretched and

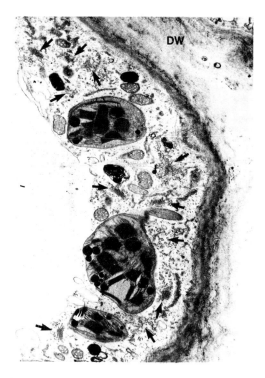

Fig. 20.10 An electron micrograph of a cell from the separation layer at the time of fracture. Note the swollen degraded wall (DW) and the disappearance of the middle lamella. The cytoplasm of the cell contains many active dictyosomes (arrows) thought to be involved in enzyme secretion. Contrary to early EM observations the vacuolar and plasma membranes appear intact. (From Sexton *et al.*, 1977.)

mechanically ruptured, though there are indications that vessel wall lysis may also be involved.

The mechanics of fracture

The earliest study of the forces involved in the rupture of abscission zones was Newton's famous observations on apple. Much later, Weisner (1871) suspected that internally-generated forces were also important since he showed that the action of both wind and gravity was insufficient to rupture the vessels of the leaves he studied. A wide variety of mechanisms has evolved to stretch the non-degraded tissues until they yield. Early anatomists

noticed that the scars on either side of a freshly-fractured zone were sometimes of differing diameters. It has been shown more recently that growth of cells on one side of the fracture line coupled with contraction on the other produces stresses at the junction that cause fracture (Wright and Osborne 1974). Another old theory that has recently been resurrected is the 'turgor mechanism'. Parenchyma cells are constrained into angular shapes partly because they are cemented together. As the intercellular cement is degraded during abscission, the cells become unglued and expand osmotically to form spheres (Sexton and Redshaw, 1981). This process will tend to enlarge the width of the separation layer and stretch tissues such as the vascular traces that bridge it (Fig. 20.9).

The biochemistry of wall breakdown

The chemistry of wall breakdown is poorly understood, interpretation of data being complicated by the fact that the separation zones contain a wide variety of cell types with walls of different chemical composition. Morré (1968) showed a 4% loss of total wall carbohydrate during abscission, the decrease being largely confined to the water-soluble fraction. This result is consistent with work on *Begonia* abscission where pectins become increasingly soluble as weakening occurs (Hänisch ten Cate, Van Netter, Dortland and Bruinsma, 1975). Electron-probe analysis and Ca autoradiography have been used to show that Ca^{2+} is lost from abscising walls (Stösser, Rasmussen and Bukovac, 1969). This result is largely ignored in the literature but may be important because removal of wall Ca^{2+} by chelators is used as a method of cell maceration. Addition of Ca^{2+} does inhibit abscission though the mechanism is complex and does not only involve wall breakdown (Poovaiah and Leopold, 1973).

In 1967 Horton and Osborne showed that an enzyme which would attack carboxymethylcellulose increased during bean leaf abscission. The enzyme is a $\beta 1:4$ glucan 4 glucan hydrolase, known by the trivial name 'cellulase'. In a flurry of publications over the following few years it was established that cellulase increased in activity as the break strength declined (Fig. 20.11) and was principally (but not exclusively) confined to the separation layer. Inhibition of protein and RNA synthesis inhibited this

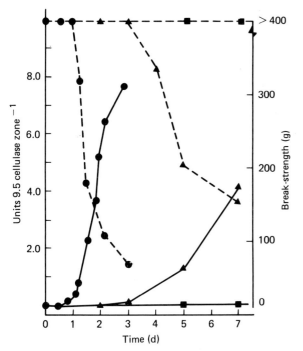

Fig. 20.11   A comparison of the rate of decline of break-strength (force necessary to rupture the abscission zone) - - - - - and 9.5 cellulase activity per abscission zone ——. In bean leaf explants exposed to 50 ppm ethylene (●), air flowing at 1.5 l min^{-1} (▲), and a hypobaric pressure of 125 mm Hg (4.61×10^4 Pa) (■) which is thought to facilitate removal of endogenous ethylene. (Sexton, Lewis and Durbin, unpublished.)

increase, and hormonal treatments which stimulated or retarded abscission had similar effects on cellulase activity (Abeles, 1969; Ratner, Goren and Monselise, 1969). Lewis and Varner (1970) showed that two cellulase isozymes were present in bean abscission zones, only one of which increased during abscission. Using D$_2$O labeling they demonstrated that this isozyme was synthesized *de novo*. The isozyme, which became known as 9.5 cellulase (pI = 9.5), was purified from bean abscission zones and antibodies were raised against it. Since these antibodies did not precipitate the other types of cellulase they could be used to quantify the 9.5 form of the enzyme (Durbin, Sexton and Lewis, 1981; Fig. 20.11). The antibody was used to show an increase in 9.5 cellulase just prior to the start of

break-strength decline. Radio-immunoassay detected no inactive precursor and immunocytochemistry and microassay confirmed high levels of activity in the separation layer (Sexton, Durbin, Lewis and Thomson, 1980). Kelly *et al.* (1982) have shown that one of the new species of mRNA that they detected early in the lag phase produced a protein on translation which was precipitable with the cellulase antibody. This would suggest transcriptional control of 9.5 cellulase production. However, translational control is also involved since 9.5 cellulase activity was not detected until 12 h after the appearance of this message.

In early experiments De la Fuente and Leopold (1969) had shown that if ethylene was removed from the atmosphere surrounding bean explants the rate of weakening slowed very rapidly. However, if the levels of cellulase were followed they continued to increase after ethylene withdrawal. Using an elegant method for washing the exocellular cellulase out of the walls, Abeles and Leather (1971) were able to demonstrate that cellulase was not secreted in the absence of ethylene. Very recently Lieberman *et al.* (1982) have employed a cytochemical method to show that cellulase is localized in the middle lamella region of the wall where breakdown is most evident. While there have been a number of studies of other species which have confirmed the observations made on bean, there are also data which cast doubt on the presumed role of cellulase in cell separation. The work of Hänisch ten Cate *et al.* (1975) on *Begonia* buds, for instance, demonstrated that abscission is complete before cellulase increases. However, Lewis' group have shown that purified abscission zone cellulase and pectinase will together cause cell separation whereas neither enzyme acts alone and that cellulase antibodies are potent inhibitors of bean abscission (Sexton *et al.*, 1980).

Cell wall cytochemists have always considered a pectinase to be the most likely candidate for an enzyme involved in cell separation. In 1968 Morré, using a bioassay, showed that a macerating enzyme (referred to as a pectinase) increased during bean abscission. Unfortunately this enzyme has never been characterized, although several years later Riov (1974) showed an exopolygalacturonase (PG) increased during citrus bud and fruit fall. Like cellulase, PG had the right characteristics to link it with abscission (Greenberg, Goren and Riov, 1975) and now pectinases have been shown to increase in several other systems. A very recent report also notes an increase in hemicellulase during mulberry leaf abscission (Hashinaga, Iwahori, Nishi and Itoo, 1981).

The organization of cell separation

The dilemma of why all the different classes of cell in the separation plane should respond to the abscission signal in a common way, while their apparently similar neighbors do not, has long preoccupied abscission physiologists. Two theories have been proposed (Sexton and Roberts, 1982). The first envisages that although these cells cannot be distinguished morphologically they may differ in some cryptic way from their neighbors. There is limited evidence to support this idea. Occasionally it has been possible to identify the separation layer as it passes through the abscission zone cortex. Polito and Lavee (1980) have demonstrated this with several staining reactions in sections of olive leaves and Wong and Osborne (1978) used endopolyploidy to identify the cells in the rather unusual separation layers of Ecballium. Nobody has yet found a characteristic which will allow the identification of the separation layer as it passes through all the tissues of the abscission zone or one which can be used in other species.

The alternative theory suggests that all cells have the ability to degrade their walls but that the hormonal signal which brings this about is limited to separation layers. If this theory of 'precise positional induction' is correct then it should be possible to manipulate the hormonal conditions and produce separation layers at unusual positions. Such 'adventitious' zones have been produced in stems, petioles and pedicels (Pierik, 1980) although no common method was involved in their induction. Unfortunately cell division and accumulation of cytoplasm precede cell separation in adventitious zones. Thus it can be argued that these processes are really inducing the formation of a separation layer which then responds by degrading its walls. This notion is supported by the observation that adventitious abscission takes longer than that in normal positions on the same plant. It has also been shown that a single abscission zone can be dissected into 30–40 small pieces without altering the plane of separation (Sexton, 1979). Such treatment would be expected to disrupt precise hormonal gradients.

It seems that evidence at present favors the existence of a cryptically differentiated separation layer. Osborne (1979) has recently used the term 'target cell' to describe such cells and it is possible that wall-degrading target cells are found elsewhere in the plant. Cells along the dehiscence lines of fruits and anthers, cells in pith cavities, resin canals, aerenchyma, lenticels and the soft flesh of fruit superficially resemble those in the separation layer.

Induction of abscission

The early botanists were aware that a variety of environmental factors could accelerate abscission. Gartner (1844) listed mineral deficiencies, drought and low light as being responsible for the shedding of flowers and fruit. He also recognized that abscission could be internally regulated and events in one part of the plant could influence an abscission zone some distance away. Pollination accelerated petal abscission and failure of embryonic development seemed to be associated with the shedding of young fruit. For details of the ways both environmental and internal factors influence the rate of natural abscission readers are referred to the excellent review by Addicott and Lyon (1973).

The idea that hormones were involved began with the observation that if the leaf blade was removed the subtending petiole was rapidly abscised. Küster (1916) showed that a small piece of lamina left attached to the petiole would stop abscission. He introduced the idea that the leaf blade produced 'chemical correlations' which inhibited the abscission zone in the petiole.

Auxin

With the demonstration that leaves produced auxins, Laibach and then Mai (1934) showed that the inhibitory effect of the lamina could be replaced by orchid pollinia rich in auxin. Two years later La Rue (1936) confirmed that purified IAA had the same effect.

The possibility that natural abscission was induced by decreasing auxin concentration was supported by the demonstration that extractable auxin levels diminished during bean leaf abscission (Shoji, Addicott and Swets, 1951). Jacobs reported that auxin transport from the lamina also decreased as leaves aged, contributing to the lowering of auxin status in the abscission zone (Jacobs, 1979). While abscission physiologists have consistently recorded reductions of auxin levels during leaf senescence (Roberts and Osborne, 1981), it is perplexing to discover that several authors propose that senescent leaves are a major source of auxin (Sheldrake, 1973).

The simple hypothesis that the induction of abscission was dependent on the auxin status of the zone was challenged by Addicott, Lynch and Carns (1955). They showed that while distal applications (lamina end) of auxin to petioles inhibited abscission, proximal applications (stem end) accelerated it. Jacobs (1955), using intact plans, confirmed that the auxin status on the stem side of the abscission zone was important. He reported that removal of the stem apex and young leaves, a major site of auxin supply, retarded abscission of debladed petioles left on the plant. Replacing the apex with IAA reinstated abscission at the original rate. To explain these data Addicott *et al.* (1955) proposed that the direction of the auxin gradient across the abscission zone was important. Auxin approaching the zone from the distal direction inhibited abscission whereas that moving from the stem accelerated the process. Since this original work, other experimenters have confirmed the effect of proximal applications, although some workers also reported that high levels applied proximally would retard abscission.

In 1950, Barlow had shown that to delay shedding, auxin had to be supplied to leaves shortly after deblading; if additions were delayed the treatment slowly lost its inhibitory effect. Using bean explants, Rubinstein and Leopold (1963) confirmed this observation and also discovered that if auxin was added later than 12 h after excision it actually accelerated abscission. They put forward the view that explants went through two stages after excision. In stage 1 auxin additions inhibited the process; in stage 2 auxin additions accelerated abscission. The length of stage 1 was thought to be related to the time taken for the IAA levels present at the end of excision to fall to a critical level when abscission could be induced. Once the induction process had started, IAA additions were no longer inhibitory.

A study of natural ageing showed that stage 1 became shorter as leaves aged. The accelerating effect of proximal applications could also be explained in terms of stages 1 and 2. It was argued that IAA applied distally moved quickly to the zone by a combination of diffusion and polar auxin transport. However, auxin applied proximally moved very slowly since diffusion was opposed by basipetal auxin transport. Thus while IAA applied distally reaches the zone in stage 1, IAA applied proximally only arrives at the zone in stage 2. If high enough concentrations of IAA were applied proximally, diffusion was faster and IAA reached the zone in stage 1. However, this would not explain the effect of the intact apex in Jacobs' (1955) experiment.

In what has become a classic paper Abeles and Rubinstein (1964) offered an explanation of why IAA could both retard and accelerate abscission. Using bean explants they confirmed an earlier observation that IAA and NAA dramatically increased ethylene production. Ethylene had long been known to accelerate abscission and they reported that if this gas was added in stage 1 it had little effect but if added in stage 2 it readily accelerated abscission. Thus, when auxin addition was delayed till stage 2 the tissue was not inhibited by auxin but responded to the ethylene induced by IAA application. On the other hand, if auxin was added in stage 1 it maintained the zone in an ethylene-insensitive condition. They showed that the accelerating effect of auxin in stage 2 could be reduced if evolved ethylene was removed by increased ventilation.

It was initially assumed that maintenance of stage 1 was in some way dependent on keeping the auxin levels above a critical threshold. In a recent illuminating paper Jaffe and Goren (1979) report that during stage 1 the ability of IAA to retard abscission is lost in parallel with a loss in its ability to stimulate H^+ efflux from the abscission zone. They speculate that these two diverse processes may become ineffective because some common component of the IAA response machinery is lost. In this connexion they showed that the ability of microsomal membranes to bind IAA also decreases. These effects of auxin have also been widely reported in fruit and flower zones; petal abscission, however, seems much less sensitive.

Ethylene

At the turn of the century physiologists were aware that some component of illuminating gas was a potent accelerator of abscission (Fitting, 1911). After the active component had been identified as ethylene, it was shown to cause abscission at concentrations as low as $100 \, nl \, l^{-1}$. For forty years the phenomenon was treated as a curious artefact until it was demonstrated that plants naturally evolve ethylene. Hall (1952) showed that young leaves evolve more ethylene than older ones. This observation seemed to preclude any natural role for ethylene in the abscission of older leaves. Using explants, Hall (1952) demonstrated that the accelerating effect of ethylene was inhibited by auxin additions, but that this could be overcome by raising the ethylene concentration. As a consequence he argued that the auxin–ethylene balance was important and that in young leaves there was sufficient auxin to inhibit ethylene's abscission-accelerating effect, while in older leaves there was not. In 1962, Burg's review brought into perspective the fact that ethylene was a common if not ubiquitous product of plant tissues and that it could be present naturally at concentrations that would stimulate ripening and abscission.

A few years later Abeles and Rubinstein (1964) suggested that IAA's ability to stimulate ethylene production explained the latter's accelerating effect in stage 2 (*see above*). Stage 1 was thought initially to be insensitive to added ethylene, while additions of the gas in stage 2 rapidly accelerated fracture. Subsequently Abeles and Holm showed that ethylene increased RNA and protein synthesis in the abscission zone in stage 2 but had no effect when added in stage 1 (*see* Abeles, 1968). The rate of break-strength decline was increased one hour after ethylene additions and reduced one hour after its removal (De la Fuente and Leopold, 1969). The increased rate of weakening was correlated with an enhanced rate of cellulase accumulation: removal of ethylene prevented cellulase secretion (Abeles and Leather, 1971). Ethylene's effect on cellulase activity has been recorded in many different abscission systems (Fig. 20.11). Polygalacturonase behaves similarly (Greenberg *et al.*, 1975).

As a result of his work, Abeles (1968) proposed the 'ageing–ethylene' hypothesis. Since Actinomycin D was an effective inhibitor of ethylene-stimulated abscission he argued that the gas induced the formation of the mRNA necessary for the production of wall-degrading enzymes. However, before ethylene could have this effect in stage 2 the tissue had to be sensitized by 'ageing'. Ageing occurred in stage 1 and involved the reduction of 'juvenility

factors' such as IAA in the abscission zone. External additions of 'juvenility factors' in stage 1 would prevent the sensitization whereas addition of ethylene increased the rate of ageing (Fig. 20.12).

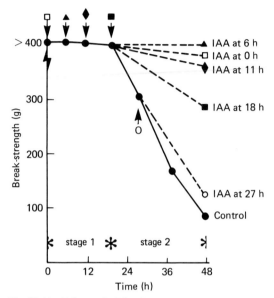

Fig. 20.12 Effects of adding IAA to bean leaf, abscission-zone explants at 0 (□), 6 (▲), 11 (◆), 18 (■), and 27 (○) h after excision. The explants were kept continuously in 50 ppm ethylene and their break-strength determined at various times. Note that even in the presence of high levels of ethylene, IAA additions at 0, 6, 11 h effectively prevent subsequent weakening. If IAA is added after 18 h it is no longer inhibitory. Stage 1, or the period when IAA additions can still prevent abscission, lasts for approximately 18 h. If this experiment has been carried out in air rather than 50 ppm ethylene, stage 1 would have been much longer. Ethylene thus reduces the duration of stage 1 and increases the rate of weakening (Fig. 20.11). Behavior of explants untreated with IAA is shown by (●).

Morgan and Beyer in a series of papers have reported that during natural abscission the ethylene concentrations reach levels which, if applied as a fumigant, induce leaf fall. They also demonstrated that these concentrations of ethylene inhibited auxin transport from the leaf and that this also declined naturally during abscission (Morgan, Ketring, Beyer and Lipe, 1972). It was possible, therefore, that ethylene acted indirectly by reducing the auxin levels in the zone. However, ethylene's ability to block auxin transport was not the only effect it exerted since DPX 184U (an auxin transport inhibitor, $3,3\alpha$ dihydro-2-(p-methoxyphenyl)-8H pyrazolo-$(5,1\alpha)$ isoindol-8-one) or exposing only the leaf blade to ethylene did not induce leaf fall. Both the inhibition of auxin transport and the presence of ethylene in the abscission zone proved essential for abscission to occur (Morgan and Durham, 1972; Beyer, 1975).

Several good correlations have been reported between ethylene levels and abscission. Young (1972) showed an increase in ethylene during *Valencia* orange fruit abscission caused by frost. Osborne's group demonstrated that ethylene production increased prior to break-strength decline in bean explants and increased naturally during the leaf senescence of several species (Jackson, Hartley and Osborne, 1973). This correlation was extended to fruit (Lipe and Morgan, 1972) and floral abscission (Swanson, Wilkins, Weiser and Klein, 1975). Another more recent example stems from work on leaflet defoliation in peanut caused by infections of *Cercospora*. Peanut varieties defoliated by the disease showed increased ethylene production prior to shedding while insensitive varieties had ethylene levels similar to the uninfected controls (Ketring and Melouk, 1982). Other authors have found no simple correlation (Marynick, 1977), but Abeles (1967) has pointed out that increasing the sensitivity of the zone with no change in ethylene concentration could be sufficient to trigger abscission.

Other evidence for ethylene's regulatory role comes from inhibition studies. Both competitive (CO_2) (Jerie, 1976) and non-competitive (Ag^+) (Beyer, 1976) inhibitors of ethylene responses delay abscission. AVG, which blocks ethylene synthesis, slows young fruit and style abscission (Einset, Lyon and Johnson, 1981). Hypobaric (low) pressures which facilitate the removal of ethylene from tissues are effective inhibitors of fruit, whole leaf and explant abscission (Morgan and Durham, 1980) (Fig. 20.11). Thus the evidence that ethylene has some natural role in the regulation of the rate of abscission is quite strong, but the question of whether it is the primary inducer or not is more contentious.

ABA and other abscission accelerators

In 1955 Osborne showed that diffusates from senescent petiole segments contained a non-volatile soluble factor which would accelerate explant abscission. Non-senescent petioles did not contain the factor and thus a regulatory role in abscission was suggested. Soon afterwards Van Stevenink demonstrated that developing pods at the base of the lupin inflorescence stimulated abscission of the apical flowers and he succeeded in extracting an abscission promotor from the young pods (Van Stevenink, 1959). A third group initially concerned with the growth of cotton fruit found a growth inhibitor–abscission accelerator which increased as abscission approached (Addicott, Carns, Lyon, Smith and McMeans, 1964). Liu and Carns (1961) crystalized the substance from mature cotton fruits and called it 'Abscisin'. Two years later a second abscission accelerator was crystalized from young cotton fruit which had a different melting point and was named 'Abscisin II' (Ohkuma, Lyon, Addicott and Smith, 1963). Both substances stimulated cotton leaf explant and fruit abscission in low concentrations. Subsequently the structure of Abscisin II was determined and the substance was renamed abscisic acid (ABA). In 1966 three separate investigations demonstrated that the lupin accelerator contained ABA (Cornforth, Milborrow, Ryback, Rothwell and Wain, 1966). Some tissues which contained Osborne's senescence factor (SF) also contained ABA and because the physiological properties of the two were similar (Chang and Jacobs, 1973) there was speculation that they were the same substance. Osborne, Jackson and Milborrow (1972) confirmed their separate identities by means of their chromatographic characteristics; in addition, their ability to induce ethylene formation differed. Dörffling, Böttger, Martin, Schmidt and Borowski (1978) made a more detailed analysis of abscission-promoting extracts from several senescent sources and showed that they contained ABA, xanthoxin and a third weakly active component which was not identified.

Conflicting conclusions have been reached from the attempts made to correlate ABA levels and abscission. Böttger (1970) showed increasing levels of an ABA-containing, abscission-accelerating fraction extracted from senescing *Coleus* leaves. Davis and Addicott (1972) provided the first correlation between ABA and cotton fruit abscission, ABA increasing during the young fruit drop and increasing again as the fruit matured. Cultivars which shed more young fruit had more ABA and the fruit that were shed contained larger amounts than those that remained firmly attached. Both petal and bean flower abscission (Bentley, Morgan, Morgan and Saad, 1975) are also associated with increased ABA levels. These and other positive correlations led Addicott (1982) to conclude that ABA's promotive role in abscission was established beyond reasonable doubt. However, Perry and Hellmers (1973) showed that while ABA increased in the leaves of two races of red maple exposed to short days, they only were shed by the northern race. More recently, other authors have also concluded that there is no simple relationship between endogenous ABA levels and whether a leaf (Peterson, Sacalis and Durkin, 1980), flower bud or fruit (Ramina and Masia, 1980; Takeda and Crane, 1980) will be shed. Morgan's group could show no correlation between the elevated levels of ABA in water-stressed leaves and whether they were abscised or not (Davenport, Jordan and Morgan, 1977). It is possible, however, to rationalize these negative correlations since IAA is also known to make tissues insensitive to ABA (Varma, 1976).

In his review Milborrow (1974) expresses surprise that amongst all the hundreds of plants sprayed with ABA more do not show an abscission response. By way of reply Addicott (1982) lists a considerable number of examples where the response does occur. ABA is much more effective when applied to explants, though some conclude that it may act by increasing ethylene production, perhaps by promoting senescence. The effect of ABA has been reduced by absorbing ethylene and Sagee, Goren and Riov (1980) have recently reported that ABA is ineffective if added in the presence of the ethylene-synthesis inhibitor AVG. While these results seem firmly to implicate ethylene in ABA action, not all observers have found ABA to promote ethylene production (Marynick, 1977). Others have shown that ABA still promotes abscission under hypobaric conditions (Rasmussen, 1974), in high concentrations of CO_2 and in the presence of saturating levels of ethylene (Craker and Abeles, 1969). ABA is thus considered by some to be directly involved in regulating wall degradation.

There is still much to be learned about abscission accelerators. The chemical identities of Abscisin 1 and SF are still unknown. Equally, there are a number of situations where abscission-accelerating activity is detected yet the nature of the accelerator is unknown. For instance, Stead and Moore (1979)

have shown that the growth of pollen tubes into the stigmatic surface of *Digitalis* generates a signal which moves down the style at a rate of 1.5 mm h^{-1} and accelerates corolla abscission.

Gibberellic acid and cytokinins

Both of these growth regulators can influence abscission, but neither is considered to act directly. In the majority of cases GA stimulates explant abscission but is less effective than ABA or ethylene (Chatterjee and Leopold 1964; Lyon and Smith, 1966). Abeles (1967) presented evidence which suggested that GA's promotive effect was mediated by enhanced ethylene production though Marynick (1977) could not confirm these observations with cotton. Perhaps the most intriguing aspect of GA action is that it acts antagonistically if added with IAA. The GA accelerating effect is reduced by IAA, and IAA's inhibiting effect is diminished by GA (Morgan, 1976).

There has been relatively little work on the role of cytokinins. When added to explants in stage 1 they will extend the duration of the latter, though less effectively than equivalent concentrations of IAA (Gorter, 1964; Chatterjee and Leopold, 1964). Most authors consider that cytokinin acts indirectly through delaying senescence or directing the movement of assimilates (Addicott, 1982). The only study of *in vivo* levels has shown them to be significantly lower in abscised fruits (Rodgers, 1981).

What induces natural abscission?

To answer the question, 'What constitutes adequate evidence that an internal factor controls abscission?', Jacobs (1962) formulated a set of six rules. The first of these was that 'there should be a parallel between the naturally occurring chemical and abscission'. A great deal of research has been dedicated to establishing such relationships. Trewavas (1982), by analogy with animal systems, has recently pointed out that growth substances (GS) probably interact with receptors (R) and that it is the levels of (GS.R) that determine the response:

$$GS + R \rightarrow (GS.R) \rightarrow response$$

Thus, not only is the concentration of the growth substance important but so too is the level of the receptor. A great deal of the abscission literature points to the fact that R could often be limiting and the plant insensitive to a hormone which, if added in other conditions, elicits a response. Since dramatic changes in sensitivity take place during abscission (*e.g.*, Stage 1 → Stage 2) the production of receptors may well prove to be the controlling factor in abscission rather than the levels of growth regulators.

Another notion which permeates the literature is that a single hormonal stimulus or master switch induces abscission. Again, this idea may well prove too simplistic. Preliminary results concerned with the formation of 9.5 cellulase indicate that both transcriptional and translational control are involved in a rather complicated control process which may well involve several switches (Kelly *et al.*, 1982).

Senescence and abscission

The senescence of fruits and leaves usually precedes abscission and as a result several authors see the former process as an essential prerequisite for the latter. Osborne (1973), for instance, puts forward the well argued case that senescence acts as a signal that induces abscission. However, there are a number of instances where senescence of the distal tissues does not occur before abscission. Some fruits do not ripen till they have fallen from the tree, and leaf abscission caused by disease, insect damage and water-stress can take place without obvious chlorophyll loss (Webster, 1973). The recent study of *Digitalis* corolla abscission induced by pollination, referred to earlier, reveals that it is not preceded by any of the more obvious symptoms of senescence (Stead, personal communication). Equally, there are examples where leaf senescence occurs in the autumn but abscission does not occur until the following spring (Addicott and Lyon, 1973). Thus, while senescence and abscission are usually coupled, the linkage can be broken and both processes can occur independently.

Duygu (1976) has listed a number of biochemical features shared by abscission and senescence and suggests that these two processes may both involve some common metabolic changes. Such conclusions would be important if abscission were shown to involve the generic biochemical features shared by all senescence processes and not just the specialized metabolism only associated with specific types of senescence (*i.e.*, wall breakdown during fruit ripen-

ing). At present the generic and specific features of senescence metabolism have been only partially unraveled. It is premature to say whether a unified biochemical definition of senescence is possible; such a proposition carries the implication that senescence involves a collection of phenomena which share elements of a common developmental program. It may do so, but on the other hand it may involve simply a collection of disparate processes, unified only to the extent that they lead to the death of the plant or some part of it but differing greatly in the mechanisms whereby the unhinging of the vital processes is achieved.

Further reading

Addicott, F. T. (1970). Plant hormones in the control of abscission. *Biol. Rev.* **45**, 485–524.

Addicott, F. T. (1982). *Abscission*, California U.P.

Kozlowski, T. T. (1973). *Shedding of Plant Parts*, Academic Press, New York.

Sexton, R. and Roberts, J. A. (1982). The cell biology of abscission. *Ann. Rev. Plant Physiol.* **33**, 133–62.

Thimann, K. V. (ed.) (1980). *Senescence in Plants*, C. R. C. Press, Boca Raton, Florida, 276 pp.

References

Abeles, F. B. (1967). Mechanism of action of abscission accelerators. *Physiol. Plant.* **20**, 442–54.

Abeles, F. B. (1968). Role of RNA and protein synthesis in abscission. *Plant Physiol.* **43**, 1577–86.

Abeles, F. B. (1969). Abscission: role of cellulase. *Plant Physiol.* **44**, 447–52.

Abeles, F. B. (1973). *Ethylene in Plant Biology*, Academic Press, New York, 302 pp.

Abeles, F. B. and Leather, G. R. (1971). Abscission: control of cellulase secretion by ethylene. *Planta* **97**, 87–91.

Abeles, F. B. and Rubinstein, B. (1964). Regulation of ethylene evolution and leaf abscission by auxin. *Plant Physiol.* **39**, 963–9.

Adams, D. O. and Yang, S. F. (1977). Methionine metabolism in apple tissue. *Plant Physiol.* **60**, 892–6.

Adams, D. O. and Yang, S. F. (1979). Ethylene biosynthesis: identification of 1-aminocyclopropane-1-carboxylic acid as an intermediate in the conversion of methionine to ethylene. *Proc. Natn. Acad. Sci. USA* **76**, 160–74.

Addicott, F. T. (1970). Plant hormones in the control of abscission. *Biol. Rev.* **45**, 485–524.

Addicott, F. T. (1982). Abscission. California U.P.

Addicott, F. T., Carns, H. R., Lyon, J. L., Smith, O. E. and McMeans, J. L. (1964). On the physiology of abscisins. In *Régulateurs Naturels de la Croissance Végetale*, ed. J. P. Nitsch, Cent. natn. Rech. Sci., Paris, pp. 687–702.

Addicott, F. T., Lynch, R. S. and Carns, H. R. (1955). Auxin gradient theory of abscission regulation. *Science* **121**, 644–5.

Addicott, F. T. and Lyon, J. L. (1973). Physiological ecology of abscission. In *Shedding of Plant Parts*, ed. T. T. Kozlowski, Academic Press, New York, pp. 85–117.

Aharoni, N., Lieberman, M. and Sisler, H. D. (1979a). Patterns of ethylene production in senescing leaves. *Plant Physiol.* **64**, 796–800.

Aharoni, N., Anderson, J. D. and Lieberman, M. (1979b). Production and action of ethylene in senescing leaf discs. *Plant Physiol.* **64**, 805–9.

Aharoni, N. and Lieberman, M. (1979). Ethylene as a regulator of senescence in tobacco leaf discs. *Plant Physiol.* **64**, 801–4.

Arditti, J. (1979). Aspects of the physiology of orchids. *Adv. Bot. Res.* **7**, 421–655.

Arney, S. E. (1947). The respiration of strawberry leaves attached to the plant. *New Phytol.* **46**, 68–76.

Barlow, H. W. B. (1950). Studies in abscission I. *J. Exp. Bot.* **1**, 264–81.

Barrett, D. H. P. and Woolhouse, H. W. (1981). Protein turnover in the primary leaves of *Phaseolus vulgaris*. *J. Exp. Bot.* **32**, 443–52.

Beevers, L. (1968). Growth regulator control of senescence in leaf discs of Nasturtium (*Tropaeolum majus*). In *Biochemistry and Physiology of Plant Growth Substances*, ed. F. Wightman and G. Setterfield, Ottawa, Runge, pp. 1417–34.

Bentley, B., Morgan, C. B., Morgan, D. G. and Saad, F. A. (1975). Plant growth substances and the effect of photoperiod on flower bud development in *Phaseolus vulgaris*. *Nature* **256**, 121–2.

Beyer, E. M. (1975). Abscission: the initial effect of ethylene is in the leaf blade. *Plant Physiol.* **55**, 322–7.

Beyer, E. M. (1976). A potent inhibitor of ethylene action in plants. *Plant Physiol.* **58**, 268–71.

Böttger, M. (1970). Die hormonale Regulation des Blattfalls bei *Coleus rehnêltianus II. Planta* **93**, 205–13.

Brady, C. J. (1981). A coordinated decline in the synthesis of subunits of ribulose bisphosphate carboxylase in ageing wheat leaves. I. Analyses of isolated protein, subunits and ribosomes. *Aust. J. Plant Physiol.* **8**, 591–603.

Brown, S. B. and Troxler, R. F. (1982). In *Bilirubin Metabolism* vol. II, eds K. P. M. Heirwegh and S. B. Brown, C. R. C. Press, Boca Raton, Florida, pp. 1–38.

Bunger, W. and Feierabend, J. (1980). Capacity for RNA synthesis in 70s ribosome-deficient plastids of heat-bleached rye leaves. *Planta* **149**, 163–9.

Burg, S. P. (1962). The physiology of ethylene formation. *Ann. Rev. Plant Physiol.* **13**, 265–302.

Butler, R. D. and Simon, E. W. (1972). Ultrastructural aspects of senescence in plants. *Adv. Gerontol. Res.* **3**, 73–129.

Callow, J. A., Callow, M. E. and Woolhouse, H. W. (1972). *In vitro* protein synthesis, ribosomal RNA synthesis and polyribosomes in senescing leaves of *Perilla*. *Cell Differentiation* **1**, 79–90.

Callow, M. E. and Woolhouse, H. W. (1973). Changes in nucleic acid metabolism in regreening leaves of *Perilla*. *J. Exp. Bot.* **24**, 285–94.

Chang, Y. P. and Jacobs, W. P. (1973). The regulation of abscission and IAA by senescence factor and ABA. *Am. J. Bot.* **60**, 10–16.

Chatterjee, S. K. and Leopold, A. C. (1964). Kinetin and gibberellin actions on abscission processes. *Plant Physiol.* **39**, 334–7.

Chibnall, A. C. (1939). *Protein Metabolism in the Plant*, New Haven, Yale Univ., 306 pp.

Colquhoun, A. J. and Hillman, J. R. (1975). Endogenous abscisic acid and the senescence of leaves of *Phaseolus vulgaris*. *Z. Pflanzenphysiol.* **76**, 326–32.

Cornforth, J. W., Milborrow, B. V., Ryback, G., Rothwell, K. and Wain, R. L. (1966). Identification of the yellow lupin growth inhibitor as (+)-abscisin II (+)-dormin. *Nature* **211**, 742–3.

Craker, L. E. and Abeles, F. B. (1969). Abscission: role of abscisic acid. *Plant Physiol.* **44**, 1144–9.

Davenport, T. L., Jordan, W. R. and Morgan, P. W. (1977). Movement and endogenous levels of abscisic acid during water stress induced abscission in cotton seedlings. *Plant Physiol.* **59**, 1165–8.

Davies, D. D. (1980). The measurement of protein turnover in plants. *Adv. Bot. Res.* **8**, 66–126.

Davis, L. A. and Addicott, F. T. (1972). Abscisic acid: correlations with abscission and with development of cotton fruit. *Plant Physiol.* **49**, 644–8.

De la Fuente, R. K. and Leopold, A. C. (1969). Kinetics of abscission in the bean leaf petiole explant. *Plant Physiol.* **44**, 251–4.

Dörffling, K., Böttger, M., Martin, D., Schmidt, V. and Borowski, D. (1978). Physiology and chemistry of substances accelerating abscission in senescent petioles and fruit stalks. *Physiol. Plant.* **43**, 292–6.

Durbin, M. L., Sexton, R. and Lewis, L. N. (1981). The use of immunological methods to study the activity of cellulase isozymes in bean leaf abscission. *Plant Cell Env.* **4**, 67–73.

Duygu, E. (1976). Biochemical and enzymological processes involved in senescence and abscission. *Bitki* **3**, 80–98.

Dyer, Y. A. and Osborne, D. J. (1971). Leaf nucleic acids. II. Metabolism during senescence and the effect of kinetin. *J. Exp. Bot.* **22**, 552–60.

Eberhardt, F. (1955). Der Atmungsverlauf alternder Blätter und reifender Fruchte. *Planta* **45**, 57–67.

Einset, J. W., Lyon, J. L. and Johnson, P. (1981). Chemical control of abscission and degreening in stored lemons. *J. Am. Soc. Hort. Sci.* **106**, 531–3.

Farkas, G. L. and Stahmann, M. A. (1966). On the nature of changes in peroxidase isoenzymes in bean leaves infected by bean mosaic virus. *Phytopathology* **56**, 669–77.

Fitting, H. (1911). Untersuchungen über die vorzeitige Entblätterung von Blüten. *Jahrb. Wiss. Bot.* **49**, 187–263.

Gärtner, C. F. (1844) quoted in Namikawa, I. (1926). *J. Coll. Agric. Hokkaido. Imp. Univ.* **17**, 63–128.

Gawadi, A. G. and Avery, G. S. (1950). Leaf abscission and the so-called abscission layer. *Am. J. Bot.* **37**, 172–80.

Gepstein, S. and Thimann, K. V. (1980). Changes in the abscisic acid content of oat leaves during senescence. *Proc. Nat. Acad. Sci.* **77**, 2050–3.

Gorter, C. J. (1964). Studies on abscission in explants of *Coleus. Physiol. Plant* **17**, 331–45.

Greenberg, J., Goren, R. and Riov, J. (1975). The role of cellulase and polygalacturonase in abscission of young and mature shamouti orange fruits. *Physiol. Plant.* **34**, 1–7.

Gregory, F. G. and Sen, G. K. (1937). Physiological studies in plant nutrition. VI. The relation of respiration rate to the carbohydrate and nitrogen metabolism of the barley leaf as determined by nitrogen and potassium deficiency. *Ann. Bot.* **1**, 521–61.

Hall, W. C. (1952). Evidence of the auxin ethylene balance hypothesis of foliar abscission. *Bot. Gaz.* **113**, 310–22.

Hänisch ten Cate, Ch. H., Van Netter, J., Dortland, J. F. and Bruinsma, J. (1975). Cell wall solubilization in pedicel abscission of begonia flower buds. *Physiol. Plant.* **33**, 276–9.

Hardwick, K., Wood, M. E. and Woolhouse, H. W. (1968). Photosynthesis and respiration in relation to leaf age in *Perilla frutescens* (L.) Britt. *New Phytol.* **67**, 79–86.

Hardwick, K. and Woolhouse, H. W. (1967). Foliar senescence in *Perilla frutescens* L. Britt. *New Phytol.* **66**, 545–52.

Hashinaga, F., Iwahori, S., Nishi, Y. and Itoo, S. (1981). Accelerated changes of cell wall degrading enzymes in abscission zones of Mulberry leaves with 2-chloroethylphosphonic acid. *Nippon Nôgeikagaku Kaishi* **55**, 1217–23.

Henry, E. W., Valdovinos, J. G. and Jensen, T. E. (1974). Peroxidases in tobacco abscission zone tissue II. *Plant Physiol.* **54**, 192–6.

Hocking, P. J. and Pate, J. S. (1977). Mobilization of minerals to developing seeds of legumes. *Ann. Bot.* **41**, 1259–78.

Horton, R. F. and Osborne, D. J. (1967). Senescence abscission and cellulase activity in *Phaseolus vulgaris. Nature* **214**, 1086–8.

Jackson, M. D., Hartley, C. B. and Osborne, D. J. (1973). Timing abscission in *Phaseolus vulgaris* by controlling ethylene production and sensitivity to ethylene. *New Phytol.* **72**, 1251–60.

Jacobs, W. P. (1955). Studies on abscission: the physiological basis of the abscission-speeding effect of intact leaves. *Am. J. Bot.* **42**, 594–604.

Jacobs, W. P. (1962). Longevity of plant organs: internal factors controlling abscission. *Ann. Rev. Plant Physiol.* **13**, 403–36.

Jacobs, W. P. (1979). *Plant Hormones and Plant Development*, Cambridge U.P., Cambridge.

Jaffe, M. T. and Goren, R. (1979). Auxin and early stages of the abscission process in citrus leaf explants. *Bot. Gaz.* **140**, 378–83.

James, W. O. (1953). *Plant Respiration*, Clarendon Press, Oxford.

Jerie, P. H. (1976). The role of ethylene in abscission of cling peach fruit. *Aust. J. Plant Physiol.* **3**, 747–54.

Kelly, P., Trewavas, A. J., Sexton, R., Durbin M. and Lewis, L. N. (1982). Regulation of abscission zone cellulase mRNA synthesis. In *Abstr. 11th Int. Conf. Plant Growth Substances*, p. 22.

Ketring, D. L. and Melouk, H. A. (1982). Ethylene production and leaflet abscission of three peanut genotypes infected with *Cercospora arachidicola. Plant Physiol.* **69**, 789–92.

Knypl, J. S. and Mazurczyk, W. (1972). Retarding effect of inhibitors of protein and RNA synthesis on chlorophyll and protein breakdown. *Biol. Plant.* **14**, 146–54.

Krizek, D. T., McIlrath, W. J. and Bergara, B. S. (1966). Photoperiodic induction of senescence in *Xanthium* plants. *Science* **151**, 95–6.

Küster, E. (1916). Beiträge zur Kenntniss des Laubfalles. *Deut. Bot. Ges.* **34**, 184–93.

La Rue, C. D. (1936). The effect of auxin on the abscission of petioles. *Proc. Nat. Acad. Sci. USA* **22**, 254–9

Lewis, L. N. and Varner, J. E. (1970). Synthesis of cellulase during abscission of *Phaseolus vulgaris* leaf explants. *Plant Physiol.* **46**, 194–9.

Lieberman, S. J., Valdovinos, J. G. and Jensen, T. E. (1982). Ultrastructural localization of cellulase in abscission zones of tobacco flower pedicels. *Bot. Gaz.* **143**, 32–40.

Lindoo, S. J. and Noodén, L. D. (1977). Studies on the behaviour of the senescence signal in Anoka Soybeans. *Plant Physiol.* **59**, 1136–40.

Lindoo, S. J. and Noodén, L. D. (1978). Correlation of cytokinins and abscisic acid with monocarpic senescence in soybeans. *Plant Cell Physiol.* **19**, 997–1006.

Lipe, J. A. and Morgan, P. W. (1972). Ethylene: role in fruit abscission and dehiscence processes. *Plant Physiol.* **50**, 759–64.

Liu, W. -C. and Carns, H. R. (1961). Isolation of abscisin, an abscission accelerating substance. *Science* **134**, 384.

Lyon, J. L. and Smith, O. E. (1966). Effects of gibberellins on abscission in cotton seedling explants. *Planta* **69**, 347–56.

Mai, G. (1934). Korrelationsuntersuchungen an entspreiteten blattstielen Mittels lebender Orchideen pollinien als Wuchsstoffquelle. *Jahr. wiss. Bot.* **79**, 681–713.

Makovetzki, S. N. and Goldschmidt, F. F. (1976). A requirement for cytoplasmic protein synthesis during chloroplast senescence in the aquatic plant *Anacharis canadensis*. *Plant Cell Physiol.* **17**, 859–62.

Marynick, M. C. (1977). Patterns of ethylene and carbon dioxide evolution during cotton explant abscission. *Plant Physiol.* **59**, 484–9.

Maunders, M. J., Brown, S. B. and Woolhouse, H. W. (1983). The appearance of chlorophyll derivatives in senescing tissue. *Phytochemistry* (in press).

McCollum, J. P. (1934). Vegetative and reproductive responses associated with fruit development in the cucumber. *Mem. Cornell Agric. Exp. Sta.* 163.

Medawar, P. B. (1957). An unsolved problem in biology. In *The Uniqueness of the Individual*, Methuen, London.

Michael, G. (1936). Über die Beziehungen zwischen Chlorophyll und Eiweissablasse in vergilbenden Laublatt von *Tropaeolum*. *Z. Bot.* **29**, 385.

Milborrow, B. V. (1974). The chemistry and physiology of abscisic acid. *Ann. Rev. Plant Physiol.* **25**, 259–307.

Mohl, H. von (1860). Über den Ablösungsprozess saftiger Pflanzenorgane. *Bot. Zeit.* **18**, 273–4.

Molisch, H. (1938). *The Longevity of Plants*, Transl. E. H. Fulling, Science Press, Lancaster Pa.

Morgan, P. W. (1976). Gibberellic acid and indole acetic acid compete in ethylene promoted abscission. *Planta* **129**, 275–6.

Morgan, P. W. and Durham, J. I. (1972). Abscission: potentiating action of auxin transport inhibitors. *Plant Physiol.* **50**, 313–18.

Morgan, P. W. and Durham, J. I. (1980). Ethylene production and leaflet abscission in Mèlia azédarach. *Plant Physiol.* **66**, 88–92.

Morgan, P. W., Ketring, D. L., Beyer, E. M. and Lipe, J. A. (1972). Functions of naturally produced ethylene in abscission, dehiscence and seed germination. In *Plant Growth Substances 1970*, ed. D. J. Carr, Springer-Verlag, Berlin, pp. 502–9.

Morré, D. J. (1968). Cell wall dissolution and enzyme secretion during leaf abscission. *Plant Physiol.* **43**, 1545–59.

Mothes, K. (1933). Die Vakuuminfiltration in Ernährungsversuch. (Dargestellt an Untersuchungen über die Assimilation des Ammoniaks). *Planta* **19**, 117–38.

Mozer, T. J. (1980). Control of protein synthesis in barley aleurone layers by the plant hormones gibberellic acid and abscisic acid. *Cell* **20**, 479–85.

Murneek, A. C. (1926). Effect of correlation between vegetative and reproductive functions in the tomato (*Lycopersicon esculentum* Mill.). *Plant Physiol.* **1**, 3–56.

Murray, B. J. and Noodén, L. D. (1981). Testing the role of nutrient drain in monocarpic senescence. *Plant Physiol.* **67**, S375.

Ness, P. J. and Woolhouse, H. W. (1980a). RNA synthesis in *Phaseolus* chloroplasts. 1. Ribonucleic acid synthesis in chloroplast preparations from *Phaseolus vulgaris* L. leaves and solubilisation of the RNA polymerase. *J. Exp. Bot.* **31**, 223–33.

Ness, P. J. and Woolhouse, H. W. (1980b). RNA synthesis in *Phaseolus* chloroplasts. II. Ribonucleic acid synthesis in chloroplasts from developing and senescing leaves. *J. Exp. Bot.* **31**, 235–45.

Noodén, L. D. and Leopold, A. C. (1978). Phytohormones and the endogenous regulation of senescence and abscission. In *Phytohormones and Related Compounds. A comprehensive treatise*, vol. II, eds D. S. Letham, P. B. Goodwin and T. J. V. Higgins, Elsevier, Amsterdam and New York.

Noodén, L. D. and Obermeyer, W. R. (1981). Changes in abscisic acid translocation during pod development and senescence in soybeans. *Biochem. Physiol. Pflanzen.* **176**, 859–68.

Noodén, L. D., Rupp, D. C. and Derman, B. D. (1978). Separation of seed development from monocarpic senescence in soybeans. *Nature* **271**, 354–6.

Ochoa, S. and de Haro, C. (1979). Regulation of protein synthesis in eukaryotes. *Ann. Rev. Biochem.* **48**, 549–80.

Ohkuma, K., Lyon, J. L., Addicott, F. T. and Smith, O. E. (1963). Abscisin II an abscission accelerating substance from young cotton fruit. *Science* **142**, 1592–3.

Osborne, D. J. (1955). Acceleration of abscission by a factor produced by senescent leaves. *Nature* **176**, 1161–3.

Osborne, D. J. (1967). Hormonal regulation of leaf senescence. *Symp. Soc. Exp. Biol.* **21**, 305–22.

Osborne, D. J. (1973). Internal factors regulating abscission. In *Shedding of Plant Parts*, ed. T. T. Kozlowski. Academic Press, N.Y., pp. 125–48.

Osborne, D. J. (1979). Target cells—new concepts for plant regulation in horticulture. *Sci. Hort.* **30**, 1–13.

Osborne, D. J. (1982). Hormones and foliar senescence. In *Growth Regulators in Plant Senescence*, Monograph 8, eds M. B. Jackson, B. Grout and I. A. Mackenzie, British Plant Growth Regulator Group, Wantage, pp. 57–83.

Osborne, D. J., Jackson, M. B. and Milborrow, B. V. (1972). Physiological properties of abscission accelerator from senescent leaves. *Nature New Biol.* **240**, 98–101.

Osborne, D. J. and Sargent, J. A. (1976). The positional differentiation of abscission zones during the development of leaves of *Sambucus nigra* and the response of cells to auxin and ethylene. *Planta* **132**, 197–204.

Paranjothy, K. and Wareing, P. F. (1971). The effect of abscisic acid, kinetin and 5–fluorouracil on ribonucleic acid and protein synthesis in senescing radish leaf discs. *Planta* **99**, 112–19.

Pearson, J. A., Thomas, K. and Thomas, H. (1978). Nucleic acids from leaves of a yellowing and a non-yellowing variety of *Festuca pratensis* Huds. *Planta* **144**, 85–7.

Perry, T. O. and Hellmers, H. (1973). Effects of abscisic acid on the growth and dormancy behaviour of different races of red maple. *Bot. Gaz.* **134**, 283–9.

Peterson, J. C., Sacalis, J. N. and Durkin, D. J. (1980). Alterations in abscisic acid content of *Ficus benjamina* leaves resulting from exposure to water stress and its relationship to leaf abscission. *J. Am. Soc. Hort. Sci.* **105**, 793–8.

Peterson, L. W. and Huffaker, R. C. (1975). Loss of ribulose 1,5-diphosphate carboxylase and increase in proteolytic activity during senescence of detached primary barley leaves. *Plant Physiol.* **55**, 1009–15.

Peterson, L. W., Kleinkopf, G. E. and Huffaker, R. C. (1973). Evidence for lack of turnover of ribulose 1,5-diphosphate carboxylase in barley leaves. *Plant Physiol.* **51**, 1042–5.

Pierik, R. L. M. (1980). Hormonal regulation of secondary abscission in pear pedicels *in vitro*. *Physiol. Plant.* **48**, 5–8.

Polito, V. S. and Lavee, S. (1980). Anatomical and histochemical aspects of ethephon induced leaf abscission in olive (*Olea europaea*). *Bot. Gaz.* **141**, 413–17.

Poovaiah, B. W. and Leopold, A. C. (1973). Inhibition of abscission by calcium. *Plant Physiol.* **51**, 848–51.

Poovaiah, B. W. and Rasmussen, H. P. (1974). Localization of dehydrogenase and acid phosphatase during the abscission of bean leaves. *Am. J. Bot.* **61**, 68–73.

Ralph, R. K. and Wojcik, S. J. (1981). Plant protein kinases and cytokinins. *Plant Sci. Lett.* **22**, 127–40.

Ramina, A. and Masia, A. (1980). Levels of extractable abscisic acid in the mesocarp and seed of persisting and abscising peach fruit. *J. Am. Soc. Hort. Sci.* **105**, 465–8.

Rasmussen, G. K. (1974). Cellulase activity in separation zones of citrus fruit treated with abscisic acid under normal and hypobaric pressures. *J. Am. Soc. Hort. Sci.* **99**, 229–31.

Ratner, A., Goren, R. and Monselise, S. P. (1969). Activity of pectin esterase and cellulase in the abscission zone of citrus leaf explants. *Plant Physiol.* **44**, 1717–23.

Richmond, A. E. and Lang, A. (1957). Effect of kinetin on protein content and survival of detached *Xanthium* leaves. *Science* **125**, 650–1.

Riov, J. (1974). A polygalacturonase from citrus leaf explants. *Plant Physiol.* **53**, 312–16.

Roberts, J. A. and Osborne, D. J. (1981). Auxin and the control of ethylene production during the development and senescence of leaves and fruits. *J. Exp. Bot.* **32**, 875–87.

Rodgers, J. P. (1981). Cotton fruit development and abscission: fluctuations in the levels of cytokinins. *J. Hort. Sci.* **56**, 99–106.

Rubinstein, B. and Leopold, A. C. (1963). Analysis of the auxin control of bean leaf abscission. *Plant Physiol.* **38**, 262–7.

Sacher, J. A. and Davies, D. D. (1974). Demonstration of *de novo* synthesis of RNase in *Rhoeo* leaf sections by deuterium oxide labelling. *Plant Cell Physiol.* **15**, 157–61.

Sagee, O., Goren, R. and Riov, J. (1980). Abscission of citrus leaf explants. Interrelationships of ABA, ethylene and hydrolytic enzymes. *Plant Physiol.* **66**, 750–3.

Sexton, R. (1976). Some ultrastructural observations on the nature of foliar abscission in *Impatiens sultani*. *Planta* **128**, 49–58.

Sexton, R. (1979). Spatial and temporal aspects of cell separation in the foliar abscission zones of *Impatiens sultani*. *Protoplasma* **99**, 53–66.

Sexton, R., Durbin, M. L., Lewis, L. N. and Thomson, W. W. (1980). Use of cellulase antibodies to study leaf abscission. *Nature* **283**, 873–4.

Sexton, R., Jamieson, G. G. C. and Allan, M. H. I. L. (1977). An ultrastructural study of abscission zone cells with special reference to the mechanism of enzyme secretion. *Protoplasma* **91**, 369–87.

Sexton, R. and Redshaw, A. J. (1981). The role of cell expansion in abscission of *Impatiens sultani* leaves. *Ann. Bot.* **48**, 745–56.

Sexton, R. and Roberts, J. A. (1982). The cell biology of abscission. *Ann. Rev. Plant Physiol.* **33**, 133–62.

Sheldrake, A. R. (1973). The production of hormones in higher plants. *Biol. Rev.* **48**, 509–59.

Shoji, K., Addicott, F. T. and Swets, W. A. (1951). Auxin in relation to leaf blade abscission. *Plant Physiol.* **26**, 189–91.

Simon, E. W. (1967). Types of leaf senescence. *Symp. Soc. Exp. Biol.* **21**, 215–30.

Simpson, E., Cooke, R. J. and Davies, D. D. (1981). Measurement of protein in leaves of *Zea mays* using [^{3}H]acetic anhydride and tritiated water. *Plant Physiol.* **67**, 1214–19.

Spiers, J. and Brady, C. J. (1981). A coordinated decline in the synthesis of subunits of ribulose bis phosphate carboxylase in ageing wheat leaves. II. Abundance of messenger RNA. *Aust. J. Plant Physiol.* **8**, 608–18.

Stead, A. D. and Moore, F. G. (1979). Studies of flower longevity in *Digitalis*. *Planta* **146**, 409–14.

Stösser, R. (1971). Localization of RNA and protein synthesis in the developing abscission layer in fruit of *Prunus cerasus* L. *Z. Pflanzenphysiol.* **64**, 328–34.

Stösser, R., Rasmussen, H. P. and Bukovac, M. J. (1969). Histochemical changes in the developing abscission layer in fruits of *Prunus cerasus*. *Planta* **86**, 151–64.

Swanson, B. T., Wilkins, H. F., Weiser, C. F. and Klein, I. (1975). Endogenous ethylene and abscisic acid relative to phytogerontology. *Plant Physiol.* **55**, 370–6.

Takeda, F. and Crane, J. C. (1980). Abscisic acid in pistachio as related to inflorescence bud abscission. *J. Am. Soc. Hort. Sci.* **105**, 573–6.

Takegami, T. and Yoshida, K. (1975). Remarkable retardation of the senescence of tobacco leaf discs by cordycepin, an inhibitor RNA polyadenylation. *Plant Cell Physiol.* **16**, 1163–6.

Tepfer, D. A. and Fosket, D. E. (1978). Hormone-mediated translational control of protein synthesis in *Elodea* leaf cells. *Developmental Biology* **62**, 486–97.

Thimann, K. V. (1978). Senescence. In *Controlling Factors in Plant Development*, eds H. Shibaska, M. Funeyor, M. Katsumi and A. Takovots, Botanical Magazine, Tokyo, Special Issue, pp. 19–43.

Thomas, H. (1975). Regulation of alanine aminotransferase in leaves of *Lolium temulentum* during senescence. *Z. Pflanzenphysiol.* **74**, 208–18.

Thomas, H. (1976). Delayed senescence in leaves treated with the protein synthesis inhibitor MDMP. *Plant Sci. Lett.* **6**, 369–77.

Thomas, H. (1977). Ultrastructure, polypeptide composition and photochemical activity of chloroplasts during foliar senescence of a non-yellowing mutant genotype of *Festuca pratensis* Huds. *Planta* **137**, 53–60.

Thomas, H. and Stoddart, J. L. (1975). Separation of chlorophyll degradation from other senescence processes in leaves of a mutant genotype of meadow fescue (*Festuca pratensis*). *Plant Physiol.* **56**, 438–41.

Thomas, H. and Stoddart, J. L. (1977). Biochemistry of leaf senescence in grasses. *Ann. Appl. Biol.* **85**, 461–3.

Thomas, H. and Stoddart, J. L. (1980). Leaf senescence. *Ann. Rev. Plant Physiol.* **31**, 83–111.

Trewavas, A. J. (1982). Growth substance sensitivity: the limiting factor in plant development. *Physiol. Plant.* **55**, 60–72.

Valdovinos, J. G. and Jensen, T. E. (1968). Fine structure of abscission zones II. *Planta* **83**, 295–302.

Valdovinos, J. G., Jensen, T. E. and Sicko, L. M. (1972). Fine structure of abscission zones IV. *Planta* **102**, 324–33.

Van Loon, L. C., Haverkort, A. J. and Lockhurst, G. J. (1978). Changes in protease activity during leaf growth and senescence. F.E.S.P.P. Abstr. **280c**, 544–5.

Van Steveninck, R. F. M. (1959). Abscission accelerators in lupins (*Lupinus luteus* L). *Nature* **183**, 1246–8.

Varma, S. K. (1976). Role of abscisic acid in the phenomena of abscission of flower buds and bolls of cotton and its reversal by other growth regulators. *Biol. Plant.* **18**, 421–8.

Von Abrams, G. J. (1974). An effect of ornithine on degradation of chlorophyll and protein in excised leaf tissue. *Z. Pflanzenphysiol.* **72**, 410–21.

Wang, T. L. and Woolhouse, H. W. (1982). Hormonal aspects of senescence in plant development. In *Growth Regulators in Plant Senescence,* Monograph 8, eds M. B. Jackson, B. Grout and I. A. Mackenzie, British Plant Growth Reg. Group, Wantage, pp. 5–25.

Wareing, P. F. and Seth, A. K. (1967). General aspects of ageing and senescence in the whole plant. *S. E. B. Symposia XXI. Aspects of the Biology of Ageing*, pp. 543–58.

Waters, S. P., Peoples, M. B., Simpson, R. J. and Dalling, M. J. (1980). Nitrogen redistribution during grain growth in wheat (*Triticum aestivum* L.). I. Peptide hydrolase activity and protein breakdown in the flag leaf, glumes and stem. *Planta* **148**, 422–8.

Webster, B. D. (1973). Anatomical and histochemical changes in leaf abscission. In *Shedding of Plant Parts*, ed. T. T. Kozlowski, Academic Press, N. Y., pp. 45–83.

Weisner, J. (1871) quoted by Facey, V. (1950). *New Phytol.* **49**, 103–16.

Whyte, P. and Luckwill, L. C. (1966). Sensitive bioassays for gibberellins based upon retardation of leaf senescence in *Rumex obtusifolius*. *Nature* **210**, 1360.

Wollgiehn, R. (1967). Nucleic acid and protein metabolism of excised leaves. *Symp. Soc. Exp. Biol.* **21**, 231–46.

Wollgiehn, R., Lerbs, S. and Munsche, D. (1976). Synthesis of ribosomal RNA in chloroplasts from tobacco leaves of different ages. *Biochem. Physiol. Pflanzen.* **170**, 381–7.

Wong, C. H. and Osborne, D. J. (1978). The ethylene induced enlargement of target cells in flower buds of *Ecballium elaterium* and their identification by the content of endoduplicated nuclear DNA. *Planta* **139**, 103–11.

Woolhouse, H. W. (1967). The nature of senescence in plants. *Symp. Soc. Exp. Biol.* **21**, 179–213.

Woolhouse, H. W. (1974). Longevity and senescence in plants. *Sci. Prog. Oxford* **61**, 223–47.

Woolhouse, H. W. (1978). Senescence processes in the life cycle of flowering plants. *Bioscience* **28**, 25–31.

Woolhouse, H. W. (1982a). Biochemical and molecular aspects of plant senescence. In *Molecular Biology of Plant Development*, eds H. Smith and D. Grierson, Blackwells, Oxford, pp. 256–87.

Woolhouse, H. W. (1982b). Hormonal control of senescence allied to reproduction in plants. In *Strategies of Plant Reproduction*, Beltsville Symposia in Agricultural Research, VI, ed. W. J. Meudt, US Dept, Agric., Beltsville, pp. 201–33. Allanheld, Osmun, Totowa.

Woolhouse, H. W. and Jenkins, G. I. (1983). Physiological responses, metabolic changes and regulation during leaf senescence. In *The Growth and Functioning of Leaves,* ed. J. Dale and F. L. Milthorpe, C.U.P., Cambridge, pp. 449–87.

Wright, M. and Osborne, D. J. (1974). Abscission in *Phaseolus vulgaris.* The positional differentiation and ethylene induced expansion growth of specialized cells. *Planta* **120,** 163–70.

Yemm, E. W. (1950). Respiration of barley plants. IV. Protein catabolism and the formation of amides in starving leaves. *Proc. Roy. Soc. Lond.* **B136,** 632–49.

Yoshida, Y. (1961). Nuclear control of chloroplast activity in *Elodea* leaf cells. *Protoplasma* **54,** 476–92.

Young, R. (1972). Relation of ethylene and cellulase activity to abscission of freeze injured citrus fruit. *J. Am. Soc. Hort. Sci.* **97,** 133–5.

Index

Index compiled by Richard Raper of Indexing Specialists, Hove